COMMUNICATION SYSTEMS AND RANDOM PROCESS THEORY

NATO ADVANCED STUDY INSTITUTES SERIES

Proceedings of the Advanced Study Institute Programme, which aims at the dissemination of advanced knowledge and the formation of contacts among scientists from different countries.

The series is published by an international board of publishers in conjunction with NATO Scientific Affairs Division

A	Life Sciences	Plenum Publishing Corporation
B	Physics	London and New York
C	Mathematical and Physical Sciences	D. Reidel Publishing Company Dordrecht and Boston
D	Behavioural and Social Sciences	Sijthoff & Noordhoff International Publishers B.V.
E	Applied Science	Alphen aan den Rijn, The Netherlands and Winchester, Mass., USA

Series E: Applied Science – No. 25

COMMUNICATION SYSTEMS AND RANDOM PROCESS THEORY

edited by

JOSEPH K. SKWIRZYNSKI

GEC-Marconi Electronics, Ltd.
Great Baddow Research Laboratories
Chelmsford, Essex, U.K.

SIJTHOFF & NOORDHOFF 1978
Alphen aan den Rijn – The Netherlands

Proceedings of the NATO Advanced Study Institute
on Communication Systems and Random Process Theory
Darlington, U.K.
August 8-20, 1977
Supported also by the National Science Foundation, USA, and by the European Research Office of the US Army

ISBN-13: 978-94-011-7579-1 e-ISBN-13: 978-94-011-7577-7
DOI: 10.1007/978-94-011-7577-7

Softcover reprint of the hardcover 1st edition 1978

PREFACE

This volume contains the complete proceedings of the second NATO Advanced Study Institute organised to relate communication theory with allied subjects, and thus to single out themes which, though peripheral at present, are gaining in importance because of recent developments in theoretical investigations by communication experts.

In 1974 we attempted to promote an interchange of ideas between specialists in Signal Processing and in Control and System Sciences *.

This time we tried to concentrate on probabilistic aspects of communication theory and practice and of the allied science of random process theory, with its novel and exciting approach to the mathematical foundations of noise phenomena.

The topics presented here have been chosen with the above in view, to enlarge ideas and to diffuse results in the two allied subjects, by providing the opportunity for gaining insight in depth into new developments. These topics fall broadly into 10 distinct parts. Each starts with two or more basic presentations of a tutorial nature, followed by one or more detailed examinations of particular aspects of the subject. In addition, some parts are concluded with reports of panel discussions organised to consider a particularly pertinent issue.

---

* "NEW DIRECTIONS IN SIGNAL PROCESSING IN COMMUNICATION AND CONTROL", Edited by J.K. Skwirzynski. NATO ASI Series E - No. 12, Noordhoff, Leyden. 1975.

The 10 parts are :

I. SOME ASPECTS OF DATA COMMUNICATION

We open with a review on Eurocon '77 Assembly and then concentrate on timing recovery problems and on the optimisation of digital transmission systems. There follows a short discussion on communication under interfering conditions such as intentional jamming.

II. MULTIPLE COMMUNICATION

We have two tutorial presentations on the status and prospects of multiplexing in a volume space (frequency X time X code). Additional topics deal with Spread Spectrum Techniques and with Probabilistic Loading of Communication Networks.

III. ALGEBRAIC CODING

This opens with a tutorial on this subject and on combinatorics and continues with discussions on the construction of new codes, on a novel mathematical approach to code generation and on soft decision in coding.

IV. PROBABILISTIC CODING

Here we have attempted to bridge the gap between deterministic aspects of the code theory with the main theme of the Institute. Following the tutorial paper there are discussions on decoding in burst-noise channels and on synchronisation recovery.

V. INTEGRATED CODING AND RATE DISTORTION THEORY

In 1974 we had interesting papers and discussions on Combined Channel and Source Coding. Now we have enlarged upon this subject and have tried to bring it up-to-date. There are three main papers on Integrated Coding, on Rate Distortion Theory and on Some Aspects of Integrated Coding. Several additional papers deal with particular aspects of these subjects. This part concludes with a report of panel discussion on 'The Practicality of Source Coding'.

VI. DISPERSIVE AND FADING CHANNELS

Here the aim is to present the status of analytic tools which are available for prediction of transmission through randomly varying channels and to correlate these with available measurement data. Specific applications deal with various situations such as optical and undersea communication. This part is closed with a report of a panel discussion on "Channel Statistics and Models".

VII. NON-STATIONARY AND NON-GAUSSIAN SIGNAL/NOISE ANALYSIS

The tutorial papers here deal with "traditional" mathematical techniques used by communication specialists. There are in addition several presentations on dealing with nonlinear channels with noise.

VIII. ADAPTIVE SIGNAL PROCESSING

The two tutorial papers are followed by a panel discussion on 'Adaptive Filtering for High Speed Modems'.

IX. STOCHASTIC CALCULUS

This part contains three tutorial papers introducing martingales, and stochastic calculus in continuous and discrete-time, followed by a paper on the 'Design Approximations for Nonlinear Filtering . The session is concluded by a report of a panel discussion on the problem: 'Are Ito Calculus and Martingale Theory Useful in Practice?'.

X. SIGNAL PROCESSING APPLICATIONS OF STOCHASTIC CALCULUS

This contains three contributions on some approaches to problem definitions and solutions using the methods of stochastic calculus.

The final part of the book contains a self contained presentation by the Communications Division of SHAPE Technical Centre in Hague. We have followed our custom of presenting to participants some work in a European establishment notable in our field.

It is hoped that this volume will be of interest to anyone who wishes to be acquainted with new developments and new possibilities for the analytic treatment of probabilistic channels, signals and systems. It is possible that we may follow the present Institute with a third one, concentrating this time on Multi-user Systems and on analytic techniques to study their performance.

The Director of this Institute was helped during two years of preparation, and then of this volume, by three colleagues, Professor K.W. Cattermole, Professor B. Picinbono and Professor W.L. Root who have not spared valuable advice on technical aspects of the program organisation. Some sessions were organised by young yet eminent specialists in their fields: Dr. P.G. Farrell was responsible for organising sessions corresponding to Parts III to V in our volume, Dr. G.A. Richards was responsible for organising session VII and Dr. M.H.A. Davis of sessions IX and X. They are thanked here for their persistence in preparing coherent and well organised Institute sessions.

My colleague, Mr. B.G. West has been responsible for the preparation of the tapes for panel discussions. In the editing of these tapes he was helped by Professor K.W.Cattermole, Dr. P.G. Farrell and Dr. Davis. They are complimented for the success of their work.

I also wish to acknowledge the assistance by Miss Marjorie Sadler and Miss Ann Williamson in administration and organisation of this Institute and Mrs. S. Carnell for performing the typing duties.

Finally, I thank Mr. G.D. Speake, Technical Director of the Marconi Research Laboratories for permission to undertake the organisation of this Institute and the publication of this volume.

J.K. SKWIRZYNSKI

Director of the Institute.

Great Baddow, 7th November, 1977.

TABLE OF CONTENTS

Part I

SOME ASPECTS OF DATA COMMUNICATION

STOCHASTIC PROBLEMS OF MULTIPLE COMMUNICATION

K.W. CATTERMOLE

Department of Electrical Engineering Science,
University of Essex, Colchester, England.

1. INTRODUCTION

Much communication theory, and perhaps especially information theory, has been expressed in terms of the properties of single channels. Yet almost all practical communication is affected by the co-existence of many channels, messages, or stations. This is, of course, very obvious in multiplex communication systems. There have been several good reviews of multiplex theory and technique, from Landon (1) to Flood (2). It is not our purpose to duplicate these - indeed, a general knowledge of such material is taken for granted - but rather to cast the net wider, taking in all forms of multiple usage. In our present context, we need also to bring out the stochastic aspects of each problem.

2. MULTIPLE-USAGE COMMUNICATIONS

The topics in multiple-usage communications here considered fall into four main categories (not without some difficulties of demarcation!):

(a) Multiplex Systems. By this term, we mean that many channels are provided by means of a permanent, or at least enduring, partition of time, frequency or space. In an ideal system of this kind, the channels might be so well isolated that no random effects were generated by their coexistence. In practice we are almost always concerned with crosstalk, intermodulation, loading statistics, etc.

(b) Multiple Access Systems. By this term, we mean that channels or signalling capacity are assigned in a coordinated way to many users whose demands are independent. Clear examples are circuit and message switching systems, demand-assigned satellite communication systems, and computers with multiple access. An example perhaps on the boundary of classes (a) and (b) is provided by speech interpolation systems. In all these cases, the capacity demand placed on the system has a random component. The signals actually carried, and their further effects, are random variables of a compound type. Impairments suffered by signals offered to such a system (distortion, delay or rejection) are further random variables.

(c) Multiple access networks. By this term we mean a set of interconnected multiple access systems with only incomplete overall coordination. Switched networks for telephony and data are the classical examples. These exhibit all the random phenomena of class (b), together with others arising from the independence of the several control centres: for example, blocking or delay of traffic at any one centre may depend on the characteristics of the overflow traffic from many others.

(d) Open media to which many users have access with little if any coordination. An example is a random-access code-division-multiplex system.

We shall proceed later to detailed discussion of some examples: but first we develop some mathematical tools for probabilistic analysis.

3. PROBABILITY DISTRIBUTION AND RANDOM PROCESSES

Many problems in multiple communications will lead us into compound processes and distributions. Consider the following examples:

(1) In a frequency-division multiplex system, with a fixed number of channels, the composite signal can be modelled as a random process, and the signal power has a probability distribution which can be derived simply from that of the individual channels.

(2) In a multiple-access system, the number of active users is again a random process with a simple distribution.

(3) In a frequency-division multiple access system, such as is common in satellite communications, both the number of active users and the signals contributed by a user can be modelled as random processes. The composite signal is therefore the superposition of a random number of random variables.

This last is an example of a compound process. We shall encounter many examples both of continuous and of discrete compound processes and distributions, and a preliminary look at their mathematical formulation may help.

Discrete probability distributions

Will be denoted by lower-case letters, and their probability generating functions by capitals: if the discrete variable X has distribution f, then

$$\text{prob. } \{X = i\} = f_i, \quad i = 0, 1, 2 \ldots$$

$$F(z) = E(z^X) = \sum_{i=0}^{\infty} f_i \, z^i \qquad (1)$$

Continuous probability densities.

Will be denoted by lower-case letters and their characteristic functions by capitals: if the continuous variable X has density f(x),

$$\text{prob. } \{x \leqslant X < x + dx\} = f(x)\, dx$$

$$F(\Theta) = E(e^{j\Theta X}) = \int_{-\infty}^{\infty} f(x)\, e^{j\Theta x}\, dx \qquad (2)$$

For non negative variables we may also use the Laplace transform

$$F(s) = E(e^{-sX}) = \int_{0}^{\infty} f(x)\, e^{-sx}\, dx \qquad (3)$$

which is closely related to the characteristic function.

We assume that the density and its Fourier transform (≡ characteristic function) exist, at least in the sense of generalised-function theory: the latter (3,4) permits us to accommodate all distributions of practical interest without worrying explicitly about continuity and convergence.

Compound discrete distributions.

Suppose that certain events can occur at random times. Let the number of events in a given interval be a random variable X with distribution g (for example, the number of noise bursts interrupting a multiplex signal). Let each event have an associated random variable W_i with distribution f (for example, the number of channels in the multiplex which are active when the jth burst occurs). What is the distribution of $Y = \sum W_j$ (the total number of interruptions of an active channel by a noise burst)?

Consider the case X = i. The total outcome of i (independent) events is the i-fold convolution f^{*i} with generating function F^i. Summing over all i with appropriate weight, we have the generating function of Y:

$$H(z) = \sum_i g_i F^i(z) = G\{F(z)\} \qquad (4)$$

From the relations between the generating function and the factorial moments of the distribution (7, Vol.I, p.250) it is easy to deduce the mean and variance:

$$E(Y) = E(X).\ E(W) \qquad (5)$$

$$V(Y) = E(X)\ V(W) + E^2(W)\ V(X) \qquad (6)$$

Note that the variance/mean ratio of Y is greater than that of W, and usually it is also greater than that of X. An important special case is the compound Poisson process, for which the events occur with uniform probability density; randomly originating traffic, random errors, etc. are often of this type. In this case,

$$\begin{aligned} G(F(z)) &= e^{-\lambda+\lambda F(z)} \\ E(Y) &= \lambda E(W) \\ V(Y) &= \lambda E(W^2) \end{aligned} \qquad (7)$$

Discrete events with a continuous random parameter.

Suppose that a parameter of a random process (a rate or a time, for example) is itself a random variable. We may wish to calculate some property of the resulting distribution. This can be done for continuous or discrete distributions: we will take first a discrete example.

Consider a multiple access system which enters a certain traffic state (e.g. fully loaded) and subsequently leaves this state. The time interval t between these two events is a random variable with a continuous density f(t). Further traffic demands can be modelled by a Poisson process with generating function

$$G(z,t) = \sum_{k=0}^{\infty} g_k\ (t)\ z^k = e^{\rho t(z-1)} \qquad (8)$$

where ρ is the mean rate of arrival. What is the distribution h of the number of arrivals occurring during the random time interval?

By taking a weighted average over all values of t it is clear that

$$h_k = \int_0^\infty g_k(t)\ f(t)dt \tag{9}$$

It is often easier to calculate the generating function

$$H(z) = \int_0^\infty G(z,t)\ f(t)\ dt \tag{10}$$

For example, let f(t) be a gamma distribution

$$f(t) = \frac{\beta^r e^{-\beta t} t^{r-1}}{(r-1)!} \tag{11}$$

(This is the waiting time to the rth event in a Poisson process, and is often useful to represent random intervals). Then

$$H(z) = \frac{\beta^r}{(r-1)!} \int_0^\infty t^{r-1}\ e^{-t(\beta + \rho - \rho z)} dt = \left(\frac{\beta}{\beta + \rho - \rho z}\right)^r \tag{12}$$

This is the generating function of a negative binomial distribution

$$h_k = \binom{-r}{k} \left(\frac{\beta}{\rho+\beta}\right)^r \left(\frac{-\rho}{\rho+\beta}\right)^k \tag{13}$$

It is notable that Wilkinson (5) has, by a completely different argument, shown that overflow traffic is approximated by a negative binomial distribution. (See section 5 below).

Continuous distributions with a discrete random parameter.

Our introductory example, namely the composite signal in a multiple-access system, falls in this category. Let the (continuous) density relating to one channel be f(x), and the (discrete) distribution of the number of active channels be g. Then clearly the composite distribution is

$$h(x) = \sum_{k=0}^{\infty} g_k\ \{f(x)\}^{*k} \tag{14}$$

and the characteristic functions or Laplace transforms are related as in equation (4) for the discrete case.

As a specific example, suppose that a channel on carrier has a random amplitude and phase, and can be modelled as a bivariate normal distribution. The power in one channel then has a negative-exponential distribution e^{-y}: hence for k active channels the total power has a gamma distribution

$$\{f(y)\}^{*k} = \frac{e^{-y}\ y^{k-1}}{(k-1)!} \qquad k = 1, 2 \ldots. \tag{15}$$

If the number of active channels has a Poisson distribution

$$g_k = \frac{e^{-\lambda}\ \lambda^k}{k!} \quad , k = 0, 1, 2 \ldots.$$

then the total power has the distribution

$$h(y) = e^{-\lambda}\delta(y) + \lambda e^{-\lambda-y} \sum_{k=o}^{\infty} \frac{(\lambda y)^k}{k!\ (k+1)!} \tag{16}$$

The series can be identified with the modified Bessel function, since

$$\frac{I_1(z)}{\frac{1}{2}z} = \sum_{k=0}^{\infty} \frac{(\frac{1}{2}z)^{2k}}{k!(k+1)!} \tag{17}$$

whence

$$h(y) = e^{-\lambda}\delta(y) + e^{-\lambda-y}\sqrt{\frac{\lambda}{y}}\, I_1\ (2\sqrt{\lambda y}\,) \tag{18}$$

We are often interested in the asymptotic behaviour of signal distributions, since this is relevant to overload or error perfor-- mance. In this case

$$h(y) \simeq \frac{\lambda^{\frac{1}{4}}\ e^{-\lambda}}{2\pi^{\frac{1}{2}}y^{\frac{3}{4}}}\quad e^{-y}\ e^{2(\lambda y)^{\frac{1}{2}}} \tag{19}$$

Comparing this with the asymptote for a fixed number of active channels equal to the mean λ, namely

$$h_{\lambda}(y) = \frac{e^{-y}\ y^{\lambda-1}}{(\lambda-1)!} \tag{20}$$

we see that (19) differs by an exponential factor, and vanishes more slowly than (20) for any λ (Fig.1). This is a natural result of the compound randomness. A further result is that the variance of the compound distribution is greater than for a fixed number of channels. Using the Laplace transform as a moment-generating function, it can be shown that if g is a Poisson distribution then the mean and variance are given by equation (7) as in the discrete case.

Compound processes and distributions are treated in some, though by no means all, well known textbooks: for example, Moran (6), Feller (7) and Parzen (8). They have also arisen in other areas of signal and communication theory, notably optical signals as discussed by Picinbono (9).

4. MULTIPLEX SYSTEMS

The predominant class of multiplex systems as defined in section 2 above or in reference (2), allot to each channel a carrier drawn from an orthogonal set of carriers. That is, we define a set of waveforms $c_i(t)$, i=1, 2 ... n such that

$$\langle c_i(t)\ c_j(t)\rangle = \delta_{ij} \begin{cases} 0, & i \neq j \\ 1, & i = j \end{cases} \qquad (21)$$

the average being taken over some appropriate period. If individual channel signals are $s_i(t)$, then the composite signal is

$$S(t) = \sum_{i=1}^{n} s_i(t)\ c_i(t) \qquad (22)$$

An individual channel signal may be recovered by means of a correlative receiver, since

$$\langle c_i(t)\ S(t)\rangle = s_i(t) \qquad (23)$$

by virtue of the orthogonality property, equation (21). Alternatively, a channel may be recovered by a filtering operation broadly defined: in the well known case of frequency-division multiplex, the filter may be a passive or active wave filter of classical type.

The orthogonal carriers are, in the commonest cases, either

(a) sinusoids of different frequency, such that the modulated channels occupy disjoint frequency bands (frequency division multiplex).

(b) pulses of common repetition frequency but interlaced in time, so that the modulated channels occupy distinct time slots (time division multiplex).

However, a variety of other orthogonal carriers have been used or proposed, including

(c) Sinusoids of common frequency but orthogonal in phase (obviously only two are available).

(d) Sinusoidal bursts with a mixture of time and frequency separation.

(e) Orthogonal polynominals such as the Hermite or Legendre polynomials.

(f) Orthogonal square-waves such as the Rademacher or Walsh functions.

Moreover, it has been recognised that if the channel signals are digital, especially binary, perfect separation by linear means is not necessary since the signals may be detected in the presence of bounded interference by non-linear means, as in regenerative repeaters. This opens the way for the use of signal sets which are approximately but not exactly orthogonal, such as

(g) Pseudo-random binary sequences.

It is clear from the definition of orthogonality that, if the signals are generated precisely and if the system is perfectly linear, isolation between channels is so complete that we need not be concerned with the joint statistics of the signals. However, there are practical impairments of several types:

(i) The carriers may not be generated with enough precision. This is a major obstacle to the use of orthogonal polynomials. It is also a problem in frequency-division multiplex, where harmonic and image frequencies must be constrained to low levels.

(ii) The modulation process may upset the orthogonality. In f.d.m., imperfect low-pass filtering of baseband signals will allow sidebands to spread and possibly overlap adjacent channels. In t.d.m. pulse position modulation must be limited in extent to avoid intrusion into adjacent time slots.

(iii) Linear dispersion in the common highway may upset the orthogonality. This is a dominant problem for analogue t.d.m. (10; 11; 12 section 2.7) and severely limits its use in transmission systems.

(iv) Nonlinear distortions in the common highway may produce a great variety of effects. This is a dominant problem in the practical design of f.d.m. systems; it gives rise to many interesting stochastic problems, some of which are discussed in detail below.

Multi-channel load statistics.

In f.d.m. transmission, many channels carrying independent signals are linearly superposed. The amplitude distribution of the composite signal is therefore the convolution of the distributions of the channel signals:

$$h(x) = f_1(x) * f_2(x) * \dots f_n(x) \tag{24}$$

Since in a telephone system we are normally dealing with a rather large number of channels, and there is a good deal of randomness in each channel signal, it is tempting to assume that h(x) is Gaussian. However, the central limit theorem is perhaps more often invoked by incantation than by proof: and there are three reasons for caution. Firstly, the individual channel statistics may be very far from Gaussian: which implies that convergences to a Gaussian sum is rather slow. Secondly, the number of channels is not everywhere large: even a 2700 - channel signal is assembled from groups of 12 and supergroups of 60 channels, and we need to know the properties of these relatively small multiplex units. Finally the properties of interest include not only central statistics such as mean and variance, but extreme statistics such as tail probabilities; in respect of the latter, convergence to the Gaussian sum is very slow indeed.

The statistics of speech signals are rather complicated. Richards (13) distinguishes

(a) the "instantaneous" amplitude distribution, which is symmetrical, with the positive half modelled fairly accurately by a negative-exponential or gamma distribution.

(b) the distribution of amplitudes on a "syllabic" time scale which can be modelled by a (different) gamma distribution, or by a Gaussian distribution with a randomised parameter.

(c) the occurrence of pauses between utterances. All the foregoing relate to speech of approximately "constant volume" (this term itself is difficult to define, as may be seen from Richards' account of volume measurement). However, in practical telephone networks we must add

(d) the distribution of mean volumes due to differences between talkers (loud or soft) and their telephone circuits (high or low attenuation). Practical measurements (13,14,15) show that this distribution is approximately log-normal: i.e. the levels expressed in a logarithmic measure such as decibels are Gaussian, typically with a standard deviation of about 5dB.

(e) the occurrence in telephone channels of non-vocal signals such as digital data or telephonic control signals.

(f) the distribution of occupancy of channels, which being accessed via a switched network exhibit the typical behaviour of smoothed traffic (see section 5 below).

It is clear that a complete account of the signal statistics in a multi-channel system would be very complicated; and it is not normally attempted. There are three main statistical problems in practical f.d.m. telephony.

(i) to estimate, and specify, the mean power of the composite signal: so that mean power limitations in transmission may be observed, and impairments which are related to mean power (such as intermodulation) may be estimated.

(ii) to estimate "peak power" (on some arbitrary but useful definition) so that peak power limitation or overload may be of sufficiently infrequent occurrence.

(iii) to appreciate the effect of departures from the nominal channel statistics around which systems are usually designed, so that the hazards of non-vocal signals may be understood and suitable constraints be specified for such signals.

The mean power is estimated from the median volume of a single channel, the variance of the lognormal distribution cited as (d) above, and the mean "activity ratio" which depends on factors (c) and (f) above. The lognormal distribution recurs in several contexts, and deserves a brief exposition: for a fuller treatment see Aitchison and Brown (16). Let Y = log X have a Gaussian (normal) distribution, with mean μ and variance σ^2. Then the rth moment of X is

$$m_r = E(X^r) = E(e^{rY}) \tag{25}$$

This is readily found by observing that the characteristic function (Fourier transform) of a Gaussian distribution is another Gaussian function:

$$\phi(\Theta) = E(e^{j\Theta Y}) = e^{j\Theta\mu-\frac{1}{2}\sigma^2\Theta^2} \tag{26}$$

Consequently,

$$m_r = \phi(-jr) = e^{r\mu+\frac{1}{2}r^2\sigma^2} \tag{27}$$

Thus the mean value of X expressed in nepers is

$$\log m_1 = \mu + \tfrac{1}{2}\sigma^2 \tag{28}$$

When reckoning in dB rather than nepers we need a numerical conversion factor of 0.1 log 10:

$$\text{mean power in dB} = 10 \log_{10} m_1 = \underline{\mu} + 0.115\, \underline{\sigma}^2 \tag{29}$$

where $\underline{\mu}$ and $\underline{\sigma}$ are in dB.

Typical practical figures are $\underline{\mu}$ = -14 dBm0, $\underline{\sigma}$ = 5dB, which gives a mean power of about -11dBm0. Mean activity ratios of about 0.35 for seized channels, or 0.25 for a large group of channels in the traffic busy hour, would reduce the mean power to about -17 dBm0. An allowance for signalling tones raises the mean to about -15.3 dBm0. It is now conventional (17) to design for a mean power per channel of -15 dBm0, hence by addition of n independent channels mean power in composite signal $= -15 + 10 \log_{10} n$ dBm0 (30)

To appreciate the significance of the peak power estimate, we must recall that all amplifiers, modulators, etc. are non-linear at least in the sense that they have an upper limit to their output power: an attempt to drive them to larger peak power will cause a sharp increase in distortion. In terrestrial multiplex systems it is conventional to design for a load capacity with a small probability of exceedance, say 10^{-4}, rather than estimate the distortions precisely. For sufficiently large numbers of channels it is reasonable to assume a near-Gaussian distribution, hence a peak/mean ratio of some 11.4 dB. However, it has been known from the time of Holbrook and Dixon (18) onwards that for smaller groups a higher ratio is observed; for example

No. of channels n	peak/mean dB
60	16.5
300	12.9
960	11.5

Further data has been published by Medhurst (19) and in CCITT documents (notably ref.17). One typical empirical rule is to allow a fixed margin of about 12 dB for peaks above a nominal loading of:

$$\begin{aligned} &-15 + 10 \log_{10} n \text{ dBm0}, \quad n \geqslant 240 \\ &-1 + 4 \log_{10} n \text{ dBm0}, \quad 12 \leqslant n < 240 \end{aligned} \qquad (31)$$

It will be seen that this nominal loading coincides approximately with the long term mean power for large groups, but not for small ones.

The general conclusion is that, for f.d.m. telephone systems, there is a well-established practice based on a partial theory and some empirical evidence. The practical performance is satisfactory: however, the impairments which it yields, and the impairments which would result from changes in practice, (such as a reduction in load capacity) are not known with any precision. A further problem is that, the practice being based on the statistics of telephone speech, the effect of mixing speech and data channels the latter having a peak/mean power ratio of some 3 to 6 dB rather than the 9 to 12 dB of speech - is potentially troublesome, and not entirely understood. It has been thought necessary to confine data channels to a mean power level comparable with speech, hence to a peak power apparently rather low, to avoid problems both of overload and intermodulation.

Nonlinearity and intermodulation in f.d.m.

Consider a smooth, instantaneous nonlinear characteristic of the form

$$y = a_1 x + a_2 x^2 + a_3 x^3 + \ldots. \tag{32}$$

This represents closely enough the type of nonlinearity at levels well within the range of a fairly linear amplifier or modulator. Consider a signal $h(t) \rightarrow H(f)$ applied to the device in question. (The notation $\rightarrow$ here implies Fourier transformation). The output of the device is

$$\sum_r a_r \{h(t)\}^r \rightarrow \sum_r a_r \{H(f)\}^{*r} \tag{33}$$

The first term in the right-hand series is, of course, the spectrum of the original signal. The remaining terms are distortion products of the 2nd, 3rd, ... rth order. Some examples of convolved spectra are shown in Figure 2.

This result is, of course, derived from consideration of a single coherent signal; and some care is required in adapting it to deal with a multiple signal, which is by definition the sum of many incoherent components. It is common both in theory and in practical testing to model the multiplex signal as band limited noise, which is defined statistically by its power spectrum. We would like a relationship between the power spectrum of an input signal, and the power spectrum of the intermodulation products. Equation (33) suggests, rather than proves, a suitable approach.

Take for example a square-law device, with input spectrum H(f) and output spectrum

$$\{H(f)\}^{*2} = \int H(f-x)\, H(x)dx \tag{34}$$

The form of the convolution integral implies that the output at frequency f is the superposition of sum frequencies generated by all input frequency pairs x, f-x: the generation of sum frequencies is of course well known from simple modulation theory. However, if the components are incoherent and random in phase, we can add their squared magnitudes. In this case, the selfconvolution of an input power spectrum yields an output power spectrum. This argument is spelt out more formally in the classical paper by Rice (20).

The exception to this argument occurs at output frequency $f = 0$, where the components add coherently. Thus the true power spectrum from a square-law device is

$$\{H(f)\}^{*2} + \{\int H(f)df\}^2 \; \delta(f) \tag{35}$$

This second term is not always negligible. For example, the autocorrelation function of the output of a square-law device includes a constant (see ref. 28 page 166) which is the Fourier transform of the impulse given here. However, in practical multiplex systems a small d.c. term is immaterial and may be ignored in calculating second-order intermodulation products. When (35) is further convoluted with H(f) to obtain third-order products, the impulse yields a multiple of H(f) which amounts to a small change of gain and can be ignored in considering third-order intermodulation. Thus the important intermodulation power spectra are of the form shown in Figure 2. The approach used here is a continuous analogue of the combinatorial method used by Bennett (21) for counting discrete intermodulation products. Bennett's method has the advantage that it can distinguish products of different types (such as 2A-B, 2A+B, A+B-C etc.) which may differ in subjective effect, and also in the cumulative effect of many sources of distortion (22).

This formulation is not suitable for nonlinear devices with amplitude or slope discontinuities, or indeed any nonlinearity for which the series (32) converges slowly. More general analytical methods are available, and have been applied for example to rectification, hard limiting, and quantizing (23-27).

A variation of the above theory may be applied to the transmission of f.d.m. signals by means of f.m. radio links. Here the effect of amplitude nonlinearities is small, but that of phase nonlinearity (i.e. delay distortion) may be substantial. Let the phase shift ϕ of a channel vary with frequency:

$$\phi(f) = 2\pi \sum_{r=0}^{\infty} a_r \, (f-f_c)^r \tag{36}$$

where f_c is the carrier frequency. Then, with a baseband signal $h(t) \rightarrow H(f)$ a quasi-stationary theory of f.m. readily shows that the demodulated output is

$$H(f) + j \, 2\pi \, f \sum_{r=2}^{\infty} a_r \, \{H(f)\}^{*r} \tag{37}$$

similar in form to the series (33) except for the high-frequency lift characteristic of f.m. noise. This may be applied to power spectra with the same reservations as expressed above. For further quasi-stationary results see Garrison (79).

This theory is entirely suitable for the smooth phase nonlinearities due to bandpass amplifiers and filters. The periodic phase variations due to reflections on feeders or multipath transmission may however correspond to a very slowly converging power series, and are best tackled by other methods (29-31).

Despite the wealth of analytical method, some problems of intermodulation due to a nonlinearity have proved intractable. This is particularly true of satellite transponders whose hard-driven travelling wave amplifiers have a rather complex form of nonlinearity in amplitude and phase. Experiment and computer simulation have yielded the most directly useful answers (32).

Limitations of present multiplex theories.

The foregoing theories and observations are useful in practice, but have two major limitations.

Firstly, they are based on the assumption of similar (sometimes Gaussian) amplitude distributions, and similar (generally uniform) spectral distributions for all channels. Problems arise when channel signals with very different statistical properties are combined. We have noted the restriction of voice-frequency data signals to a mean power level typical of speech: it is also useful, in a telephone multiplex system with a substantial data loading, to randomize the spectrum of data signals by scrambling.

Secondly, the compound random nature of much multiplex-system loading, mentioned in section 2, is not fully taken into account. Mean loading may be estimated fairly well by use of "activity factors" which are of course mean values: a more complete theory would have to consider the distribution of activity. Only with the advent of time-assignment systems has this been considered.

It would seem that there is still room for further study even in apparently well-developed topics.

Time-division multiplex.

At first glance, time-division systems evade all the statistical problems of loading and intermodulation discussed in the last two sections. The signals in the respective signals simply do not overlap, so there is no physical variable corresponding to the sum of the signals.

This is however an artefact of system design. The usual t.d.m. system employs uniformly-spaced samples independent of the signal. There have been various proposals for making the sampling interval depend in some way upon the signal, for example by "extremal sampling" (33) or by "pulse interval modulation" (34). In either case, it is claimed that more channels can be carried, by virtue of the statistical fact that peak amplitudes or rapid changes are unlikely to occur in all channels at once. These systems are not of great practical importance: most of their theoretical advantage is achieved more conveniently by time-assignment systems.

It is well known that, with conventional techniques, a coaxial cable will carry more channels by means of f.d.m./s.s.b. than t.d.m./p.c.m.. This is not, as often supposed, due to the larger bandwidth occupied by a p.c.m. channel: the author has shown (35) that it is entirely due to the use of "non-simultaneous load advantage" in the f.d.m. but not in the t.d.m. system. If we compare two systems with it, or two systems without it, the t.d.m./p.c.m. system always has the advantage.

The fact that almost all practical t.d.m. systems are digital also contributes to the evasion of the loading and intermodulation problems. However, t.d.m. digital systems have statistical problems of their own, notably in timing and framing. For example, consider a digital t.d.m. system in which a small proportion of the frame is occupied by an unvarying frame-alignment pattern. Synchronism of the receiver with the transmitter is acquired and verified by inspection of this pattern. However, the pattern is (i) embedded in a more or less random stream of digits carrying messages (ii) vulnerable to random digital errors in transmission. Recovery of synchronism requires a systematic search of the digit stream as it comes in to locate the alignment pattern. The waiting time to successful realignment is an important statistic of the multiplex system. Bylanski and Ingram (36, section 5.4) give a good account of this problem, with some further references.

Space-division multiplex.

Within the general classification used in this paper, the above is an appropriate name for the transmission of several channels on distinct conductors within a multi-pair cable. Unscreened pairs or quads exhibit some unwanted coupling. This is mainly due to capacitive unbalance, and so it rises with frequency: far end crosstalk (between pairs carrying signals in the same directions) typically at 6dB per octave, and near end crosstalk (between pairs carrying signals in opposite directions) typically at 4.5 dB per octave. Consequently, crosstalk has become a significant problem mainly with the introduction of wideband signals (such as t.d.m./p.c.m.) on multi-pair cables.

If the crosstalk loss between two cable-pairs be measured, over all possible selections of two pairs from a multi-pair cable, it is found to be far from uniform. It depends on structural features of the cable (e.g. are the two pairs in the same layer or unit, and do they have the same length of twist), and within any structural classification there is a substantial random variation. Some cable crosstalk statistics have been published (37-9) but a good deal more remains in the files of telephone administrations and cable manufacturers. Most of the author's knowledge is based on unpublished work, including his own studies in the early days of practical p.c.m. telephony circa 1959-64.

Empirically, the distribution of crosstalk losses within one structural classification is well approximated by the log-normal model, with standard deviations in the range 7 to 13 dB. When fitting a model to a limited amount of data, the extremes may show anomalies depending on the fitting technique. Gumbel (40) recommends that for best estimation of extremes, the mth of n observations should be associated with a probability m/(n+1): with this convention, all the data known to the author exhibit a rather good fit. For physical reasons, the distribution must depart from log-normal at the extremes, but the departure is not normally observable with a sample of a few hundred crosstalk exposures from a cable in good condition.

Thus with a similar signal in each of many pairs, the sum of crosstalk effects into another pair will depend on the sum of samples from a log-normal distribution. No exact analytical solution is known for the convolution of log-normals.

However, Jacobsen (41) has established that the convolution of several log-normals is fairly well modelled by a three-parameter distribution which he calls the augmented log-normal. This is defined as follows: the probability that the variate exceeds

$$W = (1-a)e^{m} + a\, e^{m-\frac{1}{2}s^{2}+ys} \qquad (41)$$

is the same as the probability that a normal random variable with zero mean and unit variance exceeds y. The first three moments of the augmented log-normal may be matched to the first three moments of the sum distribution. This process requires a solution of the cubic equation $u^3 + 3u = K$ which is tabulated by Aitchison and Brown (ref.16, Table A4).

For sufficiently large n (order of 30-100) the sum distribution is adequately modelled by the normal: but for smaller numbers, which are practically significant, the significant peak crosstalk indicated by the more precise model is greater. A numerical index which the author has found useful is the 5% point on the sum distribution: from numerical computation, this is fairly well matched by the linear approximation

$$\underline{\mu} + 1.6\underline{\sigma} + 16 \log_{10} n \text{ dB} \qquad (43)$$

where $\underline{\mu}$ and $\underline{\sigma}$ are expressed in dB.

The practical problem is a compound one, in that we need to consider a sample of cable pairs each carrying a statistically defined signal, and estimate the aggregate crosstalk into a pair.

The first published study of p.c.m. planning problems, by Cravis and Crater (43), leaned rather heavily on the normal approximation. More recent studies, such as Narayana Murthy (44) and Jacobsen (45) have depended on numerical computation or other approximations for key results.

5. MULTIPLE-ACCESS SYSTEMS

The earliest multiple-access systems, which gave rise to the first serious attempts at a stochastic theory of traffic fluctuation, were switched telephone networks. A good account of the pioneer work of A.K.Erlang is given by Brockmeyer et al (46): and an encyclopaedic coverage of work up to about 1960 by Syski (47). Useful recent textbooks are Bear (48) and Kosten (49).

A classification of multiple-access systems may be based on

(a) the disposition of a demand for service which cannot immediately be met because the service channels are all busy. In a loss system, the demand is rejected (for example, a telephone call encountering "all trunks busy"). In a delay system the demand is put into a queue and awaits service. There are two subdivisions of delay systems: either the queue consists of tokens indicating that service is demanded, or else it consists of actual messages for transmission which are held in a buffer store.

(b) the statistics of arrivals and departures, which may be simple Poisson processes (as assumed, with fair justification, in much of the established theory of telephone traffic) or something else (as in many delay systems) which usually implies greater mathematical complexity.

(c) the topology of the paths through which the messages flow. In the simplest case, a single set of inlets (users) has access to a single group of outlets (servers) via a switch or selecting mechanism. In the more complex cases, a network comprises several stages of switching with mutual dependances of various kinds: the output of one selection process is an input (maybe one of many) to another.

We shall sketch in this section

(i) a general method for dealing with a selector whose arrivals and departures are simple Poisson processes, in equilibrium.

(ii) some of the problems of compound loss systems, with simple arrivals and departures but complex topology.

(iii) some of the problems of delay systems with simple

topology but complex arrival and departures.

The simple equilibrium process.

Consider a number S of traffic sources, each making a demand at random on a group of N outlet trunks or servers. Suppose that the service or holding times have a negative exponential distribution. Then the arrival and departure processes are simple Poisson, and can be specified by a probability density independent of time. The state of the system can be represented by the number of active users either in traffic, or in a queue awaiting service: this is an integer 1, 2, 3 ... min (S,N). We denote

(i) the probability of state i by p_i.

(ii) the probability density of an upward transition from i to i+1 by a "birth coefficient" b_i.

(iii) the probability density of a downward transition from i to i-1 by a "death coefficient" d_i.

A state diagram is shown in Figure 3. It is useful to define the coefficient

$$c_i = \frac{b_0 b_1 \ldots . b_{i-1}}{d_1 d_2 \ldots . d_i} , \qquad i = 1,2 \ldots \tag{44}$$
$$c_o = 1$$

In equilibrium, the state probabilities are stationary and so the probabilities of upward and downward transitions can be equated

$$p_i \, b_i = p_{i+1} \, d_{i+1} \tag{45}$$

It follows readily that $p_i = c_i p_o$, and since $\sum p_i = 1$,

$$p_r = \frac{c_r}{\sum_{i=0}^{\min(S,N)} c_i} \tag{46}$$

This formulation covers many cases, including the following:

(a) Infinite sources and trunks: states have the Poisson distribution. Let S,N → ∞: upward probabilities $b_i = A$, a constant: downward probabilities proportional to number of active users, which in normalised units means $d_i = i$. Then,

$$p_r = \frac{A^r/r!}{\sum_{i=o}^{\infty} A^i/i!} = \frac{A^r e^{-A}}{r!} \tag{47}$$

For perfectly random traffic of this kind, the variance equals the mean (a well known property of the Poisson distribution).

(b) Loss system with infinite sources, finite trunks: states have Erlang distribution

$$p_r = \frac{A^r/r!}{\sum_{i=o}^{N} A^i/i!} \tag{48}$$

This is the usual theory of the fully available group of telephone trunks and leads to the Erlang-B formula for blocking probability

$$B(A,N) = \frac{A^N/N!}{\sum_{i=o}^{N} A^i/i!} \tag{49}$$

since blocking occurs when all N trunks are busy.

(c) Loss system with finite sources and trunks: Engset distribution (S>N). The upward probability is proportional to the number of free sources, say $b_i = \rho(S-i)$: downward probability as above. Then

$$p_r = \frac{\rho^r \binom{S}{r}}{\sum_{i=o}^{N} \rho^i \binom{S}{i}} \tag{50}$$

Blocking occurs if all trunks are busy, which is true for a fraction p_N of the time on the average (time congestion). The probability that a new demand is blocked (call congestion) is less than this: it equals the probability that all trunks are occupied by S-1 sources.

(d) Finite sources, no effective limitation on trunks (i.e. $N \geq S$). Equation (48) reduces to the binomial distribution,

$$p_r = \frac{\rho^r \binom{S}{r}}{(1+\rho)^S} = \binom{S}{r} p^r (1-p)^{S-r} \tag{51}$$

where $p = \rho/(1+\rho)$. The Engset and binomial distributions correspond to what the telephone engineer calls smooth traffic: after several stages of concentrating switches, some of the randomness has been lost, and the variance of traffic is less than the mean.

(e) Upward probabilities rising with r: negative binomial distri-- bution. In all the foregoing the upward probabilities are either constant or fall with r. However, Wilkinson (5) has argued that overflow traffic (that is, the traffic which remains after a set of first-choice circuits have accepted part of the demand from a random traffic source) may plausibly be modelled by an increasing upward probability: a new demand is more likely to arise if the traffic level is already high. Putting $b_i = \rho(k+i)$ and $d_i = i$ gives

$$p_r = \binom{-k}{r} (-\rho)^r (1-\rho)^k \qquad (52)$$

which is a negative binomial distribution. (Compare equation 13, which derives a similar distribution by a completely different argument). Wilkinson shows by comparison both with practical data and with a more rigorous theory that overflow traffic is well approximated by this distribution. Overflow traffic, unlike random or smooth traffic, has a variance exceeding its mean.

(f) A delay system may also be modelled in this way if arrivals are pure Poisson and holding times negative-exponential in distri- bution. We consider unlimited sources, a limited number of service channels N and unlimited provision for queueing. The appropriate transition densities are $b_i = A$, a constant: and $d_i = \min(i,N)$, the number of service channels in use. Then

$$p_r = \frac{A^r/r!}{\sum_{i=o}^{N-1} \frac{A^i}{i!} + \frac{A^N}{N!}\left(\frac{N}{N-A}\right)} \quad , \quad r \leqslant N$$

$$= \frac{A^r/N!\; N^{r-N}}{\sum_{i=o}^{N-1} \frac{A^i}{i!} + \frac{A^N}{N!}\left(\frac{N}{N-A}\right)} \quad , \quad r \geqslant N \qquad (53)$$

The probability that a call is delayed is the probability that all service channels are busy, namely

$$P_D = \sum_{r=N}^{\infty} p_r = \frac{\frac{A^N}{N!}\left(\frac{N}{N-A}\right)}{\sum_{i=o}^{N-1} \frac{A^i}{i!} + \frac{A^N}{N!}\left(\frac{N}{N-A}\right)} \qquad (54)$$

which is Erlang's C-formula.

Compound loss systems.

Suppose that the completion of a switched connection is dependant on finding a path through a complex network of links and switches. There are many possible paths, but since all links are accessible to random traffic from several sources it may happen that all paths are blocked. Let this event be denoted by Y. The probability P(Y) will in general be far from obvious: the number of states and transitions in the network may be so great as to render the simple equilibrium method impracticable. However, let us fix some variable X, for example the number of links busy in a particular group. The conditional probability $P(Y|X=x_i)$ may be much simpler, it may for example reduce to a known result for a single stage of switching. We then sum over all possible outcomes of X (which must of course be disjoint and exhaustive) and calculate

$$P(Y) = \sum_i P(Y|X=x_i)\ P(X=x_i) \tag{55}$$

Clearly if X is a numerical variable this is a type of compound discrete distribution as defined in section 3. We are not confined to this case: the outcomes of X may for example be topological configurations of busy links. However, we take a numerical example.

Consider a connection via two links in tandem. We designate the two stages B and C, and suppose that there are m links in each group. (Figure 4a shows a conventional trunking diagram of a 3-stage switch with m B-switches, which is one realisation of the arrangement postulated). We define probability distributions G(x) that precisely x of the C-links are busy, and H(y) that a nominated set of y B-links are busy (i.e. do not include a free link, so that by definition H(0) = 1). Then the blocking probability over the two stages is

$$P_B = \sum_{x=0}^{m} G(x)\ H(m-x) \tag{56}$$

This method was pioneered by Palm and Jacobeus (50): many specific results of this type are presented by Elldin (51), Syski (47), Bear (48) and other later authors. According to specific situations, the distributions G and H may be of Erlang, Engset and binomial types; the final blocking probability sometimes emerges as a ratio of Erlang probabilities. We present here only one example which is very simple in this context but paves the way for a theory of more complex networks. Suppose that B and C links have binomial occupancy (often valid for well-smoothed traffic) with mean values b and c respectively: then

$$G(x) = \binom{m}{x} c^x (1-c)^{m-x}$$
$$H(m-x) = b^{m-x} \tag{57}$$
$$P_B = (b+c - bc)^m$$

This result can be derived in another way. Let us draw a graph showing all possible paths by which a connection may be accomplished: this is known as a channel graph. Figure 4b shows the channel graph of the present problem: there are m parallel paths each of two links in tandem. Assuming independent occupancy of links, one path is free if both its links are free: so it is blocked with probability

$$1 - (1-b)(1-c) = b+c - bc$$

The channel graph is blocked if all parallel paths are blocked, i.e.

$$P_B = (b + c - bc)^m$$

which is identical with equation (57) above. The method may be extended quite easily to any combination of series-parallel paths.

On the same assumption of independent link occupancy we can calculate blocking probabilities in channel graphs of any form. There are two methods:

(a) Following Lee (52), enumerate the cut-sets of the channel graph, i.e. the sets of links which if deleted will cause blocking, and sum their probabilities.

(b) Define the simplest set of conditions X which reduce the channel graph to something easily calculable, and sum the conditional probabilities, as equation (55). For example, in the bridge graph of Figure 5 take as condition the centre link L free or busy: in either event the conditional channel graph is series-parallel, and one easily shows that for link occupancy p

$$P_B = p(2p-p^2)^2 + (1-p)(2p^2-p^4) \tag{58}$$

Practical switching systems may have up to seven stages of switching, with quite elaborate channel graphs. It is useful to know what form of channel graph, within certain constraints, gives the lowest blocking for a given link occupancy. The solution to this problem is due to Takagi (53,54). An introduction to Takagi's method, and a review of structural problems in switched networks, is given by Cattermole (55).

Delay systems.

One fundamental result for delay systems has already been given, namely Erlang's C-formula, equation (54). This gives the probability that a call is delayed: we would like to know the distribution of delay times. On the assumptions stated, this turns out to be a negative-exponential distribution with mean value $1/(N-A)$, for calls which are delayed. The mean length of queue is $A/(N-A)$.

The assumption of negative-exponential service time is, however, not one which is generally accepted in queueing studies. In communication systems, perhaps the commonest occurrence of a queue is for access to a control processor, whose operations may be much less random than the negative-exponential distribution implies.

As an example, we develop the theory of the single-server queue, with random arrivals, and first-in-first-out order of service, for arbitrary service time distribution $b(x)$. The significant variables are

(i) the time instant t_i when service of the ith call is completed (starting with $t_o=0$).

(ii) the number in the queue q_i at time t_i.

(iii) the service time x_i of call i.

(iv) the number of arrivals r_i during service time of call i, namely during $t_i - x_i < t^i < t_i$.

The equation for the development of the queue is

$$q_{n+1} = r_{n+1} + q_n - U(q_n)$$

where $U(x)$ is the unit-step, equal to 1 if $x>0$ and 0 otherwise. Now the r_i are independent of the q_i, and in equilibrium the distribution of q_i is stationary. So taking expectations, we can write

$$E(z^q) = E(z^{r+q-u(q)}) \tag{60}$$

Let q and r have generating functions

$$Q(z) = E(z^q) = \sum_{i=o}^{\infty} \pi_i z^i \tag{61}$$

$$R(z) = E(z^r) \tag{62}$$

Then from (60)

$$Q(z) = R(z)\ E(z^{q-U(q)}) \tag{63}$$

Now

$$E(z^{q-U(q)}) = \pi_o + \sum_{i=1}^{\infty} \pi_i z^{i-1}$$

$$= \pi_o + \frac{Q(z) - \pi_o}{z} \tag{64}$$

whence

$$Q(z) = \frac{\pi_o (1-z) R(z)}{R(z) - z} \tag{65}$$

π_o can be expressed in terms of r, since the probability of a queue existing is

$$1 - \pi_o = E[U(q)] = E(r) \tag{66}$$

the right hand equation following on taking expectations of equation (59). So we have the distribution of q in terms of the distribution of r. The latter can easily be expressed in terms of the service time distribution b(x). For the arrivals in a randomised time interval can readily be computed, along the lines of our derivation of equations (9) and (10). If the arrivals are a Poisson process with mean rate A, then

$$P(r=k) = \int_o^{\infty} \frac{(Ax)^k e^{-Ax}}{k!}\ b(x)\ dx \tag{67}$$

$$R(z) = \int_o^{\infty} e^{-Ax+Axz}\ b(x)\ dx \tag{68}$$

If we know the Laplace transform B(s) of b(x), then

$$R(z) = B\ (A-Az) \tag{69}$$

The method can be extended to give a variety of specific results, and also some general formulae of interest, for example the Pollaczek formula

$$E(w) = 1 + \frac{A\{1+V(x)\}}{2(1-A)} \tag{70}$$

relating the mean waiting time to the variance of service time.

The foregoing gives only a taste of queueing theory. For a good brief textbook see Cox and Smith (56), and for a longer treatise

Kleinrock (57). Many interesting results are available on queues with access via a switch, or with access only at discrete times: see for example Chung (58,59 and further references there given). Another important problem, if real messages rather than tokens are to be stored, is that of compound rather than simple Poisson arrivals, for which Chu (60) gives both a theory and many computed results.

Time assignment speech interpolation.

We have remarked in connection with multiplexing that speech circuits generally show an activity ratio of about 0.35. By allotting a circuit only to an active user, the sharing of C channels between n(>c) users should be possible. This was first accomplished in the TASI system used on transatlantic telephone links.

Bullington and Fraser (61) give a description of the principle and a statistical analysis of the load on a TASI system. Since at any time a definite number of users n are allotted time intervals on the c channels, the probability that too many users will demand a channel is a cumulative binomial distribution

$$P_B = \sum_{x=C}^{n} \binom{n}{x} \rho^x (1-\rho)^{n-x} \qquad (71)$$

where ρ is the activity ratio (taken for simplicity as a constant, although this too has a probability distribution: see ref. 61, Figure 1). In the numerical range of interest, this can be approximated by the normal distribution. The simple conclusion is that, for a "freeze-out" probability of 1% the number of trunks served should be bounded by

$$n\rho + \sqrt{n\rho} < c \qquad (72)$$

The problem is of course complicated by the temporal distribution of speech bursts, and by the speed limitations of speech activity detectors. However, later studies (62,63) have shown that systems dimensioned more or less in accordance with this rule give good speech quality: for example, 74 trunks can be served by 36 speech channels plus a control channel.

6. FURTHER PROBLEMS IN MULTIPLE COMMUNICATION

The last two sections have elaborated on some basic topics in multiple communication which combine long practice with extensive theory. We conclude with a brief survey of some further topics and problems.

Multiple-access networks.

The switched telephone network is the obvious example, providing a century of practical experience. There is, however, rather little traffic theory devoted to the problems of complete networks, rather than single centres. Much network dimensioning is based on the statistics of routes and centres taken singly. The problem of alternate routing has been studied (48), and efficient dimensioning of overflow routes is well established. Practical planning of multi-exchange networks is now commonly based on computer modelling more than on general theory. Some important practical problems remain unsolved. For example, what is the loss probability of a telephone call proceeding via several tandem exchanges? If the loading were known to be independent we could use the theory of compound losses in section 5. However, it is known that this does not always match practical statistics very well: the practical loss probability may be lower, which suggests a significant correlation.

Data networks of the store-and-forward type have received a good deal of study. Kleinrock (64) has established that the overall mean delay of such a network is simply related to the delay distributions at the several nodes; this work also gives a good general review of the problem.

Random-access systems.

There have been numerous proposals for the provision of communications to many users by giving them more or less unconstrained access to a common medium, and distinguishing the signals by allotment of different quasi-orthogonal functions as carriers. The intention is that, as traffic demand increases, no user is blocked or "frozen out" but each observes a gradual degradation in signal quality. The purpose is to match a particular class of application: to quote White (65),

> "Asynchronous multiplex techniques seem most applicable in cases where the nature of the service is such that the bandwidth tends to be much wider than is justified by the information to be transmitted over a single channel, and where circumstances make synchronous multiplex difficult or impossible".

This places the emphasis on radio communication between users who are either mobile, or sparsely distributed. Other authors including Leakey (66) have urged that gradual degradation is itself a desirable quality.

In such circumstances, the signals can only be quasi-orthogonal, since for all orthogonal functions other than sinusoids precise time alignment is essential to orthogonality. The signal sets proposed have usually been

(i) Potentially orthogonal codes such as Walsh functions (67), used in an asynchronous and therefore quasi-orthogonal manner.
(ii) Pseudo-random binary sequences (68) or sections thereof.
(iii) Barker codes (69), which were designed to have an impulsive correlation function.

Several accounts and analyses of such systems may be found in references (70-75).

There would seem to be considerable scope for probabilistic analysis in such systems. In particular, one would like to know the distribution of degradation (errors, noise, or whatever) in a typical channel as more channels come into use. Judge (70) estimates a mean signal/noise ratio for a certain class of codes. Titsworth (71) estimates that for the class of code he considers, there is a noise penalty of some 2dB compared with ideal orthogonal signalling. Gordon and Barrett (73), along with several other authors, consider primarily the error rate as a function of noise, with the quasi-orthogonal signals presumably contributing to the "noise". Remarkably many writers on this subject give no substantial consideration to the problem. If such systems are to be usefully employed, a good deal of study remains to be done.

Information theory of multiple usage.

The beginnings of information theory, and a good deal of the later development, apply to an isolated channel which, as we have suggested, rarely exists in practice. In recent years, however, several theorists have turned their attention to multiple usage of various kinds. We shall not treat this topic here, but merely refer to the very useful review articles by Wolf (76), Wyner (77), and van der Meulen (78).

ADDENDUM

Several other contributions to this Advanced Study Institute are relevant to the present topic. See, in particular: Wolf, Ephremides, and Segall on data networks: Haber and Ince on code-division multiple access: Bremaud on queueing theory: and Wolf on information theory of multiple usage.

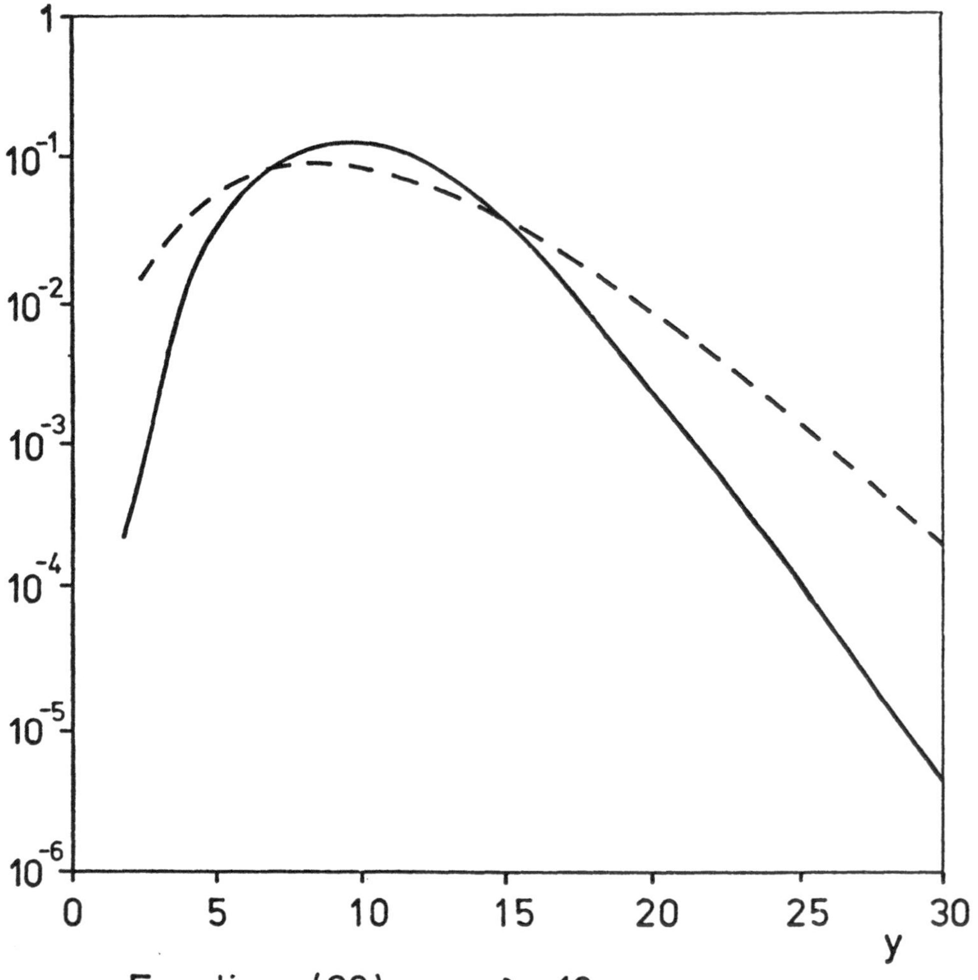

——— Equation (20) $\lambda = 10$

- - - - Equations (18 19) $\lambda = 10$

Figure 1

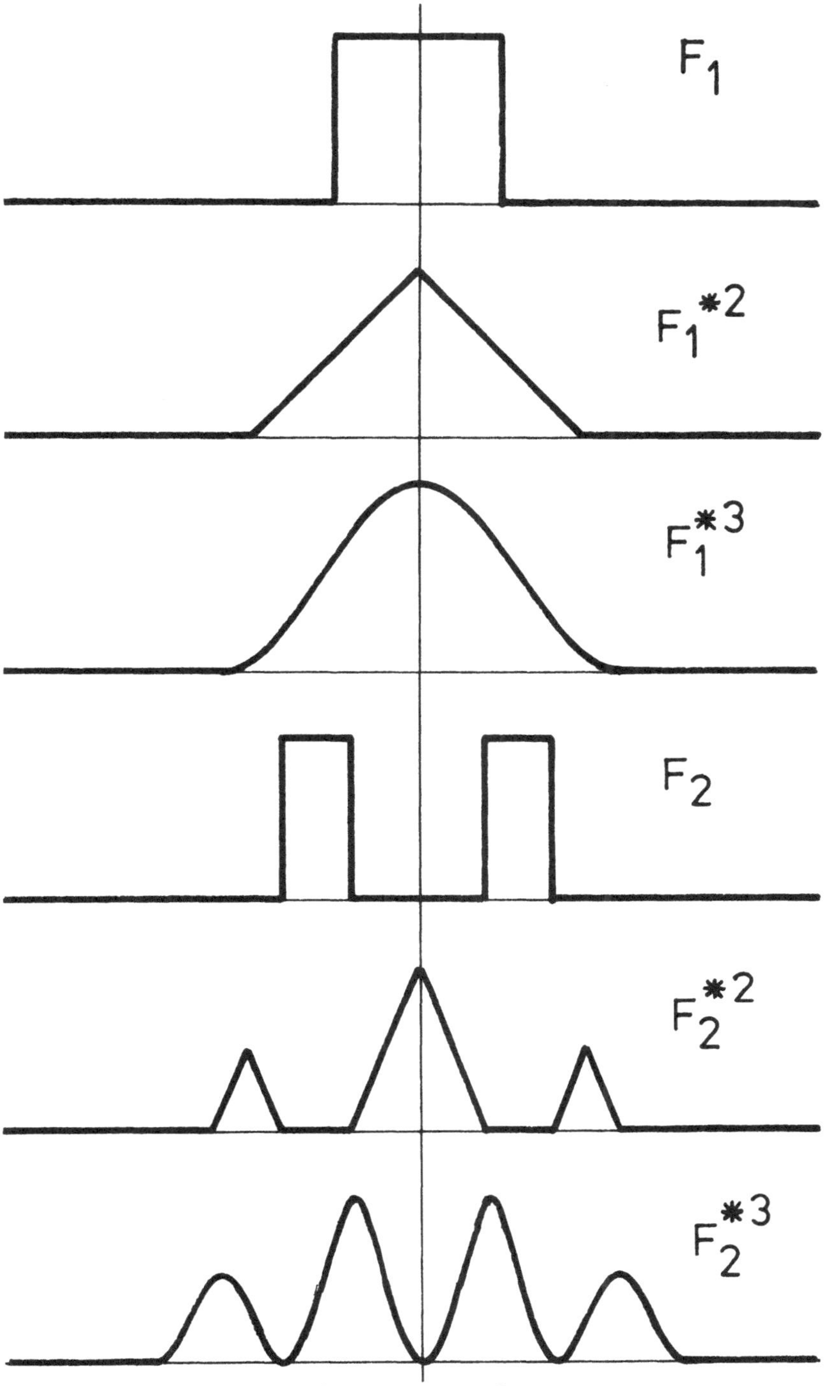

Figure 2

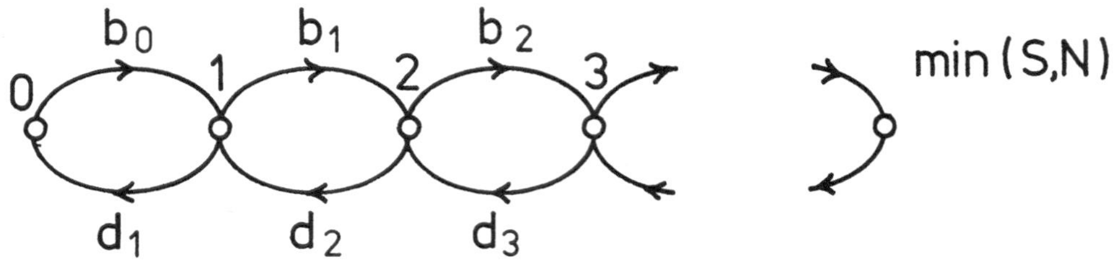

Figure 3

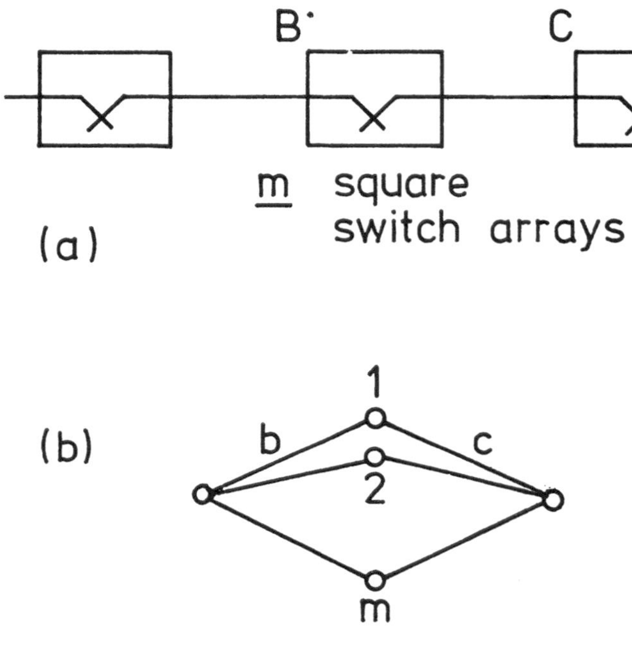

Figure 4

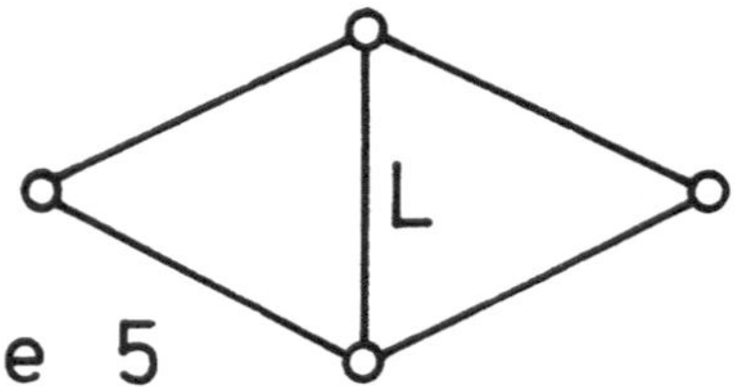

Figure 5

REFERENCES

(1) Landon, V.D.: Theoretical analysis of various systems of multiplex transmission. RCA Review, 9, 288-351 & 433-82, (1948).

(2) Flood, J.E.: Principles of multiplex communication. Nato Advanced Study Institute 1974: proceedings published as J.K. Skwirzynski (ed.). New directions in Signal Processing, Noordhoff, Leyden 1975.

(3) Lighthill, M.J.: Fourier analysis and generalised functions. Cambridge University Press, 1960.

(4) Schwartz, L.: Théorie des distributions. Hermann, Paris, 1966.

(5) Wilkinson, R.: Theories for toll traffic engineering in the U.S.A., Bell System Tech. J. 35, 421-514 (March 1956).

(6) Moran, P.A.P.: Introduction to probability theory. Oxford University Press, 1968.

(7) Feller, W.: Introduction to probability theory and its applications. Wiley, New York. Vol.I 2nd edition, 1957- Vol.II, 1966.

(8) Parzen, E.: Stochastic processes. Holden-Day, San Francisco, 1962.

(9) Picinbono, B.: Optical signals and compound stochastic processes (as ref.2).

(10) Flood, J.E. & Tillman, J.R. : Crosstalk in amplitude-modulated t.d.m. systems, Proc. IEE, 98,III, 279 (1951).

(11) Flood, J.E.: Crosstalk in t.d.m. systems using pulse position and pulse length modulation. Proc. IEE, 99, IV, 64 (1952).

(12) Cattermole, K.W.: Principles of pulse code modulation, Iliffe, London, 1970.

(13) Richards, D.L.: Telecommunication by speech. Butterworth, London, 1973.

(14) Purton, R.F.: A survey of telephone speech-signal statistics and their significance in the choice of a p.c.m. companding law. Proc. IEE, 109B, 60, (1962).

(15) McAdoo, K.L.: Speech volumes on Bell message circuits - 1960 survey. Bell System Tech. J. 42, 1999-2012 (1963).

(16) Aitchison, J. & Brown, J.A.C.: The lognormal distribution, with special reference to its uses in economies. Cambridge University Press (1957).

(17) C.C.I.T.T.: White Book Vol. III, Recommendation G223.

(18) Holbrook, B.D. & Dixon, J.T.: Load rating theory for multi-channel amplifiers. Bell System Tech. J., 17, 624-44 (1938).

(19) Medhurst, R.G.: Distortion in microwave trunk radio systems. G.E.C. Journal, 34, 75-83 (1967).

(20) Rice, S.O.: Mathematical analysis of random noise. Bell System Tech. J., 23, 282-332 (1944) and 24, 46-156 (1945). Reprinted in N. Wax (ed): Selected papers on noise and stochastic processes, Dover, New York, 1954.

(21) Bennett, W.R.: Cross-modulation requirements on multi-channel amplifiers below overload. Bell System Tech. J., 19, 587-610, (1940).
(22) Bell Telephone Laboratories: Transmission systems for communications. Third edition, 1964.
(23) Barrett, J.F. & Lampard, D.G.: An expansion for some second-order probability distributions, and its application to noise problems. Trans. I.R.E., IT-1, 10-15 (1955).
(24) Blachman, N.M.: The signal X signal, noise X noise and signal X noise output of a nonlinearity. Trans. IEEE, IT-14, 21-27, (1968).
(25) Blachman, N.M.: The uncorrelated output components of a nonlinearity. Trans. IEEE, IT-14, 250-55 (1968).
(26) Blachman, N.M.: Detectors, bandpass nonlinearities, and their optimization: inversion of the Chebyshev transform. Trans. IEEE, IT-17, 398-404, (1971).
(27) Lever, K.V. and Cattermole, K.W.: Quantizing noise spectra. Proc. IEE, 121, 945-54 (1974).
(28) Laning, J.H. & Battin, R.H.: Random processes in automatic control. McGraw-Hill, New York, 1956.
(29) Lewin, L., Muller, J.J. & Basard, R.: Phase distortion in feeders: Wireless Engineer, 27, 143, (1950).
(30) Clayton, F.M. & Bacon, J.M.: Intermodulation distortion in f.m./f.d.m. trunk radio systems in 2-path fading situations. Proc. IEE, 117, 359-68 (1970).
(31) Murphy, J.V.: Intermodulation and a.m./p.m. distortion in f.m./f.d.m. radio systems during 2-path propagation. Proc.IEE, 119, 41-7 (1972).
(32) Westcott, R.J. : Investigation of multiple f.m./f.d.m. carriers through a satellite t.w.t. operating near to saturation. Proc. IEE, 114, 726-40 (1967).
(33) Mathews, M.V.: Extremal coding for speech transmission, Trans. IRE, IT-5, 129-36 (1959).
(34) Potier, G.X.: Pulse-space modulation. Onde Electrique, 35, 159-64, (1955).
(35) Cattermole, K.W.: Problems and opportunities in digital transmission, IEEE International Conference paper, 1968.
(36) Bylanski, P. & Ingram, D.G.W.: Digital transmission systems. Peter Peregrinus, London, 1976.
(37) Hennelbarger, T.C. & Fagen, M.D.: Comparative transmission characteristics of polyethylene insulated and paper insulated communication cables. AIEE Communication and Electronics, 81, 27, (1962).
(38) Aaron, M.R.: PCM transmission in the exchange plant. Bell System Tech. J., 41, 99, (1962).
(39) Jachimowitz, L., Eager, E.S., Kolodny, L. & Robinson, D.E.: Transmission properties of polythene insulated telephone cables at voice and carrier frequencies. AIEE Communications and Electronics, 78, 618, (1959).
(40) Gumbel, E.J.: Statistics of Extremes, Columbia University Press, New York, 1958.

(41) Jacobsen, B.B.: Thermal noise in multi-section radio links. Proc. IEE, 105C, 139-150, (1958).
(42) Cramer, H.: Mathematical methods of statistics. Princeton University Press, 1946.
(43) Cravis, H. & Crater, T.V.: Engineering of T1 carrier system repeatered lines. Bell System Tech. J. 42, 431-86, (1963).
(44) Narayana Murthy, B.R.: Crosstalk loss requirements for p.c.m. transmission, Trans. IEEE, COM-24, 88-97.
(45) Jacobsen, B.B.: Cable crosstalk limits on low-capacity p.c.m. systems. Electrical Communication, 48, 98-107.
(46) Brockmeyer, E., Halstrom, H.L. & Jensen, A.: The life and works of A.K. Erlang, Copenhagen Telephone Company, 1948.
(47) Syski, R.: Introduction to congestion theory in telephone systems. Oliver & Boyd, 1960.
(48) Bear, D.: Principles of telecommunication traffic engineering, Peter Peregrinus, London, 1976.
(49) Kosten, L.: Stochastic theory of service systems. Pergamon Press, Oxford, 1973.
(50) Jacobeus, C.: Blocking computations in link systems. Ericsson Review, 3, 86-100 (1947).
(51) Elldin, A.: Automatic telephone exchanges with crossbar switches: switch calculations general survey. L.M. Ericsson, Stockholm, 3rd edition, 1967.
(52) Lee, C.Y.: Analysis of switching networks. Bell System Tech. J., 34, 1287-1315, (1955).
(53) Takagi, K.: Design of multi-stage link systems by means of optimum channel graphs. Electronics and Communications in Japan, 51A, (4), 37-46, (1968).
(54) Takagi, K.: Optimum channel graph of link system. Electronics and Communications in Japan, 54A, (8), 1-10, (1971).
(55) Cattermole, K.W.: Graph theory and the telecommunications network. Bulletin I.M.A. 11, 94-106 (1975). A revised and extended version is due for publication in R. Wilson and L. Beineke (eds.), Applications of Graph Theory, Academic Press, New York (1978).
(56) Cox, D.R. & Smith, W.L.: Queues. Methuen, London (1961).
(57) Kleinrock, L.: Queueing Systems (2 vols.) Wiley, New York, 1975.
(58) Chung, W.K.: Computer-controlled queueing systems with switching network. Proc. IEE, 122, 259-61 (1975).
(59) Chung, W.K.: Computer-controlled queueing system with general access-cycle times. Proc. IEE, 122, 262-264, (1975).
(60) Chu, W.W.: Buffer behaviour for batch Poisson arrivals and single constant output. Trans. IEEE, COM-18, 613-18 (1970).
(61) Bullington, K. & Fraser, J.M.: Engineering aspects of TASI. Bell System Tech. J., 38, 353-364, (1959).
(62) Fraser, J.M., Bullock, D.B. & Long, N.G.: Overall Characteristics of a TASI system. Bell System Tech. J., 41, 1439-54, (1962).

(63) Miedema, H. & Schachtman, M.G.: TASI quality: effect of speech detectors and interpolation. Bell System Tech. J., 41, 1455-73, (1962).
(64) Kleinrock, L.: Communication nets: stochastic message flow and delay. McGraw Hill, 1964: reprinted Dover, New York, 1972.
(65) White, W.D.: Theoretical aspects of asynchronous multiplexing, Proc. IRE, 38, 270-5, (1950).
(66) Leakey, D.: New techniques in multiplexing, GEC journal of Science and Tech. 40, 59-64, (1973).
(67) Harmuth, H.F.: Transmission of information by orthogonal functions. Springer Verlag, New York, 1972.
(68) Golomb, S.W.: Shift register sequences, Holden-Day, San Francisco, 1967.
(69) Barker, R.H.: Group synchronizing of binary digital systems. Ch.19, of Willis Jackson (ed.), Communication Theory, Butterworth, London, 1953.
(70) Judge, W.J.: Multiplexing using quasi-orthogonal binary functions. Trans. AIEE Communications and Electronics 81, 81-3, (1962).
(71) Titsworth, R.C.: A Boolean-function multiplexed telemetry system. Trans. IEEE, SET-9, 42-5, (1963).
(72) Corr, F., Crutchfield, R. & Marchese, J.: Pulsed pseudo-noise VHF radio set. IEEE Communication Convention Records, (1965), pp.143-151.
(73) Gordon, J.A. & Barrett, R.: Correlation-recovered adaptive majority multiplexing. Proc. IEE, 118, 417-422 (1971).
(74) Harmuth, H.F.: Asynchronous filters and mobile radio communication based on Walsh functions. Trans. IEEE, EMC-13, 210-18, (1971).
(75) Miller, L.S. & Thompson, W.: Use of pseudo-noise concepts in mobile communications. IEE Conference on Signal Processing Methods for Radio Telephony, (1970), pp.150-9.
(76) Wolf, J.K.: Multiple-user communications, 1973 National Telecommunications Conference Record, (2), 28E1-11.
(77) Wyner, A. : Recent results in the Shannon theory. IEEE Trans. IT-20, 2-10 (1974).
(78) van der Meulen, E.C.: A survey of multi-way channels in information theory, 1961-76, IEEE Trans. IT-23, 1-37, (1977).
(79) Garrison, G.J.: Intermodulation distortion in f.d.m./f.m. systems - a tutorial summary. IEEE Trans. COM-16, 289-303, (1968).

MULTI-USER COMMUNICATION NETWORKS*

Jack Keil Wolf

Department of Electrical and Computer Engineering
University of Massachusetts, Amherst, MA., USA

ABSTRACT. Most of the work in communications has been concerned with a single transmitter sending information to a single receiver. Satellite communication systems, computer communication networks and other systems involving multiple users, recently has caused communication system designers to focus on the problem of the simultaneous transmission of information amongst several terminals. In this paper, some of the theoretical results in multi-user communication systems are reviewed. Then some ad hoc techniques for attempting to achieve the promises of these theoretical results are considered.

1. INTRODUCTION

Most of the work in communications has focused on a model of a communications channel where one transmitter is communicating with one receiver. A startling result of information theory [1-2] which applies to this model is that noise and other disturbances on the channel does not limit the <u>reliability</u> by which digital data can be transmitted but rather only limits the <u>rate</u> at which data of arbitrarily high reliability can be transmitted. The highest rate at which such reliable data can be transmitted is known as the <u>capacity</u> of the channel. Information theory supplies us with formulae for calculating this capacity. One practical difficulty is

---

*This research was supported by the United States Air Force, Office of Scientific Research under Grant AFOSR-74-2601.

that it may be very expensive (in both cost and delay) to achieve a very low error probability in transmission.

Satellite communication systems, computer networks and other communication systems involving multiple-users, recently has caused communication system designers to focus on the problem of the simultaneous transmission of information amongst several terminals over a common communications channel. Once again information theorists have shown that many users can communicate data with arbitrarily low error probability over a noisy communications channel (where there is cross-talk as well as other disturbances) provided that the rates for the individual data streams satisfy certain inequalities. The set of rates at which simultaneous reliable transmission is possible is called the capacity region for the channel. For certain configurations of transmitters and receivers, this capacity region is known. For others the deviation of formulae for the capacity region remains an open problem.

It is the purpose of this paper to first briefly review some of the problems which are inherent to communication networks which did not appear in the single user case. Then we will focus on one of these problems—namely the random access communications problem. Next we shift to a discussion of multi-user communication channels viewed from the standpoint of an information theorist. We will note that although the problems considered by the information theorist have some similarities to those plaguing the communication network designer, there are essential differences between the two sets of problems. The remainder of the paper is concerned with ameliorating these differences.

2. COMMUNICATION NETWORKS

A communication network is a system capable of simultaneously transmitting the information of many users from various geographic locations to other geographic locations. The telephone system is one familiar example of such a network. The aim of the communication network designer is to synthesize a system whereby the many users can efficiently share the resources of the network.

We are concerned here with data communication networks where the information to be transmitted is in digital form and where we must convey this information with high reliability. The demand for data networks stems largely from communications to and from (and between) computers. Since the demand existed before special networks could be designed to satisfy this need, the switched telephone network was employed initially for this purpose. Although the switched telephone network is capable of transmitting digital data, in some cases the telephone network was used in a very inefficient manner. For example traffic between a computer terminal and a computer is usually sporadic and for long periods of time no information is being transmitted. If the terminal is connected to the computer via a link of the switched telephone network, that link (including the switching mechanisms, physical wires, etc.) cannot be used by others during the idle periods. Thus the inefficiency .

Recent interest has focused on the design of special networks tailored for the transmission of digital data. A good overview of the problems associated with the design of such networks is given in the new book by Schwartz [2]. The subject is too broad to adequately review here; instead, we give a very brief description of some of the interesting problems which are inherent to the design of data networks which did not occur for a single-user system.

The basic design problem is that of the topology of the network. That is, given the geographic locations of the transmitters and receivers and the characteristics of the traffic to be handled, how should the communication links be established between these locations such that a cost effective system results. Many possibilities exist. One could have one common communications link (such as a satellite repeater) which carries the traffic for the entire network. At the other extreme one could establish a separate link between every transmitter and every receiver. One could have links connecting only some of the nodes of the network. One could add new nodes to accommodate concentrators which act as "data smoothers". These concentrators take many low duty rate data streams and produce a single stream with a higher duty rate (i.e. fewer idle periods) which can then be more efficiently transmitted over a communications link. Not only must one consider what links should be established but one must specify the transmission rates (in bits/second) that can be supported by these links.

Once a topology has been established, new problems arise. If a message can be transmitted to a destination via two or more different paths than the routing of the messages must be considered. Flow control is concerned with strategies which prevent the network from becoming over loaded and which allow the network to recover from an overload if one were to occur.

The basic mathematical tool of analysis appears to be queueing theory. For most systems, messages cannot be instantaneously transmitted upon generation but rather must wait for service in queues. Questions of the proper lengths for queues, the average waiting time and throughput become essential parameters of the system. Since transmission channels are not noise free, the possibility exists of errors occurring in transmission, either due to extraneous noise on the channel or due to interference from other transmissions. Schemes for controlling the reliability of transmission need be considered and the effects of these schemes on the queueing delays need be established.

In the next section we consider a specific technique for transmission over a common communications channel, such as a satellite repeater, by many low duty cycle users. The analysis is not new but serves as a basis for the sections following which treat some information theoretic aspects of the multi-user communications channel.

3. A RANDOM ACCESS TECHNIQUE (THE ALOHA SYSTEM) [4]

We consider a system whereby M active users are attempting to transmit via a synchronous satellite which acts as a repeater of all signals received. We assume that each transmitter is transmitting packets of data, each packet of duration τ seconds. For simplicity we will assume that the rate of transmission of packets is the same for all active users. (This assumption is not critical to the analysis.) We assume that the statistics of the message packets at the satellite is governed by a Poisson process. That is, considering only the start times of the packets, we assume that at the satellite the probability of having exactly j message packets start in time T is given by the expression

$$P[\text{j message packets start in time T}] = \frac{(\lambda T)^j}{k!} e^{-\lambda T}, \quad j=0,1,2\ldots .$$

The parameter λ of the Poisson process is the average number of message packets per unit time. When we have both transmissions of message packets and retransmissions of message packets due to interference which occurred in the original transmission, then we will assume that the start times of the totality of the packets also are governed by a Poisson process. Later we will find a relationship between the average number of message packets and average number of message plus retransmission packets for the entire network.

We will consider two different modes of transmission called pure ALOHA and slotted ALOHA [5]. In a pure ALOHA system whenever an active user has a packet to be transmitted, he transmits it (irrespective of what the other users are doing). In a slotted ALOHA system, the time scale is segmented into slots of τ seconds duration and when a packet is available for transmission at a transmitter the transmitter waits until the beginning of the next time slot and transmits it in this slot. Figure 1 shows the situation at the transmitter for a 2 user system for both a pure ALOHA and a slotted ALOHA system. It should be noted that in both the pure

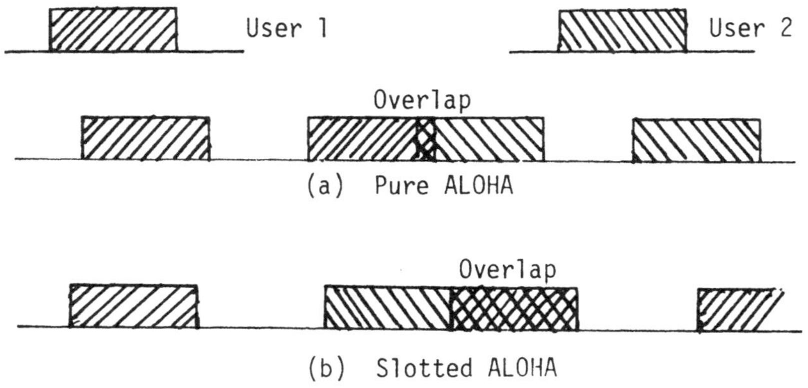

Figure 1. Packets for Pure ALOHA and Slotted ALOHA Transmission

ALOHA and the slotted ALOHA systems packets can overlap. In the pure ALOHA the overlap can be partial while in the slotted ALOHA either the entire packets overlap or they do not overlap at all. It is assumed that any packets that overlap are retransmitted by their respective transmitters. Of course, the retransmission packet can again overlap with another packet so that several retransmissions may be required in order to get the message through. It is important that the transmitters vary the delay in retransmitting packets which interfered with one another, since if they used the same delay the retransmitted packets would certainly overlap again.

We now focus on the packets as they occur at the satellite rather than at the transmitters. This allows us to characterize the entire system performance rather than the performance of a single transmitter. We consider first a pure ALOHA system.

Assume that the number of message packets received at the satellite is governed by a Poisson point process. Also assume that the number of messages plus retransmission packets received by the satellite is governed by a Poisson point process (with a different parameter). Specifically for $j = 0,1,2, \ldots$, let

$$P[j \text{ message packets start in time } T] = \frac{(rT)^j}{j!} e^{-rT}, \text{ and}$$

$$P[j \text{ message plus retransmission packets start in time } T] = \frac{(RT)^j}{j!} e^{-RT}.$$

Thus r is the average number of message packets per unit time and R is the average number of message plus retransmission packets per unit time. Let $t \triangleq R-r$ be the average number of retransmission packets per unit time and let P_t be the probability of a packet being retransmitted. Then $P_t = t/R$.

Consider a packet that starts at t_0. It will not be interfered with by any other packet if and only if no other packet arrives at the satellite with a start time in the interval $(t_0-\tau, t_0+\tau)$. The probability, P_t, that it is interfered with is $P_t = 1 - P$ [0 packets start in time 2τ] $= 1 - e^{-2R\tau}$. Substituting we find

$$R - r = (1 - e^{-2R\tau})R$$

or

$$r = Re^{-2R\tau}.$$

If packets could be packed, one after the other with no overlap and no wasted intervals we would have $R\tau = 1$. The quantity $R\tau$ is known as the channel traffic while the quantity $r\tau$ is known as the channel utilization. The relationship between these two quantities is then

$$(r\tau) = (R\tau)\ e^{-2R\tau}$$

and is plotted as one of the curves in Figure 2.

We next consider the slotted ALOHA system. In order to use the previous analysis we pretend that the transmitters transmit the signals to the satellite as in the pure ALOHA case and that the satellite puts the packets into their appropriate slots. (Of course, the slotting is actually performed by the transmitters.) Then all of the previous analyses hold except now a packet which is transmitted in the slot beginning at time t_0 will be interfered with by packets which start in the interval $(t_0 - \tau, t_0)$. (Packets which arrive after t_0 will be delayed to the next slot.) Then $P_t = 1 - e^{-R\tau}$ and our final result will be

$$(r\tau) = (R\tau)\ e^{-R\tau}.$$

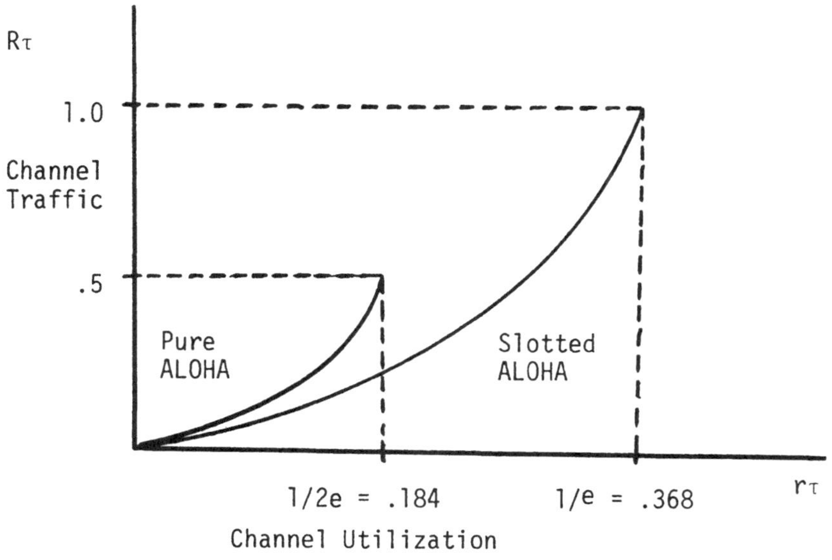

Figure 2. Performance of Pure and Slotted ALOHA

This result is also shown in Figure 2. It is seen that the maximum utilization is doubled for the slotted case over the pure case.

Metzner [6] and others have suggested a modification of the above system that increases the maximum utilization of the system. Metzner suggests that two classes of users be established with one class having much larger power than the other class. Then if a packet of high power and one of low power overlap it is assumed that the high power packet can be received without error while the low power packet need be retransmitted. All other assumptions remain the same regarding the requirement of retransmission of packets. (That is, if two high power packets overlap, both need be retransmitted. Also if two low power packets overlap, both need be retransmitted.) Metzner showed that the maximum channel utilization is now approximately 0.53 for the slotted ALOHA case (where it is 0.368 for a single power system). By using Q classes of users, all with different powers where any packet of a given power dominates all packets of less power, the total average utilization can be made to approach 1 as $Q \to \infty$. The convergence is slow and for Q = 18, 90 percent efficiency is achieved.

In the previous discussion, a basic assumption was made that if two packets collided they could not be individually resolved at the transmitter. For the case of equal powers, this assumption led to the belief that all packets involved in collisions need be retransmitted. For the unequal power case, higher power packets could survive collisions with lower power packets but one could not assume that both packets involved in a two-packet collision could survive. In a later section we will consider a model of a linear satellite repeater where the output signal is the sum of all input signals. The output signal can now take on many levels. We will show that for such a channel, the channel utilization is not upper bounded by 1. Furthermore we will give some simple codes which allow reliable transmission at channel utilizations exceeding 1.

4. MULTI-TERMINAL COMMUNICATION CHANNELS AND INFORMATION THEORY

The mathematical theory of communications, i.e., information theory, almost exclusively has been concerned with the reliable transmission of information from a single information source to a single information sink. The basic concepts for these analyses were contained in the 1948 papers of Claude Shannon [1]. Thirteen years later, Shannon [7] gave the beginnings of a theory for multi-terminal networks but this subject did not receive much attention until about 1970 when a series of papers emerged analyzing various configurations of multi-terminal communication channels and sources. Several survey articles on this subject have now appeared with extensive bibliographies [8-10].

The general multi-terminal communications channel is shown in Figure 3 where K information sources attempt to transmit digital data reliably over a common communications channel to L information sinks. The channel has P inputs which transmit the signals generated by P encoders. The channel has L outputs which serve as inputs to L decoders. The output of these decoders feed L information sinks. The box labeled T is a switch which connects certain of its inputs to certain outputs. Each decoder attempts to reproduce a specified set of the source sequences. The choice of which decoders reproduce which source sequences and the choice of the connections introduced by the switch $\underset{\sim}{T}$ result in a particular multi-terminal system to be analyzed. Some of these problems have been solved while others remain open questions at this time.

A more general problem can be formulated than that shown in Figure 3 where the decoders have as additional inputs certain outputs of the sources and/or the encoders have as additional inputs certain outputs from the channel. Indeed, the problem analyzed by Shannon in 1961 was of this more general form. Returning to Fig. 3 we assume that source i (SO_i) produces a stream of binary data (equally probable, statistically independent, binary digits) at a rate of R_i bits/time unit (i = 1,2, ..., K). The jth decoder (DEC_j) attempts to reliably reproduce certain of these data streams in order to supply this information to the jth information sink

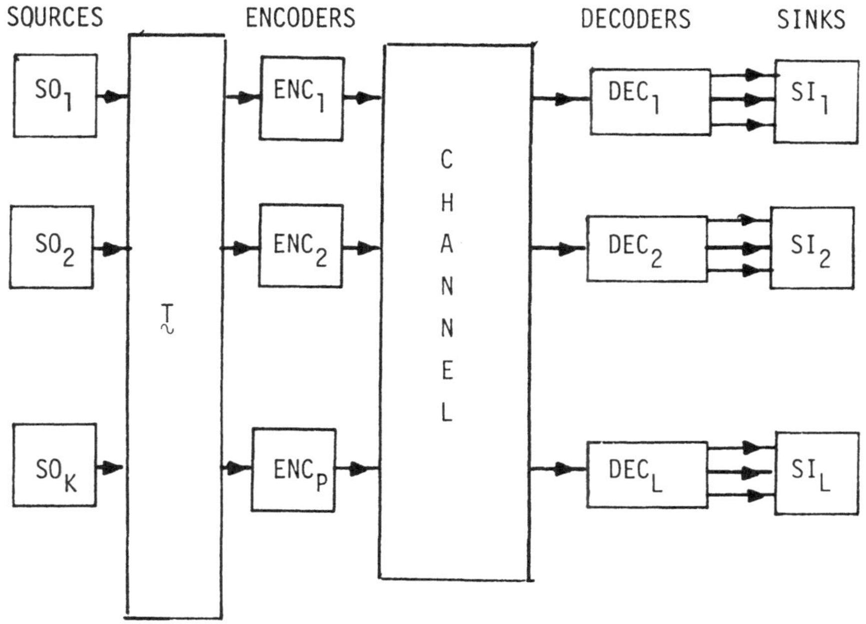

Figure 3. General Multi-Terminal Communications Channel

(SI_j, j = 1,2,...,L). A rate point $\underline{R} = (R_1,R_2,\ldots,R_K)$ is said to be achievable if and only if there exists encoders and decoders such that the probability of error in the information streams supplied to the sinks can be made as small as desired. (That is, for an achievable rate point, all the information supplied to the sinks are accurate reproductions of the information generated by the sources.) The set of all achievable rate points is called the capacity region for the system.

The capacity region is a convex region in the K-dimensional space of rate points. The boundaries of this region may be complicated curves. Thus the region is typically not just a constraint on the sum of the rates. However when we relate this work to the previously discussed random access channel, we will be interested in the largest value that the sum of the rates can achieve while in the capacity region since it is this sum that corresponds to the previously mentioned channel utilization.

We now consider a special case of the general multi-terminal communications channel as depicted in Figure 4. This system has been called the multi-access communications channel. Here there

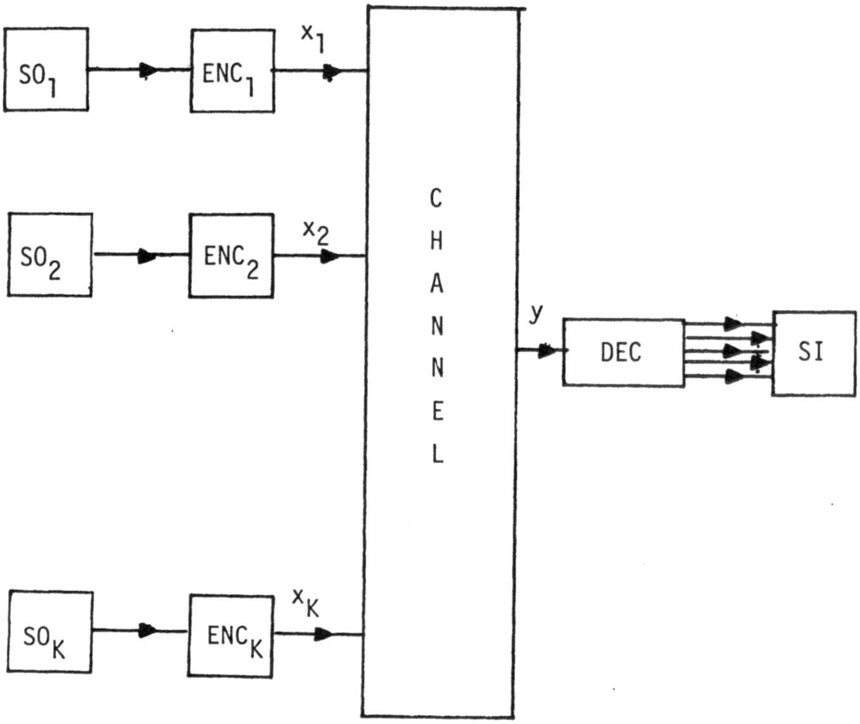

Figure 4. Multi-Access Communications Channel

are K message sources, each connected to one and only one encoder. The channel has K inputs and one output. (We will call the inputs x_1, x_2, ..., x_K and the output y.) The output of the channel is connected to a single encoder whose task is to reliably reconstruct all K message streams and furnish these streams to the information sink.

We assume that the inputs and outputs of the channel are discrete random variables and that the channel is described by the conditional probability distribution $P_{Y|X_1X_2...X_K}(y|x_1,x_2,...,x_K)$. For any joint probability on the inputs to the channel, $Q_{X_1X_2..X_K}(x_1,x_2,...,x_K)$, we can calculate the information theoretic quantities

$$I(X_1X_2,\ldots,X_K;Y) \triangleq E\left[\log_2 \frac{P_{Y|X_1\ldots X_K}(Y|X_1X_2\ldots X_K)}{P_Y(Y)}\right],$$

$$I(X_2,X_3,\ldots X_K;Y|X_1) \triangleq E\left[\log_2 \frac{P_{Y|X_1X_2\ldots X_K}(Y|X_1X_2\ldots X_K)}{P_{Y|X_1}(Y|X_1)}\right],$$

etc. Here E[ ] denotes the statistical expectation of the quantity in the brackets. The capacity region is given in terms of such quantities [11-13].

In particular, let us consider the special case where K = 2. Then, it has been shown that the rate point $\underline{R} = (R_1,R_2)$ is achievable if (but not only if)

$$0 \le R_1 \le I_Q(X_1;Y|X_2)$$

$$0 \le R_2 \le I_Q(X_2;Y|X_1)$$

$$0 \le R_1 + R_2 \le I_Q(X_1,X_2;Y)$$

where Q is taken to be any product distribution. Furthermore, the capacity region has been shown to be equal to the convex hull of the union of the regions given by the above set of inequalities where the union is taken over all possible product distributions, Q.

We now focus on a particular multi-access channel related to the channel considered in the previous section. This channel has binary inputs and an output which is the linear sum of these inputs. Thus, for this channel

$$P_{Y|X_1X_2\ldots X_K}(y|x_1x_2\ldots x_k) = \begin{cases} 1 & \text{if } y = x_1+x_2+\ldots+x_k \quad x_i\varepsilon\{0,1\} \\ 0 & \text{otherwise} \quad x_i\varepsilon\{0,1\}. \end{cases}$$

The output Y takes on values from the set $\{0,1,\ldots,K-1,K\}$. The capacity region for this channel for the case of K = 2 is the region described by the equations $0 \le R_1 \le 1$, $0 \le R_2 \le 1$, $0 \le R_1 + R_2 \le 1.5$ and is shown in Figure 5. The rates are measured in units of bits per channel use. That is, we assume that every unit of time, the channel is supplied with a pair of binary inputs, one from each encoder and instantaneously produces an output which is the sum of these inputs.

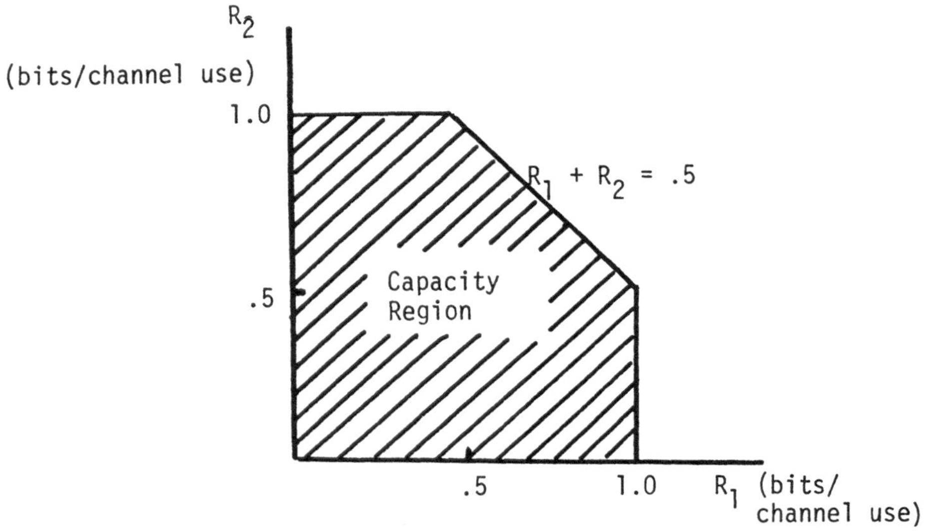

Figure 5. Capacity Region for Two Input Adder Channel

The capacity region can also be found for the case where we have K binary inputs and where the output is the linear sum of these inputs. The capacity region is somewhat more complicated since it is a region in K dimensional space. The largest sum of rates for any achievable rate point in this capacity region can be shown to be given by the equation

$$(R_1 + R_2 + \dots + R_K)_{Max} = - \sum_{j=0}^{K} \frac{\binom{K}{j}}{2^K} \log_2 \frac{\binom{K}{j}}{2^K} .$$

For large K, a good approximation to this expression is $\frac{1}{2} \log_2 (\pi e K/2)$. Some typical values of this quantity are given in the following table:

K	Max. Sum of Rates (bits/channel use)
2	1.500
3	1.815
10	2.708
20	3.208
100	4.369
200	4.869

It should be noted that a sum of rates in excess of 1 bit/channel use corresponds to a channel utilization in excess of 1. This correspondence will be discussed in a later section of this paper.

We now consider the assumptions that were made in order to derive the capacity region of the multi-access channel. The forward part of the coding theorem which established that a given rate point was achievable was based upon a block coding scheme where the encoders produced sequences of channel inputs of length N, called code words. The encoders were assumed to be synchronized with one another in that the various channel inputs were produced in synchronism by the encoders and furthermore that the blocks produced by the encoders were also synchronized. Finally it was assumed that the decoder was synchronized to the encoders both from the standpoint of block synchronization and symbol synchronization.

In a real random access system this assumed synchronism may not be present, *a priori*. Thus, it is not clear that the above results can be applied to the random access channel. In the next section we show that the assumption of synchronism amongst encoders and between the decoder and the encoders can be relaxed and that indeed a channel utilization in excess of 1 can be achieved for a random access linear sum channel without *a priori* synchronization.

5. BLOCK SYNCHRONIZATION, SYMBOL SYNCHRONIZATION, AND CODES

In this section we focus on the two-user multi-access binary-input linear sum channel whose capacity region was given in Figure 5 when block and symbol synchronism was assumed. We begin by retaining the assumption of block and symbol synchronization amongst all encoders and the decoder and we note a simple code of block length N = 2 which achieves the rate point $\underline{R} = (R_1, R_2) =$ $(.5, (\log_2 3)/2) = (.5, .7925)$ with zero error probability. Encoder 1 uses the two code words (0 0) or (1 1) while encoder 2 uses the three code words (0 0), (0 1) or (1 0). Each unique pair of code words gives a unique channel output as seen below.

	0 0	1 1
0 0	0 0	1 1
0 1	0 1	1 2
1 0	1 0	2 1

=

	Code 1
Code 2	Channel Outputs

It should be noted that for this simple code, $R_1 + R_2 = 1.2925$ bits/channel use and thus we are able to transmit more than one bit of information for each use of the channel. This is the case since the decoder is able to resolve conflicts or collisions in the binary digits transmitted. The net effect is a channel utilization in excess of 1 as previously promised. However, as noted, this scheme required synchronization between encoders and between the decoders and the encoder. We now demonstrate codes that do not require such synchronization and yet have a channel utilization in excess of 1.

We first note that if we had attempted to use the above code when the two encoders and decoder were not in block synchronism (but still in symbol synchronism) we would be unable to resolve certain conflicts. As an example of such unresolvable conflicts we note that even if encoder 1 transmits all zeros, concatenations of the code words for encoder 2 cannot be decoded by the decoder if it does not know the phasing of the blocks.

We can give an information theoretic proof that the rate point $\underline{R} = (1.0, 0.5)$ is achievable even if the two encoders and the decoder are not in word synchronization (symbol synchronization is still assumed). Let encoder 1 send an uncoded binary information stream at rate $R_1 = 1.0$ bit/channel use. Encoder 2 then sees a channel which is a binary erasure channel with erasure probability 1/2. (An erasure is synonomous with the output symbol $y = 1$.) The capacity of this channel is 1/2 bit/channel use so that encoder 2 need only use a long block code at a rate slightly less than 1/2 bit/channel use. If the decoder were in block synchronism with encoder 2, then we know that a code exists which can achieve an arbitrarily small error probability in decoding the code words of encoder 2. Once the code words for encoder 2 have been decoded, this sequence can be subtracted from the received sequence to obtain the message transmitted by encoder 1. The question remains how the decoder can obtain block synchronization with encoder 2. This can be achieved by sending a synchronization sequence prior to transmitting messages and will result in a negligible rate loss. Such a proof is only an existence proof, since although it has been shown that such codes must exist, one does not know construction methods for generating these codes.

We next give a constructive coding scheme for achieving zero error probability when neither the encoders nor the decoder are in word synchronism. Again encoder 1 uses the code words (0 0) and (1 1). Encoder 2 uses a code such that in any concatenation of the code words two ones never occur in succession. Examples of such codes and their respective rates are given in the following table:

Block Length	Code Words	Rate
2	(0 0), (1 1)	.500
3	(0 0 0), (0 0 1), (0 1 0)	.528
4	(0 1 0 0), (0 1 0 1), (0 0 0 0) (0 0 0 1), (0 0 1 0)	.580

The codes of block length N + 2 are obtained from the code words of block length N and (N + 1) by appending an initial (0) to every code word of block length (N + 1) and appending an initial (0 1) to every code word of block length N. The number of code words then form a Fibonacci sequence as N increases. Thus the limiting rate of such codes can be calculated as $N \to \infty$. Indeed this rate approaches 0.6942 so that the sum of the rates approach 1.16942.

Let us first show that the decoder can with probability 1 synchronize to the code words of encoder 1. With probability 1 the received pattern . . 0 0 1 1 0 0 . . . will occur after a sufficient delay. The two ones in this pattern could only have been caused by the code word (1 1) of encoder 1. Thus, the decoder can synchronize to encoder 1's code words.

Once the decoder is in synchronism with encoder 1 it can uniquely decompose any received sequence $\underline{Y}$ into $\underline{X}_1$ and $\underline{X}_2$ since it only sees the patterns:

$\underline{X}_1$	0 0	0 0	0 0	1 1	1 1	1 1
$\underline{X}_2$	0 0	0 1	1 0	0 0	0 1	1 0
$\underline{Y}$	0 0	0 1	1 0	1 1	1 2	2 1

Thus all code words can be decoded with zero error probability.

A simpler, very short, code exists for encoder 2 which has a rate almost as large as the limiting rate of the Fibonacci codes. Encoder 2 now has two code words (0) and (0 1). This is a variable length code of rate $R_2 = .666$ so that the sum of the rates is now $R_1 + R_2 = 1.1666$. Again we have achieved a channel utilization greater than 1 without assuming block synchronization of the encoder or decoder.

Finally we want to eliminate the requirement of symbol synchronization between the two encoders and the decoder. We again assume encoder 1 uses the code (0 0) and (1 1) while encoder 2 uses a code where no two ones can appear successively in a sequence of code words. We assume that a 1 is transmitted by sending a pulse of height E volts and duration T while a 0 is transmitted by sending no pulse for duration T. The channel remains a linear adder.

The decoder can again synchronize to encoder 1 by waiting for the waveform shown in Figure 6 which will eventually occur with probability 1. Once synchronized to encoder 1, the decoder merely

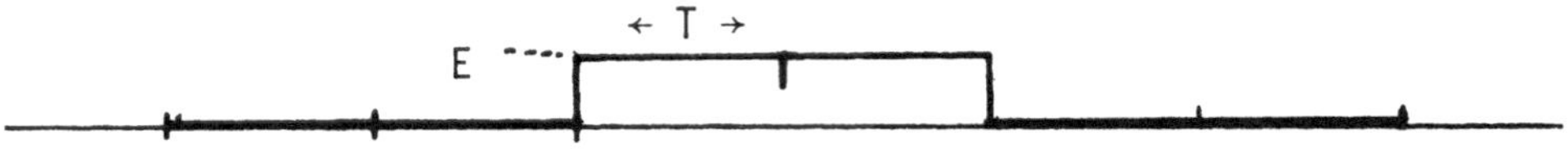

Figure 6. Output Waveform for Synchronization of Decoder to Encoder 1

looks at intervals of length 2T and notes that encoder 1 transmitted the code word (1 1) if and only if the received waveform is greater than or equal to E for this period. Thus the decoder can detect whenever encoder 1 sends the code word (1 1) and can decode the sequence sent by this encoder. Having done this, the decoder subtracts from the received waveform, the transmitted waveform of encoder 1. What remains after subtraction is the transmitted waveform of decoder 2. Thus the decoder can fully decode the information sequences of both sources without any synchronization required on the part of the encoders or the decoder.

REFERENCES

1. C. E. Shannon, "A Mathematical Theory of Communications," BSTJ, Vol. 27, pp. 379-423, 623-656, 1948.

2. R. G. Gallager, Information Theory and Reliable Communications, John Wiley, New York, 1968.

3. M. Schwartz, Computer Communication Network Design and Analysis, Prentice Hall, New Jersey, 1977.

4. N. Abramson, "The ALOHA System," in Computer Communication Networks, N. Abramson and F. Kuo, Ed., Prentice Hall, Englewood Cliffs, New Jersey, 1973, Chapter 14.

5. L. Kleinrock and S. S. Lam, "Packet-Switching in a Slotted Satellite Channel," National Computer Conference, AFIPS Conf. Proceedings, Vol. 42, pp. 703-710, AFIPS Press, 1973.

6. J. J. Metzner, "On Improving Utilizations in ALOHA Networks," IEEE Trans. on Comm. Systems, Vol. COM-24, No. 4, pp. 447-448, 1976.

7. C. Shannon, "Two Way Communication Channels," Proceedings of the Fourth Berkeley Symposium on Mathematical Statistics and Probability, Vol. 1, pp. 611-644, 1961.

8. J. K. Wolf, "Multiple-User Communications," 1973 National Telecommunications Conference Record, Part 2, pp. 28E-1-28E-11, 1973.

9. A. Wyner, "Recent Results in the Shannon Theory," IEEE Trans. on Info. Theory, Vol. IT-20, pp. 2-10, 1974.

10. E. C. van der Meulen, "A Survey of Multi-Way Channels in Information Theory, 1961-1976," IEEE Trans. on Info. Theory, Vol. IT-23, pp. 1-37, 1977.

11. R. Ahlswede, "Multi-Way Communication Channels," Second International Symposium on Information Theory, Tsakadsor, Armenia SSR, 1971.

12. H. Liao, "A Coding Theorem for Multiple Access Communications," 1972 International Symposium on Information Theory, Asilomar, California, 1972.

13. D. Slepian and J. K. Wolf, "A Coding Theorem for Multiple Access Channels with Correlated Sources," BSTJ, Vol. 52, pp. 1037-1076, 1973.

SPREAD SPECTRUM SIGNALS AND BANDWIDTH UTILIZATION

Fred Haber

Systems Engineering Dept., Moore School of Electrical Engineering, University of Pennsylvania
Philadelphia, Pennsylvania

ABSTRACT. An investigation of the bandwidth utilization of spread spectrum systems using binary code division multiple access (CDMA) is reported. Assuming a receiver sees the signals of other users as sources of random binary sequences, it is shown that bandwidth utilization is poorer by a substantial factor than it is in multiple access systems using time division, both systems being fully occupied. Consideration is also given to the case of low fractional occupancy of assigned CDMA channels, and relationships between error probability and fractional occupancy for various user populations are found. The effect of allowing multi-level base-band waveforms to be used is next investigated, and is found to result in poorer bandwidth utilization. Finally investigated is the advantage to be gained by allowing multi-dimensional base-band waveforms to be used. It is found that one can, by this method, retrieve some of the loss incurred in using spread spectrum.

1. INTRODUCTION

The use of spread-spectrum methods to obtain multiple access to a common channel carries with it a bandwidth penalty. The question of how great this cost can be in the case of code division multiple access (CDMA) is addressed. The results are expressed in terms of a merit ratio relating the number of users which can be accommodated using CDMA and the number of users accommodated in a binary time division system using the same chip duration and giving rise to approximately the same probability of error. The paper covers three cases as follows:

(1) binary base-band signals

(2) multi-amplitude base-band signals

(3) orthogonal multi-state base-band signals

For (1) above probability of error is also determined for cases where subscribers utilize their assignments only fractionally.

The transmitted signal is formed by modulating source data by a user code as described, for example, by Harris (ref. 1). Source data rate and user code rate will be denoted $1/T$ and $1/T_s$ binary digits/second, respectively. Both source signals and the user code waveform will be assumed bipolar, the latter being taken as a sequence of positive and negative unit amplitude pulses which multiplies the source waveform. It is convenient to assume that source data and user code are synchronized so that only whole chips of the user code fall into a source signal interval, T. The signal interval, T, therefore contains

$$n = T/T_s \tag{1}$$

chips so that, in effect, source data is being coded into an (n,1) code with a corresponding bandwidth increase over the base-band bandwidth by a factor of n.

If many users were to be accommodated simultaneously and if it were possible to multiplex them by time division (TDM) then, in the same bandwidth used by the spread spectrum method, one could accommodate up to n users. In the interval T, n bits of information would be transmitted, i.e., one bit per user per interval T. In practice, fewer then n users would be accommodated; some digits would be needed for acquisition and timing. The amplitude of the waveform would be chosen to give adequate transmission quality in the face of the channel noise.

The decoding process involves multiplying the incoming binary waveform by a locally stored synchronized version of the user code. By this process, the user code is removed from the binary signal and detection of the binary signal proceeds in the usual fashion. In the analyses carried out previously (ref. 1), it is assumed that user codes unfamiliar to the receiver are seen as sample functions of a bipolar random binary process. Decoding, assuming the unwanted and wanted user codes are chip synchronized, still leaves a random binary sequence which has a power spectral density at frequency f Hz given by

$$S_i(f) = A^2 T_s \left(\frac{\sin \omega T_s/2}{\omega T_s/2} \right)^2 \tag{2}$$

where A is the peak amplitude of the received waveform, $\omega = 2\pi f$, and decoding is assumed to be done with a $\pm$ unit amplitude sequence. Assuming the base-band signal bandwidth is small relative to T_s then, over the band of the decoded desired waveform, the interference power spectral density is essentially given by its value at f = o, or

$$S_i(f) = A^2 T_s \tag{2a}$$

From Eqs. (1) and (2a)

$$S_i(f) = A^2 T/n \tag{3}$$

If there are m synchronized users, each using a different user code and all arriving at the receiver with the same amplitude, the total interference power spectral density seen by a given user is

$$S_m(f) = A^2 T(m-1)/n \tag{4}$$

The total noise power spectral density seen by a given user is given by (4) plus the Gaussian noise density which we denote by N_o.

If bandwidth were available for squandering, we could make m/n small thereby making $S_m(f)$ as given by (4) small. In that event, the self-interference noise which (4) represents could be made equal to or less than N_o. However, if our interest is in getting m/n to be near unity (recall that in time division multiplexing this is what we could accomplish), then (4) will not be small and will ordinarily be greater than N_o.

2. BINARY BASE-BAND SIGNALS

Error probability in the face of Gaussian noise is determined by

$$R = \frac{\text{Signal energy per bit}}{\text{Noise power spectral density}}$$

Though the interference from other users is of the nature of a digital waveform, the detection process, which amounts to integration over the signal interval, sums over a large number of random polarity pulses arising from the unwanted user code. Such a sum, being the sum of a large number of random variables, is virtually Gaussian by the central limit theorem. The signal energy per bit is A^2T so that

$$R = \frac{A^2T}{N_o + A^2T(m-1)/n} = \frac{R_s}{1 + R_s \frac{m-1}{n}} \tag{5}$$

where $R_s = A^2T/N_o$, the ratio of received energy per bit to the two-sided noise power spectral density. The error probability P_e is determined by R; typically, for $P_e = 10^{-5}$, $R \doteq 20$ for the bipolar binary signal assumed. The maximum value of the ratio $(m-1)/n$ for a given R is obtained when $R_s = \infty$. In short, at best

$$\frac{n}{m-1} = R \tag{6}$$

This result may be interpreted as meaning that only about 1/R of the users one can accommodate by TDM are accommodated here. Thus for $P_e = 10^{-5}$ we may say that the bandwidth utilization is approximately 1/20 of what it would be with TDM.

For the case of users not mutually synchronized, the average power spectral density over the information band is (Ref. 1)

$$S_i(f) = \frac{2}{3} A^2 T_s \tag{7}$$

rather than the value given by (2a). The ratio R in this case is, at best,

$$R \doteq \frac{3}{2} \frac{n}{(m-1)} \tag{8}$$

To get R = 20, $\frac{n}{(m-1)}$ in this case needs to be about 13 so that the bandwidth utilization is about 1/13 of that obtained with TDM.

The foregoing is now reconsidered with the number of active users treated as a random variable, denoted k. Eq. (5) is rewritten.

$$R = \frac{R_s}{1 + R_s \frac{k}{n}} \tag{9}$$

Each user is assumed to be active with probability p and the user population is still m. Thus k is a binomial random variable ranging from 0 to (m-1).

The average error probability is given by

$$\overline{P}_e = \sum_{k=0}^{m-1} P_e(k)\, P(k)$$

$$= \sum_{k=0}^{m-1} \int_{\sqrt{R}}^{\infty} \frac{1}{\sqrt{2\pi}} e^{-x^2/2} dx \binom{m-1}{k} p^k (1-p)^{(m-1-k)} \tag{10}$$

where R here is given by Eq. (9) and where $P_e(k)$ is the conditional probability of error given that k interferers are present.

Trial calculations of (10) using m=n=16, values of R_s=20, 40, 80 and ∞, and p = 1/15, 1/30, and 1/45 showed $\overline{P}_e$ never better than 10^{-4}. The implication is that to get reasonable values of $\overline{P}_e$, n will have to be large. We have kept this in mind in the calculations made subsequently and now described.

We let $R_s=\infty$ so that R=n/k for k=1,2,...,(m-1). For k=0, $P_e(k)=0$ so that the k=0 term in (10) drops out. Assuming $P_e(k)$ in (10) can be replaced by its asymptotic form for large R (admissible since we're interested in small $\overline{P}_e$ and the important values of $R = \frac{n}{k}$ will be those where k is small) and that P(k) can be replaced by the Poisson approximation to the binomial (admissible if (m-1) is large and p is small) and, finally, writing

$$m = rn \tag{11}$$

we get

$$\overline{P}_e = \frac{e^{-(m-1)p}}{\sqrt{2\pi m/r}} \sum_{k=1}^{m-1} \frac{\sqrt{k}}{k!} e^{-m/2rk} [(m-1)p]^k \tag{12}$$

$\overline{P}_e$ as given by (12) is plotted in Figure 1 as a function of p for m = 101 and for r taking on values 1/2, 1, and 2.

r = 1 implies a user population equal to that obtained in TDM with slots dedicated to users, and r = 1/2 and 2 represents user population half and twice, respectively, of the TDM system with dedicated channels. From the results of Figure 1, we see that in the case of low utilization probability, say p = 0.01, implying an average of 1% of the users on at one time, $\overline{P}_e$ of 10^{-5} can be achieved even with r = 2. This case may also be interpreted as one in which the system bandwidth is half that in a dedicated TDM system. On the other hand, if r = 1/2 as many as 6% of the users

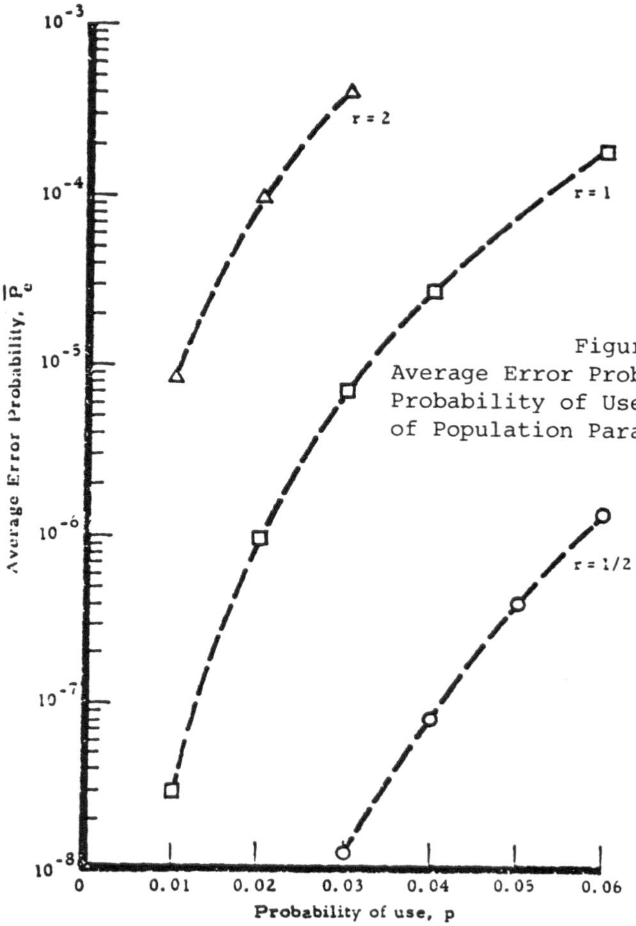

Figure 1.
Average Error Probability in CDMA vs Probability of Use for Various Values of Population Parameter, r

can be on at one time and the error probability can be around 10^{-6}, or better. In this case, however, the system bandwidth would be twice that of a dedicated TDM system.

In short, for systems involving many users each of whom make infrequent use of the system, spread spectrum code division multiplex will give a quality of service typical of TDM with dedicated channels. Signal power will be higher than for TDM and the system error probability slowly gets worse as the number of active users increases. We point out that a demand assignment system will be greatly superior though in cases of infrequent usage and that the spread spectrum advantage is that it does not require a central control making assignments.

3. MULTI-AMPLITUDE BASE-BAND SIGNALS

We now turn to the question of what the effect of base-band signal amplitude modulation coupled with spectrum spreading has on the average interference power.

Incoming multi-amplitude sequences are assumed multiplied by the bipolar unit amplitude user code to give an interference waveform as shown in Figure 2. We assume the signal takes on values $\pm a$, $\pm 3a$, ... $\pm Ka$, and to be of duration T_w, where T_w is an integral number of bit periods T. As before, T will be assumed synchronized with T_s, the chip duration.

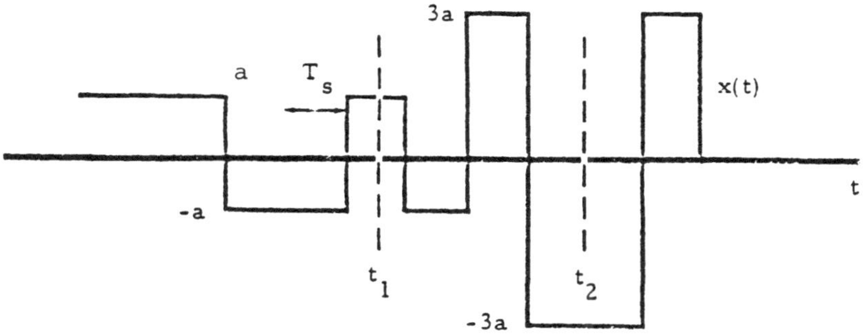

Figure 2
Typical Sample Function of a CDMA Multiple Amplitude Signal

Assuming the amplitude levels are equiprobable the autocorrelation function is found to be

$$R_x(\tau) = \frac{1}{3} K(K + 2)a^2 \left(1 - \frac{|\tau|}{T_s}\right), \quad \tau = t_2 - t_1 \leq T_s$$

$$= 0 \qquad , \text{ otherwise} \tag{13}$$

The corresponding power spectral density is

$$S_x(f) = \frac{1}{3} K(K + 2)a^2 T_s \cdot \left(\frac{\sin \omega T_s/2}{\omega T_s/2}\right)^2$$

Assume (K+1), the number of amplitude levels, is an integer power of 2; i.e. $(K+1)=2^k$. This allows k bits of information per interval T_w and we can let T_w be k times the information source bit interval T. The interference power spectral density at frequencies $|f| << 1/T_s$ is

$$S_x(f) = \frac{1}{3} K(K + 2)a^2T_s$$

The word error probability in this case is determined by the total noise and the quantity which measures the incremental energy between signal alternatives, that is, it depends on

$$a^2T_w = ka^2T$$

The power spectral density of (m-1) interferers is $\frac{1}{3} K(K+2)a^2T_s(m-1)$ so that

$$R = \frac{ka^2T}{N_0 + \frac{1}{3} K(K+2)a^2T(m-1)/n} \tag{14}$$

Again, to accommodate the largest number of users we will have to make a^2T/N_0 large so that

$$R \to \frac{3k}{K(K+2)} \frac{n}{m-1} = \frac{3 \log_2(K+1)}{K(K+2)} \frac{n}{m-1} \tag{15}$$

For any K>1 this value of R is less than n/(m-1) hence poorer than for the binary case. No advantage is therefore gained in using multiple amplitude states. Since a word error does not mean that all bits in the word are in error, the values of R required here and in the binary case differ somewhat. The difference though is not significant.

4. MULTI-DIMENSIONAL BASE-BAND SIGNALS

The failure of the multiple amplitude signal is associated with the higher amplitude states used which cause greater interference. This overcomes the advantage gained by using base-band signal words of longer duration. An alternative which may be expected to give better results is one which makes use of base-band signal waveform which in the language of vector spaces is of more than one dimension. In particular, suppose base-band waveforms are constructed using 2^k orthogonal or biorthogonal bipolar binary sequences. Such sequences can be constructed in several ways; see, for example, MacWilliams and Sloane, ref. (2) and Viterbi, ref. (3). The signal waveform so generated is next multiplied by the user code which is presumed to have a chip duration sufficiently smaller than the duration of the digits of the signal sequences so that on despreading the interference power spectral density is still essentially white and as given by (3). R is then still as given by (5) and at best (with $R_s=\infty$) is as given by (6). For multi-state waveforms of the kind being considered, however, the value of R for a given P_e is less than for a two-state waveform.

Figure 3 (obtained from Ref. 3 or 4) gives the required energy per bit/noise power spectral density for various values of k, with bit error probability fixed at 10^{-5}. Similar curves for other values of bit error probability can be obtained from the original source of the data. Orthogonal and biorthogonal waveforms are represented in Figure 3, though for values of k beyond 2 there is no important difference. We observe that by going from k=1 to k=6 a reduction by a factor of about 2.5 is obtained, from R=20 to R=8 (the biorthogonal case is being used). This implies that R needs to be

$$R = \frac{n}{m-1} = 8$$

or

$$m = \frac{n}{8} + 1$$

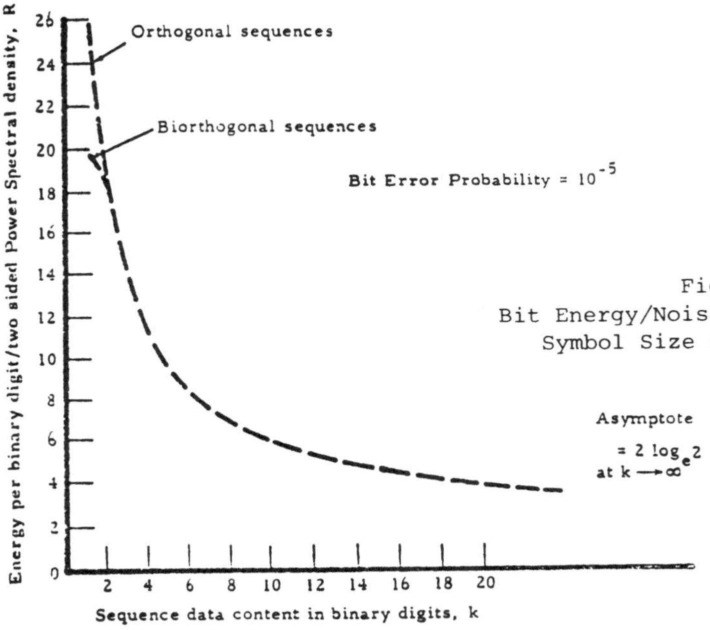

Figure 3
Bit Energy/Noise PSD Ratio vs
Symbol Size in Bits

The number of simultaneous users, m, can be about 1/8 the number of chips per bit interval, n. This is a substantial improvement over the one dimensional case where the spectrum utilization was only about 1/20. Furthermore, if we take account of the non-synchronization among users, for which the k=1 case treated initially resulted in a spectrum utilization of 1/13, we now get a spectrum utilization of about 1/5. The price one has to pay is greater complexity associated with the need for $2^6=24$ orthogonal or biorthogonal sequences. The limit of improvement obtainable is given by the asymptotic value of $R = 2 \log_e 2$ for k approaching infinity. In that event, for chip synchronized users,

$$m = \frac{n}{2\log_e 2} + 1 \doteq 0.72n$$

Finally, if users are chip unsynchronized, the interference spectral density is reduced by the factor 2/3 and accordingly

$$m = \frac{3}{2}\,\frac{n}{2\log_e 2} + 1 \doteq 1.08n$$

a limiting result which indicates the potential for rates higher than obtained in TDM.

5. ACKNOWLEDGMENT

This work was done, in part, while the author was employed by the Aerospace Corporation, Washington, D.C. during the summer of 1976.

6. REFERENCES

1. R. L. Harris, Introduction to Spread Spectrum Techniques in Spread Spectrum Communication, NATO AGARD Lecture Series No. 58, BolKesjØ Norway, 28-30 May 1973 and the Hague, Netherlands, 4-6 June 1973. Reference No. AGARD-LS-58.

2. F. J. MacWilliams and N.J.A. Sloane, Pseudo-Random Sequences and Arrays, Proceedings IEEE, Vol. 64, No. 12, Dec. 1976, pp. 1715-1729.

3. A. J. Viterbi, Principles of Coherent Communications, McGraw Hill Book Co., New York, 1966.

4. L. Baumert, M. Easterling, S. W. Golomb, A. J. Viterbi, Coding Theory and Its Applications to Communication Systems, Report No. 32-67, Jet Propulsion Laboratory, Pasadena, California, 31 March 1961.

EXTENSION OF AN ADAPTIVE DISTRIBUTED ROUTING ALGORITHM TO MIXED MEDIA NETWORKS

Anthony Ephremides
Electrical Engineering Department
University of Maryland
College Park, Maryland

ABSTRACT

By modeling the broadcast portion of a mixed media network as a fully connected point-to-point network with link capacities varying as functions of the traffic rate it is possible to extend an adaptive distributed routing algorithm that was originally developed for point-to-point ground networks. Additional modifications for improved dynamic performance at the satellite interface message processors are also included.

INTRODUCTION

Recently R. Gallager [1] developed an adaptive distributed routing algorithm which for slow changes in the user traffic rates was shown to converge to the minimum total average delay value. In that algorithm a crucial assumption was the convex dependence of the link delays only on the traffic rates in those links. Also recently the problem of routing in mixed media networks was posed by Huynh, Kobayashi, and Kuo [2]. The network there is modeled as a collection of ground networks connected among themselves by ground links as well as via satellite through satellite interface message processors (SIMP's). In this paper it is observed that for almost any protocol of multiple access, the delays at the SIMP's are increasing, convex functions of the individual and the total broadcasting traffic rates. It is then shown that Gallager's algorithm can be extended for the case where each link delay can depend on not only the traffic rate at that link, but on possibly all the traffic rates in all the links. Finally it is demonstrated how to adapt this slightly modified algorithm to the case of mixed media networks. The validity of this approach lies in the fact that the problem of quasi-static routing can be separated from that of optimum dynamic multiple access control, since for any chosen such access protocol the algorithm yields the optimum routing.

EXTENSION OF THE ALGORITHM

In this section it will be shown that Gallager's algorithm retains its optimality properties when the link delays are allowed to depend on all the link traffic rates in the network. Since many of the proofs are completely parallel to those in [1] they are summarized in the appendix.

Consider a network of n nodes, where the link connecting nodes i and k is denoted by (i, k). Let, as in [1],

$r_i(j)$ = average rate in bits/s of the external traffic entering node i and destined for node j.

$t_i(j)$ = total average rate in bits/s of traffic entering i and destined for j.

f_{ik} = total traffic rate in bits/s on the link (i, k).

$\varphi_{ik}(j)$ = percentage of the j traffic routed from node i into the link (i, k).

D_{ik} = [average delay per message on link (i, k)] x x[average number of messages per second on link (i, k)] *

D_T = $\sum_i \sum_k D_{ik}$

*Note that the quantity in the second set of brackets is equal to μf_{ik} if $\frac{1}{\mu}$ is the average length of the messages, and thus D_{ik} is the average delay weighted by the link traffic as usual.

It is assumed that D_{ik} is a function of all the f_{pq}'s, increasing, convex, and continuously differentiable in each f_{pq}.

Then the flow equations

$$t_i(j) = r_i(j) + \sum_{\ell=1}^{n} t_\ell(j)\varphi_{\ell i}(j) \tag{1}$$

$$f_{ik} = \sum_{j=1}^{n} t_i(j)\varphi_{ik}(j) \tag{2}$$

are still valid, and so is the statement, proved in [1], that Eqs. (1) have a unique solution yielding t's that are positive and continuously differentiable in the r's and the φ's.

Next consider the incremental delays due to changes in $r_i(j)$ and $\varphi_{ik}(j)$ respectively. Clearly for $i \neq j$ one has

$$\frac{\partial D_T}{\partial r_i(j)} = \sum_{k=1}^{n} \varphi_{ik}(j) \left[\frac{\partial D_{ik}}{\partial f_{ik}} + \frac{\partial D_T}{\partial r_k(j)} + \sum_{\substack{p,q \\ (p,q)\neq(i,k)}} \frac{\partial D_{pq}}{\partial f_{ik}} \right] \tag{3}$$

$$\frac{\partial D_T}{\partial \varphi_{ik}(j)} = t_i(j) \left[\frac{\partial D_{ik}}{\partial f_{ik}} + \frac{\partial D_T}{\partial r_k(j)} + \sum_{\substack{p,q \\ (p,q)\neq(i,k)}} \frac{\partial D_{pq}}{\partial f_{ik}} \right] \tag{4}$$

where $\varphi_{ik}(j)$ is taken to be zero if (i, k) is not a link and where the last term in the brackets accounts for the fact that changes in the traffic rate on link (i, k) affect directly the delays in the rest of the links.

The next proposition, whose proof can be found in the appendix, is identical to Thm. 2 in [1].

<u>Proposition 1</u>: There exist unique solutions in the variables

$\frac{\partial D_T}{\partial r_i(j)}$ of Eqs. (3). These solutions are also continuous in the variables r and φ and Eqs. (4) are true.

Next, again paralleling the development in [1], a sufficient condition is obtained for the minimization of D_T. The following statement provides that sufficient condition with the proof summarized in the appendix.

Proposition 2: A sufficient condition for the minimization of D_T over the routing variables $\varphi_{ik}(j)$ is that

$$\frac{\partial D_{ik}}{\partial f_{ik}} + \frac{\partial D_T}{\partial r_k(j)} + \sum_{\substack{p,q \\ (p,q) \neq (i,k)}} \frac{\partial D_{pq}}{\partial f_{ik}} \geq \frac{\partial D_T}{\partial r_i(j)} \quad * \tag{5}$$

with equality in Eq. (5) when $\varphi_{ik}(j) > 0$.

Note again that Eq. (5) is equivalent to

$$\alpha_{ik}(j) - \min_m \alpha_{im}(j) = 0 \tag{6}$$

when $\varphi_{ik}(j) > 0$, and where

$$\alpha_{ik}(j) \triangleq \frac{\partial D_{ik}}{\partial f_{ik}} + \frac{\partial D_T}{\partial r_k(j)} + \sum_{\substack{p,q \\ (p,q) \neq (i,k)}} \frac{\partial D_{pq}}{\partial f_{ik}} \tag{7}$$

Thus all of the motivating statements for the algorithm of [1] are valid under the new assumption that each D_{ik} may depend on the traffic rates on all the links, with a slight modifica-

---

* The left hand side can be written in a simpler form by shoving the first term inside the summation of the last term. It is left here purposely in this form in order to indicate the additional term when compared to the expressions in [1].

tion due to the additional term that accounts for the additional incremental delays $\frac{\partial D_{pq}}{\partial f_{ik}}$.

Thus the algorithm would have now the following form. For each commodity j (destination node), each node i will be computing the quantities $\alpha_{ik}(j)$ of Eq. (7), for all its neighbors k and will be decreasing those $\varphi_{ik}(j)$ for which $\alpha_{ik}(j)$ is large and increasing the $\varphi_{ik}(j)$ of the neighbor with the smallest $\alpha_{ik}(j)$, provided of course that $\varphi_{ik}(j)$ is set to zero for blocked nodes, i.e. nodes which have routing paths leading to j that include improper links (ℓ, m) with $\varphi_{\ell m}(j) \geq \eta \cdot$

$$\cdot \frac{\alpha_{\ell m}(j) - \frac{\partial D_T}{\partial r_\ell(j)}}{t_\ell(j)}$$

. The last provision of course aims at eliminating loops and the relevant discussion that can be found in [1] carries verbatim to the present case. The proof of convergence of the algorithm follows exactly the lines of argument presented in [1] and is not repeated in the appendix. The only variation in the details of the proof lie in the fact that all the partials $\frac{\partial^2 D_{ik}}{\partial f_{pq}^2}$ must be bounded instead of just D'_{ik}.

What is important to note here is the form of the quantity $\alpha_{ik}(j)$ that must be computed at each node at each updating time and for each commodity, and which is the basis for the algorithm. As can be seen from Eq. (7) there are three terms in the expression for $\alpha_{ik}(j)$. The first is the incremental delay on link (i, k) due to changes on the traffic rate f_{ik}. The second is the incremental total delay due to changes in $r_k(j)$ and the last consists of the sum of the incremental delays of the <u>other</u> links in the network due to changes in the traffic rate f_{ik}. Of course with blocking eliminated by avoiding loops the quantity $\frac{\partial D_T}{\partial r_k(j)}$ can be propagated upstream by use of Eq. (3) as explained in [1] provided the incremental link delays can be computed. These incremental delays are on the other hand the only remaining quantities for the computation of $\alpha_{ik}(j)$. Now if the functional form of D_{ik} is known these delays can be derived analytically. Such of course generally will not be the case and the algorithm is robust in that sense over the class of exact functional dependence of D_{ik} on the f's. Therefore the alternative will be to estimate directly the derivatives. Now for the estimation of $\frac{\partial D_{ik}}{\partial f_{ik}}$ one can use a variety of existing methods, see for example [3]. But it appears that there is no way to estimate $\frac{\partial D_{pq}}{\partial f_{ik}}$ for (p, q) $\neq$ (i, k). This difficulty, although a definite obstacle for the application of this version of the algorithm, need not be a hindrance when the dependence of D_{pq} on f_{ik} is not completely arbitrary. As it will be shown in the next section under most protocols of multiple access in a mixed media network this dependence is mostly symmetrical and uniform, so that as a result only the estimation of $\frac{\partial D_{ik}}{\partial f_{ik}}$ is needed, with $\alpha_{ik}(j)$ assuming the form

$$\alpha_{ik}(j) = M \frac{\partial D_{ik}}{\partial f_{ik}} + \frac{\partial D_T}{\partial r_k(j)} \tag{8}$$

where M is the number of SIMP's.

APPLICATION TO MIXED MEDIA NETWORKS

In the preceeding section it was shown that the basic idea of Gallager's algorithm is still valid when the link delay D_{ik} may be for each link a function of the traffic rates f_{pq} in other links.

Now let us consider the model of a mixed media network identical to the one considered in [2] and depicted in Fig. 1.

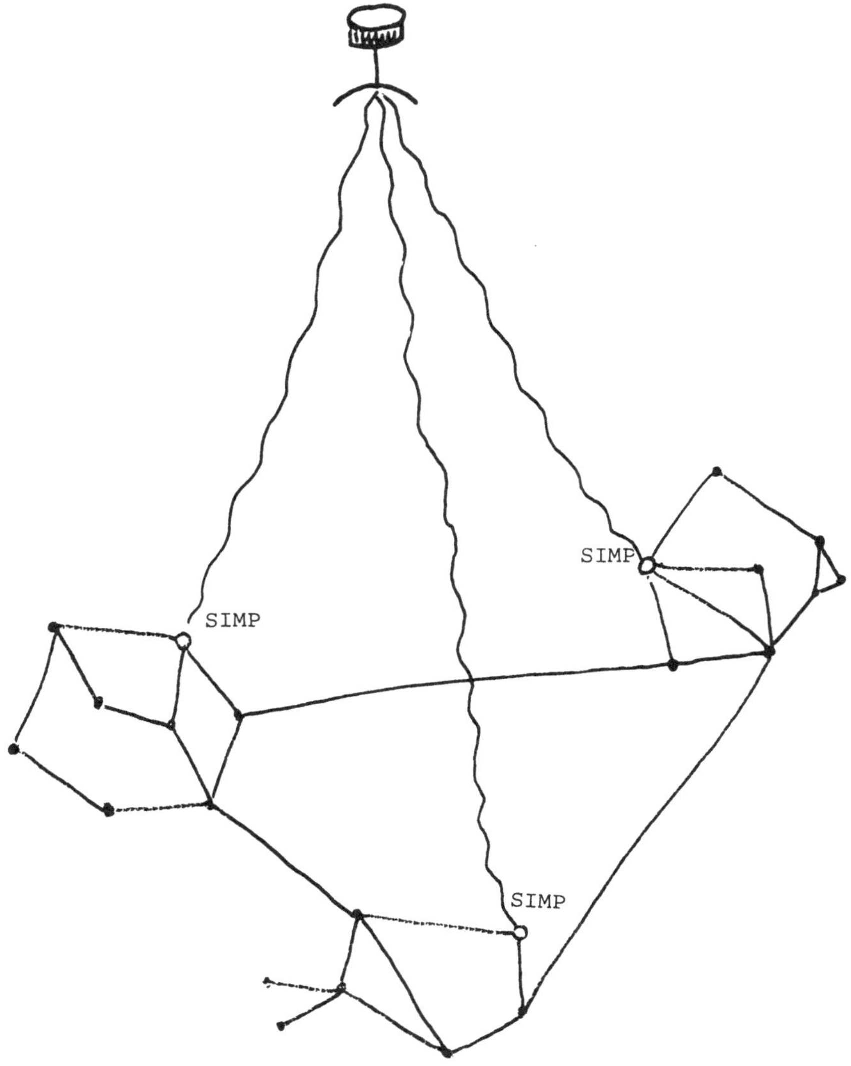

Figure 1

Suppose for the moment that the subnetwork of the SIMP's is conceptualized as a fully connected point-to-point network. Then clearly the link delay D_{ik} for any pair of SIMP's will depend on the total traffic in the broadcast medium which is commonly shared among all the SIMP's. The precise form of this dependence is not important, except that it should be convex and increasing and that the incremental delays $\frac{\partial D_{ik}}{\partial f_{pq}}$ should be computable. By stretching somewhat the realities of the situation one may think of the fully connected model and take into account the effects of the sharing of the medium by assuming that the capacity in each link is a variable quantity and a function of the total broadcasted traffic rate. Then by considering the M/D/1 result under the usual assumptions one could write

$$D_{ik} = \frac{f_{ik}}{2[C_{ik}(f) - f_{ik}]} \tag{9}$$

where $f = \Sigma f_{ik}$ and C_{ik} = the capacity of the imaginary link (i, k). Of course this is a grossly simplified assumption and it is not necessary, except for providing some insight into the motivation for the applicability of a routing algorithm such as Gallager's in the mixed media case. In fact Eq. (9) is quite similar to the heuristic formulas used in [2] on the basis of Lam's analysis [4].

In reality one need only consider the various protocols of multiple access and verify the convexity and summetry of the dependence of D_{ik} on the traffic rates. For example in any form of controlled or uncontrolled, slotted or unslotted, ALOHA mode of access the work of Lam [4] as well as others [5], done empirically or analytically, confirms the intuitively clear fact of convex dependence of delay on the traffic rate of any user. In fact even if the users do not have equal rates the incremental effect on the delay of any user's messages of variations in the rates of other users is identical to the one due to variations in his own rate. Take for example the formula used in [2] for the delay under the ALOHA mode with the average message length taken to be equal to one for simplicity:

$$D_{ik} = 0.26 + \frac{1}{C_s} + \frac{\lambda_s b(M)}{2(C_s a(M) - \lambda_s)} \cdot \frac{1}{C_s} \tag{10}$$

where C_s is the total satellite capacity, λ_s is the total traffic rate and a(M) and b(M) are heuristic increasing functions of the number of users (SIMP's) M. Or take the formula for the MASTER mode of access* where

$$D_{ik} = 0.26 + \frac{1}{C_s} + \frac{1}{2\lambda_s} \sum_{\sigma=1}^{M} \frac{\lambda_\sigma}{C_s - \lambda_\sigma} \tag{11}$$

where λ_σ is the rate of the SIMP σ traffic.

Again in both Eqs. (10) and (11) the convexity is clear and the symmetry is obvious so that indeed

$$\sum_{\substack{\text{SIMP}\\ \text{links}}} \frac{\partial D_{ik}}{\partial f_{pq}} = M \cdot \frac{\partial D_{ik}}{\partial f_{ik}} \tag{12}$$

Of course $\frac{\partial D_{ik}}{\partial f_{ik}}$ can be computed (estimated) regardless of the access mode. It should be noted that depending on the access mode actually used, the minimum average delay D_T will vary, however in all cases that minimum will be attained by using the quasi-static algorithm of Gallager with the modification proposed here. One should note also that additional minor points ought to be settled. For example for each commodity j (destination) the SIMP has the option of either routing via the ground links or via satellite, the latter alternative meaning that in the context of the fully connected model $\varphi_{ik}(j)$ is zero for all subnetworks k except for the one to which node j belongs. As a consequence a packet, once broadcasted successfully, will not have to be re-broadcasted in order to reach its destination. Also the updating information during any cycle of the algorithm should be transmitted only via the ground links to avoid additional protocol-induced collisions in the broadcast medium, or if done via satellite it should take place during pre-specified slots in an orderly way.

---

* This mode of access allows retransmission of collided packets via the ground links and therefore alters dynamically the effective routing variables φ. This effect is considered in the next section.

ADDITIONAL DYNAMIC PROCEDURES AT THE SIMP's

In the preceeding sections it was demonstrated that in mixed media networks the routing algorithm of Gallager as modified here would still be applicable. It was also noted that the separate problem of dynamic control of the access mode allows for further reducing the total delay, preventing saturation, and increasing the broadcast throughput. In that regard dynamic features of adjusting retransmission time of collided packets, of using reservation techniques, or of employing other flow control methods [6-8], can be incorporated on top of using the proposed algorithm. However when blocking of packets is used as a means of such control, or when rejected packets are rerouted via the ground links as in the MASTER mode, then the features of the algorithm are altered and the effective φ's in the ground links as well as in the satellite links are at variance relative to the values dictated by the algorithm. In such cases one should study further, either by analysis or by simulation, the performance of these procedures.

APPENDIX

A. Proof of Proposition 1

Consider Eq. (3) for one commodity j which can be taken to be the nth node. Then

$$\frac{\partial D_T}{\partial r_i} = \sum_k \varphi_{ik} \frac{\partial D_{ik}}{\partial f_{ik}} + \sum_k \sum_{p,q} \varphi_{ik} \frac{\partial D_{pq}}{\partial f_{ik}} + \sum_k \varphi_{ik} \frac{\partial D_T}{\partial r_k(j)} \qquad \text{(A. 1)}$$

By letting b_i denote the sum of the first two summations in (A. 1) and letting b denote the column vector of the b_i's, i=1, . . , n-1 and ∇D_T denote the column vector of the $\frac{\partial D_T}{\partial r_i}$'s, i=1, . . , n-1. Eq. (A. 1) can be rewritten as

$\nabla D_T = (I - \Phi)^{-1} b$ (where Φ is the (n-1)x(n-1) matrix of the φ_{ik}'s)

which by Thm. 1 of [1] has a unique solution. That solution is given by

$$\frac{\partial D_T}{\partial r_i} = \sum_\ell \frac{\partial t_\ell}{\partial r_i} b_\ell = \sum_\ell \frac{\partial t_\ell}{\partial r_i} \sum_m \varphi_{\ell m} \left[\frac{\partial D_{\ell m}}{\partial f_{\ell m}} + \sum_{\substack{p,q \\ (p,q) \neq (\ell,m)}} \frac{\partial D_{pq}}{\partial f_{\ell m}} \right] =$$

$$= \sum_\ell \sum_m \frac{\partial f_{\ell m}}{\partial r_i} \left[\frac{\partial D_{\ell m}}{\partial f_{\ell m}} + \sum_{\substack{p,q \\ (p,q) \neq (\ell,m)}} \frac{\partial D_{pq}}{\partial f_{\ell m}} \right] \qquad \text{(A. 2)}$$

where it was used that the entries in the matrix $(I - \Phi)^{-1}$ are $\frac{\partial t_\ell}{\partial r_i}$ and that $f_{\ell m} = t_\ell \varphi_{\ell m}$. A direct calculation of $\frac{\partial D_T}{\partial r_i}$ yields

$$\frac{\partial D_T}{\partial r_i} = \sum_\ell \sum_m \frac{\partial D_{\ell m}}{\partial r_i} = \sum_\ell \sum_m \left[\frac{\partial D_{\ell m}}{\partial f_{\ell m}} \cdot \frac{\partial f_{\ell m}}{\partial r_i} + \sum_{\substack{p,q \\ (p,q) \neq (\ell,m)}} \frac{\partial D_{\ell m}}{\partial f_{pq}} \cdot \frac{\partial f_{pq}}{\partial r_i} \right] \qquad \text{(A. 3)}$$

and it can be seen that the right hand sides in (A. 2), (A. 3) are equal since

$$\sum_{\ell,m}\ \sum_{\substack{p,q\\(p,q)\neq(\ell,m)}} \frac{\partial D_{\ell m}}{\partial f_{pq}}\,\frac{\partial f_{pq}}{\partial r_i} = \sum_{\ell,m}\ \sum_{\substack{p,q\\(p,q)\neq(\ell,m)}} \frac{\partial D_{pq}}{\partial f_{\ell m}}\,\frac{\partial f_{\ell m}}{\partial r_i} \tag{A.4}$$

Finally the truth of Eq. (4) is verified by direct computation and use of the above equations, that is

$$\frac{\partial D_T}{\partial \varphi_{ik}} = \sum_{\ell,m} \frac{\partial D_{\ell m}}{\partial \varphi_{ik}} = \sum_{\ell,m}\left[\frac{\partial D_{\ell m}}{\partial f_{\ell m}}\,\frac{\partial f_{\ell m}}{\partial \varphi_{ik}} + \sum_{\substack{p,q\\(p,q)\neq(\ell,m)}} \frac{\partial D_{\ell m}}{\partial f_{pq}}\,\frac{\partial f_{pq}}{\partial \varphi_{ik}}\right] =$$

$$= \sum_{\ell,m}\left[\frac{\partial D_{\ell m}}{\partial f_{\ell m}}\,\varphi_{\ell m}\,\frac{\partial t_\ell}{\partial \varphi_{ik}}\right] + \frac{\partial D_{ik}}{\partial f_{ik}}\,t_i + \sum_{\substack{\ell,m\\(\ell,m)\neq(i,k)}}$$

$$\left[\sum_{\substack{p,q\\(p,q)\neq(\ell,m)}} \frac{\partial D_{\ell m}}{\partial f_{pq}}\,\varphi_{pq}\,\frac{\partial t_p}{\partial \varphi_{ik}} + \frac{\partial D_{\ell m}}{\partial f_{ik}}\,t_i\right]$$

But $\dfrac{\partial t_\ell}{\partial \varphi_{ik}} = \dfrac{\partial t_\ell}{\partial r_k}\,t_i$ from [1] and thus

$$\frac{\partial D_T}{\partial \varphi_{ik}} = t_i\left[\sum_{\ell,m} \frac{\partial D_{\ell m}}{\partial f_{\ell m}}\,\varphi_{\ell m}\,\frac{\partial t_\ell}{\partial r_k}\right] + t_i\,\frac{\partial D_{ik}}{\partial f_{ik}} + t_i \sum_{\substack{\ell,m\\(\ell,m)\neq(i,k)}}$$

$$\left[\sum_{\substack{p,q\\(p,q)\neq(\ell,m)}} \frac{\partial D_{\ell m}}{\partial f_{pq}}\,\varphi_{pq}\,\frac{\partial t_p}{\partial r_k} + \frac{\partial D_{\ell m}}{\partial f_{ik}}\right] = t_i\left[\frac{\partial D_{ik}}{\partial f_{ik}} + \sum_{\ell,m}\left(\frac{\partial D_{\ell m}}{\partial f_{\ell m}}\cdot\frac{\partial f_{\ell m}}{\partial r_k} + \right.\right.$$

$$+ \sum_{\substack{p,q \\ (p,q)\neq(\ell,m)}} \frac{\partial D_{\ell m}}{\partial f_{pq}} \frac{\partial f_{pq}}{\partial r_k} \Bigg) + \sum_{\substack{\ell,m \\ (\ell,m)\neq(i,k)}} \frac{\partial D_{\ell m}}{\partial f_{ik}} \Bigg]$$

$$= t_i \left[\frac{\partial D_{ik}}{\partial f_{ik}} + \frac{\partial D_T}{\partial r_k} + \sum_{\substack{\ell,m \\ (\ell,m)\neq(i,k)}} \frac{\partial D_{\ell m}}{\partial f_{ik}} \right]$$

which is the same as Eq. (4).

B. Proof of Proposition 2

Again the proof is parallel to the one of Thm. 3 in [1]. Let φ be a set of routing variables satisfying Eq. (5) and φ^* any other set. Let

$$f_{ik}(\lambda) \triangleq (1-\lambda) f_{ik} + \lambda f_{ik}^{*} \quad \text{for } \lambda \in [0,1]. \qquad \text{(B. 1)}$$

Then $D_T(\lambda) = \sum_{i,k} D_{ik}(\lambda)$ = convex in λ by the convexity of D_{ik} in the f's and by (B. 1).

Thus

$$\frac{\partial D_T}{\partial \lambda}\bigg|_{\lambda=0} \leq D_T(\varphi^*) - D_T(\varphi)$$

and it is enough to show that

$$\frac{\partial D_T(\lambda)}{\partial \lambda}\bigg|_{\lambda=0} \geq 0.$$

Indeed we note that

$$\frac{\partial D_T(\lambda)}{\partial \lambda} = \sum_{i,k} \left(\frac{\partial D_{ik}}{\partial f_{ik}} \frac{\partial f_{ik}}{\partial \lambda} + \sum_{\substack{p,q \\ (p,q)\neq(i,k)}} \frac{\partial D_{ik}}{\partial f_{pq}} \frac{\partial f_{pq}}{\partial \lambda} \right)$$

$$= \sum_{i,k} \left(\frac{\partial D_{ik}}{\partial f_{ik}} (f^*_{ik} - f_{ik}) + \sum_{\substack{p,q \\ (p,q)\neq(i,k)}} \frac{\partial D_{ik}}{\partial f_{pq}} (f^*_{pq} - f_{pq}) \right) \tag{B.2}$$

From **Eq.** (5) after multiplying both sides by φ^*_{ik} (j) and summing over k we get

$$\sum_k \frac{\partial D_{ik}}{\partial f_{ik}} \varphi^*_{ik}(j) + \sum_k \sum_{\substack{p,q \\ (p,q)\neq(i,k)}} \frac{\partial D_{pq}}{\partial f_{ik}} \varphi^*_{ik}(j) \geq$$

$$\geq \frac{\partial D_T(\varphi)}{\partial r_i(j)} - \sum_k \frac{\partial D_T(\varphi)}{\partial r_k(j)} \varphi^*_{ik}(j) \tag{B.3}$$

which holds as equality if the φ^*'s are replaced by the φ's. Then multiply both sides of (B.3) by $t^*_i(j)$, sum over i, j and use **Eq.** (2) to obtain

$$\sum_{i,k} \frac{\partial D_{ik}}{\partial f_{ik}} f^*_{ik} + \sum_{i,j,k} t^*_i(j) \varphi^*_{ik}(j) \sum_{\substack{p,q \\ (p,q)\neq(i,k)}} \frac{\partial D_{pq}}{\partial f_{ik}} \geq$$

$$\geq \sum_{i,j} \frac{\partial D_T(\varphi)}{\partial r_i(j)} t^*_i(j) - \sum_{i,j,k} t^*_i(j) \varphi^*_{ik}(j) \cdot \frac{\partial D_T(\varphi)}{\partial r_k(j)} \tag{B.4}$$

Then by using **Eq.** (1) and substituting in (B.4) we get

$$\sum_{i,k} \frac{\partial D_{ik}}{\partial f_{ik}} f^*_{ik} + \sum_{i,k} f^*_{ik} \sum_{\substack{p,q \\ (p,q)\neq(i,k)}} \frac{\partial D_{pq}}{\partial f_{ik}} \geq \sum_{j,k} r_k(j) \frac{\partial D_T(\varphi)}{\partial r_k(j)} \tag{B.5}$$

But Eq. (B. 5) holds with equality if the f_{ik}'s are substituted for the f_{ik}^*'s. Thus subtracting that equality from (B. 5), and using (A. 4) and (B. 2) we get indeed that

$$\frac{\partial D_T(\lambda)}{\partial \lambda} \Big|_{\lambda=0} \geq 0$$

which completes the proof.

REFERENCES

[1] R.G. Gallager, "A Minimum Delay Routing Algorithm Using Distributed Computation", IEEE Trans. Communications, Vol. 25, No. 1, pp. 73-85, January 1977.

[2] D. Huynh, H. Kobayashi, F.F. Kuo, "Optimal Design of Mixed Media Packet Switching Networks: Routing and Capacity Assignment", IEEE Trans. Communications, Vol. 25, No. 1, pp. 158-169, January 1977.

[3] A. Segall, "The Modeling of Adaptive Routing in Data-Communication Networks", IEEE Trans. Communications, Vol. 25, No. 1, pp. 85-95, January 1977.

[4] S.S. Lam, "Packet-Switching in a Multi-access Broadcast Channel with Applications to Satellite Communication in a Computer Network", Ph.D. Dissertation, Dept. Comp. Sci. UCLA, April 1974.

[5] C.L. Heitmeyer, J.H. Kullback, J.E. Shore, "A Survey of Packet Switching Techniques for Broadcast Media", NRL Report 8035, October 1976.

[6] G. Fayolle, E. Gelenbe, J. Labetoulle, "Stability and Optimal Control of the Packet Switching Broadcast Channel", Journal of ACM, to appear, July 1977.

[7] J. Rubin, "Reservation Schemes for Dynamic Packet Access Control of Multi-Access Communications Channels" Technical Report UCLA-ENG-7712, January, 1977.

[8] A. Ephremides, "Node Level Control of Routing and Scheduling in Communication Networks", Proceedings JACC, June 1977.

Part II

MULTIPLE COMMUNICATION

FROM THEORY TO PRACTICE: ADVANCES IN COMMUNICATIONS DISCUSSED AT EUROCON '77 IN VENICE.

F.L.H.M. Stumpers Scientific Adviser (ret.)

Philips Research Laboratories, Eindhoven, the Netherlands

1. INTRODUCTION

International communications conferences stress the application more than the theory, and Eurocon '77 organized by Eurel (the Convention of National Societies of Electrical Engineers of Western Europe) and IEEE Region 8 with the special support of U.R.S.I. (the International Union of Radio Science) was no exception. On the other hand the problems of communications, statistics and random processes are so interwoven, that several subjects were treated in Venice and in Darlington and that the symposia had five speakers in common. In the opening session Reid (U.K.P.O.) gave the keynote-address:"Uncertainty and inertia:The long term demand for, and planning of new telecommunication services"[1]. There is uncertainty about the demand for such new services as facsimile, information retrieval, videophone, mobile communications and cable tv, and about supplies, costs and capabilities, even about government policy. The inertia arise from size and complexity of administration and industry, as well as from the monopolistic nature of telecommunications. The author saw the reconciliation of uncertainty and inertia in long term studies, with participation of all interested parties, as an example of which he mentioned a model study of video teleconferences. About 196 papers are reproduced in the Conference Proceedings, two books with 1357 pages. Eurocon counted five Sections, several of them with parallel sessions. We will now treat them one by one.

2. COMMUNICATION IN LARGE POWER SYSTEMS. (Chairman B. Favez, France).

The european power communication engineers are well aware of the means they can use to obtain reliable communication. Statistically telegraph lines have average bit error rates of 10^{-4} interrupted by "error bursts". The residual error probability required in telecontrol systems is 10^{-10}. To achieve this, Funk[2]) finds signal quality supervision important, as well as the use of error detecting codes. BCH codes are ineffective against synchronizing slip errors. By using block of constant length as well as well protected start and stop signals and error correction, he obtains sufficiently reliable and efficient communication. In british telecontrol systems[3]) synchronization, start, address and stop signals also get extra protection. The standardized remote terminal units of Italy's ENEL[4]) use information words of 10 useful bits and 5 control bits. ENEL has its own microwave radiolinks, standard telephone channels as back-up, and a low frequency control network for emergencies. In its data acquisition network[5]) high-speed modems (9600 bit(s) use digital transversal equalizers with (re-) training sequences shorter than 200 ms. The maximum efficiency block length is a function of bit error rate. Error rate 10^{-4}, block length 50 characters, efficiency 67.9%. Error rate 10^{-6}, block length 650 characters, efficiency 93.5%. This depends also on the modulation system. The swedish system[6]) uses message switching and adaptive routing. It has special features to test for line or interface error. In Norway's teleprotection channels, a tripsignal is sent when an alarm is required, and otherwise a guard signal. For maximum efficiency the code for the guard signal is the inverse of the code for the alarm signal. The detector is a sequential probability ratio tester, counting up when an alarm arises, protected against false alarms by a decision threshold. A french system uses two transmission channels (mutually backed-up) and 2 real-time computers, each with a second one ready to take over. The real time computers work in a conversational mode. They control the hydroelectric stations in south-west France, and supervise other equipment. - There are obvious uses for microprocessors and minicomputers in telecontrol[9,10]), and the italian power industry even has its own computer network[11]). "A power failure can have grave economic and social consequences particularly in large city centres, and thus the degree of responsability placed on control and communication reliability is very high indeed".[12)] All the predicted mishaps occurred in New-York, six weeks later, though the information on the status of the circuits was always available.

3. NEW DEVELOPMENTS IN COMMUNICATIONS. (Chairman W.J. Bray, U.K.)

The telephone network[13)], a global operating automatic machine, the largest ever devised and the most important mass communication network is estimated at an investment of $ 0,5 x 10^{12} and it serves 300 x 10^6 subscribers. Today[14)] approximately 4% of subscriber's lines in the world are connected on electronic switching systems. Within 10 years this number will represent about 50%. The operational program of a public exchange of medium capacity is stored on 6,5 . 10^6 bits among which 30% are telephonic, 30% administrative and 40% maintenance programs.
Wiest has reservations with regard to the application of large scale integration in switching systems. "It is possible to use LSI even in the switching environment of the speech path network, although this entails a considerable expenditure and usually results in a certain amount of deterioration in the transmission parameters.- Space division networks with electronic crosspoints have many disadvantages compared with those equipped with electromechanical contacts (high attenuation loss, crosstalk problems, relatively narrow transmission bandwidth). P.A.M. switching networks have the same disadvantages. Even with PCM electronic switching has a number of disadvantages especially where terminating traffic is concerned. No administration has yet introduced electronic switching into local exchanges, where particular importance is attached to transmission quality." Wiest sees the main field of application of electronic components in switching systems as system control; microcomputers are very well suited for carrying out decentralized control tasks, and in larger systems mini- or large size switching computers are employed for performing control tasks, which require overall system knowledge. Stored program central control of the No.1 Electronic Switching System was described in 1964. The full advantage of such a system could be fully realized only when suitable processors became available. The combination of digital signal processing and a time division switching network leading to a fully electronic exchange was very advantageous[15)]. The switching net-

work is a synchronous switch of the time-space-time (TST) type. A time switch module contains inlet and outlet speech memories for 512 channels. It converts the data to parallel form and it is connected to the space matrix.by highways for incoming and outgoing data. The bit rate on the highways is 4 Mbit/s. "Distributed processing" assigns all routine and time-consuming tasks to small peripheral processors, always duplicated for safety. An analogous system is described by Kevorkian[16]. His central control is composed of two computers in load sharing mode. If one computer fails, the other one is able to cope with the whole traffic. Fault detection means are implemented. The swedish Hugo system is designed for 128000 terminations. It also has the TST format, but its space switch has the form of a 10 bit parallel 10 stage circular shift register working at 80 MHz. 8 to 10 bits expansion introduces parity and check bits. The reverse concentration 10 to 8 bits is made at the transmitting end. Address memories control the relative addresses from incoming and outgoing groups of 16 time slots or subphases.

An all-electronic, four-stage, non-blocking exchange working in time-space-time multiplex was described by van Helvoort, Havermans and Greefkes[18]. It has time-division multiplexed input and output stages and two space-division multiplex stages, that as usual in such systems work also in TDM. A reduction in cross-talk is achieved by grounding the bus-bars not in use, or by using pulses with a duty cycle of 50% and grounding the bus-bars during the idle period. Non-blocking networks were studied by Clos in 1953 [19].

If a network matrix has n inputs and n outputs it is clear that a connection between a free input and a free output can always be made when there are n^2 cross-point switches. If $n = N^2$ (n and N integers) one can first take N matrices, each of N rows and 2N-1 columns, then a second group of 2N-1 matrices of N rows and N columns, and finally another group of N matrices, with N rows and 2N-1 columns. Each free output can now be reached by each free input again with $3N^2(2N-1)$ crosspoint switches, or $6n^{3/2}-3n$. If one goes to five groups instead of 3 the number is proportional to $n^{4/3}$ and so forth. Since time switching networks can be generated by an equivalence transformation on space division networks[20], the group structure can also be used in this case.

PCM equipment has to compete with the already introduced very economic carrier frequency equipment. It is therefore desirable that a new switching network can

connect both digital and conventional exchanges. This is possible in the italian PROTEO system[21]. An actual trend in the radio relay systems at frequencies below 10 GHz, and in the channels allocated to the mobile services is to use the existing radio links to carry digital information[22,23]. These authors see good possibilities in frequency shift keying systems.
The frequencies of a multitone signalling system can be detected by the application of nonrecursive digital filters to find both Fourier coefficients and hence the amplitude at a given frequency[24].
Vocoders, first described in 1929 by Homer Dudley, are sometimes used in the military sector to transmit digital speech signals over adverse narrow band channels. In the CIPHON vocoder the acoustical signal is analysed through a set of 32 channel filters. The central frequencies are exponential functions of channel order and the selectivity factor is a linear function of the frequency. The output signal of each channel is rectified, smoothed and sampled at a 300 Hz rate. Data compression gives the frequencies and amplitudes of the three predominant formant peaks. Analogously the output of a 500 Hz filter is dedicated to the analysis of melody, by rectification and analysis through a set of 32 bandpass filters in the range 80-250 Hz. Frequency (5 bits) and amplitude (4 bits) are sampled every 6,66 msec and the pitch frequency (5 bits) is renewed every 13.33 msec. Further data compression leads to 2400, 1200, or 600 bits/s, all with fair intelligibility[25]. Derived from the channel vocoder is the vocal synthesizer unit. Acousticdata (sentences, words, syllabes) are stored in the form of vocoder code samples in the speech data memory. This enables the synthesis of all numbers and one minute twenty seconds of miscellaneous sentences. The french PTT network will use this system for automatic enquiries or disnumbering, later to answer enquiries on charges, routing and for automatic consultation of the telephone directory[26].
Rothgordt[27] uses an intermediate ternary code for facsimile. A sequence of pels (picture elements) is binary coded as a number in 0's and 1's for a white runlength and as binary number in 0's and 2's for a black run length. The result may be, e.g. 110 20 1001 200 101. Now one combines these numbers in ternary pentades 11020 10012 00101 and translates the result in decimal numbers: 114 86 10. This leads to the binary octades 01110010 01010110 00001010 that are transmitted. The remaining codewords between the decimal numbers 243 and 255 are used for synchronisation,

end of line and error correction purposes. Every scan line terminates in an end of line codeword. The redundancy of the code allows in more than 90% of the cases to localise one short error burst.
The writing of data on a blackboard is a low frequency process. The position of the moving pen-point is coded, and the action of putting down or lifting the pen-point has a specially protected codeword[28]. 18 bits are necessary to fix the first pel of a trace. The first direction is chosen from 20 directions (5 bits). A relatively complicated picture, sketched in 90 seconds, requires an average data rate of 141 bits/s.
Fjällbrant[29] samples speech at 1.56 times the Nyquist rate. Afterwards he deletes either every second sample or he chooses the skipped samples in a random manner with the restrictions that not more than two samples in succession should be deleted, and an average sampling rate close to 0.5 of the original rate should be obtained. The non-uniformly distributed samples can be uniformly arranged in the transmission media. A non-recursive filter with five sets of coefficients stored in a ROM memory is used for the reconstruction network. The nonuniform case is preferred to the uniform one.
Adaptive array technology can be used to discriminate against interference and to discriminate against multipath. The communications signal needs a unique feature, e.g. a pilot signal. The array may consist of one omnidirectional antenna and three antennas with nulls in random directions. A gradient search routine may be used to find the optimum situation against interference. With multipath the reference processor is locked in time to one of the modes. The array can give 15 kB discrimination against multipath. In case of an interfering signal (white noise on the same frequency) about 28 dB processing gain was found[30].
Use of distributed microprocessors in telecommunication systems is expected to simplify some of the problems connected with central processing, e.g. reliability and availability, changeability and extensibility[31]. The establishing of a peripheral processing capacity enables the network to handle certain phases of the connection and the flow of date from user to user without the intervention of the system central unit[32].
The implementation of data modems (modular-demodulator circuits) by means of commercially available general purpose microprocessors, requires a number of signal processing methods, which simplify the arithmetical operations in both transmitter and receiver, so that

sufficient computation time can be gained to enable the use of these microprocessors in versatile data modems of bit speeds up to 9600 bits. Snijders[33] considers 4 phase modulation of two types, 8 phase modulation for 4800 bps and modified 4 phase, 4 amp modulation for 9600 bps. Methods of simplification include 1) Design of a FIR (finite impulse response) filter as an interpolating digital filter. By taking the sampling frequency for filter construction M times higher than the frequency of incoming data the repetition frequency is increased and a better filter can be made. However, since the date come in at the lower rate there are many zeroes in the convolution form requiring only N/M multiplications (if the digital filter is given by N coefficients). 2) Generate the modulated signal as a sum of filter signals, rather than by multiplication. 3) Use "table-look up" to find a filter output. (One or more ROM's contain all filter coefficients). 4) Observe that in phase modulation the signal constellations only use a limited number of values in the x,y plane, e.g. $\pm$ 1 in 4-phase modulation, $\pm$ 1, $\pm$ 3, $\pm$ 5 in 4 phase, 4 amplitude modulation. Only these weights are necessary then ("weighting accumulation"). In the data receiver a simple equalizer is used[34]. Analog modems were constructed by Artom, Paladin and Rocci[35].

4. COMMUNICATIONS AND COMPUTERS. (Chairman L. Dadda, Italy).

Packet switching means that information structured in formatted blocks is exchanged between nodes by "store and forward" techniques, assuring transmission correctness by control procedures. Next to a data field of a fixed maximum length, a header allows directly or by referring to switching node resident tables, the correct routing of information. A private packet switching network joining together data processing centres in many european countries in the European Informatics Network EIN, that had a terminal in one of the rooms of the CINI convent during the Venice Symposium. This network is the result of an agreement between France, Italy, Norway, Portugal, Sweden, Switzerland, United Kingdom, Yugoslavia and Euratom, later joined by the Netherlands and Germany[36,37,38]. In private packet switching networks the administration takes only a very limited responsability, not guaranteeing against packet loss, duplication or sequence modification. User data terminals must control

the packet flow and restore the information by high level end-to-end protocols. In EIN the line protocol deals with problems of error detection and retransmission and the identification of line and switch failures. It is responsible for ensuring that packets arrive error free at the Transport Station. Another protocol governs the interactions between Transport Stations.
Euronet is a public packet switching network for the European Economic Community[39], eventually to be integrated into a public international data network. In January 1979 the network will consts of four nodes in Paris, London, Frankfurt and Rome, linked by 48 Kbit/s circuits with remote concentrators in Amsterdam, Copenhague, Brussels, Luxembourg and Dublin, connected by 9600 bps circuits. The network design provides operational management and development facilities like: registration of data for billing and accounting, analysis of traffic of the network, program testing, fault detection and initiation of diagnostics. The availability is assured greater to be than 99.7% with a connection delay not greater than 250 ms and a packet delay of 70 ms per node.
Among the other networks discussed we mention the IIASA network of the International Institute for Applied Systems Analysis, Vienna, expected to connect Vienna, Bratislava, Berlin (GDR), Budapest, Pisa, Warsaw, Moscow, Kiev and Kiga, the Canadian Integrated Access Network (IANET), the RCPNET among educational and research organisations in Italy, the swedish ANA 30 system and CIGALE, the french packet switching network. In Cigale adaptive routing is based on a path "cost" propagation scheme, in which cost (e.g. proportional to delay) is obtained by performing a distributed algorithm in each node[40]. A conceptually simple adaptive approach is based on the queu length in every node. Positive acknowledgement of received data packets gives a sample and robust system[41].
Although a unique control point is generally advocated, this was impossible for EIN, as it was considered unacceptable for any one participant to be able to interfere with or exclude any other participant. A hierarchical distributed approach seems more acceptable in autonomous but cooperative systems. Although a star network may attain the minimum delay-capacity product, Rubin considers it undesirable for reasons of reliability[42]. He obtains reliable topological structures for hierarchical networks. Bucci[43] c.s. found local optimum allocations based on cost functions and information from other centres. Rubin strives towards a

minimum of the maximum average delay.
The european space agency has instigated a study of a simple TDMA satellite system in a packet switching data network. Open loop control seems sufficient. A bit rate of 20 Mbit/s is possible or 10 earth stations with 16000 packets of 1000 bits[44].
Encryption may be used to protect the transmitted information from unauthorized observers, to detect illegal tampering and for identity control. The addition of a random sequence, possibly with different keys for different users and changing with time is effective[45].

5. COMMUNICATIONS AND SIGNAL PROCESSING IN MEDICINE. (Chairman R.I. Magnusson, Sweden).

Myoelectric signals are used in prosthesis and electrical stimulation of the nerves can be used for sensory feedback for below elbow amputees. Electroencephalograms are analyzed in different ways (power spectra and digital filtering). Electrocardiograms are processed and maps of heart potential constructed. Other applications of signal processing treat carbon dioxide expiratory curves and ultrasonic ophtalmology, as well as chromosome autoradiographs. Contour enhancement of a radiograph by a simple photographic technique was analyzed and tried out. A microcomputer-based communication system for the non-verbal severely handicapped was described. Communications for rural health care and hospital communication systems were the subject of several papers.

6. COMMUNICATIONS IN THE DEVELOPING COUNTRIES. (Chairman B.S.Rzo, India).

Senior officials of several organizations gave a survey of their work for developing countries: Mili-I.T.U., Kirby-C.C.I.R., Fobes-Unesco, Astrain-Intelsat, Mason-United Nations Development Program.
In the more technical papers prediction and entropy of written arabic tents were estimated. New techniques for digitalization of colour t.v. signals for the proposed Arab satellite are studied. Training strategies, assistance programs and pilot plants for developing countries were described as well as educational tv programmes and cheap receivers for satellites. The agricultural real time imaging satellite system (Artiss) and the cryptographic techniques used to

prevent unauthorized observers from getting the information were discussed.
The indonesian satellite communication system and the King Faisal hospital in Riyadh were examples of cases where developing countries are in front of the western world. In the panel Mr. A.C. Clarke (Sri Lanka), probably the inventor of the geostationary satellite, showed pictures on the second hundred years of telephony.

REFERENCES

1. A.A.L.Reid: Uncertainty and inertia: the long term demand for, and planning of, new telecommunication services, vol.1, 17-26.
2. G.Funk: Comparison of data reliability and efficiency in various standard protocols for information exchange in computer teleconrol networks. vol.1, 59-65.
3. K.G.W.Bolton, D.J.Taylor, W.Ritchie, J.P.Wimbush: A large radio-scanning alarm and control scheme for an dectricity distribution network. Some design features with comments on operational experience. vol.1, 88-93.
4. F.Galli, A.Schiavi: The new ENEL data acquisition system for dispatch and remote control of the power generation and transmission system. vol.1, 120-126.
5. D.Bisci, M.Fiorina: Technical characteristics and functional performance of equipment used for operational data transmission in large power systems. vol.1, 150-159.
6. T.Cegrell, S.O.Herngren: TIDAS-The swedish state power board's computerized message switching network. vol.1, 127-136.
7. S.Finnestad, Chr. Magnus, T.Olsen, T.Teien: A new design for teleprotection equipment for use in power networks. vol.1, 160-166.
8. M.Vial, M.Lion: Comac-system. Brive-Toulouse link. vol.1, 167-171.
9. J.Bishop: The use of microprocessors in telecontrol systems. vol.1, 83-87.
10. F.Gaboudin, A.Le Roy, R.Bouafia: The use of process control computers in electric power dispatching (e.g. in Algeria). vol.1, 142-149.
11. D.Bisci, C.Muzzi: Communication problems and user's considerations in a private computer network. vol.2, 71-81.
12. H.C.A.Hankins, T.Sealy: Communications for the power industry. vol.1, 77-82.

13. G.Wiest: Large scale integration in switching systems.vol.1, 502-508.
14. R.le Corvec,J.C.de Wilde: Influence of the stored program central on the structure, defense, maintenability and administrative functions in switching systems. vol.1, 481-485.
15. A.Como: A digital SPC switching system for transit exchanges. vol.1, 344-349.
16. K.B.Kevorkian: PCM transit exchange. vol.1, 335-343.
17. J.Andersson, O.Ryden: Hugo: a proposal for a high capacity modular digital switch using high speed buses. vol.1, 330-334.
18. H.van Helvoort, G. Havermans, J. Greefkes: A simple switching network for analogue and digital signals with reduced crosstalk. vol.1, 475-480.
19. C.Clos: A study of non-blocking switching networks. B.S.T.J. 33, 406-424, 1953.
20. S.Andresen, S.R.Harrison: Towards a general class of time-division-multiplexed connecting networks. IEEE Trans.Comm. COM20, 836-846, 1972.
21. A.Bovo, R.Camiciotoli, R.Delle Donne: Integration of supervision and maintenance for PROTEO system trunk network. vol.1, 467-474.
22. D.Di Zenobio, P.Mandarini, G.F.Meucci: Digital transmission over analog terrestrial radio-links using frequencies under 10 GHz, vol.1, 249-255.
23. C.B.Dekker: A comparison of digital transmission techniques for standard fm mobile radio sets. vol.1, 243-248.
24. M.S.Buser: A digital signal processor for PCM/TDM tone encoding. vol.1, 350-355.
25. C.H.Picon, P.Deman: Survey of speech digitization techniques (non-predictive). vol.1, 292-297.
26. A.Vulmière, J.Guinand, P.Lavanant: O.S.V., a voice-synthesis unit for telecommunications systems. vol.1, 200-205.
27. U.Rothgordt: Facsimile encoding and the influence of transmission errors. vol.1, 206-211.
28. A.Kegel, J.H.Bons: On the digital processing and transmission of handwriting and sketching. vol.1, 324-329.
29. T.Fjällbrant: A method of data reduction of sampled speech signals. Vol.1, 298-304.
30. P.M.Hansen: Application of adaptive array technology to hf communications systems. vol.1,277-285.
31. S.Nestel: Microprocessors in communication systems-centralized vs. distributed processing. vol.2, 159-164.
32. C.Casalino, G.Zaffignani: The multimicroprocessor architecture in electronic exchanges. vol.2,165-169.

33. W.A.M.Snijders: Microprocessor implementation of data modems. vol.1, 556-562.
34. F.de Jager, M.Christiaens: A fast automatic equalizer for data links. Philips Tech. Rev. 37, 10-24, 1977.
35. A.Artom, G.Paladin, R.Rocci: Microcomputer-based ACE with voiceband modem for new data networks vol.1, 563-571.
36. D.L.A.Barber: EIN, a focus for the future. vol.2, 98-111.
37. G.Alirsi, G.Carrera, A.Gambaro, C.le Moli: Structure of the connection of subscriber computers to the EIN node of Milan. vol.2, 118-126.
38. K.Weaving: The design of the network control program for the ISPRA connection to EIN, vol.2, 127-133.
39. F.Arciprete, A.M.Repichini: Packet switching techniques for public and dedicated data communication networks. vol.2, 11-20.
40. J.L.Grange: Operating the CIGALE packet switching network. Concepts, techniques and results. vol.2, 44-49.
41. B.H.Pardoe, R.L.Grimsdale: The effects of node protocol and network flow control in packet-switched communication networks. vol.2, 56-60.
42. J.Rubin: On the design of reliable hierarchical computer communication networks. vol.2, 146-150.
43. G.Bucci, S.Gobinelli: A distributed approach to data base decomposition in hierarchical computer systems. vol.2, 146-150.
44. F.T.Knabe, P.A.Ljungström: A simple TDMA system in a packet switching data network. vol.2, 88-93.
45. J.Ingemarsson: The use of encryption in multiuser communication systems. vol.2, 94-97.

All references not otherwise indicated are to the two vols. Conference Proceedings, IEEE-EUREL, Venice 1977.

THE OPTIMUM TRANSMISSION SIGNAL FOR DIGITAL TRANSMISSION SYSTEMS

H. Marko

Institute of Communications Technique,
Technical University of Munich, Munich, Germany

ABSTRACT. Using reasonable assumptions, this paper will demonstrate that the rectangular pulse is the optimum elementary transmission signal for digital transmission systems. This may seem obvious; however, the introduction and use of various alternative coding systems necessitates its demonstration. Coding systems are spectral deformers of the transmitted signal, and as such include the pseudo-multilevel codes such as bipolar and duo-binary. These all lead to a reduction of the signal-to-noise ratio relative to that of the optimum transmission signal calculated in this paper. The validity of the argument and of the noise margin reduction calculations is independent of modulation method, i. e. binary or true multilevel.

1. BLOCK DIAGRAM OF A DIGITAL TRANSMISSION SYSTEM

The block diagrams of the digital transmission systems on which the following is based are shown in figure 1. System A will be described first. This system, valid for an arbitrary transmission signal, is supplied with a binary or multilevel data signal that is composed of a succession of positive and negative rectangular pulses. The time interval and the duration of these pulses is T. The permissible amplitude range at point 1 is ± 1. The signal is then converted into an arbitrarily coded transmission signal by a coding network with transfer function S_C. This network could, for example, consist of a ladder network with a time delay of T (tapped delay line). With this particular circuit it is possible to generate, among others, the various pseudo-multilevel codes (cp. [1]);

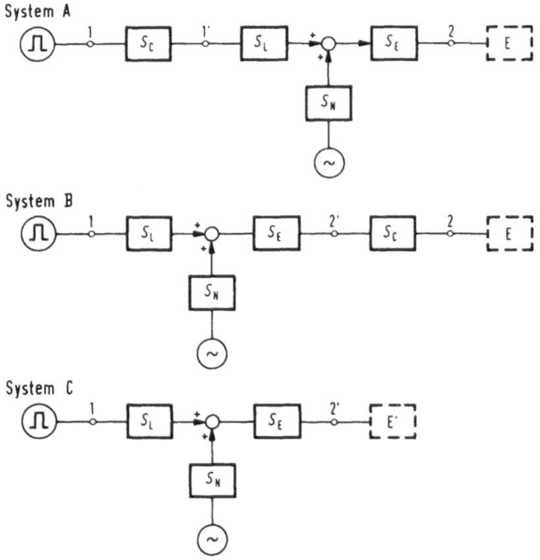

Fig. 1: System variations of digital communication systems

the preliminary modulo-2 adder required for this purpose may be ignored for our considerations. The permissible amplitude range at point 1' is also ± 1, i. e. the digital signal may not, under any circumstances nor for any coding combination, lie outside this range.

Following this is the circuit with transfer function S_L and the equalizing network with transfer function S_E, which also supplies the necessary amplification to ensure that the signal at point 2 is again in the permissible amplitude range of ± 1. (The latter is necessary only for normalization).

Connected at point 2 is the non-linear and in many cases timed receiver E, which performs the digital determination with regard to the received signal. The particular structure of this unit need not be gone into here. Customarily we would think of a timed threshold receiver that evaluates each message unit individually, rather than a correlation receiver that evaluates a number of message units simultaneously. The interference signal is coupled into the circuit in front of the equalizing network (at which point the strength of the received signal is weakest). Normally this interference is white noise, but colored noise (or similarly acting crosstalk noise from adjacent systems or both) may be taken into consideration with the S_N term. With this the noise source has the frequency independent power spectrum N_0, and the total noise power of system A at point 2 is given by:

$$N_A = N_0 \int_{-\infty}^{+\infty} |S_N|^2 \, |S_E|^2 \, df . \quad (1)$$

System B is considered next. This system is operationally similar to system A except for the connection of coding network S_C which is now on the receiver side of S_E. The signal itself is unaffected by this change. Arguments aimed at finding a suitable spectral form of the transmission signal at point 1', e. g. for the purpose of avoiding DC current considerations, may be bypassed as the signal at point 2 is precisely the same, independent of the properties of network S_L, in both systems. The composite transfer function from point 1 to point 2 for systems A and B is given by:

$$S_{12} = S_C \cdot S_L \cdot S_E . \quad (2)$$

A difference does, however, exist between system A and system B with regard to noise signals. Namely, that the total noise at point 2 of system B is given by:

$$N_B = N_0 \int_{-\infty}^{+\infty} |S_N|^2 \, |S_E|^2 \, |S_C|^2 \, df . \quad (3)$$

In the following section it will be shown that in the general case $|S_C| \leq 1$ is valid for all frequencies. It then follows that the reduction in the signal-to-noise ratio of system A with respect to system B is given by:

$$V_{AB} = 10 \log \frac{N_A}{N_B} = 10 \log \frac{\int_{-\infty}^{+\infty} |S_N|^2 \, |S_E|^2 \, df}{\int_{-\infty}^{+\infty} |S_N|^2 \, |S_E|^2 \, |S_C|^2 \, df} \geq 0 . \quad (4)$$

The transmission signal of system B is therefore to be regarded as optimum. It follows that the optimum system transmits basic binary (or multilevel) code with a rectangular pulse as the elementary signal.

The only possible further improvement that may be made to system B is the utilization of the signal present at point 2', as opposed to that at 2, through use of an appropriate receiver. This may correspond to a system C of figure 1, in which the coding network S_C has been omitted. A comparison of systems A and C shows that the noise levels are the same in either case, but that the transmission signals are different. If S_C generates a pseudo-multilevel code with p amplitude levels, then the transmission signal at point 2 of system A has the amplitude step of 2/(p-1). This is a consequence of the amplitude range being ± 1 and the amplitude steps having the same interval, as may be assumed under best-case conditions. Similarly, a transmission signal with q amplitude levels at point 2' of system C has an amplitude step of 2/(q-1). (It is also assumed that at the sampling instants there is no intersymbol interference, which is the case for an optimum S_E; cp. [2]). Therefore, the signal-to-noise-ratio reduction (for pseudo-multilevel codes) of system A compared to C may be calculated by use of:

$$V_{AC} = 20 \log \frac{p-1}{q-1} \qquad (5)$$

Commonly q = 2 and p = 3 (bipolar, duo-binary codes) giving a V_{AC} = 6 dB.

A comparison of system C with system B reveals that system C does not include the network S_C. In the case of pseudo-multilevel codes the receiver E of system B operates with p amplitude levels and therefore with p-1 thresholds. Receiver E' of system C, on the other hand, is simpler, having to differentiate between only q levels and q-1 thresholds

Prior to demonstrating that $|S_C| \leqslant 1$, which is the necessary condition for equation (4) to hold, it should be pointed out that the crosstalk interference between two type

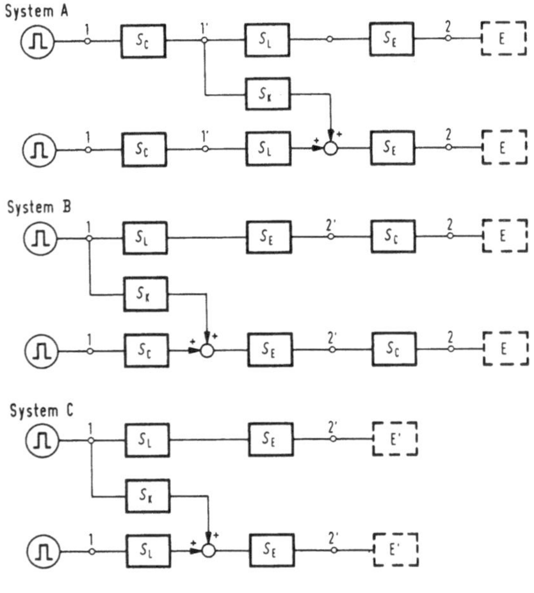

Fig.2: Crosstalk interference for the three systems

A systems is equal to that between two type B systems. As is clearly shown in figure 2, the transfer function for crosstalk interference from point 1 to point 2 is

$$S_{12N} = S_C \, S_K \, S_E \tag{6}$$

for both system A and B. S_K is the transfer function for the crosstalk coupling. System A and system B therefore have equal performance with respect to crosstalk.

In system C a certain amount of improvement is possible. At point 2' both signal and noise levels are higher than at point 2, as long as $|S_C| \leq 1$. Now if S_C generates a pseudo-multilevel code, and given the same conditions that applied previously, then the interference margin reduction for system A or B compared with system C is given by

$$V_{AC} = V_{BC} = 10 \log \frac{\int_{-\infty}^{+\infty} |S_Q|^2 \cdot |S_K|^2 |S_E|^2 |S_C|^2 \, df}{\int_{-\infty}^{+\infty} |S_Q|^2 \cdot |S_K|^2 |S_E|^2 \, df} + 20 \log \frac{p-1}{q-1} \, . \tag{7}$$

Here $|S_Q|^2$ is the signal power spectrum of the source and is, for example, given by $|S_Q|^2 = T^2 \, si^2 \, (\pi f T)$, which holds for a random binary signal with equal state probabilities.

2. PROOF OF CONDITION $|S_C| \leq 1$

This proof assumes that the amplitude ranges at points 1 and 1' of figure 1(system A) are equal. This assumption is necessary for a valid and technically realistic comparison. Points 1 and 1' are, though not expressly indicated as such on the block diagrams, the outputs of the power amplifiers. These amplifiers limit the amplitude range at both points by limiting the peak signal amplitude. Due to the desired code transparency (viz.every signal must be capable of transmission and reception), the worst-case code combinations must be used in calculating the peak signal amplitudes.

With these assumptions the proof will be carried out in two steps: first for a rectangular signal (NRZ-code) and then for an arbitrarily modified signal (e.g. RZ-code).

Figure 3 shows a rectangular signal with amplitude 1 and period T which is the elementary transmission signal of a binary or multilevel digital communications system.

A series of such signals, of either positive or negative polarity as determined by the transmission signal, are present at point 1 of figure 1. In the case of a binary code their amplitudes are ± 1, and in the multilevel case $\pm c_\mu$,

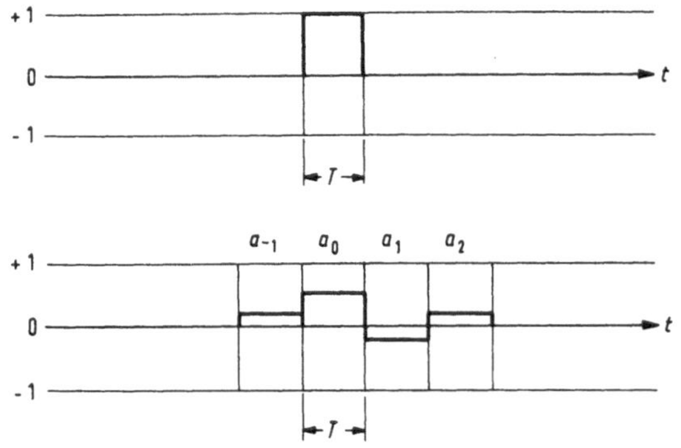

Fig.3: Elementary signal at point 1 (above) and point 1' (below) when using coding network of figure 4 ($a_{-1}=a_2$ = 1/6, a_o = 1/2, a_1 = -1/6)

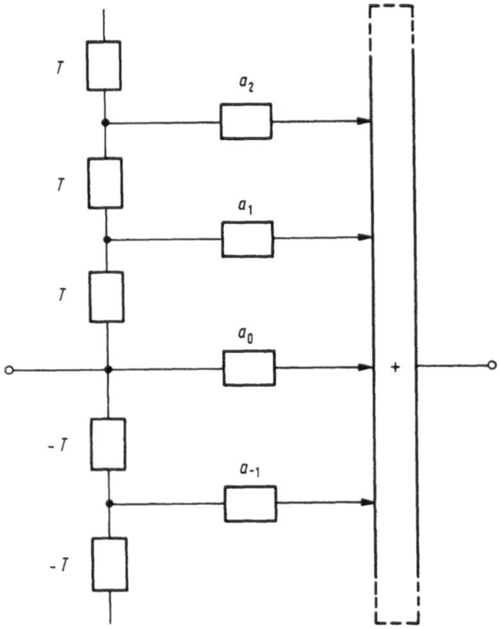

Fig.4: Coding network S_C as ladder network

where $|c_\mu| \leq 1$. This ensures that, while the available amplitude range is completely used, its limits will not be exceeded. An arbitrary signal is now constructed at point 1' of system A from time displaced rectangular pulses (see fig.3). The transfer function S_C is, therefore, as in figure 4. (The fact that this involves negative time delays is not troublesome as these may be eliminated through introduction of a time delay network T_0 in front of S_C). The coefficients a_ν may be chosen arbitrarily, bearing in mind the amplitude range limitation of $\pm$ 1 at point 1'. It must then be assumed that these fundamental signals may be superimposed as determined by the coding network. It is now apparent that due to the amplitude restriction,

$$\sum_{-\infty}^{+\infty} |a_\nu| = 1 . \qquad (8)$$

There is then in all cases at least one code that attains the amplitude limit of +1. (In the example shown in the figure this is +1 -1 +1 +1 and the peak amplitude is reached in the third time interval). Nearly any wonted signal may therefore be constructed; their amplitudes are, however, restricted by equation (8). For the first order bipolar code, for example, $a_0 = 1/2$, $a_1 = -1/2$ and all other $a_\nu = 0$. For the second order bipolar code $a_0 = 1/2$, $a_2 = -1/2$ and all other $a_\nu = 0$. And for the duo-binary code, $a_0 = 1/2$ and $a_1 = 1/2$, with all other $a_\nu = 0$.

Using figure 4, S_C is immediately seen to be given by

$$S_C = \sum_{-\infty}^{+\infty} a_\nu \, e^{-j\omega\nu T} . \qquad (9)$$

Clearly for every frequency

$$|S_C| \leqq 1 \qquad (10)$$

since the summation in eqation (9) must always be less than or equal to the summation of the absolute values of a_ν 's. It has then been shown that codes of this type (pseudo-multilevel) lead to a reduction of the noise margin, which reduction may be calculated from equations (4) or (5).

This result may now be modified and broadened if the slopes of the signal shown in figure 3 are smoothed a continuous signal results. This may be accomplished through use of a low-pass filter with transfer function S_G (see figure 5). Then

$$S_C = S_G \qquad (11)$$

for the source code and

$$S_C = S_G \sum_{\nu=-\infty}^{+\infty} a_\nu \, e^{-j\omega\nu T} \tag{12}$$

for the pseudo-multilevel code. Since the low-pass filter should not reduce the amplitude range, the following conditions must be met to avoid overshoods.

$$\begin{aligned} S_G &= 1 \text{ for } f = 0 \\ |S_G| &< 1 \text{ for } |f| > 0\,. \end{aligned} \tag{13}$$

This type of smoothed signal would therefore compare unfavorably with the rectangular pulse with regard to noise margins.

Next we consider signals with short durations, such as RZ-signals. An example is shown in figure 6. The transfer function that generates an impulse of duration ΔT from an impulse of duration T, with both impulses being equal in amplitude, is given by

$$S_I = \frac{\Delta T}{T} \cdot \frac{\mathrm{si}(\pi f \Delta T)}{\mathrm{si}(\pi f T)}\,. \tag{14}$$

(Here si (x) = sin x/x, but the physical realization of such a function will not be gone into here).

Now $|S_I|$ is not less than or equal to 1 over the complete frequency range. In fact, there are even poles of S_I at the frequency $f = \nu/T$. The frequency range is nevertheless limited by S_E to a bandwidth of $f_o = 1/2$ T. For $|f| < 1/2$ T the condition $|S_I| \leqq 1$ still holds and for $f \to 0$ $|S_I| = \Delta T/T$. From this it may be seen that the use of short duration signals (RZ-codes) also leads to a reduction of the noise margin.

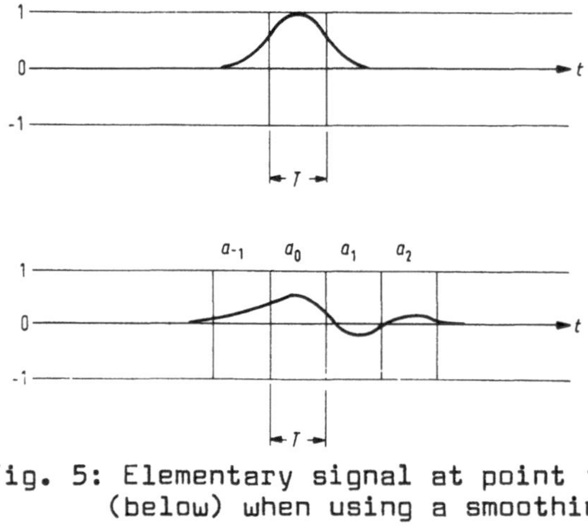

Fig. 5: Elementary signal at point 1 (above) and point 1' (below) when using a smoothing low pass filter

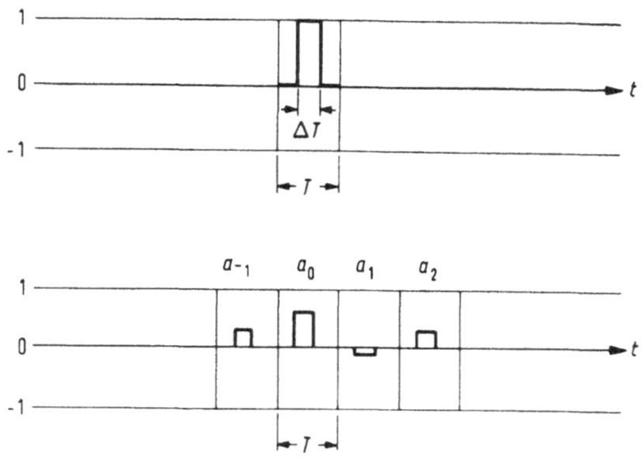

Fig.6: Elementary signal at point 1 (above) and point 1' (below) when using RZ code.

3. NOISE MARGIN REDUCTIONS FOR SOME COMMON PSEUDO-MULTI-LEVEL CODES

In this section the noise margin reduction will be calculated for common transmission codes using equation (4). The coding network S_C is used and S_N is set equal to 1 over the entire frequency band (white noise). The transmission line transfer function S_L is assumed constant and is given by

$$S_L = e^{-\alpha \sqrt{f/f_0}} . \qquad (15)$$

This indicates that the attenuation varies over the frequency range as a square root function (e.g.coaxial cable), where α is the attenuation at f_o = 1/2 T (determined by the bandwidth of the digital signal). To be able to evaluate equation (4) the following various assumptions could be made for frequency gain of the equalizer

$$\begin{aligned} S_E &= \frac{1}{S_L}, \qquad |f| < f_0 ; \\ &= 0, \qquad |f| > f_0 ; \end{aligned} \qquad (16)$$

$$\begin{aligned} S_E &= \frac{1}{S_S\, S_L}, \qquad |f| < f_0 ; \\ &= 0, \qquad |f| > f_0 ; \end{aligned} \qquad (17)$$

$$S_E = \frac{S_S\, S_L(f)}{\sum\limits_{-\infty}^{+\infty} S_S^2\, S_L^2\left(f - \frac{\nu}{T}\right)} \qquad (18)$$

$$S_E = \frac{1}{S_L} \cdot \exp\left[-\pi\left(\frac{f}{2\, f g}\right)^2\right] \qquad (19)$$

The low-pass filter function for the rectangular pulse signal

$$S_S = \mathrm{si}(\pi f T) \quad (20)$$

was used in equations (17) and (18).

An equalizer gain according to equation (16) would lead to intersymbol interference exceeding the tolerable range and is therefore not permissible. The equalizer gain of equation (17) leads to the sin x/x-function for the rectangular pulse answer at point 2' (see figure 1). Intersymbol interference is then avoided due to the fact that the zero crossings of the sin x/x-function occur at the sampling instants. This also holds if a coding network S_C is inserted. However equation (17) is not the optimum equalizer. It is shown in [2] that the optimum equalizer is given by equation (18). This also leads to a system without intersymbol interference with an elementary signal similar but not equal to the sin x/x-function.

Figure 7 shows the optimum transfer function $S = S_E\ S_L$ for various values of α, namely α = o; 2.3 N (20 dB); 4.6 N (40 dB); 6,9 N (60dB); 9,2 N (80 dB) and 11.5 N (100 dB). For large values of α S approaches S_E as given by equation (17) and as shown in figure 7 by the dashed line. This interference-free optimum system according to equation (18) is, however, difficult to realize physically and requires tight

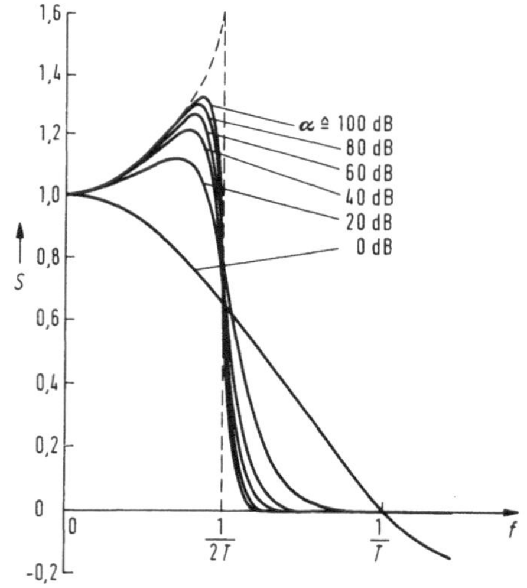

Fig.7: Optimal transfer ratio S for binary and multilevel transmission signals assuming white noise; shown for various values of transmission line attenuation (α: line attenuation at f_0 = 1/2 T).

sampling time tolerances. For this reason the analyses of [2] and [3] were based on systems that could be realized with Gaussian transfer functions for $S = S_E S_L$ (cp. eq. (19)). This allows the frequency f_g to be optimized for each value of α, with the following results (for binary coding and a receiver without quantized feedback):

$\alpha =$	0	20	40	60	80	100	dB
$f_g =$	1,5	1,0	0,78	0,7	0,66	0,62	$\cdot 1/2\,T$

Figure 8 shows the respective transfer functions.

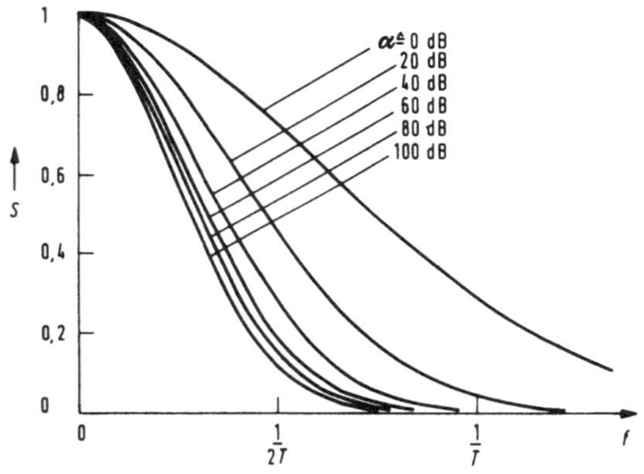

Fig. 8: Gaussian transfer function S, optimized for binary signals and white noise, for various values of transmission line attenuation (α: line attenuation at $f_0 = 1/2T$).

The noise margin reduction V_{AB}, eq. (4), was calculated for the equalizing network transfer function as given in eq. (18) (optimum interference-free system) and also as given by eq. (19) (Gaussian low-pass for $S = S_E S_L$). In doing so S_N was set equal to 1 (white noise). The results for first and second order bipolar code and for duo-binary code are shown in Figure 9. Also shown in this figure are the block diagrams for S_C and plots of S_C versus frequency.

This figure shows that for the first order bipolar code V_{AB} decreases with increasing α (transmission line attenuation) for both optimum and Gaussian low-pass filters. As V_{AC} is constant at 6 dB for all codes (see eq. (5)), system C is to be preferred in this instance. The second order bipolar and the duobinary codes, on the other hand, show a considerable increase in V_{AB} with increasing α. This is particularly true for the optimum filter and is due to the fact that S_C has a zero within the range of the Nyquist slope

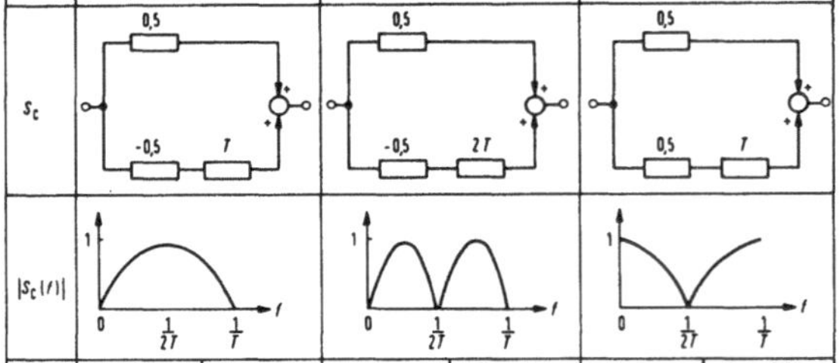

V_{AB}/dB	Bipolar-Code 1. Ordnung Opt. Filter	Bipolar-Code 1. Ordnung Gauß-T.P.	Bipolar-Code 2. Ordnung Opt. Filter	Bipolar-Code 2. Ordnung Gauß-T.P.	Duobinär-Code Opt. Filter	Duobinär-Code Gauß-T.P.
α ≐ 0	3,01	3,14	3,01	3,01	3,01	2,89
20 dB	1,00	1,75	3,80	2,80	6,88	4,80
40 dB	0,47	1,16	5,52	3,30	9,86	6,29
60 dB	0,27	0,98	7,18	3,62	12,13	6,94
80 dB	0,18	1,07	8,66	3,49	13,94	6,59
100 dB	0,12	1,14	9,96	3,31	15,44	6,38

Fig. 9: Noise margin reductions for white noise; shown for optimum and Gaussian low-pass filters (α : line attenuation at f_0 = 1/2 T)

f_0 = 1/2 T, whereby the high frequency noise is strongly attenuated. From a comparison of values for the Gaussian low-pass filter it may be seen the improvement in V_{AB} is not much over 6 dB. This indicates that for the Gaussian low-pass filter the basic binary system, system C, should be used. Comparatively, a gain of well over 6 dB may be obtained by employing an optimum filter. In this case the use of system B and duo-binary coding in the receiver is to be preferred. This type of system, while using a binary transmission code, requires the use of a ternary receiver with two thresholds. This will yield, for example, a 14 dB improvement over system A and an 8 dB improvement over system C (true binary) for an α = 80 dB. It should, however, be pointed out that the use of quantized feedback (cp. [2]) will improve the performance of system C. With feedback from a single quantized sample this improvement for large values of α, is sufficient to make system C comparable to system B.
The improvement by using quantized feedback is even more pronounced if multiple samples are fed back and system C then becomes the optimum system. In any case, system A is distinctly the worst which proves the contention that the rectangular pulse is the optimum transmission signal.

In figure 9 an S_N of 1 was assumed, which was reasonable for thermal noise (coaxial line). However, with symmetrical cables switching noise may also be present. Using

$$S_N = 1, \; |f| < f_u ; \quad S_N = 0, \; |f| > f_u \tag{21}$$

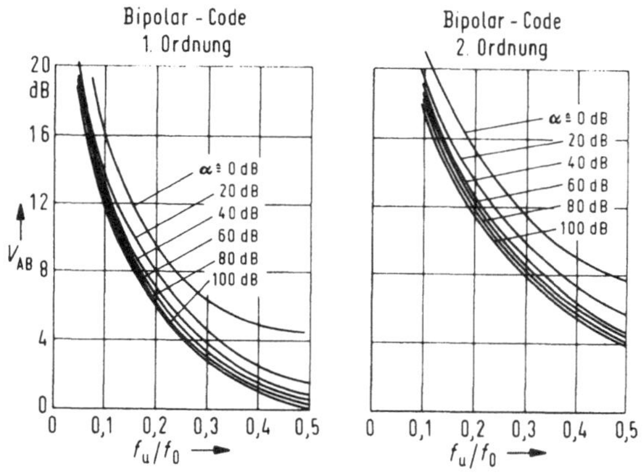

Fig. 10: Noise margin reduction V_{AB} for low frequency noise (α : line attenuation at $f_0 = 1/2\ T$)

V_{AB} is again calculated from eq. (4). The use of either the optimum filter transfer function (eq.(18)) or the Gaussian low-pass transfer function (eq.(19)) yields approximately the same result and even variations in α have a relatively minor effect. Figure 10 shows V_{AB} plotted against f_u/f_0 for the first and second order bipolar codes. At low values of f_u much better performance may be obtained from system B. Therefore, for values of $f_u/f_0 < 0.4$ for first order and $f_u/f_0 < 0.2$ for second order bipolar code, system B should be given preference over system C as V_{AC} becomes larger than 6 dB. Here again ternary coding at the receiver side results in a gain due to S_r having a zero at f = 0.

It has then been shown that system A is, in all cases, out performed by systems B and C, viz. the optimum transmission signal is a rectangular pulse and a redundancy-free code.

SUMMARY

In the USA bipolar code was introduced quite early by A.T.T. (Bell System) as the transmission code for their PCM system, T1-System.

It was then adopted with few modifications by most European countries and by Japan for their PCM systems. This is regrettable considering the inherently poor noise margin at the low frequencies present in the local communications exchanges.

Stimulated by the promise of an improved signal-to-noise, A.T.T. is now using redundancy-free binary code in their new T4-System.

Pseudo-multilevel (partial-response codes) are occasionally considered for use with long distance digital systems. Bell even introduced a 4/7-partial response code for a 1.5 Mbit/s data-under-voice channel. This makes use of the coding network S_C for second order bipolar code (see figure 9). Presently this code is also being considered for use with long range systems employing coaxial transmission lines. As the results of this paper are also valid for multilevel systems, the reduction in interference margin for the 4/7-partial response code in comparison to a quarternary code, with coding network and 7 level sampling at the receiver, may be read directly from the middle column of figure 9. For attenuation (α) between 60 and 80 dB at f_0 = 1/2 T this reduction is approximately 8 dB for the optimum filter and 3.5 dB for the Gaussian low-pass filter. In comparison to the normal quaternary system the reduction is 6 dB. Therefore, 4/7-code is not to be recommended for use with coaxial systems. Particulars concerning the advantages of binary code are considered in [2] and [4] .

* * *

The author's thanks are due to Dr. Appel and Dr. Tröndle not only for performance of the computations and design of the program used for this analysis, but also for many critical discussions and helpful suggestions.
The original paper has been published in german language [7]. The author thanks Prof. Cattermole for correcting the english version.

REFERENCES

[1] Appel, U.; Tröndle, K.: Zusammenstellung und Gruppierung verschiedener Codes für die Übertragung digitaler Signale. NTZ 23 (1970) p.11 - 16.

[2] Marko, H.: Optimale und fast optimale binäre und mehrstufige digitale Übertragungssysteme. AEÜ 28 (1974) v.10, p. 402 - 414.

[3] Marko, H.: Comparison of binary and multilevel digital transmission systems in the presence of intersymbol interference and thermal noise. Nachrichtentechn.Z. 27 (1974) v.6, p. 239 - 242.

[4] Marko, H.; Tröndle, K.: Der Grenzverstärkerabstand digitaler Übertragungssysteme für koaxiale Kabel bei gaußförmigem Übertragungsfaktor. NTZ 28 (1975) v. 5, p. 160 - 165.

[5] Marko, H.; Tröndle, K.; Söder, G.: Vergleich binärer und mehrstufiger Regenerativverstärkersysteme für koaxiale Kabel bei symmetrischer Impulsform unter Berücksichtigung der Toleranzen. NTZ 29 (1976) v. 8, p. 601 - 608.

[5] Marko, H.; Tröndle, K.; Söder, G.: Vergleich binärer und mehrstufiger Regenerativverstärkersysteme für koaxiale Kabel bei symmetrischer Impulsform unter Berücksichtigung der Toleranzen. NTZ 29 (1976) v. 8, p. 601 - 608.

[6] Marko, H.; Tröndle, K.; Söder, G.: Vergleich optimaler, binärer und mehrstufiger Regenerativverstärkersysteme mit quantisierter Rückkopplung und unsymmetrischer Impulsform. NTZ 30 (1977) v. 4, p. 316 - 323.

[7] Marko, H.: Das optimale Sendesignal (Leitungscode) für digitale Übertragungssysteme. NTZ 28 (1975) v. 1, p. 7 - 12.

TIMING RECOVERY PROBLEMS IN DATA COMMUNICATION*

L. E. Franks

Department of Electrical and Computer Engineering
University of Massachusetts
Amherst, MA 01003

INTRODUCTION

Signal formats designed for highly efficient utilization of channel capacity will exhibit the following properties: (1) no data-independent signal components; such as a residual unmodulated carrier or added periodic pilot tones. This also usually means that the modulating data signal should be a zero-mean process. (2) a bandwidth-efficient carrier modulation scheme such as vestigial-sideband (VSB) or quadrature-amplitude (QAM) modulation of a sinusoidal carrier with a multilevel baseband pulse-amplitude modulated (PAM) digital data signal. (3) sophisticated pulse design and equalization capability in order to minimize sideband bandwidths. In some systems, the actual bandwidth is only a small fraction in excess of the idealized Nyquist bandwidth for the signal format employed.

Successful demodulation of signals of this type requires highly accurate recovery of timing information. A coherent (phase-locked) reference carrier is required at the receiver to avoid quadrature distortion (VSB) or co-channel interference (QAM). In addition, an accurate reference for symbol timing must be developed at the receiver in order to avoid intersymbol interference in the sampling process. Timing recovery problems are compounded by the fact that with signals exhibiting the above properties (1-3), their periodically-varying features tend to be highly suppressed. It is the periodic (cyclostationary)

*This work has been supported in part by the National Science Foundation under Grant ENG 76-19492.

statistical properties [1] of the signal that make timing recovery, and hence self-synchronizing systems, possible.

Our objective is to determine the receiver structure and performance of various schemes to estimate the parameters τ and θ in a received carrier signal of the form, $z(t) = y(t) + u(t)$, where $u(t)$ represents white, Gaussian noise with a power spectral density of N_0 watts/Hz, and the noise-free carrier signal is

$$y(t) = \operatorname{Re}[\beta(t)\exp(j2\pi f_0 t)] \tag{1}$$

where f_0 is the carrier frequency and $\beta(t)$ is the complex envelope representation for $y(t)$. In the case of VSB/PAM, we have

$$\beta(t) = \sum_{k=-\infty}^{\infty} a_k \gamma(t - kT - \tau)\exp(j\theta) \tag{2}$$

where $\gamma(t) = g(t) + j\tilde{g}(t)$ and $\{a_k\}$ is a wide-sense stationary data sequence. We refer to $g(t)$ as the baseband pulse shape and $\tilde{g}(t)$ is related to it through a time-invariant linear transformation such that $\Gamma(f) = 0$ for $f < -\delta/2T$. This corresponds to an upper-sideband signal and the lower (vestigial) sideband occupancy is expressed as a fraction (δ) of the Nyquist bandwidth ($1/2T$).

In the QAM/PAM case, we use two data sequences and

$$\beta(t) = \exp(j\theta) \sum_k a_k g(t-kT-\tau) + jb_k h(t-kT-\tau) \tag{3}$$

where $g(t)$ and $h(t)$ are real, baseband pulses. In normal QAM, the same pulse shapes are used in both in-phase and quadrature channels, however another arrangement, known as staggered QAM (SQAM), has some important implications in timing recovery. In this case we have $h(t) = g(t - T/2)$. In both the VSB and QAM cases, we assume that the baseband pulses are bandlimited to a fraction, γ, in excess of the Nyquist bandwidth; i.e., $G(f)$, $H(f) = 0$ for $|f| > (1 + \gamma)/2T$.

The carrier phase and symbol timing parameters, θ and τ, are assumed to be unknown, non-random parameters. This is the classical situation for application of the maximum-likelihood estimation principle [2]. We propose estimator structures based on this principle, and develop expressions for the r.m.s. value of error (also called jitter) in the estimation of these parameters. Our specific goals are to relate the r.m.s. jitter to the following system parameters: (1) receiver complexity; the simplest receiver designs result from independent estimation of the carrier phase, symbol timing, and data parameters but the performance of this approach may be inadequate. Improved performance results when various strategies of joint estimation are employed. (2) Signal design; both the modulation format (VSB, QAM, or SQAM) and the baseband pulse shape have important effects on jitter performance. (3) Signal bandwidth; the behavior of r.m.s. jitter as a function of the excess bandwidth parameters, δ and γ, is of considerable interest, especially when these values are relatively small. Many previous studies of timing jitter have tended to overlook the situation of a high degree of pulse overlap that occurs in systems with small excess bandwidth. Finally, our study will emphasize the high signal-to-noise ratio performance of the parameter estimation schemes. Timing jitter has two components; that due to the additive noise and that due to the random data signal itself (sometimes called the pattern-dependent jitter). The first component tends to be quite similar in all estimation strategies, while the second component can be markedly affected by system design.

BACKGROUND: BASEBAND PAM TIMING RECOVERY

Despite the fact that baseband timing recovery has been extensively studied [3] - [7], and that we are primarily interested here in the possibilities inherent in joint parameter estimation in carrier systems, the single-parameter baseband case is an effective and simple means to illustrate the principles used in the carrier-system analysis. Accordingly, we first examine the problems associated with estimation of τ in the case that $z(t) = y(t) + u(t)$ and

$$y(t) = \sum_k a_k g(t - kT - \tau) \tag{4}$$

The likelihood function, within an arbitrary multiplicative constant, for this case is; [3]

$$L(\hat{\tau}, \{\hat{a}_k\}) = \exp\left[\frac{1}{N_0}\int_{T_0} z(t)\hat{y}(t)dt - \frac{1}{2N_0}\int_{T_0} \hat{y}^2(t)dt\right] \tag{5}$$

where $\hat{y}(t)$ carries the <u>estimates</u> ($\hat{\tau}$, $\hat{a}_k$) of the parameters in (4). We assume that the observation interval, T_0, is long compared to a symbol period, T. This allows us to make the approximation,

$$\int_{T_0} z(t) \sum_{k=-\infty}^{\infty} \hat{a}_k g(t-kT-\hat{\tau})dt$$

$$\cong \sum_{k=0}^{K-1} \hat{a}_k \int_{-\infty}^{\infty} z(t)g(t-kT-\hat{\tau})dt$$

$$= \sum_{k=0}^{K-1} \hat{a}_k q_k \qquad (6)$$

where $T_0 \cong KT$ and q_k can be interpreted [3], [4] as the sampled (at $t = KT + \hat{\tau}$) output of a matched filter, $g(-t)$, driven by the received signal, $z(t)$. We assume that the last term in the exponent of (5) is substantially independent of $\hat{\tau}$, so that a maximum-likelihood receiver determines $\hat{\tau}$ to maximize (6). The popular decision-directed or "data-aided" strategy [8] - [14] involves the joint estimation of the timing parameter $\hat{\tau}$ and the data parameters, $\hat{a}_k$, so that the output of the data detector is fed back into the timing recovery circuit which operates the sampling switch. In analyzing the jitter performance of this strategy, it is usually assumed that $\hat{a}_k = a_k$; i.e., data errors are sufficiently infrequent that they do not affect r.m.s. jitter.

If one desires not to use the data-aided strategy, then the appropriate likelihood function to use is (5) after averaging it over all the data variables. This averaging is easily performed if we assume that the data are jointly Gaussian random variables with mean, m, and covariance matrix, M. Then we get

$$L_0(\hat{\tau}) = \underset{\{a_k\}}{E} [L(\hat{\tau}, \{a_k\})]$$

$$= \text{const.} \times \int \cdots \int L(\hat{\tau}, x_i \cdots) \exp - \frac{1}{2} \sum\sum M_{ij}^{-1}(x_i-m)(x_j-m)dx_i \cdots$$

$$= \text{const.} \times \exp - \frac{1}{2} \sum_{kj}\sum Q_{kj}(q_j - \mu)(q_k - \mu) \qquad (7)$$

The integration above is performed by "completing the square" on the exponent of the integrand to form a multivariate Gaussian p.d.f. which integrates to one. The matrix Q is defined by

$$Q^{-1} = M^{-1} + \frac{1}{2N_0} \Gamma \qquad (8)$$

where

$$\Gamma = \int_{-\infty}^{\infty} g(t-kT)g(t-mT)dt \stackrel{\Delta}{=} r(kT-mT) \tag{9}$$

and $\mu = -2N_0M^{-1}m$.

The q_k values in (7) are the same sampled matched-filter outputs as in (6). The bias term μ tends to zero for high signal-to-noise ratio and is zero in any case for zero-mean data. We note that Q is a positive-definite Toeplitz matrix so that it can be factored, $Q = V'V$, and we can write the log-likelihood function as

$$\Lambda(\hat{\tau}) = \ell n\ L(\hat{\tau}) = \text{const.} \times \sum_{k=0}^{K-1} s_k^2 + \text{const.} \tag{10}$$

where (assuming m = 0),

$$s_k = \sum_j v_{k-j}q_j \tag{11}$$

From (11) we see that the test statistic can be obtained by filtering the sampled output sequence from the matched filter with a discrete-time filter having a transfer function, $V(\exp - j2\pi Tf)$. Alternatively, the matched filter can be modified so that its sampled output is the s_k sequence. From the expression (8) for Q, it is interesting to note that for high $SNR(N_0 \to 0)$, the correlation in the data sequence becomes relatively unimportant in determining the maximum-likelihood estimator, while intersymbol interference (as represented by Γ) has a major effect. Note that the elements of Γ are the samples of the time-ambiguity function [1] for the baseband pulse, g(t). If r(t) is a Nyquist pulse, $r(kT) = 0$ for $k \neq 0$, and if the data is uncorrelated, then Q is diagonal and the K-term sum of squares of the matched filter output is the appropriate test statistic.

Choosing $\hat{\tau}$ to maximize the test statistic is usually implemented by locating the zeros of the derivative with respect to $\hat{\tau}$. The mean value of this derivative is helpful to characterize the performance of the estimator. We define the following quantities:

$$w(\hat{\tau}) \stackrel{\Delta}{=} \partial\Lambda/\partial\hat{\tau}; \quad F(\hat{\tau}) \stackrel{\Delta}{=} E[w(\hat{\tau})]; \quad A \stackrel{\Delta}{=} \partial F/\partial\hat{\tau} \Big|_{\hat{\tau}=\tau} \tag{12}$$

The function $F(\cdot)$ is often referred to as the "S-curve" or "discrimination characteristic" of the estimator. If the estimation is unbiased, which is the case in our examples, then $F(\tau) = 0$, where τ is the "true" value of the timing parameter. The slope, A, of the S-curve can be regarded as the gain of the discriminator. These quantities, illustrated in Figure 1, can be used to derive a useful approximate expression [5], [7] for error variance that is suitable for the case of relatively small jitter. Let $\hat{\tau}_o$ denote the maximum-likelihood estimate $(w(\hat{\tau}_o) = 0)$ then the approximation is:

$$\mathrm{Var}(\hat{\tau}_o - \tau) = A^{-2}\mathrm{Var}\ w(\tau) \tag{13}$$

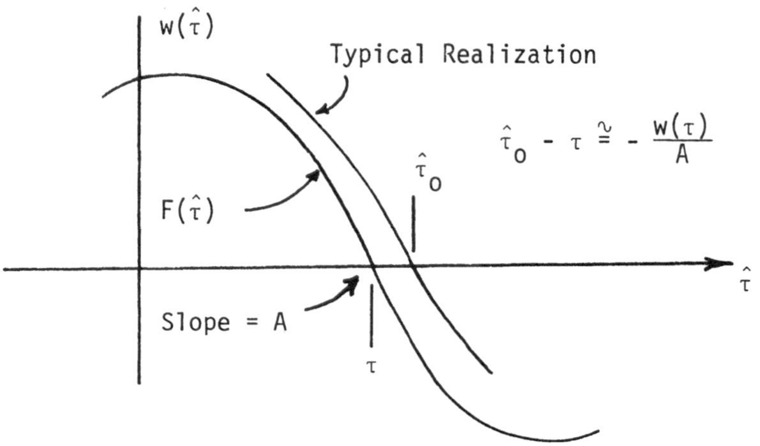

Figure 1: S-curve of maximum-likelihood estimator.

We first evaluate the jitter variance for data-aided timing recovery in the high SNR case. Substituting y(t) (4) for z(t) in (6), assuming $\hat{a}_k = a_k$, and differentiating with respect to $\hat{\tau}$, we can obtain the expression

$$w(\tau) = \sum_{k=0}^{K-1} \sum_{m=-\infty}^{\infty} a_k a_m \dot{r}(kT-mT) \tag{14}$$

where $\dot{r}(t)$ is the time derivative of r(t), the time-ambiguity function for the data pulse, g(t). Now if we assume that the data are independent, identically-distributed, zero-mean, Gaussian random variable with variance σ^2, then

$$A = \sigma^2 K\ddot{r}(0) = -(2\pi\sigma)^2 K \int f^2\ R(f)df \tag{15}$$

and we can use the relation $E[a_1a_2a_3a_4] = E[a_1a_2]E[a_3a_4] + E[a_1a_3]E[a_2a_4] + E[a_1a_4]E[a_2a_3]$ to calculate the variance of $w(\tau)$ in (14), with the result that

$$\text{Var } w(\tau) = \sigma^4 \sum_{k=0}^{K-1} \left[\sum_{m=-\infty}^{\infty} \dot{r}^2(kT-mT) - \sum_{i=0}^{K-1} \dot{r}^2(kT-iT) \right] \tag{16}$$

Notice that the variance decreases as K(the length of the observation interval) increases because more of the terms in the infinite sum in (16) are cancelled out. A frequency-domain expression for (16) is more useful and it can be obtained by an application of the Poisson Sum Formula [1].

$$\text{Var } w(\tau) = \sigma^4 \sum_{\ell=-\infty}^{\infty} \left[\frac{K}{T} H(\tfrac{\ell}{T}) - \int H(f + \tfrac{\ell}{T}) L(f) df \right] \tag{17}$$

where $H(f) = \int \dot{R}(f - \nu)\dot{R}(\nu)d\nu; \quad \dot{R}(f) = j2\pi f R(f)$

$L(f) = [K \text{ sinc}(KTf)]^2.$

The expression (17) is approximated by making a power series expansion of $H(f)$ and integrating out to the first zero of the $L(f)$ function. The result is

$$\text{Var } w(\tau) \cong \frac{-\sigma^4}{2KT(\pi T)^2} \int \sum_{\ell} \dot{R}'(f - \tfrac{\ell}{T})\dot{R}'(f)df \tag{18}$$

where the primes represent differentiation of the frequency-domain functions.

To facilitate comparision with the non-data-aided case to be treated next, we shall assume that $r(t)$ has the Nyquist shape, with $r(0) = 1$ (unit energy data pulses). Within this constraint, we can choose the rolloff shape that minimizes (18). This turns out to be the linear rolloff shown in Figure 2.

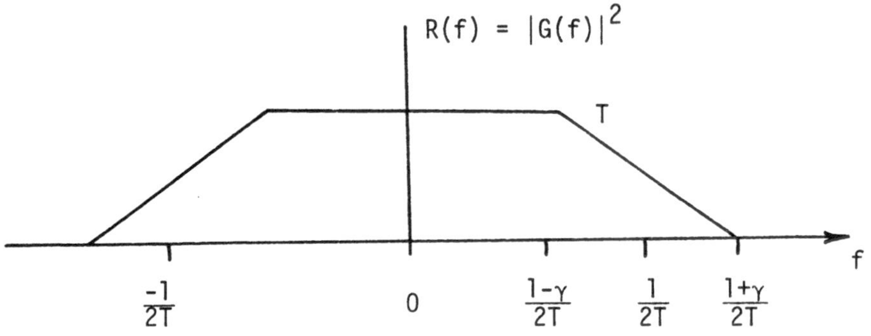

Figure 2: Spectrum of data pulse.

Using the R(f) shown in evaluating (18) and (15), we get the following simple expression for relative variance of the timing jitter in terms of excess bandwidth (γ) and length of observation interval (KT).

$$\mathrm{Var}\ \frac{\tau}{T} = \frac{18((1/\gamma) - 1)}{K^3\pi^4(1+\gamma^2)^2} \tag{19}$$

To examine the performance of independent estimation of $\hat{\tau}$ (non-data-aided), we assume that the data variables are independent and that r(t) is Nyquist. This means that $s_k = q_k$ in (10), and in the noise-free case we have

$$\Lambda(\hat{\tau}) = \frac{1}{2}\sum_{k=0}^{K-1}\left[\int_{-\infty}^{\infty} y(t)g(t-kT-\hat{\tau})dt\right]^2 \tag{20}$$

$$w(\hat{\tau}) = \sum_k \left[\int_{-\infty}^{\infty} y(t)g(t-kT-\hat{\tau})dt\right]\left[\int_{-\infty}^{\infty} y(t)\dot{g}(t-kT-\hat{\tau})dt\right] \tag{21}$$

$$w(\tau) = \sum_k\sum_m\sum_n a_m a_n r(kT-nT)\dot{r}(kT-mT) \tag{22}$$

and with the Nyquist assumption,

$$w(\tau) = \sum_k\sum_m a_k a_m \dot{r}(kT - mT) \tag{23}$$

which is identical to the expression (14) for the previous case. The mean slope (A) however is considerably different from (15) in this case.

$$A \triangleq E[\partial w(\hat{\tau})/\partial\hat{\tau}]_{\hat{\tau}=\tau}$$

$$= -\frac{K\sigma^2}{2T} \sum_{\ell} \left(\frac{2\pi\ell}{T}\right)^2 \int R\left(\frac{\ell}{T} - f\right)R(f)df \qquad (24)$$

Because the $\ell = 0$ term in the sum (24) does not appear, it is clear that A gets very small as the excess bandwidth parameter, γ, tends to zero. Obviously, data-aided performance will be much superior in the case of small excess bandwidth. Using the same shape for R(f) (Figure 2) in this case, the jitter variance expression is

$$\mathrm{Var}\ \frac{\tau}{T} = \frac{9((1/\gamma) - 1)}{2K^3\pi^4\gamma^2} \qquad (25)$$

Implementation of strategies based on the maximum-likelihood principle can take on many forms. We presume that the receiver can form the statistic, $w(\hat{\tau})$. In a "one-shot" approach, where the estimate must be available immediately after the observation interval, there could be many trial values of $\hat{\tau}$ used and the one giving the smallest w is selected. Alternatively, one value could be used (provided that it is reasonably accurate) and the estimate obtained by extrapolation using an assumed slope, A; i.e., $\hat{\tau}_0 \cong \hat{\tau} - w(\hat{\tau})/A$. Such schemes, and variants thereof [3], [5], [7], can be regarded as open-loop realizations. Closed-loop or "tracking" realizations result when the test statistic is used to control the frequency (or phase) of a clock (VCC) that actuates the sampling switch as shown in Figure 3. There are many variations on this also. The clock phase could be updated once every KT seconds, or it could be updated every T seconds. In this latter case, the K-term summing device can be eliminated since the VCC itself acts as an integrator. The loop gain can be adjusted so that the loop bandwidth [15] is the same as that provided by the K-term summer. Let τ_m denote the sampling phase in the mth symbol interval, then

$$\tau_{m+1} = \tau_m + \alpha w(\tau_m) \qquad (26)$$

where α is the VCC gain constant. The steady-state r.m.s. timing jitter and bandwidths for the open- and closed-loop schemes are approximately equal if we make $\alpha^{-1} = KA/2\pi$.

The implementation shown in Figure 3 has an equivalent

version which is perhaps more familiar. The other version [4] does not have a squarer, but has two arms with a sampler in each, configured very much like a Costas loop. The situation is completely analogous to the known fact [15] that the Costas loop is identical in behavior to a squarer followed by a conventional phase-locked loop.

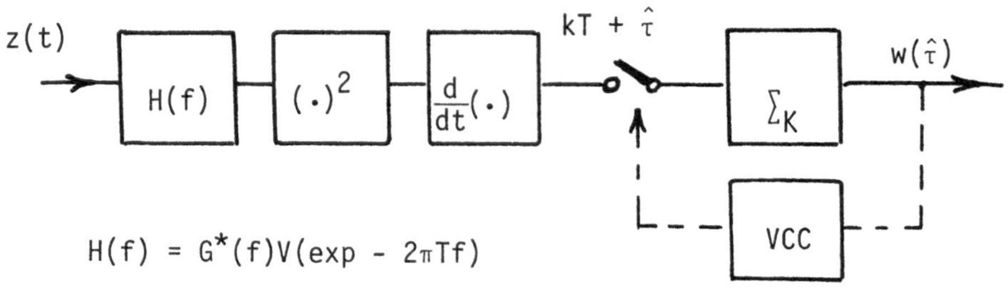

Figure 3: One implementation of maximum-likelihood baseband timing recovery (non-data-aided).

CARRIER SYSTEMS: JOINT ESTIMATION OF PARAMETERS

VSB System

For the VSB system described by (2), the log-likelihood function analogous to (6) is given by

$$\Lambda(\hat{\theta}, \hat{\tau}) = \sum_{k=0}^{K-1} \hat{a}_k q_k(\hat{\tau}, \hat{\theta}) \qquad (27)$$

where

$$q_k(\hat{\tau}, \hat{\theta}) = \mathrm{Re}[\exp(-j\hat{\theta}) \int_{-\infty}^{\infty} \alpha(t)\gamma^*(t-kT-\hat{\tau})dt] \qquad (28)$$

and $\alpha(t)$ is the complex envelope of the received signal, z(t). The receiver generates the q_k sequence by sampling the output of a coherent demodulator operating at a reference phase $\hat{\theta}$. The input to the demodulator is the response of a bandpass filter (which is the matched filter for the bandpass data pulse) to the received signal. In this case, the receiver seeks the simultaneous zero values of two statistics, $w_1(\hat{\theta}, \hat{\tau}) = \partial\Lambda/\partial\hat{\theta}$ and $w_2(\hat{\theta}, \hat{\tau}) = \partial\Lambda/\partial\hat{\tau}$. A tracking loop implementation could use these quantities to drive a VCO and VCC respectively, however, the estimates for θ and τ are, in general, coupled and for best performance the control signals should be linear combinations of w_1 and w_2 [14].

The two-parameter analog of (13) is obtained by taking the variance of each component in

$$\begin{bmatrix} \hat{\theta}_o - \theta \\ \hat{\tau}_o - \tau \end{bmatrix} \cong -A^{-1} \begin{bmatrix} w_1(\theta, \tau) \\ w_2(\theta, \tau) \end{bmatrix} \tag{29}$$

where the elements of the 2 x 2 matrix A are.

$$A_{11} = E[\partial w_1/\partial\hat{\theta}]; \quad A_{22} = E[\partial w_2/\partial\hat{\tau}]; \quad A_{21} = A_{12} = E[\partial w_1/\partial\hat{\tau}]$$

and each of these partial derivatives is evaluated at the "true" values, θ and τ. Substituting (2) for $\alpha(t)$ in the noise-free case into (28) and performing the differentiations, we get

$$w_1(\theta, \tau) = \sum_k \sum_m a_k a_m r_2(kT-mT)$$

$$w_2(\theta, \tau) = \sum_k \sum_m a_k a_m \dot{r}_1(kT-mT) \tag{30}$$

where we have defined real functions $r_1(t)$ and $r_2(t)$ such that

$$\int_{-\infty}^{\infty} [r_1(t) + jr_2(t)]\exp(-j2\pi ft)dt = |\Gamma(f)|^2 \tag{31}$$

Note that $r_1(t)$ is the time-ambiguity function for the baseband pulse. $R_1(f)$ and $R_2(f)$ are the even and odd parts, respectively, of $|\Gamma(f)|^2$ whose shape used in this analysis is sketched in Figure 4. This particular shape minimizes the variance of the expressions in (30).

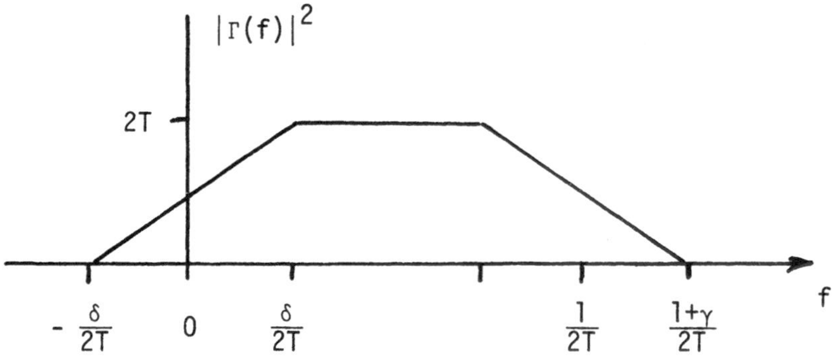

Figure 4: Spectrum of the complex envelope of VSB data pulse.

Averaging over the data, assumed i.i.d. Gaussian, we get

$$A = K\sigma^2 \begin{bmatrix} -r_1(0) & \dot{r}_2(0) \\ \dot{r}_2(0) & \ddot{r}_1(0) \end{bmatrix} \tag{32}$$

The size of the off-diagonal terms represents a high degree of coupling in the estimates. To evaluate r.m.s. jitter, the variances of w_1 and w_2 in (30) are evaluated using the same method as outlined for the baseband case in Eqns. (16)-(18). The calculated jitter values for some typical system parameter values are shown in the graphs in Figures 5 and 6.

For joint estimation of τ and θ (non-data-aided), the appropriate log-likelihood function is

$$\Lambda(\hat{\theta}, \hat{\tau}) = \frac{1}{2} \sum_{k=0}^{K-1} q_k^2(\hat{\tau}, \hat{\theta}) \tag{33}$$

where the q_k quantities are the same as those in (28). Exactly as in the baseband case, data correlation and intersymbol interference can be accommodated by modifying the bandpass matched filter at the receiver input. Henceforth we restrict consideration to the case of independent data and baseband pulses with a Nyquist time-ambiguity function. In this situation, the w_1 and w_2 functions are exactly the same as in (30). The A matrix, however, is quite different for the non-data-aided case.

$$A = -\frac{K\sigma^2}{T}\begin{bmatrix} C_o + C_1 & -\frac{\pi}{T}\ C_1 \\ -\frac{\pi}{T}\ C_1 & (\frac{\pi}{T})^2\ C_1 \end{bmatrix} \tag{34}$$

where

$$C_p = \int |\Gamma(\frac{p}{T} - f)|^2 |\Gamma(f)|^2 df \tag{35}$$

For the pulse spectrum shape shown in Figure 4, we have $C_o = 2T\delta/3$ and $C_1 = 2T\gamma/3$. This causes r.m.s. jitter in both carrier phase and symbol timing to increase very rapidly as excess bandwidth becomes small; as illustrated in the graphs in Figures 5 and 6.

QAM System

The log-likelihood function for the QAM signal (3) is given by

$$\Lambda(\hat{\theta}, \hat{\tau}) = \sum_{k=0}^{K-1} \hat{a}_k p_k(\hat{\tau}, \hat{\theta}) + \hat{b}_k q_k(\hat{\tau}, \hat{\theta}) \tag{36}$$

where

$$p_k(\hat{\tau}, \hat{\theta}) = \mathrm{Re}[\exp(-j\hat{\theta}) \int \alpha(t)g(t-kT-\hat{\tau})dt]$$

$$q_k(\hat{\tau}, \hat{\theta}) = \mathrm{Re}[-j\ \exp(-j\hat{\theta}) \int \alpha(t)h(t-kT-\hat{\tau})dt] \tag{37}$$

If we let $h(t) = g(t)$ and assume that the time-ambiguity function, $r(t)$, of these pulses is Nyquist, then

$$w_1(\theta, \tau) = 0$$

and

$$w_2(\theta, \tau) = \sum_k \sum_m (a_k a_m + b_k b_m)\dot{r}(kT-mT) \tag{38}$$

The A matrix is diagonal,

$$A = -2K\sigma^2 \begin{bmatrix} r(0) & 0 \\ 0 & \ddot{r}(0) \end{bmatrix} \tag{39}$$

so we see that jitter-free phase recovery results, and the variance of the timing jitter is exactly one-half of that given by (19) for the baseband case.

For the non-data-aided case, we use

$$\Lambda(\hat{\theta}, \hat{\tau}) = \frac{1}{2} \sum_{k=0}^{K-1} p_k^2 + q_k^2 \tag{40}$$

with p_k and q_k as given by (37). However, for the regular QAM case with $h(t) = g(t)$, we find that the mean value of Λ is independent of $\hat{\theta}$, hence phase estimation cannot be performed. Fortunately, the situation is quite different for SQAM and we find

$$w_1(\theta, \tau) = \sum_k \sum_m a_k b_m r(kT-mT- \frac{T}{2}) - b_k a_m r(kT-mT + \frac{T}{2}) \tag{41}$$

and

$$w_2(\theta, \tau) = \sum_k \sum_m (a_k a_m + b_k b_m)\dot{r}(kT-mT)$$

If we define the function

$$\rho(t) = \sum_m r^2(t-mT) \tag{42}$$

then the A matrix is diagonal with

$$\begin{aligned} A_{11} &= -2\sigma^2 K[\rho(0) - \rho(\frac{T}{2})] \\ A_{22} &= \sigma^2 K\ddot{\rho}(0) \end{aligned} \tag{43}$$

and these expressions are easily converted into the more convenient frequency-domain expressions by application of the Poisson Sum Formula.

RESULTS AND CONCLUSIONS

The r.m.s. jitter for the various estimation techniques and modulation formats has been calculated as a function of the excess bandwidth parameter, assuming $\delta = \gamma$ for simplicity in the VSB case, and the results are summarized graphically in Figures 5 and 6. A single value, $K = 10$, for the observation interval is used since all the schemes exhibit the same dependence on K. The rapid decrease in jitter as bandwidth increases is noted in all cases, but the effect is most pronounced in the non-data-aided schemes. An interesting point is that with sufficient bandwidth, the superiority of the data-aided approach is probably not great enough to warrant the increased receiver

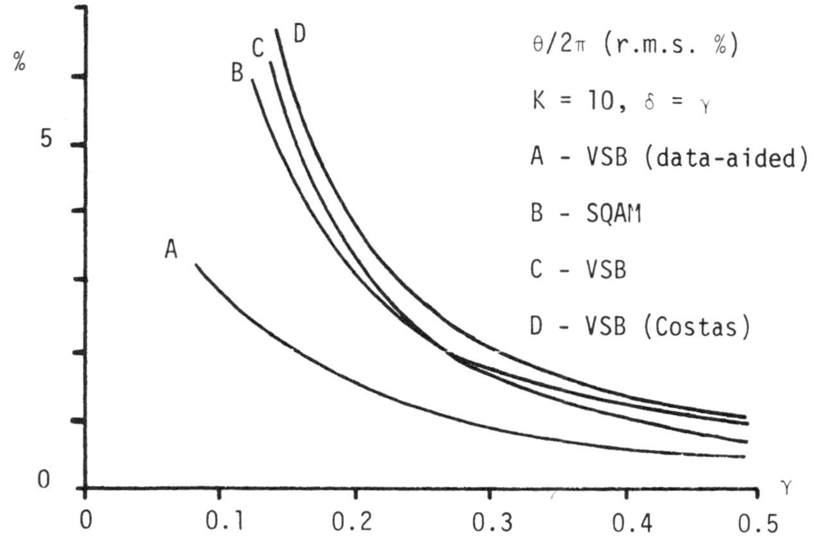

Figure 5: R.M.S. phase jitter vs. excess bandwidth.

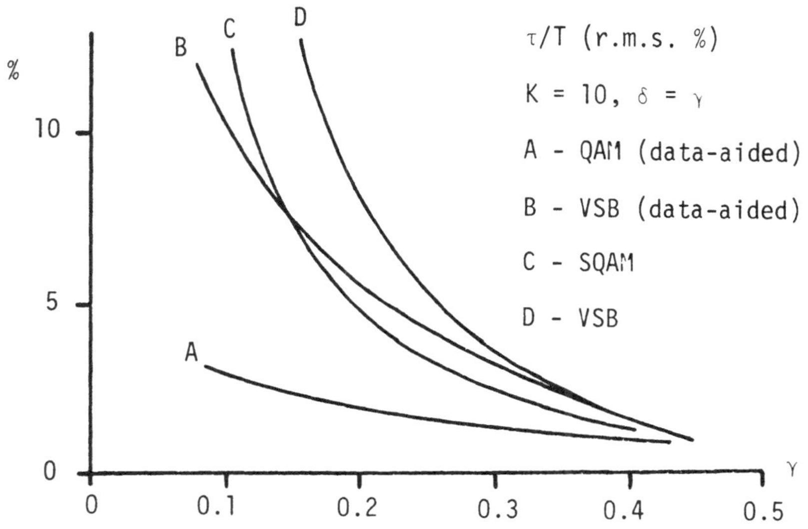

Figure 6: R.M.S. timing jitter vs. excess bandwidth.

complexity. In the VSB case, the phase recovery performance of a Costas loop is also shown for comparison and we note that its performance is only slightly poorer than joint phase and timing recovery over a wide range of bandwidth values. In comparing the system perofrmances, it should be remembered that, with the pulse shapes employed, the data-aided QAM phase recovery was jitter free, while the non-data-aided QAM system was inoperative.

REFERENCES

1. L. E. Franks, Signal Theory, Prentice-Hall, 1969.

2. H. L. Van Trees, Detection, Estimation and Modulation Theory - Part I, Wiley, 1968.

3. L. E. Franks, "Acquisition of Carrier and Timing Data - I," Signal Processing in Communications and Control, pp. 429-447, Noordhoff International, 1975.

4. R. D. Gitlin and J. Salz, "Timing Recovery in PAM Systems," Bell Sys. Tech. Jour., Vol. 50, pp. 1645-1669, May-June 1971.

5. L. E. Franks and J. P. Bubrouski, "Statistical Properties of Timing Jitter in a PAM Timing Recovery Scheme," IEEE Trans., Vol. COM-22, pp. 913-920, July 1974.

6. E. Roza, "Analysis of Phase-Locked Timing Extraction Circuits for Pulse Code Transmission," IEEE Trans., Vol. COM-22, pp. 1236-1246, September 1974.

7. L. Bellato and G. L. Cariolaro, "Time Jitter in Line Regenerators with Pattern Dependent Pulse Waveforms," Alta Frequenza, Vol. 41, pp. 800-815, November 1972.

8. H. Kobayashi, "Simultaneous Adaptive Estimation and Decision Algorithm for Carrier Modulated Data Transmission Systems," IEEE Trans., Vol. COM-19, pp. 268-280, June 1971.

9. W. C. Lindsey and M. K. Simon, "Data-Aided Carrier Tracking Loop," IEEE Trans., Vol. COM-19, pp. 157-168, April 1971.

10. G. Ungerboeck, "Adaptive Maximum Likelihood Receiver for Carrier-Modulated Data Transmission Systems," IEEE Trans., Vol. COM-22, pp. 624-636, May 1974.

11. R. Matyas and P. J. McLane, "Decision-Aided Tracking Loops for Channels with Phase Jitter and Intersymbol Interference," IEEE Trans., Vol. COM-22, pp. 1014-1023, August 1974.

12. R. Dogliotti, U. Mazzei, and U. Mengali, "Decision-Directed Carrier Synchronization for SSB Data Transmission Systems," International Communications Conference, June 1975.

13. U. Mengali, "Synchronization of QAM Signals in the Presence of ISI," IEEE Trans., Vol. AES-12, pp. 556-560, September 1976.

14. D. D. Falconer and J. Salz, "Optimal Reception of Digital Data Over the Gaussian Channel with Unknown Delay and Phase Jitter," IEEE Trans., Vol. IT-23, pp. 117-126, January 1977.

15. W. C. Lindsey and M. K. Simon, Telecommunication Systems Engineering, Prentice-Hall, 1973.

INTERFERENCE AND JAMMING PRONE COMMUNICATION

Prof. Harb. S. Hayre

W.P. Lab. - E.E. Dept.
University of Houston
Houston, TX 77004

ABSTRACT

Interference, intentional or unintentional, due to poor engineering practices or due to lack of coordination between various disciplines involved in the design and building of communication system on board air frames and ships etc., is often present, and the same is the case in many land based transmit-receive systems.

Numerous techniques to reduce interference effects on communication systems have been developed with a major portion of the efforts have been directed in the area of widebanding the signal spectrum and the use of coding. These techniques are discussed in a comparative fashion. Other methods employed to achieve this effect include the use of nulling arrays, modulation techniques, blankers, limiter, and filters etc.

Finally long range predictions are examined in view of these techniques and limitations of the state of the art.

INTRODUCTION

One may divide interference into the following basic categories:

1. From Co-located Sources:
   - Antenna Effects - Main or side lobe
   - Multipath Effects
   - Electrical Coupling
   - Overlaping of Frequency Bands
2. From Distance Sources:
   - Multipath Unintentional
   - Intentional (Jamming)
     - Low Power
     - High Power

- Signal Effects
- Environmental Mode
- Destruct Mode

In view of these, the basic intent of the designer's strategy is to minimize these by appropriate overall design or after-the-installation attempt to overcome this effect by using wideband signals in the four dimensional domain of modulation techniques, frequency, time and coding. The basic fact that broader bandwidths are required to improve signal to noise ratios in communication system, and secrecy of the mode of operation including modulation technique, code format, duration of transmission, and the center frequency of transmission leads a pratitionaer of this art to consider the use of the following signal techniques:

1. Time Hopping (TH)
2. Frequency Hopping (FH)
3. Time-Frequency Hopping (TFH)
4. Direct Sequence Systems (DS)
5. Frequency Hopping - Direct Sequence System (FH/DS)
6. Time Hopping - DS - (TH/DS)
7. Pulsed F.M. (CHIRP)

In the case of electromagnetic-Transmitter-Receiver systems area, one must design to:

1. Minimize inter-antenna interference due to colocation on a common frame such as a tower, airframes and ships
2. Obtain adaptive nulling arrays to minimize the interference by pointing a null in the direction of the incoming interference from a distance source
3. Employ adaptive processing of adaptive nulling arrays to minimize jamming effects
4. Reduce side lobes and the use of hard to detect codes such as pseudo-random codes

Listed below is a comparative signal to noise ratios (S/N) for various classical modulation systems which are employed in all these schemes to improve jamming environment-performance of communication systems; (Table 1) Zeimer & Tainter. The symbols listed are defined below:

P_T = Power Transmitted

N_o = Noise spectral density

W = Bandwidth

a = Modulation index for AM i.e. $x(t) = A_c(1+am_n(t))\cos w_c t$

k_p = Phased deviation constant i.e. $\Theta(t) = k_p\, m_n(t)$ for P.M.

m_n = PPM or PWM $\tau(t) = \tau_o + \tau_1\, m_n(t)$

D = Deviation Ratio = f_d/w

f_d = Frequency deviation (FM)

f_3 = 3db frequency of de-emphasis filter for PM

For pulsed modulation, S/N for various cases are given below:

$$(S/_N)_{PAM} = m^2(t)/N_oW$$

$(S/N)_{PWM} = (4\ \tau_1^2\ m_n^2\ \tau s\ B_T\ W/_{\tau o})\ (P_T/N_o W)$

$(S/N)_{PCM} = q^2$

$= 2^{2n}$ for binary quantization

where τ_s = Sampling period

B_T = Bandwidth of singal $\approx 1/T_R$ where T_R = rise time

TECHNICAL DISCUSSION

In case of wideband systems designed to perform satisfactorily in an interference environment, one can obtain an approximation for bandwidth W, in terms of noise power N, and signal power S from the basic Shannon's theorem for channel capacity C, (bits/sec.),

$$C = W \log_2(1 + S/N) \quad (1)$$

or $$W \cong NC/S \quad \text{for } S/N \leq 0.1 \quad (2)$$

a few other general definitions are given before a discussion of various systems is given. These are Process Gain G_p,

$$G_p = W_{RF}/R_{info} \quad (3)$$

where R_{INF} is the information rate in bits/sec. in the baseband channel. The Jamming margin M_j is defined as

$$M_j = G_p - [L_{sys} + (S/N)_{out}] \quad (4)$$

where L_{sys} = System losses

$(S/N)_{out}$ = (S/N) at the information output

Table 2 lists the process gain for various spread spectrum techniques, as well as those obtained by antenna pointing and electronic interference cancellation (Dixon, 1976). Now various techniques are briefly examined as how, in general, these are implemented and some of the associated waveforms.

DIRECT SEQUENCE SYSTEMS (DS)

Fig. 1 shows a block diagram of D.S. system and both the wave forms at different points in the system, as well as the spectrum with and without the carrier for biphase code modulated case. Basically Pulse AM and FM modulations are used in such systems although there is no bar for use of any other amplitude or phase angle modulation techniques. 180° biphase balanced phase shift keying (PSK) is usually employed in DS system because

a) suppressed carrier is difficult to detect
b) more power is available for information
c) constant envelope signal enables the maximization of transmitted power
d) biphase modulator circuitry is very simple.

The null-to-null bandwidth of the main lobe in the power spectrum is 1.2 R_c, where R_c is the code bit rate, and this bandwidth may be reduced by one half by using quadriphase shift keying (QPSK). The number of frequency sets available being a function of the length of the code or (n + 1) and their spacing being $(2n - 1)/R_c$ where n is the length of the code generator sequence and the maximal code length is (2n - 1).

FREQUENCY HOPPING (FH), FH/DS SYSTEMS & CHIRP (PULSED FM)

Figure 2 shows a FH system block diagram & the associated wave forms, for which the process gain, G_p, is equal to the number of available frequency choices (N), and the expected error rate is J/N where J is the number of jammers with power greater than or equal to signal power. In strong jammer power environment, its jamming capability is enhanced by a faster hopping rate than the jammer can keep up with, even though this basic rate is a function of information rate. M-ary frequency hopping systems reduce the CHIRP error rate shown in Fig. 3b and c although it may not be the most desirable end result in case of a repeating jammer.

In order to extend spread spectrum capability, provide multiple access and discrete address and multiplexing capability as well, direct sequence modulation (DS) is added to FH to produce FH/DS systems. The process gain of FH/DS systems is the sum of the process gains of FH and that of the DS systems, or

$$G_p(FH/DS) = 10 \log(N) + 10 \log\frac{(BW)DS}{R_{INF}} \quad (5)$$

where N is the number of channels.

CHIRP - PULSED FM

Chirp is a pulsed FM system whose frequency is swept in a given fashion such as a linearly swept signal f(t) would be

$$f(t) = A \cos(w_c t + 1/2\, \mu\, t^2) \quad (6)$$

where $\mu = dw/dt$, the frequency sweep rate. In such systems, the frequency resolution is the inverse of signal duration (sweep time), and time resolution is the inverse of the sweep frequency. This system does not use any coding as such and some of delay and bandwidth parameters of typical delay times (chirp matched filter) are shown in Fig. 3a.

TIME HOPPING (TH), TFH & TH-DS

TH systems are very simple to implement but are highly jamming susceptible, because of the carrier frequency being fixed, even though such systems force the jammer to operate continuously, thus making them easy to be detected. TH is usually used in conjunction with FH or DS. Systems requiring discrete address and random access capability often use TFH. Near and Far Transmitters are seen by such systems with considerable attenuation disparity, and processing alone cannot resolve the problem. Therefore, a synchronous system for time slots and their frequency channels is needed for multi-station operation.

TH-DS systems accomplish the time division multiplexing (TDM), otherwise not possible with DS systems alone. A system block diagram with its associated waveforms is shown in Fig. 4.

DEMODULATORS

Figures 5 and 6 show a comparison of PDM, FM & FH systems, and a block diagram with appropriate waveforms for FH demodulator, the detectors used for various systems are listed here:

Modulation Type	Detectors
Carrier FM FSK	Phase-lock detector
Clock FM FSK	Costas dector
PDM PSK	Costas plus PDM demodulator
Frequency hopping (coherent)	Phase-lock detector
Frequency hopping (noncoherent)	Envelope and integrated and dump detectors

The reader is also referred to other published material.

ANTENNAS & RADAR SYSTEMS

Friedman (1968) gives a detailed calculation of system parameters for one and two-way communication systems, including main and side lobe reception. Radar bandwidths, pulsewidths, and radar cross-sections are discussed for practical cases and very useful data is provided.

Colocation of antennas is a topic which cannot be discussed fully here for reasons of space. A rule of thumb is to locate antennas in the shadow area of other antennas if at all possible, even though it may not be practical in many applications. Failing that it is desirable to minimize antenna interference by determining composite interference pattern prior to installation. Such work has been carried out by the University of Houston Wave Propagation Lab. for a number of years. Simple computer models are not known to be adequate for such calculations. Comprehensive computer programs for this work have been in use at Ohio State University (U.S.A.) and GEC Marconi Electronics Ltd. (UK) for a few years now.

1980'S - PREDICTIONS

It is expected that computer augmentation would make future radar/communication systems less vulnerable to jamming. The time varying random frequency diversity up to 15% of the center frequency will be common, and in the case of pulsed radars, even PRF diversity may be included, even though the latter is limited by maximum range. A practical upper limit on higher frequencies up to 15GHZ will be upped to 40 to 60 GHZ or even 100 GHZ for short range systems with 1db/km O_2 & H_2O absorption. The present distribution of radar jamming spectrum of 0-20GHZ is broad enough, and any further broadening will even be more difficult to cope with. Increased use of C.W. and monopulse, leading edge tracking, STC, FTC, and other common ECCM features such as mixed frequency and PRF diversity will require new combination of presently known techniques to minimize interference effects (Hartman, 1977). Digital transmissions, controlled power, directional signals, spread spectrum, and chirp derivative transmissions will be the order of the day and EMP effects may be expected to add to this complexity.

Table 1 Summary fo Noise Performance Characteristics

SYSTEM	POSTDETECTION SNR	TRANSMISSION BANDWIDTH
Baseband	$\frac{P_T}{N_0 W}$	W
DSB with coherent demodulation	$\frac{P_T}{N_0 W}$	$2W$
SSB with coherent demodulation	$\frac{P_T}{N_0 W}$	W
AM with square-law detection	$2\left(\frac{a^2}{2+a^2}\right)^2 \frac{P_T/N_0 W}{1+(N_0 W/P_T)}$	$2W$
PM above threshold	$k_p^2 \overline{m_n^2} \frac{P_T}{N_0 W}$	$2(D+1)W$
FM above threshold (without preemphasis)	$3D^2 \overline{m_n^2} \frac{P_T}{N_0 W}$	$2(D+1)W$
FM above threshold (with preemphasis)	$\left(\frac{f_d}{f_3}\right)^2 \overline{m_n^2} \frac{P_T}{N_0 W}$	$2(D+1)W$

Table 2 Comparison of Process Gains Available for Various Techniques, Including Signal Rejection and Cancellation

System	Process Gain
Direct sequence	$\frac{BW_{RF}}{R_{info}} = TW$
Frequency hopping	$\frac{BW_{RF}}{R_{info}} = TW =$ number of frequency choices
Time hopping	$\frac{1}{\text{transmit duty cycle}}$
Chirp	Compression ratio $= \tau\, dF = TW$
Antenna rejection	Antenna gain
Electronic cancellation	Depends on accuracy of replica; sometimes 40 dB
Selective rejection	High but useful only against narrowband interference

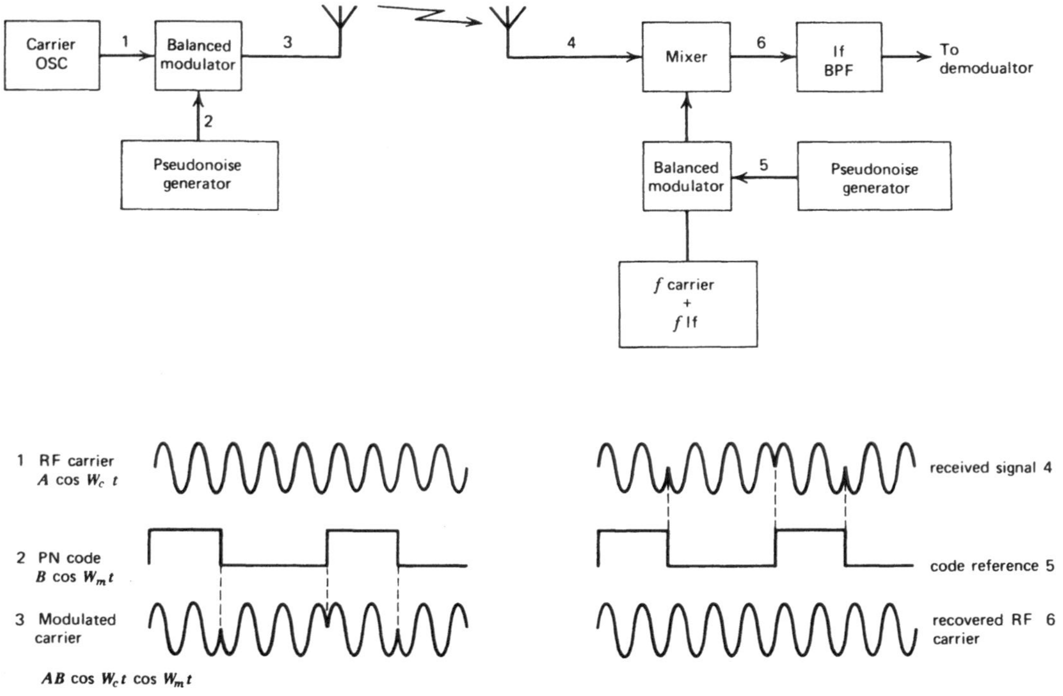

Overall direct sequence system showing waveforms

Figure 1 a.

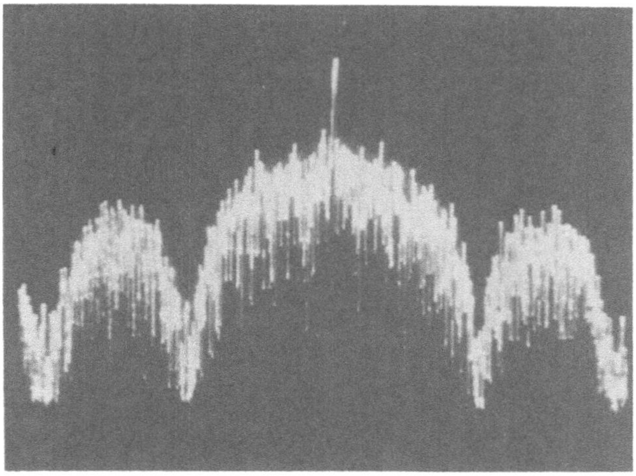

Direct sequence, unsuppressed carrier spectrum--
biphase, code-modulated

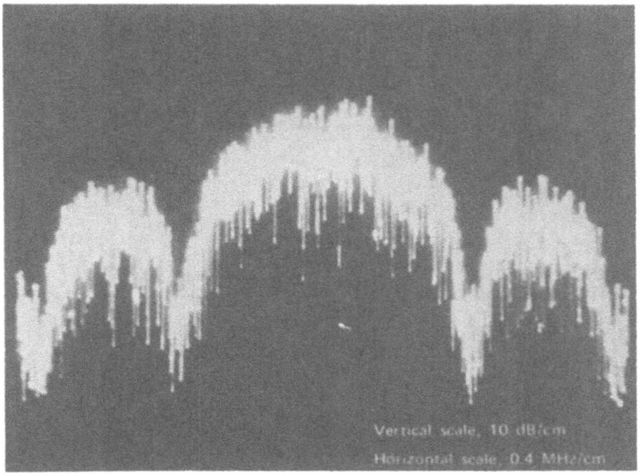

Direct sequence, suppressed carrier spectrum--
biphase, code-modulated

Figure 1 b.

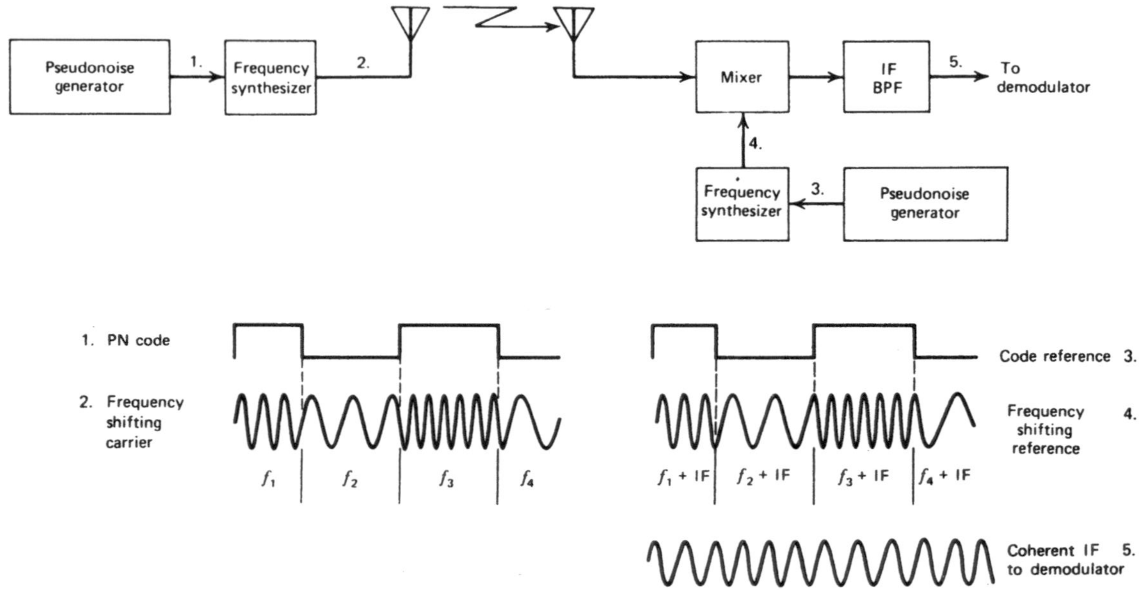

Basic frequency hopping system with waveform

Figure 2

Figure 3a Chirp filter TW product curve

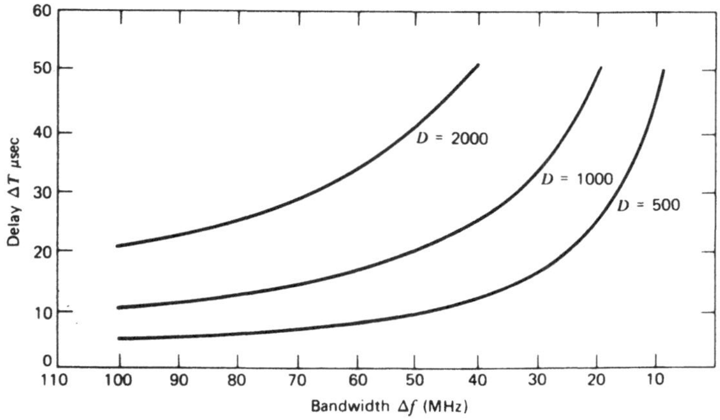

Figure 3b Number of channels required versus fraction of channels jammed(J/N) and number of jammers(FH).

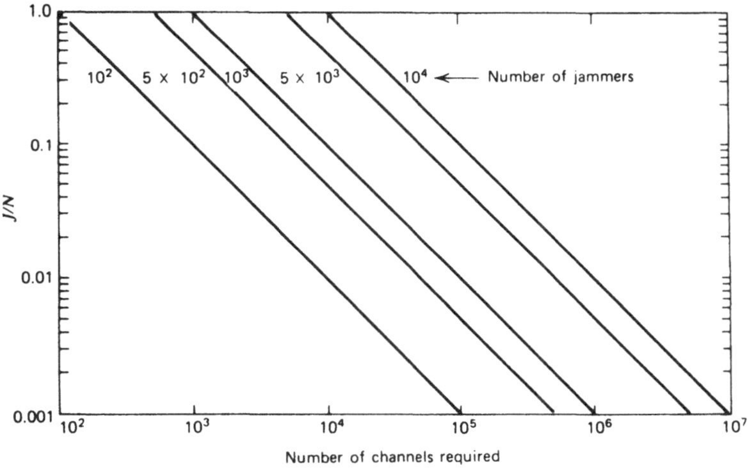

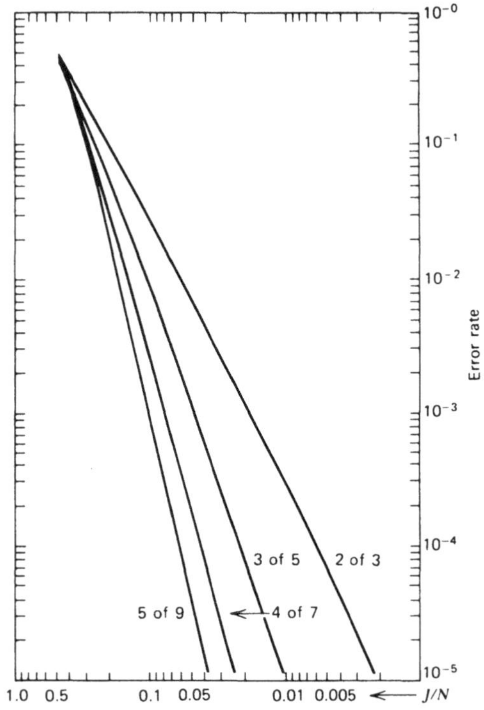

Figure 3c Error rate versus fraction of channels jammed (J/N) for various chip decision criteria in multichip transmissions (FH).

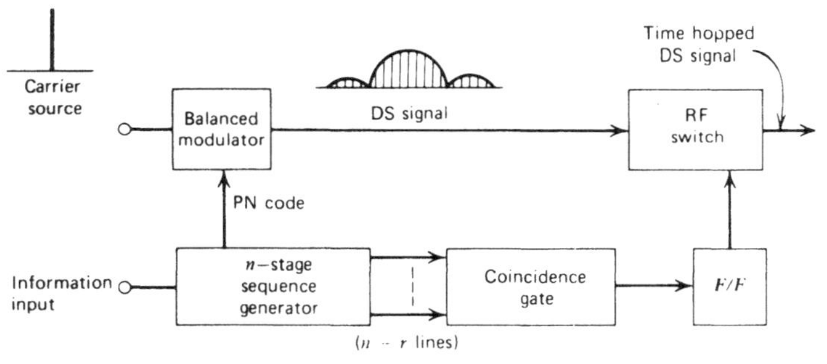

(a)

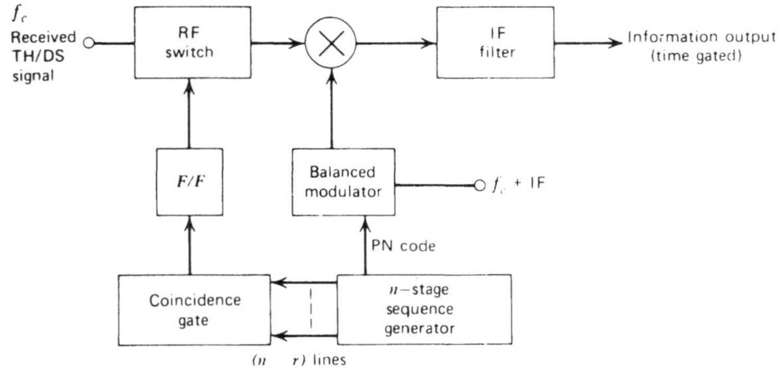

(b)

Figure 4 TH/DS system block diagrams:(a) time hopping/direct sequence transmitter;(b)TH/DS reciever

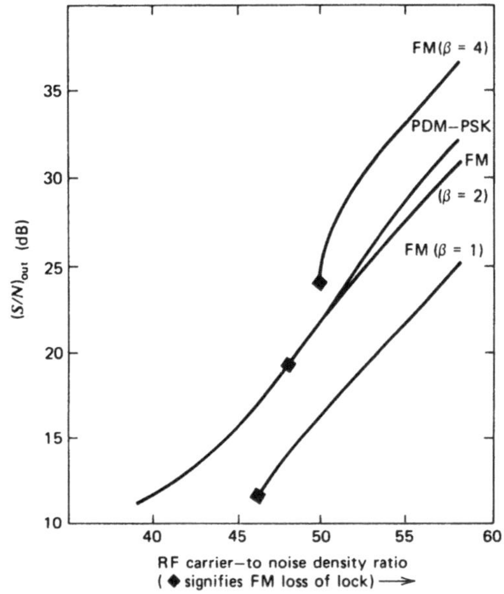

Figure 5 Performance of PDM demodulator compared with FM.

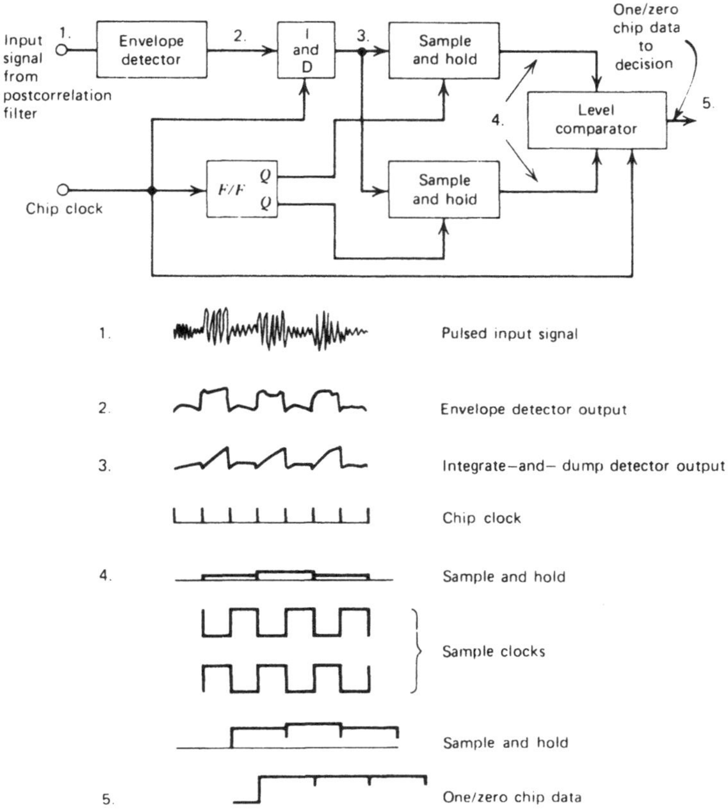

Figure 6 Frequency hopping demodulator(noncoherent).

REFERENCES

1. Abraham, L. G., Jr.; Barrow, B. B.; Cowan, W. M.; Gallant, R.M., Anti-Jam Techniques for Troposcatter Communications. Volume 3., RADC-TR-67-130
Tropospheric Scatter Multipath Tests Final Report (Tropospheric Scatter Multipath Tests for Anti-Jamming Systems)
Sylvania Electric Products, Inc., Waltham, Mass Griffiss AFB,

2. Baghdady, E. J.; Getchell, E. H.; Goff, M. E.; Gutwein, J. M.; Jam-Resistant Secure FM Command Receivers
Progress Report, 19 Apr. - 31 Oct. 1962
ADCOM, Inc., Cambridge, Mass.
NASA-CR-98088 NAS8-2668 62/10/31 351 pages

3. Baklashov, P. I., Radar Jamming and Protection of Radar Stations against Noise Joint Publications Research Service, Washington, DC (Translation of S.P.O.
Radiolokatsionnoy Tekhnika USSR, Mil. Publishing House, 1967 p. 469-524) 68/10/14 63 pages

4. Cook, D. H., Radar or ECM/ECCM System Simulation by Digital Computer Using Signal and Jamming Spectral Inputs
Electronics and Aerospace Systems Convention, Washington, D.C., Sep. 9-11, 1968, Record pp. 181-186

5. Crepeau, P. J., Topics in Naval Telecommunications Media Analysis Final Report, Oct. 1975 - July 1976
Naval Research Lab., Washington, D.C.
AD-B016496L NRL-8080, 76/12/31, 51 pages

6. Dixon, R. C., "Spread Spectrum System," John Wiley & Sons, New York, 1976

7. Ehrman, L.; Parl, S.; Pierce, J. N.; Richman, S. H., Study of Optimum Simulation Techniques for the Design and Evaluation of Anti-Jam Communication Systems Final Report, 1 Feb. 1974 - 30 June 1975, ADFAL-TR-75-186 289 pages

8. Freeman, E. R.; Sachs, H. M.; (Sachs/Freeman Associates, Inc., Hyattsville, MD.)
HF AM Signal-to Interference Ratio
International Symposium on El-ctromagnetic Compatibility, Washington, D. C., July 13-15, 1976 Record. (A77-31751 13-33) New York pp. 73-78.

9. Friedman, H. J., Jamming Susceptibility
IEEE Transactions on Aerospace and Electronic Systems, Vol. AES-4, July 1968 pp. 515-528

10. Garrett, J. C., Packet Radio Communications
Quarterly Technical Report, 1 Oct. 1973 - 31 Dec. 1976
Rockwell International Corp., Dallas, Tex. (Collins Radio Group)
AD-B008938L, REPT-523-0602121-001C3L

11. Gooding, D. J., Modulation Stuy
Final Technical Report
REPT-6070-01 N62269-76-M-2163 75/10/28 45 pages
Stein Associates, Inc., Waltham, Mass.

12. Grigorin-Riabov, V. V., Interference-Immunity of Radar Stations
Lockheed Missiles and Space Co., Palo Alto, Calif. John Crerar Libr., Chicago, Ill. 60616
(Translation from "Radiolokatsionnye Ustroistva..." S. R. Press 1970 pp. 480-520)

13. Haddad, A. H.; Fletcher, R., Pseudo-Random Coded CW Radar for Target Detection in Clutter, AD-A008964 RD-75-25
Army Missile Research, Development and Engineering Lab., Redstone Arsenal, ALA 74/08/01 20 pages

14. Hartman, Richard, "1984 - The Threat," Military Electronics & Countermeasures, July 1977, pp. 60-

15. Hayre, H. S., Multipath Communication Errors, URSI-Commission-F Conf. LaBaule, Fr., April 28 - May 5, 1977

16. Hopkins, P. M.; Simpson, R. S.; Probability of Error in Pseudo-noise /PN/-Modulation Spread Spectrum Binary Communication Systems
IEEE Transactions on Communications, Vol. COM-23, Apr. 1975, pp. 467-472

17. Jones, K. (AAED), Communications and Electronics Digest, December 1966
Air Defense Command, ENT AFB, Colo. (Directorate of Communications & Electronics)
ADCRP-100-1, No. 7 66/12/00 36 pages

18. Jones, R. R.; Schellenberg, J.; Tanski, W. J.; Moore, R. A.; Transplexing Saw Filters for ECM II -- Surface Acoustic Wave Noise Jammer Filterbanks
Westinghouse Advanced Technology Laboratories, Baltimore, MD, Microwaves, Vol. 14, Jan. 1975 pp. 68-73

19. Katyl, R. H., Moire Screens Coded with Pseudo-Random Sequence
Applied Optics, Vol. 11, Oct. 1972, pp. 2278-2285

20. Kremer, I., Effect of Modulating/Multiplicative/interference on Signal Processing in a System Consisting of a Phased Array Antenna and a Receiver
Radio Engineering and Electronic Physics, vol. 17, Sept. 1972, pp. 1451-1457
(Translation of Radiotekhnika I Electronika pp.1823-1830)

21. Kremer, I.; Karpukhin, V. I., Measurement of an Energy-Independent Parameter of a Radio Signal in the Presence of High-Level Additive and Modulating Interference
Radiotekhnika 2, Vol. 27, June 1972, pp. 11-15
Telecommunications & Radio Engineering, Vol. 27, June 1972 pp. 69-73 (Translation)

22. Kyle, Robert H., "DF Antenna System Selection," Military Electronics & Countermeasures, July 1977, pp. 20-

23. Larow, J. F.; Johnson, G. H., Simulation and Analysis of Phased Coded Waveforms Plus Noise (Interim Report)
Army Electronics Command, Fort Monmouth, N. J.
AD-A000795 ECOM-5493 73/04/00 76 pages

24. LaRue, G. D., The Evolution of Spread Spectrum Equipment for the Defense Satellite System Communications
Armed Forces Communications and Electronics Association, Annual Convention, 29th, Washington, D. C., June 3-5, 1975. Signal, Vol. 29, Aug. 1975, pp. 72-75

25. Lovrenski, V., Principles and Equipment for Jamming Radar Devices, F Stc-Ht-23-1260-73
Transl. into English from Vojnotehnicki Glasnik Yugoslavia)
AD-921112
Army Foreign Science and Technology Center, Charlottesville, VA. 73/05/13, 15 pages

26. Lueg, R. E.; Freet, R. A., An Introduction to Spread Spectrum Modulation
Tactical Air Command, Langley AFB, VA, AD-A017499 XPS-TN-75-9 75/10/24 27 pages

27. Means, R. W.; Speiser, J. M.; Whitehouse, H. J.; Image Transmission Via Spread Spectrum Techniques, AD-787502 QTR-4
Quarterly Technical Report, 2 Jan. - 1 July 1974, 150 pages

28. Meshkovskii, K. A., Coding in Communications Engineering, AD-682065
Air Force Systems Command, Wright-Patterson AFB, Ohio (Foreign Technology Div.)
(Translation of Russian Book Kodirovanie Itekhnike Svyazi 1966 pp. 1-324) 68/06/00 277 pages

29. Miamidian, L. R., Comments on Jamming Susceptibility,
IEEE Trans., Vol. AES-5, No. 3, May 1969, pp. 561-62

30. Mikenas, V. A.; Rassweiler, G. G.; Bustelo, R. A.; Payne, L. M.; Lehman, D. F., Null-Steering Feasibility Demonstration (Limited Distribution)
Final Technical Report, AD-B004502L, AFAL-TR-74-352
Harris Corp., Melbourne, Fla. (Electronic Systems Div.), 75/03/00, 193 pages

31. Parkhomenko, A., Interference and Interference Suppression in Radio Equipment
Joint Publications Research Service, Washington, D.C., JPRS-51003, 70/07/22 9 pages

32. Patterson, J. E., Method of Analysis of Radio Frequency Interference Between RF Systems (NASA DCAF E 00 3092)
Weapons Research Establishment, Salisbury (Australia).
Avail. ESRO?ESA, 114 Ave. Charles DeGaulle, 92522 Neuilly/Seine, France
Issue 1 Page 18 Category 17 WRE-TN-EC-25 67/04/00 29 pages

33. Prozorov, V. A., Protection of Radiometers from Pulse Interference
Pulkovo, Glavnaia Astronomicheskaia Observatoriia, Izvestiia, No. 188, 1972 pp. 180-183 (in Russian)

34. Ramsey, J. L., Effective Acquisition of FH TDMA Signals in Jamming
Mitre Corp., Bedford, Mass. (Limited Distribution)
AD-914652L ESD-TR-73-214 MTR-258 F19628-73-C-0001 AF Proj. 637A 73/09/00 46 pages

35. Rood, R. D., EM Tactical Communications Under Intentional Interference
M. S. Thesis, Final Report
Army Command and General Staff Coll., Fort Leavenworth, Kansas
AD-B006029L, 75/06/06, 76 pages

36. Schmidt, H. J., DDP 124 Computer Programs for the Study of Correlation Properties of Binary Sequences (Computer Programs for Study of Correlation Properties of Binary Sequences)
Air Force Systems Command, Wright-Patterson AFB, Ohio. Avionics Lab.) AD-843779 AFAL-TR-68-131 68/08/00 42 pages

37. Schreder, K. D.; Wilson, S. H., Jr.; Experimental FFH Equipment for RPV Command and Control
Quarterly Report, 1 June. - 30 Sep. 1972
Northrop Corp., Hawthorne, Calif. (Research and Technology Center)
AD-905107L NRTC-72-9R ECOM-0262-1 QR-1, DAAB07-72-C-026 ARPA Order 2164, 72/10/00 34 pages

38. Singarayar, S.; Natarajan, K.; Swamping Radar Interference
Institution of Engineers (India), Journal, Electronics and
Telecommunication
Engineering Division, Vol. 55, Apr. 1975, pp. 80-81

39. Watterson, C. C., Advantages of Wideband Modulation in Short
MF-HF Communication LINK Phase C, Part 6, Final Report, 1 Feb.
- 30 June. 1969 ESSA-TM-ERLM-ITS-188,
Institute for Telecommunication Sciences, Boulder, Colo.
AD-861429, 69/07/00, 41 pages

40. Ziemer, R. E.; Tranter, W. H., Principles of Communications
Systems, Modulation, and Noise
Houghton Mifflin Company, Boston 1976

Part III

ALGEBRAIC CODING

organised by Dr. P. G. Farrell

ALGEBRAIC CODING AND COMBINATORICS
A TUTORIAL SURVEY

Giuseppe Longo

Università di Trieste
Trieste, Italy

1. INTRODUCTION

The transmission of information over a distance is subject to disturbances and to combat their effects it is often convenient to process the information, i.e. to use error-correcting codes. The structural regularity of the codes which people have been studying has probably put too much restriction on their performance and the good codes which are known to exist have not been discovered yet (although some not too bad codes have been constructed). We shall only deal with block codes (as opposed to convolutional codes) and we assume the reader is familiar with the basic concepts and results from the theory of such codes.

Several concepts from combinatorics have been rediscovered by coding theorists, who in turn have applied combinatorial results to algebraic coding theory. What follows is a brief description of the relationships between these two disciplines which are influencing each other so much.

Let A be the alphabet over which a code $\mathcal{C}$ is defined; often A is a finite or Galois field of order q (q a prime power), GF (q) or F for short; most often q = 2 (binary case). An (n,M,d) code $\mathcal{C}$ is a code containing M codewords of length n at a minimum (Hamming) distance d apart; a linear code $\mathcal{C}$ of dimension k will be denoted by (n,k); therefore an (n,k) linear code is an (n,q^k,d) code for some d.

2. ALGEBRAIC CODING PREREQUISITES

Let F^n be the set of all n-tuples on the field F. Any element $v = v_1v_2...v_n$ of F^n can be represented by a polynomial in the indeterminates $z_1, z_2, \ldots, z_n$:

(1) $$v = v_1v_2 \ldots v_n \rightarrow z^v = z_1^{v_1} z_2^{v_2} \ldots z_n^{v_n}$$

If $z_i^2 = 1$ for all i, the set of all z^v is a multiplicative group G isomorphic to F^n.

Now consider the set QG of all formal sums

(2) $$\sum_{v \in F^n} a_v z^v, \qquad a_v \in Q, \qquad z^v \in G,$$

where Q is the field of the rational numbers. Addition and multiplication in QG are defined as follows:

(3.1) $$\sum_{v \in F^n} a_v z^v + \sum_{v \in F^n} b_v z^v = \sum_{v \in F^n} (a_v + b_v) z^v$$

(3.2) $$r. \sum_{v \in F^n} a_v z^v = \sum_{v \in F^n} r a_v z^v \qquad (r \in Q)$$

(3.3) $$\sum_{v \in F^n} a_v z^v . \sum_{w \in F^n} b_w z^w = \sum_{v \in F^n} \sum_{w \in F^n} a_v b_w z^{v+w}$$

This makes QG into a commutative algebra, and every subset of F^n (e.g. a code $\mathcal{C}$) corresponds to an element of QG (C, say):

(4) $$C = \sum_{u \in \mathcal{C}} z^u .$$

Any element of QG can be thought of as a "generalized code" with each codeword v appearing a_v times. Thus the (3,2) code $\mathcal{C} = \{000, 011, 101, 110\}$ corresponds to $C = 1 + z_2z_3 + z_1z_3 + z_1z_2$.

If $Y_i = \sum_{wt(u)=i} z^u$ is the set of vectors of weight i, namely $Y_o = 1$, $Y_1 = z_1 + z_2 + \ldots z_n$, $Y_2 = z_1 z_2 + z_1 z_3 + \ldots + z_{n-1} z_n$, etc., then the sphere of radius e around a vector v is described by $z^v(Y_o + Y_1 + \ldots Y_e)$ and consequently an (n,M,2e + 1) perfect code $\mathcal{C}$ satisfies the identity

(5) $C\,(Y_o + Y_1 + \ldots + Y_e) = \sum_{u \in F^n} z^u,$

where C is given by (4).

Let f be any mapping defined on F^n. The Hadamard (or Fourier) transform of f, $\hat{f}$, is defined as:

(6) $\hat{f}(u) = \sum_{v \in F^n} (-1)^{u.v} f(v)\ , \ u \in F^n,$

where u.v is the scalar product of u and v. The above definition is based upon the characters: given any $u \in F^n$, consider the mapping χ_u from G (cf.(1)) onto {+1, -1} defined by

(7) $\chi_u\,(z^v) = (-1)^{u.v}.$

χ_u is called a character of G and can be extended to QG: if $C = \sum_{v \in F^n} a_v z^v \in QG$, then:

(8) $\chi_u(C) = \chi_u\,(\sum_{v \in F^n} a_v z^v) = \sum_{v \in F^n} a_v\, \chi_u(z^v) = \sum_{v \in F^n} (-1)^{u.v} a_v.$

It is easy to check that for any two elements C_1, C_2 of QG one has (cf. (3.3)):

$\chi_u\,(C_1)\, \chi_u\,(C_2) = \chi_u(C_1 C_2).$

Let now $C = \sum_{v \in F^n} c_v z^v$ be an element of QG such that $M = \sum_{v \in F^n} c_v \neq 0$(e.g. a code containing M codewords).

Let $A_i = \sum_{wt(v)=i} c_v (i=0\ , \ldots, n)$ (e.g. for a code, A_i is the number of codewords of weight i), then $\sum A_i = M$, and the set $\{A_o, \ldots, A_n\}$ is called the weight distribution (W D) of G. The polynomial

$$W_c(x,y) = \sum_{v \in F^n} x^{n - wt(v)} \; y^{wt(v)}$$

$$(9) \qquad = \sum_{i=o}^{n} A_i \, x^{n-i} y^i$$

is called the weight enumerator (WE) of C.

The Hadamard transform of C is the element C' of QG defined as (cf.(6)):

$$(10) \quad C' = \frac{1}{M} \sum_{u \in F^n} \chi_u(c) z^u.$$

Let now Y_k be defined as in (5) and consider the following quantity

$$(11) \quad \chi_u(Y_k) = \sum_{wt(v)=k} (-1)^{u.v} \overset{def}{=} P_k(i)$$

where $wt(u) = i$. $P_k(i)$ is called Krawtchouk polynomial (n is fixed), and it can be seen that (11) is a particular case of the following more general definition:

$$(12) \quad P_k(x) = \sum_{j=o}^{k} (-1)^j \binom{n-x}{k-j} \binom{x}{j} \qquad (k = 0, 1, 2, \ldots).$$

If we write $C' = \sum_{u \in F^n} c'_u z^u$ and compare with (10), we find

$$c'_u = \frac{1}{M} \chi_u(C) = \frac{1}{M} \sum_{v \in F^n} (-1)^{u.v} c_v.$$

The WD $\{A'_o, A'_1, \ldots A'_n\}$ of C' is now given by

$$(13) \quad A'_k = \sum_{wt(u)=k} c'_u = \frac{1}{M} \sum_{wt(u)=k} \chi_u(C)$$

and it turns out that

$$(14)\quad A'_k = \frac{1}{M}\sum_{i=0}^{n} A_i P_k(i) \qquad (k = 0, \ldots, n),$$

where the $P_k(i)$ are defined in (11).

Now, given any $(n + 1)$-tuple $\{A_o, \ldots, A_n\}$ with $M = \sum A_i \neq 0$, the $(n + 1)$-tuple $\{A'_o, \ldots A'_n\}$ given by (14) is called its MacWilliams transform.

The set of relations expressed by (14) are equivalent to the following relationship between the weight enumerators (WE) of C and C', $W_c(x,y)$, $W_{c'}(x,y)$, which we state as a theorem

Theorem 1 (MacWilliams theorem for nonlinear codes):

$$(15)\quad W_{c'}(x,y) = \frac{1}{M} W_c(x + y, x - y).$$

If C is a linear code, then C' is its dual $C^{\perp}$, and (15) specializes to the well-known MacWilliams identities for linear codes.

Given any (linear or nonlinear) code $\mathcal{C}$, the distance distribution of $\mathcal{C}$ is the vector $\{B_o, B_1, \ldots B_n\}$, where

$$B_i = \frac{1}{M} \cdot (\text{number of ordered pairs of codewords } u, v \text{ at distance } i \text{ apart}).$$

Note that $B_o = 1$, and $\sum B_i = M$. For linear codes the distance distribution coincides with the WD. In terms of the QG algebra, let C be as in (4) and let

$$(16)\quad D = \frac{1}{M} C^2 = \sum_{w \in F^n} d_w z^w.$$

Then the WD of D is given exactly by the elements B_i where $B_i = \sum_{wt(w)=i} d_w$.

The MacWilliams transform of the distance distribution (see (13) and (14)) is $\{B'_o, B'_1 \ldots B'_n\}$ where

$$(17)\quad B'_k = \frac{1}{M} \sum_{wt(u)=k} \chi_u(D) = \frac{1}{M} \sum_{i=0}^{n} B_i\ P_k(i) \quad (k = 0, 1, \ldots n),$$

and from the properties of characters one gets:

$$(18)\quad B'_k = \frac{1}{M^2} \sum_{wt(u)=k} \chi_u (C^2) = \frac{1}{M^2} \sum_{wt(u)=k} (\chi_u(C))^2 \geqslant 0,$$

which turns out to be a useful result in the derivation of the linear programming bound.

The following properties of the A_i, B_i, A'_i, and B'_i parameters can be checked: if C is any code which contains the all-zero codeword $\underline{0}$, then

(i) if $B_i = 0$, then $A_i = 0$

(ii) if $B'_i = 0$, then $\chi_u(C) = 0$ for all u of weight i and $A'_i = 0$,

(iii) if $B'_i \neq 0$, then $\chi_u(C) \neq 0$ for some u of weight i.

Further, if C is an (n, M, d) binary code containing $\underline{0}$, let $0, \tau_1, \tau_2, \ldots, \tau_s$ be the subscripts i for which $B_1 \neq 0$ $(0 < \tau_1 < \tau_2 < \ldots < \tau_s \leqslant n)$ and let $0, \sigma_1 \ldots, \sigma_{s'}$ be the subscripts i for which $B'_i \neq 0 (0 < \sigma_1 < \sigma_2 < \ldots < \sigma_{s'} \leqslant n)$. Then $d = \tau_1$ is the minimum distance of C, s is the number of distinct nonzero distances between codewords, s' is called the external distance of C and $d' = \sigma_1$ its dual distance. It can be shown that d' is the largest number such that each (d' - 1)-subset of the coordinates in C contains all (d' - 1)-tuples an equal number of times.

3. INTRODUCTION TO COMBINATORIAL DESIGNS

Let X be a set with v elements or points (a v-set). A t-design is a collection of distinct k-subsets (called blocks) of X such that any t-subset of X is contained in exactly λ blocks. This is called a t- (v, k, λ) design.

To find a connection between t-designs and codes, we consider again the set F^n of all n-tuples over GF(q), and for any two vectors $u = (u_1, \ldots, u_n)$, $v = (v_1, \ldots, v_n)$ of F^n we say that u is covered by v if $u_i (u_i - v_i) = 0$ for all i. Then as X we take the set of coordinates and a t - (n, k, λ) design is a set $\mathcal{B}$ of k-weight vectors (blocks) such that every t-weight vector is covered by exactly λ blocks .

As an example, the nonzero words of the (4,2) ternary Hamming code, namely

0112	1011	1202	2101
0221	1120	2022	2210

which all have weight 3, form a 2- (4,3,1) design (i.e. each of the $2^2 \cdot \binom{4}{2} = 24$ ternary 4-vectors of weight 2 is covered by exactly one codeword).

From now on we shall consider the binary case only.

A 2-design is called a (balanced incomplete) block design. As an example, consider the set $X = \{1, 2, \ldots, 7\}$ and the following subsets of X (blocks): {1, 2, 4}, {2, 3, 5}, {3, 4, 6}, {4, 5, 7}, {5, 6, 1}, {6, 7, 2}, {7, 1, 3}. If each block is called line, then the corresponding block design is called projective plane of order 2. The figure illustrates this 2 - (7, 3, 1) design.

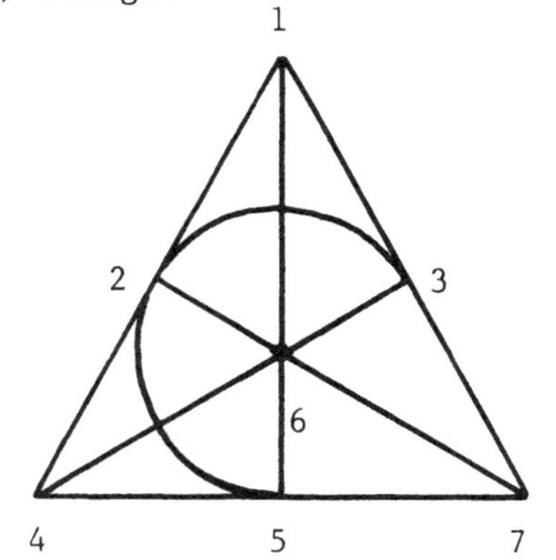

A Steiner system is a t-design with $\lambda = 1$ and a t-(v, k, 1) design is usually called an S(t, k, v). The last example therefore is an S(2, 3, 7). In particular, an $S(2, n+1, n^2 + n + 1)$ is called projective plane of order n (for n = 2 the last example is obtained again). An affine plane of order n is an $S(2, n, n^2)$. A Steiner system in which t = 2, k = 3, i.e. an S(2, 3, v) is called a Steiner triple system, SST(v).

Lemma 1 In a t-(v, k, λ) design let $P_1, \ldots, P_t$ be any t distinct points, let λ_i be the number of blocks containing $P_1, \ldots, P_i$ for $1 \leq i \leq t$ and $\lambda_o = b$ the total number of blocks. Then λ_i does not depend on the choice of $P_1, \ldots, P_i$ and actually for $0 < i < t$:

$$(19)\quad \lambda_i = \lambda \binom{v-i}{t-i} \Big/ \binom{k-i}{t-i} = \lambda \frac{(v-i)\ (v-i-1)\dots(v-t+1)}{(k-i)\ (k-i-1)\dots(k-t+1)}$$

Proof For $i = t$, equation (19) is true ($\lambda_t = \lambda$). By induction on i, if λ_{i+1} is known to be independent from $P_1, \dots, P_{i+1}$, for each block B containing $P_1, \dots, P_i$ and for each point Q distinct from $P_1, \dots, P_i$, let $z(Q,B) = 1$ if $Q \in B$, $Z(Q,B) = 0$ if $Q \notin B$. Then from the induction hypothesis and counting in two ways:

$$(20)\quad \sum_Q \sum_B z(Q,B) = \lambda_{i+1}\ (v-i) = \sum_Q \sum_B z(Q,B) = \lambda_i (k-i)$$

Q.E.D.

Corollaries (i) A $t - (v, k, \lambda)$ design is also an $i - (v, k, \lambda_i)$ design for $1 \leqslant i \leqslant t$.

(ii) In a t-(v, k, λ) design the total number of blocks is $\lambda_0 = b = \lambda \binom{v}{t} \Big/ \binom{k}{t}$ (from(19)for $i = 0$) and each point belongs to exactly $\lambda_1 = r = bk/v$ blocks (from (20) for $i = 0$).

(iii) In a 2-design $\lambda(v - 1) = r(k - 1)$ (from (20) for $t = 2$ and $i = 1$, since $\lambda_2 = \lambda$).

(iv) A necessary condition for a t-(v, k, λ) design to exist is that the numbers $\lambda \binom{v-i}{t-i} \Big/ \binom{k-i}{t-i}$ be integers for $0 \leqslant i \leqslant t$. (This condition is sometimes sufficient, but not in general.)

Let $P_1, \dots, P_k$ be the points belonging to one of the blocks in a $t - (v, k, \lambda)$ design. Consider the blocks containing $P_1, \dots, P_j$ but not $P_{j+1}, \dots, P_i$, for $0 \leqslant j \leqslant i$; if their number does not depend on the choice of $P_1, \dots, P_i$, we indicate it by λ_{ij}. These are the block intersection numbers (bin's) of the design.

Lemma 2 The λ_{ij} are constants for $i \leqslant t$. Actually $\lambda_{tt} = \lambda$ and $\lambda_{ii} = \lambda_i$ as given by (19). Further, the λ_{ij}

(if defined) satisfy the following Pascal property

(21) $\lambda_{ij} = \lambda_{i+1,\, j} + \lambda_{i+1,\, j+1}.$

Remark that in a Steiner system $(\lambda = 1)$,

$$\lambda_{tt} = \lambda_{t+1,\, t+1} = \dots = \lambda_{kk} = 1$$

and the λ_{ij} are defined for all $0 \leqslant j \leqslant i \leqslant k$.

Without entering into details, let us mention that several designs can be constructed from a given $t - (v, k, \lambda)$ design:

(a) Given a $t - (v, k, \lambda)$ design with bin's λ_{ij}, if we omit one point from all its blocks, we get two sets of new blocks which form a $(t-1) - (v-1, k, \lambda_{t,\, t-1})$ and a $(t-1) - (v-1,k-1,\lambda)$ design, with bin's $\lambda_{i+1,\, j}$ and $\lambda_{i+1,\, j+1}$, respectively.

(b) In particular, if an $S(t, k, v)$ exists, there exists an $S(t - 1, k - 1, v - 1)$.

(c) Taking the complements of all the blocks of a t-design gives the <u>complementary</u> design, which is a $t - (v, v-k, \lambda_{to})$ design with bin's $\lambda_{i,\, i-j}$.

One way of describing a $t - (v, k, \lambda)$ design with v points $P_1, \dots, P_v$ and b blocks $B_1, \dots B_b$ is to construct its $b \times v$ <u>adjacency</u> (or <u>incidence</u>) <u>matrix</u> $A = (a_{ij})$ defined by

$$a_{ij} = \begin{cases} 1 \text{ if } P_j \in B_i \\ 0 \text{ if } P_j \notin B_i \end{cases}$$

Here is the incidence matrix A of the $2 - (7, 3, 1)$ design (projective plane of order 2).

1	1	0	1	0	0	0
0	1	1	0	1	0	0
0	0	1	1	0	1	0
0	0	0	1	1	0	1
1	0	0	0	1	1	0
0	1	0	0	0	1	1
1	0	1	0	0	0	1

It can be verified that A satisfies $A^T A = (r - \lambda_2)I + \lambda_2 J$, where J is the all-one square matrix of order b. This condition reads $A A^T = (r - \lambda)I + \lambda J$ for a 2- (v, k, λ) design and it is a necessary and sufficient condition for A to be its incidence matrix.

If $b = v$ and hence $r = k$ (from Corollary (ii) to Lemma 1) then the corresponding design is called <u>symmetric</u> or <u>square</u> since its incidence matrix is square. The incidence matrix of a square 2-design need not itself be symmetric, rather it satisfies the following conditions, which are symmetric on its rows and columns:

(1) Any row contains k ones	(1') Any column contains k ones
(2) Any pair of columns both have ones in exactly λ rows	(2') Any pair of rows both have ones in exactly λ columns

Properties (2) and (1') are equivalent to

(22) $A^T A = (k - \lambda)I + \lambda J$,

whereas (1) and (2') are equivalent to

(23) $AA^T = (k - \lambda)I + \lambda J$.

On the other hand it can be proven that if A is a square (0, 1) matrix satisfying (22), then (23) also holds, and conversely.

4. HADAMARD AND CONFERENCE MATRICES

An $n\times n$ matrix H whose entries are ± 1 and such that $HH^T = nI$ is called Hadamard matrix. As an example

$$\begin{matrix} 1 & 1 \\ -1 & 1 \end{matrix} \qquad \text{and} \qquad \begin{matrix} 1 & 1 & -1 & 1 \\ 1 & -1 & -1 & -1 \\ -1 & -1 & -1 & 1 \\ 1 & -1 & 1 & 1 \end{matrix}$$

are both Hadamard matrices. For the existence of H it is necessary that $n = 1, 2$ or $n \equiv 0 \pmod 4$, but it is not known whether this is sufficient. On the other hand there is an infinite supply of Hadamard matrices, since $K = \begin{bmatrix} H & H \\ H & -H \end{bmatrix}$ is Hadamard whenever H is Hadamard.

Given a Hadamard matrix, any two columns or rows can be interchanged and any row or column can be multiplied by -1 and the matrix obtained is still Hadamard. As a consequence it is possible to reduce a H matrix in normalized form, in which its first row and column consist entirely of 1s. The second matrix above can be normalized by multiplying its third row and third column by -1.

A conference matrix (or C-matrix) C of order n is an $n \times n$ matrix with diagonal entries 0 and other entries ± 1 and such that $CC^T = (n-1)I$. Again a C-matrix can be normalized and put into the form $C = \begin{bmatrix} 0 & \underline{1} \\ \underline{1}^T & S \end{bmatrix}$, where S is a square matrix of order $(n-1)$ satisfying $SS^T = (n-1)I - J$. A construction for C-matrices due to Paley is the following:

Let $n = p^m + 1 \equiv 2 \pmod 4$, where p is an odd prime, and let $Q = (q_{ij})$ be the $p^m \times p^m$ Jacobstahl matrix, where $q_{ij} = \chi(j-i)$ and $\chi(.)$ is the Legendre symbol defined by

$$\chi(0) = 0$$

$$\chi(i) = \begin{cases} 1 & \text{if } i \text{ is a quadratic residue} \\ -1 & \text{if } i \text{ is a non-quadratic residue} \end{cases}$$

Then $C = \begin{bmatrix} 0 & \underline{1} \\ \underline{1}^T & Q \end{bmatrix}$ is a symmetric conference matrix of order n.

If C is a skew C-matrix then I+C is Hadamard, and this fact is used in the construction, due again to Paley, of Hadamard matrices of order n where n - 1 is a prime power: one starts again from the Jacobstahl matrix Q defined above and constructs the matrix $H = \begin{bmatrix} 1 & \underline{1} \\ \underline{1}^T & Q-I \end{bmatrix}$ which can be seen to be Hadamard.

If $HH^T = nI$, then also $H^T H = nI$ and therefore what is said of rows can be said of columns too.

Let H be a normalized Hadamard matrix of order $n = 4m$ and let B be the matrix obtained from H if the first row and the first column are deleted. Since the rows of H are orthogonal, any row of B contains (2m-1) +1 elements and (2m) -1 elements. Further any two rows of B_n contain +1 in exactly (m-1) positions. If now -1 is replaced by 0 in B, the matrix A obtained in this way satisfies $AA^T = mI + (m-1)J$ and therefore A^T is the incidence matrix of a 2-(4m - 1, 2m - 1, m - 1) design (cf. the property of incidence matrices for 2-designs).

5. CODES AND DESIGNS

In a code the words should be as much distant from each other as possible, and it is not surprising that designs can be used to construct codes. Thus in the incidence matrix of a 2-design, each row contains k 1's and (v-k) 0's and any two rows have 1's in the same λ columns. This implies that these rows differ in $2(k-\lambda)$ positions.

111	111	000	000
111	000	111	000
λ	$k-\lambda$	$k-\lambda$	$v-\lambda-2(k-\lambda)$

Given any Steiner system S(t, k, v) the rows of its incidence matrix form a non linear code with parameters $n=v$, $M=b=\binom{v}{t}\Big/\binom{k}{t}$, $d \geq 2\,(k-t+1)$.

111	111	000	000
111	000	111	000
$\leq t-1$	$\geq k-t+1$	$\geq k-t+1$	

Actually two blocks cannot have more than t-1 points in common (or t points would be contained in two blocks!). Every codeword in these codes has the same weight k; these are constant weight codes.

Conversely, it is sometimes possible to obtain a design from the codewords of a fixed weight in a given (n, M, d) code. For example the $(n = 2^m - 1, k = 2^m - 1 - m)$ binary Hamming code $\mathcal{H}$ with minimum distance $d = 3$ and the corresponding extended code $\mathcal{H}^*$ give rise to the following designs: the codewords of weight 3 in $\mathcal{H}$ form an $S(2, 3, 2^m - 1)$ and the codewords of weight 4 in $\mathcal{H}^*$ form an $S(3, 4, 2^m)$. (Remark that the second statement implies the first one.) More generally, if $\mathcal{C}$ is a perfect e-error-correcting code of length n (e odd) and $\mathcal{C}^*$ is the corresponding extended code, then the codewords of weight 2e + 1 in $\mathcal{C}$ form an S(e+1, 2e+1, n) and those of weight 2(e+1) in $\mathcal{C}^*$ form an S(e+2, 2e+2, n+1).

A very interesting code is the <u>extended binary Golay code</u> $\mathcal{G}$, which we shall briefly consider for the designs which can be constructed from it. This (24, 12) linear code has the following WE (cf. (9)):

$$x^{24} + 759\, x^{16}y^{8} + 2576\, x^{12}y^{12} + 759\, x^{8}y^{16} + y^{24}. \tag{23}$$

We shall refer to the codewords in $\mathcal{G}$ of weights 8 and 12 as <u>octads</u> and <u>dodecads</u>, respectively.

<u>Lemma 3</u> Any binary 24-vector v of weight 5 is covered by exactly an octad of $\mathcal{G}$.

The proof follows immediately by noting that if v were covered by two octads, their distance would be $\leqslant 6$ (against (23)). Therefore each octad covers $\binom{8}{5}$ distinct vectors of weight 5, and in fact $759 . \binom{8}{5} = \binom{24}{5}$. Lemma 3 can equivalently be stated as: The codewords of weight 8 in $\mathcal{G}$ form a Steiner system S(5, 8, 24). Therefore there exist Steiner systems S(4, 7, 23), S(3, 6, 22), and S(2, 5, 21) (this follows from property (b) of designs given in section 4). Considering the complementary design (property (c)), one obtains that the codewords of weight 16 in $\mathcal{G}$ form a 5-(24, 16, 78) design.

Another design generated by $\mathcal{G}$ is a 5-(24, 12, 48) design whose blocks are the dodecads, and there are several others.

To give now a nonbinary example, consider the <u>ternary</u> Golay code of length 11 and minimum distance 5, which contains 132 words of weight 5. As in the proof of Lemma 3, two codewords of weight 5 cannot cover the same 3-weight 11-vector (this code too is linear) and each such codeword covers $\binom{5}{3} = 10$ distinct triples. And in fact $\binom{11}{3} . 2^3 = 132 \cdot \binom{5}{3} = 1320$. Thus the weight-5 codewords in the ternary Golay code are the blocks of a ternary 3-(11, 5, 1) design.

We now consider the designs connected with binary codes derived from Hadamard matrices. Consider a 2-(4m - 1, 2m - -1, m - 1) design with square incidence matrix A constructed as in section 4. If the rows of A are considered as codewords, they give rise to a (4m - 1, 4m - 1, 2m) code. To increase the number of codewords, we take the complement $\bar{A}$ of A, which correspond to the complementary design, which is a 2-(4m-1, 2m,m) design. Again the rows of $\bar{A}$ differ in 2m places and separately the two codes A and $\bar{A}$ can correct (m-1) errors each. The number of places in which a row of A differs from a row of $\bar{A}$ equals the number of places in which a row of A does <u>not</u> differ from a row of A itself, and this is $v - 2(k - \lambda) = 2m - 1$. This is enough to correct (m - 1) errors, but now the code has 2(4m - 1) words. Each codeword now has either weight (2m - 1) or 2m and therefore the two extra words $\underline{0}$ and $\underline{1}$ can be added without decreasing the minimum distance. These 8m codewords detect (2m - 2) errors and correct (m - 1). To increase the detection capability the code can be extended by an overall check digit and thus a (4m, 8m, 2m) code is obtained. Of course one could get the same code starting from a normalized Hadamard matrix of order n, replacing +1 by 1 and -1 by 0 to get an $n \times n$

binary matrix A and considering as codewords the rows of A and those of the complementary $\bar{A}$.

6. DESIGNS FROM CODES: A GENERAL APPROACH

At the end of section 2 four parameters d, d', s, s' for a code $\mathcal{C}$ were introduced (d is the minimum distance and s the number of distinct nonzero distances between codewords of $\mathcal{C}$; if $\mathcal{C}$ is linear d' and s' have the same meaning for the dual $\mathcal{C}^{\perp}$).

Several important codes are such that $s \leqslant d'$ or $s' \leqslant d$ and as a consequence the MacWilliams identities can be solved. Further if $t \geqslant d'-s$ or $t \geqslant d-s'$, the codewords of each weight form a t-design. Finally, if the codewords of a linear code $\mathcal{C}$ of any weight form a t-design, the same happens with the codewords of any weight of the dual code $\mathcal{C}^{\perp}$. Here are several examples of such codes:

(1) The $(n, 2^{n-1}, 2)$ even-weight code (dual of the $(n, 2, n)$ repetition code), for which $d = 2$, $s = \frac{n}{2}$, $d' = n$, $s' = 1$.

(2) The $(2^m - 1, 2^m, 2^{m-1})$ simplex code (dual of the $(2^m - 1, 2^m - 1 - m)$ linear Hamming code), for which $d = 2^{m-1}$, $s = 1$, $d' = 3$, $s' = n - 4$.

(3) The $(n, n + 1, \frac{n+1}{2})$ Hadamard code obtained from an $(n+1) \times (n+1)$ Hadamard matrix deleting the first column ($n \equiv 3 \pmod 4$), for which $d = \frac{n+1}{2}$, $s = 1$, $d' = 3$, $s' = n - 4$.

(4) The $(2^m, 2^{m+1}, 2^{m-1})$ first-order Reed-Muller code (dual of the $(2^m, 2^m - m - 1)$ linear extended Hamming code), for which $d = 2^{m-1}$, $s = 2$, $d' = 4$, $s' = \frac{n}{2} - 2$.

(5) The $(24, 2^{12}, 8)$ extended Golay code (self-dual) for which $d = d' = 8$, $s = s' = 4$.

Lemma 4 If $s \leqslant d'$ the elements of the distance distribution $\{B_i\}$ can be expressed explicitly in terms of n, M and the τ_j's.

It is not essential to give this formula (or the proof of the lemma), rather it is preferable to give a series of results which lead to the main theorem of this section:

Lemma 5 If $s \leq d'$, $A_i = B_i$ for all i (sufficient condition). Remark that if $A_i \neq 0$, then $B_i \neq 0$, so the nonzero A_i values are $A_0, A_{\tau_1} \ldots, A_{\tau}$.

Lemma 6 If $s' \leq d$, $A'_i = B'_i$ for all i (sufficient condition).

Corollary If $s' \leq d$, $A_i = B_i$ for all i (actually $\{A_i\}$ and $\{B_i\}$ are the MacWilliams transforms of $\{A'_i\}$ and $\{B'_i\}$).

We now come to designs. Let u be a fixed vector in F^n, $wt(u) = t$, where $0 < t < d'$. For $\tau_i \geq t$ let $\lambda_{\tau_i}(u)$ be the number of codewords of $\mathcal{C}$ of weight τ_i which cover u. By counting the vectors of weight $t + j$ $(0 \leq j \leq d' - 1 - t)$ which cover u and are covered by a codeword of $\mathcal{C}$ and using the property of d' stated at the very end of section 2, it is seen that the $\lambda_{\tau_i}(u)$ satisfy the d'-t equations

$$(24) \quad \sum_{i=1}^{s} \binom{\tau_i - t}{j} \lambda_{\tau_i}(u) = \frac{M}{2^{t+j}} \binom{n-t}{j}, \quad (0 \leq j \leq d'-t-1).$$

If $s \leq d'$, from Lemma 5 $A_n = B_n = 0$ or 1, and we assume that A_n is known and put

$$(25) \quad \bar{s} = \begin{cases} s & \text{if } A_n = 0 \\ s-1 & \text{if } A_n = 1 \end{cases}$$

With this position eq.s (24) can be rewritten as

$$(26) \quad \sum_{i=1}^{\bar{s}} \binom{\tau_i - t}{j} \lambda_{\tau_i}(u) = \left(\frac{M}{2^{t+j}} - A_n\right) \binom{n-t}{j},$$

since the vector $\underline{1}$ covers all vectors.

Theorem 2 If $\bar{s} < d'$, then the codewords of weight τ_i in $\mathcal{C}$ form a $(d' - \bar{s}') - (n, \tau_i, \lambda_{\tau_i})$ design, provided that $\tau_i \geq d' - \bar{s}$.

Proof If $\bar{s} < d'$, t can be chosen so that $d' - t = \bar{s}$ and (24) or (26) are $\bar{s}$ equations in the $\bar{s}$ unknowns $\lambda_{\tau_i}(u)$. Then the $\lambda_{\tau_i}(u)$ do not depend on the particular u, and can be denoted by λ_{τ_i}: the

the codewords of each weight τ_i ($\geqslant d' - \bar{s}$) form a $t(= d' - \bar{s})$-design. (It is not essential to give here an expression for the λ_{τ_i}'s).

As an application of Theorem 2, consider example (1) above (even-weight code). If n is even, $\underline{1}$ is in C, $\bar{s} = s - 1 = \frac{n}{2} - 1$ and the theorem states that the codewords of any weight $2h \geqslant \frac{n}{2} + 1$ form an $(\frac{n}{2} + 1)$ - design. Remark that $\mathcal{C}$ contains <u>all</u> vectors of weight 2h and therefore is a 2h-design: this is stronger than the conclusion from Theorem 2.

The extended Golay code (example (5)) being self-dual contains $\underline{1}$, so $\bar{s} = s - 1 = 3$ and the 8-weight codewords form a 5-design, actually a 5-(24, 8, 1)-design.

To conclude this presentation, we shall illustrate the fact that also the dual code $\mathcal{C}^{\perp}$ gives t-designs when $\mathcal{C}$ does. Let $\mathcal{C}$ be an $(n, 2^k, d)$ linear code whose words of each weight $w > 0$ form a t-design, $t < d$.

Let T be any fixed set of t coordinates among the n coordinates and let $\mathcal{C}^T$ be the linear $(n-t, 2^k, d-t)$ code obtained from $\mathcal{C}$ when the coordinates of T are deleted. Let $\{A_i^T\}$ be the weight distribution of $\mathcal{C}^T$. Since the words of weight w of $\mathcal{C}$ form a t-design, they also form an i-design for $i \leqslant t$ (corollary (i) of lemma 1) and consequently the number of codewords of weight w which contain exactly i coordinates of T does not depend on the choice of T; if μ_i^w indicates this number, also the sum

$$S_j = \mu_0^j + t\mu_1^{j+1} + \binom{t}{2} \mu_2^{j+2} + \dots + \mu_t^{j+t}$$

is independent from the choice of T. It is now easy to see that $A_j^T = S_j$. If $\mathcal{C}$ is shortened to a code $\mathcal{C}_0^T$ by taking those codewords of $\mathcal{C}$ which have 0 coordinates in T and deleting those coordinates, then $\mathcal{C}_0^T$ is an $(n - t, 2^{k-t})$ linear code whose minimum distance is again d. Now shorten $\mathcal{C}^{\perp}$ in the same way to get $(\mathcal{C}^{\perp})_0^T$; obviously the dual of $\mathcal{C}^T$ contains $(\mathcal{C}^{\perp})_0^T$, but their dimensions are the same and therefore $(\mathcal{C}^T)^{\perp} = (\mathcal{C}^{\perp})_0^T$. Since the weight distribution of $\mathcal{C}^T$ does not depend on the choice of T, the same is true of its MacWilliams transform, i.e. the weight distribution of $(\mathcal{C}^{\perp})_0^T$.

Theorem 3. If $\mathcal{C}$ is an $(n, 2^k, d)$ linear code with $k > 1$, such that for each $v > 0$ its codewords of weight v form a t-design, $t < d$, then also the codewords of each weight in $\mathcal{C}^{\perp}$ form a t-design.

Proof. The proof is best given in a number of steps.

(i) Let w be the weight of a (nonzero) codeword b of $\mathcal{C}^{\perp}$. If $w = n$ the theorem is true, thus we assume $w < n$.

(ii) Suppose $w = n - t$, and choose a codeword a of weight v in $\mathcal{C}$. If $v - t$ is odd, a will be chosen in such a way as to have 1 in the t positions where b has 0 (this is possible since the codewords of weight v in C form a t-design)

$$
\begin{array}{ccccc}
 & \leftarrow\ w & \rightarrow & \leftarrow\ t & \rightarrow \\
b = & 1\ 1\ 1 & 1\ 1\ 1 & 0\ 0\ 0 & 0\ 0\ 0 \\
a = & 0\ 0\ 0 & \underbrace{1\ 1\ 1}_{v\,-\,t} & 1\ 1\ 1 & 1\ 1\ 1
\end{array}
$$

This however leads to a contradiction since then a.b would not be zero. But if v-t is even, a can be chosen in such a way as to have 1 in (t-1) of the positions where b has 0 (again this is always possible because by (21)

$$\lambda_{t-1} = \lambda_{t-1,\, t-1} = \lambda_{t-1,t} + \lambda_{t,t} \geqslant \lambda_{t,t} = \lambda_t,$$

and therefore given any t positions, those where b has 0 -(t-1) among them can always be covered by a vector of weight v in $\mathcal{C}$).

$$
\begin{array}{cccc}
\leftarrow\ w & \rightarrow & \leftarrow\ t & \rightarrow \\
1\ 1\ 1 & 1\ 1\ 1 & 0\ 0\ 0 & 0\ 0\ 0 \\
0\ 0\ 0 & \underbrace{1\ 1\ 1}_{v-t+1} & 1\ 1\ 1 & 1\ 1\ 0
\end{array}
$$

This again implies $a.b \equiv 1 \pmod 2$. So we have proved that $w \neq n - t$.

(iii) Suppose now $w > n-t$, i.e. $w = n - t + j = n - (t - j)$ for some $j > 0$. But then the same reasoning as in (ii) applies, since a t-design is also a $(t - j)$-design. From (ii) and (iii) we conclude $w < n-t$.

(iv) It is now sufficient to prove that the A'_w codewords c_i of weight w in $\mathcal{C}^\perp$ form a t-design, where $w < n-t$. Let $\underline{c}_i$ $(1 \leq i \leq A'_w)$ be the complements of these codewords and let T be again any set of t coordinate places. If (and only if) c_i is a codeword of weight w in $(\mathcal{C}^\perp)_0^T$ its complement $\underline{c}_i$ has 1 on T. Therefore the number of complements $\underline{c}_i$ having 1 on T does not depend on the choice of T (since the weight distribution of $(\mathcal{C}^\perp)_0^T$ does not). Thus those complements form a t-design.

(v) The codewords c_i of weight w form the complementary t-design with respect to the $\underline{c}_i$.

Q.E.D.

Remark We assumed $k > 1$, because if $k = 1$, $\mathcal{C}$ can give t-designs only if it is a repetition code. In this case, however, $\mathcal{C}^\perp$ is the even-weight code that gives only trivial designs, as we saw.

From Theorems 2 and 3 above we get the following:

Corollary If $\mathcal{C}$ is a linear code with parameters d, s, d', s', let $\bar{s}$ be defined by (25) and define similarly

$$\bar{s}' = \begin{cases} s' & \text{if } A'_n = 0 \\ s'-1 & \text{if } A'_n = 1. \end{cases}$$

Then if either $\bar{s} < d'$ or $\bar{s}' < d$, the codewords in $\mathcal{C}$ of weight w form a t-design, where

$$t = \max(d' - \bar{s},\ d - \bar{s}'),$$

provided $t < d$.

REFERENCES

1. E.R. Berlekamp, Algebraic Coding Theory, McGraw-Hill, 1968.

2. E.R. Berlekamp, A Survey of Algebraic Coding Theory, CISM Courses and Lectures no. 28, Springer, 1970.

3. W.W. Peterson and E.J. Weldon, Jr., Error-Correcting Codes, 2nd ed., The M.I.T. Press, 1972.

4. N.J.A. Sloane, A Short Course on Error-Correcting Codes CISM Courses and Lectures no. 188, Springer 1975.

5. N.J.A. Sloane and F.J. McWilliams, The Theory of Error-Correcting Codes, North-Holland Publishing Company, to appear shortly.

6. J.H. van Lint, Combinatorial Designs Constructed from or with Coding Theory, in "Information Theory: New Trends and Open Problems," G. Longo ed., CISM Courses and Lectures no. 219, Springer 1975.

CODES ASSOCIATED WITH FINITE GEOMETRIES

D. W. Erbach

Department of Pure Mathematics and Mathematical Statistics, Cambridge University, Cambridge, England

1. INTRODUCTION

This is an expository presentation about codes which are generated by the incidence matrix of a finite geometry. The relations between codes and finite geometries have always been inertly there, but it is only in the fairly recent past that much serious investigation has been undertaken using more or less equally results from both the coding and geometric sides. This investigation has been provoked on the coding side by the remarkable theorems of MacWilliams [7] and Gleason [5] about weight enumerators of certain very symmetric codes, and on the geometric side by the appearence of codes as carriers of interesting mathematical structures, e.g. Designs and Finite Simple Groups.

Here we wish to consider several particular examples in order to illustrate some explicit techniques of combinatorial coding theory. On the whole our knowledge of the subject is still rather fragmentary and there are many unsolved problems.

2. FINITE GEOMETRIES

A Finite Geometry is a finite set of points usually denoted by integers $\{1,2,...,n\}$ together with certain distinguished subsets of the points, called Blocks or Lines. We are interested in two special cases, Designs and Projective Planes. Though they are finite combinatorial structures, designs have some properties of ordinary geometry, and it is customary to use geometric language to speak of "collinear points" to indicate sets of points lying in a single line or block, etc.

We use standard notation:

v = no. of points ("varieties")

b = no. of blocks

r = no. of blocks containing a given point ("replications")

k = no. of points in a block

Definition: A Design is a finite geometry in which

D1: Each block contains the same number of points.

D2: Each point is contained in the same number of blocks.

D3: For some fixed integer t, each t points are contained in a unique block.

Definition: A (Finite) Projective Plane is a finite geometry in which

P1: Any two distinct lines meet in a unique point.

P2: Any two distinct points are contained in a unique line.

P3: There are four points, no three collinear.

Designs are, in the literature, often called t-designs, or block designs, or even balanced incomplete block designs (BIBD's). There were originally applied by statisticians to the design of experiments, whence the notation conventions.

The parameters v,b,r,k, and t are not independent. Easy counting arguments show that

$$vr = bk \qquad \binom{v-1}{t-1} = \binom{k-1}{t-1} r$$

and that in fact

$$\binom{v-i}{t-i} \Big/ \binom{k-i}{t-i}$$

must be an integer if $1 \leq i \leq t$. Thus designs can exist for only a few parameter combinations. The conditions above are necessary but not sufficient, and the determination of the possible parameters for which designs exist is a major research area of combinatorial mathematics. A list of smallish consistent parameter values can be found in [6], though recent investigations have settled some questions listed there as unanswered. We mention in passing that designs with t = 2 and 3 are common, with t = 4 and 5 rare, and that except for degenerate cases (e.g. b = r = 1 and k = v) not a single design is known for $t = 6$.

Projective planes are designs for which $v = b$, $k = r$, and $t = 2$. A finite projective plane has more regularity than designs generally, and in fact the following is true:

Proposition 2.1 If P is a finite projective plane, there is an integer n such that

1) $v = b = n^2 + n + 1$

2) $k = r = n + 1$.

The integer n is the Order of the plane. Proofs of this as well as the other standard facts about designs mentioned can be found in any good combinatorics text, e.g. [6].

3. THE CODE OF THE PLANE OF ORDER 2.

The following well-known collection of 7 points and lines is the simplest example of a finite projective plane. It is the plane of order 2.

$L_1 = 1\ 2\ 4$
$L_2 = 2\ 3\ 5$
$L_3 = 3\ 4\ 6$
$L_4 = 4\ 5\ 7$
$L_5 = 5\ 6\ 1$
$L_6 = 6\ 7\ 2$
$L_7 = 7\ 1\ 3$

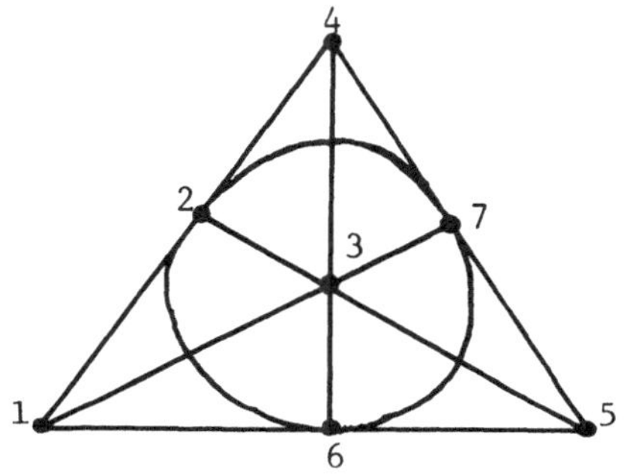

The order 2 plane generated a small interesting code C_7 whose properties we can deduce directly. We take the 7 lines of the plane and form their characteristic vectors into a matrix A, the Incidence Matrix of P

The code is the set of vectors we can generate by treating these vectors as binary numbers and adding, but without carries.

$$A = \begin{pmatrix} 1 & 1 & 0 & 1 & 0 & 0 & 0 \\ 0 & 1 & 1 & 0 & 1 & 0 & 0 \\ 0 & 0 & 1 & 1 & 0 & 1 & 0 \\ 0 & 0 & 0 & 1 & 1 & 0 & 1 \\ 1 & 0 & 0 & 0 & 1 & 1 & 0 \\ 0 & 1 & 0 & 0 & 0 & 1 & 1 \\ 1 & 0 & 1 & 0 & 0 & 0 & 1 \end{pmatrix}$$

This code evidently has length 7 and at least 7 weight 3 vectors, namely the characteristic vectors of lines of the plane. It also has the $\underline{0}$ vector which we get by adding any row to itself. Since there are 3 1's in each column, if we add all the line vectors, we get the $\underline{1}$ vector whose coordinates are all 1. Adding it to any vector gives the opposite vector which we obtain by replacing 1's by 0's and 0's by 1's. So we also have 7 weight 4 vectors which correspond to sets of 4 points, no 3 on a line in the plane. One can check that these 16 vectors are the only ones one gets by adding any number of lines from the above set. Since this is a binary code with 16 vectors, it has dimension 4.

Proposition 3.1 The plane of order 2 generates a (7,4) code whose codewords correspond to the characteristic vectors of the 7 lines, the 7 complements of lines, the whole 7 points and the empty set.

Since this code is linear, the minimum distance between codewords is the same as the minimum weight, namely 3. Thus C_7 is a 1-error correcting code. Now a Hamming Sphere of radius 1 around any word contains 8 vectors of the ambient space, so the maximum conceivable number of words in a length 7 1-error correcting code is $(2^7)/8 = 2^4 = 16$. This bound is attained by C_7 which is therefore a Perfect Code.

There are infinite degenerate families of perfect codes, an interesting infinite family, the Hamming Perfect Codes, and the two well-known exceptional examples known as Golay Codes.

4. CODES OF ODD ORDER PLANES

The order 2 plane is the smallest example of a finite projective plane. It is easy to summarize our current knowledge of planes.

Proposition 4.1 If n is a power of a prime, there is at least one plane of order n. An example can be constructed by taking a 3-dimensional vector space V over a finite field K. The "points" are 1-dimensional substaces of V, the "lines" 2-dimensional ones, and point-line incidence is subspace containment.

Proposition 4.2 (Bruck-Ryser Theorem) If $n \equiv 1$ or 2 (mod 4) a necessary condition for the existence of a plane of order n is that n be a sum of two integer squares.

There is no single n which has been shown to be the order of a plane unless it is a prime power, or not to be the order of one unless eliminated by the Bruck-Ryser Theorem.

We now show that we cannot hope to obtain interesting binary codes from planes of odd order.

<u>Proposition 4.3</u> Let P be a plane of odd order n and C_N be the binary code of length $N = n^2 + n + 1$ generated by the rows of the incidence matrix A of P. Then C_N has dimension N-1 and consists precisely of all vectors of even weight.

To see this we note that the entries of the matrix AA^t are inner products of rows of A, so

$$AA^t = \begin{pmatrix} n+1 & & & \\ & \cdot & & \underline{1} \\ & & \cdot & \\ \underline{1} & & & n+1 \end{pmatrix} \equiv \begin{pmatrix} 0 & & & \\ & 0 & & \underline{1} \\ & & \cdot & \\ \underline{1} & & & 0 \end{pmatrix} \pmod 2$$

If we add the last row to all the others, we obtain

$$\begin{pmatrix} 1 & & & & & 1 \\ & 1 & & \underline{0} & & 1 \\ & & \cdot & & & \cdot \\ & & & \cdot & & \cdot \\ & \underline{0} & & & \cdot & \cdot \\ & & & & 1 & 1 \\ 1 & 1 & \cdot & \cdot & \cdot 1 & 0 \end{pmatrix} \pmod 2$$

in which the last row is the sum of the first N-1 and in which the first N-1 rows are plainly linearly independent. So AA^t has rank (mod 2) at least N-1. It also has rank at most N-1 since rank A = rank A^t and the rank of a product is at most the rank of each factor. Further, since n is odd, the generators of C_N have even weight, hence so do all vectors of C_N since the sum of vectors of even weight also has even weight. Since C_N contains half the vectors of the underlying vector space, and only half the vectors of the space have even weight, C_N must exhaust those vectors, and so consist precisely of those vectors.

In general in order to get an interesting code, the characteristic of the underlying finite field must divide the order of the plane, so binary codes of interest arise only from planes of even order.

5. THE PLANE OF ORDER 4

Proposition 4.3 means we must consider planes of even order to find interesting binary codes. Now we consider some general properties of such codes.

It is convenient to number the points of the plane in some arbitrary fixed order and to identify words of the code with the characteristic vectors of subsets of the plane. This allows one to speak for example of "adding lines to subsets" when one really means adding their characteristic vectors (mod 2). One has the customary formula for the weight of a sum of vectors in terms of the cardinalities of the associated subsets:

$$\mathrm{wt}(v + w) = |v| + |w| - 2|v \cap w|$$

<u>Proposition 5.1</u> Let C be the code generated by the incidence matrix of a plane P of even order n. Then any line g of P has even inner product with words of C of even weight and odd inner produce with words of odd weight.

As usual, C has length $N = n^2 + n + 1$. If we adjoin to words of C an extra component and use it as a parity bit we get an Extended Code C^* in which all words have even weight. In C, lines have odd weight $n + 1$, so in the extended code they have a 1 in the parity bit. In order to have even inner product with an arbitrary word w^* of C^*, it follows that restricted to C, w has odd inner product with g if w has odd weight, and even if even.

This fact has wide ranging consequences for codes with substantial combinatorial content. To illustrate them, we can work out easily the entire structure of the (21,10) code C_{21} generated by the plane of order 4. This code, as we shall see, has rather few non-zero weights.

<u>Proposition 5.2</u> Let C_{21} be the code of the plane of order 4, and a_i the number of words of weight i in C_{21}. Then C_{21} has dimension ten and

$$a_0 = a_{21} = 1$$
$$a_5 = a_{16} = 21$$
$$a_8 = a_{13} = 210$$
$$a_9 = a_{12} = 280$$
$$a_i = 0 \quad \text{for all other } i \le 21.$$

We note first that since by 2.1 every point appears in 5 lines, the sum of all lines is the all-ones vector, so with each vector we have its opposite and $a_i = a_{21-i}$.

The minimum weight of C_{21} is 5, because given any word w of the code of even weight, each of the 5 lines through a point of w meets, by 5.1, w an even number of times, i.e. at least twice, so w has weight at least 6. But if w has odd weight, each of the 5 lines through a point not in w meets w an odd number of times, in particular at least once, and w has weight at least 5. Thus $a_k = 0$, $1 \leq i \leq 4$ and $17 \leq i \leq 20$.

By 5.1 any vector w of weight 5 meets a line in 1,3, or 5 points, but if 3, adding w to the line would give a word of weight $5 + 5 - 2 \cdot 3 = 4$ which is impossible. So the vectors of weight 5 with more than one point on a line have 5 points, hence are just the 21 lines, so $a_5 = a_{16} = 21$.

There are $\binom{21}{2} = 210$ ways to choose 2 lines which are distinct, and their sum is a word of weight 8, hence $a_8 = a_{13} \geq 210$. But in fact $a_8 = 210$, for if a word w of weight 8 meets a line in 4 points, adding the line gives a word of weight 5, which must be a line, so w is a sum of 2 lines. But it is a property of projective planes that any set of at least n + 3 points meets some line in more than 2 points, so in particular w does. Thus 5.1 implies that if w has 3 points on a line it has at least 4.

In a similar way one can show that $a_8 = a_{12} = 280 =$

$\frac{1}{4}[\binom{21}{3} - 21\binom{5}{3}]$ which is the number of ways to sum three lines

which do not go through a single point, each such sum having 4 representations.

There remains the question of words of weights 6,7,10,11,14, and 15. If we add a line to a word of even weight m, we get a word of weight m + 5, m + 1, or m - 3. If we add a line to a word of odd weight we get a word of weight m - 5, m - 1, or m + 3. Writing the possibilities out, we see that a word of any unknown weight above can only arise by adding a line to a word of one of those weights or one of the impossible weights 1 - 4 and 17 - 20. Thus we find ourselves "locked out" - since we do not start with any of those weights, we can never produce one, whence $a_i = 0$ for $i = 6,7,10,11,14,15$. Now we add the words we have already found and get a total of $1024 = 2^{10}$ so C_{21} has dimension 10 and we have verified all the assertions of the proposition.

A more complete description of the above can be found in [3]. Notice that none of these deductions required us to write down the plane of order 4 explicitly, only to know that a design of the relevant parameters exists.

The highest rate code of length 21 and minumum distance 5 has dimension 12. It can be obtained from C_{21} by adjoining words of three more weights and their complements. One takes as words of weight 6 those sets of 6 points of the plane, no three collinear. As words of weight 7 one takes subplanes of order 2. As words of weight 11 one takes words of weight 6 together with lines which do not meet them. To all of these one takes their complements. In this way one obtains a code of high rate in which every word has a precise interpretation in a suitable finite geometry.

The code C_{21} has a very remarkable extension. If we take the lines and add 3 more coordinates, all 1's, the code they generate has length 24 and only 4 non-zero weights: 8, 12, 16, 24. This code is the extended Golay code, obtained from the exceptional linear perfect Golay code by the addition of a parity bit.

The Golay code can be generated from finite geometry in another way. It was shown in 1938 in [8] that, to within relabeling points, there is a unique design with parameters $t = 5$, $v = 24$, $k = 8$, $b = 759$. This design is one of only two known cases of a design with any five points in a unique block. If one takes the characteristic vectors of the 759 blocks, only 12 are linearly independent and they generate the extended Golay code. That this exceptional code and exceptional design (and they have an equally remarkable symmetry group) should be so closely related is a tribute to the subtle mathematical architecture of the universe.

6. PLANES OF HIGHER ORDER

The codes of planes of higher order are not very well understood, partly because the length of the code goes up with the square of the order of the plane, so the codes rapidly become enormous and unwieldy.

Proposition 4.2, the Bruck-Ryser Theorem, implies that there is no plane of order 6 to yield a code.

There is a unique plane of order 8 whose code, of length 73, minimum distance 9, and dimension 28 has, to the best of the writer's knowledge, not been investigated.

The plane of order 10 however is a different matter. It is not known if there is a plane of order 10, and its possible existence is one of the most famous of all unsolved problems in combinatorial mathematics. Indeed it was the prospect of getting evidence about this question which has been the main cause of combinatorial mathematicians' thinking about coding theory. What is known is that the code has dimension 56, all weights congruent to 3 or 4 (mod 4), and almost no symmetry.

There are several planes of order 16, not all arising from constructions with finite fields, and nothing is known about their codes. As for the general questions of codes generated by designs, these too have been only slightly investigated. What is known is largely summarized in [1] and [2].

7. AN APPLICATION TO CODES FOR DETECTING DECEPTION

As this is an engineering conference we close by mentioning an application taken from [4].

Given a casino owner G (for Good guy) and B (for Bad) we imagine the following situation. B has reported takings less than they are and pocketed the difference. G wishes to install an encoder which produces a message $y = f(x,m)$ where x is the daily take and m a parameter in a coding algorithm f. The encoder punches y onto a paper tape which B posts to G. One assumes B knows x and f(.,.) but cannot change them, and y, which he can change, but does not know m. G knows y, m, and f(.,.).

In general one wishes to find codes in which B, if he changes y, has a small chance p_o escaping detection, but for which there are many possible x's.

The authors derive inequalities on p_o and x and show that optimal codes exist and can be generated from finite projective planes.

REFERENCES

1. E.F. Assmus, Jr., and H.F. Mattson, Jr., Coding and Combinatorics, SIAM Rev. 16(1974), 349-388.

2. P.J. Cameron and J.H. VanLint, Graph Theory, Coding Theory, and Block Designs, LMS Notes #19, C.U.P., Cambridge 1975.

3. D.W. Erbach, The code associated with the plane of order 4, Archiv. der Math. XXVIII(1977) 669-672.

4. E.N. Gilbert, F.J. MacWilliams, and N.J.A. Sloane, Codes which detect deception, Bell System Tech. J. 53(1974) 405-424.

5. A.M. Gleason, Weight polynomials of self-dual codes and the MacWilliams identities, Actes Congrès Internat. Math. 1970 vol. 3, Gauthiers-Villars, Paris 1971.

6. M. Hall Jr., "Combinatorial Theory", Blaisdell, Waltham Mass. 1967.

7. F.J. MacWilliams, A theorem on the distribution of weights in a systematic code, Bell System Tech. J. 42(1963) 79-84.

8. E. Witt, Uber Steinersche Systeme, Abh. Math. Sem. Univ. Hamburg 12(1938) 256-264.

SOFT-DECISION MINIMUM-DISTANCE DECODING

P. G. Farrell

Electronics Laboratories,
University of Kent at Canterbury,
Kent, England, CT2 7NT.

1. INTRODUCTION

Soft-decision (probabilistic) techniques have been applied with success to the decoding of convolutional (non-block) codes. Recent developments in semiconductor devices have now made the application of soft-decision techniques to block coding systems both possible and practical. This presentation will describe the design, implementation, and testing of such a system.

An optimum method of detection (demodulation and decoding), for a data transmission system with block coding, is coherent correlation detection of the sequence of signal elements that corresponds to each block of the code. In practice, unless the block is very short, this ideal detector is too complex to realise, because of the difficulty of generating, storing and correlating the large number of analogue signal elements required. Thus, most practical detectors consist of an analogue demodulator, possibly coherent, operating on individual signal elements, followed by a purely digital decoder operating on the blocks of digits produced by the "hard" decisions of the demodulator. This leads to an inevitable loss of performance. Instead of making a hard decision, on each received signal element, a soft-decision demodulator first decides whether it is above or below the decision threshold, and then computes a "confidence" number which specifies how far from the decision threshold it is. This number could in theory be an analogue quantity, but practical considerations require it to be quantised. Thus, the output of the demodulator is quantised, but into many more than the two regions of a hard-decision device. The confidence number can be used to assist in the decoding operation; in this way much of the performance lost by a hard-decision detector can be regained[1,2,5,6].

A data transmission system with a transparent, binary, product code; with interleaving; and with m-sequence inversion keying ASK modulation, will be described. It has variable data-rate capability, coherent detection, and full soft-decision minimum-distance decoding. The implementation is relatively simple and cheap. With simulated random errors, for a channel error rate of 2×10^{-1}, the output (decoded) error rate is better than 1×10^{-6}. Tests over practical HF channels demonstrated improvements in the error rate of 3-6 orders of magnitude, depending on the channel conditions.

2. THE SYSTEM TRANSMITTER

A block diagram of the transmitter is given in Fig. 1. Channel 1 is for binary data at 100, 67, 50 or 33 bits/sec (the faster channel) and channel 2 for data at 10, 6.7, 5 or 3.3 bits/sec (the slower channel). Digits on channel 1 are fed simultaneously into the cyclic outer encoder (a 4-stage feedback shift-register); the inner cyclic encoder (basically an 11-stage feedback shift-register); and a random-access memory (RAM) which is used for bit interleaving. The RAM is organised into four sub-frames, each consisting of 15 rows and 15 columns, as shown in Fig. 2.

Data from channel 2 is fed into the first position in each row; it is also used to invert (if it is a ONE) or non-invert (if a ZERO) the 10 information digits from channel 1 being fed into the outer and inner encoders. The channel 2 digit is also fed into the encoders, which require 11 information digits in all. Positions 2-11 in the first row are filled with channel 1 data, and then the remaining 4 positions with the parity checks derived by the outer cyclic encoder; this completes one code word row of 15 digits. Rows 1-11 are then filled with outer code words in the same way; and the final 4 rows are completed with the parity checks derived by the inner cyclic encoder (which is itself interleaved to degree 15, the code block length, in order to make addressing the RAM easier). Thus the rows of the sub-frame are outer code words, and the columns are inner code words. The remaining 3 sub-frames are filled in the same way. Digits can now be read out of the RAM. It is not possible to interleave by just reading out vertically, because this would mean that consecutive inner code word digits would not be interleaved. Instead, the RAM is read out in a diagonal pattern, in the digit order shown in the diagram: first the main diagonals in each sub-frame; then the ones immediately below, which since they have less than 15 digits, are completed with the digits in the top-right-hand-side of the sub-memories; and so on, taking one digit from each sub-frame in turn. The pattern of read-out for each sub-frame is shown in Fig. 3. In this way an interleaving factor of 60 is achieved for the inner code, and 56 for the outer code (they are different because the sub-frames are square, as the codes have the same block lengths). While the RAM is being read out, encoding continues in a second RAM, organised in the same manner; thus complete frames are encoded in each RAM alternately. More detail of the interleaving process are given in section 3.4, below.

The encoded and interleaved binary digits read out from the RAM are then used to invert or non-invert a 15-figit m-sequence. The clock rate of the m-sequence is 3 KHz for the fastest data rate, and is reduced pro-rate for the others. After sequence inversion keying, the waveform is passed through a baseband equaliser, adjusted so as to minimise symbol distortion and intersymbol interference arising in the system due to the various filter characteristics. Use was made of an 11-stage baseband equaliser previously developed.[3] The equalised sequence waveform is then amplitude-keyed onto a 1.6KHz sub-carrier, and then fed into a SSB-ASK HF transmitter with associated wide-band linear HF amplifier. The nominal transmission bandwidth of the system was 2.7KHz, as determined by the transmitter filter.

There is provision in the system for inserting a frame synchronisation sequence, so that the decoder can be correctly aligned before actual data transmission begins, and checked afterwards (see section 3.2 below).

3. THE SYSTEM RECEIVER

3.1 GENERAL

A block diagram of the receiver is given in Fig. 4. After RF and IF filtering and amplification, carrier is extracted and re-generated in a P.L.L. circuit, and applied with the IF waveform to a product detector. The output of the detector is integrated to recover the sequence-inversion-keyed waveform, and is also processed to regenerate clock waveform at the appropriate rate. Four time-constant settings in the integrator are available, corresponding to the four data rates. The variable threshold circuit adjusts the mean level of the output of the integrator to the optimum input value for the soft decision digital correlator. The sequence waveform is effectively sampled by the clock as it enters the correlator; this minimises noise and intersymbol interference effects. The correlator, by comparing the received sequence, correctly phased, with a locally generated 15-digit m-sequence, removes the sequence-inversion-keying and quantises each received encoded digit into 16 levels, represented by a four-digit binary character: a decision digit and three confidence digits. The correlator has built-in inertia which avoids mis-synchronisation by continuing in its initial state until three consecutive digits having the same re-synchronisation position are detected. The output is delayed while this consistency check is proceeding. This arrangement was modified in use; see section 3.2 below. The characters are then fed into a RAM, organised into four sub-frames holding 15 x 15 = 225 characters each, in the same way that the corresponding coded digits are read out from the transmitter RAM; so that when the receiver RAM is full, the characters in it are in exactly the same position that the corresponding coded digits were in in the transmitter RAM (see section 3.4 below). Frame synchronisation for the de-interleaving and decoding operations is established from the hard-decision digit (weight 2^3).

Decoding of the received characters can begin as soon as the RAM is full and synchronisation has been achieved; subsequent demodulated characters are fed into a second RAM, identical with the first, so that reception can continue uninterrupted. Decoding of columns 12-15 of each sub-frame is carried out first, as these columns contain checks on checks, and are required for decoding columns 1-11. The confidence digits of the rows, and of columns 1-11, are then summed, and the rows and the columns separately ranked in order of confidence. The row with highest confidence value is decoded next, followed by the column with highest confidence and so on. Rows and columns are thus decoded alternately in order of confidence value, as this minimises the probability of erroneous decoding. In this sense decoding is an adaptive process. Details are given in section 3.5 below.

Decoding is done (see also section 3.6 below) by computing the soft-decision (SD) distance between each received - possibly erroneous - row or column (consisting of 15 4-digit characters) and all the possible correct code words. The correct code words are stored in a programmable read only memory (PROM). As the distance computations are done sequentially in a systematic order, it is only necessary to store the parity checks of the code words. Also, the code is transparent, so that only half the words need be stored anyway, as the remainder are inversions of the first half. When the nearest code word is found, it is read into the RAM, via a buffer, replacing the decision digits of the appropriate row or column, the confidence digits being all reduced to zero if an error has been detected. Since only half the code book is stored, the modulo-225 threshold circuit inverts the nearest code word in the buffer if necessary.

After the first-pass decoding operation, a second pass is done. In this second operation the first eleven rows are passed again sequentially through the soft-decision distance processor, without regard for their rank. This further improves the reliability of the information digits on both channels.

Once all rows and columns have been decoded, then the corrected decision digits can be read out of the RAM. All the processing will take place in the time it takes to fill the second RAM, so that the first RAM then becomes free for storing demodulated characters while those in the second RAM are being processed, and so on.

3.2 FRAME SYNCHRONISATION

At the transmitter a 31-digit maximal-length sequence is used to provide "pseudo-data" for the framing mode. It is from this highly redundant signal that the frame timing is derived at the receiver prior to each data run. The selection of this sequence length is a compromise, giving a cross-correlation maximum which offers a high degree of immunity to severe disturbance conditions but is sufficiently short to allow an increased cross-correlation margin to be used if necessary by transmission of a number of such sequence cycles. This short framing sequence permitted the use of an asynchronous frame correlator. Framing sequence digits are not interleaved and only the hard-decision (HD) digits are used to establish frame synchronisation at the receiver.

3.3 SOFT-DECISION CORRELATOR

Two methods were used to recover soft-decision (SD) digit synchronisation from the serial data stream. The original technique, the block diagram of which is shown in Figure 5, was aimed at providing self-synchronisation at the encoded data bit level. Initially it was felt that a check for position consistency over three m-sequence cycles would give sufficient immunity to mis-synchronisation, whilst limiting data loss, in the event of incorrect timing, to an acceptable level.

It became apparent during the early transmission tests, however, that the performance of this re-synchroniser, together with that of the clock generator, was inadequate and prevented error control results being taken for even moderately poor reception conditions. The second method of synchronisation simply involved the use of one of the correlators in Fig. 5, together with a crystal controlled clock. (The transmitter was also crystal controlled). The receiver clock was manually adjusted to achieve synchronisation.

Figure 6 illustrates the SD digit format which is output by this correlator. Inversion of the confidence digits is carried out whenever the HD digit is a 0. This is done so that 111 is the highest degree of confidence for either level of data digit (HD digit).

3.4 BUFFER STORAGE FOR INTERLEAVING AND DE-INTERLEAVING

1 K-bit RAM's are used in the transmitter and receiver to provide the buffer storage necessary for interleaving and de-interleaving, and for the hierarchical (in order of confidence) decoding procedure. Two RAM chips were required at the transmitter, four times the number being required at the receiver because of

soft-decision detection. For clarity the sub-frame address input convention shown in Fig. 7 is adopted. From the total of ten address inputs on each RAM, the two remaining terminals A_o and A_5 are used to select the required sub-frame. This appears as a kind of time-division-multiplexing when reading interleaved digits from the transmitter RAM, or sorting the soft-decision digits in the receiver buffer. The 225-digit product-code word requires only 15 of the 16 possible co-ordinates on both the X and Y axes of the RAM. Hexadecimal synchronous counters and related control logic provide systematic address sequencing for read and write cycles. Operation of this circuitry will now be briefly explained.

In the transmitter: the write cycle requires that the X counter be incremented from positions 1 to 15 and then reset to scan the first row, whilst the Y is held in position 1. Subsequent rows in the sub-frame are then scanned by repeating this operation for positions 2 to 15 on the Y axis. The remainder of the frame is then filled in the same manner. The read cycle requires that complete diagonals of 15 digits be scanned, as indicated in Fig. 3. Both X and Y addressing registers increment in unison from 1 to 15 to access the leading diagonal of each sub-frame. Y then increases by one for the next cycle, and so on according to the diagram (Fig. 3) until all digits in a frame have been accessed.

In the receiver: the write circuitry is identical to that used for the read operation at the transmitter. The complicated addressing sequence for decoding is discussed in the next section.

Commutational switching is used to direct the control, input and output signals to the terminals of the appropriate memory bank in both the transmitter and receiver.

3.5 RECEIVER CONTROL LOGIC

The five stages of product decoding, shown in Fig. 8, necessitate a great deal of buffer accessing. Five thousand one hundred bit locations are accessed in decoding one frame. The decoding time for a complete frame is limited to a maximum of 4.5 seconds by the highest data rate. With parallel processing this can easily be accomplished in less than 2 seconds using a modest clock rate of 350 KHz. Fig. 8 illustrates the five stages of decoding, which are sequentially executed for each of the four sub-frames in a frame. Preparatory stages (a) and (b) compute the confidence measures for the 11 outer and the 11 inner code words on which hierarchical (ranked) decoding of these component code words will be based. Only one soft-decision (SD) digit at a time can be accessed from the buffer. An eight-bit adder with memory is used to accumulate the word confidence measure over any 15 digit word.

The third stage, (c), is the first decoding operation. No adaptivity is necessary in the decoding order for these four inner code words, but this step (to attempt correction of the errors in the outer code word parity check digits) must be completed before hierarchical decoding. Hierarchical decoding of alternate outer and inner code words is executed next, using the ranked lists compiled during the first two stages. Error corrections are automatically carried out in stages (c) and (d) by rewriting the decoded word back into the RAM in the same locations from which the predecoded word was taken. When a HD digit inversion (attempted error correction) has been made then the corresponding confidence digits are reduced to the minimum level (000).

The final part of decoding, stage (e), reprocesses the 11 outer code words given by the previous step, in an attempt to clean up the remaining errors. Hierarchical ordering of these code words offers no advantage and is therefore not used.

Accessing the component (15, 11) code words from the RAM and rewriting the decoded words demands a complex address format. The hardware solution incorporates a modulo-15 counter for bit-by-bit addressing of the row or column code word. Selection of the specific row or column is made by a second 4-bit counter. This address register plays a dual role since it is also used as the co-ordinate reference for the hierarchical selection store and is known as the code word co-ordination register (CCR). The addressing sequence for these two counters is identical for stages (a) and (b) of the decoding cycle. It is, however, necessary to interchange the X and the Y axis connections between these two stages. "One from two" data selectors are used for this purpose. The order in which inner/outer code words are accessed is known for all but the hierarchical decoding section. It naturally follows that during stages (a), (b), (c) and (e) the CCR is incremented to the next predetermined address by the reset pulse from the module-15 counter.

During each table look-up decoding in sections (c) and (d) the CCR holds the address of the code word for the rewriting cycle. Combinational logic on the outputs of the CCR are used to indicate the ends of the decoding stages and to control the axis switching logic for RAM addressing. The control logic for hierarchical decoding (stage d) alternates between outer and inner code word selection by reversing the axis address, switching after each rewrite of the decoded word. Hierarchical accessing of either inner or outer code words is achieved by first scanning the appropriate list using the CCR and choosing from the remaining code words the one with the highest total confidence. A four-bit temporary store retains the co-ordinate of the code word having maximum magnitude. At the end of the scan this is then loaded into the CCR and the corresponding value is cleared from the list. Accessing and decoding of this code word can then begin in the usual manner.

The three constituent parts of the hierarchical decoding selection logic are: the inner and outer list memories, each formed from a 16-word-by-8-bit RAM arrangement; a 7-bit full adder, and a 7-bit magnitude comparator. The word confidence magnitude (WCM) value is accumulated in the 7-bit adder as each of the 15 code word digits are read from the buffer, and the total is stored in the appropriate list location. Selection of the co-ordinate of the maximum WCM in the list being scanned is given by the magnitude comparator.

3.6 MINIMUM SOFT-DECISION DISTANCE DECODER

This decoder exhaustively compares the first half of the (15, 11) code book with the soft-decision (SD) (15, 11) code word held in the comparator store. Figure 6 illustrates the digit by digit basis on which each comparison is made. It is necessary to modify (re-invert) the confidence digit format used for the data digits stored in the buffer back to that shown in this diagram. This is achieved very simply by using exclusive-OR gates to invert the confidence digits when the hard-decision digit is a 0.

Three interacting circuits form the minimum soft-decision distance (MSDD) decoder. These are the SD distance processor, the table look-up circuit and the nearest binary-code-word selector. This latter circuit is represented in the receiver block diagram by its three component parts (see Fig. 4).

A partial schematic representation of the operation of the SD distance processor is given in Fig. 9. Here the complete SD code word is retained in the 60-bit parallel output store during the table look-up cycle. At each of the 1024 steps through the first half of the code book, whenever a 1 occurs on any of the fourteen table look-up inputs, the outputs of the corresponding SD digit is inverted. This is equivalent to inverting Fig. 4, and is necessary so that simple addition, after deweighting, of these outputs over the whole code word gives the true degree of deviation from the reference word. Deweighting was carried out using 4-bit asynchronous adders, by increasing the significance of the total for each weighting level by the appropriate number of shifts (see Fig. 10). The SD distance value is given at this output.

The table look-up information is given by the four outputs from the PROM and its 10 address inputs. When a table look-up cycle is initiated the 10 address inputs are taken through all the possible states from the all zero to the all one state using binary counters. The corresponding parity-check digits for any one state are given by the PROM outputs. Only the first half of the code book is generated, consequently the SD digit in the leading information digit position is always compared to a 0.

The purpose of the nearest-binary-code-word circuitry is to select from the entire code book the HD code word which has the minimum soft-decision distance from the word being decoded. In order to choose the code word from the second half of the code book, if that has a smaller SD distance than its counterpart from the first half, it is necessary to set a threshold on the SD distance calculator output above which the inverted table look-up code word would be taken. The maximum possible binary coded output of the SD distance calculator is 225, and so a threshold of 112.5 is used to decide from which half of the code book the word should be taken if the SD distance is the lowest value up to that point in the table look-up cycle. An additional complication is that when this threshold has been exceeded, signified by a 1 on the combinational logic output, the SD distance calculator output vector must be modulo-225 complemented before being compared against the previous minimum value. An 8-bit full added is used to subtract 225 from the input vector, operating with one's complement arithmetic. An 8-bit magnitude comparator with the same amount of storage generates a nearest-code-word signal and retains the new SD distance if this is found to be the present minimum value.

When the nearest-code-word signal occurs the present table look-up reference or its inverse is transferred to a parallel-in series out shift register. This word is complemented if the SD distance threshold has been exceeded. After an exhaustive search of the code book the nearest code word is clocked serially from the store into the HD layer of the RAM buffer. It is simultaneously compared with the HD digits of the decoded word. In the event of an error correction having been made, indicated by a dissimilarity in this comparison, the confidence digits associated with the SD digit are deemed to be unreliable and are consequently set to zero before being rewritten into the RAM's. A single TRUE/COMPLEMENT/ZEROS/ONES logic IC is used for this purpose, and for the reversion to the predecoding format for confidence digits, necessary if no error correction has been made.

4. TEST RESULTS

The performance of the system has been investigated in three ways.

4.1 DIGITAL TESTS

The digital parts of the transmitter and receiver were tested by adding (modulo-2) random errors and burst error patterns, from an experimental error generator developed at Kent[4], to the output of the sequence inversion keyer. Error rates were measured before soft-decision correlation (the channel error rate, P_e), after correlation (on the hard-decision digits), and after the first and second passes of the soft-decision decoding process (with separate counts on information and check digits). The results are shown in Fig. 11, for random errors; and in Fig. 12, for various burst lengths (the burst error densities were approximately 0.56). These tests, for convenience of measurement, were done at a digit rate of 25K bits/sec.

4.2 WHITE NOISE TESTS

The complete transmitter and receiver (without the linear HF power amplifier) were connected back-to-back, the equaliser was adjusted for optimum operation, and error rates were measured, as detailed above, with white Gaussian noise added to the transmitted signal, at sequence digit rates of 2K bit/sec and 1.5K bit/sec. These tests were done with the sub-carrier frequency adjusted to 100KHz.

The results for the 2 K bit/sec rate are given in Fig. 13, as plots of signal-to-noise ratio (db) against error rate.

4.3 FIELD TRIALS

The complete HF system was tested in three sets of trials, two over approximately 120 miles, and one over approximately 300 miles. HF frequencies of about 4.5MHz were used. Error rates as above were measured, and a chart recording was made of the input signal strength during each trial. Some of the results obtained are given in Fig. 14, which shows a plot of error rate against percentage of lowest output error-rate frames received. As a preliminary to these tests, the transmitted equaliser was re-adjusted to match, as far as possible, the actual HF transmitter characteristic.

5. CONCLUSION

It was not difficult to implement the digital parts of the system. Complexity is moderate (34 cards, or approximately a total of 418 IC chips) and the cost surprisingly low (in the region of £500). Apart from some difficulties connected with the synchronisation of the correlator (see section 3.3), arising from the unstable nature of the HF channel, the experimental system operated quite satisfactorily and reliably during the field trials.

The performance of the system in the digital and white noise tests is clearly very good, as the results (section 4.1 and 4.2) show. This is to be expected from a system with an overall coding efficiency of 3%. The improvement due to use of soft-decision decoding is over an order of magnitude at a channel error rate of 2×10^{-1} (Fig. 11); the improvement would be much greater at lower channel error rates. As might be expected, the performance of the system falls off as the length of the burst error patterns is increased. The second pass of the decoding procedure effected a considerable improvement in the burst error cases, more so than in the random error case.

Evaluation of the field trial results is not yet complete, so any conclusions based on the preliminary results given in Fig. 14 can only be rather tentative Notable features, however, are:

(a) the very high channel error rate, which hardly alters as the worse output-error-rate frames are removed from the computation;

(b) that the second decoding pass made almost no difference to the final output error-rate.

It is not clear at this stage how much improvement in performance in the field trials can be ascribed to the use of soft-decision decoding.

REFERENCES

1. P.G. Farrell & E. Munday: Variable Redundancy HF Digital Communication with Adaptive Soft-Decision Minimum-Distance Decoding: First Report on Research Study Contract AT/2099/05/ASWE, MOD (ASWE), Jan. 1975.
2. P.G. Farrell & E. Munday: Economical Practical Realisation of Minimum-Distance Soft-Decision Decoding for Data Transmission: Proceedings of the Zurich International Seminar on Digital Communications, March 1976, pp B5.1-6.
3. E. Munday: Phase Shift Keying Equalisation, M.Sc. Dissertation, University of Kent at Canterbury, 1973.
4. V.C. Rocha: Versatile Error-Control Coding Systems, Ph.D. Thesis, University of Kent at Canterbury, 1976.
5. D. Chase: A combined coding and modulation approach for communication over dispersive channels. IEEE Trans. on Comm., Vol. COM-21, No. 3, March 1973, pp 159-174.
6. E.J. Weldon: Encoding and Decoding for Binary Input, Q-ary Output Channels. Res. Rep. AFCRL Contract F19628-70-C-0082, 1970.

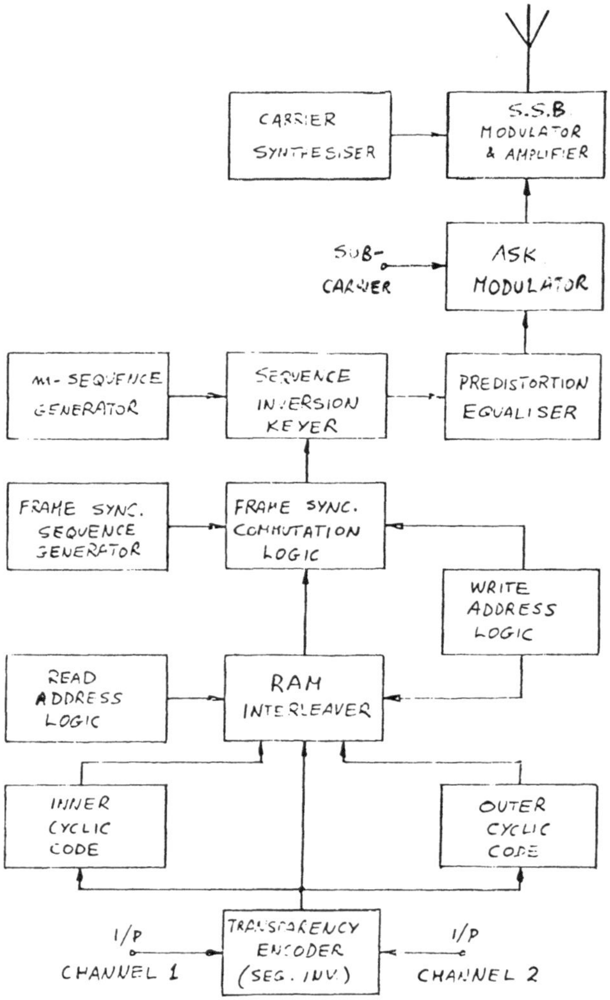

FIG. 1 : TRANSMITTER

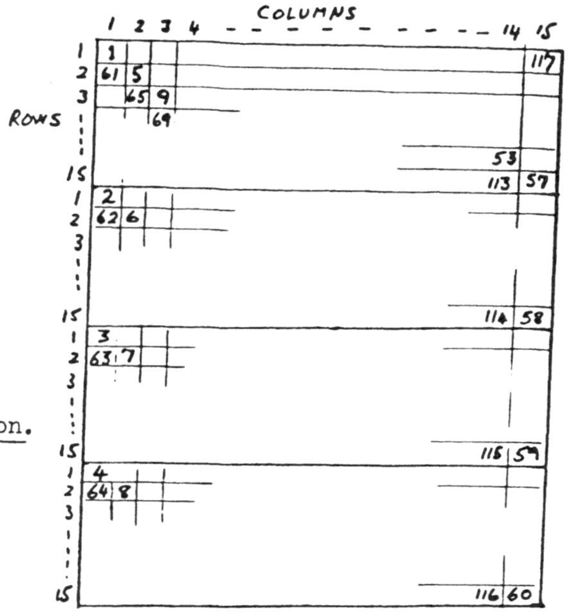

Fig. 2.

RAM:

Organisation.

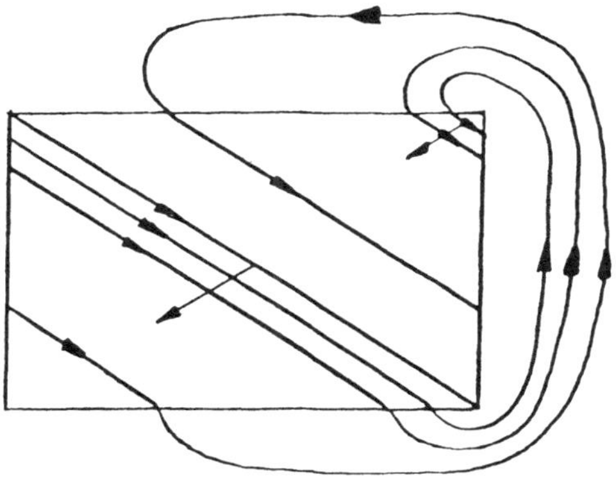

Fig. 3.

RAM:

Pattern of read-out.

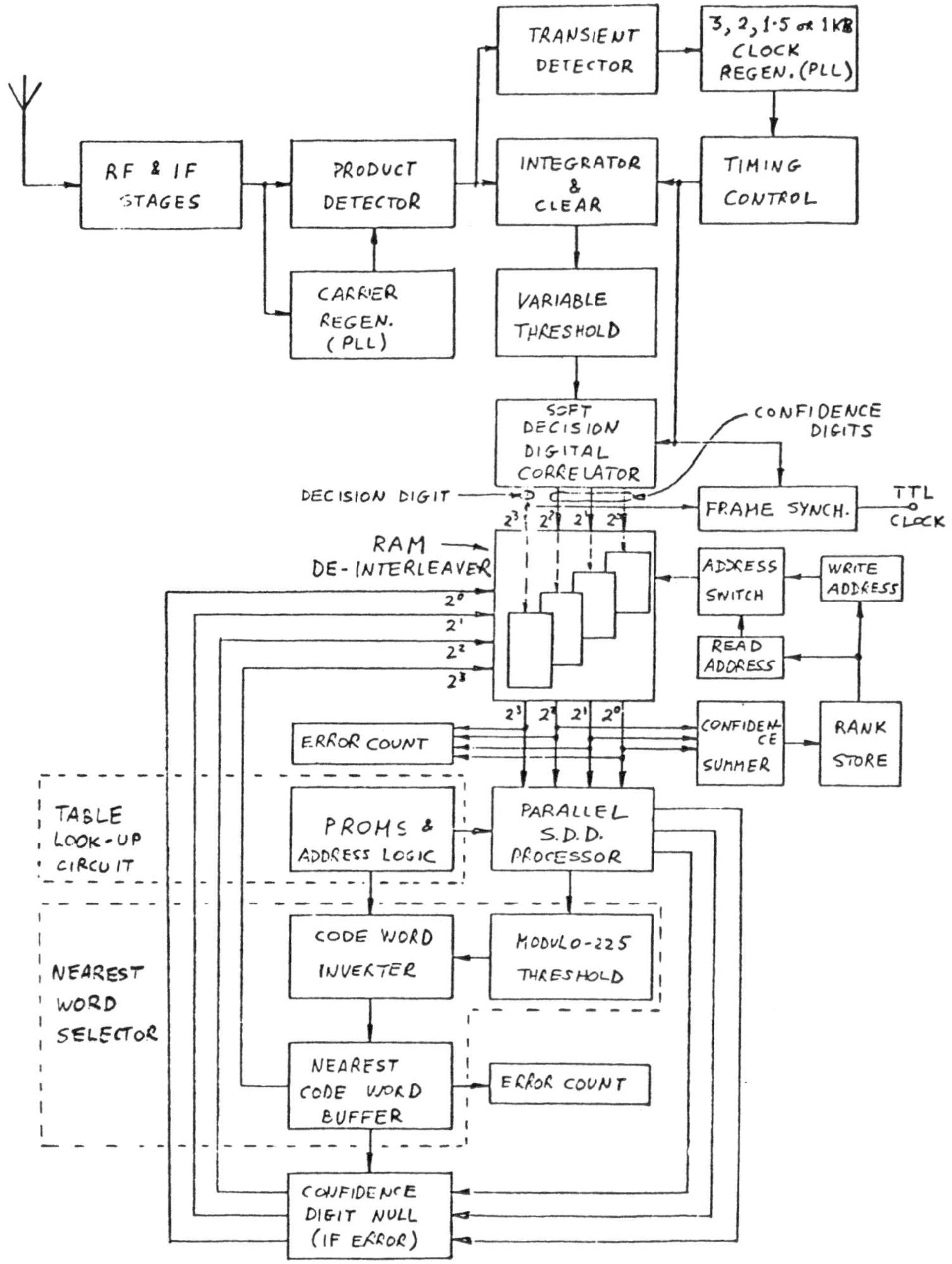

FIG. 4 : RECEIVER

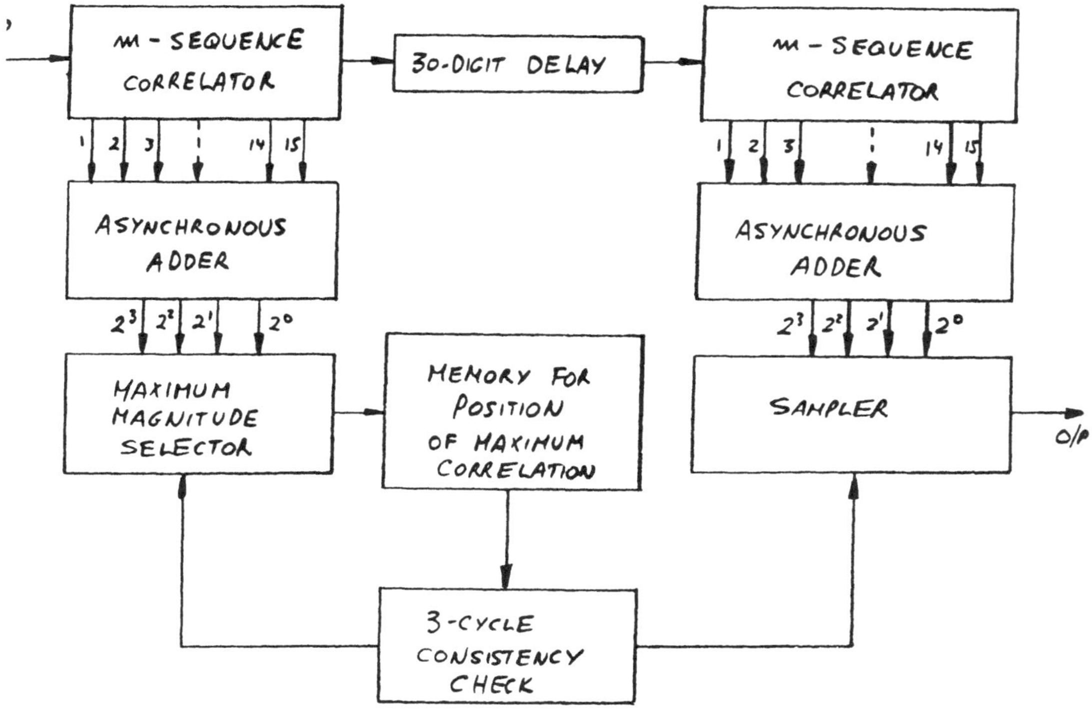

Fig. 5. Soft-Decision Digital Correlator

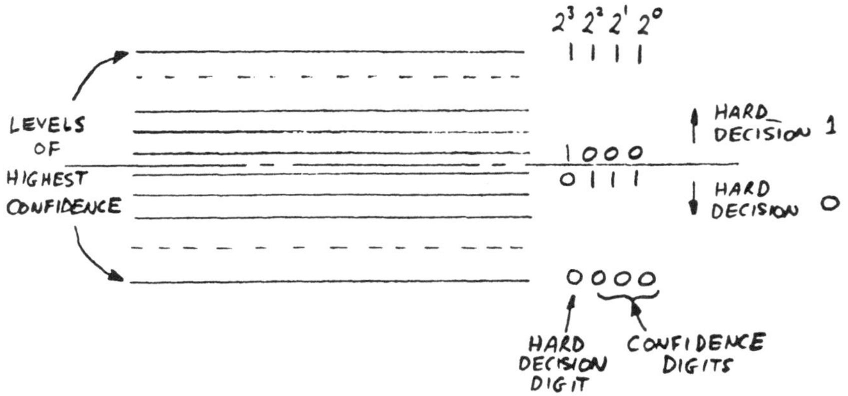

Fig. 6. Soft-Decision Digit Format

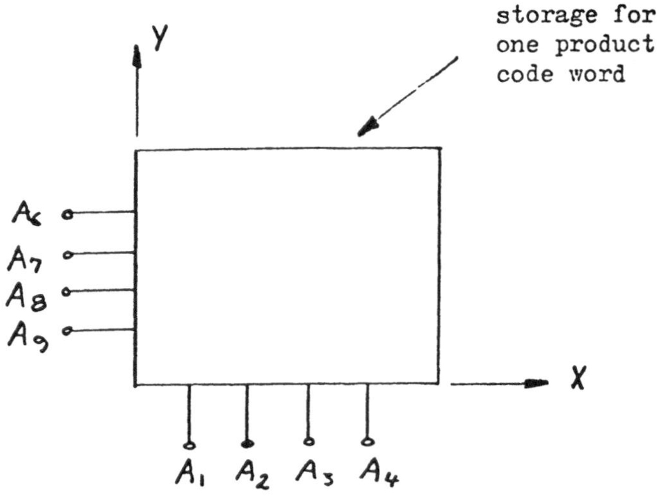

Fig. 7. R.A.M. Address Convention

Fig. 8. The Five Decoding Stages

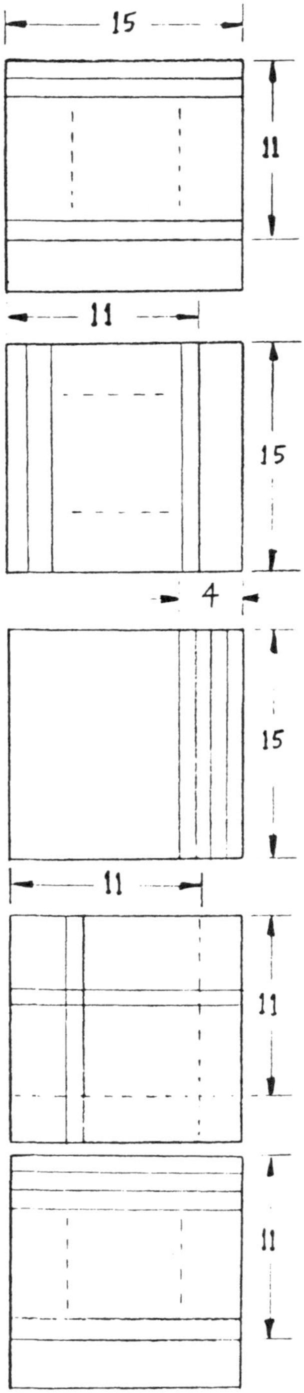

a) Confidence sum on outer (row) code words

b) Confidence sum on inner (column) code words

c) Decoding of the four inner code words whose information digits are the parity checks of the eleven outer code words

d) Hierarchical decoding of the first eleven inner/outer code words

e) Second-pass decoding of the first eleven outer code words

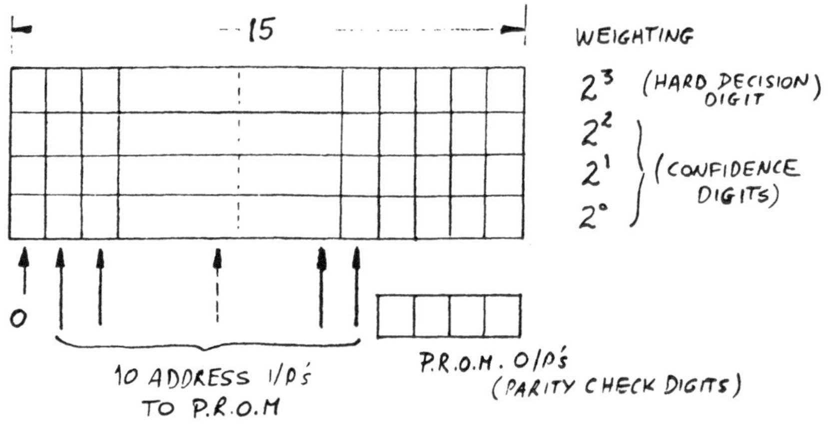

Fig. 9. Operation of the Soft-Decision-Distance Decoder

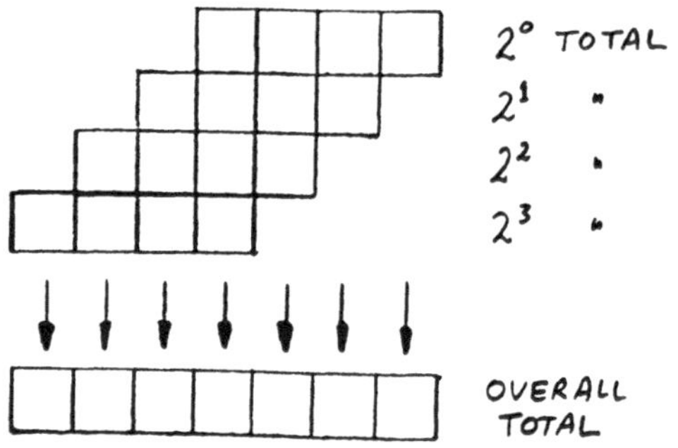

Fig. 10. Deweighting Operation

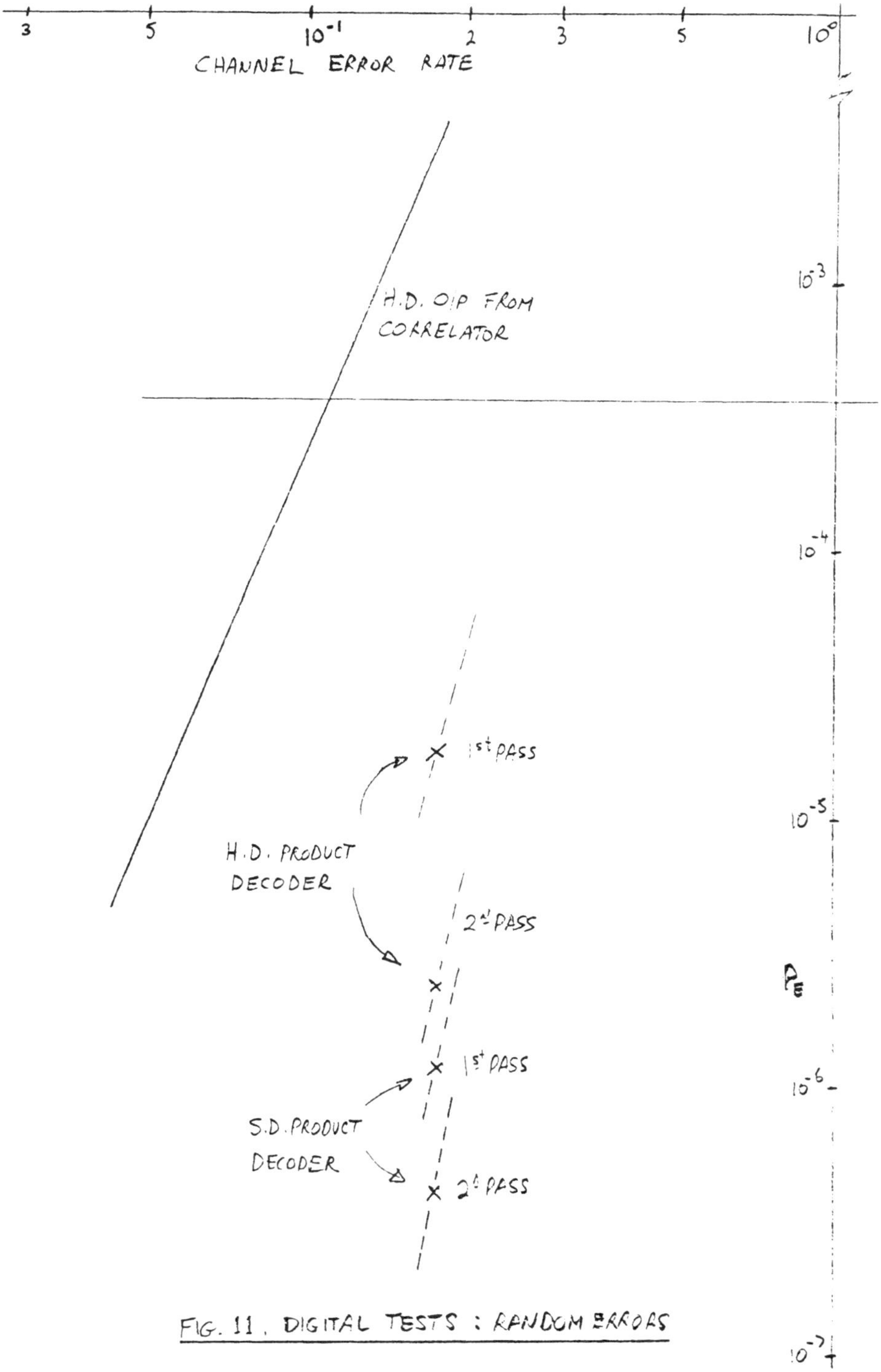

FIG. 11. DIGITAL TESTS : RANDOM ERRORS

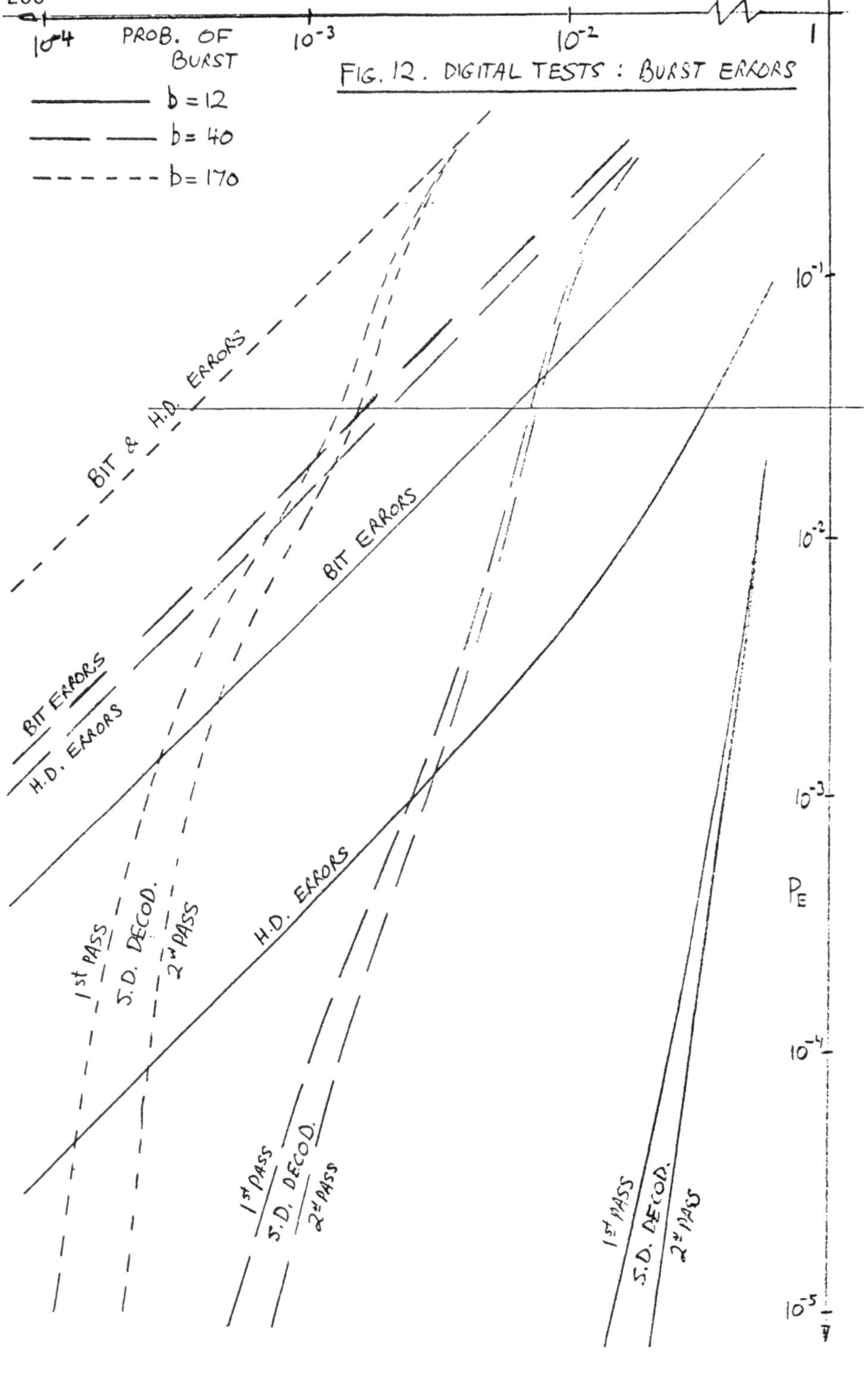

FIG. 12. DIGITAL TESTS : BURST ERRORS

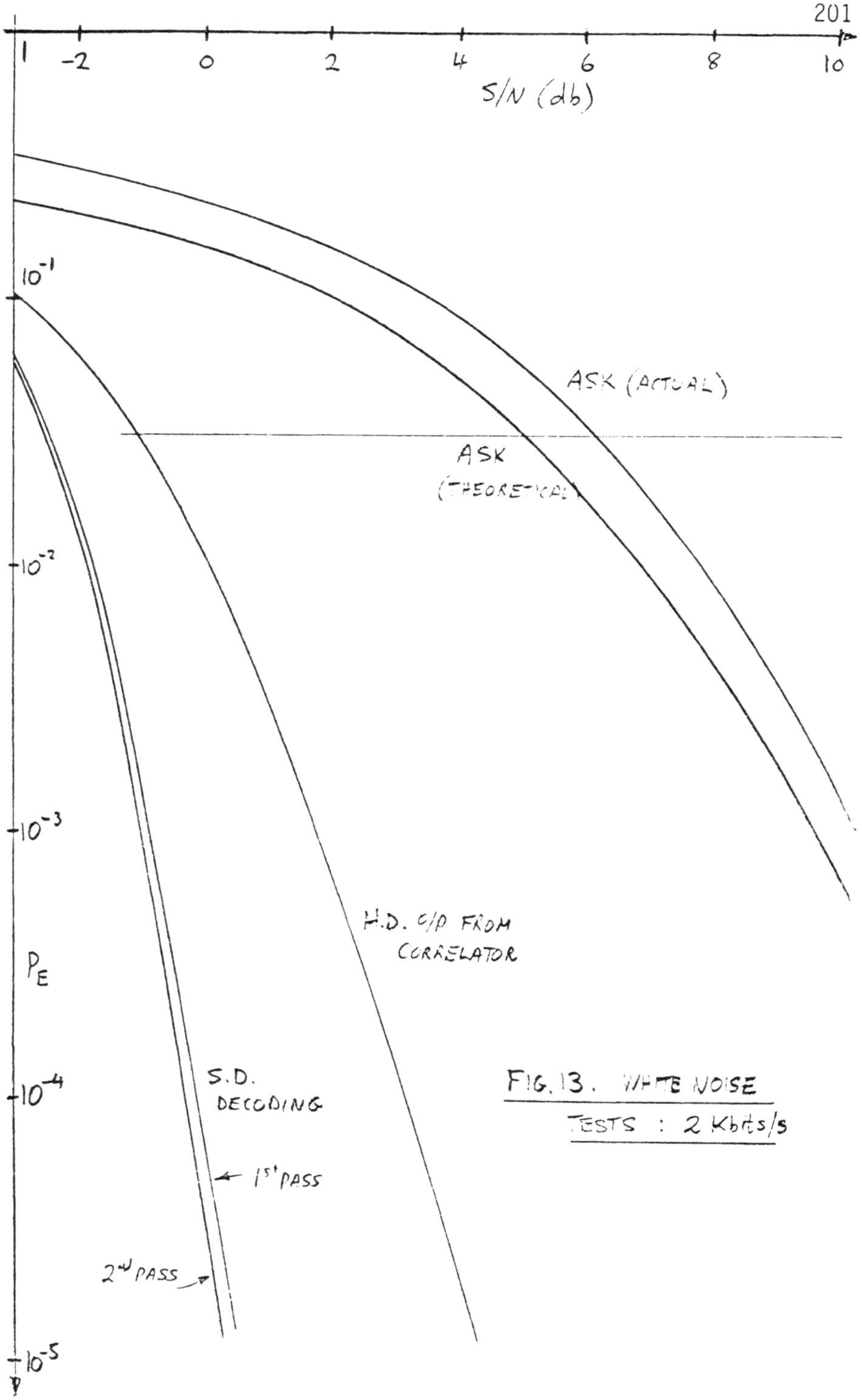

FIG. 13. WHITE NOISE TESTS : 2 Kbits/s

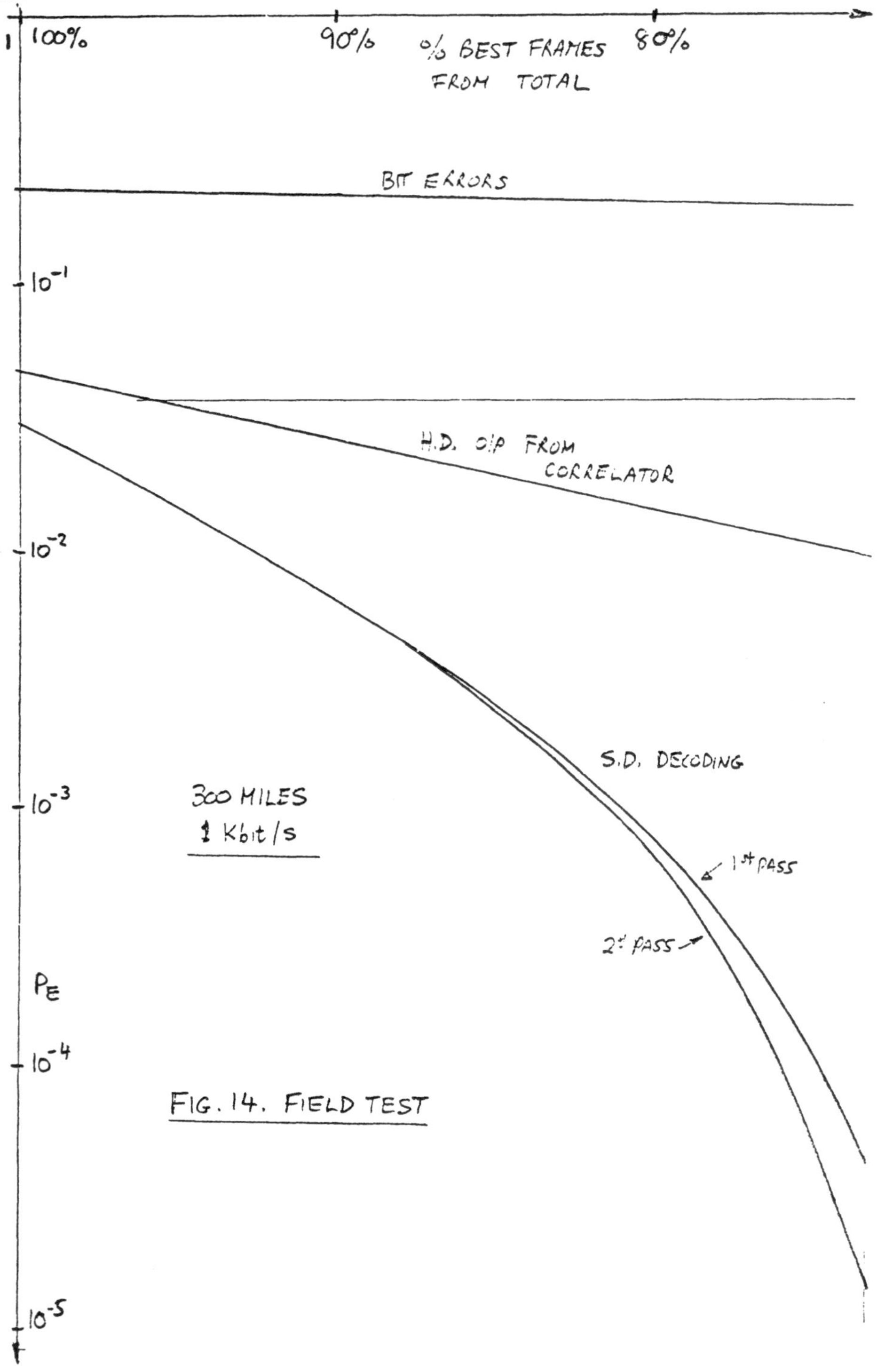

FIG. 14. FIELD TEST

ALGEBRAIC CODES CONSTRUCTED FROM OTHER ALGEBRAIC CODES: A SHORT SURVEY AND SOME RECENT RESULTS.

A. Brinton Cooper, III

U.S. Army Materiel Systems Analysis Activity
Aberdeen Proving Ground, Maryland 21005, U.S.A.

ABSTRACT. Algebraic codes can be combined in various ways to produce longer algebraic codes. Over the past twenty three years, research into such constructions has resulted in two major classes: iterated codes and "concatenated" codes. Arbitrarily long codes from these two classes provide arbitrarily small error probabilities after decoding while affording transmission rates which are greater than zero. While these rates are smaller than Shannon's channel capacity, the fact that they remain greater than zero for arbitrarily large code length is, of itself, significant.

This short tutorial survey reviews the development of each class of codes with Shannon's results for channel capacity in mind. Descendents of Elias's iterated Hamming codes and Forney's concatenated codes are presented, and some recent results are reviewed. Several constructions which do not proceed directly from these classes are mentioned.

1. INTRODUCTION

The objective of algebraic coding theory is to construct and decode error correcting codes which permit communication over a noisy channel at a rate known as channel capacity and with an arbitrarily small error probability.

An algebraic code can provide vanishing error probability with increasing block length if and only if the ratio of minimum distance to block length is bounded away from zero for arbitrarily long blocks [1].

Most known classes of algebraic codes (e.g., BCH codes) for which general construction rules are known ("constructive" codes) have minimum distances which are not so bounded. There are, however, classes of algebraic codes which are constructed from other algebraic codes and which have been shown to provide, for fixed rate, either a vanishing error probability or a lower bounded ratio of distance to length. The most significant of these classes are constructive and have constituent codes which are well known and which have manageable decoding algorithms. These important classes are the subject of this tutorial discussion paper.

2. OVERVIEW

Berlekamp [2] suggests a natural division of this topic: iterated codes and their descendents; concatenated codes and their descendents; other techniques. The first two classes are nearly mutually exclusive although, philosphically, concatenated codes can be considered a special case of iterated codes.

First, Elias's basic technique for iterating Hamming codes and its significance are considered. Improvements in Elias's bounds and a generalization of Elias's construction are reviewed, and improvements in code performance are noted.

Next, Forney's basic scheme for concatenating algebraic codes is covered. Several signficant extensions are presented which produce "constructive" codes with error correction capabilities that do not degenerate asymptotically with block length.

3. CODES DERIVED FROM ITERATED CODES

The iterated Hamming codes of Elias constitute the only known subclass of linear block codes (LBC's) having both the following properties:

1. The construction of an encoder and a non-exhaustive decoder for every code is known.

2. For every $\varepsilon>0$ and for channel error probability $<1/32$ there exists a lower bound, R_b, to the code rate, R, such that

Decoding error probability $<\varepsilon$

$R>R_b$

Following a description of the classical construction of Elias's codes improved lower bounds are presented.

3.1 Elias's Construction

Let C_i be an (n_i,k_i) algebraic code.

Suppose the output of a binary source is encoded using C_1. The output of this encoder, in turn, can be encoded using C_2. The process is called iteration, and the resulting code is an iterated code.

For decoding an iterated code, the channel output is divided into blocks of length n_1 and applied to the decoder for C_1. The output of that decoder is applied to the decoder for C_2. The purpose for iterating codes is to achieve a greater reduction in error probability than can be achieved using a single code.

This error probability reduction will not be realized unless the iteration is constructed properly. The decoder for C_1 occasionally will decode erroneously. When that happens, the number of errors per block at the decoder output will be greater than at the input. If this block is fully contained within the block of length n_2 which is input to the decoder for C_2, the latter may not be capable of correcting all the errors present. Obviously, C_2 will not be able to handle as high an input error probability as C_1 since the latter is chosen to reduce the channel error probability prior to using C_2.

If each symbol in a block of length n_1 from the decoder for C_1 would appear in a distinct block of length n_2 for decoding of the C_2, the difficulty of the previous paragraph would be avoided. This type of construction can be realized by iterating C_1 and C_2 as shown in Figure 1.

This iteration has produced an (n_1n_2,k_1k_2) code which can be iterated with another code, say C_3, in exactly the same manner. The result will be an $(n_1n_2n_3,k_1k_2k_3)$ code with rate $k_1k_2k_3/n_1n_2n_3$. Repeated such iterations of Hamming codes produce Elias's codes.

Extended Hamming codes are single error correcting, double error detecting codes with length $n=2^m$ for integer $m \geq 2$. If an

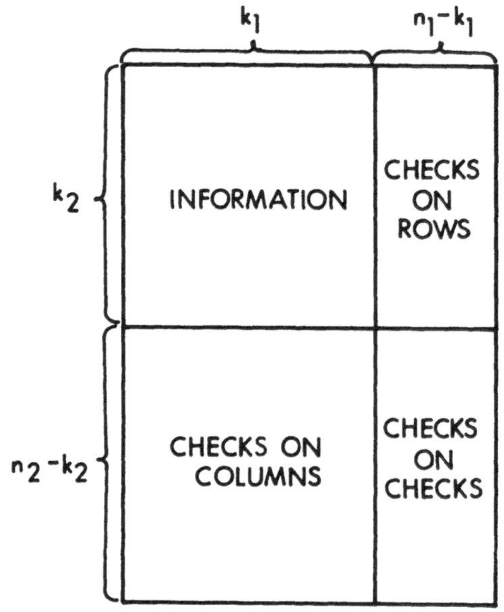

Figure 1. Structure of an Iterated Code.

even number of channel errors occurs, the decoder should take no action; the channel errors will be "passed through" to the decoder output. If three errors occur in a single block, the decoder will see a received vector which is distance one from a code word. The decoder will view this as the occurrence of a single error in the block and will change one position. This action can increase or decrease the number of errors in the block by one. The classical results for Hamming code performance assume that the number of errors always increased by one. Berlekamp's results which are presented later show how to obtain better bounds by determining when an error is added and when one is corrected.

It is well known [1] that the bit error probability produced by a Hamming single error correcting code obeys the inequality $p_1 < np^2$.

Lemma: On the binary symmetric channel with error probability p, the Hamming code with $n = 4$ always decodes to a bit error probability which is less than p. (See Figure 2).

Proof: See [3].

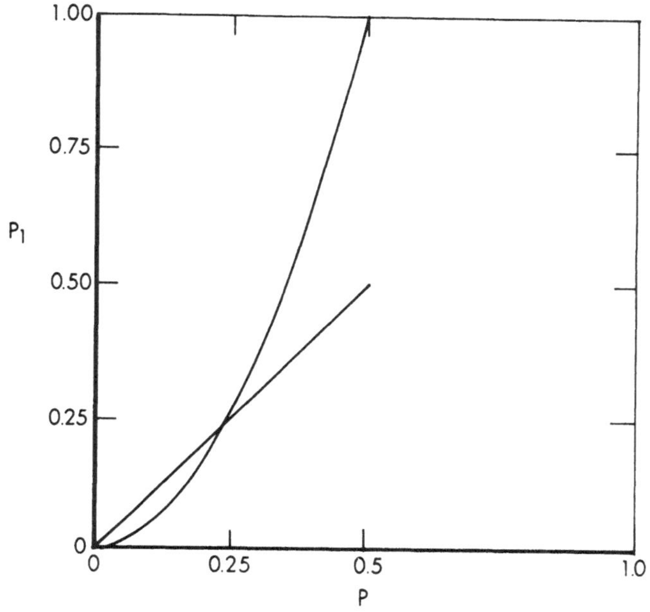

Figure 2. Two Bounds to Decoding Error Probability for the Hamming Code with n=4

Now consider an extended Hamming code of length n_1 interated with a Hamming code of length $2n_1$ on the binary symmetric channel with error probability p.

$$P_2 < (2n_1)p_1^{\,2}$$

$$< (2n_1)(n_1p^2)^2$$

$$< 2(n_1p)^4$$

Repeating this iterative process for a total of r Hamming codes and simplifying gives $P_r < (2n_1p)^{2^r}\ (1/n_1)2^{-(1+r)}$ Hence, if $2n_1p<1$, $\lim_{r\to\infty} p_r = 0$

Since the rate of an iterated code is the product of the rates of the constituents, this product is smaller than either. Yet, the following lemma holds.

Lemma: The rate of an infinitely long Elias code is lower bounded away from zero.

Proof: See [1] which shows that the rate of the iteration is underbounded by,

$$R > 1 - \frac{m+2}{2^{m-1}}$$

provided $2.2^m p < 1$. Thus, the error probability can be driven to zero while the rate is bounded away from zero.

Notice that the smallest integer for which the rate bound is positive is m=4. For m=4, n_1=16 and p < 1/32. Tighter bounds which apply to all interesting values of p are given below.

Derivations of the bounds on rate and error probability of Elias's codes utilize certain approximations which penalize the apparent performance of these codes. For noisy channels, these approximations provide a significant contribution to the upper bound on the value of channel error probability and make it appear to require more redundancy than actually necessary to drive the error probability to zero. Further, the derivation of the lower bound to the rate of an Elias code discards many terms for the sake of mathematical tractability.

In [3], these approximations are circumvented in order to tighten the lower bounds on the rates of Elias's codes. An algorithm to produce decoding error probability p_r and rate R_r for an r-stage Elias code was run, increasing r until p_r was upper bounded by 10^{-300}.

Results are displayed in Figure 3 showing R_r approaching an asymptotic value as a function of r for several values of p. The small arrows indicate the number of iterations following which $p_r < 10^{-300}$. The largest such r was 9, indicating that values for R_{30} are reasonable lower bounds to the rates of Elias's codes for any practical definition of error free decoding.

For each m, the largest value of p satisfying $2(2^m)p<1$ was found. For

$$2^{-(m+2)} \leq p < 2^{-(m+1)}$$

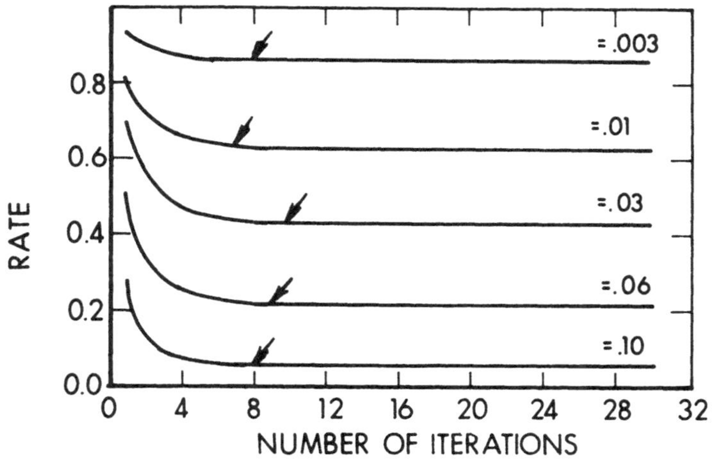

Figure 3. Rate of Elias's Codes vs Number of Iterations for Several Binary Symmetric Channels.

the values for R_{30} and for Elias's lower bound represent the bounds to code rate vs channel error probability. This set of rate bounds vs ranges of channel error probability are shown graphically in Figure 4.

The derivation of the classical upper bound to the decoding error probability of a Hamming code assumes that the occurrence of more than 3 channel errors in a block always increased the number of errors by one in the decoder [1]. It is quite likely, however, the decoder may actually decrease the number of errors by one.

Berlekamp [2] used the distribution of weights of code words in the extended Hamming codes to calculate the probabilities that the decoder would increase or decrease the number of errors in a block. This produced the largest p for which an Elias code can be constructed beginning with $n_1 = 2^m$ to provide error free decoding. Using m to determine the value of rate bound produces Berlekamp's results shown in Figure 4. Also shown are Elias's original results and tightened versions of same obtained by computing actual code rate, R_{30}, for 30 iterations. In every case, the decoded bit error probability is less than 10^{-300}.

Figure 4 shows that Berlekamp's method, coupled with the use of R_{30} to provide a reasonable estimate of the lower bound to code rate, provides the greatest lower bound known. The most significant improvements contributed by the tighter bounds were for the larger values of p, which is where the greatest improvements are needed.

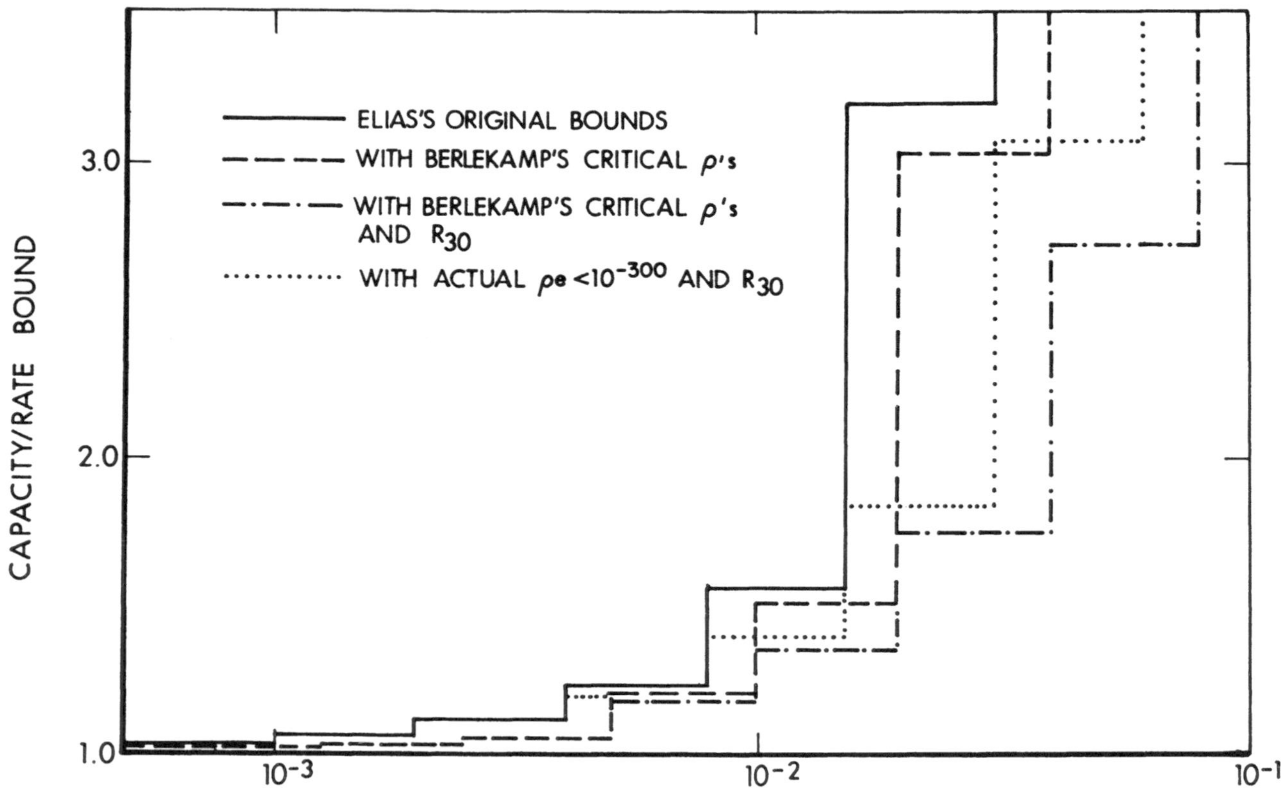

Figure 4. Improved Bounds for Elias's Codes

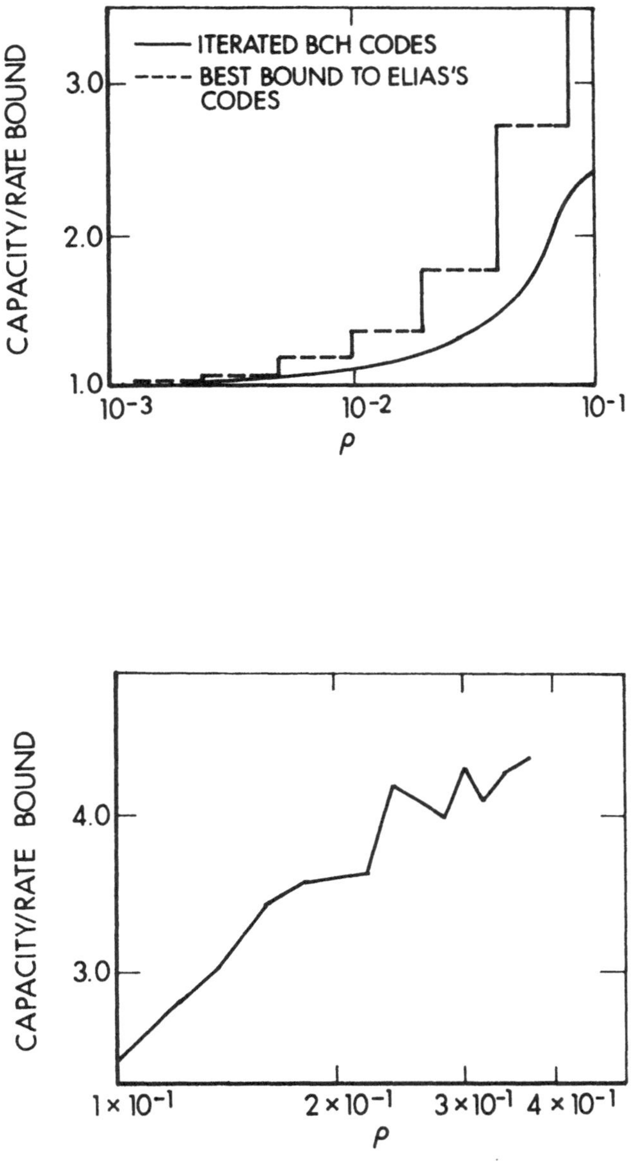

Figure 5. Iterated BCH Codes

However, none of the lower bounds to the rates of Elias's codes give any information about code performance for p>.40. This is one of the contributions to be made by the generalizations of Elias's codes discussed below.

An algorithm for producing "good" iterations at BCH codes is discussed in some detail in [3] and [4]. Basically, the algorithm begins by finding an Elias code for a very small value of p. A lemma proved in [3] and [4] shows that the rate of an Elias code approaches channel capacity as p approaches zero. For progressively noisier channels, a search algorithm searches for BCH codes to iterate with that Elias code to provide error free decoding. Thus, the iterated code for any value of p is constructed from the iterated code for some smaller value of p. Results are shown in Figure 5.

4. CODES DERIVED FROM CONCATENATED CODES

The Varsharmov-Gilbert (VG) bound [1], establishes the existence of algebraic codes for which the ratio of minimum distance to block length is bounded away from zero, for any rate, as the length becomes arbitrarily large. Peterson and Weldon [1] show that this property is sufficient to drive the error probability after decoding to arbitrarily small values. Every coding technique known (to the author) for which d/n is so bounded involves concatenated codes.

4.1 Basic Concatenated Codes.

The basic model for concatenated codes was devised by Forney [5]. Information symbols from GF(2) are treated in blocks of K_1K_2 digits, organized as K_2 vectors of K_1 symbols. Each such vector is, therefore, a symbol from $GF(2^{K_1})$. These K_2 symbols are encoded using a Reed-Solomon (RS) code of length N_2 over $GF(2^{K_1})$. This is the "outer code". Each vector of K_1 symbols is then encoded using an "inner" (N_1, K_1) code over GF(2). The resulting code has length $N_1 N_2$ and dimension K_1K_2. The study of concatenated codes involves the selection of the outer and inner codes so as to optimize the performance of the overall code.

Decoding concatenated codes reveals the prevalent error mechanisms If the inner code is decoded first, any uncorrected pattern of errors over GF(2) of length not greater than N_1 produces one symbol error in the outer code. If a sufficient number of such decoding errors occurs, the outer code will not be able to correct the overall error pattern.

Forney [5] examined the asymptotic rate and error probability of long concatenated codes when the inner code is any maximum likelihood decoded code and the outer code is an RS code. He showed that such codes exist which have rate greater than any desired value less than channel capacity and error probability less than any small, fixed number. He constructed several numerical examples.

4.2 A Class of Concatenated Codes

The goal of much coding research is to construct codes which, for arbitrary rates, meet the VG lower bound to (d/n). Successful attempts at providing non-zero lower bounds to (d/n) which fall short of the VG bound have been achieved for several variants of "concatenated" codes, which are discussed in this section. In spite of the fact that the Zyablov bound [6] shows that concatenated codes fall short of the VG bound, these codes remain attractive because of their "good" asymptotic behavior of values of (d/n). Some, in fact, achieve the Zyablov bound for large rates.

Weldon [11] has defined a sequence of codes to be an infinite set of codes (n_1, k_1), (n_2, k_2), . . ., $n_j > n_{j-1}$ where all member codes are defined by the same rules. Justesen [7] termed a sequence of codes to be constructive if it can be specified without searching. In light of these definitions, Justesen [7] showed a constructive sequence of concatenated codes. Whereas Forney's codes have decreasing error probability for fixed rate, Justesen's codes provide a non-zero, asymptotic lower bound to the ratio d/n of minimum distance to block length. The latter is achieved by selecting as the inner code an "ensemble" of codes having the property that no n-tuple appears in more than one code. Each character of the outer code is encoded by a distinct inner code, thus requiring at least N_2 inner codes. The inner codes are the so-called randomly shifted codes of Wozencraft as described by Massey [9] and constructed as follows.

For an ensemble of (N_1, K_1) codes, let $m = \max(K_1, N_1 - K_1)$ and construct a maximal length, m stage shift register sequence generator. Load the K_1 information positions into the low order stages of the register and clock it s times, $0 \leq s \leq 2^m - 1$. Select the contents of the low order $N_1 - K_1$ stages as the parity checks. The value of s is distinct for each inner code word, and m is chosen so that $N_2 \leq 2^m - 1$.

As before, the outer code is an (N_2, K_2) RS code over $GF(2^{K_1})$. Using an argument based upon the total weight of a set of distinct, non-zero, binary n-tuples, Justesen proved that, for $0<R<0.5$,

$$\liminf_{n\to\infty} (d_n/n) \geq (1-2R)H^{-1}(1/2)$$

where d_n = minimum distance of the concatenated code of length n
R = least upper bound to the rate R_n of the same code.

To achieve larger rates, the inner code is punctured by deleting several digits. The bound becomes

$$\liminf_{n\to\infty} (d_n/n) \geq (1-R/r)\, H^{-1}\, (1-r)$$

This value is maximized by differentiating with respect to r. Then r is found to be the maximum of 0.5 and the solution to:

$$R = r^2/[1+\log_2(1-H^{-1}(1-R))]$$

These bounds are shown in Figure 6.

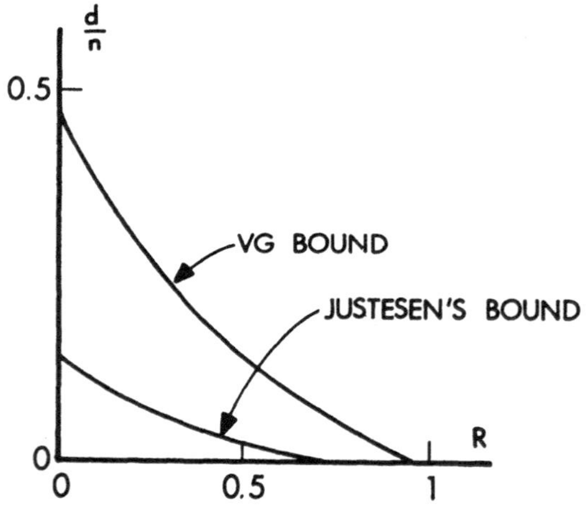

Figure 6. Justesen's Codes and the VG Bound

4.3 Improvements for Low Rates

In an effort to increase the minimum distance of Justesen's codes for low rates, Weldon [8] replaced the outer RS code over $GF(2^m)$ of length 2^m-1 with a BCH subfield subcode over $GF(2^s)$ having the same length, where m=bs and b is a positive integer. Now, instead of encoding $2^{bs}-1$ (bs)-tuples, the inner codes must encode $2^{bs}-1$ s-tuples. A more general form of the randomly shifted codes is used. These (s+bs, s) binary codes can be produced in the same manner as the (2m,m) codes but have rates <0.5 and proportionately larger minimum distances.

Generating these inner codes with a shift register as before guarantees that each non-zero vector will appear in either one code or no codes. (This is a consequence of the property that each non-zero t-tuple appear exactly once per period in the output of a maximal length t-stage sequence generator). Massey [9] showed that every such class of codes meets the VG bound.

Weldon then showed that the code rate and minimum distance are bounded by

$$R_{BCH} > (1 - d_o^*/n)^{1/b}$$

$$d_o^* = d_o + 2^{s(b-1)}$$

d_o = design distance for the BCH code

$$d/n \geq (1-R_{BCH}{}^{1/b})\ H^{-1}(b/(b+1))$$

These results are compared with Justesen's in Figure 7.

Sugiyama, et al [10] suggested, for much lower rates, an additional stage of concatenation to Justesen's codes. An (N_3, K_3) RS code over $GF(2^{K_1K_2})$ encodes a block of $K_1K_2K_3$ binary information digits. Each character in $GF(2^{K_1K_2})$ is then encoded using an (N_2,K_2) RS code over $GF(2^{K_1})$. Finally, each of these N_2N_3 characters of $GF(2^{K_1})$ is encoded with a code from the ensemble of (N_1,K_1) codes of Wozencraft.

The inner codes are produced by a shift register sequence generator having (N_1-K_1) stages, thus constraining their rates to less than 0.5. The number of codes in the inner ensemble, then, is chosen by selecting N_1 and K_1 so that $N_2N_3 = 2^{N_1-K_1} - 1$.

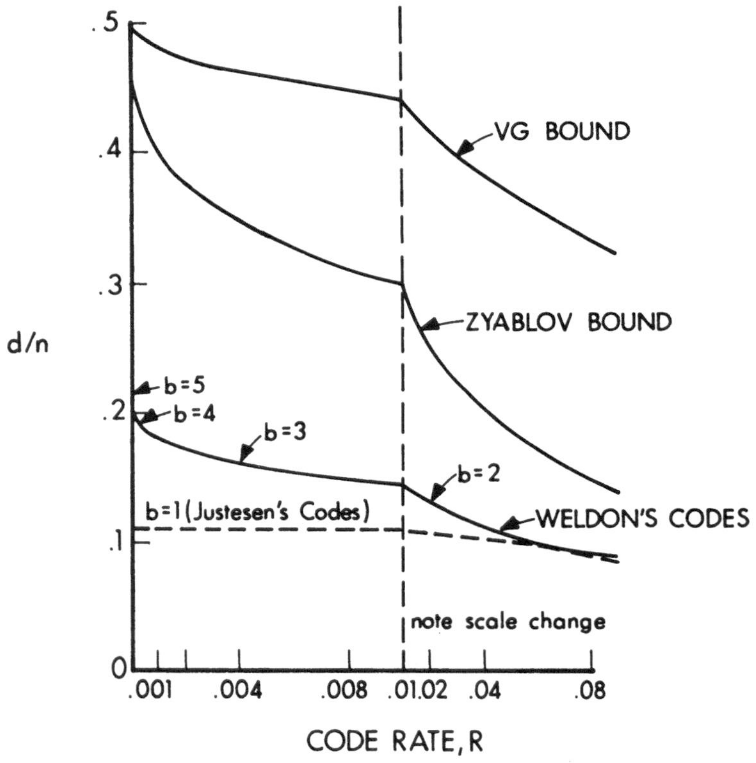

Figure 7. Weldon's Codes with BCH Outer Codes

Rates and lengths of the three classes of codes are known. The minimum distance of each RS code combined with Justesen's lemma [7] on the weight of a set of distinct binary L-tuples gives

$$\liminf_{n\to\infty} (d/n) \geq (1-R_2)(1-R_3)H^{-1}(1-R_1)$$

Setting $r = R_1R_2R_3$, the right hand side of this relation can be maximized. The results are shown in Figure 8. Weldon's bounds [8] were computed for very low rates and are shown. Details are found in [10].

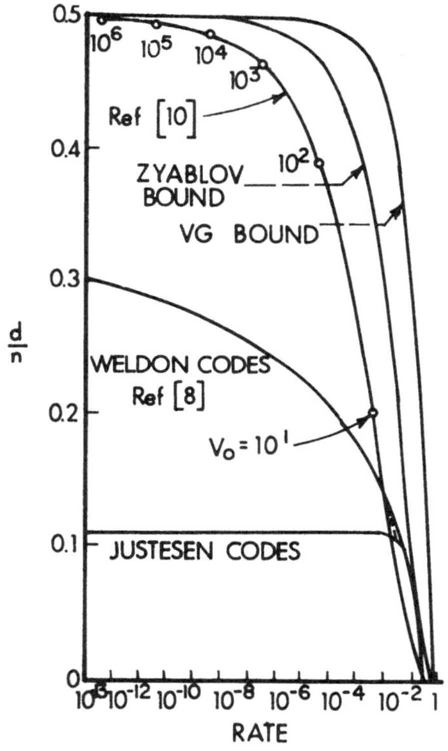

Figure 8. Bounds With Additional Concatenation

Finally, Weldon [11] devised a class of inner codes "near" the VG bound. When concatenated with RS codes, the asymptotic values of d/n for low rates are bounded above zero. Code construction uses the following.

Lemma: If an (n, k+p) code has fewer than 2^P codewords of weight less than δ, then a class of 2^P disjoint (n,k) codes can be constructed that has fewer than 2^P codewords of weight less than δ.

The proof uses a class of (k+p, k) codes, k < p, each having 2^P words and obtained by shortening Justesen's (2p,p) inner codes.

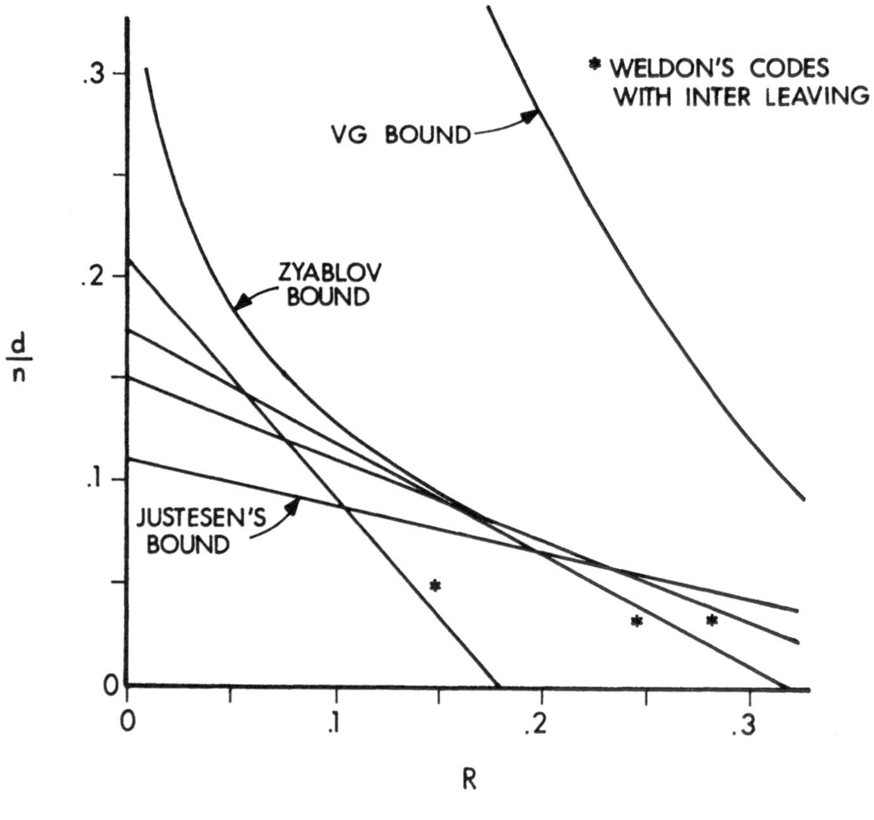

Figure 9. Performance With Interleaved Inner Codes

Thus, the problem is to construct the aforementioned (n, k+p) code C for arbitrary p. Hence, the distribution of weights (spectrum) of the words of C must be known. In order to construct sequences of fixed rate codes with known weight spectra, interleaving a given code with a known spectrum is used. Using the weight spectra of some BCH, Golay, quadratic residue, and Reed Muller codes, interleaved codes were constructed with values near the VG bound.

Of greater interest are the asymptotic results when these disjoint inner codes are concatenated with an extended RS (N,K) code having $N=2^m$, $D=N-K+1$. The concatenation operations are performed as in Justesen [7]. Rate was computed in a straightforward manner; minimum distance bounds on the inner (interleaved) code were obtained using the Chernoff bound on the tail of the probability distribution of a sum of independent, indentically distributed, random variables to provide a lower bound to the total weight of the low weight code words in the interleaved code. The results are plotted in Figure 9.

5. CONCLUSIONS

There still is no general solution to the problem of code construction to meet the VG bound. A solution involving some searching has been demonstrated to provide error free decoding on nearly any binary symmetric channel at a rate not less than 0.2 times channel capacity. Except for Weldon's work on low rate codes, Zyablov's bound to d/n for concatenated codes has not been achieved.

On the other hand, several techniques have been shown for using "bad" codes (e.g. BCH codes) to construct asymptotically "good" codes.

Further, a modest amount of searching in the design of codes produces encouraging results, the iterated BCH codes being one example. Justesen speculated that using the "best" inner code rather than the ensemble of inner codes might improve d/n. This remains an open question, however, since all concatenation schemes known to the author use varying inner codes.

Finally, none of these constructions produces "good" cyclic codes. The asymptotically best long cyclic codes [12] are a class of concatenated codes which have degenerate values of d/n, that being due to the performance of the inner codes. It still is not clear that "good" cyclic codes exist.

REFERENCES

1. W. W. Peterson and E. J. Weldon, Jr., Error-Correcting Codes, Second Edition, MIT Press, Cambridge, 1972.

2. E. R. Berlekamp, Algebraic Coding Theory, McGraw-Hill, New York, 1968.

3. A. B. Cooper III, Iterated Codes with Improved Performance, PhD Dissertation, The Johns Hopkins University, Baltimore, May 1976.

4. A. B. Cooper, III and W. C. Gore, "Iterated Codes with Improved Performance, IEEE Transactions on Information Theory, September 1977.

5. G. D. Forney, Jr., Concatenated Codes, MIT Press, Cambridge, 1966.

6. V. V. Zyablov, "On Estimation of Complexity of Construction of Binary Linear Concatenated Codes, "Problemy Peredachi Informatsy, Vol 7, No. 1, 1971.

7. J. Justesen, "A Class of Asymptotically Good Algebraic Codes," IEEE Transactions on Information Theory, September 1972.

8. E. J. Weldon, Jr., "Justesen's Construction - the Low Rate Case," IEEE Transactions on Information Theory, September 1973.

9. J. L. Massey, Threshold Decoding, MIT Press, Cambridge 1963.

10. Y. Sugiyama, et. al., "A Modification of the Constructive Asymptotically Good Codes of Justesen for Low Rates, "Information and Control, no. 25, 1974.

11. E. J. Weldon, Jr., "Some Results on the Problem of Constructing Asymptotically Good Error-Correcting Codes," IEEE Transactions on Information Theory, July 1975.

12. E. R. Berlekamp, Jr. and J. Justesen, "Some Long Cyclic Linear Binary Codes are not so Bad," IEEE Transactions on Information Theory, May 1974.

Part IV

PROBABILISTIC CODING

organised by Dr. P. G. Farrell

NONPROBABILISTIC AND PARTLY PROBABILISTIC CHANNEL CODING AND TIME VARYING CHANNELS

W.L. Root

Aerospace Engineering Department, The University of Michigan, Ann Arbor

ABSTRACT. In a single-input, single-output communication system with unknown channel the communicators have the choice of trying to identify the channel so as to know how to code effectively, of trying to devise a universal code and decoding rule which will work for the whole class of possible channels, or of trying some combination of these procedures. In particular, for some time-varying channels for which the pattern of time-variation is not known and cannot be described statistically very well, universal codes which do not depend or only partly depend on probabilities seem to be appropriate.

For the simplest case of finite-alphabet, discrete-time, memoryless but time-varying channels, where the actual mapping from input to output is unknown but is known to belong to a certain class of maps, a simple coding theory is described which yields useful answers in some cases. Under certain symmetries, rates of guaranteed error-free transmission are given; more generally, bounding inequalities are stated. This theory is purely combinatoric.

1. INTRODUCTION

These lectures touch on parts of the topic: coding for communications channels when the communicators have "incomplete knowledge" of the behavior of the channel. This circumstance of incomplete knowledge of the channel is likely to occur when its characteristics can change with time, so time-varying channels are emphasized.

The material to be discussed falls into two categories: (i) coding for (partly) probabilistic channels; (ii) coding for nonprobabilistic channels. When dealing with the first, we would say there is complete knowledge of the channel if each finite string of letters from the sending alphabet induces a probability distribution on all possible finite strings of symbols received and these distributions are known to sender and receiver. One way to describe many situations where there is less than complete knowledge of the channel in this sense is to use the concept of compound channel (in the terminology of Wolfowitz). We shall review briefly some basic theory of compound channels and tie it in with time-varying probabilistic channels. There is nothing new in this material; most of it is classical. However, the point-of-view taken here is perhaps not quite the usual one. Also it is desired to bring out some analogies with the nonprobabilistic channels.

The nonprobabilistic channels are defined entirely without the use of probabilities; they involve only mappings from inputs to outputs. Such channels are trivial in an information-theoretic sense if the communicators have full knowledge of the channel, but in general they are not if the communicators have only incomplete knowledge of the channel. This latter situation is of interest here.

Because the nonprobabilistic theory is unconventional, we go into some detail with definitions and basic facts. A few results were given in the conference paper [1], and most of these are reproduced with only minor changes here. As far as the author is aware the theory has not been carried very far. It does appear to have some promise, both in terms of direct applicability and in terms of possible combinations with probabilistic methods. It is, of course, inherently more simple than the conventional probabilistic theory.

Section 2 is devoted to definitions of channels. Section 3 contains the short review of basic results on coding for compound channels, and also some preliminary facts about rates for nonprobabilistic channels. Section 4 contains the material from [1]. A few general comments about applicability are given in the concluding Section.

2. CHANNEL DESCRIPTIONS

For now we consider only single-sender, single-receiver, discrete-time communication systems. Let X and Y be sets, finite or infinite, which represent the sending and receiving alphabets, respectively. To facilitate descriptions the terms sender and receiver are personalized; thus, for example, the sender may "choose a certain code," or the receiver may "infer a certain message was sent."

Let $(x_1, x_2, \ldots, x_n)$, $x_i \in X$, be an input sequence. Each x_i is thought of as being transmitted at an instant $t_i < t_{i+1}$. Corresponding to this input sequence is an output sequence $(y_1, y_2, \ldots, y_n)$, $y_i \in Y$, where y_i is thought of as being available to the receiver at time t_i. The y-sequence must be dependent on the x-sequence in some way, or obviously there is no information being transmitted in any reasonable sense of the word. If each y_i can depend only on $(x_1, x_2, \ldots, x_i)$ the system is <u>causal</u>; if each y_i can depend only on x_i the system is <u>memoryless</u>. We shall always assume the communication systems being discussed are causal, and unless there is indication to the contrary we assume them to be memoryless.

The term <u>channel</u> is to mean a description, as complete as possible, of what the input x-sequence and output y-sequences can be and how the y-sequences depend on the x-sequences; the description is to include the state of prior knowledge of the sender and of the receiver. It is convenient for this discussion to give a preliminary classification of channels into three types: (1) probabilistic channels, (2) partly probabilistic channels, and (3) nonprobabilistic channels.

In general, for a probabilistic channel each finite input sequence starting at $t'_o = t_i$ determines a probability distribution on input sequences starting at $t'_j \geq t'_o$ and of arbitrary length. This distribution is known to both sender and receiver. If the channel is memoryless, each $x \in X$ determines a probability distribution on Y. In case Y is a finite set $w(y|x;i)$ will denote the probability that the received symbol at t_i is y given that the transmitted symbol at t_i is x. The memoryless channel is <u>stationary</u> if $w(y|x;i) = w(y|x;j)$ for any i, j, no matter what x and y may be; then we write $w(y|x;i) = w(y|x)$ for all i. Otherwise the channel is nonstationary. The functions $w(\cdot|\cdot;i)$ or $w(\cdot|\cdot)$ are called channel probability functions, (c.p.f.), and the finite-alphabet, memoryless channels are denoted $(X, Y, w(x|y|i))$ or $(X, Y, w(x|y))$.

By a partly probabilistic channel is meant one in which probability distributions are involved but there is not a known full probabilistic description. More precisely, a particular finite input sequence either does not by itself determine a probability distribution on output sequences, or if it does, the distribution is imperfectly known to sender and receiver. Well-known examples are the "classes of channels" or "compound channels" originally considered by Blackwell, Breiman and Thomasian and by Wolfowitz (see, e.g. [2] and [3]). For use in later discussion, we make precise what will be meant by a stationary, memoryless compound channel with finite alphabets (or, usually in what follows, just stationary compound channel). With X and Y finite sets, let W be an indexed family of c.p.f.'s $w_a(\cdot|\cdot)$ for X and Y, $a \in A$. The index set A may be finite or infinite. When a message is sent the channel operates as an ordinary stationary memoryless channel (X, Y, w_a) for one of the c.p.f.'s $w_a \in W$, but neither sender nor receiver knows which w_a is operative. This concept may be modified by the assumption that one or the other of sender and receiver knows which c.p.f. is operative, but when such a case is considered it will be indicated specially. Also there is an immediate and obvious generalization to nonstationary compound channels. We note without further elucidation here that there are other kinds of partly probabilistic channels.

A nonprobabilistic channel is one for which probabilities are not involved in the description. The concept is familiar, even if the term is not. For example a binary code of block length K that corrects e or fewer errors in each block fits very well a channel that may be described as follows: each input sequence of length K is carried into an output sequence identical to the input, except that as many as e symbols may be different. A less common example is given by the following situation: X and Y are infinite sets provided with a metric space structure; for each $x \in X$ that is sent it is known that the received y will lie in a ball of radius ϵ about a point f(x), where f is a known mapping from X into Y and ϵ is a fixed positive number (see, e.g. [4]). Despite the superficial differences, these two examples are fundamentally similar, of course.

The major concern of these lectures is with nonprobabilistic channels, but of a really different character from that of these two examples. (It will be evident that the definition to be given can be easily generalized to include both of them, but this is largely irrelevant). We first define what is to be called a <u>memoryless</u> <u>primitive</u> <u>channel</u> (MPC). Let X and Y be finite sets with N_x and N_y elements respectively, and Φ a set of mappings $\phi: X \to Y$. Φ is a finite set $\{\phi_1, \ldots, \phi_M\}$. Then for each input sequence $u = (x_1, \ldots, x_n)$, the corresponding output sequence available to the receiver is

$$v = (y_1, \ldots, y_n) = (\phi_{i_1}(x_1), \phi_{i_2}(x_2), \ldots, \phi_{i_n}(x_n))$$

where each $\phi_{i_j} \in \Phi$. If the ϕ_i which is operative at some t_i must be the same as the ϕ_j which is operative at t_j, for all j, we shall say the channel is <u>time-invariant</u>; otherwise the channel is <u>time-varying.</u> Note that a channel need only have the potential for change to be called time-varying.

The class of time-varying MPC's will be further divided into subclasses as follows. If there is a positive integer K such that for some t_1 it is guaranteed that

$$\ldots, \phi_1 = \phi_2 = \ldots = \phi_K, \ \phi_{K+1} = \phi_{K+2} = \ldots = \phi_{2K}, \ldots,$$

$$\phi_{nK+1} = \ldots = \phi_{(n+1)K}, \ldots$$

all $\phi_i \in \Phi$, and there is no larger integer that satisfies this condition, then the channel is said to have <u>mesh</u> <u>size</u> K. A set of transmission times $(t_1, \ldots, t_K)$, $(t_{K+1}, \ldots, t_{2K})$, etc., for which the ϕ_i are guaranteed to be the same is called a <u>mesh.</u> Note it is not required that ϕ_{nK+1} actually be different from ϕ_{nK}, only that it be allowed to be different. Any time-varying MPC has mesh size $K \geq 1$; small K corresponds of course to a quantized version of rapid variation, large K to slow variation. Unless a specific statement is made to the contrary it will be assumed that both sender and receiver know whether the channel is time-invariant or time-varying, and if the latter, what the mesh size is. A memoryless primitive channel with mesh K is denoted $(X, Y, \Phi; K)$; if the channel is time-invariant K is replaced by ∞.

If each $\phi \in \Phi$ is injective, the channel is said to be <u>non-</u>

singular; otherwise it is singular. It is convenient also to refer to an individual $\phi \in \Phi$ as nonsingular or singular (instead of injective or noninjective, respectively).

It is clear that MPC's are in a sense one kind of simplification of compound channels, while ordinary (simple) probabilistic channels are a different kind of simplification. Other kinds of nonprobabilistic channels than MPC's obviously can be defined. When the term primitive channel, without modification, is used below it will refer generally to nonprobabilistic channels consisting of a family of maps from input sequences to output sequences.

3. RATES OF TRANSMISSION

3.1 Channels with Probabilities

It is not at all the purpose of these lectures to go into the probabilistic theory of channel coding. However we do note a few standard basic definitions and results pertaining to compound channels and to nonstationary channels. The definitions are at most slight modifications of ones in [2], [3] and [5].

Let X and Y be the sending and receiving alphabets for an otherwise arbitrary channel. For $G \geq 1$ and n a positive integer, a (G, n) code for this channel is the ordered collection

$$((u_1, B_1), (u_2, B_2), \ldots, (u_{[G]}, B_{[G]}))$$

where the u_i are elements of X^n and the B_i are disjoint subsets of Y^n. $[G]$ denotes the integer part of G, and we adopt the notation $m = [G]$ The u_i are the code words and the B_i the decoding sets of the code. A code word u_i is correctly received if its transmitted image falls in B_i. We say the code is of length n and size m.

Now let it be assumed till further notice that X and Y are finite sets and the channel is memoryless. Then the c.p.f.'s which are operative while a code word is being transmitted determine a probability measure $P(\cdot \mid u_i)$ on the subsets of Y^n. For example, if the channel is a stationary compound channel

$(X, Y, w_a(y|x), A)$ then $P(\cdot|u_i) = P_a(\cdot|u_i)$

is determined by

$$P_a(\{y_1, \ldots, y_n\} | u_i) = \prod_{k=1}^{n} w_a(y_k | x_k(u_i)) \tag{3.1}$$

where w_a is the operative cpf, and $x_k(u_i)$ is the letter in the k'th position of the word u_i.

Consider in particular the case that the channel is a stationary compound channel. The maximum error probability, ϵ_m, for the code is then defined as

$$\epsilon_m = \max_i \sup_a P_a(B_i^c | u_i) \tag{3.2}$$

For any $\epsilon \geq \epsilon_m$ we say the code is a (G, n, ϵ) code for the specified channel.

A number $\rho \geq 0$ is an <u>attainable</u> <u>rate</u> of transmission for (X, Y, w_a, A) if there is a sequence of $(e^{\rho n}, n, \epsilon(n))$ codes for the channel with $\epsilon(n) \to 0$ as $K \to \infty$. The supremum of the attainable rates is called the <u>capacity</u> of the channel and is denoted by C. If N_x is the number of elements of X, $0 \leq C \leq \log N_x$.

To state the coding theorem for stationary compound channels one introduces probability distributions q(x) on X. Let Q denote the set of all such q, and define the joint probability $P_{a,q}(x,y) = w_a(y|x)\, q(x)$ for $a \in A, q \in Q$. Define the mutual information $J_{a,q}$ on X x Y by

$$J_{a,q}(x,y) = \log \frac{P_a(x,y)}{P_a(x)\, P_a(y)} \quad \text{if } P_{a,q}(x,y) > 0 \tag{3.3}$$

$$= 0 \quad \text{if } P_{a,q}(x,y) = 0$$

Then the channel coding theorem (due to Blackwell, Breiman and Thomasian [2] and to Wolfowitz, see [3]) states that

$$C = \sup_{q \in Q} \inf_{a \in A} E_a J_{a,q}. \tag{3.4}$$

This is the theorem in its simplest form; no results are quoted here about the rate of convergence of ϵ_n to zero, and the distinction between the weak converse and strong converse is not made evident. It is pointed out in the original papers that

$$C \leq \inf_{a \in A} \left[\sup_{q \in Q} E_a J_{a,q} \right]$$

or, since the quantity in brackets is just the capacity C_a of the channel with c.p.f. w_a, that $C \leq \inf_{a \in A} C_a$. A little reflection indicates that this is what one would expect to be true. A more subtle observation (see [3]) is that C = 0, however, only if $\inf_{a \in A} C_a = 0$. Thus one has the perhaps unexpected result that if the individual channels comprising the compound channel have capacities bounded away from zero, it is always possible to communicate reliably through the compound channel at a nonzero rate.

Stationary compound channels may be defined which are different from what has been described here, perhaps more general in some respects and less general in others, but one expects a capacity formula (if one can be obtained) to be some version of Eq. (3.4). For example, in [6] compound channels described by $y = H_a x + z$, $a \in A$, are considered where the inputs and outputs, x and y, are real-valued functions of time on a finite interval of fixed length; H_a is a linear integral operator belonging to a specified class, and z is white Gaussian noise. Because the structure of the channel is known the expected value of mutual information can be computed in terms of the quantities determining the channel. The final formula for capacity under an average power constraint on input signals is

$$C = \sup_{\tilde{s}} \inf_{\tilde{h}} \int_{-\infty}^{\infty} \log \left(1 + \frac{|\tilde{h}(f)|^2 \tilde{s}(f)}{N} \right) df, \tag{3.5}$$

where $\tilde{h}$ is the Fourier transform of the kernel h of the linear integral operator H; $\tilde{s}$ is the spectral density of a real-valued stationary process which has mean zero and autocorrelation function bounded by 1 and integrable; N is the average power density of the white noise. For a precise statement of conditions, see [6] ; essentially the requirements are that the operators H_a have bounded memory and that the set of all H_a be compact in a suitable topology.

Nonstationary channels, channels with time-varying c.p.f.'s, present some additional features to be considered in a formulation of channel coding problems. Consider first a simple nonstationary channel; for such a channel the c.p.f.'s $w(y|x; i)$, $i = 1, 2, \ldots$, are known to sender and receiver. With the proviso that communication is started at a known fixed time t_{i_o} the maximum error probability for a (G, n) code is

$$\epsilon_m = \max_j P(B_i^c | u_j)$$

where P is determined by

$$P(\{y_{i_o}, \ldots, y_{i_o+n-1}\} | u_j) = \prod_{k=i_o}^{k=i_o+n-1} w(y_k | x_k(u_j); k)$$

Clearly ϵ_m depends on i_o; with an obvious modification of a previously introduced notation, we can refer to a $(G, n, \epsilon; i_o)$ code.

A kind of pathology which a nonstationary channel can have is illustrated by the following trivial example. Take $Y = X$. Set $w(b_j | a_j; 1) = 1$ for $j = 1, \ldots, N_x$ (= number of elements in X), where the a_j are the elements of X and the b_j are the elements of Y. Set $w(b_1 | a_j; i) = 1$, $j = 1, \ldots, N_x$, for $i = 2, \ldots, 11$; again set $w(b_j | a_j; i) = 1$, $j = 1, \ldots, N_x$, for $i = 12, 13, \ldots, 1011$. For the succeeding block of 10^6 t_i again set $w(b_1 | a_j; i) = 1$, $j=1, \ldots, N_x$. Continue to alternate between these two channel descriptions for successive blocks of length 10^{10}, 10^{15}, etc. With the preceding definition of attainable rate the capacity of this channel is zero, yet clearly the channel can be used very effectively for certain periods of time. The difficulty cannot entirely be defined away, but it seems better in view of this kind of situation to relax the definition of attainable rate: a number $\rho \geq 0$ is an <u>attainable rate</u> for $(X, Y, w(y|x; i))$ starting at t_{i_o} if there is a subsequence K_n of the positive integers and a sequence of codes $(e^{\rho K_n}, K_n, \epsilon(K_n); i_o)$ such that $\epsilon(K_n) \to 0$ as $n \to \infty$.

Capacity is again defined as the supremum of attainable rates. A priori, capacity appears to depend on t_{i_o}; in fact it does not as may readily be proved (see [5]) - and plausibly inferred from the fact that capacity is an asymptotic property. Let $J_q(i)$ be the mutual information for the time t_i, defined by making the obvious modifications in Eq. (3.3). Let

$$S_n = \frac{1}{n} \sum_{i=1}^{n=1} \sup_q E(J_q(i)). \tag{3.6}$$

Then the capacity of the time varying channel is:

$$C = \limsup_{n \to \infty} S_n. \tag{3.7}$$

See, e.g. [5] or [7] .

Although capacity is not a function of starting time, the maximum probability of error for any particular code is, in general. This situation suggests the problem of finding the maximum rate that can be achieved when the starting time is not known to the communicators, so that the choice of codes cannot depend on starting time. But this problem is equivalent to finding the maximum rate for a nonstationary compound channel with fixed starting time for which the set of c. p. f. sequences is exactly the set of all time - translates of the original c. p. f. sequence. That is, $w_n(y|x; i) = w(y|x; i+n)$. See [5] and [7] .

The nonstationary compound channels arrived at in this way consist of at most countably many individual channels, so they are a special case of the general class of nonstationary compound channels. However, we do not pursue the subject of coding for nonstationary compound memoryless channels here (see, e.g., [5] , [8] , [9] , [10] , [11]), except to note in the following paragraph a condition for nonzero capacity.

The "worst" kind of nonstationary compound channel is that for which the c. p. f. can change arbitrarily within the prescribed set of c. p. f.'s from one symbol transmission to the next. The question arises as to when effective communication is possible at all under these circumstances. An answer is given by Kiefer and Wolfowitz [10] for the case the set of c. p. f.'s is finite. Let X and Y be finite alphabets with $N_x = d$ and $N_y = b$ elements respectively; let $A = \{1, 2, \ldots, M\}$. Consider the nonstationary compound channel with c. p. f.'s $w_a(x|y)$, $a \in A$, for which the operative c. p. f. for each symbol transmission may be any one of the w_a. For each $x \in X$ let T(x) be the smallest convex body which contains the M points of b-dimensional space: $(w_a(y_1|x), w_a(y_2|x), \ldots, w_a(y_b|x))$, $a = 1, 2, \ldots, M$. Then a necessary and sufficient condition that there be an attainable rate greater than zero when neither sender nor receiver knows the channel sequence is that at least two of the convex bodies $T(x_1)$, ..., $T(x_d)$ be disjoint.

Before going on to a consideration of rates in primitive channels we make two remarks.

(1) Suppose a stationary compound channel has nonzero capacity, and a code c is a "good" code for this channel. We define the <u>rate</u> r of any code of length n and size m to be

$$r = \frac{1}{n} \log m.$$

For c to be a good code means that its rate is fairly near channel capacity and the maximum error probability is small. Now, although by definition the operative c.p.f. is to remain forever the same, in fact if it were to change instantaneously between code word transmissions to another c.p.f. within the prescribed set, then the code would perform as well as it is guaranteed to perform for the stationary compound channel. Thus, for a particular - and admittedly artificial - mode of time variation, the code would provide satisfactory communication. This simple and old observation may have practical importance for slowly varying channels. It is mentioned here because it points up a situation analogous to one to be discussed for primitive channels.

(2) If a real-life communication system can be satisfactorily modeled as a stationary compound channel with nonzero capacity, then, in principle at least, codes can be devised for use with this system which will result in satisfactory communication. However, there is the question: why do it this way? Why not just identify the channel and then use this resulting information? There are two possibilities we wish to consider: (i) there is no feedback channel, so that although the receiver may identify the channel, the identification is not available to the sender. (ii) there is a feedback channel (which for simplicity is assumed to be error-free and capable of transmission at a high nominal rate), so that the identification is available to the sender. We also want to consider long term and short term effects.

The answer to the question for case (i) is largely given by the fact that the capacity of a stationary compound channel is the same whether or not the receiver knows the operative c.p.f., as long as the sender does not know it (see [3]). This result is intuitively clear because the c. p. f. can always be estimated as closely as desired by using a fixed number $n_T < n$ of known symbol transmissions as test signals at the start. Thus, any code of length n_c may be replaced by one of length $n_T + n_c$ prefixed with the n_T test signals and followed by the original code. As $n_c \to \infty$ the rates become the same, and with the new code the receiver knows the c. p. f. to within some error depending on n_T.

In case (ii), where there is a feedback channel, there is a clear advantage in first identifying the channel if the channel is truly stationary. With identification the channel capacity will be at least as great as $\inf_a C_a$ whereas the capacity of the compound channel never exceeds this value.

However, this comment does not seem to settle the question completely. Suppose the real-life channel is slowly varying, so that remark (1) may apply. Then the identification would have to be repeated from time to time, and the relative lengths of the time interval required for identification and the time interval for the longest usable code would have to be considered. It is not known to the author how this would work out. Again, a similar situation will appear for certain primitive channels, but there the identification is a simple matter.

3.2 Primitive Channels

The definitions of a (G, n) code and of code rate r given above work perfectly well for a primitive channel. The other definitions are obviously not applicable to primitive channels since they involve probabilities. To recover the concept of attainable rate for primitive channels one must first establish some criterion of satisfactory communication; one can then investigate the rates of <u>acceptable</u> codes, codes that meet this requirement. Suppose the channel has some sort of appropriate stationarity property so that a code with rate r that is acceptable for one block of n symbols is also acceptable for succeeding blocks of n symbols. Then the total possible number of sequences of length kn available for satisfactory communication is at least e^{knr}, $k = 1, 2, \ldots$; so the code rate has the significance of an "attainable rate".

One reasonable criterion of acceptability for codes for primitive channels is that they be guaranteed error-free. There is the usual complication caused by the fact we are dealing (in general) with time-varying channels, so a little care must be taken with definitions. A code of length n is guaranteed error free (g.e.f.) at time t_1 if the decoding yields the correct code word no matter which code word was sent and no matter which of the set of possible channel maps $(x_1, \ldots, x_n) \to (y_1, \ldots, y_n)$ is operative. If we denote by Ψ the set of possible maps $\psi: X^n \to Y^n$ for the time block (t_1, t_n), this may be rephrased by saying that for each code word $u_j \in X^n$ and corresponding decoding set $B_j \subset Y^n$, the total image $\Psi(u_j)$ must be contained in B_j. It is useful to introduce another term: $u_1 = (x_1, \ldots, x_n)$ and $u_2 = (x_1', \ldots, x_n')$ are distinguishable if $\Psi(u_1)$ and $\Psi(u_2)$ are disjoint subsets of Y^n. Any pair of code words from a guaranteed-error-free code must be distinguishable.

We now specialize to MPC's. Consider $(X, Y, \Phi; K)$, where the transmission times are numbered so that $(t_1, \ldots, t_K)$ is a mesh. Let $R_n(i)$ denote the supremum of rates for codes of length n guaranteed error-free at t_i. Since a code that is guaranteed error-free at t_i is also guaranteed error-free at $t_{i+\ell K}$, ℓ any integer, it is only necessary to consider $R_n(i)$ for $i = 1, \ldots, K$. Put

$$R(i) = \limsup_{n\to\infty} R_n(i), \quad i = 1, \ldots, K,$$

$$= \limsup_{n\to\infty} \left[\sup_c \frac{1}{n} \log m(c)\right] \qquad (3.8)$$

where c is any code guaranteed-error-free at t_i. It is easy to show that all the R(i) are the same, since i only ranges from 1 to K. Hence we can define a constant $C = R(i) = \limsup_{n\to\infty} R_n(1)$; C is called the capacity of $(X, Y, \Phi; K)$.

Consider codes of length n = K, guaranteed error-free at t_1. Each of these codes has size $m \leq N_x^K$; let μ denote the largest of these m, and let c be one of the codes of size μ. Thus

$$R_K(1) = r(c) = \frac{1}{K} \log \mu \qquad (3.9)$$

Proposition 1. $C = R_K(1) = r(c)$. (see Errata)

Proof: The code c^{ℓ} formed by repeating c for the first ℓ meshes has μ^{ℓ} code words and rate $r(c^{\ell}) = r(c)$. Hence

$$C = \limsup_{n\to\infty} R_n(1) \geq \limsup_{\ell\to\infty} R_{\ell K}(1) \geq r(c).$$

To obtain the opposite inequality, let c' be any code of length $n = \ell K + k$, $0 \leq k < K$, g.e.f. at t_1. For any code word in c', let $u|(i,j)$ denote the restriction of u, when started at t_1, to $(t_i, t_{i+1}, \ldots, t_j)$. Consider

$$u_i|(1,K),\ u_i|(K+1, 2K), \ldots, u_i|((\ell-1)K, \ell K),\ u_i|(\ell K, \ell K+k),\ i=1, 2,$$

where u_1, u_2 are code words in c'. Since c' is g.e.f., u_1 and u_2 are distinguishable. This can happen only if at least one of the pairs of restrictions is distinguishable. The maximum number of distinguishable code words for any mesh is μ; let μ_1 be the maximum number of distinguishable code words for the partial mesh at the end. Then the maximum number of code words there can be in c' is $\mu^{\ell} \cdot \mu_1$. It follows easily that $\mu_1 \leq \mu$, so we have:

$$r(c') \leq \frac{1}{\ell K+k} \log(\mu^{\ell} \cdot \mu_1)$$

$$\leq \frac{\ell+1}{\ell K+k} \log \mu.$$

Since as $\ell \to \infty$ this is arbitrarily close to $r(c) = \frac{1}{K} \log \mu$,

$$C = \limsup_{n\to\infty} R_n(1) \leq r(c). \quad |||$$

Note that the argument used involving restrictions of the code words is only valid when the code words are partitioned according to the meshes.

It may seem that the requirement of guaranteed-error-free decoding for codes for primitive channels is unduly stringent. In the opinion of the author it is not, because it is expected that in application there will always be some approximation or some probabilistic consideration involved. Thus one would not be in the position of asking for completely error-free communication in practice.

4. MEMORYLESS PRIMITIVE CHANNELS

4.1 Preliminaries

We consider coding for channels $(X, Y, \Phi; K)$ where K may be any positive integer or infinity, and where X and Y are finite sets. By virture of Proposition 1, when K is finite it is sufficient to restrict attention to g.e.f. codes of length $n = K$ which are designed to operate during one mesh. As before, μ will denote the maximum possible number of code words for such a code. The notation $\mu(K)$ will indicate dependence of μ on K when X, Y, Φ are fixed; obvious analogous notations will also be used. A number of trivial observations are collected as Proposition 2.

Proposition 2. (a) Let X, Y and K be fixed and $\Phi \subset \Phi'$. Then $\mu(\Phi) \geq \mu(\Phi')$.
(b) Let $X \subset X'$ ($N_x \leq N'_x$) and let Y and K be fixed. Let Φ consist exactly of the restrictions to X of the maps $\Phi' : X' \rightarrow Y$ in Φ'. Then $\mu(X, Y, \Phi) \leq \mu(X', Y, \Phi')$.
(c) $\mu \leq N_y^K$. |||

Consider now the case $N_x = N_y = N$, and identify the elements of Y with those of X by an arbitrary but fixed bijection. With this identification of elements any map $\phi \in \Phi$ can be thought of as a map from X into itself; if ϕ is injective it is necessarily bijective and hence has an inverse ϕ^{-1}. As above, Ψ is the collection of maps ψ given by $\psi(u) = (\phi(x_1), \ldots, \phi(x_K))$, $\phi \in \Phi$.

Proposition 3. With $N_x = N_y = N$, if Φ is a group then:
(a) The total images $\Psi(u)$, $u \in X^K$, form the cells of a partition of $Y^K = X^K$. The cells are denoted B_i.
(b) $\mu \in B_i$ iff $\Psi(u) = B_i$.
(c) μ = the number of cells B_i in the partition.

Proof: First we note that if Φ is a group of mappings of X onto itself, Ψ is a group of mappings of X^K onto itself. Now, suppose for some $\psi_1, \psi_2 \in \Psi$ and some u_1 and u_2 that $\psi_1(u_1) = \psi_2(u_2)$. Let ψ_3 be any map in Ψ. Then

$$\psi_3(u_1) = \psi_3 \psi_1^{-1} \psi_2(u_2) = \psi_4(u_2)$$

for some $\psi_4 \in \Psi$. Thus $\Psi(u_1) = \Psi(u_2)$, and we see that the total images of elements in X^K either coincide or are disjoint. Since

$\Psi(X^K) = Y^K$ the assertion (a) follows. Assertion (b) follows because Ψ contains the identity mapping.

Clearly μ is not less than the number of cells in the partition, because a g.e.f. code can be constructed by taking one u from each cell. If μ exceeded this number there would be some code for which two code words u_i and u_j would have to be in the same cell. Then for some ψ_1, ψ_2 it would follow that $\psi_1(u_i) = \psi_2(u_j)$ and u_i and u_j would not be distinguishable. Thus (c) is proved. |||

This situation in which Φ is a group is the cleanest mathematically and easiest to handle. When Φ is not a group but the channel is nonsingular (each ϕ in Φ is a bijection), the above result gives bounds for μ by the following obvious corollary.

Proposition 4. If $N_x = N_y$ and $(X, Y, \Phi; K)$ is nonsingular then $\mu \geq$ the number of cells in the partition of X^K determined by any group containing Φ, and $\mu \leq$ the number of cells in the partition determined by any group contained in Φ. |||

4.2 Channels for which $N_x = N_y$ and Φ is the Symmetric Group

Perhaps the most important special case of the situation covered by Prop. 3 is that in which Φ is the group of all permutations of N letters. We shall refer to a channel $(X, Y, \Phi; K)$ with $N_x = N_y = N$ and Φ the symmetric group on N letters as a full nonsingular channel, and use the special notation $\nu(N, K) = \mu$.

Proposition 5. For a full nonsingular channel $(X, Y, \Phi; K)$,

$$\nu(N, K) = \sum_{i=1}^{\ell} S(i, K) \tag{4.1}$$

where $\ell = \min(N, K)$ and the $S(i, K)$ are Stirling numbers of the second kind. (Note that $S(i, K)$ may be defined as the number of ways the set of integers $\{1, 2, \ldots, K\}$ can be partitioned into i unordered classes, $1 \leq i \leq K$. See [12])

Proof: A formal proof is not given. It is easier to understand what is going on by looking at an example; once the example is completed it will be clear how a general proof goes. Take N = 4,

$K = 5$, and write $X = Y = \{a, b, c, d\}$. Consider the words $u_1 = (a\ b\ c\ d\ a)$, $u_2 = (c\ d\ a\ b\ c)$ and $u_3 = (a\ a\ b\ c\ d)$. With

$$\phi_1 = \begin{pmatrix} a\ b\ c\ d \\ a\ b\ c\ d \end{pmatrix}, \quad \phi_2 = \begin{pmatrix} a\ b\ c\ d \\ c\ d\ a\ b \end{pmatrix}$$

one has $\phi_1(u_1) = \phi_2(u_2) = (a\ b\ c\ d\ a)$. Thus u_1 and u_2 belong to the same B_i (or u_1 is equivalent to u_2, $u_1 \sim u_2$). On the other hand, for any ϕ and ϕ' $\phi(u_1)$ must have the same first and fifth letters with the other letters all different, while $\phi'(u_3)$ must have the same first and second letters, with the others different. Thus $\phi(u_1)$ can never be the same as $\phi'(u_3)$, and u_1 is not equivalent to u_3. In fact, $u \sim u_1$ if and only if it has the pattern (15) (2) (3) (4). (Note that the ordering within parentheses and the ordering of parenthesized terms are both irrelevant; thus (15) (2) (3) (4) = (2) (4) (51) (3), for example.) Similarly, a word will be equivalent to u_3 if and only if it has the pattern (12) (3) (4) (5). But (15) (2) (3) (4) and (12) (3) (4) (5) are partitions of the set $\{1, 2, 3, 4, 5\}$. In general, there is a 1:1 correspondence between the patterns of words of length 5 containing 4 letters and the partitions of $\{1, 2, 3, 4, 5\}$ into 4 classes. Each pattern represents one cell of the partition of X^5 determined by Ψ (one B_i), so there are $S(4, 5)$ cells of this partition which contain words with 4 distinct letters. We may pick one word of each pattern (one from each cell) to start the list of code words.

Now consider words using exactly 3 letters, e.g. $u_4 =$ b b a c a. The pattern of u_4 is (12) (35) (4). Clearly no word with 3 letters can be equivalent to a word with 4 letters. It follows as above that there are $S(3, 5)$ patterns with 3 letters. We may pick $S(3, 5)$ new words, one with each of these patterns, to add to the code list.

Similarly we may pick an additional $S(2, 5)$ words with exactly 2 letters, and finally $S(1, 5) = 1$ code word with exactly 1 letter. All the possible patterns are accounted for; there is a 1:1 correspondence between patterns and cells B_i in the partition of X^5, so by Proposition 2 it follows that $\nu(4, 5) = S(1, 5) + S(2, 5) + S(3, 5) + S(4, 5)$. |||

A formula due to Stirling is, for $N \leq K$,

$$S(N, K) = \frac{1}{N!} \sum_{n=0}^{N} (-1)^{N-n} \binom{N}{n} n^K. \tag{4.2}$$

From this, or from recurrence formulas (see, e.g., [12]), one has for the example discussed above: $S(1,5) = 1$, $S(2,5) = 15$, $S(3,5) = 25$, $S(4,5) = 10$ and $\nu(4,5) = 51$. Since $S(5,5) = 1$, one has $\nu(N,5) = \nu(5,5) = 52$ for $N \geq 5$.

The sum of the numbers of elements in the cells B_i must be N^K, hence, for $N \leq K$

$$N^K = \sum_{i=1}^{N} N(N-1)\dots(N-i+1)\, S(i,K). \tag{4.3}$$

Proposition 6. (a) For $N \leq K$, $\nu(N,K) \geq \frac{N^K}{N!}$.
(b) For fixed N, $\nu(N,K)$ is asymptotic to $\frac{N^K}{N!}$ as $K \to \infty$.

Proof: The assertion (a) follows from Eq. (4.3). From Eq. (4.2) one sees that

$$S(N,K) \cdot \frac{N!}{N^K} = 1 + o(1) \text{ as } K \to \infty.$$

Then (b) follows from Prop. 5. |||

A simple consequence of the preceding calculations is:

Proposition 7. If $(X, Y, \Phi; \infty)$ has $N_x = N_y = N$ and is nonsingular, its capacity is log N.

Proof: Since $R_n \leq N^n$, $n = 1, 2, \dots$, $C \leq \log N$. For arbitrary n, an optimum code for a full nonsingular channel with mesh size K = n is g.e.f. for the time-invariant channel in question, no matter what Φ may be. For such a code $\mu = \nu(N,n)$. Then,

$$R_n \geq \frac{1}{n} \log \nu\ (N,n) \geq \log N - \frac{1}{n} \log N!\ ,\ n \geq N.$$

Thus $C = \limsup_{n \to \infty} R_n \geq \log N$. |||

4.3 Singular Channels with $N_x = N_y$.

Recall that a MPC is singular if it contains at least one singular map, i.e. at least one map that is not injective. Once singular maps are permitted, the coding problems become very much more complicated than in the case of full nonsingular channels. Only preliminary results are presented here.

One conclusion can be drawn immediately:

Proposition 8. Let $(X, Y, \Phi ; K)$, $(X, Y, \Phi'; K)$ be respectively nonsingular and singular channels with the same X, Y and K and with $N_x = N_y$. Then for K sufficiently large the capacity C of the nonsingular channel exceeds the capacity C' of the singular channel.

Proof: Suppose Φ'' consists of one singular map that carries the N letters of X into N' symbols in Y, $N' \leq N-1$. Then $\mu'' = \mu''(\Phi'', K) \leq (N-1)^K$. Hence, if Φ' is any collection of maps containing a singular map, the corresponding $\mu' = \mu'(\Phi', K) \leq (N-1)^K$ by Proposition 1.

On the other hand $\mu = \mu(\Phi, K) \geq \nu(N, K) \geq \frac{N^K}{N!}$ Hence

$$\frac{\mu}{\mu'} \geq \frac{\nu(N, K)}{\mu'} \geq \left(\frac{N}{N-1}\right)^K \cdot \frac{1}{N!} \cdot \quad |||$$

Now, for purposes of illustration consider, two code words: $u_1 = (a\ b\ d\ a\ c\ a\ )$, $u_2 = (a\ b\ b\ a\ c\ d)$. The pattern of u_1 is (146) (2) (3) (5). This pattern corresponds to a partition of $\{1, 2, \ldots, 6\}$ of type 3+1+1+1, or 1+3+1+1, etc., since the order does not matter (see [12] for a discussion of the combinatorics relating to the types of partitions): we will say simply the pattern is of type 3+1+1+1. u_2 has a pattern (14) (23) (5) (6), of type 2+2+1+1. A nonsingular map ϕ preserves patterns; however a singular ϕ does not. For example if $\phi = \begin{pmatrix} a & b & c & d \\ d & d & a & b \end{pmatrix}$ then $\phi(u_1) =$ (d d b d a d), which has pattern (1246) (3) (5) of type 4+1+1. Let Φ be the set of all maps which carry 4 letters into 3 letters. Then the images under Φ of any u containing 4 letters will exhibit new patterns and types; there will in fact be $\binom{4}{2} = 6$ new patterns of various types. We illustrate for u_1:

$$\Phi[\ (146)(2)(3)(5)] \to \{(1246)(3)(5),\ (1346)(2)(5),\ (1456)(2)(3),$$
$$(146)(25)(3),\ (146)(35)(2),\ (146)(23)(5)\}$$

With this Φ, the complete set of image patterns of u depends only on the pattern of u; for arbitrary Φ this is not true of course. Note that the sets of image patterns of u_1 and u_2 happen to intersect. We will call a pattern with m classes, an m-pattern. The

Φ as prescribed carries each 4-pattern into a set of 3-patterns; furthermore each 3-pattern is an image of 4-patterns (for example, a 3+2+2 pattern is the image of five 4-patterns). When this Φ acts on a word with fewer than 4 letters, the situation is a little different. For example,

$$\Phi[(1234)(5)(6)] \rightarrow \{(1234)(5)(6), (12345)(6), (12346)(5), (1234)(56)\}$$

The original pattern is repeated in the image set because some of the maps in Φ carry the 3 letters of u into 3 distinct letters.

This discussion generalizes in an obvious way to arbitrary N and K. Let Φ_p denote the class of all maps from N letters into N-p letters, p = 0, 1, . . . , N-1. Then the discussion also generalizes directly to classes Φ which are a Φ_p or a union of Φ_p's. It does not generalize to arbitrary classes Φ because the image patterns for a word u will in general depend on u and not just on the pattern of u. To keep things simple we consider here only channels for which $\Phi = \Phi_1$ or $\Phi = \Phi_o \cup \Phi_1$.

An <u>(N, K) - graph</u> is described as follows. On one horizontal line list all N-patterns for words of length K. On a horizontal line below this list all (N-1)-patterns for words of length K. On a horizontal line below this, all (N-2) patterns. Keep this up until the last line has the single entry (1, 2, . . . , K). These entries are the vertices of the graph. From each vertex (N-pattern) in the top line draw a line to each vertex ((N-1)-pattern) in the second line which is an image of this N-pattern under Φ_1. Similarly, connect vertices in the second line ((N-1)-patterns) to their images in the third line under Φ_1. The only image of an (N-1)-pattern that is an (N-1)-pattern is itself, so there are no horizontal connecting lines. Keep up this procedure; finally all vertices on the next-to-bottom line will connect to the single vertex at the bottom.

The N-patterns, (N-1)-patterns, . . . , 1-pattern of length K are a partially ordered set under the ordering given by this directed graph. A <u>1-tree</u> is a vertex and all the vertices of the line below to which it is connected; the vertex on the higher line is the <u>base vertex</u> for the tree. The following assertions follow immediately.

Proposition 9. (a) Consider the channel $(X, Y, \Phi_1; K)$ with $N_x = N_y = N$ and Φ_1 as defined above. Then $\mu_1 = \mu_1(\Phi_1)$ is equal to the largest number of disjoint 1-trees that can be imbedded in the (N-K) graph. A biggest g.e.f. code is given by any set of code words whose patterns are those of the base vertices in such an imbedding.
(b) Let μ_{01} be the μ for the channel $(X, Y, \Phi_0 \cup \Phi_1; K)$. Then $\mu_{01} = \mu_1$, and a biggest code for the channel of (a) is also a biggest code for this channel. |||

Example. For the channel $(X, Y, \Phi_0 \cup \Phi_1; K)$ with $N = 4$, $K = 5$, the following sets of patterns each give biggest error-free codes:

(i) (12)(34)(5), (13)(45)(2), (14)(25)(3), (15)(23)(4), (24)(35)(1), (12345)

(ii) (13)(2)(4)(5), (12)(34)(5), (14)(25)(3), (15)(23)(4), (24)(35)(1), (12345).

4.4 Remarks

(1) For channels $(X, Y, \Phi; K)$ with $N_x = N_y$ and Φ any of certain subgroups of the symmetric group, calculations similar in spirit to those for the full nonsingular channel can be made without difficulty. Some of these channels would appear to be of interest.

(2) For singular channels of the type included in Proposition 8 an upper bound on μ is given in [1]. The author does not know a good lower bound, nor an asymptotic expression for μ.

(3) Various kinds of channels can be defined to which the material of this Section applies. For example, let X be a finite set and Y be a metric space. Let Φ be a collection of maps ϕ from X into subsets of Y with the following properties: (i) for some fixed $b > 0$, $\phi(x)$ is a ball in Y of radius b; (ii) $\phi(x')$ is disjoint from $\phi(x'')$, $x' \neq x''$. If the channel has mesh size K, the sort of coding used for a full nonsingular channel applies, and $\mu \geq \nu(N_x, K)$. This formulation allows for the interpretation that $\phi(x)$ is really the center of the image ball and random noise of magnitude not exceeding b is added.

A modification of the description of the maps ϕ would convert this to a singular channel.

(4) This comment relates to the remark (2) in the previous Section. Suppose the channel is a full nonsingular MPC with $K \geq N$. Suppose a procedure is adopted in which first the channel mapping is identified for the receiver, and from then on the sender is allowed to use the "full code" of all possible words - which the receiver can correctly decode because the map ϕ is known to him. Then the number of code words of length K available for g.e.f. reception is $N^{K-(N-1)}$. But

$$\nu\,(N,K) \geq \frac{N^K}{N!} \geq \frac{N^K}{N^{N-1}}$$

with equality (for both inequalities) when and only when $N = 2$. Thus the universal code gives a strictly greater rate for any $N > 2$ than does the procedure with identification first. One sees that this latter procedure comes nowhere near achieving capacity when N is large and K is not much greater than N.

5. COMMENTS ON APPLICABILITY

It appears that the primitive channel concept and the coding results for particular primitive channels should prove useful for various real-life single-sender, single-receiver channels with the following characteristics. (i) Additive noise at the receiver is a relatively minor source of the distortion caused by the channel (but it need not be negligible; see the remark (3) of the preceding Section). (ii) The major factors causing channel distortion are difficult to characterize statistically; or more fundamentally, any statistical characterization of these factors is unrealistic because of a lack of statistical regularity in the underlying mechanisms. This is related to: (iii) the channel is time-varying in an unpredictable way. Of course if a channel is undergoing unpredictable time-variation at a very rapid rate, no coding can be very effective.

The material should also prove useful for single-sender, multiple-receiver channels that satisfy the conditions mentioned. The universal nature of acceptable codes makes them as valid for simultaneous communication over different individual channels as for communication over a single channel, provided the set of channel mappings is the same for each.

As indicated in more than one place above, a universal

code procedure is to be weighed against the possibility of channel identification when a channel is not adequately known. This comment applies to probabilistic channels as well as non-probabilistic ones. It would appear that the advantage might lie either way, depending on a number of circumstances.

The restriction imposed on the mode of time-variation for the primitive channels discussed is unrealistic, of course. If a time-varying primitive analogue channel is converted to a finite-alphabet, discrete-time channel by quantization, one can choose a nominal mesh size K for the discrete channel and code for it on the basis of its characteristics and the nominal K. If K is small enough the code should work with only occasional errors; increasing K will result in a higher superficial rate of transmission, but also in a higher frequency of errors. It should be possible to analyze various situations of this sort on a probabilistic basis, but the author has not yet attempted this. The resulting over-all channel characterization would then be partly probabilistic. One thing to be noted is that absolute knowledge of the image $\phi(x)$ of a letter is never necessary for the kind of coding-decoding described in Section 4; it is only necessary for the receiver to be able to make comparisons within one code word.

Again, an analogous comment applies to the use of the compound channel concept for slowly time-varying probabilistic channels; see the remark (1) in Section 3.

REFERENCES

1. W.L. Root, Primitive Coding Theory for Certain Classes of Discrete Channels, Proc. Fourteenth Annual Allerton Conf. on Circuit and System Theory p. 684, 1976.
2. D. Blackwell, L. Breiman and A.J. Thomasia, The Capacity of a Class of Channels, Ann Math Stat, 30, pp. 1229-1241, 1959.
3. J. Wolfowitz, Coding Theorems of Information Theory, Springer-Verlag, Berlin, 1961.
4. W.L. Root, Estimation of ϵ -Capacity for Certain Linear Communication Channels, IEEE Trans. on Info Theory, 14, pp. 361-369, 1968.
5. J.S. Kaufman, Capacity of Classes of Time-varying Memoryless Channels, Ph.D. Dissertation, Univ. of Michigan,

Ann Arbor, 1970.

6. P. P. Varaiya and W. L. Root, Capacity of Classes of Gaussian Channels, Info and Control, 6, pp. 1350-1393, 1968.
7. K. Jacobs, Almost Periodic Channels, Colloquium Aarhus Univ., 1962.
8. R. Ahlswede, Beitrage zur Shannonschen Informationstheorie in Falle nichstationarer Kanale, Z. Wahrs Verw. Geb., 10, pp. 1-42, 1968.
9. R. Ahlswede and J. Wolfowitz, The Capacity of a Channel with Arbitrarily Varying Channel Probability Functions and Binary Output Alphabet, Lecture Notes in Math 89, Springer-Verlag Berlin, 1969.
10. J. Kiefer and J. Wolfowitz, Channels with Arbitrarily Varying Probability Functions, Info and Control, 5, pp. 44-54, 1962.
11. J. Ziv, The Capacity of the General Time Time-discrete Channel with Finite Alphabet, Info and Control, 12, pp. 233-251, 1969.
12. C. Berge, Principles of Combinatorics, Academic Press, New York, 1971.

ERRATA

1. Proposition 1 is incorrect in the generality in which it is stated. The trivial direct half of the Proposition, $C \geq R_K(1)$ is true of course, but the converse is false. The Proposition is true in the case that $N_x = N_y$ and Φ is a group; in this case the results of Proposition 3 are sufficient to validate the argument used for the converse.

The error was pointed out to the author by J. L. Massey with a simple counter example.

2. Reference 6 should be: SIAM J. Appl. Math., v 16 n 6, pp. 1350-1393, 1968.

SEQUENTIAL DECODING FOR BURST-ERROR-CHANNELS

J. Hagenauer

DFVLR, Institut für Nachrichtentechnik
D-8031 Oberpfaffenhofen, W-Germany

ABSTRACT. Sequential decoding of convolutionally encoded data cannot be used directly on channels with additional bursts of errors. A modification of sequential decoding is proposed which goes beyond simple interleaving. It uses the channel error statistics of a Markov model and a burst-tracking method for better performance. Simulation results with high rate codes and modified Fano- and Stack-decoders show the feasibility of the proposed method.

1. INTRODUCTION

Sequential decoding of convolutionally encoded data has been used very successfully on memoryless channels. However sequential decoding cannot be used directly on burst noise channels, because clusters of errors cause a sequential decoder to do an enormous amount of searching, which results in time or storage overflow. One common method of transforming bursty channels into non-bursty channels is by the process of interleaving the data before transmission over the channel - and deinterleaving before decoding. If the degree of interleaving is chosen to be much longer than the average length of a burst, the clustered errors of the same burst appear as separate single error. The usual method now considers the channel as memoryless and treats the spreaded errors of one burst as uncorrelated. However, mere interleaving is an information destroying process and it is obvious that one should use the correlation between error digits in a burst to get better performance of the decoder.

In this paper we propose in addition to interleaving a modification of sequential decoding, using a method to keep track of the bursts during decoding. The basic idea is to let the decoder utilize as much information about the channel state as it is available at the current state of decoding from the already decoded data. A convenient way to pass this information to the decoder is via the likelihood metric, which controls the search of the decoder. This controllability of the decoding process via likelihood metric gives probabilistic decoding a high flexibility.

In chapter 2 we give the definitions of the digital system and establish the Markov model for the burst channel. Chapter 3 gives the basic concepts and the likelihood calculations whereas chapter 4 deals with the practical implementation of the method for the Fano algorithm. In chapter 5 we show some results of a computer simulation of the proposed method at a coderate R = 4/5.

2. DIGITAL SYSTEM AND MARKOV BURST CHANNEL MODEL

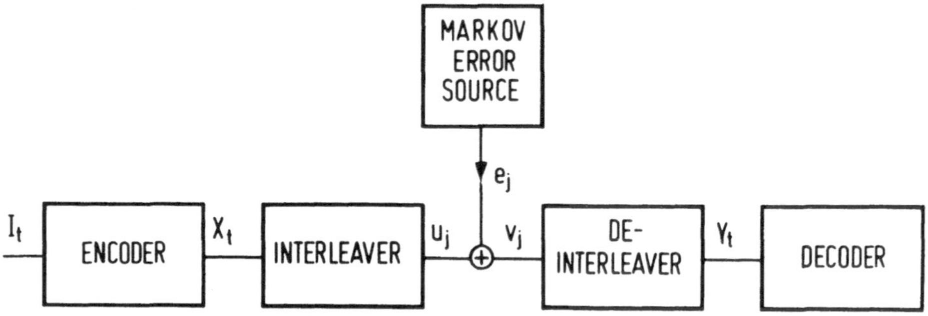

Fig. 1 - Block diagram of the digital system

Figure 1 shows the block diagram of the binary digital system. The stream of binary information digits is delivered to the encoder in blocks of K bits

$$I_t = (i_t^{(1)}, i_t^{(2)}, \ldots, i_t^{(K)}),$$

where t = 0, 1, 2, 3 is the discrete time.

The convolutional encoder with memory length M produces a subblock of length N bits

$$X_t = (x_t^{(1)}, x_t^{(2)}, \dots x_t^{(N)}).$$

In the following a systematic high rate code with $K = N - 1$ will be used, but this is no restriction for the use of other convolutional codes.

The interleaver can be thought of being a matrix with q rows and $r_o \cdot N$ columns as shown in fig. 3, each element containing one bit, which is filled with subblocks X_t row by row. When the interleaver is filled up with $q \cdot r_o$ subblocks, i.e. $q \cdot r_o \cdot N$ bits, its content is transmitted over the channel column by column. The bit stream u_j is disturbed by the additive channel noise e_j. The channel output

$$v_j = u_j \oplus e_j$$

is written into the deinterleaver matrix column by column. Initially it is assumed, that the demodulator makes hard decisions and passes no reliability information to the decoder. In moving forward the decoder will take the subblocks from the deinterleaver row by row.

The error pattern of the channel is thought to be generated by the Markov source of fig. 2. This model of a burst channel is a slight modification of the models proposed by Gilbert [3] and Elliott [4] and is a special case of Fritchman's partitioned Markov chain [5]. The four distinct channel states are denoted by

$$S_j = (b_j, c_j) \qquad b_j, c_j = 0,1. \tag{1}$$

There are two different modes of the error source, the good mode ($b_j = 0$) and the burst or bad mode ($b_j = 1$). Errors can occur in either mode ($c_j = 1$). Most of the time the channel remains at state (0, 0). With probability p_S the channel switches to the burst mode and with probability p_L it switches back to the good mode. A burst always starts and ends with an error. Being in one state of the good mode, the probability, that the next bit will be in error is $p_{GS} = p_G + p_S$, in the burst mode the corresponding probability is p_B, which is sometimes named burst severity. The model is especially meaningful, if the two modes are clearly distinct, i.e. if

$$p_{GS} \ll p_B \text{ and } p_L \ll 1 - p_B. \tag{2}$$

The state transition probabilities of the Markov models

$$p\,(l, m | l', m') = \Pr\ \{S_{j+1} = (l, m) | S_j = (l', m')\},$$

$$l, m, l', m' = 0, 1 \tag{3}$$

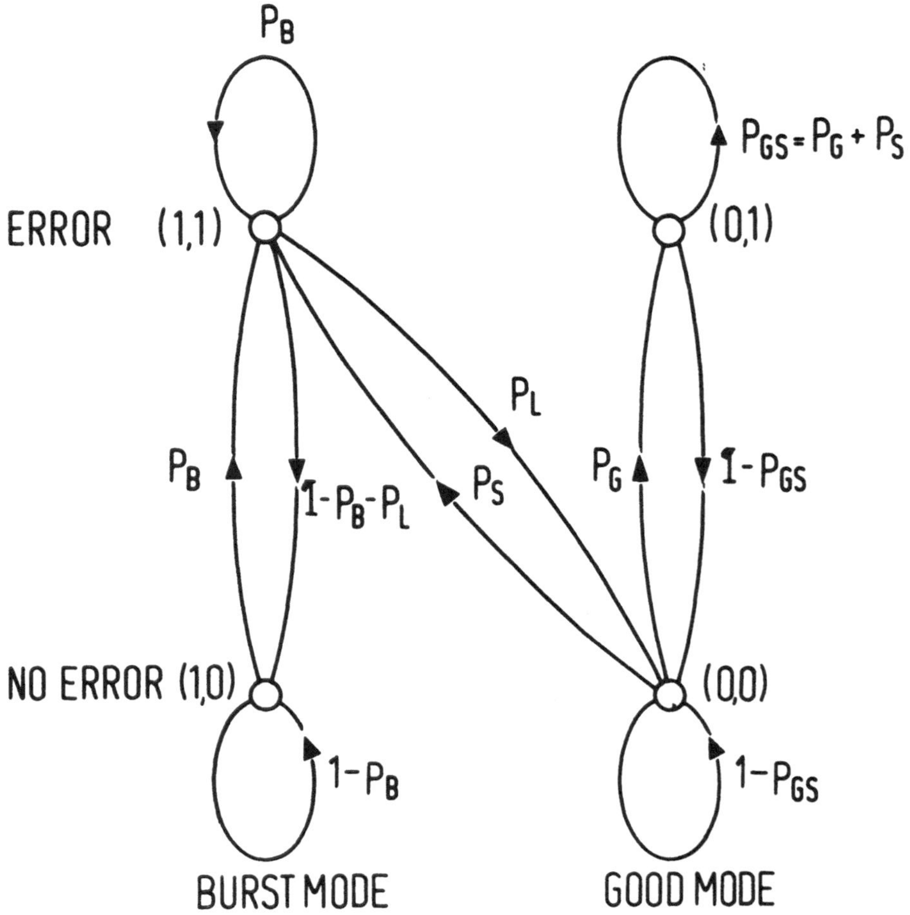

Fig. 2 - State transition diagram of the Markov error source

can be looked up in fig. 2 and lead to the transition equations

$$\begin{bmatrix} \Pr\{S_{j+1} = (1,1)\} \\ \Pr\{S_{j+1} = (1,0)\} \\ \Pr\{S_{j+1} = (0,1)\} \\ \Pr\{S_{j+1} = (0,0)\} \end{bmatrix} = \begin{bmatrix} p_B & p_B & 0 & p_S \\ 1-p_B-p_L & 1-p_B & 0 & 0 \\ 0 & 0 & p_{GS} & p_G \\ p_L & 0 & 1-p_{GS} & 1-p_{GS} \end{bmatrix} \begin{bmatrix} \Pr\{S_j\} = (1,1) \\ \Pr\{S_j\} = (1,0) \\ \Pr\{S_j\} = (0,1) \\ \Pr\{S_j\} = (0,0) \end{bmatrix} \tag{4}$$

These equations can be used to calculate the stationary probabilities Pr $\{b_j, c_j\}$, which will be used later. From the stationary probability one also obtains the average length of a burst to be equal to

$$\frac{1}{p_L} \quad \frac{(1 - p_L)}{p_B} \quad . \tag{5}$$

3. THE GENERAL METHOD

There exist several variations of sequential decoding algorithm. The reader's familiarity with either the stack algorithm or the Fano algorithm is assumed. For reference of these two algorithms see Jelinek [1], [6] or Gallager [2]. All sequential decoding algorithms attempt to follow a path in the code tree, whose metric increases in the long run. The metric associated with each path is the so called Fano metric. It measures the weighted distance between the code word $\tilde{X}_0, \tilde{X}_1, \ldots, \tilde{X}_t$ of the current path and the distorted received code word $Y_0, Y_1, \ldots Y_t$. The likelihood metric L_t can be calculated recursively

$$L_t = L_{t-1} + \lambda_t \ , \ t = 0, 1, 2, \ldots , \tag{7}$$

where the likelihood increment is

$$\lambda_t = \log \frac{\Pr \{Y_t | \tilde{X}_t \}}{\Pr\{ Y_t \}} -NR,$$

assuming that all subblocks X_t are equally likely. If interleaving is large enough to produce equally likely Y_t, one has

$$\Pr \{Y_t\} = 2^{-N}$$

and with a code of rate $R = (N-1)/N$

$$\lambda_t = 1 + \sum_{r=1}^{N} \log \Pr \{y_t^{(r)} | \tilde{x}_t^{(r)}\}. \tag{8}$$

Pr $\{\cdot\}$ is the probability, that the r-th bit of the t-th subblock in the interleaver was changed during its transmission over the channel. Assume that this bit was transmitted with index $j + 1$, which can be calculated using t, q, r_o, N.

$$\Pr\{y_t^{(r)} \mid \tilde{x}_t^{(r)}\} = \begin{cases} \Pr\{e_{j+1} = 1\} & \text{if } \tilde{x}_t^{(r)} \neq y_t^{(r)} \\ \Pr\{e_{j+1} = 0\} & \text{if } \tilde{x}_t^{(r)} = y_t^{(r)} \end{cases} \tag{9}$$

In the case of a BSC the error probability would be the crossover probability. In our case we have to determine, whether the channel was in the burst mode or in the good mode at the preceding bit with index j to get the error probability for the actual bit with index j + 1.

Let $p_{burst}(j)$ be the probability, that the Markov error source at "time" j was in the burst mode. From the Markov model we have

$$p_{burst}(j) = \Pr\{(b_j = 1)\} = \Pr\{b_j = 1,\ c_j = 1\} + \Pr\{b_j = 1,\ c_j = 0\}$$

and

$$\Pr\{e_{j+1} = 1\} = p_B\, p_{burst}(j) + p_{GS}\,(1 - p_{burst}(j)). \tag{10}$$

If there is no information obtainable about the state of the channel at "time" j, the best thing to do is to treat the interleaved channel as a BSC and to use for p_{burst} (j) the stationary probability

$$\overline{p_{burst}} = \overline{\Pr\{b_j = 1,\ c_j = 1\}} + \overline{\Pr\{b_j = 1,\ c_j = 0\}}. \tag{11}$$

With this value the error probability (10) is independent of j and the same probability is used for all binary digits.

However, the information about the already decoded subblocks can be used to trace and identify bursts in the deinterleaver matrix, in order to obtain a better estimate for p_{burst} (j). Consider the situation depicted in Fig. 3. The rows of the deinterleaver matrix are filled with the subblocks Y_t. The matrix was filled from the channel column by column. Thus a possible burst of length 2q would extend from A to B for instance. The sequential decoder moves forwards in horizontal direction from one subblock to the other. Occasionally it moves backwards to correct wrong path assumptions. Assume the decoder is located at the subblock with index t - 1 and is about to investigate for the first time the subblock with index t. In order to obtain the likelihood increment λ_t it has to evaluate eqs. (9) and (10), which requires p_{burst} (j) for all N bits of the subblock. Assume the number of columns Nr_o to be sufficiently large, so that the decoder will never backup more than r_o subblocks. (We shall see later on, that this assumption can be dropped somewhat).

Assuming no undectectable decoding error, it can be concluded, that all the subblocks with index less than $t - r_o$ are successfully decoded and the decoder knows the error pattern in this area. In fig. 3 some errors in the dashed column above subblock y_t would indicate with high probability, that the channel is in the burst mode for e_j.

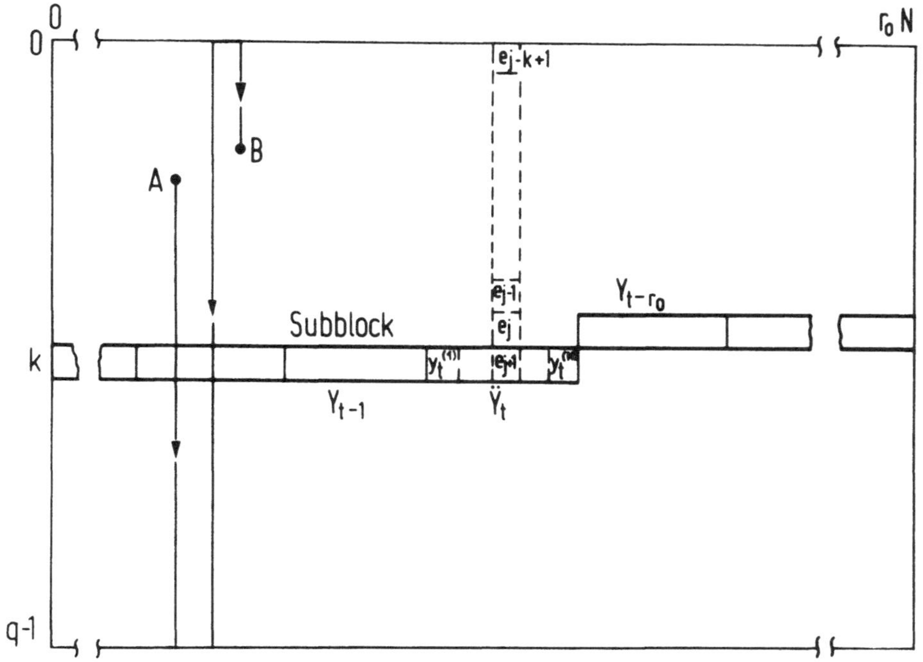

Fig. 3 - Deinterleaver matrix with possible error patterns

Using the known error pattern

$$e_j, e_{j-1}, e_{j-2}, \ldots, e_{j-k+1},$$

one is able to obtain an estimate for $p_{burst}(j)$ as the a posteriori probability (APP)

$$\beta_j = \Pr\{b_j = 1 | c_j = e_j, c_{j-1} = e_{j-1}, \ldots c_{j-k+1} = e_{j-k+1}\} . \quad (12)$$

β_j can be calculated recursively using the Markov property of the channel:

$$\beta_j = \frac{p(1, e_j \mid 1, e_{j-1}) \; \beta_{j-1} + p(1, e_j \mid 0, e_{j-1})(1 - \beta_{j-1})}{\sum_{l=0}^{1} p(l, e_j \mid 1, e_{j-1}) \; \beta_{j-1} + p(l, e_j \mid 0, e_{j-1})(1 - \beta_{j-1})} . \tag{13}$$

Since there is no preceding error pattern available in the first row, the best value for β_{j-k+1} is the a priori known stationary probability (11)

$$\beta_{j-k+1} = \overline{P_{burst}} \quad . \tag{14}$$

Now we are able to outline the operations of the decoder for calculating the likelihood increment:

1) The decoder stores the $r_o \cdot N$ a posteriori probabilities β_j; one for each bit in a row. The starting values are given by (14) and (11).

2) When the decoder reaches for the first time the subblock Y_t, it has to update the N probabilities β_{j-1} for the subblock. For each bit it has to check, whether the bits in the two lines above were in error and to calculate β_j

3) Using the APP β_j the decoder has to calculate

$$\Pr\{ e_{j+1} = 1 \} = p_B \, \beta_j + p_{GS} \, (1 - \beta_j) \tag{15}$$

to get finally with (9) the desired value λ_t from (8).

It should be noted, that step 2) is not necessary, while the decoder moves backwards and again forwards along positions, which he had already reached before. The number of updating steps equals the number of subblocks to be decoded. However, if the decoder occasionally backs up more than r_o subblocks, it may get wrong estimates, because the error pattern used for updating may have been wrong. This results in possibly wrong indications of bursts, which cause the decoder to perform more wrong searches, but it does not harm its ability to find the correct path finally, provided that $\Pr\{\text{backup} > p_o\}$ is sufficiently small.

Since the discussed updating operations and likelihood calculations are fairly complicated to be implemented in a real decoder, it is desirable to have simple suboptimal approximations, which will be discussed in the next chapter.

4. PRACTICAL IMPLEMENTATION INTO THE FANO-ALGORITHM

A way to avoid the calculation of the likelihood metric at each decoding step, is the use of precalculated tables. In order to get a finite number of tables, it is necessary to quantize β_j before calculating (15). Let the quantization of β_j be in Q steps, including the values 0 and 1. Let the code of rate $(N-1)/N$ be systematic, with 2^{N-1} branches leaving each node and with each subblock containing N-1 information bits and one parity bit. While extending a node, first the syndrome for the new subblock is calculated. The 2^{N-1} possible branches are characterized by the alterations of the received information bits, i.e.,

$$\tilde{x}_t^{(r)} = y_t^{(r)} \oplus z^{(r)} \qquad r = 1, \ldots, N-1 \tag{16}$$

Thus $z^{(r)}$, $r = 1 \ldots N-1$, is the tentative error pattern of the first N-1 digits of y_t. From the syndrome the possible error in $y_t^{(N)}$ can be determined. Thus depending on the syndrome bit, we are able to calculate two tables, each containing 2^{N-1} entries with the error pattern and the associated likelihood values. The tables should be ordered with the highest likelihood value on the top. Since each of the N bits of Y_t has assigned a quantized β_j out of Q distinct values, we get a total of $2\,Q^N$ tables.

The tables can be calculated and ordered before decoding starts, because they depend only on p_B, p_{GS} and the fixed quantization values of β_j. The decoder merely has to determine the table address, using the syndrome bit and the quantization level of β_j.

The calculation and updating of the APP β_j can be considerably simplified, if we assume for the channel parameters

$$p_{GS} \ll p_B \text{ and } p_L \ll 1-p_B$$

which we mentioned already as the condition (2) of having two sufficiently distinct modes on the channel. Such a channel is termed a "dense-burst" channel. In this case evaluation of (13) shows, that for practical purposes β_j depends almost solely on e_j and not on e_{j-1}. Furthermore, if $e_j = 1$, β_j is approximately 1 for all β_{j-1} except the case $\beta_{j-1} \approx 0$, where it is equal to p_S/p_{GS}. For $e_j = 0$ the tradeoff between β_j and β_{j-1} is not far from being linear.

These facts suggest, to use in this case a simplified burst tracking method: A simple counter $d_j = 0, 1, \ldots D-1$ called the <u>burst indicator</u> counts up and down, according to the rules

$$d_j = \begin{cases} D-1 & \text{if } e_j = 1 \text{ and } d_{j-1} \neq 0 \\ [D \cdot p_S/p_{GS}] & \text{if } e_j = 1 \text{ and } d_{j-1} = 0 \\ d_{j-1} - 1 & \text{if } e_j = 0 \text{ and } d_{j-1} \neq 0 \\ 0 & \text{if } e_j = 0 \text{ and } d_{j-1} = 0 \end{cases} \qquad (17)$$

The initial value for the first row in the deinterleaver is $[\overline{p_{burst}} \cdot D]$. The setting of the burstindicators d_j can be included into the channel model, which the decoder uses. The states $S_j = (b_j, c_j)$ of figure 2 can be thought of being split up into D substates (b_j, c_j, d_j).

From the transition matrix of the extended Markov model the stationary probabilities of the 4D mutually exclusive states can be calculated. Of special interest is the stationary probability

$$\overline{Pr} \; \{b_j = 1 \mid d_j = d\} \qquad (18)$$

which is the APP of being in a burst, when the burst indicator d_j is at certain level d. The burst indicators d = 0, 1, ... D-1 can be arranged into Q groups according to their probabilities (18) as mentioned above. In the case Q = 2 a threshold separates the high valued burst indicators which indicates burst mode, from the low valued burst indicators, which indicate good mode.

The discussed methods have been implemented into the stack and Fano algorithm. Only the modified Fano algorithm will be described here, because it is more easily implemented in hardware.

Fano Algorithm:

The Fano-type of decoder for decoding an (N-1)/N code stores for each of the r_o subblocks of a row the following information

Syndrome bit,
Extension number n_e $\quad 1 \leq n_e \leq 2^{N-1}$,
Error pattern of the subblock in the line above ,
N burst indicators.

Two pointers mark the current position of the decoder and the maximum position it ever reached. The three likelihood values L_{t-1}, L_t, L_{t+1} are stored together with the current threshold and are changed as the decoder moves forwards and backwards. In moving backwards the decoder uses the stored syndrome bit and the indicators to get the table address, and n_e to get the table entry.

When moving forward the decoder calculates the new syndrome bit and changes n_e. When it moves beyond the old maximum position, it has to update the error pattern in the line above and to set the burst indicators using this error pattern. The other rules for moving the decoder are essentially the same as for the classical Fano decoder (see for example [1]).

A remark about the threshold setting seems to be appropriate:
The operation of the decoder depends on the proper setting of the threshold, which is changed in steps of Δ T. Investigations show that a good choice for ΔT is the value of a likelihood drop caused by a single unindicated error.
The classical Fano decoder without burst indication investigates all "accessible" paths, whose likelihood lies above the current threshold T. Thus errors cause the decoder to drop the threshold and to reinvestigate the "accessible" path with likelihood metric above the lowered threshold. In our case we have indicated digits, which are in error with relatively high probability and which will generate only a slight drop of the likelihood. Consequently in most cases L_t remains above the threshold and the decoder keeps moving forwards. The indications of some digits determine an "easily accessible" part of the tree, which the decoder will investigate first, probably without dropping the threshold. Thus the "backup and run forward again " move to drop the threshold, which is characteristic for the Fano decoder, is avoided in decoding burst errors. Only if the search in the "easily accessible" subtree is unsuccessful, the decoder drops the threshold, thus allowing unindicated errors to be decoded.

Finally we point out a few additions to the proposed method, which can improve performance considerably.

1. As mentioned before, all digits in the first row of the deinterleaver are treated equally and no burst is indicated. However, there may be wrap-around-bursts, causing errors in the first row. It was found during simulation, that a considerable part of the effort to decode a frame was spent in decoding the first row. To avoid this, a preamble should be used, which involves transmitting a known sequence prior to transmitting a block of encoded data. Typically the preamble should be a few rows long and is used to initialize the first indicators. With the preamble most of the wrap-around-bursts are detected and bursty bits in the first row are indicated.

   In many transmission systems a preamble is used to provide some sort of synchronization pattern after transmitting q digits. The very same synchronization digits could be used as preamble, which would keep the overall redundancy to minimum.

2. Any available information from the analog channel or the demodulator, which indicate a burst with some probability, i.e. signal dropout or loss of synchronization can be marked in the deinterleaver. During decoding these markings can be used to set a bias for the burst indicators, in order to 'tell' the decoder about a higher probability of bursts.

3. Other ways of interleaving can be used to spread long wrap-around-bursts over distinct subblocks. As an example the columns of the interleaver can be sent over the channel as the sequence 0, N, 2N, ...$(r_o-1)N$; 1,N + 1, ... Normally it will be easier for the decoder to decode two indicated bits contained in two subblocks, rather than in one subblock.

5. SIMULATION RESULTS

The binary system of Fig. 1 was simulated in PL/1 on an IBM 370/168 computer. A high rate systematic convolutional code with rate R = 4/5 (N = 5) and memory M = 25 (total constraint length 104) was used. The code was constructed by computer search with the aim to achieve a rapidly growing column distance function [7], [8], which enables the decoder to reject wrong paths quickly. Simulating the channel conditions, a Markov error source generated error patterns with the parameters

$$p_G = p_S = 10^{-4}, \; p_B = 0.49, \quad p_L = 3.92\cdot 10^{-2}. \tag{19}$$

These parameters yield an average burst length of 50 bits. The interleaver as well as the deinterleaver have r_o = 50 subblocks per row, meaning 250 columns, and the number of rows was varied between q = 50 and q = 1,000 i.e. 1 to 20 times the average burst length. In order to compare frames of the same size, one frame consisted of Γ = 2495 subblocks each containing N = 5 bits, followed by a tail of M = 25 parity bits, which is required to terminate the tree search. The M parity bits are generated by the encoder using M known (i.e., all zero) information blocks, which need not to be transmitted. Thus a frame occupied 50 rows of the deinterleaver and the number of frames per interleaver varied between 1 and 20.
A preamble of q_{pre} = 5 rows was used for some runs. In this case each block of q bits was preceded by a known (i.e., all zero) pattern of 5 bits before sending it over the channel.

It was found that the quantization of the burst indicators with Q >2 resulted only in a slight reduction of the decoding effort.

Therefore Q = 2 was used, requiring 64 pre-calculated tables with the likelihood values. Furthermore a D = 8 level burst indicator was found to be sufficient, with setting rules as described in Chapter 4.

The amount of "work" to decode a frame was monitored. For the Fano-type decoder, the number of decoder moves (forward or backward), was counted. We define the relative additional work

$$w = \frac{\text{number of moves to decode a frame}}{\text{number of moves to decode an error free frame}} - 1$$

$$w = \frac{\text{additional number of moves}}{\Gamma + M} \quad . \qquad (20)$$

The number of moves is identical with the number of nodes the decoder visited. Of special interest are the average $\bar{w}$ and the work distribution Pr $\{ w > w_o \}$, which, as a function of w_o on a log-log scale, shows the characteristic Pareto behaviour.

For each of the distributions shown in the figures between 360 and 500 frames were decoded, meaning between $4.5 \cdot 10^6$ and $6.25 \cdot 10^6$ transmitted binary digits. As the figures show, in most cases no undetectable errors were recorded. It is well known, that the main problem with sequential decoding is not to reach low error rates, but to avoid buffer or time overflow. A few frames of the Fano decoder are registered as erasure frames. An erasure frame occurs, when the decoder backs up more than r_o subblocks, because in this case the memory with the path information is empty.

Figure 4 shows the influence of the interleaving constant q. Including the preamble interleaving was varied between 1.1 and 20.1 times the average burst length. The average work is only a few percent more than the work in the error free case and is not as important as the work distribution. Concluding from the shown upper part of the distribution, interleaving of 3 to 5 times the average burst length seems to be sufficient.

Figure 5 shows the influence of the preamble, which is especially observable, when interleaving is small. Both distributions were obtained by decoding the same frames, but in case B) the preamble information was not passed to the decoder.

In figure 6 we show the advantages of the proposed method, when compared with the usual method, which treats the interleaved channel as a BSC.

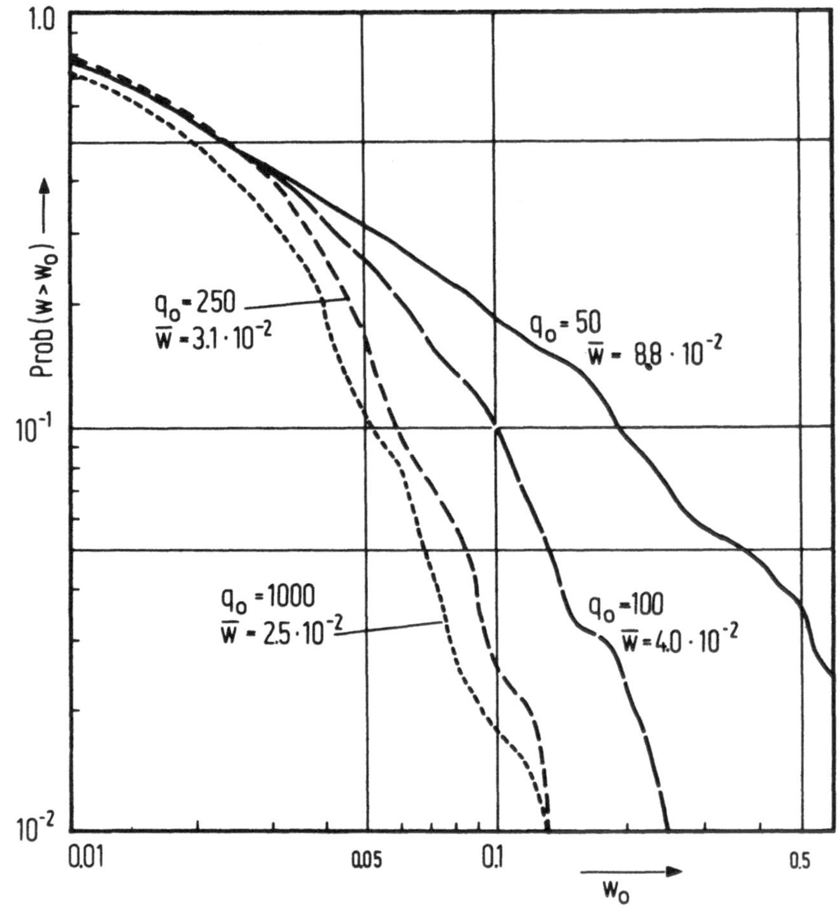

Fig. 4 - Fano-Decoder:
Distribution Prob $(w > w_0)$, eq. (20) of the relative additional work to decode a frame. Coderate 4/5; frame size 50 x (5 x 50) bits, Interleaver: 5 x 50 columns, q_0 rows, 5 rows preamble. Channel statistics as in eq. (19), approx. 450 frames or $5.6 \cdot 10^6$ bits per curve; no decoding error, 1 erasure frame at $q_0 = 50$

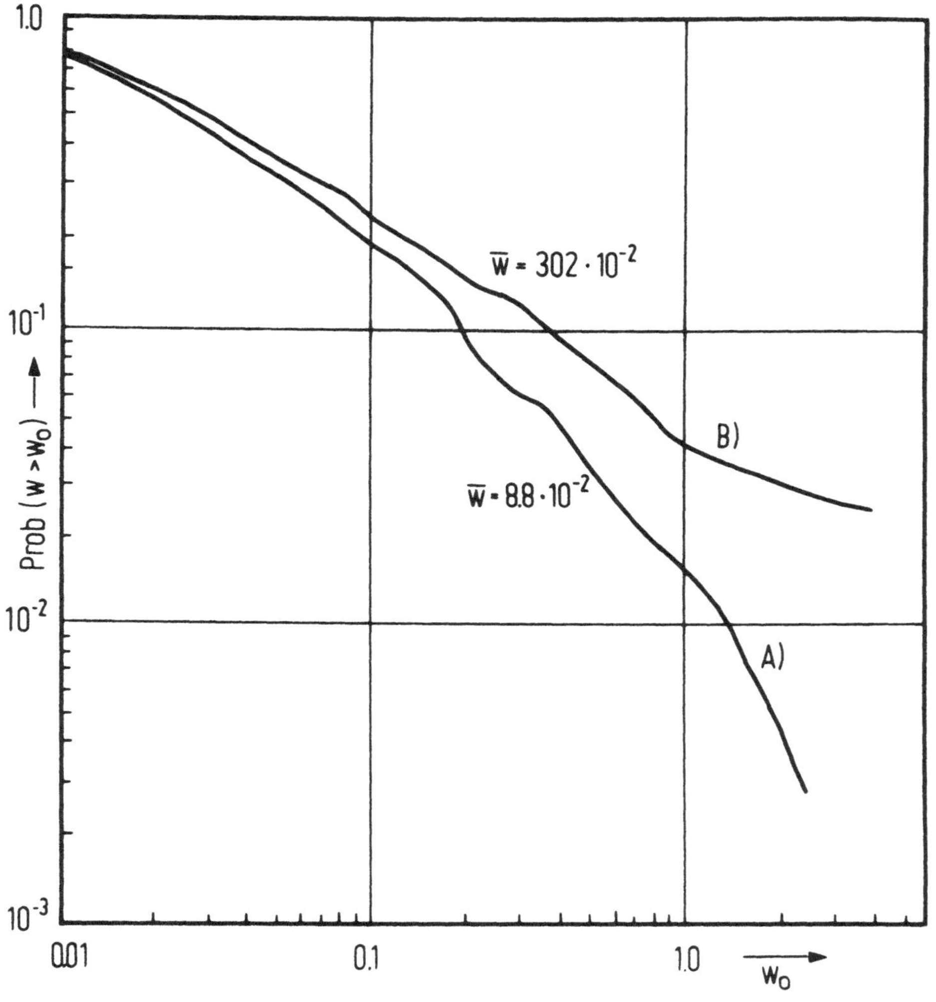

Fig. 5 - Fano-Decoder:
Distribution Prob($w > w_0$), Parameters as in fig. 4,
$q_0 = 50$: A) 5 lines preamble uses: 1 erasure frame, no error
B) no preamble used: 7 erasure frames, 7 errors.

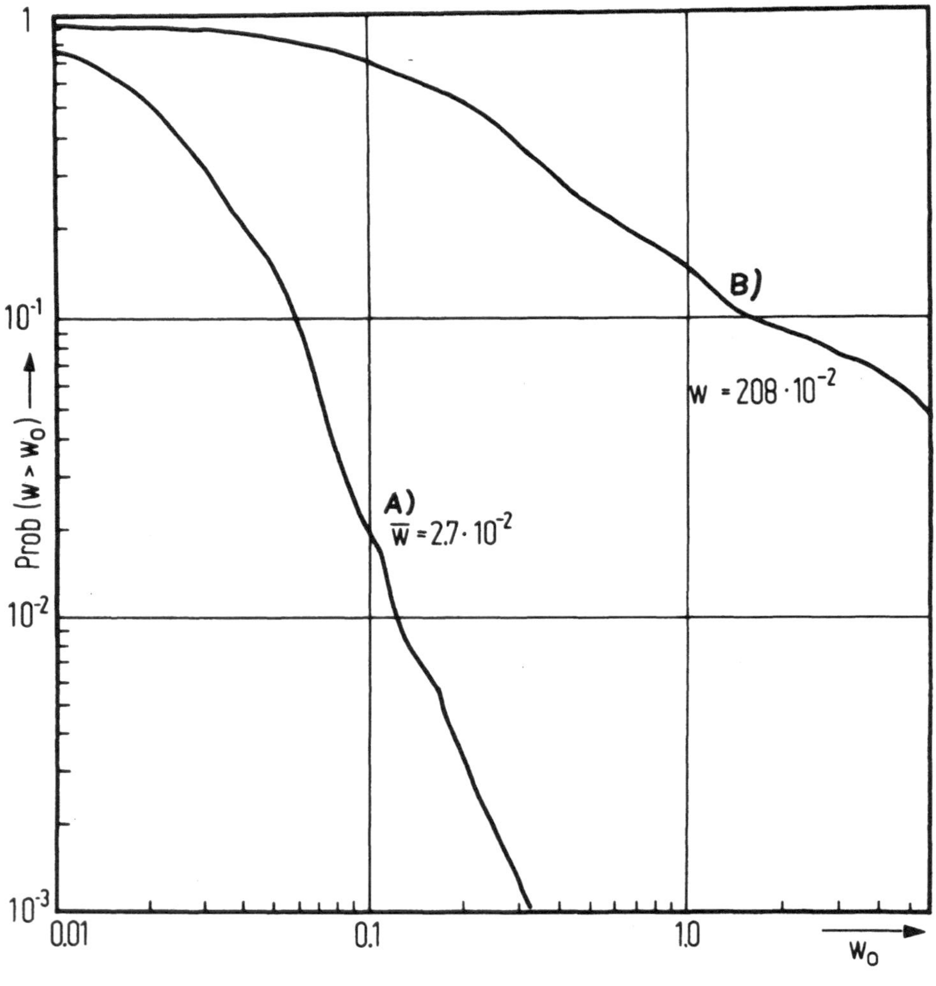

Fig. 6 - Fano-Decoder:
Distribution Prob$(w > w_0)$ same parameters and error patterns as in fig. 4, $q_0 = 100$, columns transmitted as the sequence 0, N, 2N, ..., 1, N+1, ...

A) proposed method with burst indicators; no erasures, no errors
B) simple interleaving without burst indicators; 21 erasure frames, 20 errors

ACKNOWLEDGEMENT

This work has been done at the IBM Th. J. Watson Research Center, Yorktown Heights, NY. The author is grateful to Drs. L. Bahl, J. Cocke and C.D. Cullum for their contributions and helpful discussions.

REFERENCES

[1] F. Jelinek, Probabilistic Information Theory, McGraw Hill, New York, 1968.

[2] R.G. Gallager, Information Theory and Reliable Communication, John Wiley, New York, 1968.

[3] E.N. Gilbert, Capacity of a Burst-Noise Channel, B.S.T.J., 39, 1960, pp. 1253-1265.

[4] E.O. Elliot, Estimates of Error Rates for Codes on Burst-Noise Channels, B.S.T.J., 42, 1963, pp. 1977-1998.

[5] B.B. Fritchman, A channel characterization using partitioned Markov Chains, IEEE Trans. on Inf. Theory, Vol. IT-13, pp. 221-227, Apr. 1967.

[6] F. Jelinek, Fast Sequential Decoding Algorithm using a Stack, IBM J. Res. Develop., vol. 13, pp. 675-685, Nov. 1968.

[7] P.R. Chevillat and D.J. Costello, Jr., Distance and Computation in Sequential Decoding, IEEE Trans. on Comm., Vol. COM-24, April 1976, pp. 440-447.

[8] J. Hagenauer, High Rate Convolutional Codes with Good Distance Profiles, to appear at IEEE Trans. on Inf. Th. Vol. IT-23, Sept. 1977.

SYNCHRONIZATION RECOVERY AND ERROR CORRECTING CODES

J.C. Prabhakar

Department of Electrical Engineering, Texas Tech University, Lubbock, Texas 79409 USA

ABSTRACT. This work presents a technique to construct code dictionaries capable of correcting single amplitude or single synchronization errors in any block of length equal to code length (n). An M-derivative of the code vector is defined. Both the code word and its derivatives are said to belong to a particular differential group formed under a defined criterion. Maximal differential groups (MDG) have n + 1 members. Bounds on the number of MDGs are presented. Code dictionaries are constructed from minimum valued members from each MDG. The transmitted code consists of a code vector followed by its M-derivative.

The decoding procedure is discussed in a number of sequential steps to compare the received sequences with stored codes of the dictionary and to account for the following situations. (1) Minimum distance of zero, (2) Minimum distance of one and (3) Minimum distance of more than one.

1. INTRODUCTION

A typical binary data communication system consists of a source, a source encoder, channel encoder, and a modulator transmitting messages via a communication channel (corrupted with noise) followed by a demodulator, channel decoder, receiver decoder and the receiver, all connected in cascade. These auxilliaries are needed to offset the disruptive effects of the channel. The noisy channel may cause amplitude errors in which a zero may be received as a one and vice-versa. Coding schemes to correct such errors have been investigated at great length. One of the best known among these is the class of cyclic (group) error-correcting codes due to Bose-Chaudhury and Hocquenghem (BCH) which can correct an arbitrary number of amplitude errors with a specified low probability of error within the framework of Shannon's Second Theorem.

The noise in the channel may cause another serious kind of error, i.e., synchronization error. The fact is that the individual symbols have physical meaning to the receiver only when considered together with certain other symbols of the sequence. The sequence must be correctly grouped into "words" or "blocks" in order for the receiver to understand the message. A deletion (insertion) of one or more symbols from any block will cause a forward (backward) drift in the following symbols of that and of the following blocks. When the error is such that the receiver incorrectly groups symbols into words, a synchronization error is said to have occurred.

1.1 Literature review

The previous work in the field of synchronization error correction may be divided into two categories, codes of variable length and codes of the fixed length. Codes of first variety with the property of correcting synchronization errors have been designed by Gilbert and Moore (1) and , Neuman (2), (3). Neuman has in fact proposed codes which can correct the synchronization errors or the amplitude errors. However, no specific proposal for correcting the synchronization error is given. Variable length codes capable of detecting and correcting synchronization errors have also been presented by T.R. Hatcher (4) and R.A. Scholtz (5). Fixed length synchronizable codes have been discussed by Golomb, Gordon, Welch (6), Gilbert (7), and Eastman (8). Golomb has developed "Comma Free Codes" with the property that any code word received under de-synchronized conditions would not be a code in the defined code dictionary. Gilbert has presented comma free codes by beginning each code word with a particular "synchronizing prefix," while ensuring that the sequence of any genuine code word does not have the prefix at any place within it, except at the beginning of each code word. Eastman and Even (9) have developed codes capable of detecting synchronization errors. Sellers (10) has given a class of codes which detects and corrects single and multiple adjacent synchronization errors and, in addition, corrects a burst of amplitude errors surrounding the position of synchronization error. Jeffery D. Ullman (11) has developed codes capable of correcting single synchronization error, which have redundancy close to the minimum possible. L. Calabi and W.E. Hartnett (12) and, R.C. Bose and J.C. Caldwell (13) have also designed fixed length codes, capable of correcting synchronization errors.

1.2 Problem statement

The synchronization errors are more serious than the substitution errors because the latter affect only the code word in which the error has occurred. A deletion or insertion of a digit in a code word, on the other hand, shifts the following words one digit forward or backward causing all these to be received erroneously. In this work, code dictionaries are so designed as to have an intrinsic property of detection and correction of single synchronization errors. This is achieved even if there exists, in addition, at most one additive error in any block of length equal to the code word length (n).

2. SYNCHRONIZATION RECOVERY AND ERROR CORRECTING CODES

Let us define the M-derivative of any code $V = V_1V_2V_3 \ldots. V_n$ as $V^1 = V_1^1V_2^1V_3^1 \ldots. V_n^1$ where $V_i^1 = V_i \oplus V_{i+1} \oplus \ldots. \oplus V_n \oplus V_1^1 \oplus V_2^1 \oplus \ldots \oplus V_{i-1}^1$ the operator $\oplus$ being the mod 2 adder. Thus the M-derivative of 01011 is 10101.

<u>Theorem 1.</u> Distinct code vectors have distinct M-derivatives.

Proof (By contradiction): Let $V^1 = V_1^1V_2^1V_3^1 \ldots. V_n^1$ be the M-derivative of two distinct code words $V = V_1V_2 \ldots. V_n$ and $U = U_1U_2 \ldots. U_n$. Then

$$V_i^1 = V_i \oplus V_{i+1} \oplus V_{i+2} \oplus \ldots. \oplus V_n \oplus V_1^1 \oplus V_2^1 \ldots. \oplus V_{i-1}^1$$
$$= U_i \oplus U_{i+1} \oplus U_{i+2} \oplus \ldots. \oplus U_n \oplus V_1^1 \oplus V_2^1 \ldots. \oplus V_{i-1}^1$$

Since this is true for all i, for i = n this results in $U_n = V_n$. Also i = n-1 leads to $U_{n-1} = V_{n-1}$. Thus $U_i = V_i$ for all i (by induction) which contradicts our assumption.

<u>Theorem 2.</u> (If $V = V^i$ for any i and any V, then V_s^j are said to belong to the same differential group, $j < i$). Each differential group has at most (n + 1) elements where n is the length of the code.

Proof: Let V be any code vector, such that

$$V = V_1V_2V_3 \text{ ------------------------ } V_n$$

the M-derivative of V, U, will be

$$V^1 = U = U_1U_2U_3 \text{ ------------------ } U_n$$

From the definition of the M-derivative,

$$U_1 = V_1 \oplus V_2 \oplus V_3 \oplus V_4 \text{ -----------} \oplus V_n$$

$$U_2 = (V_2 \oplus V_3 \oplus V_4 \text{ ---------------} \oplus V_n) \oplus V_1 \oplus (V_2 \oplus V_3 \oplus \text{-------------------------} \oplus V_n).$$

But since $\oplus$ is addition mod 2, 1 + 1 = 0 + 0 = 0, one gets

$$U_2 = V_1$$

Similarly,

$$U_3 = (V_3 \oplus V_4 \oplus V_5 \text{ ---------------} \oplus V_n) \oplus U_1 \oplus U_2.$$

Substitution for U_1 and U_2 yields $U_3 = V_2$.

Proceeding on the same lines as above it can be shown that,

$$U_n = V_{n-1}$$

Thus $V^1 = U = U_1V_1V_2V_3 \text{ ---------------- } V_{n-1}$

and $V^2 = U^1 = W = W_1W_2W_3 \text{ ------------- } W_n$

$$= W_1U_1V_1V_2V_3 \text{ --------- } V_{n-2}$$

where $W_1 = U_1 \oplus V_1 \oplus V_2 \oplus V_3 \oplus \text{---------} \oplus V_{n-1}$

Substituting for U_1 one gets,

$$W_1 = V_1 \oplus V_2 \oplus V_3 \oplus V_4 \text{ ----------} \oplus V_n \oplus V_1 \oplus V_2 \oplus \text{---} \oplus V_{n-1} \oplus V_1 = V_n.$$

So, $V^2 = V_n U_1 V_1 V_2 V_3$ --------------- V_{n-2}

$V^3 = V_{n-1} V_n U_1 V_1 V_2$ ------------- V_{n-3}.

So, a general formula for r-th M-derivative is found as,

$$V^r = V_{n-r+2} V_{n-r+3} \text{-----} V_n U_1 V_1 V_2 \text{-----} V_{n-r}.$$

So the n-th M-derivative will be,

$$V^n = V_2 V_3 V_4 \text{------} V_n U_1$$

and the (n + 1)st M-derivative will be

$$V^{n+1} = V_1 V_2 V_3 V_4 \text{----------} V_n.$$

But this is the code vector itself, hence $V = V^{n+1}$ and one has a maximal differential group (MDG).

Theorem 3. For odd n, the code vector having identical sequence of (n-1)/2 bits about the central bit (0 or 1) does not belong to MDG. The proof is easy to construct and is omitted to conserve space.

Corollary 1: For odd n, the number of MDGs, D(n) is bounded by

$$D(n) \leq \frac{2^n - 2^{(n+1)/2}}{(n+1)}.$$

Proof: The number of code vectors having identical sequences of (n-1)/2 bits about the central bit which is fixed at 0 is $2^{(n-1)/2}$. Also the number of code vectors with identical sequences about the central bit, when it is fixed at 1, is $2^{(n-1)/2}$. Hence the total number of the code vectors which are not in the maximal differential group will be $2^{(n+1)/2}$. The total number of n-tuples is 2^n. Thus the maximum number of code vectors which can belong to the maximal differential group will be $2^n - 2^{(n+1)/2}$, and since each maximal group has n+1 members, the proof directly follows.

Corollary 2. For even values of n, the number of maximal groups, D(n), is bounded by,

$$D(n) \leq \frac{2^n-1}{n+1} .$$

The sequence 00 ----------- 00 cannot form a MDG. Hence.

Table I gives the number of maximal differential groups for different values of n (code vector length).

Table II shows a code dictionary formed by selecting one minimum valued element from each of the Maximal Differential Groups. If the number of digits in the message sequence is seven, the number of MDGs will be 14 (Table I). Each message code is followed by its M-derivative making a 14-digit code word to be transmitted.

TABLE I

NUMBER OF MDGs FOR DIFFERENT CODE VECTOR LENGTHS

n	D(n)
2	1
3	1
4	3
5	4
6	9
7	14
8	28
9	48
10	93
11	165
12	315
13	585

TABLE II

A CODE WORD DICTIONARY OF LENGTH 14

0	0	0	0	0	0	1	1	0	0	0	0	0	0
0	0	0	0	0	1	0	1	0	0	0	0	0	1
0	0	0	0	1	0	0	1	0	0	0	0	1	0
0	0	0	0	1	1	1	1	0	0	0	0	1	1
0	0	0	1	0	1	1	1	0	0	0	1	0	1
0	0	0	1	1	0	1	1	0	0	0	1	1	0
0	0	0	1	1	1	0	1	0	0	0	1	1	1
0	0	1	0	0	1	1	1	0	0	1	0	0	1
0	0	1	0	1	0	1	1	0	0	1	0	1	0
0	0	1	0	1	1	0	1	0	0	1	0	1	1
0	0	1	1	0	1	0	1	0	0	1	1	0	1
0	0	1	1	1	1	1	1	0	0	1	1	1	1
0	1	0	1	1	1	1	1	0	1	0	1	1	1
0	1	1	0	1	1	1	1	0	1	1	0	1	1

Code Transmitted

2.1 Encoding - decoding procedures

The encoder simply finds the M-derivative of the information sequence and transmits it immediately following the information sequence. The errors in the received stream of symbols or sequence of symbols are assumed separated, at least, by a distance equal to the code word length, say n. The code dictionary is stored in the memory of the decoder. When the receiver starts receiving the message, the decoder shall synchronize it with the transmitter right from the first complete and correct code word received.

Assuming that the receiver is synchronized with the transmitter, the decoding procedure is discussed in the following steps.

(1) Compare the received sequence of first n symbols with the codes in the dictionary. If the minimum distance is zero, go to step 2; if one, go to step 3; if more than one, go to step 4.

(2) There are three possibilities. (a) No error. (b) An insertion of identical bit in last run. (c) A deletion of a bit from the last run and the first symbol of the next code word is identical to the symbols in the last run.

In any of the above cases the received code word is assumed to be correct. Since it has been assumed that the code word is correct, in the case of (b) and (c), the error is shifted to the next code word and if no other error occurs in the next code word this shifted error will be corrected as a single error in the next code word. But if another error occurs in the next code word then the error does not fall under single amplitude or single synchronization error and so will be considered separately.

(3) There are three possibilities. (a) Single amplitude error (No sync. error). (b) An insertion of a bit in the last run which is different from the bits in that run. (c) A deletion of a bit from the last run and when the first bit of the next code word is different from the bits in the last run.

In any of the above cases the received code word is corrected for single amplitude error. Since it has been assumed that a single amplitude error occurred, in the case of (b) and (c), the error is shifted to the next code word and if no error occurs in the next code word this shifted error will be corrected as single synchronization error in the next code word. But if another error occurs in the next code word then the error would not fall under single amplitude or single synchronization error and so will be treated separately.

(4) Check the sequence of redundant bits excluding the last bit i.e., a sequence of (n/2 - 1) bits. If this sequence is the prefix of any code word in the dictionary then go to step 5, otherwise go to step 6.

(5) The single deletion error has occurred in the first half of the code word, and the correct code word is that for which the prefix is the (n/2 - 1) sequence of redundant bits (of the received code word). Compare the received code word with the correct code word and insert a bit before the position where the received and the correct code words differ. The inserted bit has to be the differing bit in the correct code word.

(6) Check whether n/2 bit prefix of the received code word is a prefix of any code word in the code dictionary. If it is, go to step 7; if not, go to step 8.

(7) Check for single deletion error in the second half of the received code word, by comparing the received code word with the code word in the dictionary for which the condition in step 6 is satisfied. Insert a bit in the received code word before the place where the received code word and the code word in the dictionary differ. The inserted bit has to be the differing bit of the correct code word. Now compare the corrected code word with the correct code word in the dictionary. If they are identical the synchronization error has been corrected; if not, the error was not a single deletion in the second half of the received code word. Restore the original received code word and go to step 8.

(8) Check (n/2 - 1) bit sequence starting from the third redundant bit of the received code word (the last bit of this sequence will be the first bit of the next code word). If this sequence is a prefix of any code word in the dictionary, go to step 9; if not, go to step 10.

(9) A single insertion error in the first half of the received code word is detected. Delete a bit from the received code word by comparing it with the code word in the dictionary which satisfied step 8. The deleted bit has to be the first bit in the received code word which differs from a bit in the correct code word in the dictionary.

(10) Check n/2 bit prefix of the received code word. If it is a prefix of any code word in the dictionary, go to step 11; if not, go to step 12.

(11) A single insertion error in the second half of the received code word. Delete a bit from the received code word by comparing it with the code word in the dictionary which satisfies step 10. The deleted bit has to be the first bit in the received code word which differs from a bit in the correct code word in the dictionary.

(12) The received code word has two errors, a synchronization error at the first place which has been transferred from the previous code word, and one error in the second half of the received code word.

Check n/2 bit sequence, starting from the second bit of the received code word. If it is a prefix of any code word in the dictionary, go to step 13; if not, go to step 14.

(13) The first bit has been inserted, delete it and go to step 1 to correct the second error.

(14) First bit has been deleted. Since every code word in the dictionary begins with 0, insert a 0 at the first place and go to step 1 to correct the second error.

This completes the single error detecting and correcting procedures.

REFERENCES

1. Gilbert, E. N. and Moore, E. F. "Variable length binary encodings." Bell Systems Technical Journal, vol. 38, pp. 933-967, July 1959.
2. Neumann, P. G. "Efficient error-limiting variable-length codes." IRE Trans. on Information Theory, vol. IT-8, pp. 292-304, July 1962.
3. Neumann, P. G. "Error limiting coding using information-lossless sequential machines." IEEE Trans. on Information Theory, vol. IT-10, pp. 108-115, April 1964.
4. Hatcher, T. R. "On a Family of Error-Correcting and Synchronizable Codes." IEEE Trans. on Information Theory, vol. IT-15, pp. 620-623, September 1969.
5. Scholtz, R. A. "Codes with Synchronization Capability." IEEE Trans. on Information Theory, vol. IT-12, pp. 135-142, April 1966.
6. Golomb, S. W., Gordon, B., and Welch, L. R. "Comma free Codes." Canadian J. Math., vol. 10, pp. 202-209, 1958.
7. Gilbert, E. N. "Synchronization of Binary Messages." IRE Trans. on Information Theory, vol. IT-6, pp. 470-477, September 1960.
8. Eastman, W. L. "On the construction of Comma free Codes." IEEE Trans. on Information Theory, vol. IT-11, pp. 263-267, April 1965.
9. Eastman, W. L. and Even, S. "On synchronizable and PSK-synchronizable block codes." IEEE Trans. on Information Theory, vol. IT-10, pp. 351-356, October 1964.
10. Sellers, F. F., Jr. "Bit loss and gain correction codes." IRE Trans. on Information Theory, vol. IT-8, pp. 35-38, January 1962.
11. Ullman, J. D. "Near-optimal, single-synchronization-error-correcting codes." IEEE Trans. on Information Theory, vol. IT-12, pp. 418-424, October 1966.
12. Calabi, L. and Hartnett, W. E. "A family of codes for the correction of substitution and synchronization errors." IEEE Trans. on Information Theory, vol. IT-15, pp. 102-105, January 1969.
13. Bose, R. C. and Caldwell, J. C. "Synchronizable Error Correcting Codes." Information and Control, vol. 10, pp. 616-630, 1967.

Part V

INTEGRATED CODING AND RATE DISTORTION THEORY

organised by Dr. P. G. Farrell

JOINT SOURCE AND CHANNEL CODING*

James L. Massey

Visiting Professor of Electrical Engineering
and Computer Science
Massachusetts Institute of Technology
Cambridge, Massachusetts U.S.A.

ABSTRACT. The advantages and disadvantages of combining the functions of source coding ("data compression") and channel coding ("error correction") into a single coding unit are considered. Particular attention is given to linear encoders, both for sources and for channels, because their ease of implementation makes their use desirable in practice. It is shown that, without loss of optimality, a joint source/channel linear encoder may be used when the goal is the distortionless reproduction of the source at the destination. On the other hand, it is shown that in general there is an inherent and significant loss of optimality if a joint source/channel linear encoder is used when the goal is relaxed to reproduction of the source within some specified non-negligible distortion.

1. INTRODUCTION

Our aim in this tutorial paper is to treat the separability of the two basic coding functions that arise in communications, namely source coding and channel coding, first in the general case and then in the important practical case when these functions are both linear. We shall find that the desirability of joint linear source/channel coding is closely (and, to us, surprisingly) linked to the degree of fidelity specified in the reconstruction of the source at the destination.

---

*This research was supported by the Office of Naval Research under Contract ONR-N00014-64-C-1183.

The model of a communications system with separate source and channel coding is shown in Fig. 1.

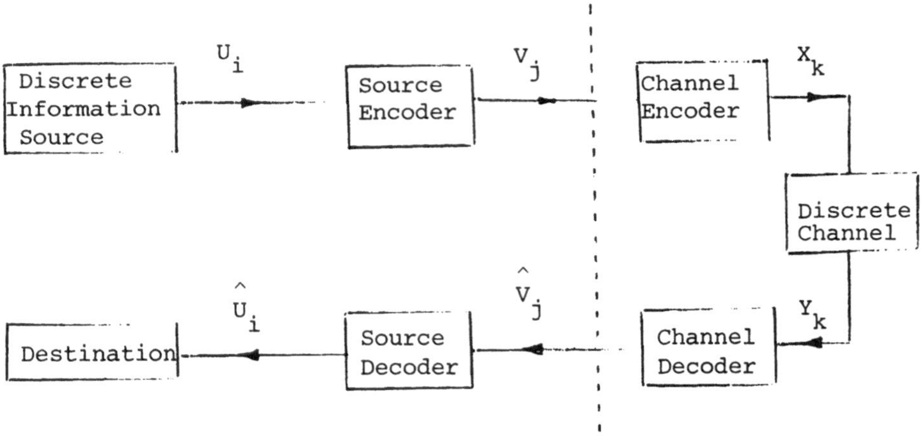

Fig. 1 A Digital Communications System with Separate Source and Channel Coding

It will be noted that there are three different subscripts on the various symbols shown in Fig. 1, namely, i, j, and k. We use this artifice to distinguish between sequences that may not be equi-numerous over a long time inverval. For instance, there may be more source output digits per second, say, than encoded source digits per second--in fact, we hope that there are many more so that the source encoder is doing well its task of "data compression". Also for instance, there may be fewer encoded source digits per second than encoded channel digits per second--we may be forced into this situation by the need to insert redundancy into the channel input digits so that the channel decoder can do well its task of "error correction".

Roughly speaking, we may use the terms "source coding", "data compression", and "redundancy removal" as synonymous. Again roughly speaking, we may use the terms "channel coding", "error correction", and "redundancy insertion" as synonymous. A wag might accuse the International Brotherhood of Information Theorists of featherbedding: it provides jobs for those who take out redundancy and jobs for those who put redundancy back in, at least when source coding and channel coding are performed separately as shown in Fig. 1. But it is a serious question to ask whether one box, a "joint source/channel encoder" as shown in Fig. 2, couldn't do a better job (or at least do the same job more economically) than does the tandem combination of the "source encoder" and "channel encoder" boxes in Fig. 1. As we shall soon be seeing, this simple question has a rather complicated answer.

In fact, one of the important results in Shannon's celebrated 1948 paper[1] was his demonstration that the source and channel coding functions are fundamentally separable in the sense that, without loss of efficiency in the use of a given channel to transmit a given source

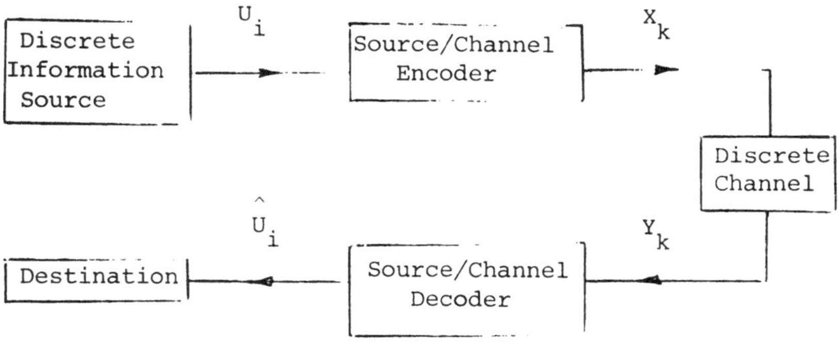

Fig. 2 A Digital Communications System with Joint Source/Channel Coding

with some specified fidelity to a destination, these two coding subsystems can be designed entirely independently. One can always design an optimum system by combining (1) a source encoder which has been designed to transform (at least, approximately) the source output into a stream of independent binary digits, each equally likely to be a 0 or a 1, and (2) a channel encoder which has been designed quite independently of the actual statistics for its input binary digits (i.e., has been designed for use with a maximum-likelihood decoder). Fano[2] has aptly commented on the significance of this fundamental separability: it means that those parts of the communications system to the right of the dashed line in Fig. 1 can always be designed, with no loss of optimality, as a system to transmit binary digits reliably. Binary digits are a kind of standard interface between the source coding world and the channel coding world, and one pays no surtax in efficiency for crossing at this interface.

As characteristic as the generality of the above-stated separability result of Shannon is the fact that his 1948 paper gives little clue as to how complex an efficient communications system becomes when the source and channel coding functions are separated as in Fig. 1. With tongue-in-cheek, we now assert:

Theorem 1: For a given efficiency (measured in number of source letters transmitted per use of the channel and fidelity (measured in the quality of the source reproduction at the destination) achievable by separate source and channel coding for a given source and a given channel, there always exists a joint source/channel coding scheme for the same source and channel that is at least as efficient, that gives at least as much fidelity, and is no more complex than the separate coding system.

Proof: Let Fig. 1 be a diagram of the hypothesized separate system. Then, in Fig. 1, draw a large box to enclose the "source encoder" and "channel encoder". Draw a second such box to enclose the "channel decoder" and "source decoder". Call the first new box the "source/channel encoder" and call the second new box the "source/channel decoder". You have just constructed a joint source/channel coding system that satisfies the assertion in the theorem. (Naturally, you might be able to build a simpler joint system that works at least as well; in fact, you might be able to build a far simpler system!)

Its triviality not withstanding, Theorem 1 does illuminate the chief attractive feature of joint source/channel coding, namely, the possible reduction in complexity compared to a similarly-performing system with separate source and channel coding. We will pursue this point further, but not without first giving a caveat: the reduction in complexity is purchased by a loss in flexibility! If one opts for a jointly coded system, he can no longer easily adapt his system later to a different source; in the separately designed system, one could continue to use the same channel coding subsystem, changing only the source encoder to the source encoder matched to the new source. Telephone companies worldwide are beginning to experience how painful this loss of flexibility can be. Most telephone systems were originally designed as a joint source/channel coding system (even if the designers were unawares that they were doing "coding") for transmitting the voice source over a narrowband channel. As more and more of their customers are changing from voice sources to data sources, the telephone companies are madly scrambling to adapt their communications brontosaurus to its new environment.

2. DEFINITIONS AND PRELIMINARIES

So that we can begin to speak more precisely as engineers should, we state here a few definitions.

A binary memoryless source (BMS) with parameter q is a device whose output is a sequence $U_1, U_2, U_3, \ldots$ of statistically independent, binary-valued random variables such that

$$P(U_i = 1) = 1 - P(U_i = 0) = q, \quad \text{all } i.$$

This is the only source that we shall consider hereafter; it is general enough for all our purposes even if it is a realistic model of only few actual information sources. When $q = 1/2$, the BMS is called the binary symmetric source (BSS); this very special type of BMS will play a key role in what follows. In fact, the goal of the source/encoder in Fig. 1 is to make its output a good approximation to the output of a BSS.

A binary symmetric channel (BSC) with cross-over probability p is memoryless channel which accepts binary digits at its input and emits binary digits at its output according to the following conditional probabilities:

$$P(Y = 1 \mid X = 0) = P(Y = 0 \mid X = 1) = p$$

$$P(Y = 1 \mid X = 1) = P(Y = 0 \mid X = 0) = 1 - p.$$

Again, although the BSC is a realistic model for only a few actual discrete channels, it is general enough for our purposes.

Next, we recall some well-known results from information theory[1,2,3,4].

Let $h(x) = - x \log_2 x - (1 - x) \log_2 (1 - x)$ (where $0 \leq x \leq 1$) be the usual binary entropy function. Then the entropy (or "rate") of the BMS is given by

$$H(U) = h(q) \quad \text{bits/letter}$$

where "letter" means a binary digit emitted by the source. According to Shannon's Noiseless Coding Theorem, H(U) is the lower limit of rate, measured in encoded binary digits per source letter, for a source encoder such that the source output sequence can be reconstructed from the encoder output with an arbitrarily-small specified per-digit error probability. Equivalently, 1/H(U) is the upper limit of compression, measured in source letters per encoded binary digit, which can be achieved by coding schemes which convert the source output into a stream of binary digits from which the source output can be reconstructed with an arbitrarily-small specified per-digit error probability.

The capacity of the BSC is given by

$$C = 1 - h(p) \quad \text{bits/use,}$$

where a "use" means the transmission of a single binary digit through the channel. According to Shannon's Noisy Coding Theorem, C is the upper limit of the rate of binary digits from a BSS (which we can think of as being the output of the source encoder in Fig. 1) per channel use for a channel encoder such that there is a channel decoder which delivers the BSS digits with an arbitrarily-small specified per-digit error probability.

A very fundamental characterization of an information source is that given by its rate-distortion function. The rate-distortion function of the BMS is given by

$$R(D) = \begin{cases} h(q) - h(D) \quad \text{bits/letter}, & 0 \leq D \leq \min(q,1-q) \\ 0, & D > \min(q,1-q) \end{cases}$$

where D is the Hamming distortion defined by

$$D = \lim_{n\to\infty} \frac{1}{n} \sum_{i=1}^{n} P(\hat{U}_i \neq U_i),$$

i.e., D is the per-digit error probability in the source reconstruction. According to Shannon's Theorem for Coding Relative to a Fidelity Criterion, R(D) is the lower limit of rate, measured in binary digits per source letter, for a source encoder such that the source output sequence can be reconstructed from the encoder output with a distortion of D or less.

3. LINEAR CODING

We now consider the special case of linear coding, both linear source coding and linear channel coding. We begin with the latter because the relevant theory[5] is more widely known.

A [block] linear (N, K) binary channel encoder is specified by a K x N binary matrix G, of rank K, in the manner that

$$\underline{X} = \underline{V}\, G \tag{1}$$

where $\underline{V} = [V_1, V_2, \ldots V_K]$ is the information (row) vector, and $\underline{X} = [X_1, X_2, \ldots X_N]$ is the codeword. The operations in (1), and hereafter for all matrices and vectors, are in the finite field GF(2), i.e., in modulo-two arithmetic. The code rate is R = K/N bits/use.

It is well-known[2,3] that linear channel coding is sufficiently general to attain the performance promised by the Noisy Coding Theorem (although we hasten to add that it is only the encoder which is linear; a good channel decoder is always nonlinear!). That is, for a given $\varepsilon > 0$ and a given R such that R < C, there exists, for sufficiently large N, linear (N, K) encoders and appropriate decoders such that

$$\frac{1}{N} P(\hat{\underline{X}} \neq \underline{X}) \leq \varepsilon$$

when this channel coding system is used on a BSC of capacity C, regardless of the source statistics. In fact, it is known that no other type of coding can give a significantly smaller decoding error probability. Add to this the simplicity with which a linear encoder can be implemented and you will see why no one seriously proposes the use of other than linear channel encoders.

For the given G, one can always find an (N-K) x N matrix H, of rank N-K, such that

$$G\,H^T = 0 \qquad (2)$$

where the superscript T denotes "transpose". Moreover, a given vector $\underline{X}$ is a codeword if and only if

$$\underline{X}\,H^T = \underline{0}.$$

If one writes the vector $\underline{Y} = [Y_1, Y_2, \ldots Y_N]$ received over the BSC as $\underline{Y} = \underline{X} + \underline{E}$, where $\underline{E} = [E_1, E_2, \ldots E_N]$ is the error pattern, then it follows from (2) that

$$\underline{S} \triangleq \underline{Y}\,H^T = \underline{E}\,H^T. \qquad (3)$$

The (row) vector $\underline{S} = [S_1, S_2, \ldots S_{N-K}]$ is consequently called the syndrome because it depends only on the errror pattern $\underline{E}$ that has infected the codeword in its passage through the BSC.

It is a well-known fact in coding theory that, without loss of optimality, the decoder for a linear code can always be built in the manner shown in Fig. 3 such that the decoder first forms the syndrome and then estimates the error pattern solely from this syndrome. One should not be misled by Fig. 3; the leftmost and

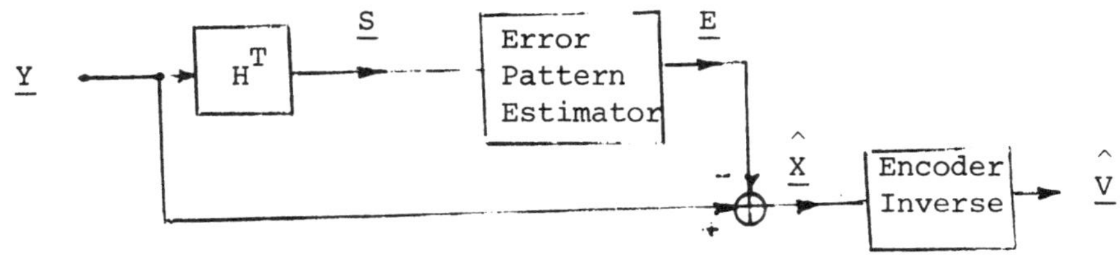

Fig. 3 A Syndrome Decoder for a Linear Code

rightmost boxes therein are linear devices and easy to implement, but the box labelled "error pattern estimator" may be unimaginably difficult to implement for very long and powerful codes.

We now turn to the description of linear source coding. A [block] linear (N, K) source encoder is specified by an (N-K) x N binary matrix H, of rank N-K, in the manner that

$$\underline{V} = \underline{U}\, H^T \tag{4}$$

where $\underline{U} = [U_1, U_2, \ldots U_N]$ is the source message, and $\underline{V} = [V_1, V_2, \ldots V_{N-K}]$ is the encoded version of the source message. (We shall place the subscript c or s on K, N, H and G whenever the context does not make it clear whether we are specifying the channel encoder or the source encoder, respectively.) Thus, the <u>compression ratio</u> of a linear (N, K) source encoder is

$$\beta \triangleq N/(N-K).$$

The <u>rate</u> of this linear source coding scheme is

$$R_L \triangleq 1/\beta = 1 - K/N.$$

The reason for our choosing the above notation for linear source encoding is the interpretation that we now wish to make. We first make the key observation that the error pattern $\underline{E}$ of the BSC is statistically identical to the output vector $\underline{U}$ of a BMS with parameter q equal to p. Thus, we are always free to consider that a linear source encoder treats the output of the BMS as an "error pattern" and forms the "syndrome" of this error pattern, according to (4), which syndrome is then the encoded version of the source message. Hence, we can <u>always</u> consider linear source coding conceptually as shown in Fig. 4 where the source decoder is an "error pattern estimator". This interpretation of linear source coding appeared first in the literature in the work of Ohnsorge[6] and has been rather fully developed by Ancheta[7].

Fig. 4 The Syndrome-Source-Coding Interpretation of Linear Source Coding

4. JOINT LINEAR SOURCE/CHANNEL CODING--THE DISTORTIONLESS CASE

We now consider linear source encoding when the goal is reproduction of the source with a negligibly small (but non-zero) probability ε of digit error, so-called "distortionless coding".

Consider a BMS with parameter q where, for convenience with no real loss of generality, we take $0 \leq q \leq 1/2$. For the BSC with crossover probability p equal to q, we know there is a linear channel coding scheme (G_c, H_c) such that, for any given $\delta > 0$, it has

$$R \geq C - \delta = 1 - h(q) - \delta$$

and achieves per-digit error probability ε or less in the estimated codeword $\underline{\hat{X}} = \underline{\hat{U}}\, G_c$. For this channel coding scheme, the per-digit error probability in the vector $\underline{\hat{E}}$ of Fig. 3 coincides with that in the vector $\underline{\hat{X}}$. Thus, if we use these same two matrices as the G_s and H_s of the source coding scheme of Fig. 4, it follows that the per-digit error probability of the reconstruction $\underline{\hat{U}}$ is again the same, i.e., is or less. (Here we assume that the source coding ε scheme uses the same error pattern estimator as did the channel devices that have no effect on the latter's operation. If D is the per-digit

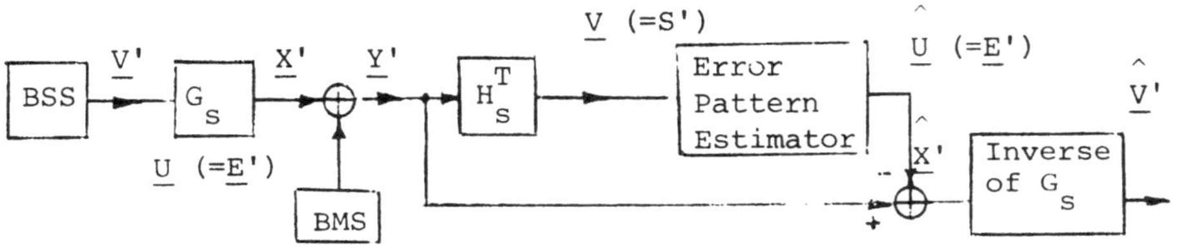

error probability in $\underline{\hat{U}}$ for the linear source coding scheme, we see from Fig. 5 that it is also the per-digit error probability in $\hat{X}'$. Now, as is well-known in coding theory, given H_s, one can always choose G_s such that G_s has an identity matrix in some K of its columns. But then $\underline{\hat{V}}'$ is just the vector composed of the K digits in these K positions of $\hat{X}'$. It follows that the per-digit error probability in $\underline{\hat{V}}'$ is at most (N/K)D. But, since this is also the fidelity with which the BSS (not the BMS!) in Figure 5 is being transmitted through the BSC created by considering the output of the BMS to be an error pattern $\underline{E}$, and since K digits of the BSS are being transmitted with N uses of this BSC with capacity $C = 1 - h(q)$, it follows from the properties of the rate-distortion function of the BSS that

$$\frac{N[1 - h(q)]}{K} \geq R_{BSS} \left(\frac{N}{K} D\right) = 1 - h\left(\frac{N}{K} D\right)$$

or, equivalently,

$$h\left(\frac{N}{K} D\right) \geq 1 - \frac{N}{K} [1 - h(q)]. \qquad (5)$$

We can put (5) into more revealing form in terms of

$$R_L = \frac{1}{\beta} = 1 - \frac{K}{N}.$$

Then (5) becomes

$$D \geq (1 - R_L) \; h^{-1} \left[\frac{h(q)-R_L}{1-R_L}\right] \qquad (6)$$

coding scheme.) The compression ratio achieved is

$$\beta = \frac{N}{N-K} = \frac{1}{1-R} \geq \frac{1}{h(q)+\delta} = \frac{1}{H(U)+\delta}$$

which is arbitrarily close to the upper limit of achievable compression ratios, $1/H(U)$, established by the Noiseless Coding Theorem. Thus, as has been observed by Hellman[8] and Ancheta[7], linear source encoding entails no loss of optimality when the goal is distortionless reproduction of the source.

But we now recall that linear channel coding never entails a loss of optimality. Moreover, if we have

$$N_s - K_s = K_c$$

(which can always be achieved simply by redefining the block lengths, if necessary, to be integer multiples of the original block lengths), then we can write for the tandem combination of the two linear systems

$$\underline{X} = \underline{V}\, G_c = \underline{U}\, H_s^T\, G_c.$$

It follows then that we can consider $A \triangleq H_s^T G_c$ to be the defining matrix of a linear joint source/channel encoder which operates as

$$\underline{X} = \underline{U}\, A.$$

It follows, as first observed by Hellman[8], that joint linear source/channel encoding entails no loss of optimality when the goal is distortionless reproduction of the source. Moreover, the implementation of the matrix $A = H_s^T G_c$ cannot avoid being far simpler in general that the separate implementation of the matrices H_s^T and G_c.

Example: Suppose that we are to transmit, with negligibly small distortion, a BMS with q = .10 through a BSC with p = .10. Since h(.10) = 0.47, it follows that a compression ratio of 1/h(.10) = 2.13 can be approached, and that a channel coding rate of C = 1 - h(.10) = .53 can be approached. Thus, an overall efficiency of (2.13)x(0.53) = 1.13 source letters per channel use can be approached arbitrarily closely with joint source/channel linear coding, and no larger overall efficiency can be obtained by any distortionless coding scheme. In particular, for suitably large K, we can find an R = 1/2 linear channel encoder specified by

$$G_c = [I_K \; P]$$

(where P is some K x K binary matrix) and a $\beta = 2$ linear source encoder

$$H_s = [P^T \; I_K]$$

such that the overall distortion is smaller than the specified small amount. But then

$$A = H_s^T G_c = \begin{bmatrix} P & P^2 \\ I & P \end{bmatrix}$$

describes a linear joint source/channel encoder which has overall efficiency $\beta R = 1$, quite close to the theoretical limit. Moreover, we see that A can be implemented quite straightforwardly from a device which implements only P, whereas implementation of G_c and H_s would each require implementation of P in separate source and channel coding. It is interesting to note that A is an N x N matrix, but that its rank is only N/2; this lack of full rank appears to be fundamental for useful linear joint source/channel encoders.

We conclude that joint linear source/channel coding is a highly attractive approach when the goal is the distortionless reproduction of the source.

5. JOINT LINEAR SOURCE/CHANNEL CODING--THE NON-NEGLIGIBLE DISTORTION CASE

With many actual data sources (e.g., with facsimile), one is often content to accept non-negligible distortion D in the source reproduction (e.g., D = 1/10). The rate-distortion function of the source specifies how such a relaxed demand on the fidelity of reconstruction can be translated into more efficient use of the channel, i.e., fewer uses of the channel for each source letter.

Following recent work by Ancheta[9], we now show that, for a given D (non-negligibly) greater than zero, the performance of linear source coding is bounded in general strictly below the compression ratio 1/R(D) which Shannon has shown can be approached arbitrarily closely by some sort of source coding.

The key (and clever) idea in Ancheta's proof that linear source encoding for non-negligible distortion in inherently suboptimal was his exploitation of the fact that a linear source encoder "cannot see" a vector which lies in the null space of the matrix H_s^T, i.e., its output is zero for any vector which could be the output of the linear device which implements the matrix G_s. Consider then the situation shown in Fig. 5, where we have merely supplemented the source coding system of Fig. 4 by adding some where $h^{-1}(\cdot)$ is the inverse (made unique by restricting its values to be between 0 and 1/2) of the binary entropy function.

The significance of (6) can perhaps be most easily seen by its specialization to the BSS, i.e., to q = 1/2. Then h(q) = 1 and (6) simplifies to

$$D \geq (1 - R_L)/2. \tag{7}$$

In Fig. 6a, we have plotted both the bound (7) on the attainable distortion D of a linear source coding scheme of rate R_L for the BSS, together with the rate-distortion function R(D) = 1 - h(D) of the BSS. This figure clearly illustrates how far away from optimal a linear source coding system must be when a non-negligible D is specified. For example, with D = .11, R(D) = .50 but R_L = .78. Thus the linear scheme can have at best $\beta = 1/R_L = 1.28$, compared to the compression ratio 1/R(D) = 2 that can be approached by more general source coding schemes.

A similar interpretation can be made from Fig. 6b where we have shown the rate-distortion function R(D) for the general BMS and also the corresponding bound on R_L from (6).

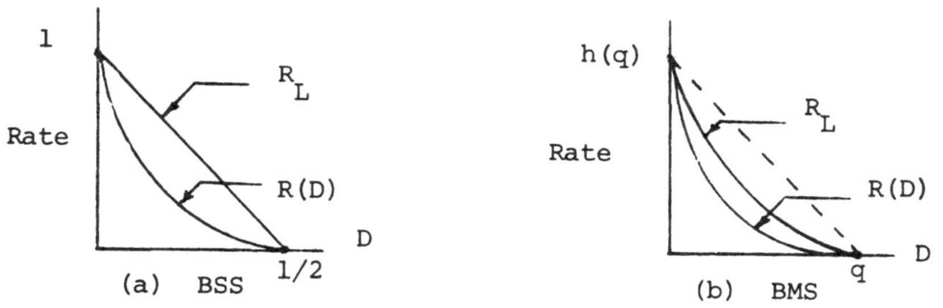

Fig. 6 Bounds on the Achievable Rate R_L with Linear Source Coding

Ancheta[9] actually has a lot more to say about the non-optimality of linear source coding with non-neglible distortion, but we shall leave the rest for him to tell in his own publications, except to mention his conjecture that the achievable R_L is actually more strictly bounded away from R(D) according to the dashed line shown in Fig. 6b.

We now give a simple argument to show that the inherent lack of optimality of linear source coding in the non-negligible distortion case implies in general an inherent lack of optimality for linear joint source/channel encoding in the non-negligible distortion case.

Suppose that the $N_s \times N_c$ matrix A describes a linear joint source/channel encoder, for a BMS and BSC, which achieves distortion D (where D is not negligibly small). Suppose that A has rank r. Then one can always find an $r \times N_s$ matrix H_s of rank r and $r \times N_c$ matrix G_c of rank r such that $A = H_s^T G_c$. Thus, we can consider the matrix H_s^T as describing a linear source encoder and the matrix G_c as describing a linear channel encoder; the original linear joint source/channel encoder is equivalent to separate encoding with these derived linear encoders.

Let D' be the best obtainable distortion when the BMS is reconstructed directly from the output of the linear source encoder H_s^T. It follows that $D' \geq D$, because the best service which the channel encoder G_c can provide is to permit perfect transmission of the source encoder output to the best source reconstructor. Hence, the rate R_L of the linear source encoder must satisfy (6) for the given distortion D.

The overall efficiency of the linear separate source/channel coding system (and hence also of the entirely equivalent original linear joint coding system) is $\beta R_c = R_c/R_L \leq 1/R_L$ source letters per channel use, where the inequality follows from the fact that

$R_c \leq 1$. On the other hand, there exist coding systems whose overall efficiency approaches C/R(D) source letters per channel use, where C is the capacity of the BSC and R(D) is the rate-distortion function of the BMS. Thus, when, for a given D, the bound (6) specifies an R_L such that $R_L > R(D)/C$, then there is an inherent loss of optimality when linear joint source/channel encoding is used. In other words, when the bound (6) gives an R_L which exceeds R(D) by a factor of more than 1/C, then linear joint source/channel encoding is sub-optimum.

Example: Consider the BSS together with the BSC having p = .10, and suppose that D = 1/4 is specified. Then, R(D) = h(1/2)-h(1/4)= .19. From (6), we find $R_L = .50$. Thus, R_L is (.50)/(.19) = 2.63 times as great as R(D). But 1/C = 1.89. Because 2.63 > 1.89, it follows that a linear joint source/channel coding system must be sub-optimum. To put it another way, any such linear joint coding system has an efficiency of at most $1/R_L = 2$, whereas there exist more general coding systems whose efficiency approaches C/R(D) = 2.79 source letters per channel use.

We should point out in closing that a joint linear source/channel coding system can sometimes "accidently" be optimal when R_L, as given by (6), exceeds R(D) by a factor of only 1/C or less. In the above example, if we had taken D = .10 rather than D = 1/4, we would have found $R_L = .80$ and R(D) = .53 so that $R_L/R(D) = 1.51 < 1/C = 1.89$. C/R(D)=1 is the maximum approachable efficency. But the "straight wire" encoder, which merely transmits the BSS output directly over the channel, has efficiency 1 and distortion D = .10. We can consider this trivial but optimum coding scheme as the linear joint source/channel coding scheme with A = 1. [The reason for this accidental optimality is that the given BSC happens to be the appropriate "forward channel" for the given distortion D and the BSS, cf. Berger[4]]

References

1. Shannon, C.E., A Mathematical Theory of Communication, Bell System Technical Journal, 27, 379, 1948.

2. Fano, R.M., Transmission of Information, M.I.T. Press, Cambridge, Massachusetts, 1961. (page 3)

3. Gallager, R.G., Information Theory and Reliable Communication, John Wiley and Sons, New York, 1968.

4. Berger, T., Rate Distortion Theory: A Mathematical Basis for Data Compression, Prentice-Hall, Englewood Cliffs, New Jersey, 1971.

5. Peterson, W.W., Error-Correcting Codes, M.I.T. Press, Cambridge, Massachusetts, 1961.

6. Ohnsorge, H., Data Compression System for the Transmission of Digitalized Signals, Proceedings IEEE International Conference on Communications, 485, 1973.

7. Ancheta, T.C., Jr., Syndrome-Source-Coding and its Universal Generalization, IEEE Transactions on Information Theory, 22, 432, 1976.

8. Hellman, M.E., Convolutional Source Encoding, IEEE Transactions on Information Theory, 21, 651, 1975.

9. Ancheta, T.C., Jr., Bounds and Techniques for Linear Source Coding, Ph.D. Thesis, Department of Electrical Engineering, University of Notre Dame, Notre Dame, Indiana, August, 1977.

RATE DISTORTION THEORY: THE INFLUENCES OF THE SOURCE AND DISTORTION MEASURE IN STRUCTURING SOURCE CODING

David J. Sakrison

Department of Electrical Engineering and Computer Sciences
and the Electronics Research Laboratory
University of California, Berkeley, California 94720

INTRODUCTION: WHAT'S SO IN TRANSMITTING A RANDOM SOURCE OVER A NOISY CHANNEL.

This section outlines the main results that pertain in trying to transmit a source over a channel; the objective is to give the reader some perspective of the problem and then outline in the context of this perspective those topics that are covered in depth in the remaining sections. In order to focus on the concepts involved rather than expending our effort on building up mathematical machinery, we will focus on the case in which the source output is a sequence of random variables $U_1, U_2, \ldots$ and the channel inputs and outputs are sequences of random variables $X_1, X_2, \ldots$ and $Y_1, Y_2, \ldots$ respectively. We assume that the reader has a working knowledge of such quantities as differential (integral) entropy and average mutual information for continuous-valued random variables, including the properties and basic inequalities pertaining to these quantities. The reader who is not totally comfortable with these ideas is referred to Shannon [1, parts I and III] or chapter two of Gallager [2]. All of the concepts used can be extended to the situations of more immediate practical interest in which the source output and the channel input and output are waveforms (random processes).

A <u>source</u> is a mathematical abstraction of some information producing device, such as a speaker of instrument. The behavior of a source is indeterminate (or we would not be interested in it) and we thus consider the source output to be random. In the simplest case which we consider initially, we take the source output to be a sequence of random variables $U_1, U_2, \ldots$. We may or may not know the probability distribution of the source, but in any case we use α to denote the source distribution. For simplicity, we assume initially that the source output is a sequence of independent identically distributed (i.i.d.) random variables. Physically the source is characterized by limited accuracy, either due to limited accuracy of the measuring instrument (e.g. a microphone) or the final end user (e.g. the human auditory system). Mathematically this is represented by a distortion measure: if u is the source output and $\tilde{u}$ the output presented to the user at the end of a transmission system, $d(u,\tilde{u})$ is a non-negative function which expresses the corresponding amount of distortion relative to the users criterion of fidelity. Mathematically a source is thus defined by two quantities: a) its probability distribution α (determined by the physics governing the source); b) a distortion measure $d(\_,\_)$ (which is determined by the sensitivity or requirements of the user).

Let us now introduce a hypothetical transition distribution for the system output $\tilde{u}$ presented to the user given that the source output is u; we label such a distribution by γ and denote its density function by $q_\gamma(\tilde{u}|u)$. The labels α and γ are used to indicate a dependence on the source and transition distributions. Given the source distribution α with density function $p_\alpha(u)$ and any transition distribution γ we can calculate the average distortion $E_{\alpha\gamma}\{d(U,\tilde{U})\}$. The value of average distortion may be compared against various values of a dummy variable d^*, which can represent various choices of acceptable distortion level. For a given source α, one can thus define an acceptable set of transition distributions as

$$\Gamma_\alpha(d^*) = \{\gamma: E_{\alpha\gamma}\{d(U,\tilde{U})\} \leq d^*\} \tag{1}$$

The average mutual information between U and $\tilde{U}$ depends on both α and γ and we denote it by $I_{\alpha\gamma}(U;\tilde{U})$. In terms of the above, we define the rate distortion function of a source α to be

$$R_\alpha(d^*) = \inf_{\gamma \in \Gamma_\alpha(d^*)} I_{\alpha\gamma}(U;\tilde{U}) \tag{2}$$

This function is monotone decreasing and convex $\cup$; its significance will be explained shortly.

For a channel there is an exact dual to the above quantities. A channel is characterized by: a) a transition (noise) probability distribution b) constraints on quantities such as input power and bandwidth. The capacity C of a channel, which is a function of constraints on the average input power and bandwidth, is defined as the sup of the average mutual information between the channel input X and channel output Y over all input distributions on X satisfying the given constraints on the input power and bandwidth.

The problem of transmitting a source over a channel is completely characterized by the rate distortion function of the source and the capacity of the channel. Part of this is stated in the

Converse to the Coding Theorem: Given any source with rate distortion function $R_\alpha(d^*)$ and any channel of capacity C, there does not exist any (coding or modulation) method of transmitting the source output U to a user over the channel with average distortion less than d*, where d* is the solution of

$$R_\alpha(d^*) = C \tag{3}$$

The reader should note the generality of this statement. It says that whatever source (speech, video, census data) we have in mind, we need only characterize it by its rate distortion function; similarly, whatever channel (telephone line, satellite link, etc.) we have may be independently characterized by its capacity. Also, the statement of what we cannot do contains no qualification as to the complexity or nature of the transmission method, be it PCM, FM, coded PCM or whatever. This converse is proven by juggling of a number of Information Theoretic inequalities, most of which have some intuitive interpretation. We do not present these here since some of the details are not too enlightening; the interested reader should consult Gallager [2, pp. 449] or Sakrison [3, pp. 143], or Berger [4, sect. 3.2].

The converse Theorem by itself would not be tremendously exciting; what makes it so is that Information Theory also tells us that the average transmission distortion can be made to be arbitrarily close to the solution of Eq. (3). To establish this positive result, we split the problem into two subproblems: channel coding and source coding and consider somewhat specific schemes for each.

For any integer N let us consider blocking N channel inputs $x_1, x_2, \ldots, x_N$ together and denoting the block by $\underline{x}$; similarly with channel outputs $\underline{y}$, source outputs $\underline{u}$, and system outputs (user inputs) $\tilde{\underline{u}}$. The channel coding Theorem may be stated as follows.

Channel Coding Theorem: Given any channel of capacity C and any (small positive numbers) ε and δ, there exists an N, a set of $Q=2^{(C-\delta)N}$ (Q an integer) input vectors (N-tuples $\underline{x}_1, \underline{x}_2, \ldots, x_Q$) and a partition of the space of all possible output vectors into Q sets $\mathcal{Y}_1, \ldots, \mathcal{Y}_Q$, such that if the channel decoder declares that the input was $\underline{x}_q$ whenever $\underline{y} \in \mathcal{Y}_q$, then the probability of a transmission error is less than ε. (This states that the input vector $\underline{x}_q$ nearly always causes an output $\underline{y} \in \mathcal{Y}_q$).

For a proof and complete discussion of this theorem the reader is referred to Gallager [2, Chap. 5]. It states, in essence, that digital information can be transmitted over a channel of capacity C bits/sec. at a rate of C-δ bits/sec. with an arbitrarily small probability of error, if we are willing to make the block length N sufficiently long.

There is a dual Source Coding Theorem as follows:

Source Coding Theorem: Given a source with rate distortion function $R_\alpha(\cdot)$ an average distortion level d*, and any small positive numbers ε and δ; we then consider selecting a block length L, a partition $\mathcal{U}_1, \mathcal{U}_2, \ldots, \mathcal{U}_M$ of the space of all possible source output vectors $\underline{u}$ with

$$M = 2^{[R_\alpha(d^*)+\delta]L} \qquad \text{(integer M)} \tag{4}$$

and a set of corresponding code or reproduction (approximation) vectors $\underline{u}_1, \underline{u}_2, \ldots, \underline{u}_M$, with the following property. Let the encoder function so that whenever the source output $\underline{u} \in \mathcal{U}_m$ the decoder presents the user with the reproduction vector $\underline{u}_m$. Then for L sufficiently large it is possible to select these quantities such that the average distortion satisfies

$$E_\alpha\{d(\underline{U},\underline{u}_m)\} \leq d^* + \varepsilon \tag{5}$$

This theorem states what can be done to minimize the average encoding distortion if we are willing to collect a block of L source outputs together (L may have to be large for Eq. (5) to be satisfied) and encode them for simultaneous transmission (using a block of N channel symbols). In interpreting this theorem it is necessary to specify how we measure distortion over the block. The mathematically most tractable way, and the one assumed in the above source coding theorem, is to average the distortion over all symbols in the block

$$d(\underline{u},\underline{\tilde{u}}) = (1/L) \sum_{\ell=1}^{L} d(u_\ell,\tilde{u}_\ell) \tag{6}$$

Let us now see what these two theorems say about how we can transmit a source over a channel of capacity C at close to the performance given by the Converse to the Coding Theorem. The Channel Coding Theorem states that by using only certain blocks of N channel inputs (with N sufficiently large) we can transmit information over the channel very reliably (probability of error < ε) at a rate C-δ which is very close to the channel capacity of C bits/sec. The Source Coding Theorem states what can be done by multidimensional (L) quantization of the source for large L. The space of L source symbols is partitioned up into M subsets and a standard quantized source symbol $\underline{u}_m$ is used to represent any L-tuple of source outputs $\underline{u}$ that falls in $\mathcal{U}_m$. Note that the source space is not quantized dimension-by-dimension as would be done by an a-d converter; the performance of Eq. (5) is predicated on a multidimensional quantization procedure which is much more gruesome to implement. If this coding or multidimensional quantization is done at a rate which generates $R_\alpha(d^*)+\delta$ bits/sec., then the average coding or quantization distortion can be made arbitrarily close to d*. Combining the two coding procedures gives us a way to come arbitrarily close to the performance given by $R_\alpha(d^*) = C$. There is one thing that needs attention in this argument; the effect of infrequent channel errors must contribute only a small amount of average distortion. This will be guaranteed if the following (practically very reasonable) condition is met: for any finite u'

$$E\{d(U,u')\} < \infty \tag{7}$$

The two coding theorems establish that the performance $R_\alpha(d^*) = C$ can be achieved by a PCM type of solution as shown in Fig. 1; quantizing the source and transmitting the resulting digital information over the channel.† This two step

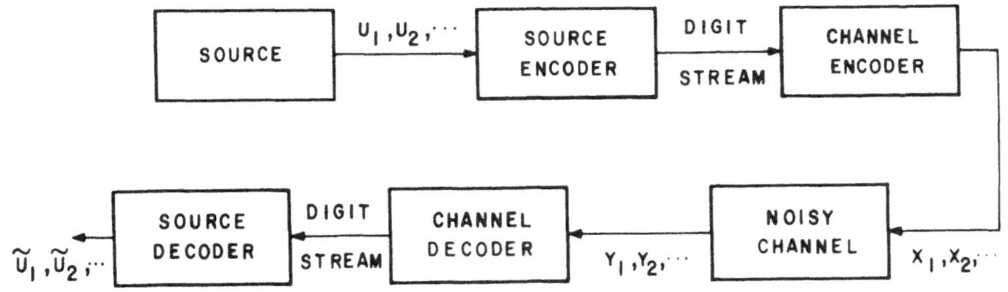

Figure 1 The Source Communication Problem Broken into a Source Coding Problem and Channel Coding Problem.

solution of the problem is complex, but has some advantages. It means that a person can focus on a smaller problem and the solution to that problem can be used generally. That is, the person faced with the problem of efficient transmission of speech can concentrate on how to find an efficient digital representation of speech without worrying about what type of channels will be used for transmission; his encoder can be used in conjuction with a number of different channels. Put another way, if there are N_s sources that might ultimately be used with N_c channels, we need solve only N_s+N_c problems and not N_cN_s. Another advantage is that information from quite different sources (speech and weather maps)

†This is not to imply that a one-step procedure might not work. Transmission of a Gaussian Random Process of bandwidth W over a Gaussian Channel of Bandwidth W is achieved at $R_\alpha(d^*) = C$ by simple AM. However, this is the only known example of direct efficient encoding. Other forms of modulation, such as FM, fall below the $R_\alpha(d^*) = C$ bound.

can easily be multiplexed onto a common channel by multiplexing the two digit streams.

In the following sections we take a closer look at the encoding or multidimensional quantization or digitization of sources to see what makes it work. We then consider different ways in which the source coding problem may be broken up and part of it recombined with the channel coding problem. We conclude with an example of this procedure.

A LOOK AT THE SOURCE CODING THEOREM

In this section we take a brief look at how one establishes the source coding theorem. Our goal is not to provide a rigorous proof, very readable proofs exist in several places: Gallager [2, sec. 9.3], Sakrison [3, pp. 148], Berger [4, sects. 3.1,3.2], but rather to examine it with the goal of understanding what makes the proof work and how that affects the way in which the coding problem may be modularized. We begin by examining the special case of a Gaussian Source with a square-error criterion from a geometric approach and then give brief consideration to the more general case.

Coding the Gaussian Source

We consider the Gaussian Source with a square-error criterion because the Gaussian Distribution has an L-dimensional probability distribution which admits of an easy description and because the square-error criterion corresponds to the usual Euclidean L-dimensional norm. Our distortion measure for a block of L random variables $\underline{u} = u_1,\ldots,u_L$ is simply

$$d(\underline{u},\underline{\tilde{u}}) = (1/L)\sum_{\ell=1}^{L} (u_\ell - \tilde{u}_\ell)^2$$

which is 1/L times the usual definition of the Euclidean norm of $\underline{u}-\underline{\tilde{u}}$. To avoid carrying this factor of 1/L along, we define the norm to include this factor

$$\|\underline{u},\underline{\tilde{u}}\|^2 = (1/L)\sum_{\ell=1}^{L} (u_\ell - \tilde{u}_\ell)^2 = d(\underline{u},\underline{\tilde{u}}) \tag{8}$$

Now consider the probability distribution for a block of L source variables $\underline{U} = U_1,\ldots,U_L$ under the assumption that the source outputs are i.i.d. Gaussian Random Variables with zero mean and variance σ^2. The probability density function is given by

$$f_{\underline{U}}(\underline{u}) = \prod_{\ell=1}^{L} \frac{1}{\sqrt{2\pi}\,\sigma} \exp\left[-\frac{u_\ell^2}{2\sigma^2}\right] = \frac{\exp[-L\|u\|^2/2\sigma^2]}{(2\pi\sigma^2)^{L/2}} \tag{9}$$

The distribution is spherically symmetric. Moreover, $E\{\|\underline{U}\|^2\} = \sigma^2$ and as L becomes large $\mathrm{var}\{\|\underline{U}\|\} \to 0$ as 1/L. These statements are easily shown; the interested reader is referred to Sakrison [3, sec. 6.1, or 5]. They imply that nearly all of the vectors $\underline{u}$ lie near the surface of an L-dimensional sphere of radius σ and that, from a mean-square error standpoint, we introduce negligible mean-square error in taking a vector to lie on the surface of the sphere. Specifically, if $\underline{u}_p$ denotes the projection of a vector $\underline{u}$ onto the sphere

$$\underline{u}_p = \underline{u}\,\frac{\sigma}{\|\underline{u}\|}$$

then for any vector $\underline{u}'$

$$E\{\|\underline{U}-\underline{u}'\|^2\} \leq E\{\|\underline{U}_p-\underline{u}'\|^2\} + \text{const.}/L \tag{10}$$

Thus in our discussions we will always consider only those source points $\underline{u}$ lying on the surface of the sphere $\|\underline{u}\| = \sigma$ or within a thin spherical shell about this surface.

Now let us consider the task of encoding. Our problem is to map all source points $\underline{u}$ in L-dimensional Euclidean space, E^L, into one of the M source or representation vectors $\underline{u}_m$. Given a set of M code vectors $\underline{u}_m$, m = 1,...,M, the optimum way of partitioning or quantizing E^L is as follows

$$\begin{aligned}\mathcal{U}_m &= \{\underline{u}\text{: the quantizer output is } \underline{u}_m\} \\ &= \{\underline{u}\text{: } d(\underline{u},\underline{u}_m) = \|\underline{u}-\underline{u}_m\| \leq \|\underline{u}-\underline{u}_j\|,\ j = 1,\dots,M\}\end{aligned} \tag{11}$$

Such an encoding partition is depicted for two dimensions in Fig. 2. The goal of encoding is to minimize, M, the number of

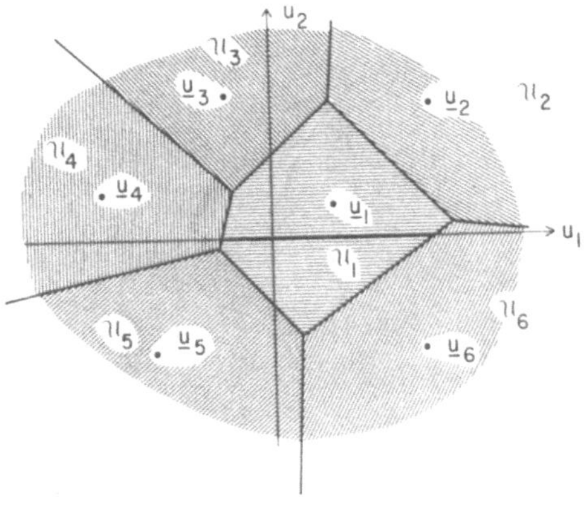

Figure 2 L-dimensional (L=2) quantization or encoding of source outputs.

code vectors $\underline{u}_m$, consistent with the constraint

$$E\{\|\underline{U}-\underline{u}_m\|^2\} \leq d^* \tag{12}$$

in which $m = m(\underline{U})$ is the index of the code vector closest to $\underline{U}$.

We now wish to focus on how the problem description, specifically the source distribution and the choice of a distortion measure, affect the solution. At this point, the reader can appreciate our choice of the Gaussian Random Variable source and mean-square-error distortion measure. The properties of the source distribution and the distortion constraint (12) can simply be interpreted as follows:

a) the source distribution requires that our encoding partition cover the surface of the L-dimensional sphere of radius σ and

b) the distortion criterion requires that this covering be such that the square of the Euclidean length of the average "quantization" or encoding error be d*.

Let us thus consider the most effective way to position points to meet this constraint and minimize the number of points required to cover the message sphere (sphere of radius σ). We first consider a single code vector $\underline{u}_m$ and the shape of its representation set $\mathcal{U}_m$. This set will of course depend on other points as determined by (11), but let us start by considering what the most desirable shape might be. This would be the set that intersected the largest possible amount of surface mass of a sphere of radius σ (to minimize the number of code points required) consistent with a fixed value of moment-of-inertia (quantization distortion) of this set. Figure 3 shows a code vector $\underline{u}_m$ with

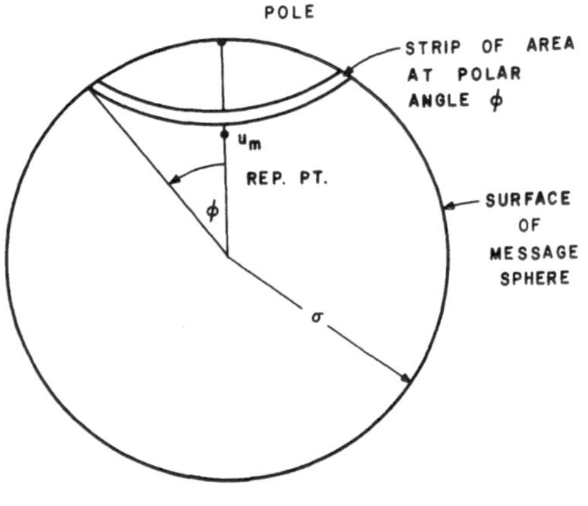

Figure 3 Geometry indicating the shape of the optimum representation region.

a line from the origin drawn through this vector to a pole. It is clear that distance of a point increases monotonically with the polar angle. The intersection of the optimum shaped $\mathcal{U}_m$ with the message sphere is thus a symmetric polar cap (a chunk of area outside the cap could be spread uniformly around the cap, reducing the quantization error while leaving the area covered fixed). Next, one might ask at what radius the code point $\underline{u}_m$ should be located to have an average distortion d*. For the dimension L being large, virtually all of the area of the polar cap is at its edge, therefore the distance from the code point to the edge of the cap should be $\sqrt{d^*}$. Since this half-chord from $\underline{u}_m$ to the edge of the cap is perpendicular to the polar line, the code point should be at radius $\sqrt{\sigma^2 - d^*}$.

We can now put a lower bound on the number M of the code points required to cover the message sphere; it is

$$M \geq \frac{\text{surface area of L-dimensional message sphere radius } \sigma}{\text{area polar cap of chord length } 2\sqrt{d^*}} \tag{13}$$

The ratio is a lower bound because it is not clear that it is possible to configure a set of code points so that the resulting $\mathcal{U}_m$ actually have the effectiveness of the polar cap. By using formulae and bounds for the surface areas in inequality Eq. (13), one can obtain the bound [3, sec. 6.1; 5]

$$R_{gauss}(d^*) \geq (1/L)\log_2 M \geq (1/2)\log_2 \sigma^2/d^* \tag{14}$$

$1/L \log_2 M$ being the rate in bits/random variable that it takes to describe which of the M code vectors to use to approximate $\underline{u}$. The right hand side of Eq. (14) is in fact the rate distortion function.

Now let us consider how to show that the rate on the right hand side of Eq. (14) is attainable. To do this, one uses a random coding argument, the classic method for establishing a coding theorem. We consider a random ensemble of codes; each code consists of M code vectors $\underline{u}_m$ drawn independently according to a uniform distribution on a sphere of radius $\sqrt{\sigma^2 - d^*}$ (equivalently the components of the code vectors are independent Gaussian random variables of variance $\sigma^2 - d^*$). One then asks what is the probability over the ensemble of codes that some source vector $\underline{u}$ on the surface of the message sphere will not be within $(1+\varepsilon)\sqrt{d^*}$ of one of the M code vectors in a set. From Fig. 4 one sees that this probability of the distortion being greater than d* for M independently chosen code vectors is

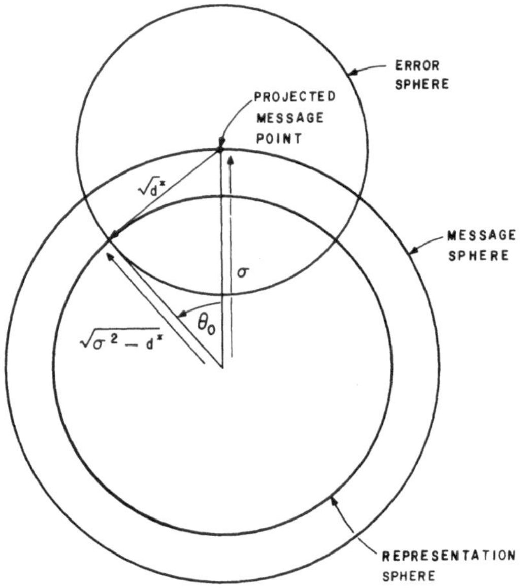

Figure 4 Figure illustrating situation used for calculating P_{d*}.

$$P_{d*} = \left[1 - \frac{\text{area polar cap radius } \sqrt{\sigma^2 - d*} \text{ angle } \theta_0}{\text{area sphere radius } \sqrt{\sigma^2 - d*}} \right]^M \qquad (15)$$

$$\sin \theta_0 = \sqrt{d*}/\sigma$$

If one uses appropriate bounds for the L-dimensional areas involved and takes $(1/L)\log_2 M$ slightly greater than the right-hand-side of (14), one finds [3,6] that this probability goes to zero as L becomes large. This establishes that a set of code points with rate slightly greater than $R_{gauss}(d*) = 1/2 \log_2 \sigma^2/d*$ can be found that achieves nearly the covering efficiency of the polar cap.

This is, in outline form, a proof of the coding theorem and its converse for an i.i.d. Gaussian Random Variable source. Let us pause a moment to note how the source and distortion measure determined the character of the optimum coding solution. First, the <u>probability distribution of the source</u> dictated not only that we need only cover the <u>surface</u> of an L-dimensional sphere of <u>radius σ</u>, but also that <u>all</u> points on that sphere were equiprobable and we needed to cover essentially all of that sphere. Second, <u>the distortion measure</u> determined that the polar cap was the best distribution of source points about a code vector. We remark as a generalization that this conditional distribution would result from an L-dimensional probability density proportional to

$$\exp[-\|\underline{u}-\underline{u}_m\|^2/2d^*] = \exp[-d(\underline{u},\underline{u}_m)/2d^*]$$
$$= \prod_{\ell=1}^{L} \exp[-d(u_\ell,u_{m,\ell})/2d^*] \tag{16}$$

Third, where we sprinkled or distributed the code vectors (uniformly on the sphere of radius $\sqrt{\sigma^2-d^*}$) depended on <u>both</u> the source distribution and the distortion measure.

The above was an overview of the coding theorem and what makes it work for a Gaussian source under a squared-error criterion. Let us next comment briefly on the situation that applies to the general case, specifically how a set of optimum code vectors is chosen. For a given source α, let γ' denote the transition distribution that achieves the minimum in Eq. (2). Then we have a joint distribution on u, $\tilde{u}$

$$f_{\alpha\gamma'}(u,\tilde{u}) = f_\alpha(u)f_{\gamma'}(\tilde{u}|u) \tag{17}$$

which determines a marginal distribution on $\tilde{u}$ which depends on <u>both</u> α and γ'

$$f_{\alpha\gamma'}(\tilde{u}) = \int_{-\infty}^{\infty} du\, f_{\alpha\gamma'}(u,\tilde{u}) \tag{18}$$

In the general case, an optimum code can be determined by drawing L-dimensional code vectors $\underline{u}_m$ according to the distribution

$$f_{\alpha\gamma'}(\underline{\tilde{u}}) = \prod_{\ell=1}^{L} f_{\alpha\gamma'}(\tilde{u}_\ell) \tag{19}$$

The essence of the coding theorem is unchanged; what is required is more general methods of bounding P_{d^*}; the interested reader is referred to Gallager [2, sec. 9.3], Berger [4, sects. 3.1,3.2], or Sakrison [3, pp. 148].

The optimization problem of Eq. (2) is a difficult one mathematically; further, the fact that the optimizing γ depends on both the source and distortion measure greatly complicates things as does the fact that the distribution used to generate code vectors depends on both α and γ'. As a prelude to modularizing the coding problem and perhaps simplifying it, let us consider the special case in which the distortion measure is a function not of both u and $\tilde{u}$, but only of the error or difference $u-\tilde{u}$, $d(u-\tilde{u})$ (termed then a <u>difference distortion</u> measure). For this case there is a useful bound which is determined by a much simpler optimization problem. The expression for the rate distortion function can be manipulated as follows

$$R_\alpha(d^*) = \inf_{\gamma \in \Gamma(d^*)} I_{\alpha\gamma}(U;\tilde{U})$$

$$= \inf_{\gamma \in \Gamma(d^*)} [H_\alpha(U) - H_{\alpha\gamma}(U|\tilde{U})]$$

$$= H_\alpha(U) - \sup_{\gamma:\ E_{\alpha\gamma}\{d(U-\tilde{U})\} \leq d^*} H_{\alpha\gamma}(U-\tilde{U}|\tilde{U})$$

$$\geq H_\alpha(U) - \sup_{\gamma:\ E_{\alpha\gamma}\{d(U-\tilde{U})\} \leq d^*} H_{\alpha\gamma}(U-\tilde{U}) \tag{20}$$

Now the maximization on the right-hand-side of Eq. (20) does not depend on α or U; specifically, letting $Z = U-\tilde{U}$, we have

$$R_\alpha(d^*) \geq R_{\alpha L}(d^*) \triangleq H_\alpha(U) - \sup_{E\{d(Z)\} \leq d^*} H(Z) \tag{21}$$

This is a much simpler optimization problem; the maximizing distribution is easily determined by variational methods to be of the form

$$p_Z(z) = c_1 \exp[-d(z)/c_2] \tag{22}$$

with c_2 picked such that $E\{d(Z)\} = d^*$ and c_1 such that the density has mass one. The right-hand-side of (21) is known as the Shannon Lower Bound. Berger [4, sec. 4.3.1] has a much more elegant derivation which yields the following very useful result:

$$R_\alpha(d^*) = R_{\alpha L}(d^*) \tag{23}$$

if and only if the source variable U can be expressed as the sum of two independent variables, $U = W+Z$, with Z having the distribution of Eq. (22).†

This result is extremely useful for two reasons. First the equality given in Berger's result does pertain in many situations: for most α and d( ) this condition will be met for all values of d* less than some critical value. Although this critical value is typically small compared to $E\{d(U)\}$, the region of practical interest is when $E\{d(U-\tilde{U})\} << E\{d(U)\}$, i.e. when the "power" of the distortion is small compared to the "power" of the source. Secondly, when equality applies, the optimizing density of Eq. (22) depends only on d( ) and not on α and can be found explicitly.

---

†Note that as $d^* \to 0$ this condition becomes weaker and weaker; thus $R_\alpha(d^*) \to R_{\alpha L}(d^*)$ as $d^* \to 0$ for a broad class of distributions and distortion measures.

This means that our optimum quantizer can be designed only from knowledge of d( ) and need not involve knowledge of the source distribution α (since the optimizing γ for $R_\alpha(d^*)$ is the optimizing γ for Eq. (20)). Proper functioning of the encoder does depend on α, but this portion of the problem can be broken off and solved elsewhere, perhaps in conjunction with channel encoding. How this separation should be effected we consider in the next section by means of two examples.

DESIGNING CODERS OR QUANTIZERS WHEN THE SOURCE DISTRIBUTION IS ONLY PARTLY KNOWN

In this section we explore by means of two examples several related facets of the source coding problem: how the effects of the source distribution and distortion measure may be separated out in the encoder design, how to proceed when the source distribution is only partially known, and how this can lead to modularization of the problem.

Encoding a nongaussian source of known variance

Let us reconsider the coding problem of the previous section in which the source variables $U_1, U_2, \ldots$ were i.i.d. random variables of zero mean and variance σ^2 and the distortion measure is square-error. Now however, let us assume that the distribution is not Gaussian and that nothing other than the mean and variance are known to the designer of the encoder; this is a fairly realistic situation.

Let us start by observing what happens to the distribution of a block $\underline{U}$ of L source variables when the source is nongaussian. The i.i.d. property and the law of large numbers dictate that, for large L, nearly all the probability mass still has to lie in a shell about the L-dimensional sphere of radius σ; i.e. inequality (10) still holds. What changes is that a nongaussian source will not have a spherically symmetric distribution and hence nearly all the probability is confined to some proper subset of the spherical surface.† Our coding vectors thus need to cover only that subset, hence we need less of them and the Gaussian source is the worst source (requires the largest rate) of all sources with given variance [1, part V, 3, pp. 131,5,6].

Let us now ask to what extent we need to find a new solution to the coding problem. We will focus on the case in which the

---

†This follows from the "typical" set property of a long block of independent random variables, where the probability density is nearly constant over a set of probability close to one; this is the continuous counterpart of Shannon's Theorem 3 [1].

Rate Distortion function equals the Shannon Lower Bound (Eq. (23) holds) or when the two rates are nearly the same. Equality holds when U can be represented as $U = U' + U_g$ in which U_g is a Gaussian random variable of variance d^* (see Eq. (22)). This will hold for d^* sufficiently small. What is the geometric interpretation of this? Consider a code point $\underline{u}_m$ as in Figure (3); if the quantization region $\mathcal{U}_m$ still has source points $\underline{u}$ distributed in a spherical cap about $\underline{u}_m$, then this corresponds to the components of $U-\underline{u}_m$ having an i.i.d. Gaussian distribution of variance d^*. This situation pertains when U has (or nearly has) a uniform distribution on the surface of the sphere within the spherical cap. We note that this will not happen for all values of d^*, since the distribution of U is no longer spherically symmetric or uniform over the surface of the sphere of radius σ. However, if $\sqrt{d^*}$ is sufficiently small that the density of U is nearly uniform within a sphere of radius $\sqrt{d^*}$, then the distribution of coding error is (nearly) the same as for the Gaussian Source and the Rate Distortion function (nearly) equals the Shannon Lower Bound.

If in this situation we use a set of M code vectors chosen for the Gaussian source (picked according to a uniform distribution on a sphere of radius $\sqrt{\sigma^2-d^*}$) then the coding errors will still be distributed nearly in the optimum manner: a uniform distribution on the spherical cap. What has changed from the Gaussian situation is that these caps or coding regions $\mathcal{U}_m$ contain or cover different amounts of source probability mass and hence have different probabilities of being generated by the encoder. (For L large they will essentially fall into two groups: one group, containing nearly all the source mass, will contain caps that contain nearly equal amounts of source mass; the other group will collectively cover only a negligible amount of source mass.)

The aware reader should grasp (or pause to grasp) at this point the implication that the above facts have as to what extent we need a new solution to the coding problem for the nongaussian case. If we already have a good code for the Gaussian Source we may use it as the first portion (the quantizer) of an encoder for any distribution with the same variance. What remains to be done is to design a digital code which takes into account the fact that the code point indices m which are to be transmitted occur with unequal probability. There are many methods for efficiently encoding discrete random variables, among them Huffman coding [2, section 3.2] and Lexographic codes [7]. We will not elaborate on this problem of encoding a discrete source variable (the index m) since it is one of the oldest topics in Information Theory, and because other papers in this volume consider how this discrete source encoding may be combined with channel coding.

The conclusion of the above is as follows. Consider encoding a source under a difference distortion measure when the distortion d* is sufficiently small that the Rate Distortion function (nearly) equals the Shannon Lower Bound. Then the distribution of coding errors is determined only by the distortion measure (does not depend on the source distribution), is the one that optimizes Eq. (21), and is of the form of Eq. (22). Further, if the source distribution is known only up to its second moments, then an optimum encoder may be built by designing a code for the worst source distirbution (a Gaussian Distribution which is symmetric and uniform on the surface determined by the second-order constraint) and using that code as a quantizer. The digital output of this quantizer (the indices m of the code points $\underline{u}_m$) can be encoded using discrete coding methods. The result of this two-step procedure can be made efficient, i.e. the overall rate can be made arbitrarily close to that given by R(d*). We have given only a plausibility argument for this statement in the case in which U is a random variable; a mathematically tight argument can be made in the more general case [6].† The efficiency of the quantizer portion of the encoder can be measured by the entropy of the encoding error, $H(\underline{U}-\underline{u}_m)$, relative, to $\max_{E\{d(Z)\} \leq d^*} H(Z)$. Similarly the efficiency of the digital part of the encoder is measured by the rate of the encoder relative to the entropy of the output of the quantizer, H(M).

Encoding of a Gaussian Source of Unknown Variance

The above observations are the principal contributions of this paper. Let us briefly explore by means of a second example another facet of encoding when the source is not completely known.

This time we assume for simplicity that $U_1, U_2, \ldots$ is a sequence of i.i.d. zero-mean Gaussian random variables whose variance, σ^2, is known only to lie within some interval $[\sigma_\ell^2, \sigma_u^2]$. How do we go about encoding the U_i in this situation? Since the Rate Distortion function $R_{gauss}(d^*) = 1/2 \log_2(\sigma^2/d^*)$ depends on σ^2, we must ask what our constraint is; we could continue to demand that $E\{d(U,\tilde{U})\} = d^*$ and ask to minimize the transmission rate, or fix the transmission rate and seek to minimize $E\{d(U,\tilde{U})\}$. The corresponding Distortion Rate function is the inverse of the Rate Distortion Function. Either approach

---

†This can be noted from the remark on page 305 of [6] together with the fact that when the Rate Distortion function equals the Shannon Lower Bound then the γ in $\Gamma_\alpha(d^*)$ that minimizes $I_{\alpha\gamma}(U;\tilde{U})$ is the γ that minimizes the right hand side of Eq. (21)

is valid; which one we adopt depends on the problem environment. In real-time transmission of a source (such as moving images in TV format) over a given channel we have no choice but to fix the rate and settle for the minimum d* consistent with the source distribution. In relaying images back from Mars, we may instead demand a d* comparable to the scanning optics and use whatever transmission period is determined by the source rate, $R_\alpha(d^*)$. (In the preceeding section we implicitly adopted this fixed d* approach).

In the example at hand, the solution to the fixed rate situation is simple. One builds a good encoder (quantizer) for some reference level σ_r^2 and a corresponding d* that yields the required rate. For a source with unknown σ^2, we simply collect a long block of variables, estimate σ^2, scale the variables by σ_r/σ, encode, and rescale the decoded output $\tilde{U}$ at the receiver by σ/σ_r. We leave it to the reader to verify that this simple procedure is optimum.

If we fix the distortion and seek to minimize the rate, the situation is somewhat more complex, since the code for a source of variance σ^2 should have $(\sigma/\sqrt{d^*})^L$ code points distributed on a sphere of radius $\sqrt{\sigma^2-d^*}$. The code thus depends in a non-trivial way on σ^2. One approach would be to generate K codes for K values of $\sigma^2, \sigma_k^2, k = 1,\ldots,K$, in the interval $[\sigma_\ell^2,\sigma_u^2]$, combine all the points from these codes into one union code and pick the approximation $\underline{u}_m$ to $\underline{u}$ from this union code. This method will not yield the rate $R_{gauss}(d^*) = 1/2 \log_2 \sigma^2/d^*$ if we encode a source output $\underline{u}$ into the closest $\underline{u}_m$; most of these $\underline{u}_m$ would lie on a sphere of radius larger than $\sqrt{\sigma^2-d^*}$ and would hence yield a rate larger than $1/2 \log_2 \sigma^2/d^*$. The correct solution is to note that in optimum encoding at average distortion level d* the block error $\|\underline{u}-\underline{u}_m\|$ is nearly always close to d*. A correct coding algorithm would thus be based on a strategy such as picking the first $\underline{u}_m$, m = 1,2,... such that $|d(\underline{u}-\underline{u}_m)-d^*| < \varepsilon$, where ε is small compared to d*.

APPLICATION TO ACTUAL SOURCES

The preceeding discussion has been in the context of a source whose output was a scalar-valued random variable. While mathematically and conceptually easy to work with, such a source does not model those physical sources of greatest interest. In this section we direct our discussion to more closely model sources of practical interest. We first extend some of the ideas to multidimensional sources and then conclude by describing an image encoding algorithm that embodies many of the ideas we have discussed.

Multidimensional Sources

Most physical sources are not described directly by scalar random variables. One source of practical interest might be speech, in which a block of source output might be a waveform u(t) defined on a T second interval, $[t_0, t_0+T]$. This source would be presented by a random process U(t). A second source of practical interest would be an image source in which a block of source output might be a log-intensity function u(x,y) defined on a raster $0 \leq x \leq X$, $0 \leq y \leq Y$, or a sequence of such functions defined on this raster.

Let us consider how to represent such a source in order to apply Rate Distortion Theory. For ease of description we will consider speech, but our remarks apply to image representation simply by considering functions of two variables (or three variables for moving (TV) images). Suppose then we initially take a block as a sample u(t), $t \in [0,T]$, with integral-square error distortion measure

$$d(u,\tilde{u}) = \int_0^T [u(t)-\tilde{u}(t)]^2 dt \tag{24}$$

A more useful distortion measure might be frequency-weighted-integral-square-error, in which v(t) and $\tilde{v}(t)$ are respectively u(t) and $\tilde{u}(t)$ filtered (or preemphasized) with a frequency weighting that represents the sensitivity of the listener and

$$d(u,\tilde{u}) = \int_0^T [v(t)-\tilde{v}(t)]^2 dt \tag{25}$$

Let us immediately make a considerable mathematical simplification. In Rate-Distortion Theory there is always one potential for simplification that should not be overlooked. The theory is constructed on the realization that we are only interested in an approximate description of the source; thus we should always look to see if by adding some small amount, ε, of average distortion we can simplify the problem. If so, all results we obtain for the simplified version of the problem apply to the real one since ε can be made arbitrarily small. In the case at hand, we note that for the distortion measure Eq. (25) we can describe any waveform (of finite energy) to within ε by a finite dimensional vector, either a vector whose components are samples of U(t) or a finite number of Fourier Coefficients of U(t) in some appropriate orthonormal expansion. We will do this since it allows us to use ordinary multidimensional calculus, probability densities, and differential (integral) entropy. A set of exponential Fourier coefficients would be particularly appropriate if we are using a frequency-weighted error criterion, for the frequency weighting then applies in a direct manner. We thus take our source output to be an n-dimensional vector $\underline{U}$. The reader who takes comfort in specific associations should associate the components of $\underline{U}$ with the first n sine-cosine expansion coefficients on $[0,T]$; it is conceivable that a sample of T seconds contains sufficient statistical regularity that this will also serve as a block for coding purposes; if not, we might form a block by simultaneously encoding a number of such vectors.

By Parseval's Theorem, the distortion criterion of Eq. (24) (or (25) if the components are frequency weighted) becomes simply the n-dimensional Euclidean norm

$$d(\underline{u},\underline{\tilde{u}}) = \|\underline{u}-\underline{\tilde{u}}\|^2 = \sum_{j=1}^{n} [u_j-\tilde{u}_j]^2 \tag{26}$$

A rotational (orthogonal) transformation leaves the distances of vectors (and hence distortion) unchanged; we may thus consider any transformed version of $\underline{u}$ that may be convenient. It will make life much simpler if we assume that $\underline{U}$ has been transformed so that

$$E\{U_i U_j\} = \sigma_i^2\, \delta_{ij} \qquad \delta_{ij} = \begin{cases} 1 & i=j \\ 0 & i\neq j \end{cases}$$
$$\sigma_i^2 \geq \sigma_{i+1}^2 \tag{27}$$
$$i,j = 1,2,\ldots,n$$

(i.e. that $\underline{U}$ represents the Karhunen-Loève representation of the source). If the U_j are Fourier sine-cosine coefficients, then the U_j are already approximately uncorrelated for T long compared to the "correlation time" of the correlation function of U(t) [8].

We now simply state some useful results from Rate Distortion Theory. First is the Rate Distortion Function for a Gaussian source under the Euclidean norm distortion measure. Let a source output be a sequence of i.i.d. zero-mean Gaussian <u>vectors</u> with covariances given by Eq. (27). Then the Rate Distortion Function $R(d^*)$ is easily described parametrically in terms of the parameter μ. Let $j(\mu)$ denote the largest j such that $\sigma_j^2 > \mu$; then the Rate Distortion Function per vector is given by

$$d^*(\mu) = \mu\, j(\mu) + \sum_{j=j(\mu)+1}^{n} \sigma_j^2 = \sum_{j=1}^{n} \min(\mu,\sigma_j^2) \tag{28}$$

$$R(\mu) = (1/2)\sum_{j=1}^{j(\mu)} \log_2(\sigma_j^2/\mu) \tag{29}$$

These two equations can be interpreted with the help of Figure 5; none of the components U_j for which $\sigma_j^2 \leq \mu$ are transmitted and contribute the error under the tail in the figure; the remaining

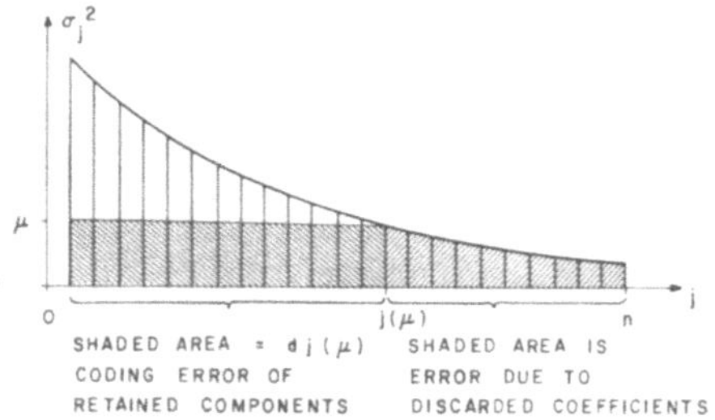

Figure 5 Interpretation of Error Terms in Expression for the RDF of a Gaussian Source.

components are all transmitted with distortion μ, requiring a rate of $1/2 \log_2 \sigma_j^2/\mu$ for the jth component. This rate can also be expressed

$$R(\mu) = H(\underline{U}') - H_{g\mu}(\underline{Z}') \tag{30}$$

in which $\underline{U}'$ is the vector of the first $j(\mu)$ components of $\underline{U}$ and $\underline{Z}'$ is a $j(\mu)$ dimensional error vector of i.i.d. Gaussian random variables of variance μ. The derivation of the Rate Distortion function for a scalar Gaussian source follows neatly from Berger's variational formulation of the calculation of the R(d*) function [4, pp. 99]; the multidimensional case follows using matrix properties and simple multidimensional optimization arguments [4, sect. 4.5,2].

The scalar random variables U_j can be grouped together and encoded; since they are Gaussian and uncorrelated they are independent and the remarks made earlier would apply to encoding a group of the U_j with near-equal variance. In fact a block of Gaussian variables of different variances can be efficiently encoded or simultaneously quantized [2, Theo. 9.7.1]. Further, only 0.25 bit per coefficient is lost by quantizing the coefficients individually (as with an A-D converter) and then Huffman encoding the quantized variables [9].

We can now easily give an explicit expression for the Shannon Lower Bound in the multidimensional mean-square-error case. The covariance of $\underline{U}$ is still that given by Eq. (27), but $\underline{U}$ is no longer presumed Gaussian. Let $\underline{U}''$ denote the components of $\underline{U}$ not in $\underline{U}'$. Then if $\underline{U}'$ and $\underline{U}''$ are independent, the bound is given by Eqs. (28) and (30). If not, $H(\underline{U}')$ must be changed to $H(\underline{U}',\underline{U}'')$ and the entropy for a Gaussian vector with the same covariance as $\underline{U}''$

$$H_g(\underline{U}'') = (1/2) \sum_{j>j(\mu)} \log_2(2\pi e \lambda_j) \qquad (31)$$

must be subtracted from (30).
Equality holds in the bound under the following conditions. Let $\underline{U}'$ represent the first $j(\mu)$ coefficients of $\underline{U}$ and $\underline{U}''$ the last $n-j(\mu)$. Then if U' and $\underline{U}''$ are independent and $\underline{U}'$ can be written as $\underline{U}' = \underline{W}'+\underline{Z}'$ with $\underline{W}'$ and $\underline{Z}'$ independent and $\underline{Z}'$ Gaussian with i.i.d. components of variance μ, equality holds. For small d*, and hence small μ, this is a reasonable model for many sources of practical interest. This statement is somewhat well known, its derivation being an extension of some results of Berger [2, pp. 90-93]. We leave it as an exercise for the reader to reconcile Eqs. (29 and 30) with Eq. (21).

Example: An Image Encoding Algorithm

We now describe briefly an image encoding algorithm which exploits many of the concepts so far discussed. To begin with, we must consider what distortion measure is appropriate for evaluating image degradation. A number of psychophysical experiments have been done which measure the detectability of various patterns on a uniform field and superimposed on an image. These are directly applicable to determining a distortion measure d(,) at threshold levels of distortion (where errors are just barely noticable). We simply state some conclusions of this work; for a detailed summary of this topic the reader is referred to Sakrison [10]. In detecting threshold image stimuli, a human observer is sensitive to contrast, rather than absolute intensity. The sensitivity of an observer to gratings of different spatial frequencies is a function of spatial frequency, being a maximum at about 4 cycles per degree (of arc subtended in the observer's field of view) and falling off rapidly for higher spatial frequencies and slowly for lower spatial frequencies. Further, patterns consisting of a number of spatial frequency components are seemingly detected independently; i.e. the pattern is at threshold when one component is at threshold, independent of the level of the others. Each component is detected by an incoherent detector with a bandwidth of about ± one-half its center frequency and with a nonlinearity which is close to a 6th power law at low levels and nearly linear at higher levels [10]. One could abstract from this description a numerical distortion criterion and it would not resemble a weighted-integral-square of the log-intensity error $u(x,y)-\tilde{u}(x,y)$; i.e., one could find error patterns which would be small under one criterion and large under the other. However, the entropy of the error distribution that is optimal (has maximum entropy) under the weighted-integral-square error criterion is not much different than the entropy of the error distribution that is optimal under the criterion appropriate to the human observer [10]. For this reason, and because of its overwhelming analytic tractability, we consider here weighted-integral-square error of the log intensity.

Next we consider the source distribution. A useful starting point is to consider a log-intensity profile across a horizontal scan line as shown in Figure 6. This function is clearly not a

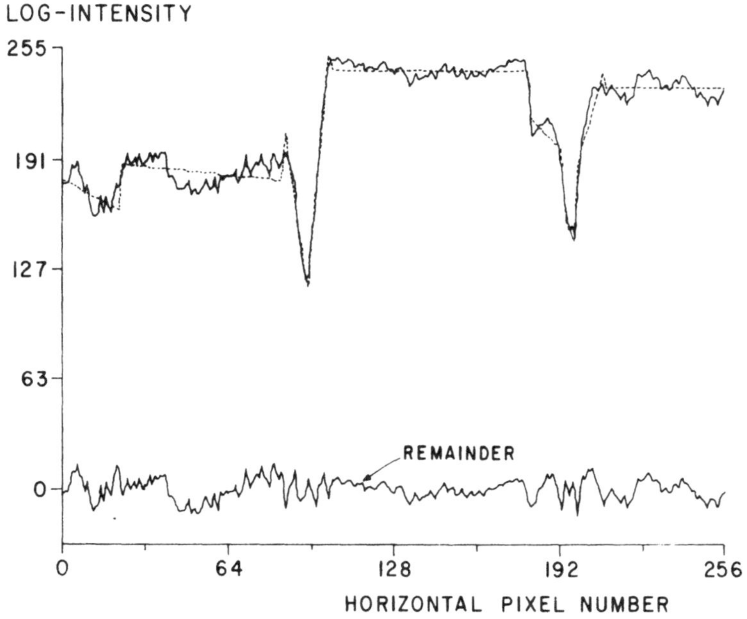

Figure 6 Horizontal Scan Lines of the original log-intensity u(x,y) (solid line), straight-line fit to u(x,y) or discontinuous component $u_d(x,y)$ (dashed line), and remainder image $r(x,y) = u(x,y)-u_d(x,y)$ (lower solid line).

sample function from a stationary Gaussian process; the abrupt changes cause a very nongaussian density function for the difference of two samples taken about 4 pixels apart. Thus a Gaussian source model is inappropriate. However, we might consider the log-intensity to be the sum of two components, a discontinuous one, shown as the dashed line in Figure 6, and a continuous remainder component which is obtained by subtracting this discontinuous component from the original log-intensity. This remainder component shown in Figure 6 rather resembles a sample function from a Gaussian process. If the source is reasonably represented by $U(x,y) = U_d(x,y) + R(x,y)$ in which U_d represents the discontinuous component, R is a Gaussian Random Field, and U_d and R are independent, then the preceeding discussion of the Shannon Lower Bound implies that the image source can be encoded efficiently by separately encoding U_d and R and that the

Rate Distortion Function is given by the distortion value of Eq. (28) and a rate per vector of

$$R_L(\mu) = H(\underline{U}') - H_{g\mu}(\underline{Z}) = H(\underline{U}_d') + H(\underline{R}') - H_{g\mu}(\underline{Z}) \qquad (32)$$

This argument overlooks one point: the Gaussian vector $\underline{R}'$ is presumed to contain $\underline{Z}$; i.e. all the components of $\underline{R}'$ must be $\geq \mu$. Since the $(j(\mu)+1)$th component of $\underline{U}$, which is the sum of the corresponding components of $\underline{U}_d$ and $\underline{R}$, has variance $\leq \mu$, this is not a reasonable assumption. An analysis can be made which eliminates the need for this assumption, which leaves the rate of Eq. (32) unchanged, and which adds to the right-hand-side of Eq. (28) a distortion term representing the error involved in estimating $\underline{U}_d''$ from $\underline{U}_d'$.

We have in fact implemented a code based on this model on a mini-computer [11]. The component U_d is quantized to one pixel in position and coarsely in log-intensity. It is described in terms of a differential description (along a line and from line-to-line) of the discrete parameters that describe its breakpoints. The component R(x,y) is encoded (quantized) using a near-optimum code [9] for the Gaussian Source: it is expanded in a Fourier series, the low r.m.s. coefficients discarded, and the high r.m.s. coefficients quantized to the same level. The breakpoint parameters and quantized coefficients then need to be encoded using discrete coding methods to reduce their data rate. This last step could be incorporated with error-correcting encoding for the channel.

REFERENCES

1. C.E. Shannon, The Mathematical Theory of Communication; Illini Books; Urbana; 1964. Also available in Key Papers in the Development of Information Theory; D. Slepian, Ed; IEEE Press, 1973.
2. R.G. Gallager, Information Theory and Reliable Communication, J. Wiley; New York; 1968.
3. D.J. Sakrison, Notes on Analog Communication, Van Nostrand-Reinhold; New York; 1970.
4. T. Berger, Rate Distortion Theory, A Mathematical Basis for Data Compression, Prentice-Hall, 1971.
5. D.J. Sakrison, "A Geometric Treatment of the Problem of Source Encoding a Gaussian Random Variable," IEEE Trans. on Info. Theory, Vol. IT-14, pp. 481-486, September 1968; also available in Benchmark Papers in Electrical Engineering and Computer Science, Vol. 14, Data Compression, L. Davisson and R. Gray Eds., Dowden, Hutchison, and Ross, pp. 96-101, Stroudsberg, Pa.
6. D.J. Sakrison, "Worst Sources and Robust Codes for Difference Distortion Measures," IEEE Trans. on Info. Theory, Vol. IT-21,

pp. 301-309, 1975.

7. T.J. Lynch, "Sequence time coding for data compression," Proc. IEEE (Lett.), Vol. 54, pp. 1490-1491, Oct. 1966; and L.D. Davisson, Comments on "Sequence time coding for data compression," Proc. IEEE (Lett.), Vol. 54, p. 2010, Dec. 1966.
8. W.L. Root and T.S. Pitcher, "On the Fourier-Series Expansion of Random Functions," Ann. of Math. Stat., Vol. 26, pp. 313-318, June 1955.
9. T.J. Goblick, Jr. and J.L. Holsinger, "Analog Source Digitization: A Comparison of Theory and Practice," IEEE Trans. on Info. Theory, Vol. IT-13, pp. 323-326, Apr., 1967.
10. D.J. Sakrison, "On the Role of the Observer and a Distortion Measure in Image Transmission," IEEE Trans. on Comm., special issue on image bandwidth compression, to appear in summer 77.
11. J.K. Yan and D.J. Sakrison, "Encoding of Images based on a Two-Component Source Model," IEEE Trans. on Comm., special issue on image bandwidth compression, to appear summer 77.

ACKNOWLEDGEMENT

Research sponsored by the Joint Services Electronics Program Contract F44620-76-C-0100 and the National Science Foundation Grant ENG75-10063. Grateful thanks are due to Dimitri Anastasiou for reading the manuscript and offering many useful comments and to Cindy Tast for typing the manuscript.

ÇONVOLUTIONAL-CODES FOR INTEGRATED CODING*

J.P.M. Schalkwijk

Department of Electrical Engineering, University of Technology, Eindhoven, The Netherlands°

Lectures presented at the NATO Advanced Study Institute on Communication Systems and Random Process Theory, August 8-20,1977, Darlington, England.

These two lectures concern the relation between complexity and performance in (binary) convolutional coding. In particular, we study the hardware requirements of the syndrome decoder for channel coding, as this syndrome implementation of the maximum likelihood (ML) decoding algorithm appears most economical. The core part of the syndrome decoder for channel coding is formed by a syndrome former, and an error sequence estimator. The syndrome former can be used as a source encoder, and the error sequence estimator as the corresponding source decoder.

The first lecture concerns the state space of the syndrome former. Alternative definitions of "state" are considered. The source- and sink-tuples are defined. We mention the concept of "metric-equivalence" that will be further explored in the second lecture.

The second lecture is on metric functions and error sequence estimation. Metric equations are introduced, as is the concept of a "survivor". We define a class $\Gamma_{n,h,\ell}$ of syndrome formers. For this class of syndrome formers the number of metric equivalence classes, $N_{n,h,\ell}$ is exponentially less than the number of states.

---

* This research was supported in part by the U.S. Navy under Contract N00123-76-C-0842.

° Present address.

It is shown that in the implementation of the ML error sequence estimator one metric/path-register combination per metric equivalence class suffices. Thus, compared with one metric/path-register combination per state in the traditional implementation one obtains an exponential saving in hardware! In sequential estimation of the error sequence, i.e. in Fano-,and Stack-decoding, one also obtains major savings [6] in the number of computations and in the required storage by exploiting the state space symmetries of $\Gamma_{n,h,\ell}$.

I. SYNDROME SEQUENCE FORMING

For a state space analysis it is convenient to represent an n-input, 1-output syndrome former by an n-tuple (A,B,C,...,D) of binary polynomials, see Fig. 1. Note that an equivalent syndrome former can be realized having only h memory cells. The circuit of Fig. 1 with nh memory cells has been chosen for mathematical convenience, i.e. an n-input, (n-k)-output syndrome former can then be considered as a set of n-k syndrome formers as in Fig. 1 that have the nh memory cells in common. If these n-k syndrome formers are coherent [6], the hardware saving state space symmetries of $\Gamma_{n,h,\ell}$ can be preserved.

Obviously, one single noise vector in the sequence $\ldots,[e_{1,-1}, e_{2,-1},\ldots,e_{n,-1}]^T, [e_{10},e_{20},\ldots,e_{n0}]^T, [e_{11},e_{21},\ldots,e_{n1}]^T,\ldots$ can at most influence h+1 successive syndrome digits. We define the "physical state" of the system to be the nh-dimensional binary vector representing the contents of all shift register stages in Fig. 1. Every noise vector that enters the system causes a transition of its physical state and gives rise to a binary

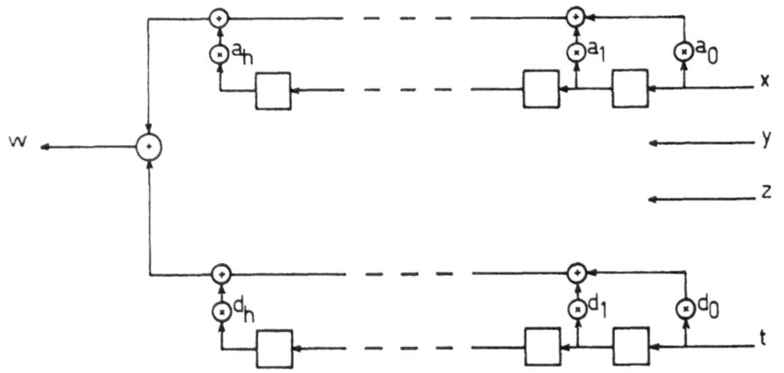

Fig. 1. The syndrome former for a rate-(n-1)/n convolutional code.

syndrome digit. The phenomenom occurs that two different initial physical states are syndrome-indistinguishable, i.e. that under every noise vector sequence $[e_{10},e_{20},\ldots,e_{n0}]^T$, $[e_{11},e_{21},\ldots,e_{n1}]^T,\ldots$ their syndrome sequences are identical. It is left to the reader [3,4] to prove that this natural concept of syndrome indistinguishability is exactly the same as the following equivalence relation: Two physical states are called equivalent if their difference has a sequence of syndrome digits identically zero under a sequence of noise vectors identically zero. In fact, we may restrict ourselves in this definition to sequences of zero-noise vectors of length h, since all following zero-noise vectors simply must yield zero-syndrome digits.

The equivalence classes of the above equivalence relation will be called "abstract states", or briefly "states" of the system. There are several equivalent state descriptions. In ref. [3] Schalkwijk and Vinck use the contents of the bottom register D of the syndrome former, Fig. 1, as a description of the state. Forney [4] uses the zero-noise syndrome sequence to represent the state. In the present paper we opt for this latter description.

We are now ready to introduce some convenient notation: States (given by their zero-noise syndrome sequence) are denoted by lower case greek letters with a subscript, e.g.

$$\sigma_1 \triangleq [s_1,s_2,s_3,\ldots,s_{h-2},s_{h-1},s_h],\ \text{and its left shifts}$$

$$\sigma_2 \triangleq [s_2,s_3,s_4,\ldots,s_{h-1},s_h,0]\ ,$$

$$\sigma_3 \triangleq [s_3,s_4,s_5,\ldots,s_h,0,0]\quad,\ \text{and so on.}$$

Occasionally, i.e. if sufficiently many terminating components $s_h, s_{h-1},\ldots$ vanish, we also write the right shifts, e.g.

$$\sigma_0 \triangleq [0,s_1,s_2,\ldots,s_{h-3},s_{h-2},s_{h-1}]\ \text{if}\ s_h = 0,$$

$$\sigma_{-1} \triangleq [0,0,s_1,\ldots,s_{h-4},s_{h-3},s_{h-2}]\ \ \text{if}\ s_h = s_{h-1} = 0.$$

Finally, we introduce the symbols α_1, β_1, $\gamma_1,\ldots,\delta_1$ to denote the generator states of the system, i.e.

$$\alpha_1 \triangleq [a_1,a_2,\ldots,a_h]\ ,$$

$$\beta_1 \triangleq [b_1,b_2,\ldots,b_h]\ ,$$

$$\gamma_1 \triangleq [c_1,c_2,\ldots,c_h]\ ,$$

$$\vdots$$

$$\delta_1 \triangleq [d_1,d_2,\ldots,d_h]\ .$$

Without loss of generality we assume $a_h=1$. This assumption is justified by the definition of h and implies that the state space has dimension h.

The output of the syndrome former, see Fig. 1, at time t and the state time t+1 are completely determined by the state σ_1 and the input $[e_{1t},e_{2t},\ldots,e_{nt}]^T$ at time t, t=...,-1,0,+1,... . As the syndrome former is supposed to be time invariant there is no purpose in retaining the subscript t in the state space analysis. Thus, we denote the syndrome former input by $[x,y,z,\ldots,t]^T$. The corresponding state transition and the output ω are given by

$$\begin{array}{l} \qquad [x,y,z,\ldots,t]^T \\ \qquad\quad \downarrow \\ \sigma_1 \longmapsto \sigma_2 + x\alpha_1 + y\beta_1 + z\gamma_1 + \ldots + t\delta_1 \\ \qquad \searrow \\ \qquad\quad \omega = s_1 + xa_0 + yb_0 + zc_0 + \ldots + td_0 \ . \end{array} \tag{1}$$

Now consider the linear subspace spanned by the generators α_1, β_1, $\gamma_1,\ldots,\delta_1$. If the subspace has dimension q then according to (1) each state σ_1 has exactly 2^q state transition images. Again by (1), these images from a coset of the linear subspace $L[\alpha_1,\beta_1,\gamma_1,\ldots,\delta_1]$. This coset will be called the "sink-tuple" of σ_1.

The linear subspace $L[\alpha_1,\beta_1,\gamma_1,\ldots,\delta_1]$ is identical to the linear subspace $L[\alpha_1,\beta_1 + b_h\alpha_1, \gamma_1 + c_h\alpha_1,\ldots, \delta_1 + d_h\alpha_1]$. However, as $a_h=1$, the vectors $\beta_1 + b_h\alpha_1$, $\gamma_1 + c_h\alpha_1,\ldots$, $\delta_1 + d_h\alpha_1$ have a rightmost coordinate equal to 0. Thus, these vectors have a right shift. Furthermore,

$$\operatorname{rank}\begin{bmatrix} a_1 & ,\ldots, & a_{h-1} & ,1 \\ b_1+b_ha_1 & ,\ldots, & b_{h-1}+b_ha_{h-1} & ,0 \\ \vdots & & & \vdots \\ d_1+d_ha_1 & ,\ldots, & d_{h-1}+d_ha_{h-1} & ,0 \end{bmatrix} = \operatorname{rank}\begin{bmatrix} 1,0 & ,\ldots, & 0 \\ 0,b_1+b_ha_1 & ,\ldots, & b_{h-1}+b_ha_{h-1} \\ \vdots & & \vdots \\ 0,d_1+d_ha_1 & ,\ldots, & d_{h-1}+d_ha_{h-1} \end{bmatrix}$$

Define,

$$\varepsilon_1 \triangleq [1,0,0,\ldots,0] \ ,$$

a row vector of length h. Then

$$\dim L[\alpha_1,\beta_1,\gamma_1,\ldots,\delta_1]=\dim L[\varepsilon_1,(\beta+b_h\alpha)_0,(\gamma+c_h\alpha)_0,\ldots,(\delta+d_h\alpha)_0] \ .$$

Each state has at least one preimage. If $\tau_1 = [s_1, s_2, \ldots, s_{h-1}, 0]$, then $\tau_0 = [0, s_1, \ldots, s_{h-2}, s_{h-1}]$ is a preimage under $[x, y, z, \ldots, t] = [0, 0, 0, \ldots, 0]$. If $\tau_1 = [s_1, s_2, \ldots, s_{h-1}, 1]$, then $(\tau+\alpha)_0$ is a preimage under $[x, y, z, \ldots, t] = [1, 0, 0, \ldots, 0]$. But, if a state τ_1 has a preimage then is has at least 2^q preimages, i.e. all the states in the coset of $L[\varepsilon_1, (\beta+b_h\alpha)_0, (\gamma+c_h\alpha)_0, \ldots, (\delta+d_h\alpha)_0]$ that contains the particular preimage. We now have the following results. Each state σ_1 has exactly 2^q images, i.e. the sink-tuple of σ_1. On the other hand, each state τ_1 has at least 2^q preimages, i.e. the above mentioned coset of $L[\varepsilon_1, (\beta+b_h\alpha)_0, (\gamma+c_h\alpha)_0, \ldots, (\delta+d_h\alpha)_0]$. Hence, we conclude that τ_1 has exactly 2^q preimages that constitute the "source-tuple" of τ_1. It is easily verified that each element σ_1 of a source-tuple has the same sink-tuple.

It is this source/sink-tuple description of the state space that will play an important role in the remainder of the paper. Hence, to make things more concrete, we give a specific example for the syndrome former of Fig. 2.
We have

$\alpha_1 = [1\ 1\ 0\ 1]_2 = 13$ $\qquad\qquad \varepsilon_1 = [1\ 0\ 0\ 0]_2 = 8$

$\beta_1 = [1\ 0\ 0\ 1]_2 = 9$ $\qquad\qquad (\alpha+\beta)_0 = [0\ 0\ 1\ 0]_2 = 2$

Source-tuples		Sink-tuples
{0, 2, 8,10}	I →	{0, 4, 9,13}
{1, 3, 9,11}	II →	{2, 6,11,15}
{4, 6,12,14}	III →	{1, 5, 8,12}
{5, 7,13,15}	IV →	{3, 7,10,14}

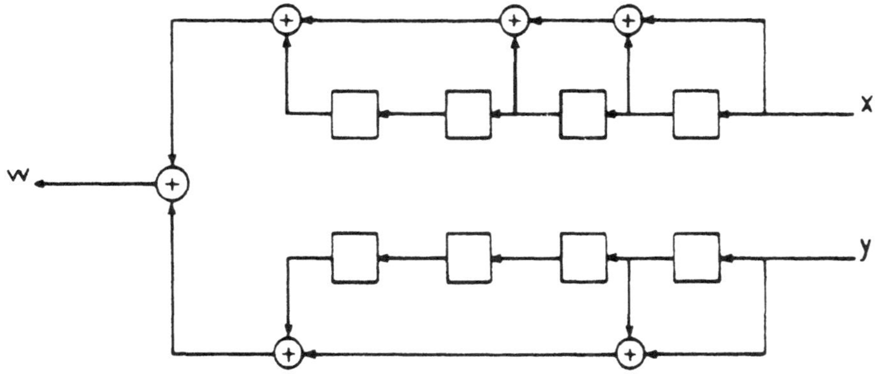

Fig. 2. The syndrome former for a rate-½ convolutional code.

Fig. 3 shows a partition of the state space in source/sink-tuples.

sink-tuples \ source-tuples	I	II	IV	III
I	0	9	13	4
II	2	11	15	6
IV	10	3	7	14
III	8	1	5	12

Fig. 3. State space partition in source/sink-tuples.

Anticipating on Section II the states in Fig. 3 have been geometrically arranged in such a way that also the metric equivalence classes {0}, {4}, {8}, {12}, {9,13}, {6,14}, {1,5}, {2,10}, and {3,7,11,15} are easily distinguishable. Two states that are in the same metric equivalence class have the same metric value [5].

II. ERROR SEQUENCE ESTIMATION

As the estimation algorithm to be described in this section is similar to Viterbi's [5], we can be very brief. To find the required state sequence Viterbi introduces a "metric function". A metric function is defined as a nonnegative integer-valued function on the states. With every state transition we now associate the Hamming weight W_H of its noise vector $[x,y,z,\dots,t]^T$.

PROBLEM: Given a metric function f and a syndrome digit ω, find a metric function g which is statewise minimal, and for every state is consistent with at least one of the values of f on its preimages under syndrome digit ω, increased by the weight of its corresponding state transition.

The solution to this problem expresses g in terms of f and ω, and can be formulated in terms of the source/sink-tuples of Section I. In fact, the values of g on a sink-tuple T_i are completely determined by the values of f on the corresponding source-tuple S_i, and by the syndrome digit ω. The equations that express g in terms of f and ω are called "metric equations". They have the form

$$g(\tau_1)=\min\{f(\sigma_1)+W_H([x,y,z,\dots,t]^T)\,|\,\sigma_1 \xrightarrow[\omega]{[x,y,z,\dots,t]^T} \tau_1\} \,. \quad (2)$$

The particular preimage σ_1 in (2) that realizes the minimum is called the "survivor". When there are more preimages for which the minimum in (2) is achieved, one could flip a multi-coin to determine the survivor. However, we will shortly discover that a judicious choice of the survivor among the candidate preimages offers the possibility of significant savings in decoder hardware. The construction of (2) can be repeated, i.e. starting with a metric function f_0, given a syndrome sequence $\omega_1,\omega_2,\omega_3,\ldots$ one can form a sequence of metric functions $f_1,f_2,f_3,\ldots$ iteratively by means of the metric equations:

$$f_0 \xmapsto{\omega_1} f_1 \xmapsto{\omega_2} f_2 \xmapsto{\omega_3} f_3 \longmapsto \ldots \quad .$$

The metric function f_s, whose value $f_s(\sigma_1)$ at an arbitrary state σ_1, equals the Hamming weight of the lightest path from the zero-state to σ_1 under an all zero syndrome sequence, $\omega_1,\omega_2,\omega_3,\ldots =$ $= 0,0,0,\ldots$, is called the "stable metric function". It has the property

$$f_s \xmapsto{\omega=0} f_s \ .$$

Without further ado we now introduce the class $\Gamma_{n,n,\ell}$ of rate-(n-1)/n binary convolutional codes (A,B,C,...,D), i.e. in terms of their syndrome formers, that exhibits state space symmetries that allow for an exponential reduction of decoder hardware. To wit $(A,B,C,\ldots,D) \in \Gamma_{n,h,\ell}$ if and only if

$$a_h = 1 \tag{3a}$$

$$a_j = b_j \ ; \ 0 \le j \le \ell-1 \tag{3b}$$

$$a_j = b_j \ ; \ h-\ell+1 \le j \le h \tag{3c}$$

$$C,\ldots,D \text{ all have degree } \le h-\ell \tag{3d}$$

$$\gcd(A,B,C,\ldots,D) = 1 \tag{3e}$$

$$L[\varepsilon_1,(\alpha+\beta)_0,\gamma_0,\ldots,\delta_0] \cap L[(\alpha+\beta)_1,\ldots,(\alpha+\beta)_{\ell-1}] = \{0\} \ . \tag{3f}$$

The code $A(X) = 1+X+X^2+X^4$, $B(X) = 1+X+X^4$ of Fig. 2 is an element of $\Gamma_{2,4,2}$. As a consequence of (3) we have

$$\Gamma_{n,h,1} \supset \Gamma_{n,h,2} \supset \Gamma_{n,h,3} \supset \ldots \quad . \tag{4}$$

If condition (3e) is satisfied, then it follows from the invariant factor theorem [4] that the n-tuple $(\Lambda,B,C,\ldots,D)$ is a set of syndrome polynomials for some non-catastrophic rate-$(n-1)/n$ convolutional code (in fact, for a class of such codes).

Assume $\Gamma_{n,h,\ell} \neq \phi$. For $(A,B,C,\ldots,D) \in \Gamma_{n,h,\ell}$ an "ℓ-singleton state" is defined to be a state the last ℓ components of which vanish . Linear combinations and left shifts of ℓ-singleton states are ℓ-singleton states, too. For every state ϕ_1 the states $\phi_i (i \geq \ell+1)$ are ℓ-singleton states. We have the following lemma, the proof of which is left to the reader.

LEMMA 1: For every state σ_1 there exists a unique ℓ-singleton state $\phi_{\ell+1}$ and a unique index set $I \subset \{1,2,\ldots,\ell\}$ such that

$$\sigma_1 = \phi_{\ell+1} + \sum_{i \in I} \alpha_i \ . \tag{5}$$

Using this lemma we now associate with the state σ_1 the set $[\sigma_1]^{(\ell)}$ defined by

$$[\sigma_1]^{(\ell)} = \{\phi_{\ell+1} + \sum_{i \in I} [\alpha_i + r_i (\alpha+\beta)_i] \,|\, r_i \in \{0,1\} \text{ for all } i\} \ .$$

We shall prove the following theorem:

THEOREM 2: The collection of all sets $[\sigma_1]^{(\ell)}$ forms a partition of the state space.

PROOF: Obviously the union of all $[\sigma_1]^{(\ell)}$ is equal to the state space. So, we only have to prove that $[\sigma_1]^{(\ell)} = [\sigma_1']^{(\ell)}$ whenever $[\sigma_1]^{(\ell)} \cap [\sigma_1']^{(\ell)} \neq \phi$. Let us assume that $[\sigma_1]^{(\ell)} \cap [\sigma_1']^{(\ell)} \neq \phi$. Then there exist r_i and s_i such that

$$\phi_{\ell+1} + \sum_{i \in I} [\alpha_i + r_i (\alpha+\beta)_i] = \phi'_{\ell+1} + \sum_{i \in I'} [\alpha_i + s_i (\alpha+\beta)_i] \quad ,$$

or

$$\phi_{\ell+1} - \phi'_{\ell+1} + \sum_{i \in I} r_i (\alpha+\beta)_i - \sum_{i \in I'} s_i (\alpha+\beta)_i = \sum_{i \in I \Delta I'} \alpha_i \quad .$$

Now the LSH of above equation is an ℓ-singleton state by (3c) so that the symmetric difference $I \Delta I'$ must be empty, in other words $I=I'$. Therefore we get

$$\phi'_{\ell+1} - \phi_{\ell+1} = \sum_{i \in I} (r_i - s_i)(\alpha+\beta)_i \quad ,$$

i.e. $\phi'_{\ell+1}$ and $\phi_{\ell+1}$ differ by some linear combination of $\{(\alpha+\beta)_i | \, | i \in I\}$. But then we must have $[\sigma_1]^{(\ell)} = [\sigma_1']^{(\ell)}$, since in the construction of these classes all linear combinations of

$\{(\alpha+\beta)_1 | i\epsilon I\}$ are involved.

Q.E.D.

COROLLARY: Based on the partition of the state space according to Theorem 2 an equivalence relation $R_{n,h,\ell}$ can be defined, where two states σ_1 and σ_1' are called $R_{n,h,\ell}$-equivalent iff $[\sigma_1]^{(\ell)} = [\sigma_1']^{(\ell)}$.

The one-element equivalence classes of $R_{n,h,\ell}$ consist of exactly one ℓ-singleton state. An example are the states 0,4,8, and 12 in Fig. 3. The number $N_{n,h,\ell}$ of $R_{n,h,\ell}$-equivalence classes can be found as follows. First, take $I \subset \{1,2,\ldots,\ell\}$ in (5) fixed, and let j denote the cardinality of I. The last ℓ components of an ℓ-singleton state are zero. Hence, there are $2^{h-\ell}$ ℓ-singleton states. Now 2^j of these $2^{h-\ell}$ ℓ-singleton states correspond to the same $R_{n,h,\ell}$-equivalence class, i.e. all ℓ-singleton states differing by a linear combination of $\{(\alpha+\beta)_i | i\epsilon I\}$. Hence, there are $2^{h-\ell-j}$ $R_{n,h,\ell}$-equivalence classes for each I of cardinality j. Thus

$$N_{n,h,\ell} = \sum_{j=0}^{\ell} \binom{\ell}{j} 2^{h-\ell-j} = 2^{h-2\ell} 3^{\ell} \quad . \tag{6}$$

THEOREM 3: Let $(A,B,C,\ldots,D) \epsilon \Gamma_{n,h,\ell}$, and assume that $1 \leq \ell' \leq \ell$. Then every $R_{n,h,\ell}$-equivalence class of $(A,B,C,\ldots,D)$ is a union of $R_{n,h,\ell'}$-equivalence classes of $(A,B,C,\ldots,D)$, cf (4).

PROOF: Let σ_1 and τ_1 be $R_{n,h,\ell'}$-equivalent states of $(A,B,C,\ldots,D)$. Then we may write for some $r_i \epsilon \{0,1\}$, $i \epsilon I' \subset \subset \{1,2,\ldots,\ell'\}$:

$$\sigma_1 = \phi_{\ell'+1} + \sum_{i\epsilon I'} \alpha_i \quad ,$$

$$\tau_1 = \phi_{\ell'+1} + \sum_{i\epsilon I'} [\alpha_i + r_i(\alpha+\beta)_i] \quad .$$

On the other hand, for some $I'' \subset \{\ell'+1,\ldots,\ell\}$ and some $\Psi_{\ell+1}$ we have

$$\phi_{\ell'+1} = \Psi_{\ell+1} + \sum_{i\epsilon I''} \alpha_i \quad .$$

Letting $I = I' \cup I''$, $r_i=0$ for $i\epsilon I''$ we now obtain

$$\sigma_1 = \Psi_{\ell+1} + \sum_{i\epsilon I} \alpha_i$$

$$\tau_1 = \Psi_{\ell+1} + \sum_{i\epsilon I} [\alpha_i + r_i(\alpha+\beta)_i] ,$$

i.e. σ_1 and τ_1 are $R_{n,h,\ell}$-equivalent Q.E.D.

In Fig. 3 we exhibit the $R_{2,4,2}$-equivalence classes for the $A(X) = 1+X+X^2+X^4$, $B(X) = 1+X+X^4$ code. We claimed that any two states within the same equivalence class have the same metric value irrespective of the noise vector sequence input. We are now ready to prove this result.

THEOREM 4: Assume that $(A,B,C,...,D) \epsilon \Gamma_{n,h,\ell}$. Let f_0 be any starting metric function, and let $\omega_1,\omega_2,\omega_3,...$ be any syndrome sequence. Then every iterate f_u is constant on the $R_{n,h,u}$-equivalence classes of $(A,B,C,...,D)$, $1 \leq u \leq \ell$.

PROOF: The proof is by induction on u. Consider the two $R_{n,h,1}$-equivalent states $\phi_2+\alpha_1$, and $\phi_2+\beta_1$. Obviously they belong to the same sink-tuple. We list their preimages, corresponding noise vectors and symdrome digits according to (1).

Preimage	$\phi_2+\alpha_1$	$\phi_2+\beta_1$
	Noise;Syndrome	Noise;Syndrome
$\phi_1 \quad +z\gamma_0+...+t\delta_0$	$[1,0,z,...,t]^T;\omega_1$	$[0,1,z,...,t]^T;\omega_1$
$\phi_1 \quad +(\alpha+\beta)_0+z\gamma_0+...+t\delta_0$	$[0,1,z,...,t]^T;\omega_1$	$[1,0,z,...,t]^T;\omega_1$
$\phi_1+\varepsilon_1 \quad +z\gamma_0+...+t\delta_0$	$[1,0,z,...,t]^T;\bar{\omega}_1$	$[0,1,z,...,t]^T;\bar{\omega}_1$
$\phi_1+\varepsilon_1+(\alpha+\beta)_0+z\gamma_0+...+t\delta_0$	$[0,1,z,...,t]^T;\bar{\omega}_1$	$[1,0,z,...,t]^T;\bar{\omega}_1$

We see that on every line, i.e. for every preimage the syndrome bits and the Hamming weights of the state transitions to $\phi_2+\alpha_1$, and $\phi_2+\beta_1$ are identical. Hence, $f_1(\phi_2+\alpha_1) = f_1(\phi_2+\beta_1)$ for every f_0 and every ω_1. This proves the assertion for u=1. Now let us assume that the statement is true for a fixed u, $1 \leq u \leq \ell-1$. Let f_0 be any starting metric function and let $\omega_1,\omega_2,\omega_3,...$ be any syndrome sequence. Then, by our induction hypothesis, f_u is constant on the $R_{n,h,u}$-equivalence classes. Let χ_1 and χ_1' be any pair of $R_{n,h,u}$-equivalent states. Then there is a state Ψ_{u+1} and an index set $I \subset \{1,2,...,u\}$ such that for some $r_i \epsilon \{0,1\}$

$$\chi_1 = \Psi_{u+1} + \sum_{i\in I} \alpha_i \quad ,$$
$$\chi'_1 = \Psi_{u+1} + \sum_{i\in I} [\alpha_i + r_i(\alpha+\beta)_i] .$$

We now consider the cosets S and S' of $L[\varepsilon_1,(\alpha+\beta)_0,\gamma_0,\ldots,\delta_0]$ to which χ_1 and χ'_1 belong, respectively, and compare them element wise. The states

$$\chi_1 + p\varepsilon_1 + q(\alpha+\beta)_0 + r\gamma_0 + \ldots + s\delta_0 \quad ,$$

and

$$\chi'_1 + p\varepsilon_1 + q(\alpha+\beta)_0 + r\gamma_0 + \ldots + s\delta_0$$

are obviously $R_{n,h,u}$-equivalent for all $p,q,r,\ldots,s \in \{0,1\}$, since by the definition of ε_1 and by (3c,d) the last u components of $p\varepsilon_1 + q(\alpha+\beta)_0 + r\gamma_0 + \ldots + s\delta_0$ vanish. Furthermore, by (3b) we have

$$\sum_{i\in I} a_i = \sum_{i\in I} [a_i + r_i(a_i+b_i)] \quad .$$

Hence, by (1) the preimages

$$\chi_1 + p\varepsilon_1 + q(\alpha+\beta)_0 + r\gamma_0 + \ldots + s\delta_0 \quad ,$$

and

$$\chi'_1 + p\varepsilon_1 + q(\alpha+\beta)_0 + r\gamma_0 + \ldots + s\delta_0$$

give rise to identical syndrome digits in response to an input vector $[x,y,z,\ldots,t]^T$. These arguments together, however, imply that the values of f_{u+1} on the corresponding state transition images are equal and, hence, f_{u+1} is constant on the $R_{n,h,u+1}$-equivalence classes of $(A,B,C,\ldots,D)$. Q.E.D.

Theorem 4 proves that one needs only one metric register for each $R_{n,h,\ell}$-equivalence class. We will now show that, except for the last $\ell-1$ stages, the same is true for the path registers. Let $(A,B,C,\ldots,D) \in \Gamma_{n,h,\ell}$. Condition (3f), where $\{(\alpha+\beta)_1,(\alpha+\beta)_2,\ldots,(\alpha+\beta)_{\ell-1}\} = \{\underline{0}\}$ for $\ell=1$, implies that a coset of $L[\varepsilon_1,(\alpha+\beta)_0,\gamma_0,\ldots,\delta_0]$ and a coset of $L[(\alpha+\beta)_1,(\alpha+\beta)_2,\ldots,(\alpha+\beta)_{\ell-1}]$ can have at most one element in common, i.e.

<u>LEMMA 5</u>: No two distinct $R_{n,h,\ell-1}$-equivalent states can belong to the same source tuple.

On the other hand, from the proof of Theorem 4 it follows that whenever χ_1 and χ_1' are $R_{n,h,\ell-1}$-equivalent, then the same holds for the states

$$\chi_1 + p\varepsilon_1 + q(\alpha+\beta)_0 + r\gamma_0 + \ldots + s\delta_0 \quad ,$$

and

$$\chi_1' + p\varepsilon_1 + q(\alpha+\beta)_0 + r\gamma_0 + \ldots + s\delta_0 \; , \quad p,q,r,\ldots,s \in \{0,1\} \; ,$$

that form the source-tuples containing χ_1 and χ_1'. These results lead to a natural equivalence between source-tuples. Two source-tuples are said to be equivalent if they contain a pair of $R_{n,h,\ell-1}$-equivalent states. It is left to the reader to prove that this relation is an equivalence relation. The unique and natural one-to-one correspondence between the states of two equivalent source-tuples, that is induced by the intersection with $R_{n,h,\ell-1}$-equivalence classes is, by the proof of Theorem 4, consistent with the algebraic difference structure of the source-tuples. Hence, in view of Theorem 4, we see that for the m-th iterate f_m, $m \geq \ell-1$, of any metric function f_0 under any syndrome sequence $\omega_1, \omega_2, \omega_3, \ldots$ the values of f_m on the corresponding states of two equivalent source-tuples are identical.

Given two successive iterates f_{j-1} and f_j, $j \geq \ell$, of a metric function f_0, linked by the syndrome digit ω_j,

$$f_{j-1} \xrightarrow{\;\omega_j\;} f_j \quad .$$

In Viterbi decoding [5] one determines for each state τ_1 a survivor σ_1, such that

$$f_j(\tau_1) = f_{j-1}(\sigma_1) + W_H([x,y,z,\ldots,t]^T) \quad ,$$

subject to (2). Survivors of a state τ_1 in the sink-tuple T_i always belong to the corresponding source-tuple S_i, see Section I. However, as discussed in this Section, there are situations in which more than one survivor may be chosen, i.e. when two or more σ_1's in (2) achieve the minimum. In this case, one has a choice of two possible strategies that result in the same decoded error rate by transmission over a binary symmetric channel (BSC), i.e. (i) flip a (multi) coin, or (ii) decide for every tie-pattern once and for ever which survivor shall be taken. We shall use the second strategy, that according to the properties of equivalent source-tuples can be realized in the following way: Whenever two source-tuples S_i and S_i' are equivalent (and, hence, have statewise identical f_{j-1}-values) then let for the respective sink-tuples T_i and T_i' statewise corresponding survivors be chosen in such a way that $R_{n,h,1}$-equivalent states get the same survivor. Given a sequence of metric function iterates

$$f_0 \overset{\omega_1}{\longmapsto} f_1 \overset{\omega_2}{\longmapsto} f_2 \longmapsto \dots \longmapsto f_{j-2} \overset{\omega_{j-1}}{\longmapsto} f_{j-1} \overset{\omega_j}{\longmapsto} f_j, \quad j \geq \ell \;,$$

then for every state σ_1 a sequence of successive survivors can be constructed

$$\sigma_1^{(-j)} \longleftarrow \sigma_1^{(-j+1)} \longleftarrow \sigma_1^{(-j+2)} \longleftarrow \dots \longleftarrow \sigma_1^{(-2)} \longleftarrow \sigma_1^{(-1)} \longleftarrow \sigma_1 \;,$$

and the following theorem holds.

THEOREM 6: If σ_1 and η_1 are distinct $R_{n,h,m}$-equivalent states then $\sigma_1^{(-m)} = \eta_1^{(-m)}$, $m=1,2,\dots,\ell$.

PROOF: The proof is by induction on m. For m=1 the assertion is part of our assumption above. Now assume that the statement is true for m=u, u fixed, $1 \leq u \leq \ell-1$, and let σ_1 and η_1 be two $R_{n,h,u+1}$-equivalent states, that are not $R_{n,h,u}$-equivalent (otherwise $\sigma_1^{(-u)} = \eta_1^{(-u)}$ and, hence, immediately $\sigma_1^{(-u-1)} = \eta_1^{(-u-1)}$). Then we may write

$$\sigma_1 = \phi_{u+2} + \alpha_{u+1} + \sum_{i \in I} \alpha_i$$

$$\eta_1 = \phi_{u+2} + \beta_{u+1} + \sum_{i \in I} [\alpha_i + r_i (\alpha+\beta)_i] \;,$$

where $I \subset \{1,2,\dots,u\}$. It is easy to find preimages $\tilde{\sigma}_1$ and $\tilde{\eta}_1$ of σ_1 and η_1 respectively, viz.

$$\tilde{\sigma}_1 = \phi_{u+1} + \alpha_u + \sum_{i \in I \setminus \{1\}} \alpha_{i-1} \;,$$

$$\tilde{\eta}_1 = \phi_{u+1} + \beta_u + \sum_{i \in I \setminus \{1\}} [\alpha_{i-1} + r_i (\alpha+\beta)_{i-1}] \;.$$

Obviously, $\tilde{\sigma}_1$ and $\tilde{\eta}_1$ are $R_{n,h,u}$-equivalent and, hence, by Theorem 3 also $R_{n,h,\ell-1}$-equivalent. Therefore, the source-tuples containing $\tilde{\sigma}_1$ and $\tilde{\eta}_1$ are equivalent. Furthermore, we observe that

$$\sigma_1 = \tilde{\sigma}_2 + \begin{cases} 0 & \text{if} \quad i \notin I \\ \alpha_1 & \text{if} \quad i \in I \end{cases} ,$$

$$\eta_1 = \tilde{\eta}_2 + \begin{cases} 0 & \text{if} \quad i \notin I \\ \alpha_1 + r_1(\alpha+\beta)_1 & \text{if} \quad i \in I \end{cases} .$$

Hence, because of the assumption made above, the survivors $\sigma_1^{(-1)}$ and $\eta_1^{(-1)}$ are corresponding states, i.e. $R_{n,h,\ell-1}$-equivalent states. The algebraic difference structure of equivalent source-tuples is identical, hence,

$$\tilde{\sigma}_1 - \sigma_1^{(-1)} = \tilde{\eta}_1 - \eta_1^{(-1)} \in L[\varepsilon_1,(\alpha+\beta)_0,\gamma_0,\ldots,\delta_0] .$$

So, $\tilde{\sigma}_1 - \sigma_1^{(-1)} = \tilde{\eta}_1 - \eta_1^{(-1)}$ is a u-singleton state. Hence, $\sigma_1^{(-1)}$ and $\eta_1^{(-1)}$ are $R_{n,h,u}$-equivalent and therefore, by the induction hypothesis, $\sigma_1^{(-u-1)} = \eta_1^{(-u-1)}$. Q.E.D.

Theorem 6 shows that except perhaps for the last $\ell-1$ stages, $R_{n,h,\ell}$-equivalent states have the same path register contents irrespective of the noise vector sequence input. Thus, roughly speaking, one needs only one path register for each $R_{n,h,\ell}$-equivalence class of states. By Theorem 4 one only needs one metric register per $R_{n,h,\ell}$-equivalence class. Hence, the complexity [1] of a syndrome decoder for a code $(A,B,C,\ldots,D) \in \Gamma_{n,h,\ell}$ is proportional to the number $N_{n,h,\ell}$ of $R_{n,h,\ell}$-equivalence classes, i.e. by (6) the complexity is proportional to $2^{h-2\ell}3^{\ell}$. As an example take a code in $\Gamma_{2,2\ell,\ell}$, i.e. a rate-$\frac{1}{2}$ code with

$$A(X) = X^{2\ell} + A_{2\ell-1}X^{2\ell-1} + \ldots + A_1X + 1 ,$$

$$B(X) = A(X) + X^{\ell} .$$

The syndrome decoder for such a code has complexity proportional to $3^{\ell} = (\sqrt{3})^h$. The classical Viterbi decoder [5] for the same code has complexity 2^h, hence, by exploiting the state space symmetry we achieve an exponential saving in hardware.

III. CONCLUSIONS

These lectures describe a class $\Gamma_{n,h,\ell}$ of syndrome formers that exhibit certain state space symmetries. These symmetries can be exploited to obtain an exponential reduction of the complexity of the corresponding ML estimator. Our results can be extended [6] to the class $\Gamma^{(n-k)}_{n,h,\ell}$ of n-input, (n-k)-output syndrome formers. Preliminary simulation results obtained by A.J.P. de Paepe, W.J.H.M. Lippmann, and A.J. Vinck indicate that the state space formalism can also be used to advantage in sequential decoding. In Fano-, and Stack-decoding one obtains major savings in the number of computations and in the required storage by exploiting the state space symmetries of $\Gamma^{(n-k)}_{n,h,\ell}$.

Finally, the distance properties of $\Gamma^{(n-k)}_{n,h,\ell}$ are such that these results are very useful for both source-, and channel-coding.

ACKNOWLEDGEMENT

The author wants to thank his coworkers K.A. Post, A.J. Vinck, A.J.P. de Paepe, A.P.C. van Schendel, and W.J.H.M. Lippmann for many fruitful discussions on the subject of syndrome decoding. Thanks are also due to Mrs. G.H. Driever-van Hulsen for the accurate typing of the manuscript.

REFERENCES

[1] J.P.M. Schalkwijk and A.J. Vinck, "Syndrome decoding of convolutional codes", IEEE Trans. Commun. (Corresp.), vol. COM-23, pp. 789-792, July 1975.

[2] J.P.M. Schalkwijk, "Symmetries of the state diagram of the syndrome former of a binary rate-½ convolutional code", Lecture Notes, CISM Udine Summer School on Coding, Udine, Italy, September 2-12, 1975.

[3] J.P.M. Schalkwijk and A.J. Vinck, "Syndrome decoding of binary rate-½ convolutional codes", IEEE Trans. Commun., vol. COM-24, pp. 977-985, September 1976.

[4] G.D. Forney, Jr., "Convolutional codes I: Algebraic Structure", IEEE Trans. Inform. Theory, vol. IT-16, pp. 720-738, November 1970; also, correction appears in vol. IT-17, p. 360, May 1971.

[5] A.J. Viterbi, "Convolutional codes and their performance in communication systems", IEEE Trans. Commun. Technol. (Special Issue on Error Correcting Codes - Part II), vol. COM-19, pp. 751-772, October 1971.

[6] J.P.M. Schalkwijk, A.J. Vinck, and K.A. Post, "Syndrome Decoding of Binary Rate-k/n Convolutional Codes", University of Technology, Eindhoven, The Netherlands, TH-report 77-E-73, ISBN 6144 073 4, March 1977.

ROBUST LOW BIT RATE VOICE ENCODING

J.B. Anderson

Dept. of Elec. Eng., and Communications Research Lab., McMaster Univ., Hamilton, Ont., Canada

ABSTRACT. We review the principles of a new type of robust low bit rate voice encoding called tree encoding, and trace its origin in the theory of source encoding. We discuss improved tree (or trellis) codes, tree search algorithms, and the performance of actual hardware. These waveform encoders achieve telephone quality speech at 16Kbits/s or below, with white noise during speech sounds and no idle channel noise. The output symbol stream is simple and resistant to channel errors.

1. INTRODUCTION

Virtually all waveform speech encoders function sequentially. Included are such well-known examples as delta-modulation, PCM, differential PCM (DPCM), and their adaptive variants. In these schemes, a sequence of decisions is made uniformly in time, each decision according to the same set of rules. DPCM, for instance, predicts using a fixed structure and then quantizes the prediction error. The possible decision outcomes of such a waveform encoder can be graphically listed on a code tree like that of Fig. 1. On each branch lies an analog (*i.e.*, high precision) number $\hat{x}_t$ representing an outcome at time t, and b branches stem from each node. One speech sample x_t arrives for each tree level. An output bit rate of $\log_2 b$ bits/sample will specify any one path through the tree.

Contemporaneous with the development of these traditional speech digitizers was the evolution of the theory of rate-distortion. In 1958, Shannon published theorems giving the least rate R at which block source codes could operate, and still achieve an average

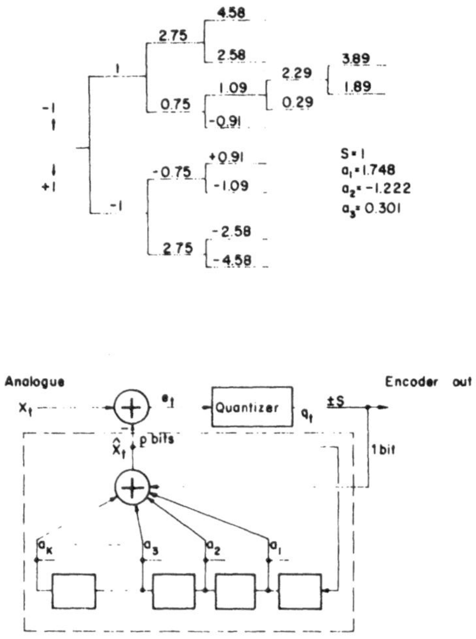

Fig. 1: Circuit and code tree for DPCM, based on K-stage recursive prediction. Example has 2 branches out of each node; hardware constructed has 4.

distortion D. There were few applications for these block codes, however, and further progress had to await the proof by Jelinek [14] in 1969 that codes with a tree structure satisfied the same relation between R and D. It was still not known how to use a tree code, except by an impossible, exhaustive search among all words.

To encode speech (or any other source), a tree encoder must exhibit various tentative code words, among the possibilities defined by the code tree, compare the respective speech samples, and release an output that denotes its final decision, all in an efficient way. Heretofore, waveform digitizers have developed only a single path through the code tree, although coding theory would suggest that several paths be carried forward in parallel and a choice made among them only after their encoding performance is known.

Multi-path encoding methods which are sequential in nature, and therefore use a code tree, are called sequential coding procedures. These first appeared in the late 1950's in the work of Fano and Wozencraft on sequential decoders for error correction. The application of these methods to analog-to-digital conversion and other theoretical source coding problems did not occur until

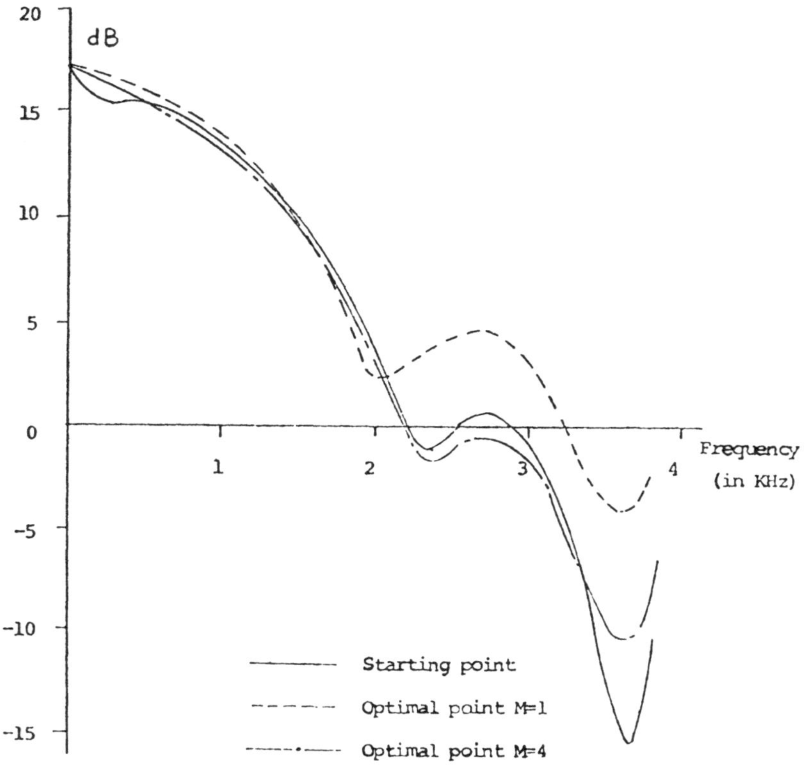

Fig. 2: Magnitude spectra of optimized code generating filters at M=1 and 4 tree paths searched; solid line is the starting point for the optimization. Length K = 4; 2-sec typical sentence of male speaker.

the early 1970's [1]-[4]; Anderson and Bodie [5] applied sequential encoding to the practical problem of speech digitization. Outside of coding theory, similar procedures have developed under different names. The term "delayed encoder" has been used, since an encoder pursuing more than one code tree path of necessity delays its choice among them. The earliest "tree" speech digitizer known to the author in the open literature was the delayed encoder of Cutler [6].

In its full generality, the tree coding problem divides into two subproblems, the choice of which outcomes $\hat{x}_t$ will be listed in the tree, the code design, and the choice of a procedure to consider some but not all of the tree paths, the design of the search algorithm.

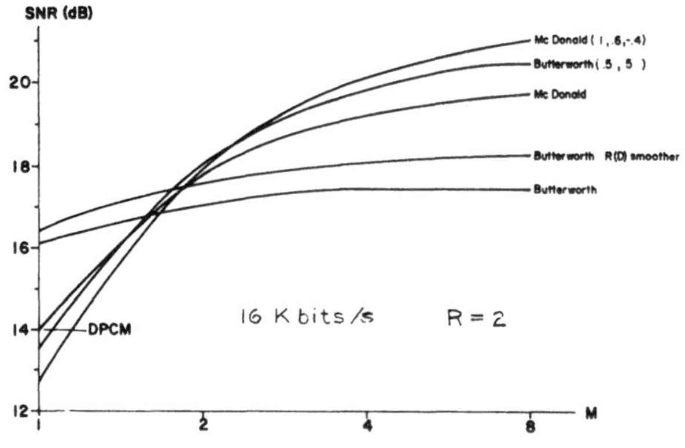

Fig. 3 : Overall encoding noise in dB SNR vs. searching intensity M, for several codes (labels refer to code names). Earlier devices use a single-path search (M=1).

2. NEW SPEECH TREE CODES

It appears that the best non-adaptive code trees in the sense of square-error are generated by simple digital filter structures, and in fact, by the convolutional structure well known in other areas of coding.

Among the traditional waveform methods which are non-adaptive, the best code trees are those of DPCM. The tree in Fig. 1 is for DPCM, a circuit for which appears in the figure; the code design is based on least mean-square prediction. A close look will show that the code letters are actually generated by a recursive digital filter (in the dashed box), whose inputs are the quantizer outputs. If one considers these b-level quantizer outputs to be b-ary symbols, the resulting sequence defines which path is taken through the code tree; this sequence $\tilde{q}$ is therefore called a path map. Although DPCM is recursively generated, it has been found ([5], Sect. IV) that an equivalent transversal generating filter with the same b-level input is more efficient in terms of both arithmetic and storage, when used with a multi-path search algorithm. Such a filter has infinitely many tap coefficients, but these decay rapidly, and may be truncated without loss. The transversal realization may be defined in terms of the convolution of the b-level $\tilde{q}$ and a (K+1)-length "generator" sequence $\tilde{c} = (c_o, c_1, \ldots, c_K)$,

$$\hat{x}_t = \sum_{i=o}^{K} c_i q_{t-i} \tag{1}$$

This is simply a convolutional code, identical in form to the familiar error-correcting convolutional codes, except for its mode of arithmetic.

When a multi-path tree search is used, several classes of codes are known to perform better than DPCM codes. Convolutional generators $\underset{\sim}{c}$ of the same length derived from rate-distortion theory can do better, and "smoothed" generators built up out of a lowpass filter in concatenation with the DPCM filter can do even better ([5], Sects. III, IV). Beyond DPCM-based codes, Uddenfeldt and Zetterberg [7] have studied single- and double-integration delta modulation codes for tree coding, with and without adaptation.

Nonetheless, these codes have shortcomings. All are based originally on a recursive structure with infinite memory, and not a priori on a finite convolution like (1). Applied to a non-stationary source like speech, their design procedures become heuristic; even universal coding theory, meant to apply to a source which exhibits one of a number of stationary modes, will not yield accurate answers, since in speech the modes last too short a time. Finally, code derivations always assume either a single-path or a completely exhaustive search of the code tree, and not the moderate searching done by a practical encoder, even though it is known that a code designed for one degree of searching generally will not perform well for another.

In work by Anderson and Law [8], a purely empirical optimization procedure was devised to find good, short-length codes of the form (1). A software copy of the actual encoder tests the code against speech, so that the final code will be matched to the encoder, and a composite speech record (several sentences, several speakers) insures that the code is not tuned to a specific record. Starting from some generator $\underset{\sim}{c}$, the procedure perturbs each coefficient of $\underset{\sim}{c}$ by δ and tests the new code, attempting to find a better generator on the "sphere" of radius δ. If it does, this becomes the centre for another attempt. The scheme amounts to hill-climbing, but instead of calculating a gradient, the direction of greatest improvement is found by testing the 2(K+1) nearest neighbors.

The outcome of this procedure is a code somewhat better (1-4 dB in long-term SNR) than those found by other means. A useful way to portray codes is through the magnitude spectrum of their generating filters, and Fig. 2 shows the spectra of the best codes found at 1 and 4 paths searched. More intensively searched codes require fewer high frequency components; Uddenfeldt and Zetterberg view this peculiarity of searching as enhanced ability to follow transients. A second conclusion of the optimization was that good, non-adaptive codes need not have generators of more than 5 coefficients.

3. TREE SEARCH ALGORITHMS

Beyond the rudimentary single path search, many effective search algorithms are known. These differ from each other in how paths are kept or dropped.

Some algorithms are efficient for a low investment of complexity, while others require high complexity and stringent performance criteria before their greater power appears. It is known, for instance, that the stack algorithm [3], the 2-cycle algorithm [1], and an algorithm due to Gallager [2], all must visit asymptotically about $\exp(C(D-\Delta)^{-1/2})$ nodes to encode each digit of the independent-letter binary source with Hamming distortion measure (see [9]). Here, $\Delta=\Delta(R)$ is Shannon's optimal distortion for encoders of rate R, and D is the average per-digit distortion achieved. On the other hand, a procedure called the M-algorithm visits about $\exp(C(D-\Delta)^{-3/4})$ nodes, a larger number, but for speech encoded with a "loose" D, a squared-error distortion and a small-scale search, the M-algorithm does much better than its asymptotically more powerful brothers.

In practical search algorithms, factors other than the number of nodes visited affect the effort required to find a path with a certain expected distortion. Whether or not the algorithm allows backtracking, how much it utilizes sorting or stacking, and the data structures it requires all influence its efficiency. Unfortunately, the interplay of these factors is not well understood.

For a practical, modestly complex encoder working with speech, the most effective algorithm is apparently the M-algorithm (originally described in [10]; see discussion in [3]). This scheme can be viewed as a highly truncated Viterbi algorithm. It pursues a fixed number M of paths at any given tree level; at each level, all bM branches are extended out of the M paths saved from the previous level (time t-1), and only the best M of these are saved for the next level. The computational heart of this procedure is a bubble sorter, which continuously orders the paths by distortion. Finite storage requires a constraint on the length of path map symbols kept, called the "memory length". The algorithm is synchronous, performing identical steps after each speech sample, an important practical advantage. The effect of the M-algorithm search on several different codes is shown in Fig. 3.

Very powerful algorithms, such as the stack algorithm, have been used to encode speech (Mohan [11]), but these very complex schemes gain 1 or 2 dB SNR at most. The full Viterbi algorithm can be used as well (see Becker and Viterbi [12] for a linear predictive coder application). However, research has shown [8] that a waveform digitizer requires about 1000 states in its trellis, but that only 4-8 of these are worth scrutiny at once. Thus the advantage of the truncated-Viterbi/M-algorithm, with its bubble sort among a few paths, over the full Viterbi algorithm, with its simpler but too numerous computations.

4. PRACTICAL HARDWARE AND ITS PERFORMANCE

An M-algorithm sequential speech encoder/decoder with M=4 and a bit rate of 16 Kbits/s has been constructed in TTL logic by the author and Ho [13]. Later prototypes should require about 80 chips, if none are custom made. The device samples speech at 8000 Hz, and has a code tree with b=4 branches out of each node, numbers which seem to be best for telephone-quality reproduction. Since two bits are needed to specify each path branch, and one branch corresponds to each speech sample, the total output rate is 16 Kbits/s. The distortion measure is square-error.

Tests have reproduced natural and intelligible speech of telephone quality at the full 4KHz bandwidth, with a certain quantity of white background noise during speech sounds, but no idle channel noise. Sounds with a high number of zero-crossings seem particularly well reproduced. White, uncorrelated encoding noise is characteristic of multi-path searching encoders; studies have shown [8] that as M increases, the noise spectrum subsides and rapidly flattens, and that its autocorrelation tends to an impulse. The long-term SNR at M=4 (including silences) is about 18 dB, rising to 22-24 dB during voiced sounds, an improvement of 4-8 dB over DPCM at the same rate. This is about the performance of adaptive DPCM.

As a practical encoding scheme, an important advantage of tree coding lies in its output bit stream. Successive 4-level output symbols are uniformly and quasi-independently distributed, for easy channel modulation. Symbol synch only is needed to decode the symbol stream, there being no words or sensitive special symbols. Finally, short-length codes like those in 2., whose use is only feasible with tree searching, lead to high error resistance (since an erroneous channel symbol affects only K+1 code word letters). Tests have shown that noise due to channel errors does not obscure speech until error rates rise above 0.01.

5. FUTURE PROSPECTS

As with any other scheme, the practicality of tree coding will take time to determine. Applications to mobile and tactical speech communication seem most likely. Other, non-waveform, methods certainly perform at much lower bit rates, but have much higher circuit complexity and require better data channels. Compared with the best waveform methods, tree encoding has roughly similar complexity, but of a very different type; the future of all methods depends to some degree on how well their complexity can be realized in LSI circuitry.

For speech communication, certain issues in the design of codes remain to be settled. It is not clear that square-error, without frequency weighting, is the best measure of goodness for the tree search, or that a white noise spectrum is desirable. Perhaps codes should be designed to concentrate noise in high-energy bandwidths of speech, where it will be less perceptible. More study is also needed on the effect of tree searching on adaptive codes, and on codes similar to those of delta-modulation. These codes have a long transversal realization, and so cannot be generated this way, but a short recursive structure. Their resistance to channel errors is very high because disturbances tend to be forced out of band.

From the theoretician's point of view, tree encoding, with its white, uncorrelated noise, and its rapid convergence toward the rate-distortion function, represents a powerful technique. For new codes, we can hope for a new theory of universal coding, aimed at sources composed of many stationary modes, each present for a short time. For tree searching algorithms, the future should bring better understanding of how structure affects performance.

REFERENCES

1. J.B. Anderson and F. Jelinek, "A 2-cycle algorithm for source coding with a fidelity criterion," IEEE Trans. Inform. Theory, vol. IT-19, pp. 77-92, Jan. 1973.
2. R.G. Gallager, "Tree encoding for sources with distortion measure," IEEE Trans. Inform. Theory, vol. IT-20, pp. 65-76, Jan. 1974.
3. J.B. Anderson, "A stack algorithm for source coding with a fidelity criterion," IEEE Trans. Inform. Theory, vol. IT-20, pp. 211-226, Mar. 1974.
4. A.J. Viterbi and J.K. Omura, "Trellis encoding of memoryless discrete-time sources with a fidelity criterion," IEEE Trans. Inform. Theory, vol. IT-20, pp. 325-332, May 1974.
5. J.B. Anderson and J.B. Bodie, "Tree encoding of speech," IEEE Trans. Inform. Theory, vol. IT-21, pp. 379-387, July 1975.
6. C.C. Cutler, "Delayed encoding: Stabilizer for adaptive coders," IEEE Trans. Commun. Technol., vol. COM-19, pp. 898-906, Dec. 1971.
7. J. Uddenfeldt and L.H. Zetterberg, "Algorithms for delayed encoding in delta modulation with speech-like signals," IEEE Trans. Commun., vol. COM-24, pp. 652-658, June 1976.

   --------------------------------, "Adaptive delta modulation with delayed decision," IEEE Trans. Commun., vol. COM-22, pp. 1195-1198, Sept. 1974.

8. J.B. Anderson and C.-W. Law, "Real-number convolutional codes for speech-like quasi-stationary sources," to appear, IEEE Trans. Inform. Theory.

9. J.B. Anderson, "Asymptotic computation of certain sequential algorithms for source coding with a fidelity criterion," IEEE Trans. Inform. Theory, vol. IT-22, pp. 82-83, Jan. 1976.

10. F. Jelinek and J.B. Anderson, "Instrumentable tree encoding of information sources," IEEE Trans. Inform. Theory, vol. IT-17, pp. 118-119, Jan. 1971.

11. S. Mohan, Ph.D. Research, McMaster Univ., 1976.

12. D.W. Becker and A.J. Viterbi, "Speech digitization and compression by adaptive predictive coding with delayed decision, NTC-75 Conf. Record, pp. 46-18 to 46-23, New Orleans, LA., Dec. 1975.

13. J.B. Anderson and C.-W.P. Ho, "Architecture and construction of a hardware sequential encoder for speech," to appear, IEEE Trans. Commun.

14. F. Jelinek, "Tree encoding of memoryless time-disctete sources with a fidelity criterion," IEEE Trans. Information Theory, vol. IT-15, pp. 584-590, Sept. 1969.

INTEGRATED DATA COMMUNICATION SYSTEMS WITH DATA COMPRESSION AND ERROR CORRECTING CODES

G.Benelli,V.Cappellini,E.Del Re

Istituto di Elettronica,Facoltà di Ingegneria,
Università di Firenze,Firenze,Italy

ABSTRACT.Two data transformations- data compression and error control coding - are considered for high efficiency integrated data communication systems.Some methods of data compression and error correcting coding are in particular described.Comparison results giving the compression ratio and r.m.s. error for some data communication systems using in different way data compression and channel coding are shown.A new integrated compression-coding stategy - called COSYDAI - is also proposed,in which a suitable synchronization is added to compressed data ,to obtain error detection and error correction capabilities.

1.INTRODUCTION

High efficiency communication systems are required in the next future to solve the problems connected to the large amount of data to be transmitted from one place to another through noisy communication channels with severely limited bandwidths.One important data transformation,which can be performed to increase the efficiency of a communication system is the data compression,which reduces the amount of not useful or redundant data.In the literature the data compression techniques are often considered on error-free channels.Neverthless the compressed data are more sensible to the channel errors than the normal data,because at the receiver the signal is reconstructed using a smaller number of bits.For this it can be useful and often necessary to use together data

---

This work has been supported by the Italian National Research Council (CNR) under contr.n. 76.00439.07.

compression the channel coding,which protects the data against disturbances,interferences and noise in the communication channel.
In this paper the performances,obtained through a computer simulation, of communication systems using data compression with or without error correcting codes are presented in the case of channels with memory.As data compression methods using prediction or interpolation and spline functions are mainly considered,while as error control coding binoid codes and generalized Hamming codes for burst error correction are considered. After a new integrated compression-coding strategy,called COSYDAI,is also proposed.For transmission channel with memory,this strategy uses in a suitable way the synchronization information,necessary in the compressed data, to detect and in many cases to correct the channel errors.The COSYDAI strategy permits a higher efficiency than other communication systems using both data compression and channel coding.

2.SOME DATA COMPRESSION AND ERROR CORRECTING CODES TO BE USED IN THE INTEGRATED SYSTEM

Let us consider now two methods of data compression,one with prediction,the other with spline functions,which seem well suitable for applications in integrated digital communication systems having also error control coding transformations.
The zero order prediction (ZOP) algorithm with floating aperture is based on the relation [1]

$$y_{pi} = y_{i-1} \pm \Delta \tag{1}$$

where y_{pi} is the predicted sample at the time $t = t_i$, y_{pi} is the preceeding sample and Δ is the aperture (or tolerance).
The predicted sample y_{pi} is compared with the actual sample y_i and if the resulting difference is in the allowable error tolerance $\pm\Delta$,the actual sample y_i is discarded.Otherwise y_i is considered one of non-redundant sample set to be transmitted.
A simple adaptive modification of this algorithm,we are proposing, can be realized considering two error tolerance values : a smaller value Δ_1 for those parts of the signal in which a higher precision is required and a greater value Δ_2 for those parts in which a lower precision is allowed.The two band tolerances are choosen according to the actual value of the samples.
This algorithm can result in a good performance especially with those signal that the most part of the time are in the amplitude range where a lower precision is required.This compression algorithm has been tested applying it to two types of signals : to an original (not-preprocessed)signal and to a lowpass digitally prefiltered signal to eliminate the high frequency components not of interest of the signal spectrum.A net improvement in the algorithm efficiency resulted in case of prefiltered signal compression.Generally a suitable signal lowpass filtering is desiderable in the

application of this simple compression method,to obtain an overall greater efficiency.

The adaptive compression algorithm using spline functions [2] is based on the approximation of the original signal by means of suitably selected linear segments.In this method it is supposed that the available signal is affected by addittive-type noise and its variance is known or may be evaluated as,for instance,when some time intervals exist in which the signal is absent [2, 3]. The available data samples are of the form

$$y_i = s_i + n_i \qquad (2)$$

where s_i are the signal samples and n_i the samples of the additive noise,which is supposed to have a Gaussian distribution with zero mean and constant variance.The compression algorithm is based on the approximation of the signal sample s_i by linear or first-order spline functions : suitable time intervals are selected so that in each of them a linear function is found to approximate closely the signal samples.The number and the length of the intervals and the parameters defining the linear approximation in each of them are determined as follow.The noise variance σ_{n_0} is preliminarily evaluated from n_0 consecutive samples of a signal-free time interval.The compression algorithm at n-th step considers n samples of the available signal y_i and determines the best linear approximation $\hat{y}_i$ to them by the least-squares criterion.The mean σ_n^2 of the squares $(y_i-\hat{y}_i)^2$ is evaluated,which can be considered an estimator of the noise variance.Therefore the quantity $\sigma_n^2/\sigma_{n_0}^2$ is an applicable statistics to decide whether the samples y_i are significantly apart from the linear approximation $\hat{y}_i$.This decision must be carried out according to the statistical criterion related to the random variable $\sigma_n^2/\sigma_{n_0}^2$ which has a Fisher distribution with $(n-1,n_0-1)$ degrees of freedom.If the approximation is statistically justified,a new linear approximation $\hat{y}_i$ of n+1 samples y_i (the preceding n samples plus the next one) is evaluated and statistically verified; if the approximation of n samples is not statistically justified,the approximation at the previous step is considered valid,the interval length n-1 and the parameters defining the linear approximation in this interval are transmitted and the procedure starts again for the evaluation of the length and the best linear approximation of a following interval.

It must be pointed out that this method,having the peculiar characteristic of involving the noise level in the data approximation procedure,performs a reduction of noise components (in particular for high frequency noise it performs a sort of smoothing).Therefore in some cases it is not necessary to filter the signal before the application of the compression algorithm based on the spline function approximation.The signal prefiltering,indeed,results in general useful in the application of prediction algorithms to obtain a higher efficiency.Finally it is to point out that in any considered data compression method the synchronization data must

be inserted.Two pratical approaches on this line are :
a)counting of the data taken out before the considered sample ;
b)counting of the sample position in a frame. The first synchronization method permits a higher compression ratio,but in general is more sensible to the channel errors.
In the following we introduce shortly some classes for burst error correction,which have in general simple encoding and decoding operations (particulary when a general purpose computer is used) and which are useful in the integrated systems,using data compression and error correcting codes.Firstly we describe a class of codes, called "Generalized Hamming codes" for burst error correction,obtained through a suitable generalization of the Hamming codes for single error correction [4].The Hamming codes for single error correction defined in a Galois field GF(q) have a parity-check matrix H_1 with m rows and $(q^m-1)/(q-1)$ columns.To obtain a code able to correct all the bursts of length b or less we consider a parity-check matrix H obtained from H_1 by substituting the elements of H_1 with the element itself multiplied by I_b (the bxb identity matrix).These codes have a very simple and fast decoding algorithm and contain,like particular cases,many other subclasses of codes with interesting error detection and correction capabilities.For example we can obtain,with m = 2, the Samoilenko binoid codes[5] which are optimum respect to the Reiger bound[6].We can obtain also some other new classes of binoid codes with a higher code-rate respect the Samoilenko codes and with some error detection capabilities together with error correction capacity [7].Another subclass is some binary burst error correcting codes,which can be decoded in a simple way both by a ge ral computer or by a decoder of the Meggit type[4].

3.A NEW INTEGRATED COMPRESSION-CODING STRATEGY

In this section a new integrated compression-coding startegy is proposed,in which the necessary time synchronization information (t.i.) in data compression algorithms is utilized in a suitable way to detect and correct channel errors.This method,called COSYDAI, (compression with synchronization control and data interpolation) presents,like it is shown in the next section,an higher efficiency than other communication systems using data compression and error correcting codes when the transmission channel is with memory. The compression data algorithms,described in the previous section, are very sensible to the errors in the time information bits and in the most significant bits of the samples.The first time information method (section 2) permits a higher compression ratio CR than the second method.Neverthless it is more sensitive to the channel errors.In the second methos,in fact, it is possible to detect some errors in the time information.In fact the t.i. sequence in the second method is strictly increasing.If an error changes a number t_i to a value t'_i such that the sequence of the t_i is not increasing,then the error can be detected.If $t'_i > t_{i+2}$ or $t'_i < t_{i-2}$ **the i-th received** time information number t'_i is,with high probability,in error.In these cases,in our simulated systems,we replace t'_i by a value equal to the mean between the preceeding and the following t.i. received number.This value is generally enough near to that exact.

Neverthless inthe case of burst-type errors,this method is not very good (as shown in section 4).In fact several consecutive t.i. are often corrupted by the bursts and therefore the necessary conditions to modify the value of that wrong are not verified and many new errors can be introduced using the previous correction of erroneous t.i.Moreover also the samples between the erroneous t.i. are often altered by the channel noise,for the nature of the burst errors.

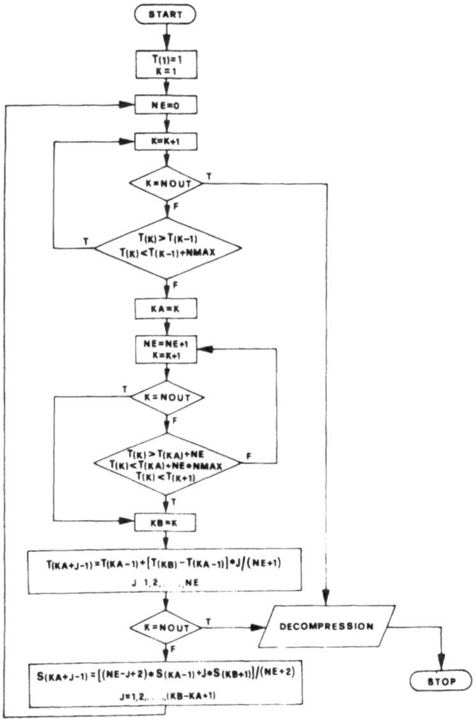

Fig.1- Block-diagram of COSYDAI system

In the COSYDAI strategy the second time information method is utilized.Also the data compression is made so that a fixed number of samples greater than NMAX is not eliminated consecutively.To the receiver,firstly, the t.i. succession is examined to see if the sequence of t.i. is increasing and obeys the previous restrictions. Until this is verified,nearly certainly any error is occurred and therefore nothing is modified.When some consecutive t.i. are detected in error following the previous criterion,they are replaced by new t.i. equidistant between themselves and with values included between the last exact which precedes them and the first we find again exact after that wrong.The values of the samples relative to the mistaken t.i. and that immediately following are modified because they are generally in error:the values that are now assigned to those wrong samples are obtained making a weighed mean of the values of the exact samples which precede them and of that

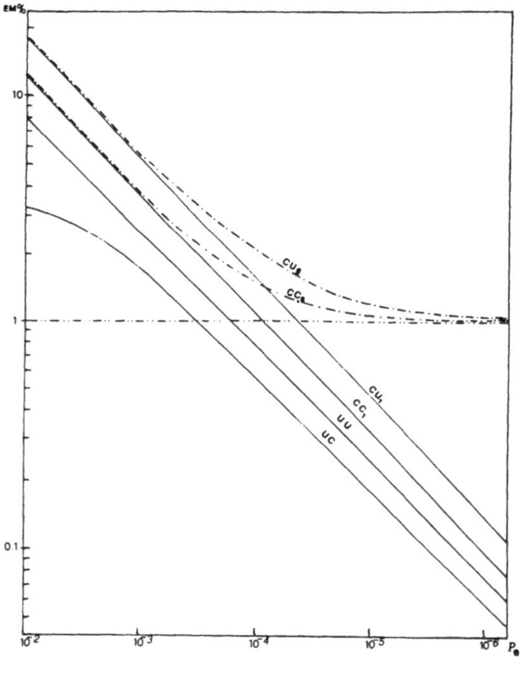

Fig.2-Simulation results regarding some integrated data communication systems

which follow them.The weights are given to the exact samples are inversely proportional to the distance between the sample to replace and that exact.The block-diagram of COSYDAI algorithm is shown in Fig.1.

4.IMPLEMENTATION OF SYSTEMS USING DATA COMPRESSION AND ERROR CORRECTING CODES

In this section the results obtained by a computer simulation of some systems using data compression with or without channel coding are presented.
To characterize the performance of these systems using data compression,we have simulated four structures : uncompressed-uncoded (UU),compressed-uncoded (CU),uncompressed-coded(UC) and compressed-coded (CC) structure.For all these cases,we have obtained the compression ratio CR and the r.m.s. error EM (as percentual of the full scale).

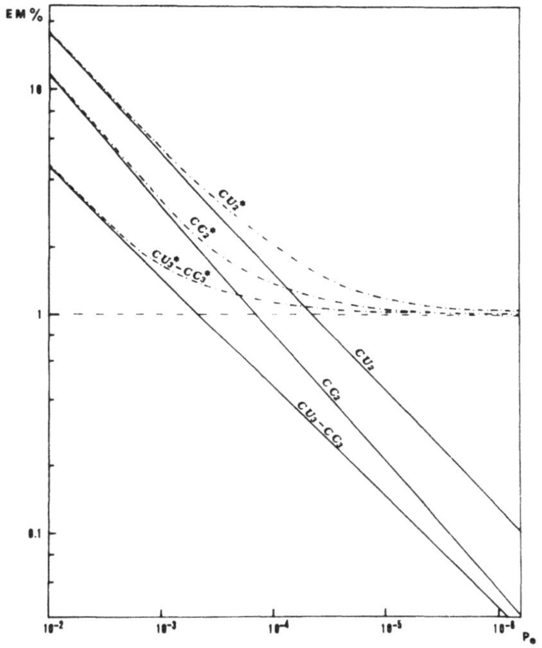

Fig.3-Simulation results regarding some other integrated data communication systems.

In the systems CU and CC,EM value includes the distortion introduced by the compression error and channel noise.To show the different influence of these two error types,we have also computed the r.m.s. error due only to the channel noise;it is obtained through the comparison between the reconstructed data at the receiver from the compressed vector with and without channel errors. In the following,the results considering only the distortion introduced by the channel noise are denoted without asterisk.
The channel noise is simulated by the Gilbert channel model,using a Markov chain with two states [8],which describes approximately the behaviour of some channels with memory,like telephone channels.The utilized code is a Samoilenko binoid code (90,80) defined in a Galois field GF(31).In the binary transmission,as we have simulated,the code is of the type (450,400) and it is able to correct all the bursts of length 21 bits or less.

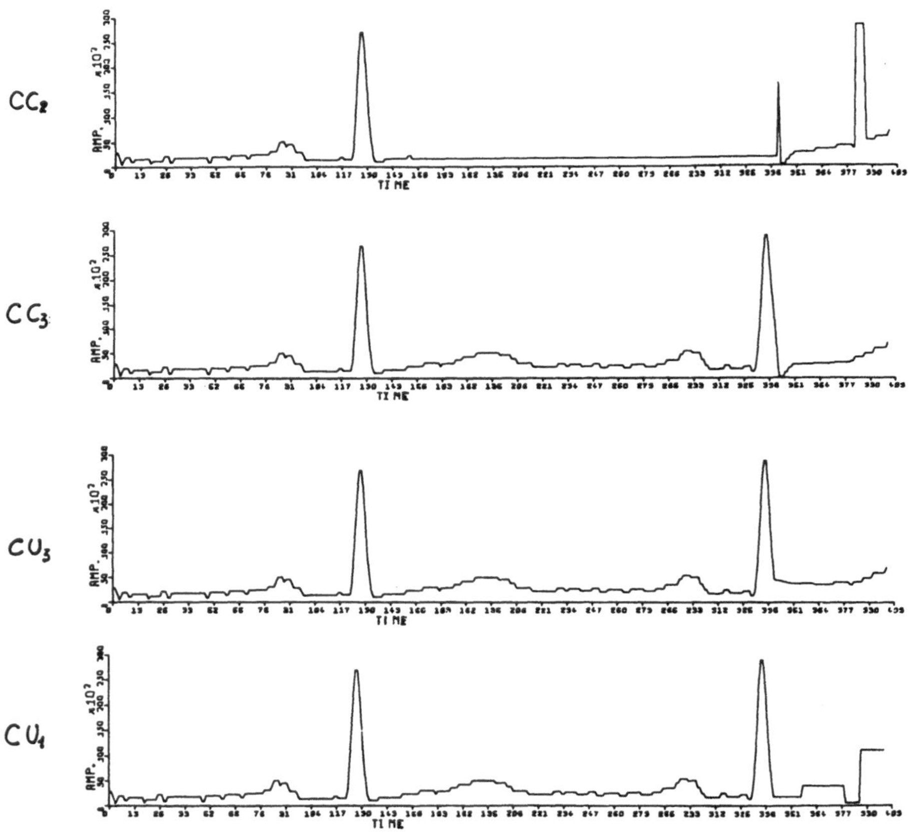

Fig.4-Examples of reconstructed ECG analog waveforms for several transformations :reconstructed ECG with CU_1,CU_3,CC_3(closely similar to the original signal),CC_2.

In our simulated system,the utilized signal,which is an electrocardiogram (ECG),is firstly processed using a low-pass filter and after compressed by a ZOP algorithm with floating aperture.For a tolerance Δ= 2.18% respect to the full scale,we have obtained an r.m.s. error EM = 1%.It is clear that in the cases CU and CC it is impossible to go below this value.

The obtained r.m.s. error EM,expressed like percentual of the full scale,versus the channel error probability P_e is shown in Fig.2 and 3 for all the simulated systems.In these figures we have denoted with the index 1 the system using the first t.i. method, with the index 2 the second t.i. method and with the index 3 the modified system of Fig.1. In the Table 1 the compression ratio CR for the different systems is shown.

For high P_e the EM error is principally determined by the channel noise,while the influence of the compression distortion is negligeable.By reducing P_e,the importance of the compression errors becames more and more high.

It is important to note that using the second method for time information (CU_2) ,the improvement is very low respect to the case CU_1.At the same time the CR value is reduced to 2.6.In the case CC_2,particulary for low P_e,we have an improvement for EM respect to CC_1,but the compression ratio is reduced to 2.2.

The third system (CU_3) shows a very high efficiency;in fact the r.m.s. error due only to the channel error,is lower respect to the case UU and UC and to the other cases.This follows from the possibility to identify in many cases the errors in the time synchronization and from this to reduce also the influence of the errors in the most important bits of the samples.

The case CC_3 generally has an efficiency similar to CU_3.In fact when an uncorrectable burst happens,the code can correct some errors in the time information symbols and some information necessary for the identification of the error positions is lost.

	SYSTEM							
	UU	UC	CU_1	CU_2	CU_3	CC_1	CC_2	CC_3
CR	1	0.89	3.07	2.6	2.6	2.73	2.16	2.16

TABLE 1-Compression ratio values for some simulated data communication systems.

In this case a higher EM can results.Fig.4 shows an example of reconstructed ECG anolog waveforms for several transformations above considered and considering a particular channel error structure.

From these results it is clear the great importance of the time synchronization in the transmission of compressed data on a noisy channel.By a properly choose of the synchronization information, in fact,it is possible to reduce the influence of the channel errors.Naturally to reduce for all the systems the error EM,some code,able to correct longer bursts,can be used.This naturally reduces also the compression ratio.Therefore for the systems CC, it is necessary with reference to the considered signal and channel, a compromise between the EM and CR values.

REFERENCES

1. J.E.Medlin,"Sampled-Data Prediction for Telemetry Bandwidth Compression",IEEE Trans.Space Elect.Telem.,vol AES-3,n.5,p.784-795,September 1967.
2. V.V.Shakin,P.Breuer,"Adaptive Least-Squares Spline Fitting the Vectorial Signal",Proceed.Florence Conf.on Digital Signal Proc. September 1975.
3. G.Benelli,C.Bianciardi,V.Cappellini,E.Del Re,"High Efficiency Digital Communication using Data Compression and Error Correcting Coding",Proceed.Europ.Conf.on Electrotechnics,EUROCON'77, Venice,May 1977.
4. G.Benelli,C.Bianciardi,V.Cappellini,"Generalized Hamming Codes for Burst-Error-Correction",Alta Frequenza,vol.44,November 1975.
5. S.I.Samoilenko,"Binoid Error Correction Codes",IEEE Trans.Inf. Theory,vol-IT-19,January 1973.
6. W.W.Peterson,E.J.WeldonJr.,"Error Correcting Codes",Cambridge Mass.,M.I.T. Press 1972.
7. G.Benelli,C.Bianciardi,V.Cappellini,"Some Burst -Error-Correction Codes",IEEE Trans.Inf.Theory,vol IT-21,November 1975.
8. E.N.Gilbert,"Capacity of Burst-Noisy-Channel",Bell System Tech.Journal,vol 39,September 1960.

SOURCE ENCODING WITH KALMAN FILTERS

Emmanuel N.Protonotarios and Christos A.Mourikis*

School of Electrical Engineering, National Technical University of Athens, Athens-Greece

ABSTRACT. Efficient digital encoding of analog sources requires knowledge of the statistics of the signals. The digital representation of signals offers the advantages of raggedness, easy regeneration and encryption, uniform format for all types of signals, as well as possible combination of transmission and switching functions. The inherent disadvantage is the need for increased transmission bandwidth. This is relieved to some degree with the use of redundancy reduction encoding schemes. Two types of redundancy are generally contained in a signal. One is due to the high correlation between signal samples. The second component of the redundancy is due to the nonuniform probability density of the samples to be quantized. In the present paper we deal with the first kind of redundancy and we explore the application of the discrete Kalman filter in a predictive coding scheme with the general structure of differential pulse code modulation (DPCM). Every sample s(nT) of the signal (where T is the sampling period, and $\underline{n}$ a nonnegative integer) can be thaught of as containging a predictable part $\hat{s}(nT)$ (on the basis of previous information), which does not carry any new information and therefore does not have to be transmitted, and an unpredicable part $e(nT) = s(nT) - \hat{s}(nT)$, which is called the "innovation" of the sample; this contains all the new information in that sample Only a digitized version of the innovation need be transmitted. In a predictive DPCM systeme this is done by applying a predictor in the feedback path, where prediction of the incomming sample

---

* Now at the E.N.S.T., Paris, France.

is based on quantized versions of previous samples. In the present paper we propose the use of a discrete Kalman filter as a predictor. This has the advantage that for the estimation of the "predictable" part of the incoming sample we effectively use all previous information available, since the Kalman filter is of the recursive type (as opposed to FIR type). Moreover, the use of the Kalman algorithm guarantees near optimality for the prediction. If the quantized optimum prediction error samples are entropy-coded the DPCM coder performs very near Shannon's rate-distortion function which constitutes an unattainable uper bound for the performance of such systems. For stationary signals the complete structure and all predictor parameters are given explicitly. In recent years a great deal of attention has been given to the predictive coding of speech signals. Because of the non-stationary nature of these signals an adaptive predictor should be used, which is readjusted periodically. In this case the predictor parameters change and have to be transmitted to the receiver. These parameters could be computed with another discrete Kalman filter algorithm.

1. INTRODUCTION

Differential pulse code modulation (DPCM) is an effective redundancy reduction scheme that provides a relatively simple way for source encoding of analog bandlimited signals. For this reason DPCM finds increasing use in the digital transmission of both voice and television signals, especially in the case of the video telephone (e.g. O'Neal 1966 [1], Protonotarios 1967 [2]). Samples of a bandlimited signal with a nonuniform spectral density are correlated; therefore these samples contain redundancy that can be removed to some degree by DPCM, encoding. Every sample $s(nT)$ of the signal (where T is the sampling period, and n a nonnegative integer) can be thaught of as containging a predictable part $\hat{s}(nT)$ (on the basis of previous information) which does not carry any new information and therefore does not have to be transmitted, since it can be reconstructed at the receiver. The difference, $e(nT) = s(nT) - \hat{s}(nT)$, is the unpredictable part and is called the "innovation" of the sample; this contains all the new information in that sample. Only a digitized version of the innovation need be transmitted. In DPCM this is exactly done by applying a predictor in a feed-back path, as shown in Fig.1. The prediction of the incoming sample is based on quatized versions of previous samples.

The important feature of the arrangement of Fig.1 is that by supplying a feedback path, quantization error accumulation is avoided. The input at the predictor is,

$$\hat{s}(nT)+e(nT)+q(nT) = \hat{s}(nT)+\{s(nT)-\hat{s}(nT)\}+q(nT) = s(nT)+q(nT)$$

where $q(nT)$ is the quantizing noise corresponding to the n-th

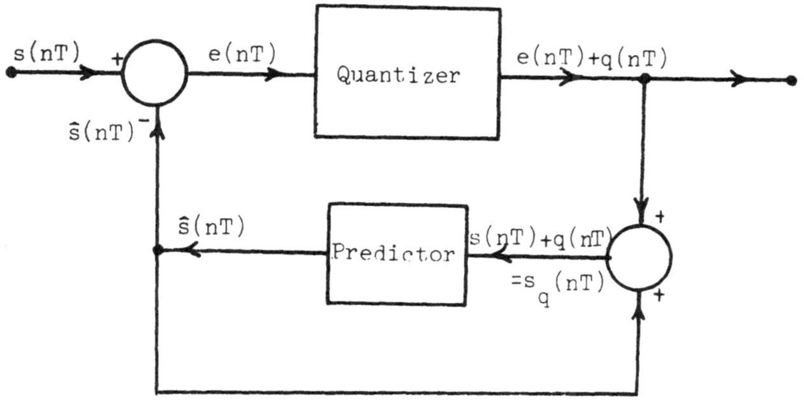

Fig. 1. DPCM Coder

sample. Most authors (e.g. Atal et al 1970 [3], Jayant 1974 [4], Noll 1975 [5], Gibson et al 1974 [6], Cohn and Melsa 1975 [7], use a finite implulse response (FIR) digital filter as a predictor, with a transfer function (in the z-domain) of the form

$$P(z) = \sum_{i=1}^{N} \alpha_i \, z^{-i} \qquad (1)$$

where the α_i are computed in terms of the correlations of the samples with a minimum mean square error (MMSE) criterion.

Recursive predictors of the Kalman filter type have been used by some authors in recent years [8-10]. In the present paper we also use a discrete Kalman filter as a predictor. This has the advantage that for the estimation of the "predictable" part of the incoming sample we effectively use all previous information available, since the Kalman filter is of the recursive type (as opposed to FIR type). Moreover, the use of the Kalman algorithm guarantees near optimality for the prediction. If the quantized optimum prediction error samples are entropy-coded it has been found (O'Neal 1976 [11]) that the DPCM coder performs within 1.53 dB of Shannon's rate-distortion function which constitutes an unattainable upper bound for the performance of such systems.

Our purpose is to give explicit results for the dynamical structure of the predictor and to assess its performance in comparison with FIR type nonrecursive predictors. We should note that Gibson et al [6] make use of a Kalman filter algorithm in order to estimate the parameters α_i of the predictor in form (1). We, on the other hand, estimate directly the signal samples with the Kalman filter, as previous authors [8-10] have done.

2. THE KALMAN FILTER

Let a discrete-time dynamical system be descrebed by the following state equations

$$\underline{x}(n+1) = \Phi.\underline{x}(n)+\Gamma.\underline{u}(n) \quad (2)$$

$$\underline{y}(n) = H.\underline{x}(n)+\underline{v}(n) \quad (3)$$

where $\underline{u}(n)$, $\underline{v}(n)$ are discreted white gaussian noises, uncorrelated with each other, as well as with the initial state vector $\underline{x}(0)$. Then, the Kalman filter equation states that the optimum estimator (in the minimum variance sense) of $\underline{x}(n)$, given the observations $\underline{y}(1),\dots,\underline{y}(n)$, is given by:

$$\hat{\underline{x}}(n|n) = \Phi.\hat{\underline{x}}(n-1|n-1)+K(n)\{\underline{y}(n)-H\Phi.\hat{\underline{x}}(n-1|n-1)\} \quad (4)$$

where the gain matrix K(n) is computed separately from known algo rithms 12 . Moreover, it is easy to prove that the best estimate of the next sample (optimal prediction) is given by:

$$\hat{\underline{y}}(n+1|n) = H\hat{\underline{x}}(n+1|n) \quad \text{with} \quad \hat{\underline{x}}(n+1|n) = \Phi\hat{\underline{x}}(n|n) \quad (5)$$

Relation (5) may also be written as

$$\hat{\underline{x}}(n+1|n) = \Phi\{I-K(n)H\}\underline{x}(n|n-1)+\Phi K(n)\underline{y}(n) \quad (6a)$$

$$= \Phi\hat{\underline{x}}(n|n-1)+\Phi\overline{K}(n)\{\underline{y}(n)-H\hat{\underline{x}}(n|n-1)\} \quad (6b)$$

The last equations will be used presently in order to obtain the optimal DPCM predictor.

3. APPLICATION TO DPCM

Application of the Kalman filter to DPCM is straightforward. We consider the signal to be DPCM-coded, s(n), as the output of the system descrebed by (2), according to the equation, s(n) $H\underline{x}(n)$. As mentioned before the input to the box labeled Predictor in Fig.1, is not s(n), but s(n) plus a quantizing noise component, q(n). Thus (3) should read in this case

$$s_q(n) = H\underline{x}(n)+q(n) \quad (7)$$

Combining equations (5) and (6) we obtain

$$\hat{s}(n+1|n) = H\Phi\{\hat{\underline{x}}(n|n+1)+K(n)(s(n)-\hat{s}(n|n-1))\} \quad (8)$$

Which is a time domain expression of the optimum predictor, requiring the estimation of the complete state vector $\hat{\underline{x}}(n|n-1)$,although we only need the estimation of s(n).

It is more interesting to arrive at a frequency (z-domain) characterisation of the optimum predictor, and then come back to the time domain in an obvious way. To accomplish this we shall consider that our system operates in steady state conditions and that, particularly, the Kalman gain K(n) has reached its steady state value K.

We define the following z-transforms, $\hat{X}_1(z) = Z\{\hat{x}(n|n-1)\}$ $S_q(z) = Z\{s_q(n)\}$ $\hat{S}(z) = Z\{\hat{s}(n|n-1)\}=Z\{s_q(n|n-1)\}$ (9)

The optimum predictor transfer function can be defined from the equation

$$\hat{S}(x) = P(z)\ S_q(z) \quad (10)$$

From (5) we have:

$$\hat{S} = H\,\hat{X}_1 \qquad (11)$$

and, $z\hat{\underline{X}}_1 = \Phi(I-KH)\hat{\underline{X}}_1 + \Phi K S_q$ or $\hat{\underline{X}}_1 = (zI-\Phi+\Phi KH)^{-1}\Phi K\,S_q$ (12)

Combining (11) and (12) and comparing with (10) we obtain the desired result:

$$P(z) = H(zI-\Phi+\Phi KH)^{-1}\Phi K \qquad (13)$$

We may easily arrive at other equivalent forms of P(z) by starting from different expressions for $\hat{s}(n+1|n)$. We cite here one more form:

$$P(z) = H\Phi(zI-\Phi+KH\Phi)^{-1}K$$

It is worth noting that further evaluation of expression (13) although including some symbol manipulation, is readily done by a digital machine and results in P(z) being expressed in the form:

$$P(z) = \sum_{i=1}^{p} \beta_i z^{-i} / \{1 + \sum_{j=1}^{p} \mu_j\, z^{-j}\} \qquad (14)$$

Transposing (14) to the discrete time domain (z^{-1} = unit delay), we have

$$\hat{s}(n|n-1) = \sum_{i=1}^{p} \beta_i s_q(n-i) - \sum_{j=1}^{p} \mu_j \hat{s}(n-j|n-j-1) \qquad (15)$$

In the above we have not made any particular assumption regarding the form of Φ. However we may sometimes consider that the signal s(nT) consists of samples of a continuous signal s(t), (with a rational power spectral density) whose evolution is described by the differential equation

$$\frac{d^p s}{dt^p} + \alpha_{p-1}\frac{d^{p-1}s}{dt^{p-1}} + \ldots + \alpha_1 \frac{ds}{dt} + \alpha_o s = bw(t) \qquad (16)$$

In this case $\Phi = \exp(FT)$ where F is the system matrix corresponding to the coefficients $\{\alpha_i\}$. Assuming T "small enough" (probably a very restrictive assumption) we may ignore terms of order T^2, T^3 etc. thus putting $\exp(FT) \cong I+FT$. After some lengthy algebraic manipulations, we arrive at explicit approximate forms of P(z). These present the advantage that, being simple, their parameters may be easily modified in real-time adaptation schemes. However their accuracy and efficiency is questionable.

Some remarks regarding the derivation and use of the optimum predictor should be made now.

1) As pointed out previously, the Kalman filter requires that both generating and observation noises be white and gaussian. As far as generating noise is concerned, we may consider this to be the case. But observation noise, i.e. quantizing error, is neither white nor strictly gaussian. However its p.d.f. approximates a norman curve reasonably well, in the case of optimal nonuniform quantizers, the better so as the number of quantizing levels is increased {see Arnstein, 1975 [13]}.Its spectrum falls off at high frequencies, but we cannot remedy this if we want to remain in the linear domain.

2) The Kalman filter operates optimaly if at every time instant nT the oppropriate gain K(n) is used. Using the steady state gain K we should expect a certain deterioration of its performance. Also, this approach cannot be applied to non-stationary signals where no steady state exists in the strict sence. It may be applied however on segments of such signals, considered as locally stationary.

3) Obviously the overall performance directly depends upon the accuracy of the signal generating model which was assumed known and exact. In general this may not be the case and then a preliminary stage of system identification should be carried out. This however is outside the scope of the present paper.

4. SIMULATION RESULTS

Simulations were conducted on a CDC series 6000 computer, using 1st and 2nd order artificial stationary signals. These cases are simple enough to allow at least some degree of analytical handling and the derivation of reference theoretical results. On the other hand they represent reasonable and useful approximations to higher order signals, for which the procedure is essentially the same. Uniform quatizers were used throughout, and the stepsize was computed accordign to Max's formulas {Max, 1960 [14]} so as to be approximately optimum. The Kalman predictor was compared primarily with the usual n-tap predictor (Eq.(1)) with n = 1 or 2 in the cases examined.

In order to compute the Kalman gain, an estimate of the quantizing noise variance σ_q^2 is needed, and as such the approximate value $\Delta^2/12$ was used, i.e. overload noise was neglected. Finaly, a reference value computed approximately was the prediction error variance, σ_e^2, derived from the classical theory of nonrecursive linear prediction. Three criteria were used to compare the performance of the predictors.

1) The overall SNR defined as $SNR=10\log_{10}\{\Sigma s^2(n)/\Sigma(s(n)-s_q(n))^2\}$

2) The variance σ_e^2 of the quantizer input sequence (which should be as small as possible) and

3) The normalized autocorrelation C_1 of this same sequence, which should ideally be zero, indicating that all of the predictable part of the samples has been effectively extracted.

For reasons stemming mainly form quantizer-signal mismatching, the above three creteria do not always simultaneously favour the same predictor.

4.1. First order signal

Signal model, $s(n+1) = \rho s(n) + \sqrt{1-\rho^2}\ u(n)$, $o \leq \rho \leq 1$

$u(n)$ = white gaussian noise with $E\{u\}=0$, $E\{u^2\}=1$

Predictor expressions

1. One-tap $\hat{s}_1(n+1|n) = \rho \hat{s}_q(n)$
2. Kalman $\hat{s}_\kappa(n+1|n) = K\rho\, s_q(n) - (K-1)\rho \hat{s}_\kappa(n|n-1)$

with

$$K = \frac{(1+\lambda)(1-\rho^2)}{2\rho^2}\left\{\left[1+\frac{4\rho^2\lambda}{(1+\lambda)^2(1-\rho^2)}\right]^{1/2}-1\right\}$$

where $\lambda = 1/\sigma_q^2 \cong 12/\Delta^2$; it is worth noting that $\lim K = 1$ as $\sigma_q \to 0$, in which case $\hat{s}_\kappa = \hat{s}_1$. Also, generally it is found that $K > 0.95$ so that no dramatic difference can be expected between the performance of the two predictors. Indeed the simulation verified this competely. The differences noted were insignificant, more so as the number of quantizer levels M was increased resulting in a finer quantizing, i.e. as $\sigma_q^2 \to 0$.

4.2 Second-Order Signal

Signal model, $\underline{x}(n+1) = \Phi \underline{x}(n) + \Gamma u(n) = \begin{bmatrix} \varphi_{11} & \varphi_{12} \\ \varphi_{21} & \varphi_{22} \end{bmatrix} x(n) + \begin{matrix} \gamma_1 \\ \gamma_2 \end{matrix}\; u(n)$

$s(n) = H\underline{x}(n) = [1\ 0]\underline{x}(n)$; $u(n)$ as before.

Predictor expressions

1. Two-tap: $\hat{s}_2(n+2|n+1) = A_1\, s_q(n+1) + A_2\, s_q(n)$

   $A_1 = \rho_1(1-\rho_2)/(1-\rho_1^2)$, $A_2 = (\rho_2-\rho_1^2)/(1-\rho_1^2)$

   $\rho_1 = E\{s(n+1)s(n)\}/E\{s^2(n)\}$, $\rho_2 = E\{s(n+2)s(n)\}/E\{s^2(n)\}$

2. Kalman: $\hat{s}_\kappa(n+2|n+1) = (K_1\varphi_{11}+K_2\varphi_{12})s_q(n+1) - K_1(\varphi_{11}\varphi_{22}-\varphi_{12}\varphi_{21})s_q(n) -$

   $-\{(K_1-1)\varphi_{11}+K_2\varphi_{12}-\varphi_{22}\}\hat{s}_\kappa(n+1|n) + (K_1-1)(\varphi_{11}\varphi_{22}-\varphi_{12}\varphi_{21})\hat{s}_\kappa(n|n-1)\}$

K_1 and K_2 are the steady state Kalman gain components. They can be easily-computed in the case of ideal observations ($\sigma_q^2 = 0$). Two sub-cases were examined. In the first the Kalman predictor degenerates into the classical 2-tap one as $\sigma_q^2 \to 0$ (a situation analogous to that pertaining to the 1st order signal). For this to happen the condition $\gamma_1(\gamma_2\varphi_{12}-\gamma_1\varphi_{22}) = 0$ must hold. In the second case, the Kalman filter is always different from the classical one, i.e. always of the recursive type. As expected, improvement in performance is more pronounced in the latter case, which was also simulated in the absence of any quatizer, thus establishing the superiority of the Kalman predictor. Because of space limitations we present here (in Table I) some representative computer simulation results only for this last case.

(α)	$\hat{s}_{\varkappa}$	$\hat{s}_{\varkappa}$
$\sigma^2 e/\sigma^2 s$	0.2659	0.3076
C_1	0.0020	0.2378

(β)	$\hat{s}_{\varkappa}$	$\hat{s}_2$
$\sigma^2 e/\sigma^2 s$	0.2778	0.3143
C_1	0.0202	0.2275
SNR	25.11	24.54

Table I. Performance criteria for 2^d-order signal, 2^d sub-case. (α) no quantizer (β) 4-bit quantizer specific values $\varphi_{11}=0.8$, $\varphi_{12}=1$, $\varphi_{21}=0.5$, $\varphi_{12}=0$; $\gamma_1=0.7$, $\gamma_2=0.4$.

It is obvious form the above tables that the Kalman predictor achieves a definite improvement over the 2-tap one. Specifically the prediction error sequence is substantially less correlated, tending to the optimal value of zero correlation. It also has a smaller variance, thus allowing more efficient coding and resulting in some gain on the over-all SNR.

5. CONCLUDING REMARKS

The method can be exterded for DPCM encoding of nongaussian signals that are generated by Marcovian schemes of the form,

$$\underline{x}(n+1) = \underline{f}\{\underline{x}(n)\}+q.\underline{w}(n) \quad ; \quad s(n) = h\{\underline{x}(n)\}$$

In this case a nonlinear predictor of the extended Kalman variety or other may be applied.

In recent years a great deal of attention has been given to speech signals. Because of the nonstationary nature of these signals an adaptive predictor should be used, which is readjusted periodically. In this case the predictor parameters change and have to be transmitted to the receiver. These parameters could be computed with another discrete Kalman filter algorithm.

REFERENCES

1. O'Neal, Jr, J.B. (1966), Predictive Quantizing Systems (DPCM) for Transmission of Television Signals, B.S.T.J., 45, 689.
2. Protonotarios E.N. (1967), Slope Overload Noise in Differential Pulse code Modulation Systems, B.S.T.J., 46, 2119.
3. Atal, B.S., Schroeder, M.R. (1970), Adaptive Predictive Coding of Speech Signals, B.S.T.J., 48, 1973.

4. Jayant, N.S. (1974),"Digital Coding of Speech Waveforms PCM, DPCM, and DM Quantizers,"Proc.IEEE, 62, 611.
5. Noll, P. (1975),"A Comparative Study of Various Quantization Schemes for Speech Encoding,"B.S.T.J., 54, 1597.
6. Gibson, J.D., et al (1974),"Sequentially Adaptive Prediction and Coding of Speech Signals,"IEEE Trans. Comm.COM-22, 1789.
7. Cohn, D.L., Melsa, J.L., (1975),"The Residual Encoder-An Improved ADPCM System for Speech Digitization,"IEEE Trans. Comm., COM-23, 935.
8. Pirani,G. and Scagliola, G. (1976), "Performance Analysis of DPCM Speech-Transmission Systems Using Kalman Predictors," 1976 IEEE International Conference on Acoustics, Speech, and Signal Processing, April 12-14, 1976, Philadelphia, Pa.
9. Gunn, J.G. and Sage, A.P. (1973), "Speech Data Rate Reduction", IEEE Trans. on Aerospace and Electronic Systems, Vol. AES-9, n.2, pp. 130-150, March 1973.
10. Irwin, J.D. and O'Neal, Jr., J.B. (1968), "The Design of Optimum DPCM Encoding Systems via the Kalman Predictor," American Automatic Control Council, June 26-28, 1968.
11. O'Neal, Jr, J.B. (1976)"Differential Pulse Code Modulation (PCM) With Entropy Conding," IEEE Trans.Info.Th.,IT-22, 169.
12. Schweppe, F.C. (1973)"Uncertain Dynamic Systems," (Englewood Cliffs, N.J. Prentice Hall).
13. Arnstein, D.S. (1975), "Quantization Error in Predictive Coders," IEEE Trans. Comm., COM-23, 423.
14. Max, J. (1960), "Quantizing for Minimum Distortion,"IRE Trans. Info. Th.,IT 6,7.

Panel Discussion
on

THE PRACTICALITY OF SOURCE CODING *

Chairman:	Prof. J. Massey	(M.I.T. Cambridge, U.S.A.)
Panel:	Prof. D. Sakrison	(Univ. of California, Berkeley, U.S.A.)
	Prof. J.P.M. Schalkwijk	(Eindhoven Univ., Holland)
	Prof. E.N.Protonotarios	(Univ. of Athens, Greece)
Contributors:	Prof. J.B. Anderson	(M[c]Master Univ., Hamilton, Canada)
	Dr. P.G. Farrell	(Univ. of Kent, U.K.)
	Prof. D.L. Snyder	(Washington Univ., St. Louis, U.S.A.)

Professor Massey began the Panel Discussion by briefly describing the two standard techniques for source coding : Huffman coding and Tunstall coding. The Huffman technique encodes a fixed length string of source symbols into variable length binary codewords in such a way as to minimise the average codeword length. The Tunstall method encodes a variable number of source symbols into a fixed length codeword. Professor Massey gave simple examples of each method, and indicated that both were optimum for the parameters considered. Though Tunstall coding has the advantage of fixed length codewords, there are in practice ways of overcoming the problem of variable length Huffman coding; for example, guarding against overflow whilst waiting for a long codeword to clear. Huffman encoding systems have been implemented by NASA (for spacecraft picture data) and by Bell Northern, Canada (for telephone accounting data).

---

* The Editor wishes to thank Dr. P.G. Farrell for the preparation of this report.

Professor Sakrison continued the discussion by stating that there was no doubt that source coding was practical. The more interesting question was about the way in which source coding would change and improve; he thought that the biggest impact would come from the advent of cheap LSI technology. Professor Sakrison described the implementation of a still image encoding scheme, using a method previously described at the Institute [see Professor Sakrison's contribution on Rate Distortion Theory: The Influence of the Source and Distortion Measure in Structuring Source Coding] of separating an image into two components: a discontinuous component and a continuous remainder component obtained by subtracting the discontinuous component from the original image data. The separation is very nearly optimum if the remainder is something like a stationary Gaussian process, and an efficient method of coding is to encode the two components separately. The continuous component is a straight line fit, line-by-line, to a 6 bit-per-pixel representation with a maximum error criterion rather than a mean square error criterion because it is essential to pick out the edges of objects accurately. This is the computationally lengthy part of the process, but it is simple enough to do with a microprocessor. The dynamic range of the remainder component is quite small (4 bits-per-pixel is more than adequate), and it is proposed to use a CCD device to do the required transform-encoding. Thus a feasible method of transmitting stills over a 19.2 Kbit/sec channel could be available in about two years. In order to take advantage of the two-dimensional structure of the data, differential line-to-line transmission was used. The scheme achieved about 0.8 bit-per-pixel.

Professor Schalkwijk then described enumerative source coding schemes. In an enumerative scheme, source output sequences are listed in lexicographical order by means of an index number. Instead of transmitting a particular sequence, the lexicographical index, suitably coded, is transmitted instead. The main part of the encoding process is to compute the index corresponding to a given sequence. This can be done, without needing to store a dictionary (as for Huffman coding), with a complexity approximately proportional to the source sequence length. The inverse, decoding, process is no more complex. This makes it possible to consider using channel coding with enumerative source coding as a more efficient scheme than that using separate channel coding and syndrome source coding. Enumerative source coding is asymptotically optimum; the method can be modified to make it optimum even for relatively short sequence length.

Professor Protonotarios spoke on practical aspects of speech encoding. This is an area where redundancy reduction schemes have had a lot of success, but much also remains to be done. About 50 bits/sec is basically all that is required to characterise speech; this is a lower bound. An upper bound is the 64 Kbit/sec required for sampled and quantised telephone quality speech - clearly there is a lot of room between the bounds. There are two main kinds of redundancy reducing schemes. One class of methods tries to reduce the time waveform itself: in this category are differential pulse code modulation (DPCM), delta modulation, adaptive delta modulation, and all kinds of adaptive DPCM. Adaption must be used, because of the non-stationary nature of speech. About 10 Kbit/sec can be achieved. More sophisticated methods, on the other hand, take account of the spectrum of the waveform as well. In this category are the analysis and synthesis methods and the so called vocoders. By using an unvoiced formant vocoder, about 600 bits/sec or less can be achieved. This is a considerable achievement, but still leaves a lot of room for improvement. In discussion it was pointed out that the lower limit of 50 bit/sec was related to the number of phonemes in a language and the rate at which a human being can utter these.

Professor Snyder described an application of Huffman encoding to enable the efficient storage of electro-cardiogram data. The electro-cardiogram waveform was sampled and digitised, and then second differences were taken to decorrelate it into essentially independent 10-bit data words. The thirteen most probable 10-bit words, and a fourteenth symbol representing the pooled collection of the 1011 remaining words, were encoded using Huffman's technique. When one of the pooled words occurred, it was transmitted with the fourteenth Huffman code preceding it as a prefix. The thirteen most probable words were transmitted as Huffman codes only. This technique resulted in a compression of something like 30%, which permitted the storage of 24 hours of electro-cardiogram recording on a single digital tape in the technology of about 5 years ago.

Professor Anderson elaborated on the hardware aspects of the tree encoding voice coding scheme he had previously described at the Institute [Robust Low Bit Rate Voice Encoding]. The scheme uses an M-algorithm; that is, 16 paths in the possible encoding tree are reduced to M=4 at each step. The bit rate of the machine constructed was 16 Kbit/sec, the bandwidth about 3.8 KHz, and the quality of the speech something like toll network (switched network) telephone quality. The main components of the hardware are the path sorter, which selects the four best paths from up to 16 offered to it; and the code generator, which is basically a transversal digital filter realised as a read-only memory.

Dr. Farrell then described a Huffman encoder for 63 6-bit data words, 3 occurring with relatively high probabilities (above 0.3), and all the rest quite low (less than 0.05). This gave 3 Huffman codes of length 2, 4 of length 7, and the rest of length 8. A compression ratio of about 1.5 was achieved. The hardware operated at 1 Mbit/sec, and required about 400 gates, equivalent to some 100 standard IC's. About 2/3 of the complexity was concerned entirely with the generation of the 8-bit Huffman codes, which makes the modification described above by Professor Snyder seem of interest as a way of reducing the complexity. The number of input data words was chosen not to be a power of 2, and the word probabilities were deliberately skewed, so as to avoid any regular structures in the Huffman encoding. Thus a good idea of the implementation complexity could be achieved for a source without too much simplifying structure.

In general discussion it was pointed out that enumerative coding could be used for a source with memory, thus perhaps avoiding the necessity for transforming a source with memory into a memoryless source before, say, Huffman or Tunstall coding (in the case of a digital source), or Karhunen-Loëve or Fourier expansion (in the case of an analogue source). Such pre-processing might also destroy the characteristic redundancy of the source, thus making the source more difficult to encode. It was also pointed out that, at least in theory, differential line-to-line transmission as mentioned by Professor Sakrison above was unnecessary; the second line can be encoded without reference to the first, provided the first line is available at the decoder when decoding the second. The two lines are a pair of correlated sources, each of which can efficiently be decoded with side information. The Lempel-Ziv source coding scheme was then mentioned. In the case of a stationary, ergodic source, certain data patterns will repeat occasionally. When this occurs, instead of transmitting the pattern again, a pointer to its location and length in the past is sent instead. The scheme does very badly when used for non-stationary sources like speech or video waveforms, however.

Part VI

DISPERSIVE AND FADING CHANNELS

REVIEW OF METHODS FOR DESCRIPTION OF RANDOM TRANSMISSION CHANNELS: APPLICATIONS TO IDENTIFICATION AND OPTIMISATION OF DETECTION AND TRANSMISSION PROCESSES.

J.L. LACOUME - G. JOURDAIN

Centre d'Etude des Phénomènes Aléatoires et Géophysiques (Equipe de Recherche Associée au CNRS), B.P. 15 - 38040 GRENOBLE-CEDEX

ABSTRACT :

The aim of this work is to present theoretical means of description of linear systems, whether stationary or not, in connection with practical methods of identification. In the different approaches used to represent channel propagation, we will focus our interest on the systemic approach leading to the representation of the system by a stochastic function : impulse response in the stationary case ; functions of two variables in the non stationary case. These representations will be characterized by their moments in the external approach or by a structural, KALMAN type, representation in the internal approach. This study will be performed for two-points-linking channels and for spatial channels (extended sources and receivers). Finally , we shall present some applications issued from laboratory simulations or from experimentations in natural media.

INTRODUCTION.

The aim of this lecture is to present the means utilized in order to represent the input/output relation in propagation channels and to define the methods of measurement of the channel characteristics (identification).

The input is the signal emitted by a system of communication or of remote sensing. The output is the signal received. From the mathematical point of view, the channel is considered as an application of a set of input signals (noted $|e\rangle$) into a set of observable or output signals (noted $|s\rangle$). The input-output relation is :

$$|s\rangle = F|e\rangle$$

where F is an operator representating the propagation channel.

1. DIFFERENT CHANNEL STRUCTURES.

The essential features characterizing the channel are :

1/ Linearity or non-linearity,

2/ Extended or not extended sources and receiving system.

3/ Deterministic or stochastic signals and system.

4/ Stationarity in time and space or non-stationarity.

We shall only consider linear systems with respect to the input, i.e. there will always be

$$F[\alpha_1|e\rangle + \alpha_2|e\rangle] = \alpha_1 F[|e\rangle] + \alpha_2 F[|e_2\rangle]$$

$$|e_1\rangle \ , \quad |e_2\rangle \in S_e$$

$$\alpha_1 \quad , \ \alpha_2 \quad \in \mathbb{C}$$

In the two first parts of the lecture, we shall consider only two-point-linking propagation channels : the input or output is only a function of the time : this case will be called biponctual channel. In the third part, we shall consider the case called spatial channel, in which the source and the receiver are spatially extended.

The signal and the channel will be generally stochastic ones, the deterministic case being a particular case of the stochastic one.

Finally, the stationarity will be supposed in some cases, but we shall generally consider the non-stationary case. The considered cases are represented in Figure 1.

2. THE DIFFERENT APPROACHES OF THE DESCRIPTION OF A PROPAGATION CHANNEL.

We shall distinguish : the systemic approach, divided into external approach and internal approach, and the physical approach. The experimentator's aim is to determine the input output relationship. This can be done by considering the channel as a "black box" and trying only to determine the properties of the operator F representing the system : this is the systemic external approach.

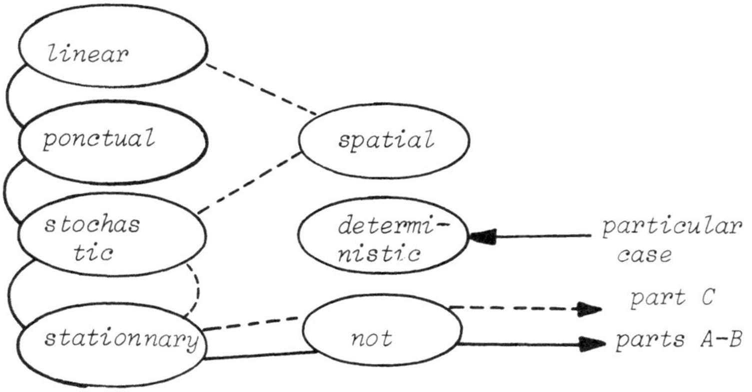

FIGURE 1

In this approach a two-point-linking stationary system (TPLSS) is described by the impulse response or equivalently by the transfert function. In the systemic internal approach, one tries to obtain a structural representation of the impulse response. With the physical approach the experimentator tries to determine the input/output relationship, taking into account the physical phenomena in the propagation medium. For example :

- ray theory, when the scale irregularities in the medium are larger than the wavelength ;
- mode theory when the medium is bounded.

It is important to notice that in actual cases the three approaches can be combined, in order to obtain the most economical description and to take into account all the data. In the following we shall mainly be concerned with systemic approaches and we shall consider :

A) The external description of two-point-linking non stationary random systems ;

B) The internal description of two-point-linking non stationary random systems ;

C) The external description of spatial stationary and non stationary systems.

Finally, we shall present in the part D applications to submarine acoustic propagation.

A) EXTERNAL DESCRIPTION OF TWO-POINT-LINKING NON STATIONARY RANDOM SYSTEMS.

1. GENERALITIES.

In this kind of approach we are concerned only with the functional relationship between input and output.

The simplest description is the bitemporal representation. In that description the input/output relationship is :

$$s(t) = \int_{\xi} g(t,\xi)\ e(t-\xi)\,d\xi \tag{1}$$

where : $s(t)$, $e(t)$ are deterministic or stochastic functions representing the output and the input,

ξ is the delay between input and output,

g is generally a stochastic function. g is the bitemporal representation of the system.

The DOPPLER-delay representation is obtained by FOURIER transform.

Let be

$$U(\nu,\xi) = TF_t[g(t,\xi)]$$

TF_t : FOURIER transform on the variable t ,

Then :

$$s(t) = \int_{\nu}\int_{\xi} U(\nu,\xi)\ e^{2i\pi\nu t}\ e(t-\xi)\,d\nu\,d\xi \tag{2}$$

It is interesting to write the relation (2) with notations using operators.

Let be

F_ν = frequency-shifting operator (ν = frequency shift)

T_ξ = time-translation operator (ξ = time translation)

Formally we can write (2)

$$|s\rangle = \int_{\nu}\int_{\xi} U(\nu,\xi)\ F_\nu T_\xi |e\rangle\,d\nu\,d\xi$$

We can see on this expression that the output $|s\rangle$ can be considered as a sum of copies of the time delayed and frequency shifted input. $U(\nu,\xi)$ represents the contribution of each of these copies after propagation in the channel. Then $U(\nu,\xi)$ is the

DOPPLER (frequency shifting) - delay (time delay) representation of the channel.

This representation can be induced by physical considerations. If in the channel there are a lot of moving scatterers, each of them generates a particular term in the integral (3) (Fig. 2). With this "physical" interpretation, we see that the DOPPLER-delay representation is an approximation only valid when the scatterers' speed is small compared to the one of the signal in the medium and when the bandwidth of the signal is small. Under these conditions, the DOPPLER effect

$$\frac{\Delta f}{f} \sim \frac{\Delta f}{f_C} = \frac{V_{sc}}{V_{sig}}$$

(f_C : central frequency of the signal), is a frequency shift.

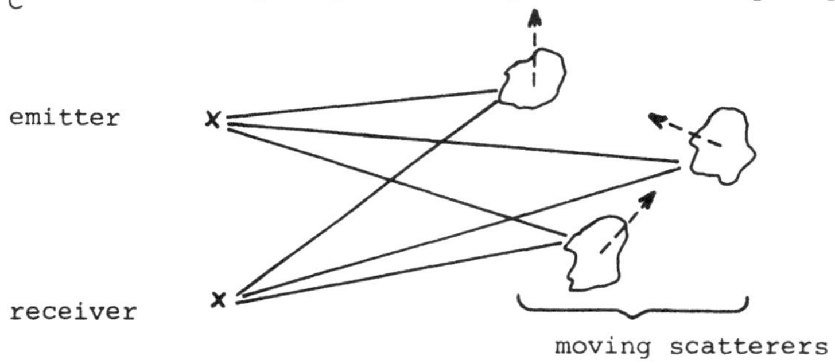

FIGURE 2

In the general case, the actual DOPPLER effect of each scatterer is a time-compression and, under these conditions, the time-compression - delay representation will be a more realistic one. In this model :

$$|s\rangle = \iint_{\lambda\,\xi} G(\lambda,\xi)\; C_\lambda T_\xi |e\rangle \, d\lambda d\xi \qquad (4)$$

C_λ : operator of time compression [5]

$$|y\rangle = C_\lambda |x\rangle \Leftrightarrow y(t) = x(\lambda t)$$

This representation is useful in submarine acoustics when considering large bandwidth signals.

For this kind of models the "representation function" ($U(\nu,\xi)$ or $G(\lambda,\xi)$) is generally a stochastic function.

In order to characterize the channel, we must perform measurements of this function. We shall limit ourselves to the study of the first two moments. This study is sufficient in the gaussian case.

2. FIRST ORDER CHARACTERIZATION. THE AMBIGUITY FUNCTION.

Let $E[A] = EA$ be the expectation of the stochastic element A.

$$\chi_{se}(v,\chi) = \langle s|F_v T_\chi|e\rangle = \int_t s(t)\, e^{-2i\pi vt} e^*(t-\chi)\, dt$$

be the interambiguity function of $|s\rangle$ and $|e\rangle$ and

$$E\chi_{se}(v,\chi) = E[\chi_{se}(v,\chi)]$$

be the expectation of interambiguity of $|s\rangle$ and $|e\rangle$.

One can see easily that

$$\chi_{se}(v,\chi) = \int_\nu\int_\xi U(\nu,\xi)\ \chi_{ee}(v-\nu,\chi-\xi)\, d\nu d\xi \qquad (5)$$

and, if the signals and the medium are statistically independent :

$$E\chi_{se}(v,\chi) = \int_\nu\int_\xi E\ U(\nu,\xi)\ E\chi_{ee}(v-\nu,\chi-\xi)\, d\nu d\xi \qquad (6)$$

This relation gives a means of measurement of the expectation of the DOPPLER-delay representation. Knowing the expectation of ambiguity of the signal $E\chi_{ee}$, one must perform a deconvolution of the expectation of interambiguity output/input, using $E\chi_{ee}$ as kernel.

In order to simplify the process, it is worthwhile to use signals with a "good" ambiguity function. The ideal case would be

$$E\chi_e(v,\chi) = \delta(v)\ \delta(\chi)$$

but, unfortunately, this is known to be impossible and what we can obtain is only a signal with an ambiguity function close to the origin of the time-frequency plane.

If we can suppose that $E\ U(\nu,\xi)$ is almost constant on the domain where $E\chi_e(v-\nu,\chi-\xi) \neq 0$.

$$EU(v,\chi) = \frac{E\chi_{s,e}(v,\chi)}{\int_\nu\int_\xi E\chi_{ee}(\nu,\xi)\, d\nu d\xi} \qquad (7)$$

With this kind of signal the measurement of $EU(v,\chi)$ is obtained directly through the measurement of $E\chi_{se}(v,\chi)$.

This is the theoretical and practical way of measuring $EU(v,\chi)$. In order to perform such measurements, we have to :

- use "performant" input signals ($\chi_{ee}(v,\chi)$ close to the origin);
- calculate the expectation of the output/input interambiguity

The same process can be applied to other channel models. For example, in the time compression model, the same process is valid, with expectations of the compression ambiguity being defined by :

$$\Psi_{se}(\lambda,\xi) = \langle s|C_\lambda T_\xi|e\rangle$$

3. SECOND ORDER CHARACTERIZATION. THE SCATTERING FUNCTION.

Let be :

$$U(\nu,\xi) = EU(\nu,\xi) + \tilde{U}(\nu,\xi)$$

$$|s\rangle = |s| + |\tilde{s}\rangle$$

Now we shall study $|\tilde{s}\rangle$ issued from $\tilde{U}$:

$$\tilde{s}(t) = \iint_{\nu\,\xi} \tilde{U}(\nu,\xi)\, e^{2i\pi\nu t}\, e(t-\xi)\,d\nu d\xi$$

We first present the simplifications leading to the description of the medium, at the second order, by means of the scattering function and the measurement methods used for this function. Thereafter, we present generalizations of this concept.

3.1 Wide sense stationary uncorrelated scatterers channels (WSSUS). The scattering function.

Making new the hypothesis of independence between the input signal and the medium

$$E[\tilde{s}(t)] = 0$$

and the covariance of $\tilde{s}$:

$$\Gamma_{\tilde{s}}(t,\tau) = E[\tilde{s}(t)\tilde{s}^*(t-\tau)]$$

is :

$$\Gamma_s(t,\tau) = \iiiint \Gamma_{\tilde{U}}(\nu,\nu',\xi,\xi')\, e^{2i\pi[(\nu-\nu')t+\nu'\tau]}\, E[e(t-\xi)\,e^*(t-\tau-\xi')]\,d\nu d\nu' d\xi d\xi' \tag{8}$$

In the general case, the medium is characterized by the function of 4 variables :

$$\Gamma_{\tilde{U}}(\nu,\nu',\xi,\xi') = E[\tilde{U}(\nu,\xi)\ \tilde{U}^*(\nu',\xi')]$$

The classical simplifications are :

- the wide sense stationary channels (W.S.S. channels) : we impose that, when the input signal is stationary, the output signal is also stationary. It can be seen on (8) that this implies :

  $$\Gamma_{\tilde{U}}(\nu,\nu',\xi,\xi') = \Gamma_{\tilde{U}}(\nu,\xi,\xi')\ \delta(\nu-\nu')$$

  Coming back to the physical interpretation, we can say that the signals scattered at two different frequencies are uncorrelated.

- Uncorrelated scatterers channel (U.S channel) : this is realized when :

  $$\Gamma_{\tilde{U}}(\nu,\nu',\xi,\xi') = \Gamma_{\tilde{U}}(\nu,\nu',\xi)\ \delta(\xi-\xi')$$

  in this situation the signals scattered by two different scatterers are uncorrelated.

- WSSUS channel : it is the combination of the two preceding hypotheses. In that case :

  $$\Gamma_{\tilde{U}}(\nu,\nu',\xi,\xi') = sc(\nu,\xi)\ \delta(\nu-\nu')\ \delta(\xi-\xi')$$

  In this situation the second order properties of the medium are described by the function of two variables $sc(\nu,\xi)$, called the scattering function. Then, for a stationary input signal :

  $$\Gamma_s(\tau) = \iint sc(\nu,\xi)\ e^{2i\pi\nu\tau}\Gamma_e(\tau)\,d\nu d\xi \qquad (9)$$

4. TEST OF WSSUS HYPOTHESIS.

The experimentator who tries to characterize at the second order a propagation channel and to utilize a WSSUS model has to perform :

- first the test of the WSSUS hypothesis ;

- thereafter, if the test is positive, the measurement of the scattering function.

4.1 Test of wide sense stationarity.

This can be done by using stationary inputs and testing the

stationarity of the output.

4.2 Test of WSSUS with pure frequencies.

Taking : $e(t) = e^{2i\pi f_\circ t}$, and calculating :

$$\gamma_s(t,f) = TF_\tau[\Gamma_s(t,\tau)]$$

we obtain :

$$\gamma_s(t,f) = \int \Gamma_{\tilde{U}}(\nu, f-f_\circ, \xi, \xi')\, e^{2i\pi(\nu - f_\circ + f)t}\, e^{2i\pi f_\circ(\xi'-\xi)}\, d\nu d\xi d\xi' \qquad (10)$$

In the general case $\gamma_s(t,f)$ is a function of : t, f and $f_\circ$.
We see on (10) that :

- the WSS hypothesis $\Leftrightarrow \gamma_s(t,f)$ independent of t ;
- the WSSUS hypothesis $\Leftrightarrow \gamma_s(t,f) = \ell(\Delta f)$ with $\Delta f = f - f_\circ$.

We have here a means of making a test of the WSS and/or WSSUS hypothesis.

When the WSSUS hypothesis is verified, (10) writes :

$$\gamma_s(f) = \ell(\Delta f) = \int sc(\Delta f, \xi)\, d\xi \qquad (11)$$

This shows us an important property : in a WSSUS channel the frequency spreading of monochromatic signals is independent of the input frequency.

Combined with the stationarity of the output, this is a characteristic property of the WSSUS channel, very useful for testing the WSSUS hypothesis.

Starting from the relation (11), we can define the frequency spreading of a WSSUS channel B by :

$$B^{-1} = \frac{\int \left| \int sc(\nu,\xi)\, d\xi \right|^2 d\nu}{\left[\iint sc(\nu,\xi)\, d\nu d\xi \right]^2} = \frac{\int \gamma_s^2(\Delta f)\, d\Delta f}{\left[\int \gamma_s(\Delta f)\, d\Delta f \right]^2} \qquad (12)$$

B can be measured using the output associated to a monochromatic input. Similarly, we can define a time spreading by :

$$\Delta^{-1} = \frac{\int \left| \int sc(\nu,\xi)\, d\nu \right|^2 d\xi}{\left[\iint sc(\nu,\xi)\, d\nu d\xi \right]^2} \qquad (13)$$

The cross correlation of two different frequencies f_1 and f_2 such as :

$$\Delta f = f_1 - f_2$$

is :

$$\Gamma_{s_1 s_2}(\Delta f, 0) = \iint sc(\nu,\xi)\, e^{2i\pi\xi\Delta f}\, d\xi\, d\nu \tag{14}$$

and thus :

$$\Delta^{-1} = \frac{\int \left|\Gamma_{s_1 s_2}(\Delta f, 0)\right|^2 d\Delta f}{\left[\Gamma_{s_1 s_2}(0,0)\right]^2} \tag{15}$$

This gives us a means of measuring the time spreading Δ .

5. MEASUREMENT OF THE SCATTERING FUNCTION.

When the WSSUS hypothesis is valid, with once more the hypothesis of statistical independence between input and medium, it comes

$$E\left[\left|\chi_{\tilde{s}e}(v,\chi)\right|^2\right] = \int_\nu\int_\xi sc(\nu,\xi) E\left|\chi_{ee}(v-\nu,\chi-\xi)\right|^2 d\nu\, d\xi \tag{16}$$

(16) gives us, like in the first order case, a means of measurement of $sc(\nu,\xi)$. With "performant signals", i.e. signals for which $\left|\chi_{ee}(v,\chi)\right|^2$ is close to the origin in the time-frequency domain :

$$sc(v,\chi) \sim \frac{E\left[\left|\chi_{\tilde{s}e}(v,\chi)\right|^2\right]}{\int_\nu\int_\xi E\left|\chi_{ee}(\nu,\xi)\right|^2 d\nu\, d\xi} \tag{17}$$

This is a means of measuring $sc(\nu,\chi)$.

6. GENERALIZATIONS : FREQUENCY VARYING MODELS.

One major fact of the WSSUS channel is the independence between the frequency spreading and the input frequency. In actual situations this condition is not always fulfilled and the WSSUS hypothesis is then invalid. Different solutions have been proposed for taking into account these situations. We will first examine the reverberation model, giving a frequency spreading proportional to the input frequency and then we shall present shortly more general approaches.

6.1 The reverberation model.

As indicated in § 1, for this model (developed at the CEPHAG laboratory) the output signal is :

$$|s\rangle = \iint G(\lambda,\xi)\; C_\lambda T_\xi \;|e\rangle \; d\lambda d\xi \qquad (18)$$

T_ξ : time delay

C_λ : time compression

In this case, we define a WSSUS medium by :

$$[E\,\tilde{G}(\lambda,\xi)\;\tilde{G}^*(\lambda',\xi')] = \widehat{sc}(\lambda,\xi)\;\delta(\lambda-\lambda')\,\delta(\xi-\xi')$$

As in § 4, when taking a monochromatic input signal, we will have output signals with the power spectral density :

$$\gamma_s(f) = \frac{1}{f_o}\int \widehat{sc}\left(\frac{f}{f_o},\xi\right) d\xi \qquad (19)$$

where f_o is the input frequency. We can see on (19), that in this case the frequency spreading of the output is proportional to the input frequency.

For physical reasons, we think that this kind of model is well adapted to submarine acoustic propagation.

6.2 The generalized diffusion function : measurements were made on aerial acoustic propagation. In this context a frequency spreading

$B = A\, f_o^{\,1,8}$ was found.

This kind of frequency spreading cannot be explained by a WSSUS or a reverberation WSSUS model and [9] propose a generalized diffusion function. The idea is roughly, the following : observing experimentally the frequency spreading versus input frequency, the experimentator will find some functions. It may be expected that in some input frequency ranges the frequency spreading will be constant (or nearly constant). Locally, with respect to the frequency input, the medium looks like a WSSUS medium. Thus it is proposed to define a scattering function varying with the input frequency.

Similarly, in order to overcome non-stationarity in the wide sense, one can propose to limit in time the validity of the WSS hypothesis.

On this basis, one can describe the medium by

$sc(\nu,\xi,f,t)$

f and t being the time and the frequency of the input. This leads to a "local WSSUS hypothesis" in the time-frequency plane.

It should be noticed that the function of 4 variables sc is apparently of the same complexity as the complete description given by $\Gamma_{\tilde{U}}(\nu,\nu',\xi,\xi')$. However, it is expected that the description is more practical because the variations of sc with respect to f and t are slow.

B) INTERNAL APPROACH OF TWO-POINT-LINKING NON STATIONNARY RANDOM SYSTEMS

1. GENERAL PURPOSE.

Up till now, we have studied the external approach of the channel, i.e. the channel was always characterized by its bitemporal response (or by one of its FOURIER transforms) in the deterministic case, and by its statistical properties (mean value and covariance) in the random case. Let us now precise what is the internal approach.

1.1 Internal approach of processes and channels.

In the simpler case, when the channel is a deterministic and non time varying one, instead of characterizing it by $g(\xi)$, it is usually supposed that the channel can be described by a differential system of a given range.Thus, the data of the impulsional response are replaced by those of the different terms of the system. The same holds true for the internal representation of a random process :instead of speaking in terms of covariance, one rather assumes that the FOURIER transform of this covariance is a rational function, i.e. that the random process originates in the filtering of a white noise process by this system. Such modelization has a double advantage :

- data reduction for the characterization : some parameters (depending on the range of the system) are sufficient to replace the "continuous" graph $g(\xi)$;

- for the further optimal processings : they generally depend on the form of the output process covariance ; a structure, identical to the one of the characterization, will be utilized and generally the realization will be easier, owing to the reduced number of data. Thus the technique of optimal filtering of

KALMAN is substituted, in the practical cases, to the WIENER method (where the solutions are given in the form of integral equations).

1.2 For the randomly time varying channel (RTVC) internal approach we could imagine to use the internal approach of a deterministic channel, and to consider that some parameters of the system are randomized.

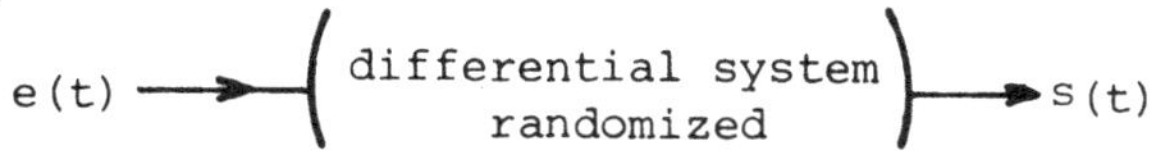

But in this case, the output statistical properties would be quite inaccessible. So the above method is not studied in the following.

In the present case, we shall consider the internal approach of the bitemporal random response $g(t,\xi)$, characterizing the channel. This case is thus quite similar to the one of representation of a random process, however, there is a fundamental difference : the process is here a function of two variables t and ξ. For this approach one needs of course the excitation of a system by a noise process ; the statistical properties of this noise will be connected to the wanted properties of the covariance function of $g(t, \xi)$. Of course the modelisation of $g(t, \xi)$ does not include the transmission process i.e. the input signal $e(t)$ will still have to be filtered by $g(t, \xi)$.

Internal approach of a RTVC

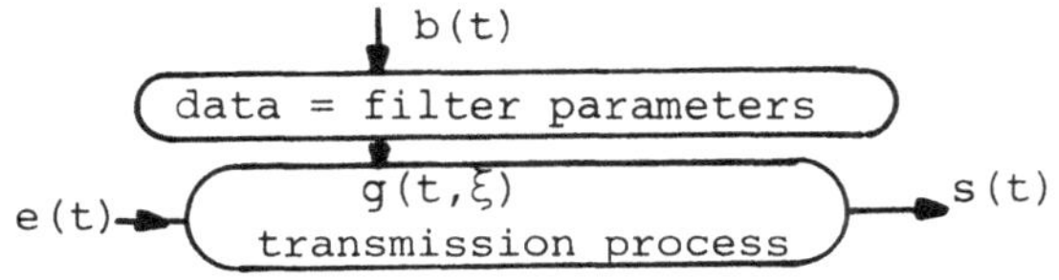

It is worthy of notice that the channel thus approached is necessarily a random one, and we have to deal with $\tilde{g}(t,\xi)$, always centered in the following. We shall always omit the symbol $\sim$ in this part B, for more convenient formulas.

1.3 Some authors, when speaking about internal approach of fluctuating channels, are concerned with the internal approach of the output process $s(t)$. It is obvious that in this case :

- we have simply to deal with the representation of a random process, as recalled above ;
- each time the input signal $e(t)$ changes, $s(t)$ changes, thus the internal representation must be changed.

We shall not deal with this problem in the following.

1.4 Finally there is another important point : this approach is not valid for the reverberating model, as this model has no direct equivalent as to the impulsional response.

Let us now discuss the internal approach of :

first, general RTVC, or doubly-spread channels ; we shall also study especially WSSUS channels ;

thereafter, degenerated "DOPPLER only" or "delay only" channels.

2. INTERNAL REPRESENTATION OF THE BITEMPORAL RESPONSE OF A RTVC

We refer for a large part to the works of KURTH J. [8] and V. TREES as to the internal representation of a random process with two variables. They refer to a distributed parameter state variable model, presented by TSAFESTAS and NIGHTINGALE and give a complex formulation of it (cof. bibliography of [8].

2.1 The fundamental ideas of this representation are the "separation" of the different aspects implied in $g(t,\xi)$:

1/ the random aspect : is obtained by exciting the input of the differential system by a noise ; the system itself is deterministic - this is the classical case of the representation of a random function and gives the variation of g versus the variable t.

2/ the "time varying" aspect : is obtained by the parametrisation versus the variable ξ , of the above mentioned classical representation, as well for the input noise as for the system.

3/ As for the different statistical hypothesis which are desired for the covariance of $g(t,\xi)$, the system is provided with a stationary or non-stationary structure ; and the adequate data are given for the covariance of the input noise as well as for the initial conditions.

Finally, as usual for this type of problem, the exciting noise is assumed to be a gaussian one and, as the filtering is a linear one, this generates a gaussian impulsional response $g(t,\xi)$.

2.2 The equations in the general case.

$g(t,\xi)$ is supposed to be the output of a distributed parameter linear system driven by a noise process $b(t,\xi)$. The system is described by the partial differential linear equation

$$\frac{\partial x(t,\xi)}{\partial t} = \underline{F}(t,\xi)\ \underline{x}(t,\xi) + \underline{A}(t,\xi)\ \underline{b}(t,\xi)$$

$$g(t,\xi) = \underline{C}(t,\xi)\ \underline{x}(t,\xi) \tag{20}$$

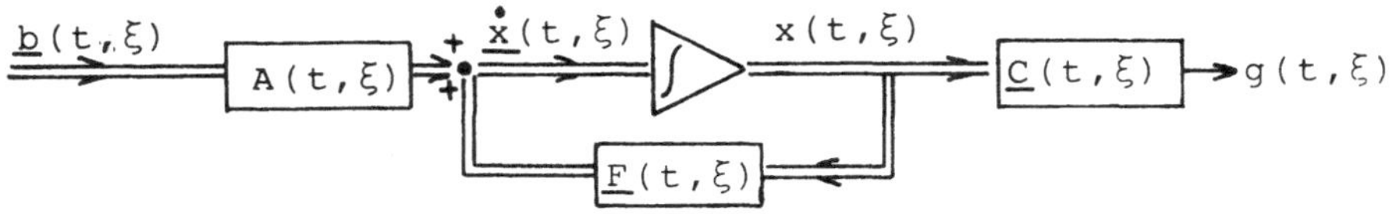

This partial differential equation is an ordinary differential equation with ξ as a parameter. The n-dimensional vector process $\underline{x}(t,\xi)$ is the state of the distributed system at the time t. The gain matrices $\underline{A}$, $\underline{F}$ and $\underline{C}$ are deterministic functions. The p-dimensional input process $\underline{b}(t,\xi)$ is centered, temporally white, with a covariance

$$\left.\begin{array}{l} E\{\underline{b}(t,\xi)\ \underline{b}^{+}(t'\xi')\} = \underline{Q}(\xi,\xi';t)\delta(t-t') \\ E\{\underline{b}(t,\xi)\ b^{T}(t',\xi')\} = 0 \\ E\{\underline{b}(t,\xi)\} = 0 \qquad\qquad + = T^{*} \end{array}\right) \tag{21}$$

Let us <u>interpret the ξ - parametrisation</u> of the excitation noise and the system matrices by the discretisation of the ξ-space : it means that, with each delay ξ_i (let us recall that ξ corresponds to a delay between the date t and the impulsion emission instant) one must dispose of a white noise $b(t,\xi_i)$ - the statistical properties of which (such as the power) are a function of ξ_i; and a system which is also a function of ξ.
The given results are the following ones:

The covariance of $g(t,\xi)$ is

$$\Gamma_g(t,t';\xi,\xi') \underline{\Delta} E\{g(t,\xi)g^{*}(t',\xi')\} = \underline{C}(t,\xi)\underline{\Gamma}x(t,t';\xi,\xi')\underline{C}^{+}(t',\xi') \tag{22}$$

where the 4-variable covariance of the state vector is expressed in terms of $\underline{\Gamma}x(t,t;\xi,\xi')$ and the system transition matrix $\phi(t,t';\xi)$, as :

$$\underline{\Gamma}x(t,t';\xi,\xi') = \left(\begin{array}{ll} \underline{\phi}(t,t';\xi)\underline{\Gamma}x(t',t';\xi,\xi') & t > t' \\ \Gamma_x(t,t;\xi,\xi')\phi^{+}(t,t';\xi') & t < t' \end{array} \right) \tag{23}$$

On the other hand, $\Gamma_x(t,t;\xi,\xi')$ is solution of :

$$\frac{\partial \underline{\Gamma}_x(t,\xi;t,\xi')}{\partial t} = \underline{F}(t,\xi)\underline{\Gamma}_x(t,t;\xi,\xi') + \underline{\Gamma}_x(t,t;\xi,\xi')\underline{F}^+(\xi',t) + \underline{A}(t,\xi)\underline{Q}(t,\xi,\xi')\underline{A}^+(\xi',t) \qquad (24)$$

with the given initial condition :

$$\underline{\Gamma}_x(t_o,t_o;\xi,\xi') = \underline{\underline{P}}_o(\xi,\xi') \qquad (25)$$

Finally we recall that $\phi(t,t',\xi)$ is solution of

$$\frac{\partial \underline{\phi}(t,t',\xi)}{\partial t} = \underline{F}(t,\xi)\underline{\phi}(t,t',\xi) \quad \text{with} \quad \underline{\phi}(t,t,\xi) = \underline{I} \qquad (26)$$

So the formal solution exists, but it is obviously quite impossible to be carried out in the general case. Particularly one cannot easily imagine such 3-variable matrices, as $\underline{Q}(t,\xi,\xi')$, which however must be done a priori. It is the same problem with the various gain matrices, all parametrised and time varying.

However, KURTH has proposed the most general internal model here above (which can lead to a 4-variable covariance of $g(t,\xi)$).

NB : In our opinion - but this should be verified - it is perhaps possible to simplify the above model without loosing its generality, since both excitation noise and system are non-stationary. If only one of these two quantities is stationary, $\underline{x}(t,\xi)$ will still be non stationary and so will $g(t,\xi)$.

2.3 Study of the simplifying hypothesis on the covariance of $g(t,\xi)$.

Let us suppose that the covariance of $g(t,\xi)$ responds to one of the simplifying hypothesis studied above : WSS, US or WSSUS. The corresponding internal approach will be simplified.

We are to study below particularly the WSSUS case ; the WSS-case and the US-case are studied in another work [4].

a) The WSS-hypothesis involves that $\Gamma_g(t,t',\xi,\xi')$ will be stationary. This will entail the following simplifications :

- with regard to the system : the gain matrices $\underline{A}$, $\underline{C}$ and $\underline{F}$ are stationary, i.e. they are no longer functions of t ;
- with regard to the input noise : $\underline{b}(t,\xi)$ must be stationary.

b) The US hypothesis involves that $\Gamma_g(t,t';\xi,\xi')$ will be uncorrelated versus ξ and ξ' ; for this purpose it is sufficient that the noise processes $\underline{b}(t,\xi)$ are supposed to be uncorrelated versus ξ; this property will be still true for the system output process, since $\underline{b}(t,\xi)$ are the only random quantities.

c) Finally the internal approach of a WSSUS channel uses both the above mentioned simplifications. It follows that :

- The gain matrices of the system are not function of t .
- The input noise process is stationary and uncorrelated versus ξ

$$E\{b(t,\xi)\, b^{+}(t',\xi')\} = \underline{Q}(\xi)\,\delta(\xi-\xi')\,\delta(t-t') \tag{27}$$

Thus, the internal modelization of a WSSUS channel involves a set of random processes $b(t,\xi_i)$ which are all white, stationary and uncorrelated. This can be interpreted in the following way : for each delay ξ_i , scatterers are equivalent to generating white stationary noise, the power of which is a function of ξ_i , $Q(\xi_i)$; the further spectral form of this noise is connected to the scatterers (the system parameters are functions of ξ_i). Saying that the scatterers are uncorrelated, is equivalent to admitting that the white noises are all uncorrelated.
The distributed parameter state-variable model for the WSSUS channel, as shown below, is :

$$\frac{\partial\, \underline{x}(t,\xi)}{\partial t} = \underline{F}(\xi)\,\underline{x}(t,\xi) + \underline{A}(\xi)\,\underline{b}(t,\xi)$$
$$g(t,\xi) = \underline{C}(\xi)\,\underline{x}(t,\xi) \tag{28}$$

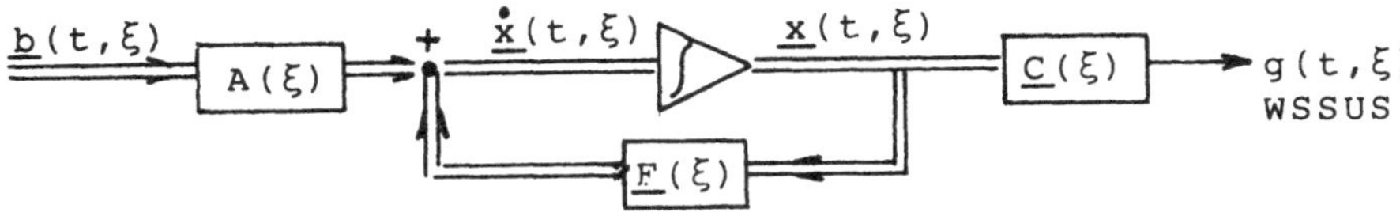

The covariance of $b(t,\xi)$ is given by (27) and

$$E\{\underline{b}(t,\xi)\,\underline{b}^{T}(t',\xi')\} = 0 \tag{29}$$

Let us now see how the different data in the fig. above are connected to the scattering function of the WSSUS channel. We recall that this function is the FOURIER transform versus τ of $\Gamma_1(\tau,\xi)$ given by :

$$\Gamma_g(t,t';\xi,\xi') \equiv \Gamma_1(t-t',\xi)\,\delta(\xi-\xi') \equiv \Gamma_1(\tau,\xi)\,\delta(\xi-\xi') \tag{30}$$

$\Gamma_1(\tau,\xi)$ is carried out by inspection of (22) and the above hypothesis :

$$\Gamma_1(\tau,\xi) = \underline{C}(\xi)\,\underline{\Gamma}_x(\tau,\xi)\,\underline{C}^{+}(\xi') \tag{31}$$

$\underline{\Gamma}_x(\tau,\xi)$ is given by the equations 23 to 26, which become :

$$\underline{\Gamma}_x(\tau,\xi) = \underline{\phi}(\tau,\xi)\ \underline{\Gamma}_x(0,\xi) \qquad \tau > 0$$
$$= \underline{\Gamma}_x(0,\xi)\ \underline{\phi}^+(\xi,-\tau) \qquad \tau < 0 \qquad (32)$$

where the transition matrix, now stationary, is solution of

$$\underline{\phi}(0,\xi) = I$$
$$\frac{\partial \underline{\phi}}{\partial t}(t,\xi) = F(\xi)\ \underline{\phi}(t,\xi) \qquad (33)$$

and $\Gamma_x(0,\xi)$ is solution of

$$\underline{0} = \underline{F}(\xi)\ \underline{\Gamma}_x(0,\xi) + \underline{\Gamma}_x(0,\xi)\ \underline{F}^+(\xi) + \underline{A}(\xi)\ \underline{Q}(\xi)\ A^+(\xi) \qquad (34)$$

As an example of the WSSUS channel model, KURTH [8] considers a first order system for (28).

$$\underline{F}(\xi) = -k(\xi) = -k_r(\xi) - j\,k_i(\xi) \quad , \quad k_r(\xi) > 0$$
$$\underline{A}(\xi) = \underline{C}(\xi) = 1 \qquad (35)$$
$$\underline{Q}(\xi) = Q(\xi) \qquad \geq 0$$

From equations (31) to (34), it follows that :

$$\Gamma_1(\tau,\xi) = \frac{Q(\xi)}{2k_r(\xi)} \exp - \{k_r(\xi)|\tau| - jk_i(\xi)\tau\} \qquad (36)$$

and the scattering function is the FOURIER transform of (36) versus τ :

$$sc(\nu,\xi) = \frac{Q(\xi)}{[2\pi\nu + k_i(\xi)]^2 + k_r^2(\xi)} \qquad (37)$$

Thus, from (37) it follows that, with a first order system, not all possible forms of scattering functions can be obtained. Particularly :

- (37) is not factorisable into $f_1(\nu)\ f_2(\xi)$;

- (37) shows that a firstorder system can represent a scattering function which, as a function of ν is a one pole spectrum, centered at

$$f_o = -\frac{k_i(\xi)}{2\pi}$$

and 3 dB points $\pm \frac{k_r(\xi)}{2\pi}$ about f_o. As a function of ν , the form of $sc(\nu,\xi)$ is quite fixed. On the other hand, versus ξ, (37) permits a considerable flexibility, since the two quantities $Q(\xi)$ and $k_r(\xi)$ are arbitrary as long as they remain > 0 . Particularly multimodal functions, versus ξ, can be obtained. KURTH gives the following example :

$$Q(\xi)=\begin{cases} 1-\cos\frac{2\pi\xi}{L} & \begin{cases} 0\le\xi\le\frac{L}{4} \\ \frac{3L}{4}\le\xi\le L \end{cases} \\ 2+\cos\frac{\pi\xi}{L} & \frac{L}{4}\le\xi\le\frac{3L}{4} \\ 0 & \text{elsewhere} \end{cases}$$

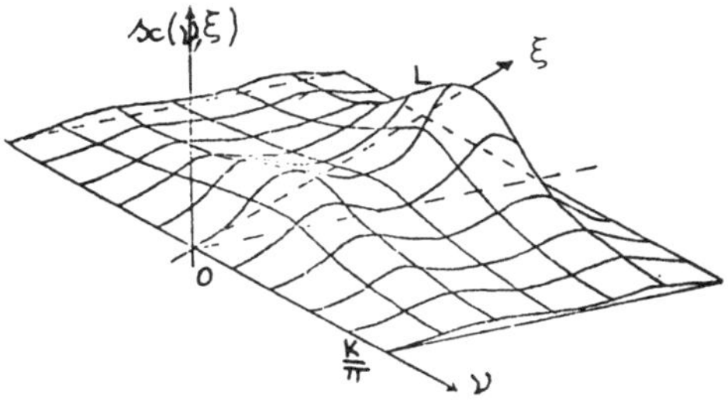

$$k_r(\xi) = k(1-\frac{\xi}{2L}) - j\,\frac{3k\xi}{5\pi L}$$

NB : Let us notice that if the system is a real first order one, $\underline{F}$, $\underline{A}$ and $\underline{C}$ are real scalars and the equations from (31) to (34) become :

$$\Gamma_1(\tau,\xi)=C^2(\xi)\Gamma_X(\tau,\xi) \quad \text{with} \begin{cases} \Gamma_x(\tau,\xi) = \Gamma_x(0,\xi)\, e^{|\tau|F(\xi)} & F<0 \\ \Gamma_x(0,\xi) = -\frac{A^2(\xi)}{F^2(\xi)}\, Q(\xi) & Q>0 \end{cases}$$

It follows that the scattering function is :

$$sc(\nu,\xi) = \frac{C^2(\xi)A^2(\xi)Q(\xi)}{F^2(\xi) + 4\pi^2\nu^2} \qquad (38)$$

Consequently :

- $sc(\nu,\xi)$ is necessarily odd versus ν , with a Lorentzian form
- the individual forms of C , A and Q as functions of ξ are not important. It is only their product that occurs in (38). Particularly, we often will incline to choose A=C=1 (as in the above example) and the form of the numerator of (38) versus ξ will be simply the one given by the power $Q(\xi)$ of the input noise.

If a scattering function, multimodal in the ν direction is wanted, the differential system must be at least of second order [4] gives an example of this case.

3. INTERNAL REPRESENTATION OF DEGENERATE CHANNELS (EITHER DOPPLER-SPREAD ONLY, OR DELAY-SPREAD ONLY).

Another particular case, quite different from those studied in the previous section, is when the transmission channel garbles the emission signal in only one way (either DOPPLER or delay). KURTH gives some results of internal representation in these two cases, the important points of which we simply present below :

3.1 DOPPLER-spread only channel.

In this case $g(t,\xi)=g(t)\delta(\xi)$ (39)
where $g(t)$ is always a random process. The received signal is, with (1) :

$$s(t) = e(t)g(t) \tag{40}$$

This kind of transmission is usually called modulation of $e(t)$ by $g(t)$, or temporal fading.

3.11 The internal approach of $g(t)$ is much simpler than in the general case, since the response is now a function of t only ; therefore this case may be obviously considered as a particular one of the case studied in § 2 ; so to have the representation of $g(t)$, it is sufficient to fix ξ in (20) to a given value (f.i.o.) Then only one noise $b(t)$ and one system (F,A,C) are to be used.

3.12 Moreover, with (40), the output process $s(t)$ will have the same representation, with adding only a multiplicative function $e(t)$ in the observation equation

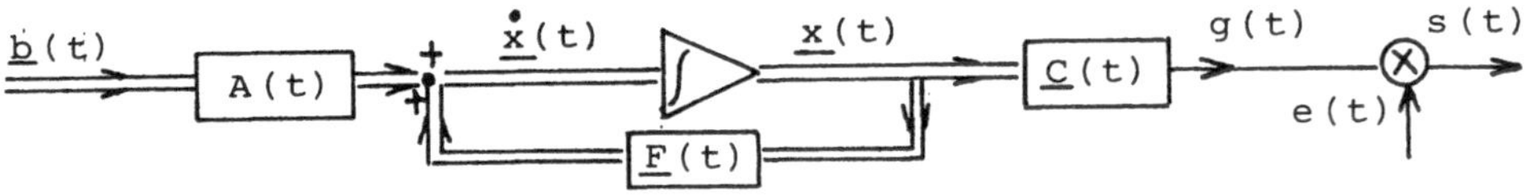

(Internal representation of a DOPPLER-only channel and output process)

3.2 Delay spread only channel or stationnary channel.

3.21 It is an absolutely different case when the channel is delay spread only, since such a channel is non time varying, and $g(t,\xi)$ is no longer a function of t. The differential representation versus t of the previous section cannot be used here. The only possibility is representing directly the output process $s(t)$, which, as mentioned in the introduction, is different from the above study. KURTH developed in this case an internal representation of $s(t)$, by using duality concepts between a delay spread channel and a DOPPLER spread channel ; this involves infinite duration signals $e(t)$. Moreover, if the channel is a delay spread and "US" one i.e. if $E\{g(\xi)g^*(\xi')\}=S(\xi)\delta(\xi-\xi')$, the internal approach involves $S(\xi)$ to be a rational function versus ξ (and so with infinite spreading). This is very difficult to be considered practically, because the scatterers should be able to generate infinite delays !

3.22 Let us recall, however, that there is another possibility of having an internal approach for a random non-time varying channel : it is obtained, as mentioned in the introduction of the part B, by supposing that the FOURIER transform (F.T.) of $g(\xi)$ exists and is a rational function, and randomizing some parameters of the differential system representing this F.T.

4. CONLUSION.

In this section, a general internal representation of a randomly time varying channel is given. Let us recall all requirements connected to this approach: set of white input noises, set of differential systems, besides the implicit hypothesis that a state variable model exists for $g(t,\xi)$,$\forall$ ξ. Another kind of difficulty is that internal representations connected to degenerate channels cannot always be carried out from the general model. Finally, one must not forget that, in the WSSUS case for instance, if any king of scattering functions is desired, the range of the differential system may be high enough.

An original example of identification of a natural channel by an internal model is given in Part D. It is obtained from experimental measure of the scattering function.

This kind of approach seems to be applied with success in optimal detection and communication processing. To our knowledge, only theoretical studies of these processings exist up till now. Particularly, KURTH [8] shew that in the case of WSSUS doubly spread channel internal model, a realizable structure of optimal detection can be carried out. With a very simple emission signal, he obtained its theoretical performances in detection and communication.

C) SPATIAL CHANNELS.

In this case the source and the receiver are extended in space. The input and output signals are stochastic (or deterministic) functions of time and position. This situation has been largely studied in many areas of research : radar, sonar, radio-astronomy, optics, and leads to a great deal of problems. We shall here limit ourselves to an extension to the spatial case of the formalism presented in the part A. Contrarily to the common approach, we shall not introduce simplifications leading to a pure spatial case or to a pure temporal one.

1. EXTENSION OF THE FORMALISM TO SPATIAL CASES.

In the following, the sources are located in a part S of space and the receivers (or the receiving system) in the part R.

The channel is linear

Then, as in A :

$$s(t,y) = \int_{\xi}\int_{(S)} g(t,\xi,y,z)\ e(t-\xi,z)\,d\xi dz \qquad (41)$$

(Notice that $s(t,y)$ and $e(t,z)$ are surfacic or volumic densities of the signal amplitude).

$g(t,\xi,y,z)$ is the biponctual, bitemporal representation of the channel. Generally, g will be a stochastic function of the 4 variables t,ξ, y, z.

As in the case A , we can introduce the bipunctual DOPPLER-delay representation of the channel by

$$U(\nu,\xi,y,z) = TF_t\ g(t,\xi,y,z)$$

In the stationary case, the bipunctual impulse response is :

$$g(r,\xi,y,z) = g_S(\xi,y,z)$$

and the bipunctual DOPPLER delay representation

$$U(\nu,\xi,y,z) = \delta(\nu)\ g_S(\xi,y,z)$$

2. CHARACTERIZATION OF THE OUTPUT AND THE INPUT-OUTPUT RELATION :

As in case A, we can use interambiguity functions χ_{eS} and expectation of interambiguity $E\chi_{eS}$. We shall also use the spectral function

$$\xi_{eS}(v,f,z,y) = TF_{\chi} E\chi_{eS}(v,\chi,\xi,n)$$

In the stationary case

$$\xi_{es}(v,f,z,y) = \delta(v)\ \gamma_{es}(\nu,z,y) \qquad (42)$$

where γ_{es} is a cross-spectral density.

Finally, in the stationary case an important role is played by the coherence function or normalized cross-spectral density :

$$C_{es}(\nu,z,y) = \frac{\gamma_{es}(\nu,z,y)}{\sqrt{\gamma_{ee}(\nu,z,z)\gamma_{ss}(\nu,y,y)}} \tag{43}$$

We shall say :

- "spatially incoherent input", if

$$C_{ee}(\nu,z_1,z_2) = 1 \qquad z_1 = z_2$$
$$= 0 \qquad z_1 \neq z_2$$

- "input totally coherent in space" if

$$|C_{ee}(\nu,z_1,z_2)| = 1 \qquad \forall, z_1 \text{ and } z_2$$

3. IDENTIFICATION OF THE FIRST MOMENT OF THE DOPPLER DELAY BIPUNCTUAL REPRESENTATION.

It is easily seen that (as medium and signal are statistically independent)

$$E\chi_{se}(v,\chi,y_1,z_2) = \iiint_{\nu\ \xi\ S} E[U(\nu,\xi,y_1,z)] \tag{44}$$

$$E\chi_{ee}[v-\nu,\chi-\xi,z\ ,z_2]d\nu d\xi dz$$

The "performant input" will be such that

$$\chi_{ee}[v,\chi,z_1,z_2] = F(z_2)\delta(v)\delta(\chi)\delta(z) \tag{45}$$

This can be only approximately obtained and then (44) gives

$$E[U(\nu,\xi,y_1,z_2)] \tag{46}$$

Then the first order identification of the medium is obtained through the measurement of the expectation of interambiguity input/output.

The condition (45) shows us that the performant inputs are then orthogonal codes with a expectation of ambiguity close to the origin.

In the stationary case, the performant signals are spatially decorrelated white noises.

4. SECOND ORDER STUDY.

At the second order, let us write :

$U = EU + \tilde{U}$
$|e\rangle = E|e\rangle + |\tilde{e}\rangle$

Then we have 3 different cases :

1/ $\tilde{U} = |\tilde{e}\rangle = 0$: signal and medium deterministic.

2/ $\tilde{U} = 0$; $|\tilde{e}\rangle \neq 0$: stochastic signal, deterministic medium

3/ $\tilde{U} \neq 0$: stochastic medium.

The case 1 is completely treated in 3. Case 2 is generally considered in classical optic. Case 3 is the general case.

We shall deal hereafter only with some considerations on the case 2 in a stationary medium, leading to the ZERNICKE - van CITTER theorem. We shall conclude with some indications on the spatial scattering function.

4.1 Medium deterministic, stationary stochastic inputs.

$$\gamma_{ss}(\nu,y_1,y_2) = \iint g_S(\xi_1,y_1,z_1,)\, g_S^*(\xi_2,y_2,z_2)\, e^{2i\pi(\xi_2-\xi_1)\nu}\, \gamma_{ee}(\nu,z_1,z_2)\, d\xi_1 d\xi_2 dz_1 dz_2 \qquad (37)$$

give the transformation law of the cross-spectral density. This equation looks like a double HUYGHENS integral and classical works [1] have shown that the propagation law of the cross-spectral density is identical to the propagation law of the amplitude of a monochromatic wave.

Interesting results are obtained :

- with totally coherent inputs ; then :

  $$\gamma_{SS}(\nu,y_1,y_2) = \sqrt{\gamma_{SS}(\nu,y_1,y_1)\gamma_{SS}(\nu,y_2,y_2)}$$

  the output is <u>totally coherent</u>.

- with incoherent input we get the ZERNICKE - van CITTER theorem : the coherency between two points of the output reproduces the figure of diffraction of the source centered on one point of measurement. This shows us that, using large bandwith signals with two sensors is equivalent to using an array of sensors and monochromatic signals : this is called time-space equivalence.

4.2 Introduction of the spatial scattering function.

Going back to A, let be $\tilde{H}(t,f)=TF_{\xi}[\tilde{g}(t,\xi)]$ the transfert function of the medium.

One can see easily that the scattering function exists when

$$E[\tilde{H}(t,f)\ \tilde{H}(t+\Delta t,f+\Delta f)] = F(\Delta t,\Delta f) \quad \text{and that :}$$

$$F(\Delta t,\Delta f) = TF_{\nu}^{-1}\ TF_{\xi} sc[\nu,\xi]$$

In the spatial case, we can then introduce the spatial scattering function through the spatial transfert function. Let be

$$H(t,f,y) = \int_{(S)} TF_{\xi}[g(t,\xi,y,z)]dz \quad (S) \text{ close to the origin.}$$

We say that the spatial scattering function exists when :

$$E[H(t,f,y)\ H(t+\Delta t,f+\Delta f,y+\Delta y)] = F[\Delta t,\Delta f,\Delta y]$$

and then :

$$SCS(\nu,\xi,K) = TF_{\Delta t}\ TF_{\Delta f}^{-1}\ TF_{\Delta y}^{-1}\ F(\Delta t,\Delta f,\Delta y).$$

D) APPLICATIONS.

In this section, some examples of experiments of characterization of random varying channels are given, where the technics presented in the two first parts are used. These experiments concern :

a) on the one hand, simulated channels - so as to define the validity of the proposed methods ;

b) on the other hand, real channels, and the examples exposed below will be essentially taken from submarine acoustics.

1 IDENTIFICATION OF THE FIRST ORDER CHANNEL CHARACTERIZATION.

This is the application of the § A2 method and this experiment has been made with an electronic simulator of reverberation model. For more details, see (6).

1.1 Electronic simulation of reverberation model.

We recall that for the reverberation model the input-output

relation is given by (18). In this model , the original operation is the time compression, which is easily obtained by electronics (by modifying a memory reading clock); this explains the choice of the technics used for the simulation of the equation. This simulation is made by means of the "best" (in a certain sense (see [5]) discretisation of (18) :

$$s(t) : \sum^{N_k} \sum^{N_j} e(k\Delta\lambda t - j\Delta\xi)\, G(k\Delta\lambda, j\Delta\xi)\, \Delta\xi\Delta\lambda \qquad (38)$$

So the channel is characterized by the set of the $N_k \times N_j$ random variables (r.v.) G_{kj} , the statistical properties of which can be fixed. In this experiment, we are concerned with the <u>estimation</u> of the <u>mean value</u> of the channel ; so the mean value of the set of r.v. will be given. Then, the simulator is composed of :

- a processing for generating the weighted delayed-dopplerized elementary echoe . This elementary echoe uses the data G_{kj} which is provided by means of a perforated band.
- a memory for cumulating these elementaries echoes. Finally, it is the size of the memory which limits the dimensions $N_k \times N_j$.

1.2 Input signals.

This simulator can work either with monochromatic signals, or with wide-band signals. In the identification experiment of $EG(\lambda, \xi)$, the latter are used. They are binary signals, $\pm$ 1, called PSK(t) [5] . They can be used, either alone for realizing a low pass signal, or for the binary phase modulation of a carrier frequency ν_o.

Let $T = K\theta$ be the signal duration

θ = the digit duration

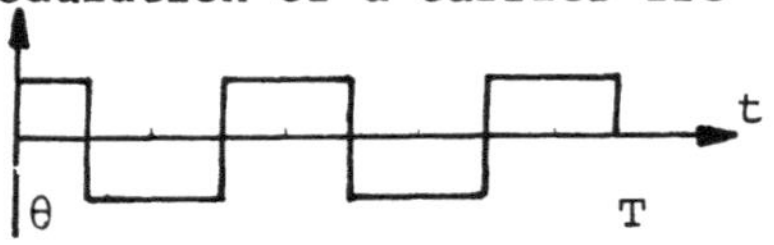

Fig. 3

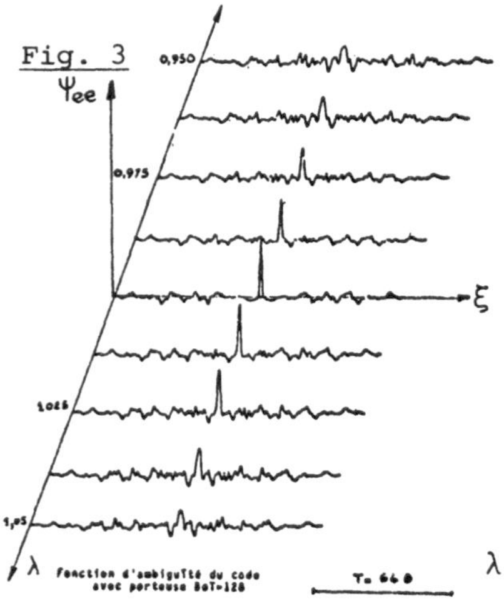

Fig. 4

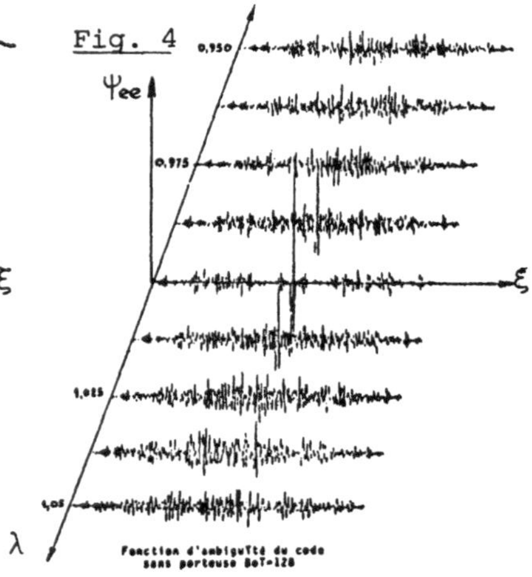

We recall that the time resolution of this signal is about $\pm\ \theta$ (so, its bandwidth is about $2/\theta$). The product WT of the signal is $2K$ (128 in our example).

For the mean value identification of the reverberation model, the used signals are completely characterized by their compression ambiguity functions which are given in the Fig. 3 and 4 above.

1.3 Identification of the mean value $EG(\lambda,\xi)$.

The reverberation simulator is driven by the signals PSK(t) and the mean value of the input-output ambiguity is carried out. We simulated various mean values for the set of r.v. and, obviously, many statistical realizations for each value (32 in our case). The obtained results are given :

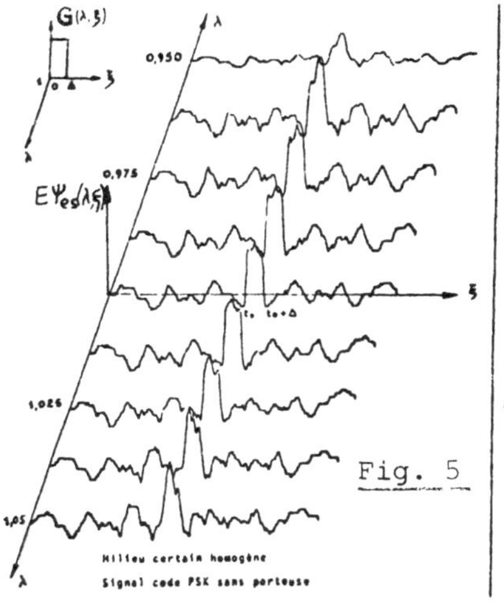

Fig. 5

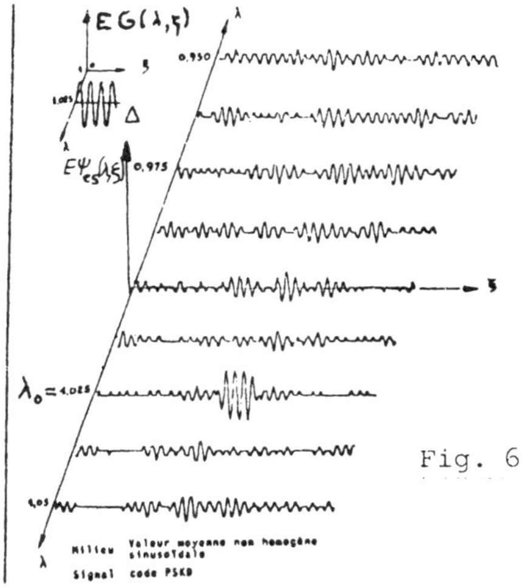

Fig. 6

a) for the identification of a deterministic, non time varying, low pass channel (fig. 5). The signal is low pass (PSK alone). All the schemes are drawn with the same scale. The $G(\lambda,\xi)$ form is well recovered from the $\lambda=1$ cross section. On the other side, as the signal ambiguity function does not really look like a "double DIRAC function", this form is recovered also from the other cross sections while distorted.

b) for the identification of a random channel, with a time varying mean value $EG(\lambda,\xi)$ which has a sinusoïdal form for $\lambda=\lambda_o$ and a Δ-duration. The mean value of the input-output ambiguity may really lead to recover $EG(\lambda,\xi)$ (fig. 6).

2. SECOND ORDER CHARACTERIZATION-EXPERIMENTS IN SUBMARINE ACOUSTIC CHANNEL

The given applications on natural channels are taken from

submarine acoustics. Deep-water and long range propagation is concerned. Neither the emitter nor the receiver are in the deep acoustic channel axis (SOFAR). The emitter is immobile and the receiver is mobile.

With the properly chosen signals and the processings stated before, the emitter-receiver channel is to be characterized by means of one of the TVRC or reverberation models and its statistical properties are to be estimated.

In this kind of propagation, the channel mean value is usually supposed to be near zero ; so, we particularly study here the measures about the 2nd order characterization (after verifying the above assumption about the mean value).

2.1 Monochromatic input signals. Frequency spreading and coherence measurements.

The channel has been driven by a set of monochromatic continuous waves , covering almost the whole frequency band of interest in submarine acoustics (frequencies spaced of 100 Hz up to 2 kHz). The received signals allow to compare and estimate :

a) the channel frequency spreadings $B_{\nu i}$, related to each emitted wave ν , by means of spectral analysis. In our experience, we have found the frequency spreading B was almost constant over the frequency band of interest and very little (< 0.5 Hz). So, the WSSUS hypothesis is not cancelled. This means also that the time fading is about a few seconds.

b) the correlations between the powers related to near frequencies to estimate the frequency coherence (i.e. the inverse of the time spreading). This method is valid if the signals spaced in frequency of more than the above quantity B Hz, are still correlated (this is our case for $B\Delta \ll 1$). So, in this experiment, signals $Y_{\nu i}(t)$ coming from each frequency ν_i , are separate by pass band filters, and we study $P_i(t) = E\ Y^2_{\nu i}(t)$. The correlation between the different $P_i(t)$ is viewed fig. 7 : on one side the frequencies are 1 Hz-spaced and the correlation is shown very high ; on the other side, they are 75 Hz-spaced and the correlation is near zero. For this kind of channel, we have found that the frequency coherence was about 50 Hz. The other meaning of this is that the elementary channel time spreading is about 0.02 sec.

2.2 PSK(t) input signals. Scattering function estimation.

We recall that, if the channel is driven by high time and frequency resolution signals, the input-output ambiguity leads to

- the channel mean value, by means of the ambiguity mean value
- the channel scattering function, by means of the ambiguity variance

Fig. 7 : Evolution of $P_i(t)$ for différent νi

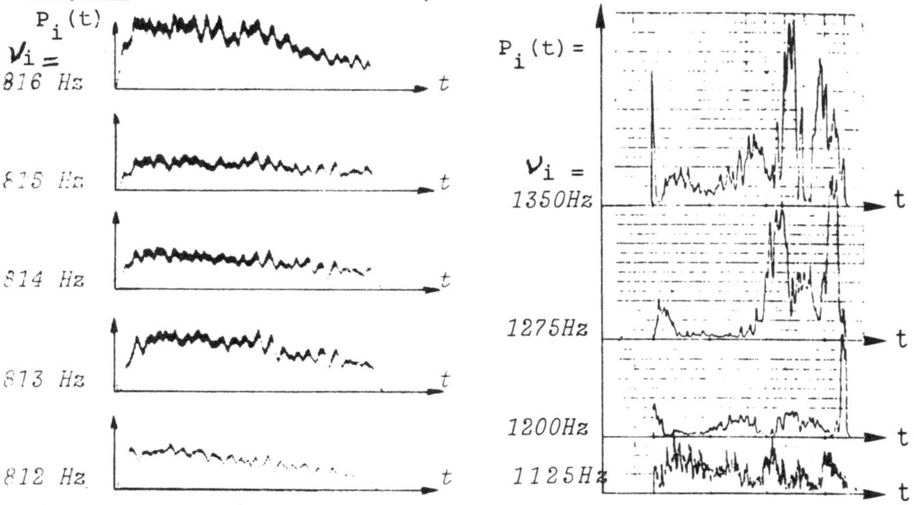

We have emitted the kind of signals stated before - high product WT ($\sim$ 400) PSK signals, which modulate the carrier frequency ν_0 = 800 Hz. Moreover, the ambiguity method has been matched to the experimental requirements of the processing, particularly by using the low-pass components of the PSK modulated signals (see [7]).

So it has been shown, first, that in the studied case, statistical averaging makes the mean value of the channel approaching zero.

Fig. 8

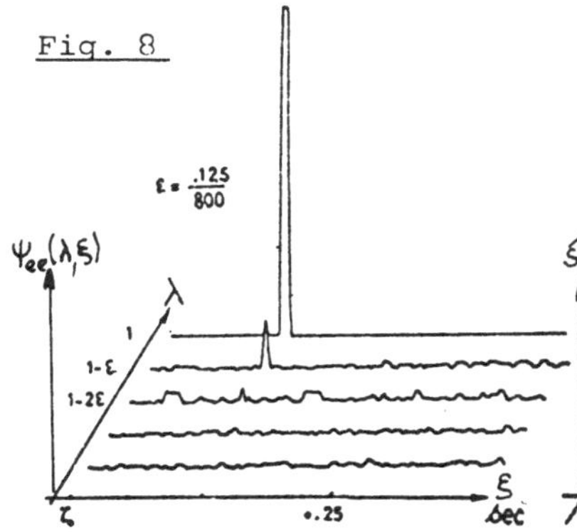

Fig. 9

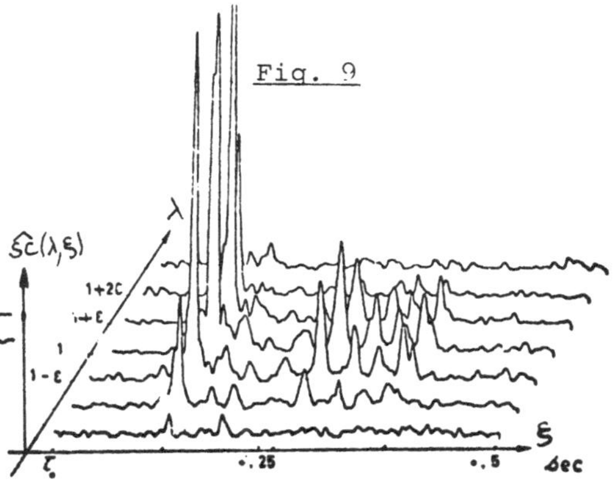

The compression ambiguity function of the used PSK signal is given in fig. 8 . A form of a scattering function obtained by this method is given fig. 9 . We notice 3 or 4 "multiple paths" : the channel time spreading, as estimated above, (§2.1b) took into account the basis of a path. As to the DOPPLER axis, we always obtain a weak frequency spreading, about $2\varepsilon\nu_0 = \pm\ 0.125$ Hz, which still agrees well with the above estimations (§2.1a).

We have obtained here a more detailed analysis of the signal energy diffusion, due to the channel, over the time-frequency space.

3. CHANNEL INTERNAL APPROACH, WITH THE ABOVE SCATTERING FUNCTION.

As an application of part B, we can see what kind of internal model can lead to a scattering function, the form of which is given on fig. 9 and obtained by experiment in a natural channel.

This identification is detailed in [4] and we could admit that in this case a first order differential model is sufficient. The conclusions of this identification are the following :

Fig. 9 has been approximated by fig. 10 opposite. So the bitemporal response of the natural channel characterized by the scattering function given by fig. 9 can be the output of a first order differential internal model, like the scheme

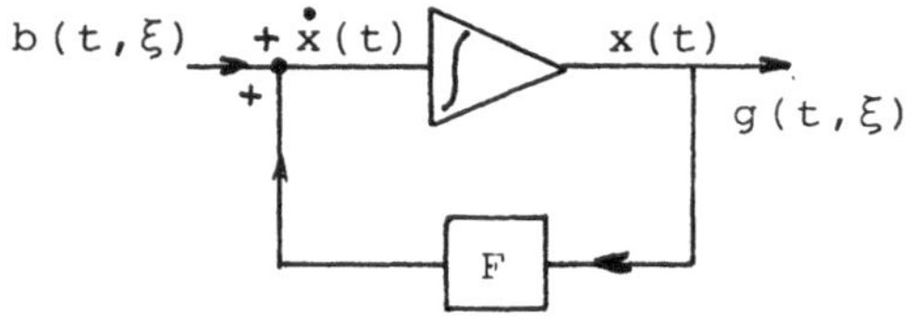

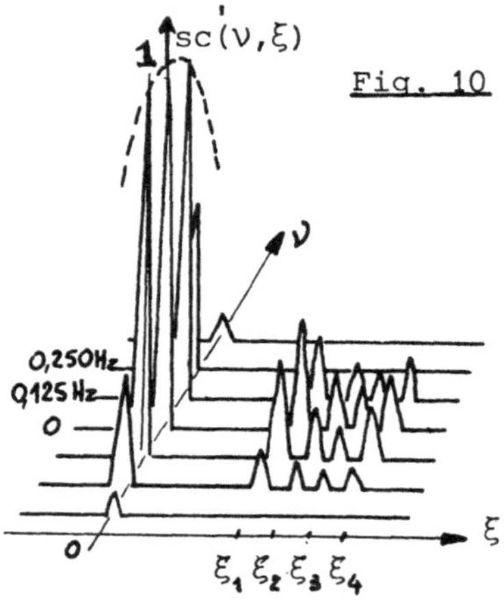

Fig. 10

where :

$b(t,\xi)$ is a white noise process, ξ—decorrelated, and with power

$$Q(\xi) = \frac{1}{F^2} \sum_{i=0}^{4} \alpha_i \Lambda_\Delta(\xi - \xi_i) \quad \text{where} \quad \Lambda_\Delta(b) = \begin{cases} 1 - b/a & b \leq a \\ 0 & b \geq a \end{cases}$$

The quantities occuring in this formula are directly connected to fig. 9 :

$F^2 \sim (3\varepsilon\nu_o\pi)^2 \sim 1.5\ \text{Hz}^2 \quad ; \quad \Delta \neq 0.015$ sec.

$\alpha_o = 1 \; ; \; \alpha_1 = 1/4 \; ; \; \alpha_2 = 1/8 \; ; \; \alpha_3 = 1/9 \; ; \; \alpha_4 = 1/8$

$\xi_o = 0 \; ; \; \xi_1 = 148\text{ms} \; ; \; \xi_2\ 187\text{ms} \; ; \; \xi_3 = 214\text{ms} \; ; \; \xi_4 = 241\text{ms}$

CONCLUSION

In this lecture we have presented some formal means used to describe non stationary random channels of propagation.

The topic is so large that it was impossible to cover all the purpose. Our constant point of view was to follow closely the applications and to give the theoretical means allowing to interpret identification experiments and to define the signals best adapted to the channel. The applications of that formalism are numerous. We have limit ourself to the submarine acoustic propagation in which we have a direct knowledge.

We think that one of the most important points was to point out the important role played by the ambiguity function in order to test the channel or to characterize the properties of signals.

We have also given some elements on the internal representation of the channel in the stationary and the non-stationary cases. This kind of representation is not always simpler than the external description. However it shall be very useful when ones tries to define the optimal processors in detection or communication.

In the spatial case, essentially in stationary situations, an important role is played by the cross-spectral density functions.

At the end we shall recall that we have limit ourselves to the study of the two first stochastic moments. This study is then only sufficient in the gaussian case.

REFERENCES

(1) BLANC-LAPIERE - A.P. DUMONTET - Revue d'optique 34 (1955).

(2) BORN-WOLF - Principles of optics-Pergamon Press (1959).

(3) FRANKS L. - Signal Theory-Prentice Hall (1969).

(4) FONTAINE - Rapport CEPHAG (1977).

(5) JOURDAIN G. - Filtres linéaires, aléatoires et non stationnaires-Modèles simulations et applications. Thèse de Doctorat d'Etat (1976).

(6) JOURDAIN G. - G. REVOL - Simulation d'échos diffus. Identification de la valeur moyenne. GRETSI.NICE (1975).

(7) JOURDAIN G. - J.Y. JOURDAIN - Méthodes et techniques d'estimation de la fonction de diffusion - GRETSI.NICE (1977).

(8) KURTH - Distributed parameters state variables models for randomly non time varying systems (Sc. Dr M.I.T. 1969).

(9) LAVAL R. - Cohérence spatio-temporelle et fonction de diffusion généralisée GRETSI - NICE (1977).

OTHER GENERAL REFERENCES ON THE CHARACTERISATION OF RANDOM CHANNELS

KAILATH, Measurements on time variant communication channels. IEEE Trans. ITT, Sept. 1962, pp. 229-236.

KAILATH, Sampling models for linear time variant filters. Tech. Rep. 352 Res. Lab. of Electronics, MIT, May 25, 1959.

BELLO, P.A., Characterisation of randomly time variant filters. MIT Res. Lab. Electr. Cambridge, Mass. Rept. No. 352, May 1959.

PRICE, R., GREEN, P.F., Signal processing in radar astronomy and communication via fluctuating multipath media. MIT Lincoln Laboratory, Lexington, Mass. Rept. No. 234, Oct. 6, 1960 (71 ref.).

ELLINTHORPE, A.W., NUTTAL, A.M., Theoretical and empirical results on the characterisation of undersea acoustic channels. 1st annual communication IEEE Convention Boulder, Juin 1965.

PERFORMANCE CAPABILITIES OF THE VITERBI ALGORITHM FOR COMBATTING INTERSYMBOL INTERFERENCE ON FADING MULTIPATH CHANNELS*

John G. Proakis

Department of Electrical Engineering, Northeastern University, Boston, Mass. 02115, U.S.A.°

ABSTRACT. Tight upper and lower bounds are obtained on the probability of error for maximum-likelihood sequence estimation of binary data transmitted over D independently Rayleigh fading multipath channels.

1. INTRODUCTION

This paper is concerned with maximum likelihood sequence estimation (MLSE) of a binary data sequence that is transmitted over a Rayleigh fading, multipath channel. The objective is to derive the error rate performance for the case in which the binary signal is transmitted on D independently Rayleigh fading, multipath channels. Thus, D represents the amount of explicit diversity in the received signal.

Figure 1 gives a baseband model for the digital communication system under consideration. A binary data sequence $\{I_k\}$ consisting of statistically independent bits is the input to the transmitter. The bit rate is 1/T where T is the duration of the signaling interval. The total transmitted signal is

$$s(t) = \sum_{n=o}^{\infty} I_n g(t-nT) \qquad (1)$$

* This work has been supported in part by the National Science Foundation under Grant GK-26329.

° Present address.

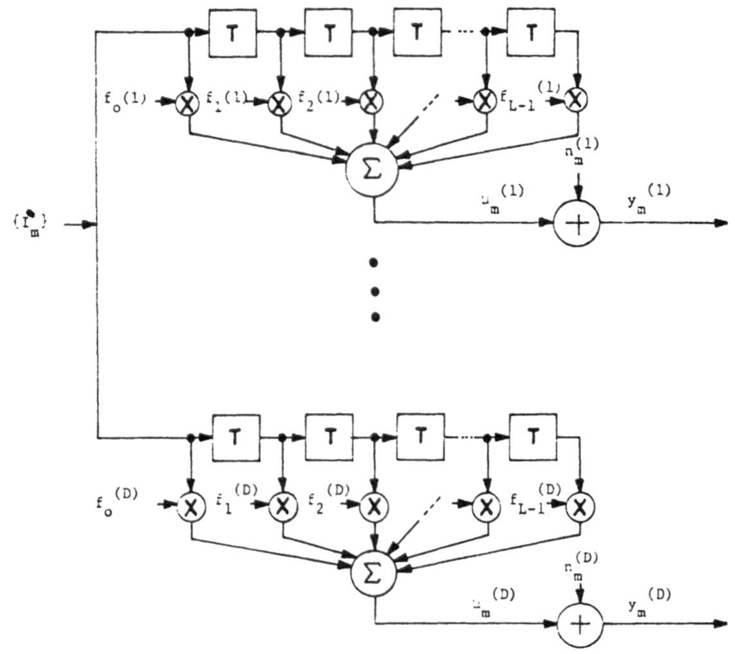

Figure 2: Equivalent Discrete-Time Model for Digital Communication System with Diversity

where g(t) represents the impulse response of the transmitting filter.

The signal s(t) is transmitted over D independent Rayleigh fading, multipath channels as shown in Figure 1. Each channel is modeled as a linear time-variant filter in which the time variations are much slower than the bit rate. The slow fading assumption is appropriate for a number of channels. For example, the bit rate on a troposheric scatter channel might be in the range of 10^6 to 10^7 bits per second, while the channel impulse response changes at a rate of 1 to 5 Hz.

The signaling rate and, hence, the signal bandwidth is assumed to exceed the coherence bandwidth (roughly equal to the reciprocal of the multipath spread) of each channel, so that the channels are time dispersive. In other words, the multipath is resolvable with a resolution equal to the reciprocal of twice the signal bandwidth. In effect, the time-dispersive nature of the channel gives rise to intersymbol interference [1].

In addition to the channel distortion resulting from the time-variant multipath, the signal on each channel is corrupted by an additive noise which is modeled as a zero mean, white Gaussian noise process with two-sided spectral density No/2. The additive noises on the D diversity channels are assumed to be mutually statistically independent.

The received signal on each channel is passed through a linear filter which, from an optimum detection viewpoint, is matched to the channel-distorted version of the transmitted signal pulse g(t). (In practice, a linear time-invariant filter matched to the signal pulse g(t) is usually employed.) For the purpose of deriving the error rate performance of the optimum MLSE receiver, it is assumed that the D channels are used over an interval of time say NT, where N is an integer, which is much greater than the time dispersion of the channels, say LT, where L is also an integer, but much smaller than the reciprocal of the fading bandwidth (coherence time) of the channels. Then, over the interval NT the channel may be characterized as a time-invariant channel having a time dispersion LT. It should be noted that the channel characteristics in an interval of duration NT are statistically independent of the channel characteristics in any other interval of the same duration provided that the intervals are separated by the reciprocal of the fading bandwidth of the channels. Consequently, in the derivation of the error rate performance it is implicitly assumed that one observes the performance of the optimum MLSE receiver through non-overlapping time windows of duration NT, so that the error rate is an average over the Rayleigh fading statistics of the channel.

The output of each receiving filter is sampled periodically at a rate 1/T samples per second and the samples are further processed in accordance with the MLSE technique. Since the multi-channel digital communication system from the input at the transmitter to the output of the samplers is linear, having a discrete-time input and D discrete-time outputs at the same rate of 1/T samples per second, this part of the system can be represented as a parallel bank of D equivalent discrete-time filters, which for all practical purposes are assumed to have finite duration responses. In general, the noise sequence corrupting the signal at the output of each discrete-time channel is colored. For purposes of simplifying the analysis a noise-whitening, discrete-time filter is assumed to follow each sampler, as described by Forney [2] for the case of a time-invariant, time-dispersive channel.

Upon incorporating a noise-whitening filter in each of the D diversity channels, we obtain an equivalent discrete-time model of the communication system that is illustrated in Figure 2. That is, each channel including transmitter and receiver filters is modeled as a linear, discrete-time filter having a finite duration impulse response. The tap coefficients of the jth equivalent discrete-time filter are denoted as $f_k^{(j)}$, k = 0,1,...,L-1 and j = 1,2,...,D. Consequently, the outputs of these filters are expressed as

$$y_m^{(j)} = \sum_{k=0}^{L-1} f_k^{(j)} I_{m-k} + n_m^{(j)}, \quad m = 1,2,...; \quad j = 1,2,...,D \qquad (2)$$

In this equivalent discrete-time channel model, which is consistent with the previous work of Kailath [3], the tap coefficients $\{f_k^{(j)}\}$ are complex-valued, mutually statistically independent Gaussian random variables having zero mean and variance

$$\sigma_{kj}^2 = \frac{1}{2} E|f_k^{(j)}|^2, \quad k = 0,1,...,L-1; \quad j = 1,2,...,D. \qquad (3)$$

Two special cases of (3) are considered. In one, the D channels are assumed to have identical statistical characteristics, i.e., the variances σ_{kj}^2 are independent of j,

$$\sigma_k^2 = \frac{1}{2} E|f_k^{(j)}|^2, \quad k = 0,1,...,L-1 \qquad (4)$$

and in the other, all the variances are assumed to be equal, i.e., σ_{kj}^2 is independent of both k and j. Since $f_k^{(j)}$ is zero mean, complex-valued Gaussian random variable, $|f_k^{(j)}|$ is a Rayleigh distributed random variable. The tap gain coefficients $\{f_k^{(j)}\}$ are statistically independent from one interval of duration NT to another.

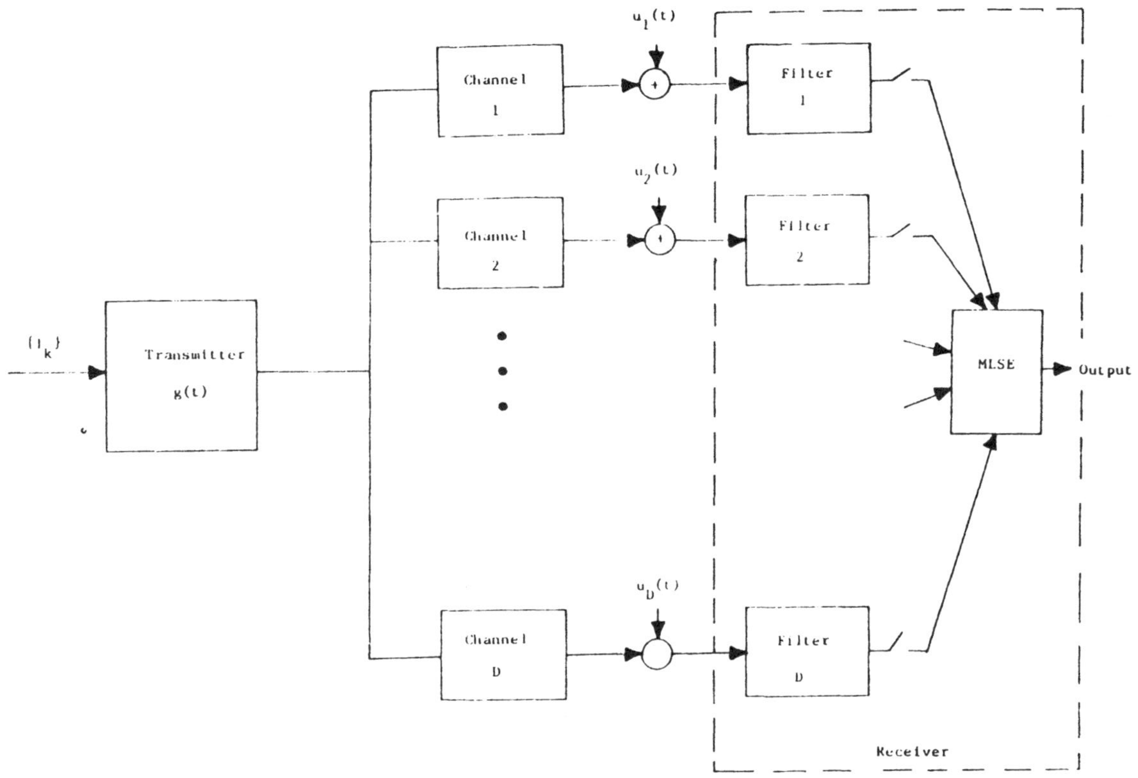

Figure 1: Baseband Model for Digital Communication System with Diversity

2. OPTIMUM MLSE FOR DIVERSITY RECEPTION

In this section, the structure and the nature of the metric computations of the optimum MLSE receiver for diversity operation are briefly described. The starting point is (2), which is the expression for the received signals on the D diversity channels. The channel tap gain coefficients are treated as known constants in the detection of the information sequence. As indicated above, these channel parameters actually vary slowly relative to the symbol interval. Consequently, the $\{f_k^{(j)}\}$ may easily be estimated by a technique described by Magee and Proakis [4]. For purposes of the following derivation, it is convenient to assume that the estimates of these channel parameters are perfect.

In the MLSE technique one observes the D received signal sequences $\{y_m^{(j)}\}$, $j = 1,2,\ldots,D$, and $m = 1,2,\ldots,N$ where $N \gg L$, and from these observables one selects the information sequence that maximizes the joint, conditional probability density function

$$p(\underline{y}_N|\underline{I}_N) \prod_{j=1}^{D} \prod_{m=1}^{N} p(y_m^{(j)}|I_m, I_{m-1}, \ldots, I_{m-L+1}) \tag{5}$$

where the $\{y_m^{(j)}\}$ are complex-valued, Gaussian variables with mean

$$\mu_m^{(j)} = \sum_{k=0}^{L-1} f_k^{(j)} \, I_{m-k} \tag{6}$$

and variance $\sigma^2 = E|n_m|^2 = N_o$, which is independent of m. Thus, we seek the sequence that maximizes the joint probability density

$$p(\underline{y}_N|\underline{I}_N) = \left(\frac{1}{2\pi\sigma^2}\right)^{DN} \exp\left\{-\frac{1}{2\sigma^2}\sum_{j=1}^{D}\sum_{m=1}^{N} |y_m^{(j)} - \mu_m^{(j)}|^2\right\} \tag{7}$$

over all possible 2^N sequences. The Viterbi algorithm is a computationally efficient technique for implementing the optimum MLSE receiver [2].

Note that maximizing $p(\underline{y}_N|\underline{I}_N)$ over $\underline{I}_N$ is equivalent to selecting the sequence that minimizes the term in the exponent, which is a measure of the distance between the received signal and any of the possible transmitted sequences. Thus, the sequence resulting in the minimum distance is the most probable sequence. In turn, the minimum distance criterion is equivalent to selecting the sequence that minimizes the index

$$J(\underline{I}_N) = \sum_{m=1}^{N} \mathrm{Re}\{ \sum_{j=1}^{D} [y_m^{(j)} - \frac{1}{2} \mu_m^{(j)}] \mu_m^{*(j)} \} \qquad (8)$$

where Re(x) denotes the real part of the quantity x. Consequently, at each stage of the optimum MLSE technique using the Viterbi algorithm (VA), the following quantities are computed and used to update the metrics:

$$U_m = \mathrm{Re}\{ \sum_{j=1}^{D} [y_m^{(j)} - \frac{1}{2} \mu_m^{(j)}] \mu_m^{*(j)} \}, \quad m = 1,2,... \qquad (9)$$

The metric computation given above may be viewed as a cross correlation operation between the possible signal terms $[\mu_m^{*(j)}]$ and the received signals $y_m^{(j)}$, j = 1,2,...,D, with a corresponding bias term $\frac{1}{2}|\mu_m^{(j)}|^2$ subtracted, followed by a diversity combiner which simply adds the results from the cross correlation. Thus, the computation of the metrics for diversity reception is a straightforward extension of the computations performed in single channel operation.

3. PERFORMANCE OF MLSE ON RAYLEIGH FADING MULTIPATH CHANNELS

In this section we derive the average probability of error achieved by the MLSE technique, implemented by means of the Viterbi algorithm, for data transmitted via binary phase-shift-keying (PSK) on a Rayleigh fading multipath channel with diversity. Our starting point for this derivation is the result on the error rate performance of MLSE for a time-invariant channel, previously derived by Forney [2]. The expression for the upper bound on the probability of a bit error in binary signaling is

$$P_e \leq \sum_{E \varepsilon W} K(E) \ \mathrm{erfc} \sqrt{\frac{d^2(E)}{N_o}} \qquad (10)$$

where

$$\mathrm{erfc}\ (x) = \frac{2}{\sqrt{\pi}} \int_x^{\infty} \exp(-t)^2 \ dt \qquad (11)$$

In the above expression $d^2(E)$ is the Euclidean weight of an error event, W is the set of all possible error events starting at some given instant in time and K(E) is a constant independent of $d^2(E)$

and N_o. The ratio $d^2(E)/N_o$ is the signal-to-noise ratio (SNR) parameter which we denote as $\gamma(E) = \gamma$. For convenience, we also let $d^2 = d^2(E)$. For a time-invariant channel, the sum in the upper bound on P_e is dominated by those error events with the minimum Euclidean weight d^2_{min}. That is

$$P_e \simeq K \, \mathrm{erfc} \sqrt{\frac{d^2_{min}}{N_o}} \tag{12}$$

where K is another constant independent of SNR.

For the moment, let us consider the case of single channel transmission, i.e. no diversity. Let $\underline{f}$ denote the (L x 1) vector of channel coefficients. Then, the Euclidean weight of an error event is given by the quadratic form [5]

$$d^2 = \underline{f}_t^* \, \underline{A} \, \underline{f} \tag{13}$$

where $\underline{A}$ is an (L x L) positive definite matrix which depends upon the error event. For example, if the error event has length one, then $\underline{A}$ is the identity matrix. If the error event is of length two, say of the form $1 - z^{-1}$, where z^{-1} denotes a unit of delay, then

$$\underline{A} = \begin{bmatrix} 2 & -1 & 0 & 0 & \cdots & & 0 \\ -1 & 2 & -1 & 0 & \cdots & & 0 \\ 0 & -1 & 2 & -1 & 0 & \cdots & 0 \\ \vdots & & & & & & \\ \vdots & & & & & & \\ \vdots & & & & & & \\ 0 & \cdots & & & 0 & -1 & 2 \end{bmatrix} \tag{14}$$

and if it is of the form $1 - z^{-2}$, then

$$\underline{A} = \begin{bmatrix} 2 & 0 & -1 & 0 & \cdots & & 0 \\ 0 & 2 & 0 & -1 & 0 & \cdots & 0 \\ -1 & 0 & 2 & 0 & -1 & 0 \cdots & 0 \\ \vdots & & & & & & \\ \vdots & & & & & & \\ 0 & \cdots & & & -1 & 0 & 2 \end{bmatrix} \tag{15}$$

For a fading multipath channel, the vector $\underline{f}$ of channel coefficients is random. Consequently, for any error event, d^2 is a random variable. Therefore, the expression for the error probability given by Eq. (10) may be viewed as the probability of error conditional on a given channel characteristic. In order to determine the performance of MLSE on a Rayleigh fading channel one must average the expression given in Eq. (10) over the statistics of the channel coefficients. Toward this end, we first compute the probability density function for the normalized random variable

$$\gamma = \frac{d^2}{N_o} = \frac{1}{N_o} \underline{f}^*{}_t \, \underline{A} \, \underline{f} \, , \tag{16}$$

which applies to single channel transmission.

For the Rayleigh fading channel, the vector $\underline{f}$ of channel coefficients is zero-mean Gaussian. The components of $\underline{f}$ are all mutually statistically independent with variances

$$\sigma_k^2 = \frac{1}{2} E \, |f_k|^2 \tag{17}$$

Thus, the covariance matrix of $\underline{f}$,

$$\underline{R} = \frac{1}{2} E \, [\underline{f}^* \, \underline{f}_t] \, , \tag{18}$$

is diagonal. It follows that for any fixed error event, γ is a quadratic form in complex-valued, zero-mean statistically independent Gaussian variables. Hence, the characteristic function of the normalized quadratic form given in Eq. (16) is [6]

$$\phi(jv) = \prod_{k=o}^{L-1} \frac{1}{(1 - j2v\zeta_k/N_o)} \tag{19}$$

where $\{\zeta_k\}$ are the eigenvalues of the matrix $\underline{A} \, \underline{R}$. It is convenient to define $\bar{\gamma}_k = 2\zeta_k/N_o$ so that

$$\phi(jv) = \prod_{k=o}^{L-1} \frac{1}{(1 - jv\bar{\gamma}_k)} \tag{20}$$

An interesting special case and the one we shall focus our attention on is where the $\{f_k\}$ have equal variance. Then

$$\underline{R} = \sigma_f^2 \, \underline{I} \tag{21}$$

where $\underline{I}$ is the identity matrix. In this case $\bar{\gamma}_k = 2\sigma_f^2\lambda_k/N_o$ where $\{\lambda_k\}$ are the eigenvalues of $\underline{A}$. Then, the characteristic function given by Eq. (20) becomes

$$\phi(jv) = \prod_{k=o}^{L-1} \frac{1}{(1 - jv\bar{\gamma}\lambda_k)} \tag{22}$$

where $\bar{\gamma}$ is defined as

$$\bar{\gamma} = \frac{2\sigma_f^2}{N_o} = \frac{E|f_k|^2}{N_o} \tag{23}$$

and represents the average signal-to-noise ratio per tap weight of the equivalent discrete-time channel. If, in addition, the error event of length one is considered, $\underline{A}$ becomes the identity matrix and, hence, all λ_k are equal to unity. The corresponding characteristic function for this case is

$$\phi(jv) = \frac{1}{(1 - jv\bar{\gamma})^L} \tag{24}$$

The probability density function for γ is obtained from the forms of the characteristic function given above. The expression that applies to the forms of the characteristic function given by Eqs. (20) and (22) is

$$p(\gamma) = \begin{cases} \sum_{k=o}^{L-1} (\pi_k/\bar{\gamma}_k)\exp(-\gamma/\bar{\gamma}_k) , & \gamma \geq 0 \\ 0 , & \gamma < 0 \end{cases} \tag{25}$$

where

$$\pi_k = \prod_{\substack{i=o \\ i\neq k}}^{L-1} (1 - \frac{\gamma_i}{\bar{\gamma}_k})^{-1} \tag{26}$$

The form for $p(\gamma)$ given by Eq. (25) is based on the assumption that the $\{\bar{\gamma}_k\}$ are distinct. On the other extreme we consider the case in which all the $\{\bar{\gamma}_k\}$ are equal. In fact, Eq. (24) represents a special case of this condition, for which $\lambda_k = 1$ for all k. Inversion of the characteristic function given by Eq. (24) leads to the probability density function

$$p(\gamma)= \begin{cases} \dfrac{1}{(L-1)!\bar{\gamma}^L} \gamma^{L-1} \exp(-\gamma/\bar{\gamma}) & , \gamma \geq 0 \\ 0 & , \gamma < 0 \end{cases} \tag{27}$$

Now, it is a straightforward computation to average the expression for P_e, given by Eq. (10), over the fading. Thus we obtain

$$\begin{aligned} \bar{P}_e &= \int_o^\infty P_e \, p(\gamma) d\gamma \\ &\leq \sum_{E\varepsilon W} K(E) \int_o^\infty \operatorname{erfc} \sqrt{\gamma} \, p(\gamma) \, d\gamma \\ &< \sum_{E\varepsilon W} K(E) \sum_{k=o}^{L-1} \pi_k \left[1 - \sqrt{\frac{\bar{\gamma}_k}{1+\bar{\gamma}_k}} \right] \end{aligned} \tag{28}$$

for the case in which the $\{\bar{\gamma}_k\}$ are distinct. We use the symbol $\bar{P}_e(E)$ to denote the inner sum in Eq. (28), i.e.,

$$\bar{P}_e(E) = \sum_{k=o}^{L-1} \pi_k \left[1 - \sqrt{\frac{\bar{\gamma}_k}{1+\bar{\gamma}_k}} \right] \tag{29}$$

so that

$$\bar{P}_e \leq \sum_{E\varepsilon W} K(E) \bar{P}_e(E) \tag{30}$$

It is interesting to note that an error event of length one, for which $\underline{A}$ is a diagonal matrix, and hence, $\lambda_k = 1$ for all k, contributes the term

$$\bar{P}_1 = \frac{1}{2} \sum_{k=o}^{L-1} \pi_k \left[1 - \sqrt{\frac{\bar{\gamma}_k}{1+\bar{\gamma}_k}} \right] \tag{31}$$

to the sum in Eq. (30), where now $\bar{\gamma}_k$ reduces to $\bar{\gamma}_k = E|f_k|^2/N_o$ and represents the received SNR from the kth tap of the equivalent discrete-time channel model. But the expression is just the average probability error for the optimum "one-shot" receiver or, equivalently, for a maximal-ratio combiner in an L-diversity receiver operating over a non-dispersive (frequency non-selective) channel, i.e., a channel with no intersymbol interference. Thus, Eq. (31) represents a lower bound on the performance of MLSE for single channel transmission in which the channel dispersion is of length L. Eqs. (30) and (31) also indicate that the MLSE takes advantage of the "implicit" diversity [7-9] provided by the time dispersion in the fading multipath channel. That is, a channel having time dispersion of length L provides L^{th}-order "implicit" diversity. We refer to this diversity as "implicit" to distinguish it from explicit diversity which is obtained by transmitting the same signal on several channels that are independently fading.

For an error event of length one and for $\bar{\gamma}_k = \bar{\gamma}$ for all k, the conditional error probability is

$$P_e = \frac{1}{2} \operatorname{erfc} \sqrt{\gamma} \tag{32}$$

where $\gamma = \underline{f}_t^* \underline{f}/N_o$ and $p(\gamma)$ is given by Eq. (27). The average of P_e over γ results in the following expression for the error probability:

$$\bar{P}_2 = \left[\frac{1 - \sqrt{\dfrac{\bar{\gamma}}{1 + \bar{\gamma}}}}{2}\right]^L \sum_{k=o}^{L-1} \binom{L - 1 + k}{k} \left[\frac{1 + \sqrt{\dfrac{\bar{\gamma}}{1 + \bar{\gamma}}}}{2}\right]^k \tag{33}$$

where

$$\binom{n}{k} = \frac{n!}{k!\,(n - k)!} \tag{34}$$

This form for $\bar{P}_2$ is just the expression for the average probability of error for the optimum "one-shot" receiver or, equivalently, for a maximal-ratio combiner in an L-diversity receiver for a non-dispersive channel in the case where all diversity branches have identical received SNR, denoted as $\bar{\gamma}$.

The above expressions on the probability of error lead to an interesting upper and lower bound on the probability of error for MLSE in the region of high SNR. First of all, for $\bar{\gamma} >> 1$, Eq. (33) is well-approximated by the expression

$$\bar{P}_2 \simeq \binom{2L-1}{L} \frac{1}{(4\bar{\gamma})^L} \tag{35}$$

This expression represents a lower bound on the performance of MLSE since it is the asymptotic performance of the optimum one-shot receiver.

On the other hand, for $\bar{\gamma}_k >> 1$, $\bar{P}_e(E)$ given by Eq. (29) is well-approximated as

$$\bar{P}_e \simeq 2 \binom{2L-1}{L} \frac{1}{4^L \prod_{k=o}^{L-1} \bar{\gamma}_k} \tag{36}$$

Eq. (36) provides an interesting result especially for the case in which the channel coefficients $\{f_k\}$ have equal variance. That is, when $E|f_k|^2 = 2\sigma_f^2$ for all k,

$$\prod_{k=o}^{L-1} \bar{\gamma}_k = \left(\frac{2\sigma_f^2}{N_o}\right) \prod_{k=o}^{L-1} \lambda_k \left(\frac{2\sigma_f^2}{N_o}\right) |\underline{A}| = (\bar{\gamma})^L \; |\underline{A}| \tag{37}$$

where $|\underline{A}|$ is the determinant of the matrix $\underline{A}$ in the quadratic form. Thus, we have the important result that the average probability error for MLSE on a Rayleigh fading channel characterized by the equal variance coefficients $\{f_k\}$ is bounded from above and below at high SNR as follows:

$$\binom{2L-1}{L} \frac{1}{(4\bar{\gamma})^L} \leq \bar{P}_e \leq K_u \binom{2L-1}{L} \frac{1}{(4\bar{\gamma})^L} \tag{38}$$

where the constant K_u is independent of SNR and is defined as

$$K_u = 2 \sum_{E \epsilon W} K(E)/|A| \tag{39}$$

Of course, the value of $|\underline{A}|$ depends on the error event assumed.

The bounds given in Eq. (38) illustrate the excellent performance, relative to the ideal, achieved with MLSE on a fading channel that introduces intersymbol interference. The MLSE technique, in effect, performs in a manner similar to an optimum diversity combiner.

The results given above for the performance of MLSE on a Rayleigh fading, time-dispersive channel can be generalized to include diversity operation. With D order diversity the expression for the conditional error probability given by Eq. (10) still holds, but, now, d^2 is defined as a sum of quadratic forms

$$d^2 = \sum_{j=1}^{D} \underline{f}_t^{*(j)} \underline{A}\, \underline{f}^{(j)} = \sum_{j=1}^{D} d_j^2 \tag{40}$$

where $\underline{f}^{(j)}$ denotes the vector of channel tap coefficients for the $j^{\underline{th}}$ channel. It is assumed that the vectors $\underline{f}^{(j)}$ are mutually statistically independent. Hence, the $\{d_j^2\}$ are jointly statistically independent. Also, each vector $\underline{f}^{(j)}$ has zero mean and covariance matrix

$$\underline{R}_j = \frac{1}{2} E[\underline{f}^{*(j)} \underline{f}_t^{(j)}] \tag{41}$$

It is now a simple matter to modify the characteristic functions for the normalized variable $\gamma = d^2/N_o$, where d^2 is given by Eq. (40). For example, in the general case where the eigenvalues of the matrices $\underline{A}\,\underline{R}_j$ are distinct, the characteristic function of the random variable γ is

$$\phi(jv) = \prod_{k=o}^{LD-1} \frac{1}{1 - j2v\zeta_k/N_o} \tag{42}$$

where $\{\zeta_k\}$ are the LD distinct eigenvalues of $\underline{A}\,\underline{R}_j$, $j = 1,2,\ldots,D$. The other case of special interest is that for which the covariance matrices $\{\underline{R}_j\}$ are identical and proportional to the identity matrix, i.e.,

$$\underline{R}_j = \sigma_f^2\, \underline{I}\,, \quad j = 1,2,..,D. \tag{43}$$

Then, the characteristic function of γ is

$$\phi(jv) = \prod_{k=o}^{L-1} \frac{1}{[1 - jv\bar{\gamma}\lambda_k]^D} \tag{44}$$

where $\{\lambda_k\}$ are the eigenvalues of $\underline{A}$. If, in addition, we consider the special case of an error event length one, $\underline{A}$ is an identity matrix, and, since $\lambda_k = 1$ for all k, Eq. (44) becomes

$$\phi(jv) = \frac{1}{(1 - jv\bar{\gamma})^{DL}} \quad . \tag{45}$$

It should be noted that Eq. (42) is similar to Eq. (19) and Eq. (45) is similar to Eq. (24). Consequently, the inversion of the characteristic function given by Eq. (42) leads to the probability density function

$$\begin{cases} \sum_{k=o}^{LD-1} (\pi_k/\bar{\gamma}_k) \exp(-\gamma/\bar{\gamma}_k), & \gamma \geq 0 \\ 0, & \gamma < 0 \end{cases} \tag{46}$$

where $\bar{\gamma}_k = 2\zeta_k/N_o$, while the inversion of the characteristic function given by Eq. (45) leads to the probability density function

$$p(\gamma) = \begin{cases} \frac{1}{(LD-1)!\bar{\gamma}^{LD}} \gamma^{LD-1} \exp(-\gamma/\bar{\gamma}), & \gamma \geq 0 \\ 0, & \gamma < 0 \end{cases} \tag{47}$$

where $\bar{\gamma} = 2\sigma_f^2/N_o$.

Now, the average of the conditional error probability given by Eq. (10) over the fading statistics of the channel specified by the density function $p(\gamma)$ given in Eq. (46) leads to the upper bound

$$\bar{P}_e \leq \sum_{E \varepsilon W} K(E)\bar{P}_e(E)$$

$$\bar{P}_e(E) = \sum_{k=o}^{LD-1} \pi_k \left[1 - \sqrt{\frac{\bar{\gamma}_k}{1+\bar{\gamma}_k}}\right] \tag{48}$$

For $\bar{\gamma}_k \gg 1$, $\bar{P}_e(E)$ is well-approximated as follows:

$$\bar{P}_e(E) \simeq 2\binom{2LD-1}{LD} \frac{1}{4^{LD}\prod_{k=o}^{LD-1}\bar{\gamma}_k} \tag{49}$$

Let us focus our attention on the important special case in which the channel tap coefficient $\{f_k^{(j)}\}$ have identical variances. That is, when $E|f_k^{(j)}|^2 = 2\sigma_f^2$ for $k = 0,1,..,L-1$ and $j = 1,2,...,D$ we have

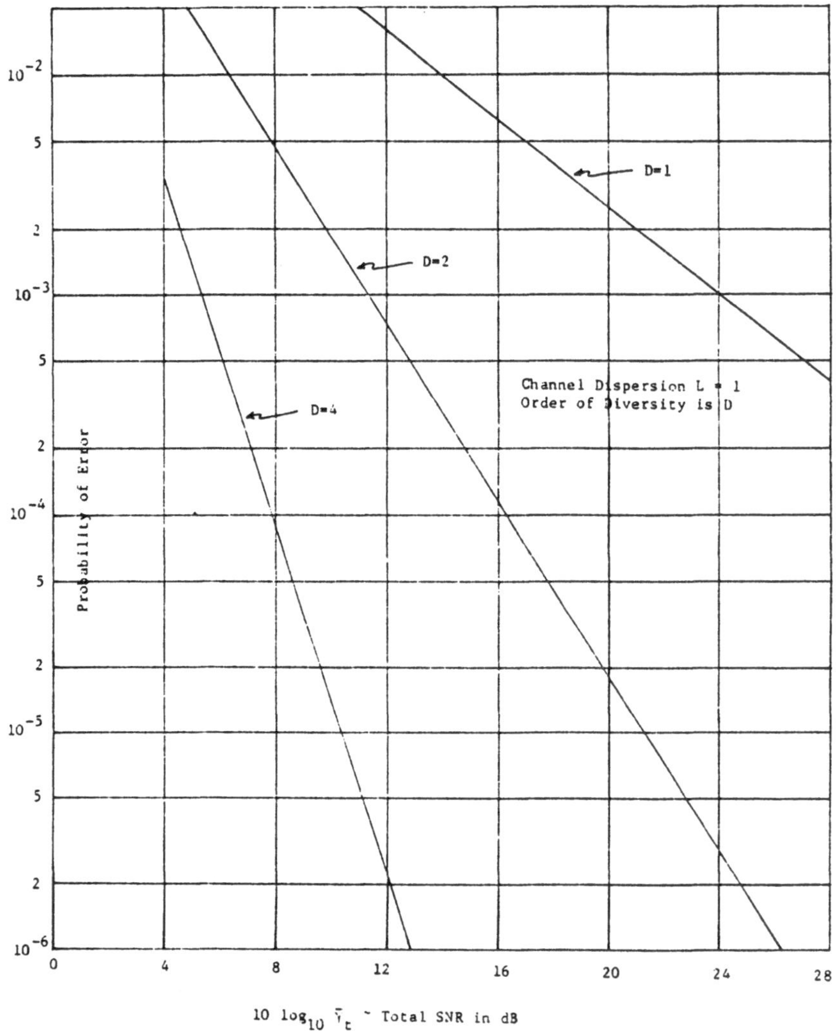

Figure 3: Upper Bound on the Performance of MLSE on Independent Rayleigh Fading Channels: L = 1

$$\prod_{k=o}^{LD-1} \bar{\gamma}_k = \left(\frac{2\sigma_f^2}{N_o}\right)^{LD} \prod_{k=o}^{LD-1} \lambda_k = \bar{\gamma}^{LD} \, |\underline{A}|^{D} \tag{50}$$

Therefore, the upper bound on the performance of MLSE for this case is

$$\bar{P}_e \leq K_{uD} \binom{2LD-1}{LD} \frac{1}{(4\bar{\gamma})^{LD}} \tag{51}$$

where

$$K_{uD} = 2 \sum_{E \varepsilon W} K(E)/|A|^D \tag{52}$$

The corresponding lower bound for this case is simply the performance of the optimum one-shot receiver. This error probability is obtained by averaging the conditional error probability given in Eq. (32) over the channel fading statistics specified by $p(\gamma)$ given in Eq. (47). The result is

$$\bar{P}_2 = \left[\frac{1 - \sqrt{\frac{\bar{\gamma}}{1+\bar{\gamma}}}}{2}\right]^{LD} \sum_{k=o}^{LD-1} \binom{LD-1+k}{k} \left[\frac{1 + \sqrt{\frac{\bar{\gamma}}{1+\bar{\gamma}}}}{2}\right]^{k} \tag{53}$$

Finally, $\bar{P}$ can be approximated as

$$\bar{P}_2 \simeq \binom{2LD-1}{LD} \frac{1}{(4\bar{\gamma})^{LD}} \tag{54}$$

Since $\bar{P}_e \geq \bar{P}_2$, it follows that the upper and lower bounds differ only by the constant K_{uD}, which is independent of the signal-to-noise ratio paramter $\bar{\gamma}$. These bounds illustrate the excellent performance capability of MLSE of binary PSK signals on a Rayleigh fading multipath channel. It is also apparent that the channel time dispersion of length L results in an increase of the order diversity by a factor L.

Some performance characteristics for MLSE of binary PSK signals are illustrated in Figures 3 through 5. In these graphs the upper bound on the probability of error is plotted as a function of the normalized SNR $10\log_{10}\bar{\gamma}_t$ where $\bar{\gamma}_t = L\bar{\gamma}$ (total SNR per channel), with the order diversity D as a parameter. The constant K_{uD} in the expression for the upper bound is in the range of 3.5 to 1.0, with the higher value corresponding to the case D = 1 and L = 4. A small number of points obtained by Monte Carlo simulation performed on a digital computer are also shown on these graphs for purposes of comparison. In spite of the fact that the simulation results were obtained by averaging over a relatively small number (30 to 100) of channel fades there is good agreement between simulation and theoretical results.

The graphs of the probability error for MLSE discussed above have been replotted in Figure 6 and 7 to illustrate the effect of

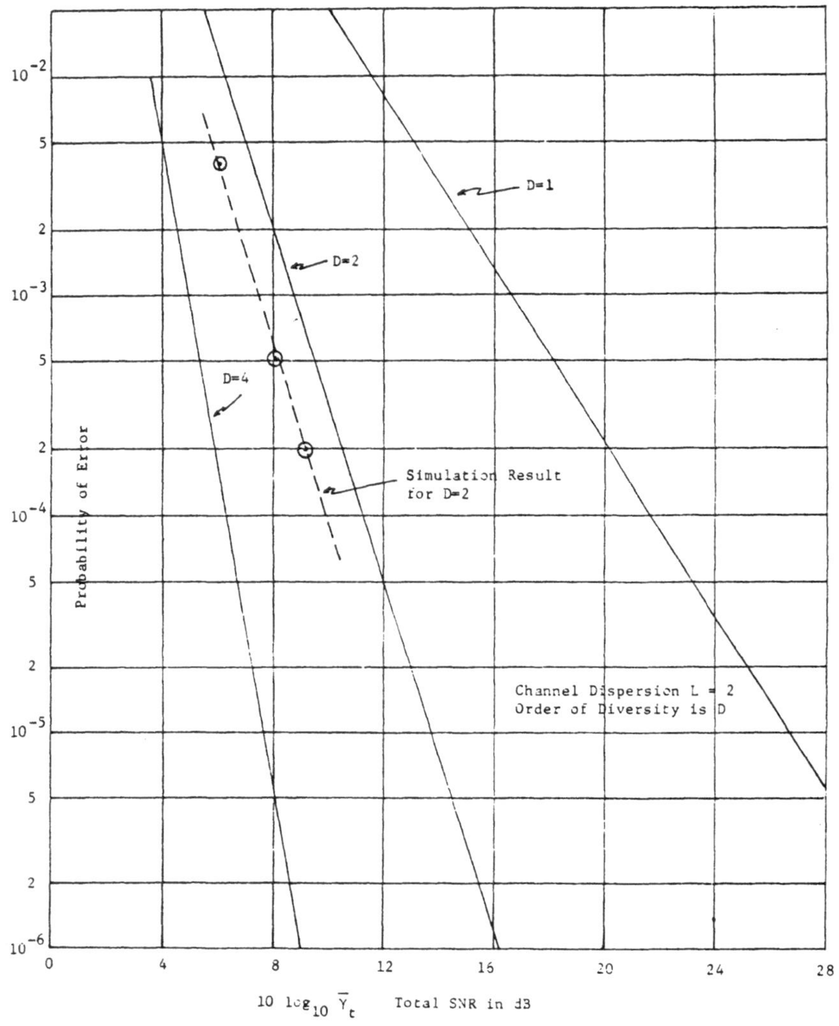

Figure 4: Upper Bound on the Performance of MLSE on Independent Rayleigh Fading Channels: L = 2

channel dispersion on performance. In these graphs, the order of diversity D is held constant and L is varied. The graphs illustrate the gain in performance of MLSE resulting from an increase in the "implicit" diversity of order L. It should also be noted that the relative gain achieved by an increase in the time dispersion (as measured by L) diminishes with an increase in the order of diversity D. Furthermore, as L increases the relative gain in performance achieved by increasing the order of the (explicit) diversity D decreases. Thus, one can trade off implicit diversity for explicity diversity.

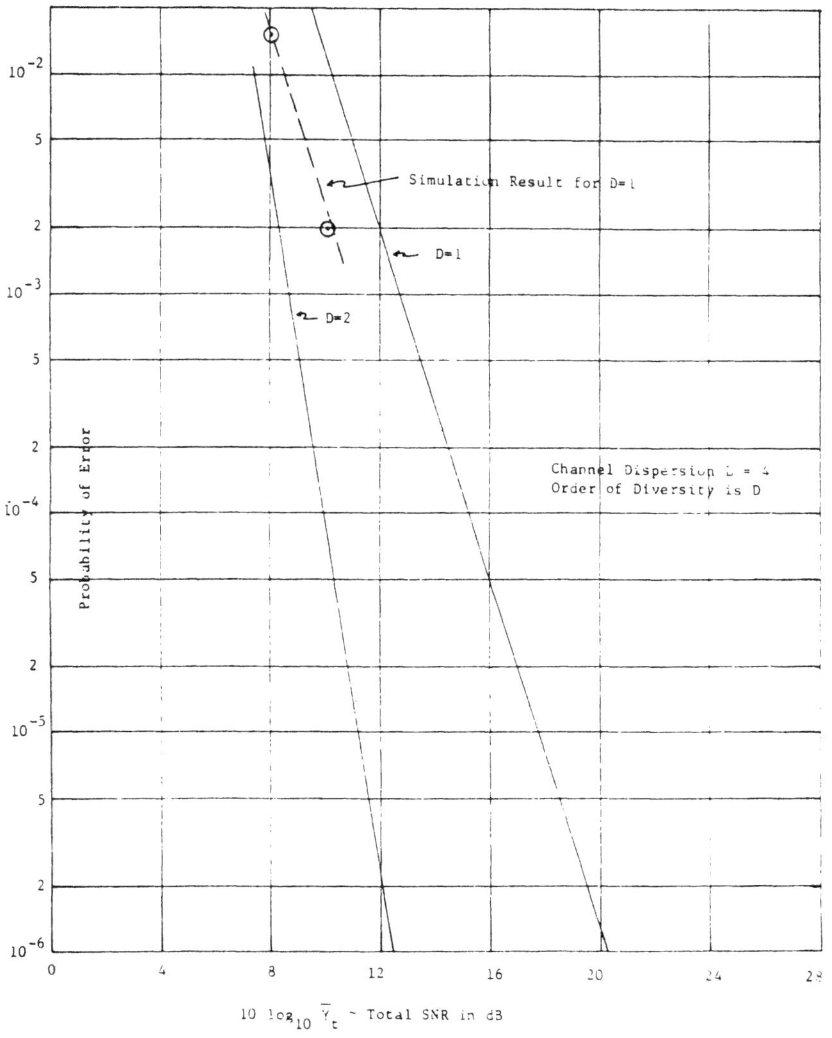

Figure 5: Upper Bound on the Performance of MLSE on Independent Rayleigh Fading Channels: L = 4

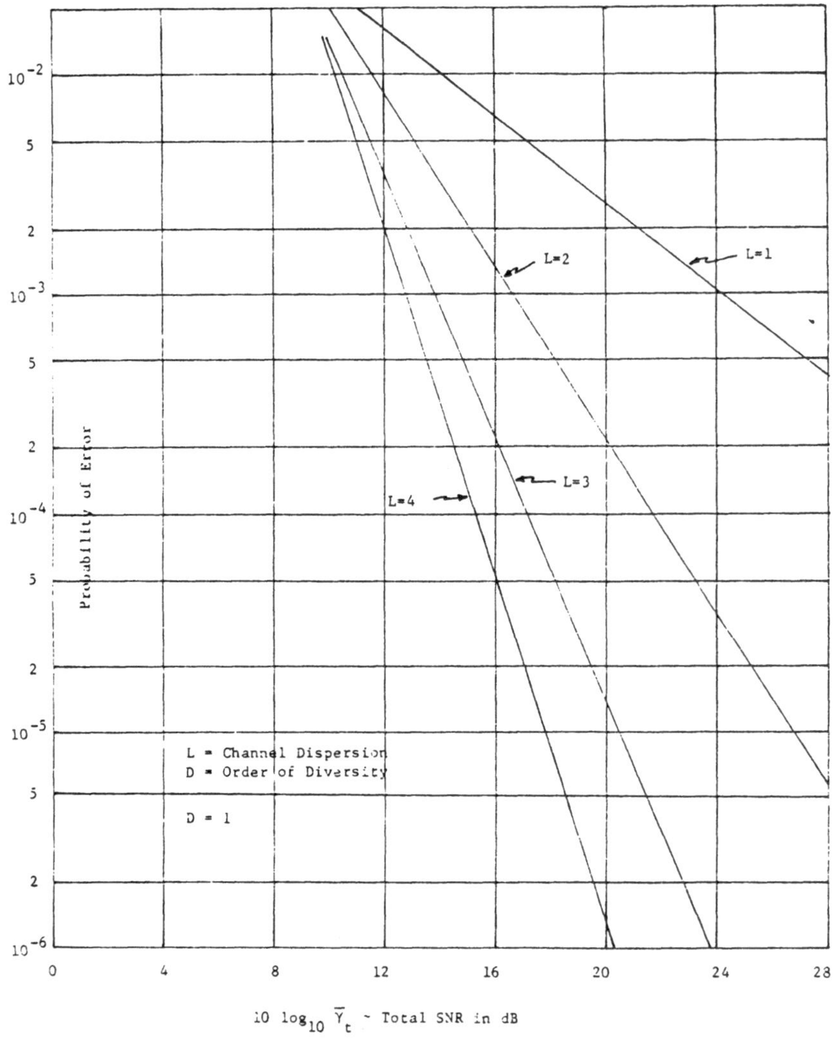

Figure 6: Upper Bound on the Performance of MLSE on Independent Rayleigh Fading Channels: D = 1

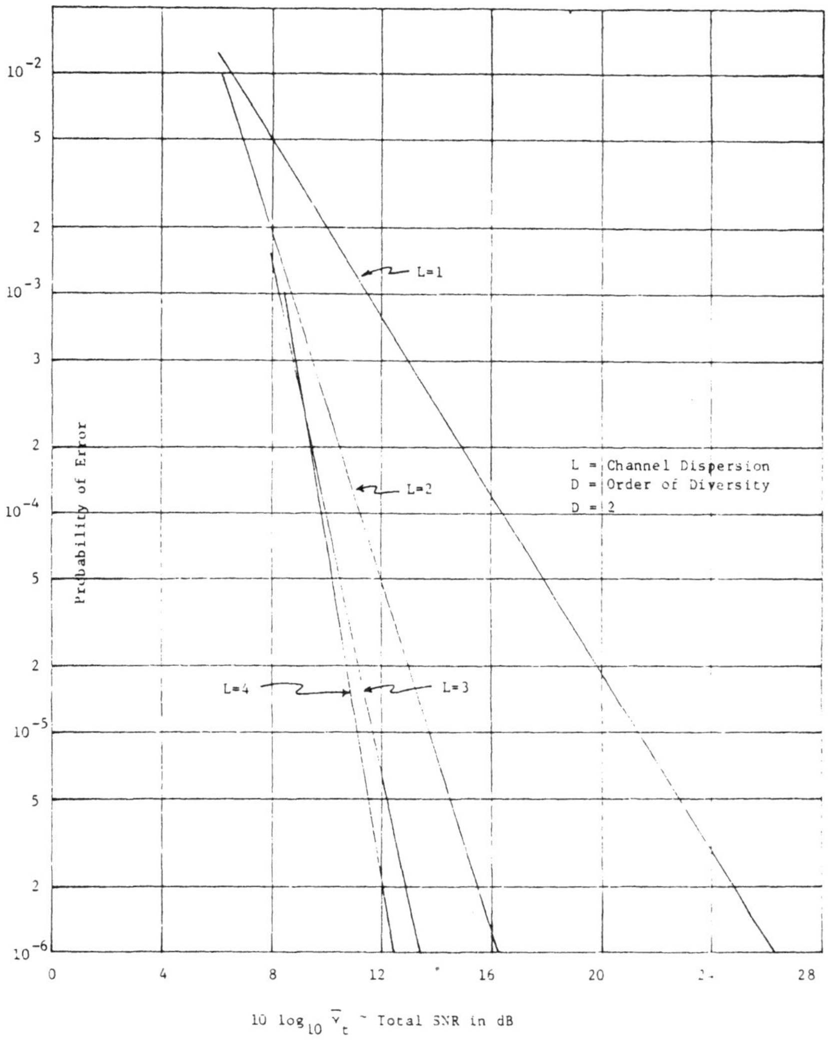

Figure 7: Upper Bound on the Performance of MLSE on Independent Rayleigh Fading Channels: D = 2

REFERENCES

1. J.G. Proakis, "Advances in Equalization for Intersymbol Interference" in Advances in Communication Systems, A.V. Balakrishnan and A.J. Viterbi, Eds., New York: Academic Press, 1975.
2. G.D. Forney, Jr., "Maximum-Likelihood Sequence Estimation for Digital Sequences in the Presence of Intersymbol Interference", IEEE Trans. Inform. Theory, vol. IT-18, pp. 363-378, May 1972.
3. T. Kailath, "Sampling Models for Linear Time-variant Filters", Res. Lab of Electronics, Massachusetts Institute of Technology, Cambridge, Mass., Tech. Report 362, May 1959.
4. F.R. Magee, Jr., and J.G. Proakis, "Adaptive Maximum Likelihood Sequence Estimation for Digital Signaling in the Presence of Intersymbol Interference", IEEE Trans. Inform. Theory (Correspondence), vol. IT-19, pp. 120-124, January 1973.
5. F.R. Magee, Jr., and J.G. Proakis, "An Estimate of the Upper Bound on Error Probability for Maximum-likelihood Sequence Estimation on Channels having a Finte-duration Pulse Response", IEEE Trans. Inform. Theory (Correspondence), vol. IT-19, pp. 699-702, September 1973.
6. M. Schwartz, W.R. Bennett, and S. Stein, Communciations Systems and Techniques, New York: McGraw-Hill, 1966, Appendix B, pp. 590-595.
7. P. Monsen, "Equalization of the Slow Fading Channel", IEEE Trans. Commun., vol. COM-22, pp. 1064-1075, August 1974
8. M.M. Goutman, "Intersymbol Interference as a Natural Code," IEEE Trans. Commun., vol. COM-20, pp. 1033-1037, October 1972.
9. A.A. Giordano, J.H. Lindholm, T.A. Schonoff, and J.G. Proakis, "Error Rate Performance Comparison of MLSE and Decision-Feedback Equalizer on Rayleigh Fading Multipath Channels", Proc. 1975 IEEE Inter. Conf. Communications, San Francisco, CA, June 1975.

CHANNEL EVALUATION TECHNIQUES FOR DISPERSIVE COMMUNICATIONS PATHS

M. Darnell

Communications Division, SHAPE Technical Centre,
The Hague, Netherlands.

ABSTRACT. The first part of the paper is concerned with a definition of the term "Channel Evaluation", together with a discussion of the inputs required before a channel evaluation algorithm can be implemented. In principle, channel evaluation techniques can be applied to any dispersive path; however, attention is confined mainly to the HF (2-30 MHz) band, since it is in this area that such techniques have been most widely applied to date. A survey of existing types of HF channel evaluation systems is given. The performance of one particular system is then considered in detail and test data obtained over various HF propagation paths is presented. Examination of these data allows the performance of an HF system incorporating this real-time channel evaluation technique to be compared with that of an HF system relying on normal HF frequency prediction methods for frequency selection.

1. INTRODUCTION

Real-time Channel Evaluation (RTCE) techniques have been developed to enable a communicator operating over a time-varying propagation path, with time and/or frequency dispersion, to make more effective use of such a path. In essence, RTCE is a modelling process which allows the communication system control strategies employed to be adaptive in response to the current state of the propagation medium as described by an appropriate quantitative path model.

Attention is confined here to electromagnetic communication systems operating in the High Frequency (HF) band covering the radio frequency range 2-30 MHz, since it is in this band that RTCE has so far been most widely applied. However, the technique can, in principle, be extended to other dispersive communication media.

2. THE HF PROPAGATION PATH

2.1 Simplified path description

Figure 1 is a simplified representation of a typical HF skywave propagation path in which the ionosphere is shown comprising four ionized layers - the D-layer, the E-layer, the F1-layer and the F2-layer, at the mean heights indicated. The signal is shown refracted back to earth via single-hop E, F1 and F2 modes, together with double-hop F1 and F2 modes. In general, as a signal passes through the D-layer, energy absorption takes place, the degree of absorption being inversely proportional to the square of the radio frequency.

Clearly, for the example illustrated in Figure 1, a short transmitted pulse can have five components when detected at the receiver, each component itself exhibiting time dispersion which causes a broadening of the corresponding received pulse. This latter form of mode time dispersion is due to the fact that the transmitted energy illuminates the ionospheric layers over a relatively large volume and not at a single point; thus, the total energy of a given received component can be seen as being made up of the energy from a large number of elemental returns, each with a slightly different propagation time, integrated over the appropriate volume of the ionospheric layer.

Frequency dispersion is normally not more than a small fraction of a Hz but, under very disturbed conditions, frequency shifts of several Hz have been observed to persist for several minutes. In practice, time dispersion due to multipath effects is the dominant feature.

For a 2000 Km HF path, at the receiver the approximate propagation time values for the single-hop E-layer mode, t_{1E}, the single-hop F1-layer mode, t_{1F1}, the double-hop F1-layer mode, t_{2F1}, the single-hop F2-layer mode, t_{1F2}, and the double-hop F2-layer mode, t_{2F2}, are given by:

$$
\left.\begin{array}{lll}
\text{E-Layer} & : & t_{1E} = 6.76 \text{ ms} \\
\text{F1-Layer} & : & t_{1F1} = 6.89 \text{ ms} \\
\text{F1-Layer} & : & t_{2F2} = 7.28 \text{ ms} \\
\text{F2-Layer} & : & t_{1F2} = 7.37 \text{ ms} \\
\text{F2-Layer} & : & t_{2F2} = 8.70 \text{ ms}
\end{array}\right\} \quad (1)
$$

It is seen that the maximum time dispersion, i.e., the delay between reception of the 1E and 2F2 modes, is approximately 2 ms. In terms of the performance of, say, an HF digital communication system, the effect of this dispersion is to cause intersymbol interference and thus limit the channel capacity achievable with conventional signal design and processing schemes.

2.2 Propagation prediction techniques

The potential complexity of the HF propagation medium having been indicated, attention is now turned to techniques for predicting the parameters of HF propagation paths. In general, two types of prediction are required:

(a) For day-to-day circuit operation, a relatively short-term prediction of the available frequency range is necessary;

(b) For system planning, long-term predictions, preferably covering a complete sunspot cycle, are required.

The most important skywave modes are single or multiple hops from the E, F1 and F2 layers, as shown in Fig. 1; of these, the F2 layer modes are by far the most significant. For each layer, there is a highest, or "critical", frequency, f_o, at which a vertically-incident signal will be refracted. Above f_o, signals will penetrate the layer and fail to be refracted. For obliquely-incident signals, the maximum usable frequency (MUF) is given by:

$$MUF = K f_o \sec \phi \qquad (2)$$

where K is a correction factor which allows for the effects of the curvature of the earth and the vertical distribution of ionization in the layer, and ϕ is the angle of incidence of the signal on the layer.

For case (a) above, the parameters of an HF circuit which can be predicted with reasonable accuracy are the critical frequencies for the E and F1-layers, f_oE and f_oF1, as a function of sunspot number, S, and solar zenith angle, Θ,; these quantities can be computed from empirical relationships [1] such as:

$$f_oE \simeq 0.9 \left((180 + 1.44S)\cos \Theta\right)^{\frac{1}{4}} \qquad (3)$$

and $$f_oF1 \simeq (4.3 + 0.01S)\cos^{\frac{1}{5}}\Theta \qquad (4)$$

The critical frequency for the F2 layer , f_oF2, cannot be determined from a simple analytical expression since it depends upon many factors. Accordingly, smoothed predictions of the value of f_oF2 as function of time of day are prepared, normally on a month-by-month basis, making extensive use of the past data for similar values of S.

On average, f_oF2 predictions are reasonably accurate. However, for a given circuit on a particular day, the predicted values of f_oF2 can be considerably in error.

The optimum working frequency (FOT) is defined to be the fequency for which there is a 90% probability that the actual MUF will be greater.

$$FOT = 0.85 \text{ (Predicted MUF)} \tag{5}$$

In the frequency selection process, it is important to predict the MUF for a circuit: it is similarly essential to have an indication of the lowest usable frequency (LUF). The LUF depends upon the noise level at the receiver, the path loss due to ionospheric absorption, and the transmitter power and antenna gains available to overcome these two effects.

To illustrate a typical output from a frequency prediction computation, Fig. 2 shows FOT and LUF plotted as a function of time of day for an arbitrary propagation path. The frequencies f_{DAY} and f_{NIGHT} are selected to provide a reasonable fit between the predicted FOT and LUF contours.

Considering case (b) above where long-term predictions are required to aid the dimensioning of an HF communication system at the planning stage: again critical frequencies for the various layers must be computed, but this time for a complete sunspot cycle. Also, the heights of the reflecting layers and the dominant propagation modes must be determined in order to specify the antenna radiation patterns required.

In the HF band, atomospheric and galactic noise is normally the dominant source of the noise at quiet receiving locations. Extensive measurements of this type of noise have been made for many years and are normally tabulated in terms of geographical location, time of day, season, frequency and receiver bandwidth (see, for example, [2]).

3. REAL-TIME CHANNEL EVALUATION (RTCE)

3.1 Definition of RTCE

The following statement is proposed as a definition of RTCE:

"Real-time channel evaluation is the term used to describe the processes of measuring appropriate parameters of a set of communication channels in real-time and of employing the data thus obtained to describe quantitatively the states of those channels and hence their relative capabilities for passing a given class, or classes, of communication traffic."

Several important points arise from the above definition:

(a) The process described by the definition is essentially one of producing simplified numerical models of the channels.

(b) The parameters measured must be such as to yield a model appropriate to the class, or classes, of traffic it is required to pass over the channel.

(c) "Real-time" implies that the measured parameter values must be updated at intervals which are less than the effective response time of the communication system to control inputs.

(d) The relative capabilities of the channels to pass a given class, or classes, of communication traffic must be expressed in terms which are meaningful to the communicator and system controller, e.g., an error rate for digital data and level of intelligibility for speech.

3.2 A generalized RTCE algorithm

It is assumed that the state of the HF channel as observed at the receiver can be characterized by a set of n measurable parameters, which are functions of both frequency f and time t,

$$v_1(f, t),\ v_2(f, t),\ \ldots\ldots\ldots,\ v_n(f, t) \qquad (6)$$

Let this set of parameters be expressed as a vector $[V\ (f, t)]$:

$$[V\ (f,t)] \equiv \begin{bmatrix} v_1(f, t) \\ v_2(f, t) \\ \vdots \\ v_n(f, t) \end{bmatrix} \qquad (7)$$

RTCE will, in general, necessitate sampling the values of selected members of the parameter set at intervals of, say, T seconds for each of m separate channels; thus, at the kth sampling instant given by

$$t = kT \quad (k \text{ integer}) \tag{8}$$

and for the ith frequency channel for which

$$f = f_i \quad (i \text{ integer and } 1 \leqslant i \leqslant m) \tag{9}$$

expression (7) becomes:

$$\left[\overline{V}(kT)\right]\Big|_{f_i} \equiv \left.\begin{bmatrix} v_1(kT) \\ v_2(kT) \\ \vdots \\ v_n(kT) \end{bmatrix}\right|_{f_i} \tag{10}$$

An RTCE coefficient matrix $[A]$ is now defined as:

$$[A] \equiv [a_1 \; a_2 \; .... \; a_n] \tag{11}$$

and an RTCE confidence level, $E_A(kT)\,|_{f_i}$, by

$$E_A(kT)\Big|_{f_i} = [A]\;[V(kT)]\Big|_{f_i} \tag{12}$$

Hence

$$E_A(kT)\Big|_{f_i} = (a_1v_1(kT) + a_2v_2(kT) + ... + a_nv_n(kT))\Big|_{f_i} \tag{13}$$

For practical RTCE schemes, the coefficient matrix normally takes the form:

$$[A] = [0 \; 0 \; 0 \; 0 \; .... \; 0 \; a_j \; 0 \; .... \; 0 \; 0 \; 0 \; 0] \tag{14}$$

$$(j \text{ integer and } 1 \leqslant j \leqslant n)$$

which implies that, for any given RTCE technique, the evaluation of the set of channels will be based upon measurements of a single parameter . Thus $E_A(kT)|_{f_i}$ reduces to:

$$E_A(kT)\Big|_{f_i} = a_j v_j(kT)\Big|_{f_i} \tag{15}$$

In addition to the actual values of the set of measurable parameters $[V\ (f,\ t)]$, the values of the first derivatives of those parameters with respect to time may also be incorporated into an RTCE algorithm. The set of corresponding first derivatives is again expressed as a column vector, $[\dot{V}(kT)]|_{f_i}$, where:

$$[\dot{V}(kT)]\Big|_{f_i} \equiv \begin{bmatrix} \dot{v}_1(kT) \\ \dot{v}_2(kT) \\ \vdots \\ \dot{v}_n(kT) \end{bmatrix}\Bigg|_{f_i} \tag{16}$$

An RTCE coefficient row matrix $[B]$ for the above set of derivatives is defined as:

$$[B] \equiv [b_1\ b_2\ \ldots\ldots\ldots\ b_n] \tag{17}$$

and an RTCE confidence level, $E_B(kT)|_{f_i}$, by

$$E_B(kT)\Big|_{f_i} = [B]\ [\dot{V}(kT)]\Big|_{f_i} \tag{18}$$

For practical RTCE schemes, $[B]$ will normally take the form:

$$[B] = [0\ 0\ \ldots\ldots\ 0\ b_1\ 0\ \ldots\ldots\ 0\ 0] \tag{19}$$

(1 integer and $1 \leqslant 1 \leqslant n$)

and therefore

$$E_B(kT)\Big|_{f_i} = b_1 \, \dot{v}_1(kT)\Big|_{f_i} \tag{20}$$

Some smoothing of the measured parameters will normally be necessary prior to sampling, so that the values obtained at the sampling instants will be average values computed over the previous T′ seconds, where

$$T' \lesssim T \tag{21}$$

Hence, using average values, expressions (12) and (18) can be rewritten respectively as:

$$\bar{E}_A(kT)\Big|_{f_i} = [A] \; [\bar{V}(kT)]\Big|_{f_i} \tag{22}$$

and

$$\bar{E}_B(kT)\Big|_{f_i} = [B] \; [\bar{\dot{V}}(kT)]\Big|_{f_i} \tag{23}$$

An overall figure of merit, $\overline{FM}(kT)|_{f_i}$, will be defined as being an arbitrary function of the confidence levels $\bar{E}_A(kT)|_{f_i}$ and $\bar{E}_B(kT)|_{f_i}$:

$$\overline{FM}(kT)\Big|_{f_i} \equiv F(\bar{E}_A(kT), \bar{E}_B(kT))\Big|_{f_i} \tag{24}$$

$$= F\,([A] \; [\bar{V}(kT)], \; [B] \; [\bar{\dot{V}}(kT)])\Big|_{f_i} \tag{25}$$

Using a simple linear interpolation for the values of the derivatives:

$$\overline{FM}(kT)\Big|_{f_i} = F\,([A] \; [\bar{V}(kT)], \; \frac{[B]}{T} \; [\bar{V}(kT) - \bar{V}((k-1)T)])\Big|_{f_i} \tag{26}$$

The form of the coefficient matrices [A] and [B] and of the function F ( ) will depend upon the nature of the particular RTCE algorithm employed in a given situation. However, in general, it can be stated that their form will be influenced by the objective(s) of the communication system and by any constraints placed upon system operation by equipment characteristics.

The function of the RTCE algorithm is to compute for the set of m channels the values of

$$\overline{FM}(kT)|_{f_i} \quad \text{for } 1 \leq i \leq m \tag{27}$$

and select the value of i for which $\overline{FM}(kT)|_{f_i}$ is a maximum. It may also be required to rank the other (m-1) channels in order of FM, and to produce a decision on which channels should be employed for communication taking account of operational and equipment constraints, and to identify possible standby channels.

Figure 3 illustrates the RTCE algorithm described above in diagrammatic form.

4. PRACTICAL RTCE SYSTEMS

There are many different forms of RTCE systems and many different measurable parameters on which particular RTCE schemes can be based. Examples of time - and frequency - dependent parameters which might be measured at the RTCE receiver, for each of the m alternative channels, are:

(a) Signal amplitude.

(b) Signal phase (absolute or differential).

(c) Frequency offset.

(d) Signal propagation time.

(e) Noise level.

(f) Channel unit impulse response.

(g) Signal-to-noise ratic.

(h) Baseband spectrum.

(i) Number of received data errors in a given time interval.

(j) Level of speech intelligibility for received signal.

(k) Telegraph distortion factor.

(l) Number of repeats requested in a given time interval for an automatic repeat request (ARQ) system.

Examples of individual techniques which have been developed for RTCE purposes will now be described.

4.1 Ionospheric pulse sounding

The most widely used form of RTCE is known as ionospheric sounding. The simplest implementation is a pulse sounder employing time - and frequency - synchronized transmission and reception. A high power sounding transmitter radiates short pulses in time sequence on preset frequency channels spaced over part, or the whole, of the HF band. The transmission schedule can be varied as required by a system program and is locked to a master clock, which itself can be locked to an external standard time transmission such as MSF or WWV. At the receiving site, the sounder receiver is synchronized with the same standard time transmission. Alternatively, the need for external synchronization can be avoided by providing atomic time/frequency standards at both transmitting and receiving sites.

When the transmitted pulse x(t) is of short duration, the sounder or RTCE receiver essentially measures the linear unit impulse response function for each channel. The received signal y(t) is given by the convolution integral:

$$y\,(t) = \int_{-\infty}^{\infty} h(u)\; x(t - u)\; du \,\Big|_{f_i \quad 1 \leqslant i \leqslant m} \tag{28}$$

where h(u) is the unit impulse response function, u being a time variable. If x(t) is an approximate impulse, then y(t) is obviously approximately proportional to the impulse response function h(t).

The output data from a sounder is usually presented in the form of a display termed an "ionogram". This is essentially a 2-dimensional projection of the raster of impulse responses for the set of m channels, as is shown in Fig. 4. The upper diagram of Fig. 4 shows impulse response samples taken from the complete m-channel raster; the lower diagram is a projection of the raster, or ionogram showing the propagation mode structure in the propagation delay (τ_p) frequency plane. Once the complete ionogram has been obtained it can, for example, be used to select frequency ranges where single mode propagation exists - as indicated in Fig. 4. Alternatively, if no regions of single mode propagation exist, a figure of merit for say the ith channel might be derived by considering the energy of the individual propagation modes. Consider the typical averaged q-mode impulse response for the ith frequency channel shown in Fig. 5. The first operation is to determine the position of the strongest propagation, i.e. the position of $h(\tau_p)_{max}\big|_{f_i}$. Let this correspond to the kth propagation mode. The energy of the kth mode, ε_k, is given by:

$$\varepsilon_k = \int_{\tau_{k-1}}^{\tau_k} \left[\overline{h(\tau_p)}\right]^2 d\,\tau_p \;\Big|_{f_i} \qquad (29)$$

where limits τ_k and τ_{k-1} are set by the single mode time dispersion which, in this case, is assumed to be similar for all modes.

The total energy for the other (q - 1) modes, ε_T, is:

$$\varepsilon_T = \sum_{j=1}^{q} \int_{\tau_{j-1}}^{\tau_j} \left[\overline{h(\tau_p)}\right]^2 d\,\tau_p \;\Big|_{f_i} \qquad (30)$$

(j integer ≠ k)

In practice, it may be difficult to specify the integration limits accurately for the weaker modes, particularly if adjacent modes have a small delay separation. Thus, the value of ε_T can be obtained more conveniently from the expression:

$$\varepsilon_T = \left(\int_{\tau_o}^{\tau_q} \left[\overline{h(\tau_p)}\right]^2 d\,\tau_p - \int_{\tau_{k-1}}^{\tau_k} \left[\overline{h(\tau_p)}\right]^2 d\,\tau_p\right) \Big|_{f_i} \qquad (31)$$

One figure of merit which could then be used to specify how nearly propagation in the ith channel approximates to single-mode conditions is therefore:

$$FM = \varepsilon_k \,/\, \varepsilon_T \;\Big|_{f_i} \qquad (32)$$

The higher the value of $\varepsilon_k / \varepsilon_T \big|_{f_i}$, the more nearly is propagation effectively single mode.

4.2 Modulated pulse sounding

A most important variant of the basic pulse sounding system is one in which the individual pulses on each frequency are modulated by a digital waveform. This confers two important advantages:

(a) it allows pulse compression coding to be applied (as in some radar systems) in order to improve the delay resolution properties of the system without having to shorten the basic pulse length, and hence increase the peak transmitted power to maintain the same transmitted energy.

(b) Small quantities of data can be transmitted with the RTCE signal, via pulse modulation coding, e.g. describing noise levels in assigned channels.

Several forms of pulse compression codes are available in practice; these include Barker codes [3], Huffman sequences [4], binary sequences of length >13 with autocorrelation functions (acf's) approximating to an impulse [5] and complementary sequences [6]. For all these types of code, impulse response evaluation is accomplished by computing the input-output cross-correlation function (ccf). For a linear system with input x(t) output y(t), the input-output ccf, $\phi_{xy}(\tau)$ is given by [7]:

$$\phi_{xy}(\tau) = \frac{1}{T'} \int_{-T'/2}^{+T'/2} x(t)\, y(t+\tau)\, dt \tag{33}$$

$$= \int_{-\infty}^{\infty} h(u)\, \phi_{xx}(\tau - u)\, du \tag{34}$$

Where τ and u are time variables, T' is the correlation interval and $\phi_{xx}(\tau)$ is the input acf. From expression (34), it is seen that if the input acf is impulsive, then $\phi_{xy}(\tau)$ will be proportional to the system impulse response function.

There are many individual pulse and modulated pulse sounders in operation all over the world but the most ambitious scheme yet proposed has been the US concept of a Common User Radio Transmission System (CURTS), in which a network of fixed sounding transmitters is provided giving complete area coverage for all users with compatible sounding receivers [8].

4.3 Chirp sounding

Another form of ionospheric sounding, fundamentally different in principle from pulse sounding, is termed "chirp" sounding. The technique derives its name from the frequency swept signal used as the sounding transmission; the sweep is typically linear, but can also be logarithmic in form. Again, a synchronised transmitter and receiver are required. At the receiver, an identical local oscillator sweep is mixed with the received chirp signal and the difference in frequency between the two signals measured [9].

Let the frequencies of the transmitted and receiver local oscillator signals, $f_T(t)$ and $f_{LO}(t)$, be given by:

$$\left[f_o + t\,\frac{df}{dt}\right]\Bigg|_{t=0}^{T'} \qquad (35)$$

Where f_o is the frequency at which the chirp commences, $\frac{df}{dt}$ is the sweep rate in Hz/sec and T' is the duration of the sweep. When i propagation modes are present, the propagation times $\tau_1, \tau_2, \ldots.. \tau_i$, the frequency components of the received signal, $f_R(t)$, will have the form:

$$\left[f_o + (t - \tau_1)\frac{df}{dt}\right]\Bigg|_{t=\tau_1}^{T'+\tau_1}, \left[f_o + (t - \tau_2)\frac{df}{dt}\right]\Bigg|_{t=\tau_2}^{T'+\tau_2}, \ldots \qquad (36)$$

$$\ldots, \left[f_o + (t - \tau_i)\frac{df}{dt}\right]\Bigg|_{t=\tau_i}^{T'+\tau_i}$$

After mixing $f_R(t)$ with the local oscillator signal, $f_{LO}(t)$, the frequency components of the difference signal, $(f_{LO}(t) - f_R(t))$, are:

$$\tau_1\frac{df}{dt}, \tau_2\frac{df}{dt}, \ldots\ldots\ldots\ldots, \tau_i\frac{df}{dt} \qquad (37)$$

Hence, propagation delays are translated directly into frequency offsets. In a linear mixing process, the relative amplitudes of the individual sweeps received via the different propagation modes will be preserved. Thus, if the output of the mixer is displayed on a spectrum analyser, a propagation mode profile equivalent to the channel impulse response will result. Consequently, by presenting a projection of the spectrum analyser output as a function of sweep frequency, ionograms can again be formed.

4.4 Oblique incidence, vertical incidence and backscatter sounding

All three ionospheric sounding techniques described previously, pulse, modulated pulse and chirp can be applied practically in different ways.

4.4.1 Oblique incidence sounding. By oblique incidence is meant operation between a sounding transmitter and receiver which are geographically separated so that the transmissions impinge on the ionospheric layers obliquely, as is shown in Fig. 1.

4.4.2 Vertical incidence sounding. Vertical incidence sounding is employed, as the name implies, to determine the propagation mode structure immediately above the sounding transmitter; consequently, the sounding receiver must be co-sited with the transmitter.

4.4.3 Backscatter sounding. Backscatter sounding is similar in nature to vertical incidence sounding in that it can be carried out from a single site. However, the transmitted energy must now be radiated obliquely, rather than vertically. The received signal is now due to that energy which has propagated via the ionosphere over a path away from the transmitter and been reflected from the ground back to the receiver via a similar ionospheric reflection in the reverse sense.

4.5 Channel evaluation and calling (CHEC) system

The channel evaluation and calling (CHEC) system has been developed in Canada and is described in [10]. CHEC is designed primarily for a scenario where one or more mobiles are required to pass traffic to a base station. On each of the m assigned channels (where m would normally be <20), the CHEC base transmitter emits a signal of several seconds duration which comprises a selective calling code, data on the average noise level at the base station in that channel, and a CW section. At the remote receiver specified by a selective call address, the base station noise levels

$$\overline{n(t)}\Big|_{f_i} \qquad 1 \leqslant i \leqslant m$$

are decoded for the sub-set of k channels actually propagating to the mobile

$$\overline{n(t)}\Big|_{f_j} \qquad \text{where } j \text{ can take } k \text{ distinct values in the range 1 to } m \text{ and } k \leqslant m \tag{38}$$

The sub-set of corresponding received average signal strengths

$$\overline{A(t)}\Big|_{f_j} \tag{39}$$

are also measured at the mobile using the CW section of the CHEC transmission. By making allowance for the relative gain characteristics of the transmitting and receiving antennas at mobile and base, together with the available transmitter powers at the two sites, and assuming propagation reciprocity, a special-purpose processor at the mobile computes the predicted average signal-to-noise ratio $\bar{\rho}_{BASE}(t)$ at the base due to its own transmission for each of the k propagating channels. Thus

$$\bar{\rho}_{BASE}(t)\Big|_{f_j} = \left[G \cdot \frac{\bar{A}(t)}{n(t)} \right]\Bigg|_{f_j} \qquad (40)$$

Where G is a frequency-dependent factor computed from the antenna gain and transmitter power characteristics for base and mobile. The optimum channel for mobile-to-base communication is then given by the value of j for which

$$\bar{\rho}_{BASE}(t)\Big|_{f_j} \quad \text{is a maximum} \qquad (41)$$

Experiments with prototype equipments have shown that the use of the CHEC technique yields a significant improvement in channel availability.

4.6 Pilot tone RTCE

The principles and performance of the Pilot Tone RTCE technique are described in [11]. The basis of the technique is that, at the RTCE receiver, a low-level CW pilot tone is detected in a narrow bandpass filter and its phase perturbations analysed. This gives data regarding the state of the channel and, using derived relationships, the performance of a specified form of communication signal using the same channel can then be predicted. Specifically, the phase of the pilot tone is sampled at regular intervals, τ_s, typically 10 ms, and the phase at the current sampling instant α_n compared with the phase measured and stored for the previous sample α_{n-1}. If the difference in phase between the two samples exceeds a certain preset threshold value α_T, then a "phase error" is indicated, i.e.

$$| \alpha_n - \alpha_{n-1} | > \alpha_T \qquad (42)$$

for phase errors.

It is evident that the sampling interval must be synchronized with an intergal multiple of the period of the pilot tone in order that, under ideal conditions when no perturbation is present, sampling will always occur at the same point in the pilot tone cycle.

The parameter chosen to indicate the state of the channel is the average number of phase errors occurring in a defined measurement interval T′, say 200 seconds, $\overline{N_{PE}}$. Therefore

$$\left[\overline{N_{PE}}\right]_t^{t+T'} = \left[\sum_{n=1}^{T'/\tau_S} K_n\right]_t^{t+T'} \quad (43)$$

where

$$K_n = 1; \text{ for } \left|\alpha_n - \alpha_{n-1}\right| > \alpha_T$$

$$\text{and} \quad K_n = 0; \text{ for } \left|\alpha_n - \alpha_{n-1}\right| \leqslant \alpha_T \quad (44)$$

For the practical tests described in [11], a low-level pilot tone was frequency multiplexed with a frequency-exchange-keyed 50 baud binary data signal; the baseband spectrum of this arrangement is shown in Fig. 6.

By suitable calibration, the number of data bit errors occurring in the given measurement interval T′for the binary data channel, $\overline{N_E}$, can be related to the pilot tone phase error count $_{PE}$ over the period T′. The theoretical relationships for steady signal, flat fading and frequency-selective fading conditions are shown as solid lines in Fig. 7. The points shown superimposed on these theoretical relationships indicate the typical scatter obtained from an experimental run in which corresponding values of $\overline{N_E}$ and $\overline{N_{PE}}$ are actually measured.

The major conclusion which can be drawn from the results of the tests carried out is that, for the majority of HF path conditions, the data error rate which would be experienced with a particular modulation scheme over any given path can be forecast with reasonable accuracy via simple phase measurements on a low-level pilot tone.

4.7 RTCE by error counting

A simple form of RTCE is to probe the m channels to be evaluated using a test signal which has essentially the same parameters as the traffic signal to be transmitted over the path. Also, it is convenient practically if this test signal has a digital format so that errors can be counted, rather than having to make on-line subjective assessments of non-digital quantities, e.g., speech intelligibility.

For all normal forms of HF traffic, it is possible to carry out reliable RTCE via a count of errors occurring in an interval of T', $N_E(t)$, which, in the case of analogue or digital speech channels, can be related to a predicted level of speech intelligibility, I(t), i.e.

$$I(t) = f(N_E(t)) \tag{45}$$

4.8 In-band RTCE

All the RTCE techniques discussed previously have been developed primarily for multiple channel situations. A technique which is designed specifically for in-band RTCE, or evaluation of sub-channels within a nominal 3 kHz HF assigned channel, is now described [12].

At the receiver, a real-time spectrum analyser monitors the distribution of noise energy for all the sub-channels within the 3 kHz bandwidth using a set of bandpass filters, each having the same bandwidth as one sub-channel. Regions of low noise are identified and indicated to the transmitter site in order that its data tones can be steered to the corresponding sub-channels.

4.9 Simultaneous RTCE and traffic transmission for a single channel

In addition to the problem of evaluating the states of alternative channels or sub-channels, RTCE may be required to evaluate the state of a given communication channel or sub-channel whilst it is actually passing traffic, in order to assess, say, when a frequency change is necessary. With some forms of traffic, it is possible to derive the RTCE data from the normal operating signals; other types of traffic may require some modification of the signal generation and processing procedures to allow the necessary information to be extracted. The possibilities for this form of RTCE are best described with the aid of specific examples.

4.9.1 Soft-decision data. Any data which can be extracted in order to assign a confidence level to a demodulation or decoding decision can, in principle, be employed for RTCE purposes. For example, in a modem employing phase modulation, the phase margin between the received signal phase, $\alpha_R(t)$ and an ideal phase reference signal, $\alpha_o(t)$, could be used to determine an RTCE figure of merit FM thus:

$$FM \propto |(\alpha_R(t) - \alpha_o(t))| \tag{46}$$

Alternatively, the FM could be taken as being proportional to the amplitude, A(t), of the received signal, or to a composite function of amplitude and phase margin:

$$FM \propto f\left[A(t), |(\alpha_R(t) - \alpha_o(t))|\right] \tag{47}$$

Soft-decision data of this type has been incorporated into an HF modem known as CODEM which employs a combination of modulation and coding techniques to enhance data transmission reliability [13].

Thus, if a phase margin threshold, α_T, and an amplitude threshold, A_T, are specified, RTCE information could, for example, be extracted in a given measurement period, T', by noting the number of occasions that

$$|(\alpha_R(t) - \alpha_o(t))| > \alpha_T$$

$$\text{and} \qquad A(t) < A_T \tag{48}$$

and relating the counts obtained to a defined channel acceptability criterion.

4.9.2 Automatic-repeat-request (ARQ) systems. An ARQ communication system typically formats the data to be transmitted into labelled blocks of a given length. The blocks are transmitted sequentially until EDC or soft-decision information at the receiver indicates that a certain block has been decoded erroneously. An ARQ signal is then sent to the transmitter via a feedback link (which need only have a relatively low capacity) requesting a repeat transmission of the corrupted block. Clearly, the number of repeats requested in a given time interval T', $N_R(t)$, will be a measure of the state of the channel and can be employed for RTCE purposes.

4.9.3 RTCE via simple modifications to the traffic signal. In certain situations it may be impractical to obtain the required RTCE information from the basic traffic signal, possibly because soft-decision parameters are not accessible or because the received data stream is encrypted. In the latter case, an additional signal generation/processing operation can be incorporated in order to specifically provide RTCE data; Figure 8 illustrates such an arrangement. It is assumed that cryptographic considerations limit access to the elements of the communication systems except for the region shown. If a special EDC system (shaded units) is introduced into this region, it can be used to format the encrypted data into, say, arbitrary codeword blocks prior to transmission. At the receiving site the received codeword blocks can be decoded to yield the original encrypted data stream. However the EDC algorithm can also be of a form which indicates, say, the number of single or multiple errors being detected and corrected per codeword and hence the state of the channel.

5. RESULTS OF PRACTICAL TESTS USING A SIMPLE RTCE SYSTEM

The system of RTCE chosen for the tests was based upon simple error counting techniques as described in section 4.7; a block diagram of the practical arrangement is shown in Fig. 9.

5.1 RTCE system description

Referring to Fig. 9: at the transmitter site the block labelled "RTCE Signal Generation" represents a memory unit which stores a library of baseband channel evaluation signals. These recorded signals take the form of modulator outputs in response to repetitive binary test sequences for digital data and to standard words and phrases for voice traffic. Currently, the following types of baseband test signal are available:

(a) Low-rate binary frequency-exchange-keyed telegraphy (75 bit/s).

(b) Combined SSB LINCOMPEX speech and telegraphy as in (a) above.

(c) Medium-speed (1200 - 2400 bit/s) digital data using a modem with 16 parallel sub-channels, each employing 4-phase DPSK modulation.

(d) Medium-speed (1200 - 2400 bit/s) digital data using a simple serial 4-phase DPSK modem.

The frequency to which the synthesized transmitter drive unit is set at any time is determined by the timing control unit, which steps the transmitter through the set of channel frequencies f_1 to f_m, according to a preset time schedule.

At the receiver the frequency to which the receiver is tuned is varied by a timing control unit, in synchronism with the transmission schedule. It should be noted that exact synchronism is not necessary and a relative accuracy in timing between transmitter and receiver of a few seconds is adequate, providing that the timing errors are not cumulative. Error counts occurring during transmitted data block lengths of 10000 bits were employed during the tests.

5.2 Format of tests

The RTCE system shown in Fig. 9 was tested over two HF paths:

(a) Oslo (Norway) to The Hague ($\simeq$ 1100 km) in June 1976.

(b) Cornwall (UK) to The Hague ($\simeq$ 700 km) in October/November 1976.

The tests described here are of two of the type transmissions listed in Section 5.1, viz FEK telegraphy and medium-speed data (parallel sub-channel modem). The circuit availability using frequency selection based upon off-line frequency prediction techniques was compared with that obtained using frequency selection based upon RTCE data. For each path, 10 possible assigned frequency channels, f_1 to f_{10}, were available. Over 24-hour periods, the signal quality, in terms of an error count for digital data and a level of intelligibility for analogue voice, was sampled for a 5-minute interval on each frequency in each hour. Half a minute was allowed for frequency changes between the sampling intervals.

5.3 Frequency selection algorithms

5.3.1 2-Frequency selection. As illustrated in Fig. 2, it is usually possible to select one "day" and one "night" frequency which will provide a reasonable fit between the predicted FOT and LUF characteristics. It is assumed for the purpose of the tests that this is the mode of operation which would be employed in order to minimize frequency changes.

5.3.2 Selection via RTCE. For RTCE using digital test signals, the average Figure of Merit for the ith channel, $FM(kT)|f_i$, at the kth sampling instant ($t = kT$) is defined to be:

$$\overline{FM}(kT)\,\big|_{f_i} = \text{Constant}\,.\,\overline{N}_E\,(kT)\,\big|_{f_i} \qquad 1 \leqslant i \leqslant m \qquad (49)$$

Where $\overline{N}_E$ (kT) is the number of errors accumulated in the period $T'(\leqslant T)$ prior to the kth sampling instant. For analogue voice signals, the Figure of Merit is defined as:

$$\overline{FM}(kT)\,\Big|_{f_i} = \text{Constant} \,.\, \overline{I}(kT)\,\Big|_{f_i} \qquad 1 \leqslant i \leqslant m \tag{50}$$

Where $\overline{I}(kT)$ is the average speech intelligibility level measured over the period T' prior to the kth sampling instant.

Normally, the optimum frequency at the kth sampling instant would be given by the value of i for which $\overline{FM}(kT)|_{f_i}$ is a maximum. However, the situation may arise where two or more channels of the m-channel set have the same maximum Figure of Merit, i.e.

$$\overline{FM}(kT)\,\Big|_{f_a} = \overline{FM}(kT)\,\Big|_{f_b} = \ldots\ldots = \overline{FM}(kT)\,\Big|_{f_c} \qquad a \neq b \neq \ldots \neq c \tag{51}$$

If any of the channels, f_a, f_b,...,f_c, are the same as the optimum channel for the previous sampling instant (t = (k-1) T), then that channel is selected as being optimum for the kth sampling instant also. The rationale for this selection is that it minimises the number of frequency changes which have to be made. Should f_a, f_b, ..., f_c not be the same as the optimum channel for the previous sampling instant, the value of i corresponding to the highest frequency channel is chosen as being optimum on the basis that, at the higher frequencies, the probability of encountering single mode propagation and lower noise levels will, in general, be higher.

Again, in order to minimise the number of frequency changes, the concept of an acceptable range for the Figure of Merit can be introduced. If it is determined that a given traffic transmission type will have acceptable quality for

$$\overline{FM}(kT)\,\Big|_{f_i} \geqslant K \tag{52}$$

Where K is a constant, then any channel for which expression (52) holds can be employed for transmission. Therefore, if

$$\overline{FM}(kT)\,\Big|_{f_a} \geqslant \overline{FM}(kT)\,\Big|_{f_b} \geqslant \ldots\ldots \geqslant \overline{FM}(kT)\,\Big|_{f_c} \geqslant K \tag{53}$$

any of the channels f_a, f_b,..,f_c can be selected and still yield acceptable communications traffic quality. Thus, if any of the channels f_a, f_b,..,f_c are the same as the optimum channel for the $(k-1)^{th}$ sampling then that channel is selected again as being optimum for the kth sampling instant also.

In the following sets of results, the RTCE selection algorithm is limited to determining the value of i for which $\overline{FM}(kT)|_{f_i}$ is a maximum.

5.4 Practical results

Practical results for the above two frequency selection algorithms, applied to two different types of traffic transmission, are now described.

5.4.1 Binary frequency-exchange-keyed (FEK) telegraphy. The sampled error counts, $\bar{N}_E(kT)$ were averaged over blocks of length 10000 data bits, measured for a 75 bit/s binary FEK telegraphy signal transmitted at a power of 150 W over the Oslo - The Hague and Cornwall - The Hague paths respectively.

The circuit availability is defined over a 24-hour period as:

$$\text{Availability} = \frac{\left[\begin{array}{l}\text{Number of hours for which } \bar{N}_E(kT) \leqslant N_T \\ \text{for a block of N data bits}\end{array}\right]}{24} \times 100\%$$

Where N_T is a variable threshold value for the number of errors above which the channel is considered unusable.

For example, Figs. 10 and 11 show diagramatically the distribution of circuit availability over representative 24-hour periods for the two paths. Shaded areas indicate periods where $N_T = 100$, i.e. the average BER $\leqslant 10^{-2}$. Actual frequency values for the channels f_1 to f_{10} are shown to the nearest 0.1 MHz in each case. The averaging interval for each error count was 5 minutes.

Figures 12 and 13 show the empirical probabilities of achieving a given BER over each of the two paths for both methods of frequency selection. The "random" characteristic is derived from the case when a random selection of frequency is made from the set of 10 available. The advantage of RTCE is clearly demonstrated.

5.4.2 Medium-speed data using parallel sub-channel modem. The best known technique for transmitting medium-speed digital data, in the range 1200 to 2400 bit/s, over HF paths is to employ a modem comprising multiple, low-rate parallel sub-channels within an overall 3 kHz channel bandwidth. The modem employed in the tests was of a form similar to KINEPLEX [14]; it had 16 parallel 4-phase DPSK sub-channels, each keyed at 75 bauds. At a data rate of 1200 bit/s, the modem could operate in a dual in-band frequency diversity mode in which the data was repeated on each of the two sets of 8 sub-channels. It is this 1200 bit/s, dual diversity, mode (abbreviated to 1200(X 2)) that has been used in the tests.

The sampled error counts, $\overline{N}_E(kT)$, were averaged over blocks of 10,000 data bits, in the 1200 (X 2) bit/s mode for the Oslo - The Hague and Cornwall - The Hague paths at an average transmitter power of 250 W.

Again, as an example, Figs. 14 and 15 show the corresponding availability time distributions for N_T = 100, i.e. BER $\leq 10^{-2}$.

Figures 16 and 17 show the empirical probabilities of achieving a given BER over the two paths for 2-Frequency, RTCE and Random selection algorithms.

5.4.3 Discussion. The results of the tests described in this section indicate, without exception, that for the two types of traffic considered, a substantial percentage increase in circuit availability, on average about 45%,can be achieved by the use of RTCE rather than 2-Frequency selection techniques.

6. CONCLUSIONS

6.1 General

The rationale for the development of RTCE techniques is simple: significant improvements in the use of the HF propagation medium can be achieved only if a communicator, or HF system controller, using a specific path at a given time has access to real-time data on the path parameters - rather than having to rely on frequency predictions which can be subject to appreciable errors.

The experimental work described in this paper has served to indicate the improvement in HF circuit availability (≃ 45%) achievable by the application of a relatively simple RTCE technique compared with that achievable using 2-Frequency operation derived from off-line frequency predictions.

As a result of the studies carried out by STC to date on the topic of RTCE, it is concluded that five particular techniques show promise of economic implementation:

(a) Simple error counting of the type used in the practical tests described in Section 5.

(b) Pilot tone phase error counting as described in Section 4.6. This would have application to the monitoring of the state of encrypted channels (Section 4.9.3). Pilot tones could also be employed for simple signal strength measurement.

(c) The use of EDC systems for monitoring the state of encrypted and clear channels (Section 4.9.3).

(d) Passive monitoring of noise/interference levels in alternative assigned channels.

(e) The use of in-band RTCE to determine the optimum sub-channel within a nominal 3 kHz assigned channel (Section 4.8).

6.2 Potential advantages of RTCE

The potential advantages accruing from the use of channel evaluation techniques can be summarized as follows:

(a) Off-line frequency predictions can be dispensed with; exact real-time data on actual propagation conditions are always available.

(b) The effect of man-made noise/interference can be measured and specified quantitatively.

(c) The facility for real-time, on-line measurement of propagation and interference allows the use of relatively transient propagation modes, e.g., via "Sporadic E Layer" propagation.

(d) Channel evaluation allows a more efficient use of the frequency spectrum by tending to select frequency channels higher than those which would be chosen via prediction techniques. Thus, spectrum congestion is reduced.

(e) Channel evaluation will provide a means of automatically selecting the best frequency and simultaneously indicating preferred stand-by channels.

(f) Transmitter radiated power can be minimized, consistent with providing an acceptable quality of received traffic.

(g) In the longer term, channel evaluation data can be used to adapt other parameters of a communication system, apart from frequency, optimally for the prevailing path conditions.

REFERENCES

1. K. Davies, Ionospheric Radio Propagation, Monograph 80, National Bureau of Standards, Washington, D.C., 1 April 1965.
2. CCIR Report 322 (1964), World Distributions and Characteristics of Atmospheric Noise.
3. R.H. Barker, Group Synchronizing of Binary Digital Systems, in "Communication Theory", Butterworth, London, 1953, pp 273 - 287.
4. D.C. Coll & J.R. Storey, Ionospheric Sounding Using Coded Pulse Signals, Rad. Sci. J. of Research, Vol. 68D(10), 1964, pp 1155 - 1159.
5. H.B. Mann, (Editor), Error Correcting Codes, Wiley, 1968, pp 195 - 225.
6. M. Darnell, Channel Estimation Techniques for HF Communications, AGARD Conference Proceedings No. 173, on Radio Systems and the Ionosphere, Paper 16, Athens, May 1975.
7. Y.W. Lee, Statistical Theory of Communication, Wiley, 1960, pp 323 - 351.
8. S.E. Probst, The CURTS Concept and Current Status of Development, in "Ionospheric Radio Communications", Plenum, 1968, pp 370 - 379.
9. G.H. Barry & R.B. Fenwick, Extra Terrestrial and Ionospheric Sounding with Synthesized Frequency Sweeps, Hewlett-Packard J., Vol. 16(11), 1965, pp 8 - 12.
10. E.E.Stevens, The CHEC Sounding System, in "Ionospheric Radio Communications", Plenum, 1968, pp 359-369.
11. J.A.Betts & M. Darnell, Real-Time HF Channel Estimation by Phase Measurement on Low-Level Pilot Tones, AGARD Conference Proceedings No. 173 on "Radio Systems and the Ionosphere", Paper 18, Athens, May 1975.
12. PYE/RAE Combine for FSK Demodulator, Communications International, Vol. 1, No. 1, 1974, p 10.
13. D.Chase, A Combined Coding and Modulation Approach for Communication over Dispersive Channels, IEEE Trans., Vol. COM-21, No. 3, March 1973, pp 159-174.
14. R.R.Mosier & R.G. Clabaugh, Kineplex, A Bandwidth-Efficient Binary Transmission System, Trans. AIEE (Comm & Electronics), Vol. 34, Jan. 1958, pp 723-727.

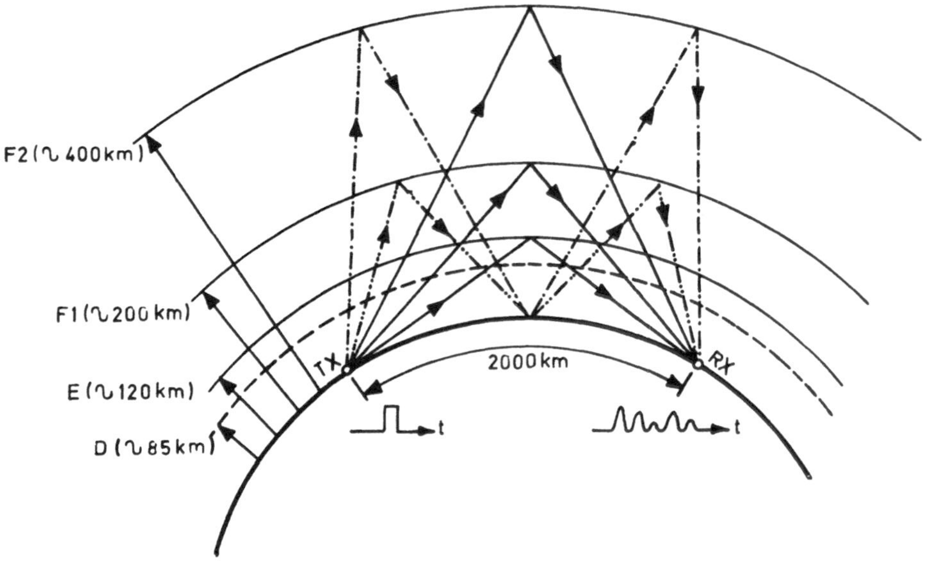

Fig. 1. Simplified HF propagation path

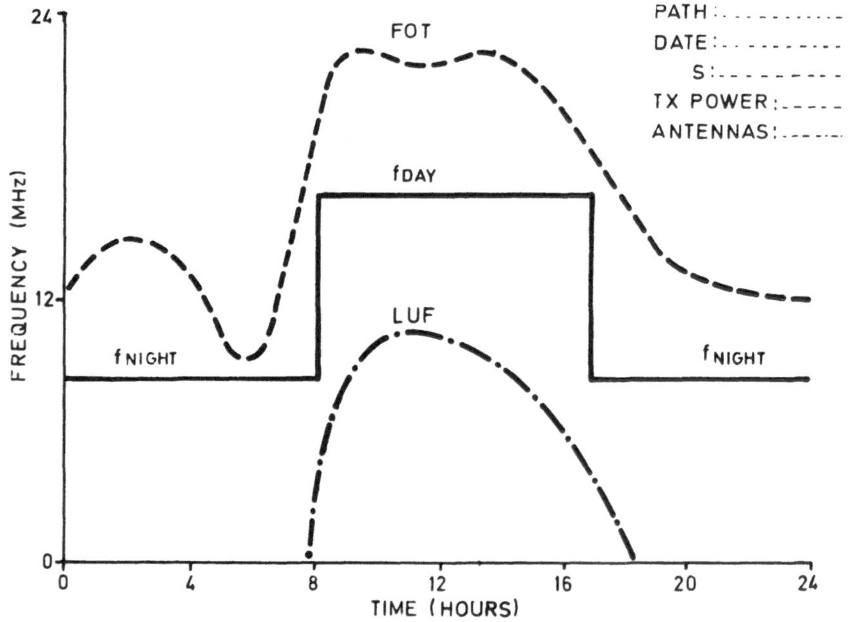

Fig. 2. Example of frequency prediction output

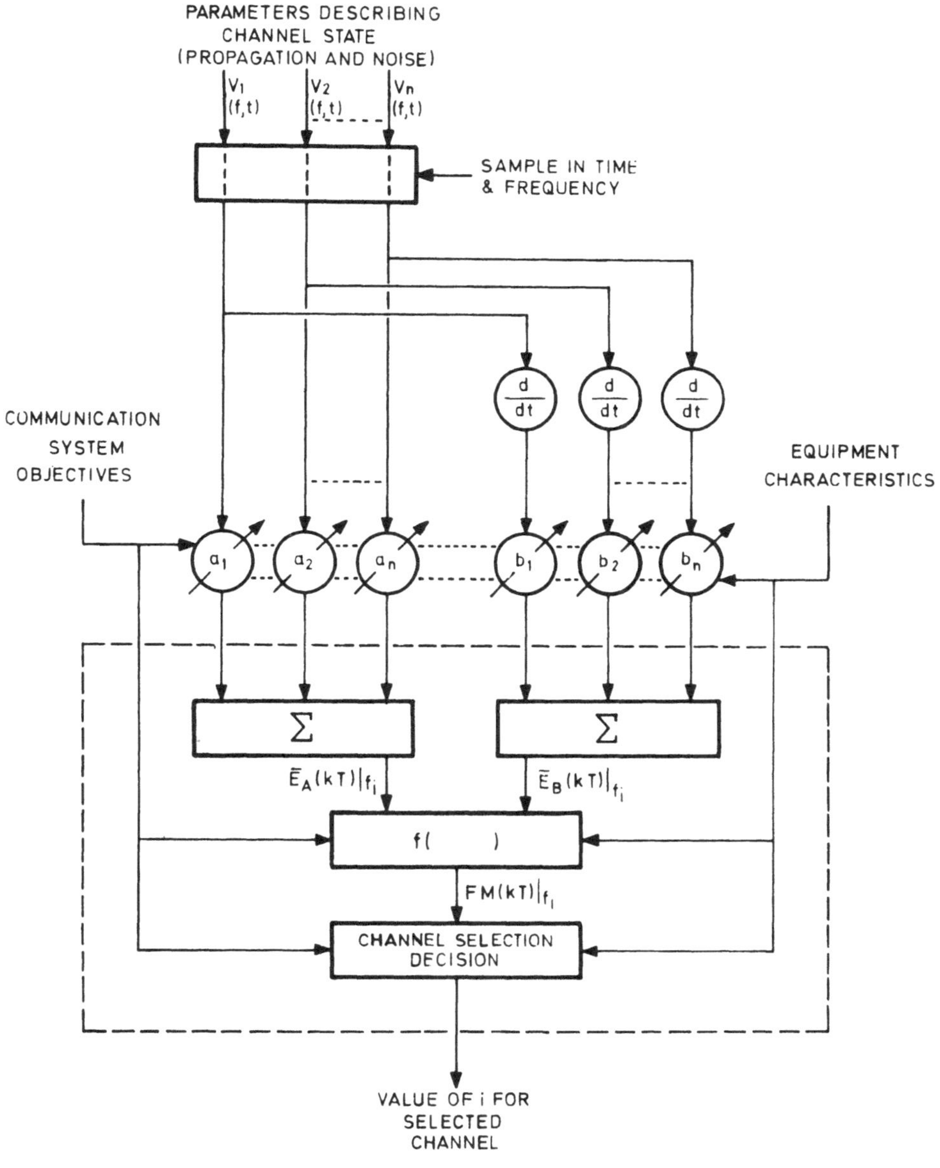

Fig. 3. Generalised RTCE algorithm

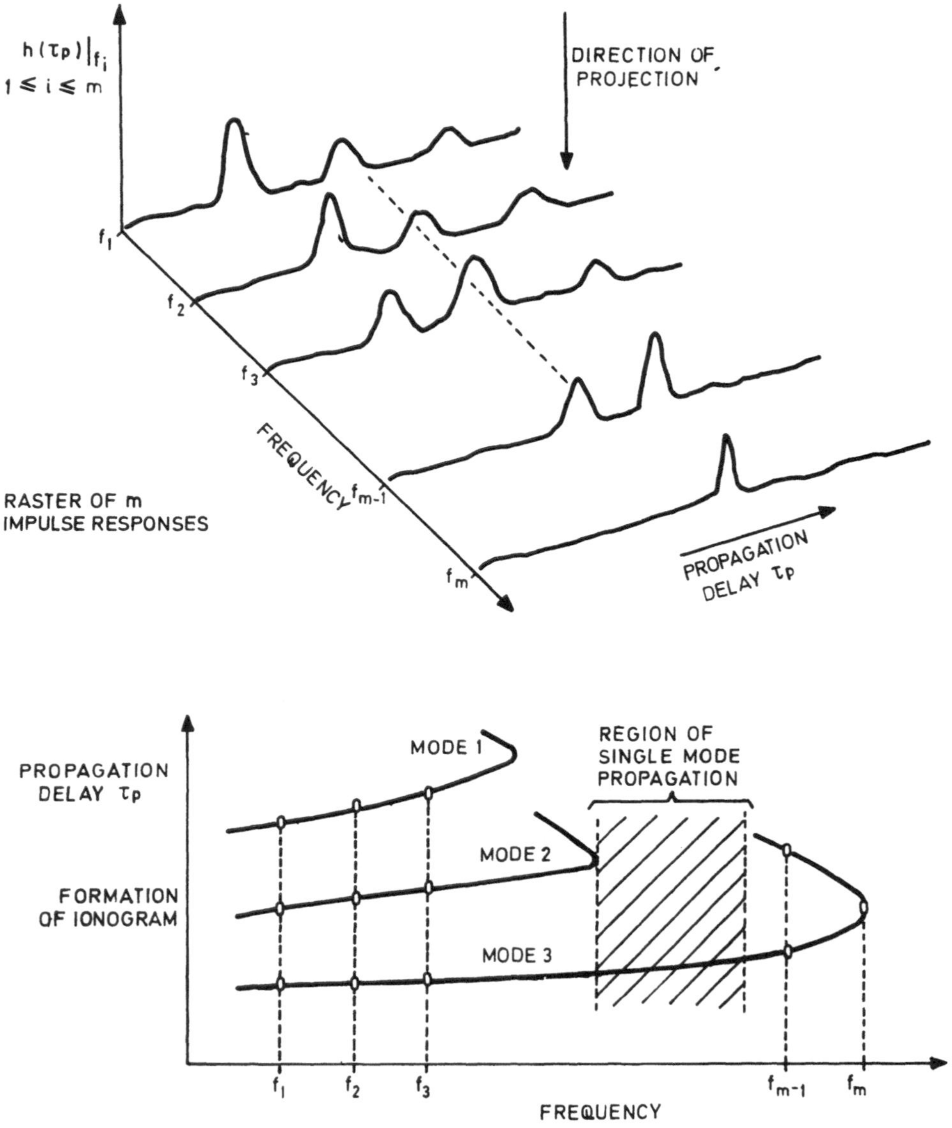

Fig. 4. Derivation of ionogram from m-channel impulse response raster.

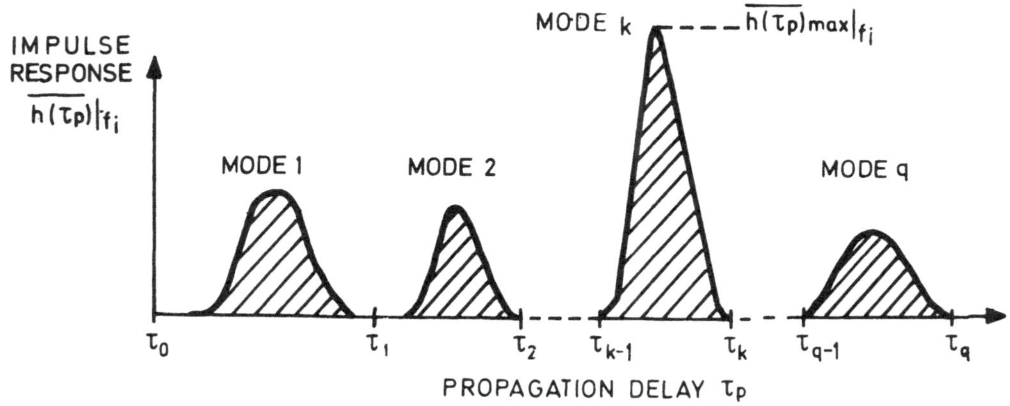

Fig. 5. q-Mode impulse response

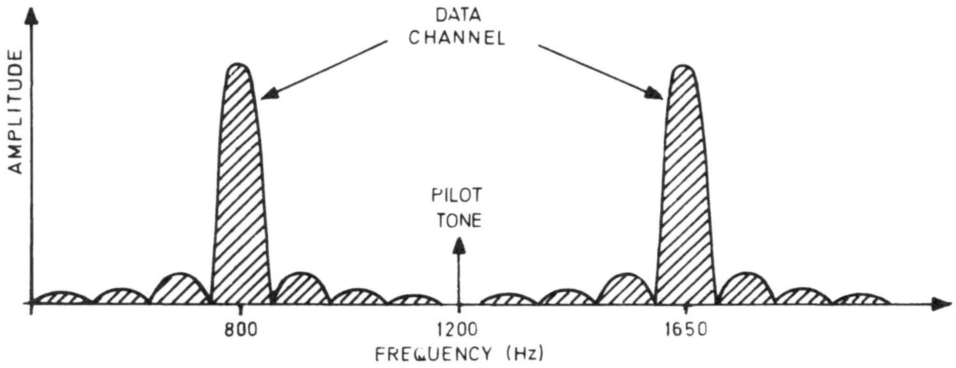

Fig. 6. Baseband spectrum of data channel with pilot tone inserted.

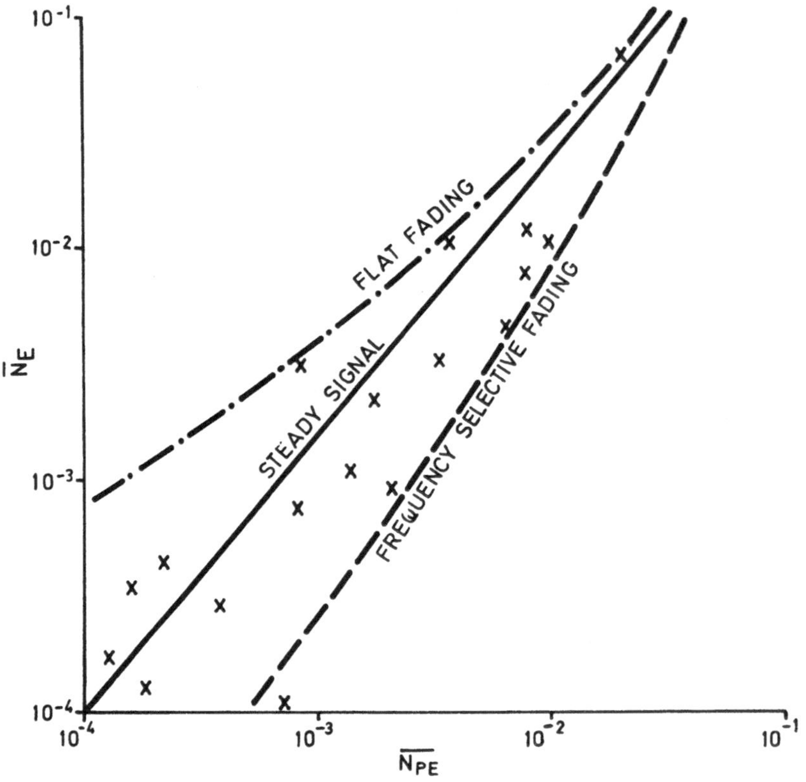

Fig. 7. Theoretical and empirical relationships between N_{PE} and BER.

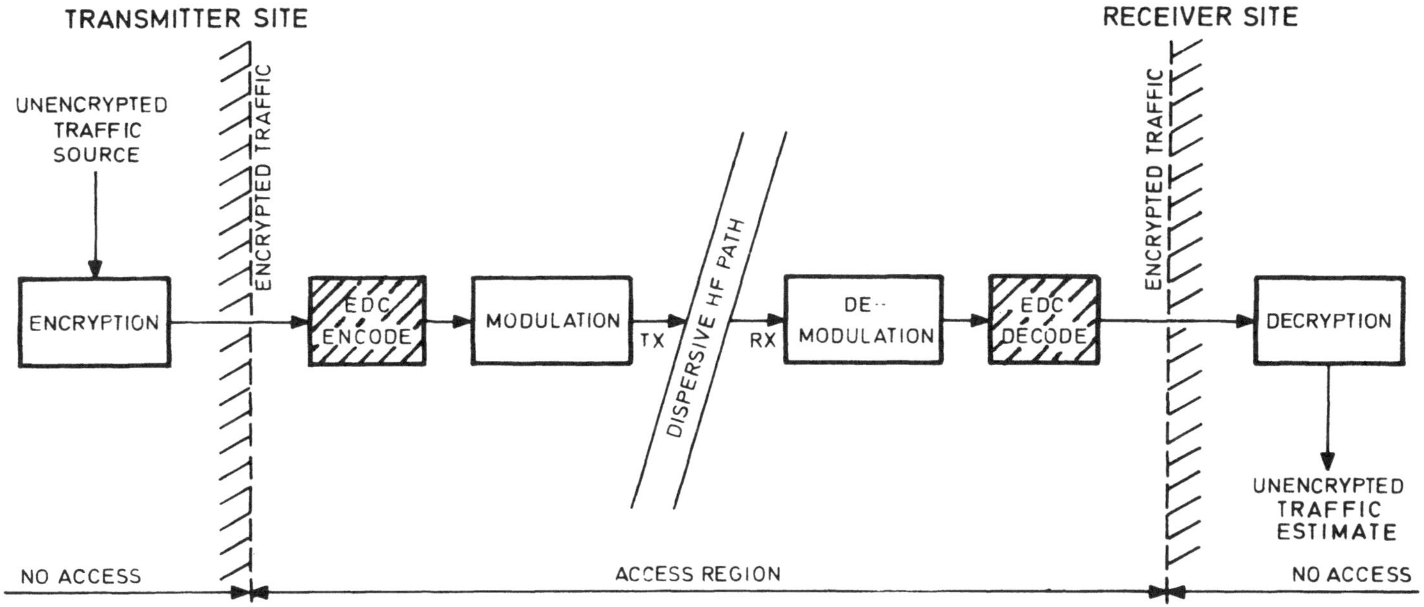

Fig. 8. The use of an auxiliary EDC system for RTCE with encrypted traffic

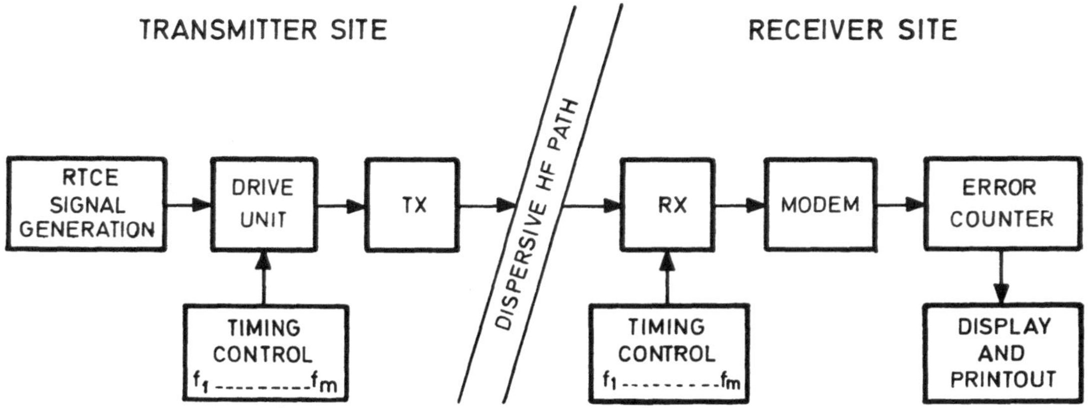

Fig. 9. Block diagram of simple RTCE system

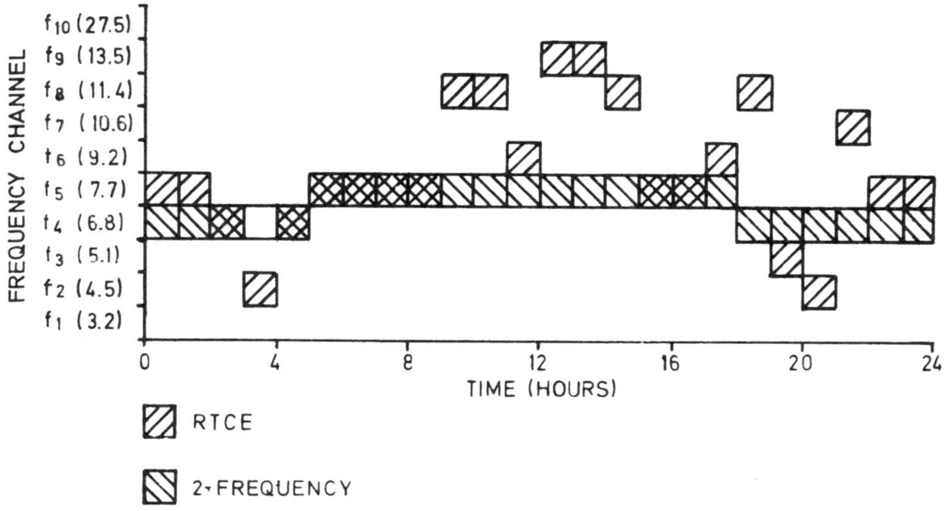

Fig. 10. Comparison of circuit availabilities for 2-frequency and RTCE frequency selection: Binary FEK telegraphy at 75 bits/sec and Tx power of 150 W.
Oslo - The Hague, 23-6-76
(Shaded Areas indicate BER $\leqslant 10^{-2}$)

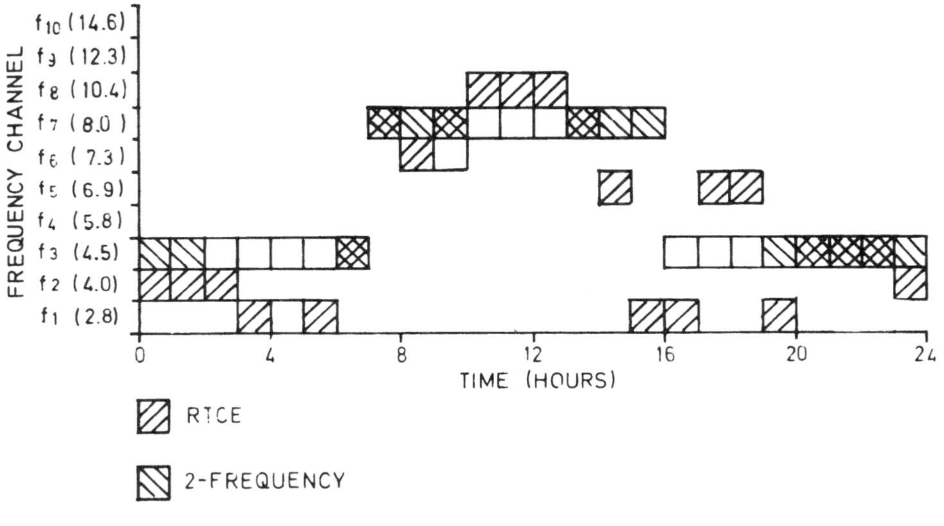

Fig. 11. Comparison of circuit availabilities for 2-frequency and RTCE frequency selection: Binary FEK Telegraphy at 75 bits/sec and Tx power of 150 W.
Cornwall - The Hague, 26-10-76
(Shaded Areas indicated BER $\leqslant 10^{-2}$)

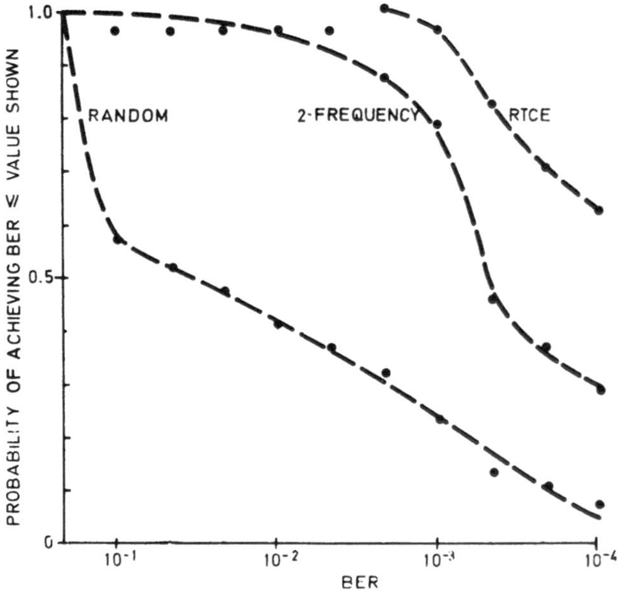

Fig. 12. Effect of different frequency selection algorithms upon the probability of achieving a given BER: Binary FEK telegraphy at 75 bits/sec and Tx power of 150 W. Oslo - The Hague, 23-6-76

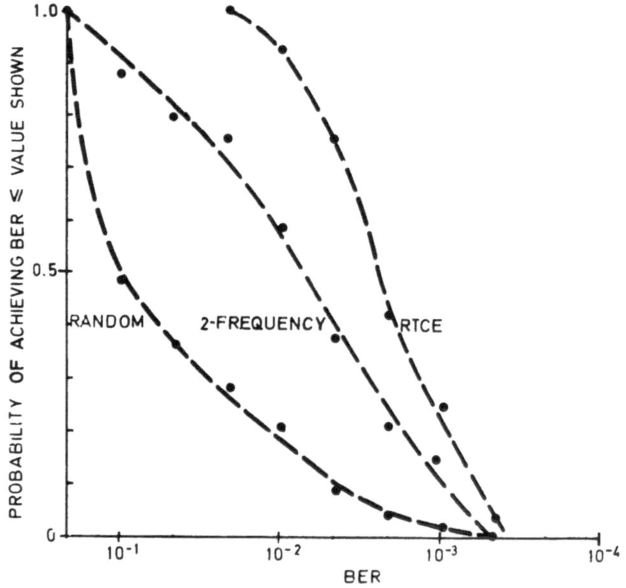

Fig. 13. Effect of different frequency selection algorithms upon the probability of achieving a given BER: Binary FEK telegraphy at 75 bits/sec and Tx power of 150 W. Cornwall - The Hague, 26-10-76.

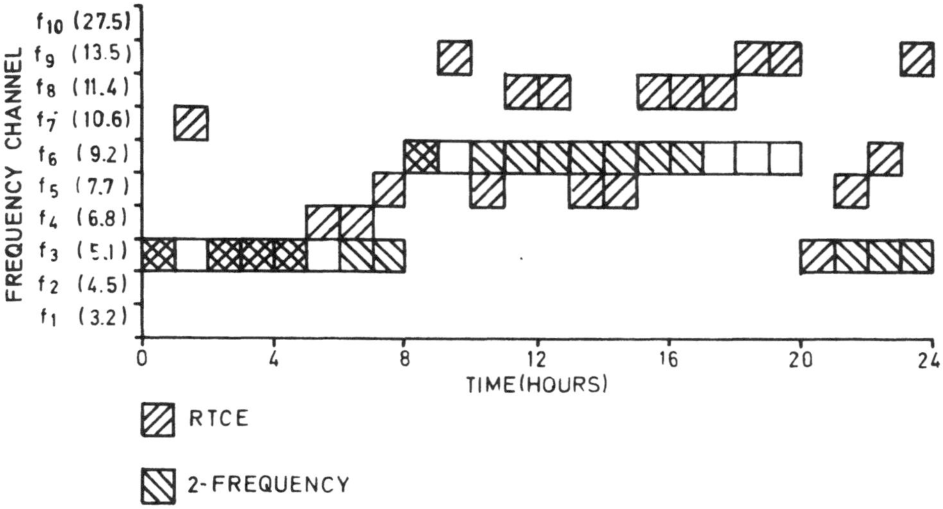

Fig. 14. Comparison of circuit availabilities for 2-frequency and RTCE frequency selection: medium-speed data (parallel sub-channel modem) at 1200 (x2) bits/sec and average Tx power of 250 W. Oslo - The Hague, 22-6-76.
(Shaded Areas indicate BER $\leqslant 10^{-2}$)

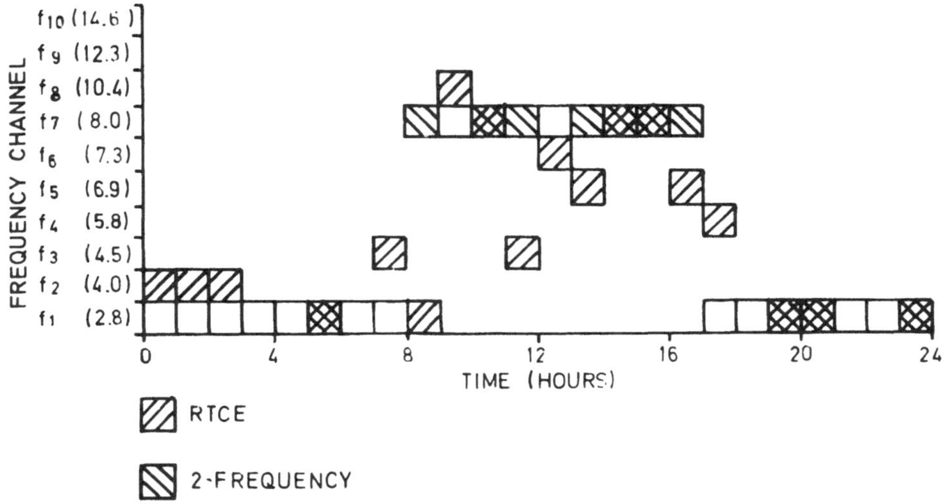

Fig. 15. Comparison of circuit availabilities for 2-frequency and RTCE frequency selection: medium-speed data (parallel sub-channel modem) at 1200 (x2) bits/sec and average Tx power of 250 W.
Cornwall - The Hague, 10-11-76.
(Shaded Areas indicate BER $\leqslant 10^{-2}$)

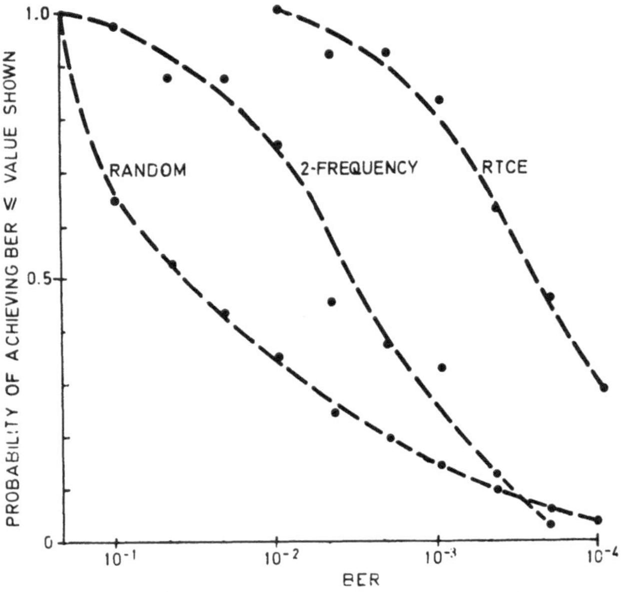

Fig. 16 Effect of different frequency selection algorithms upon the probability of achieving a given BER: medium-speed data (parallel sub-channel modem) at 1200(x2) bits/sec and average Tx power of 250 W. Oslo - The Hague, 22-6-76.

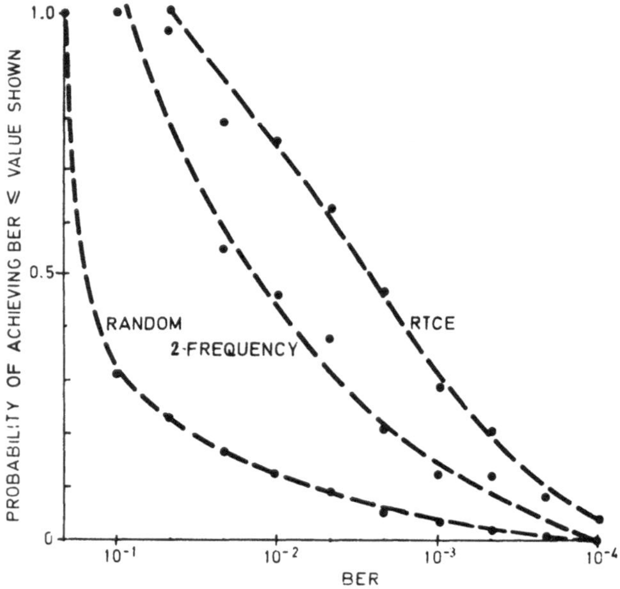

Fig. 17. Effect of different frequency selection algorithms upon the probability of achieving a given BER: medium-speed data (parallel sub-channel modem) at 1200(x2) bits/sec and average Tx power of 250 W. Cornwall - The Hague, 10-11-76.

SOME RESULTS IN SIMULTANEOUS DETECTION AND ESTIMATION

Jurgen O. Gobien

U. S. Air Force Institute of Technology

INTRODUCTION

We will consider statistical hypothesis-testing problems in which either or both hypotheses are composite. The object is to optimally estimate the uncertain parameters while detecting which hypothesis is true; the viewpoint is Bayesian, so that the parameters are treated as random variables. A certain amount of freedom in choosing a-priori densities will be assumed. We shall see that the optimal detection statistic is readily decomposed in a fashion which explicitly includes the a-posteriori densities of the parameters. Two applications of the decomposition are presented as examples. The first is detection of a narrowband signal of unknown phase in white Gaussian noise; its solution is well known, and the problem serves as a straightforward illustration. The second is detection of a known signal in Gauss-Markov noise of unknown spectral parameters. The solution is more complicated and is believed to be original.

THE DETECTOR/ESTIMATOR RELATIONSHIP

In the most general case to be considered, we observe an element of a preprobability space $y\varepsilon(\mathcal{Y},\mathcal{A})$. Under statistical hypothesis H_o, observations are generated in accordance with some member of the family of probability measures $\{P_\eta,\eta\varepsilon\mathcal{N}\}$. Under H_1 the same is true for a different family $\{P_\theta,\theta\varepsilon\Theta\}$. The problem is to make a forced choice between H_o and H_1 while simultaneously estimating parameter θ or η as appropriate. If $\mathcal{Y}$ is finite-dimensional and the measures are continuous, then the problem is specified by giving conditional densities

$$H_o : y \sim f_o(y|\eta) \qquad y\varepsilon\mathcal{Y},\eta\varepsilon\mathcal{N}$$

$$H_1 : y \sim f_1(y|\theta) \qquad y\varepsilon\mathcal{Y},\theta\varepsilon\Theta \tag{1}$$

Notation is simililfied by using a subscript to condition on hypothesis and letting the arguments indicate random variables. The parameters θ and η are considered random with a-priori densities $f_1(\theta)$ and $f_o(\eta)$ respectively.

Estimation. We will start by treating the estimation problem conditioned on a hypothesis (say, H_1). The most general Bayesian "estimator" is the a-posteriori density

$$f_1(\theta|y) = \begin{cases} \dfrac{f_1(y|\theta)\ f_1(\theta)}{f_1(y)} \\ K_1(y)\ f_1(y|\theta)\ f_1(\theta) \end{cases} \tag{2}$$

from which any point estimate is easily made. Suppose one can factor

$$f_1(y|\theta) = g_1[t_1(y),\theta]\ G_1(y) \tag{3}$$

so that $t_1(y)$ is a sufficient statistic (presumably of smaller dimension than y) for estimating θ. This can for example be done if the conditional density is of the exponential class. The probability density obtained by normalizing the first factor of (3),

$$f_1^c(\theta;\gamma) = K_1(\gamma)\ g_1[\gamma,\theta] \tag{4}$$

reproduces in functional form if used an an a-priori density in Bayes' rule (2) and is called a conjugate prior density [1, p. 50]. γ is a parameter which indexes the family of conjugate prior densities.

Suppose the a-priori density is chosen to be $f_1^c(\theta;\gamma^o)$; then the a-posteriori density is $f_1^c(\theta;\gamma^1)$, where parameter γ^1 depends only on γ^0 and the sufficient statistic $t_1(y)$. Since this density is completely characterized by the finite-dimensional parameter γ^1, it is tractable to retain it in its entirety rather than computing a point estimate. This eliminates the necessity of specifying an optimality criterion for estimation.

If the observation is a continuous-parameter random process $\{y_t, t\varepsilon T\}$, then one can in similar fashion find an a-posteriori density based on a finite number of time samples or generalized Fourier coefficients of y_t. Under suitable conditions, it is possible to obtain a limiting form for this density as the number of samples grows without bound. This is particularly true for Gaussian processes where conditions for singularity of measures and closed form expressions for Radon-Nikodym derivatives are well known. Suppose that the family of measures P_θ is dominated by P_* and that y is a finite-dimensional vector obtained from y_t as above. Then Bayes' rule (2) is unchanged if written

$$f_1(\theta|y) = K(y)\,\frac{f_1(y|\theta)}{f_*(y)}\,f_1(\theta) \tag{5}$$

If the number of components of y is allowed to grow in a suitable fashion, then the middle term of (5) converges to the Radon-Nikodym derivative dP_θ/dP_* and the a-posteriori density becomes

$$f_1(\theta|y_t) = K(y_t)\,\frac{dP_\theta}{dP_*}\,f_1(\theta) \tag{6}$$

Detection. Suppose the estimation problem for each composite hypotheses has been solved by finding the a-posteriori density as outlined above. For a wide range of criteria, the optimal detection statistic is the marginal likelihood ratio and the test is

$$\Lambda(y) \triangleq \frac{f_1(y)}{f_0(y)} \underset{D_0}{\overset{D_1}{\gtrless}} \eta \tag{7}$$

where η is a threshold, $f_1(y)$ and $f_0(y)$ are marginal densities, and the notation is self explanatory. Using Bayes' rule (2) and its analog for H_0, the marginal densities in (7) may be rewritten

$$\Lambda(y) = \frac{f_1(\theta)}{f_1(\theta|y)}\,\frac{f_o(\eta|y)}{f_o(\eta)}\,\Lambda(y|\theta,\eta) \tag{8}$$

where

$$\Lambda(y|\theta,\eta) = \frac{f_1(y|\theta)}{f_0(y|\eta)} \tag{9}$$

Thus, the composite-hypotheses detection statistic is explicitly written in terms of solutions to the two estimation problems (conditioned on H_0 and H_1) and a parameter-known (simple-hypotheses) likelihood ratio. Since (8) cannot be a function of θ or η, its right-hand side may be evaluated at arbitrary values of those parameters.

This decomposition is of particular interest in problems which admit sufficient statistics for estimation, and is discussed in detail in [2] and [3]. In many problems of interest, a limiting form for each term of (8) is known; this in turn yields a limiting form $\Lambda(y_t)$ for the composite-hypothesis detection statistic.

Rather than pursue a general development, we will now illustrate the concepts and procedures by solving two examples.

EXAMPLE 1.

Consider the problem of detecting a narrowband signal of unknown phase in white Gaussian noise while simultaneously estimating its phase:

$$\begin{aligned} H_0 &: y_t = w_t \\ & \qquad\qquad 0 \le t \le T \\ H_1 &: y_t = s(t,\theta) + w_t \end{aligned} \tag{10}$$

where w_t has spectral height $N_0/2$ and where

$$s(t,\theta) = s_I(t) \cos\theta - s_Q(t) \sin\theta \tag{11}$$

The phase θ has a-priori pdf $f_1(\theta)$ which will shortly be chosen to be of the conjugate form. This example is chosen as an illustration because its separate estimation and detection solutions are well known. All limiting (continuous observation) forms are, of course, purely formal. It should however be noted that entirely analogous results can be rigorously obtained using colored Gaussian noise in place of white.

Estimation. We will consider only the limiting form of the a-posteriori density. The dominating measure P_* is taken to be the H_0 measure on y_t. Proceeding formally through the limiting procedure of (5) and (6) yields an a-posteriori pdf (see, e.g., [4, p. 274])

$$f_1(\theta|y_t) = K_1(y_t) \exp\left\{\frac{2}{N_0} \int_0^T y_t\, s(t,\theta)dt - \frac{1}{N_0} \int_0^T s^2(t,\theta)dt\right\} f_1(\theta) \tag{12}$$

The second integral in the exponent is assumed independent of θ. Using (11) and the factorization criterion (3), we find

$$g_1[\theta, \underset{\sim}{t}(y_t)] = \exp\left\{\frac{2}{N_0}\left(\cos\theta \int_0^T y_t s_I(t)dt - \sin\theta \int_0^T y_t s_Q(t)dt\right)\right\} \tag{13}$$

where the sufficient statistic is

$$\underset{\sim}{t}(y_t) = \begin{bmatrix} t_1(y_t) \\ \\ t_2(y_t) \end{bmatrix} = \begin{bmatrix} \frac{2}{N_0} \int_0^T y_t s_I(t)dt \\ \\ \frac{2}{N_0} \int_0^T y_t s_Q(t)dt \end{bmatrix} \tag{14}$$

By inspection of (13), the natural conjugate density is

$$f_1^c(\theta;\underset{\sim}{\gamma}) = \begin{cases} K(\underset{\sim}{\gamma}) \exp\left[\gamma_1 \cos\theta - \gamma_2 \sin\theta\right] \\ K(r) \exp\left[r\cos(\theta-\phi)\right] \end{cases} \tag{15}$$

where r and ϕ are polar coordinates of the parameter $\underset{\sim}{\gamma}$, and where the normalizing constant is

$$K^{-1}(\underset{\sim}{\gamma}) = 2\pi\, I_0[r(\underset{\sim}{\gamma})] \tag{16}$$

This density arises often in phase estimation (see, e.g., [5, p. 90]). It is centered at ϕ and its dispersion is inversely related to r.

Suppose the a-priori density is conjugate with parameter $\underset{\sim}{\gamma}^0$. Using (11) in Bayes' rule (12), we see by inspection that the a-posteriori density has the same form but is indexed by

$$\underset{\sim}{\gamma}^1 - \underset{\sim}{\gamma}^0 + \underset{\sim}{t}(y_t) \tag{17}$$

Thus, computing the a-posteriori density is merely a matter of "updating" $\underset{\sim}{\gamma}$ as described by (17). Figures 1 and 2 illustrate results of a typical computer simulation in which this procudure was applied recursively, each time using the preceding a-posteriori density as a prior. Note that $r = 0$ represents a uniform phase density and $r = \infty$ a delta function at ϕ. The "true" value of θ was $\pi/4$ and the prior was arbitrarily chosen to have $r = 1$ and $\phi = 0$.

<u>Detection</u>. Since H_0 is simple, the limiting form of (8) is

$$\Lambda(y_t) = \frac{f_1(\theta)}{f_1(\theta|y_t)} \Lambda(y_t|\theta) \tag{18}$$

We'll assume that $f_1(\theta)$ is conjugate with parameter $\underset{\sim}{\gamma}^0$, see (15). Then $f_1(\theta|y_t)$ is conjugate with parameter $\underset{\sim}{\gamma}^1$ given by (17). The "signal known exactly" likelihood ratio is well known to be

$$\Lambda[y_t|\theta] = \exp\left\{\frac{2}{N_0} \int_0^T s(t,\theta)\, y_t dt - \frac{1}{N_0} \int_0^T s^2(t,\theta) dt\right\} \tag{19}$$

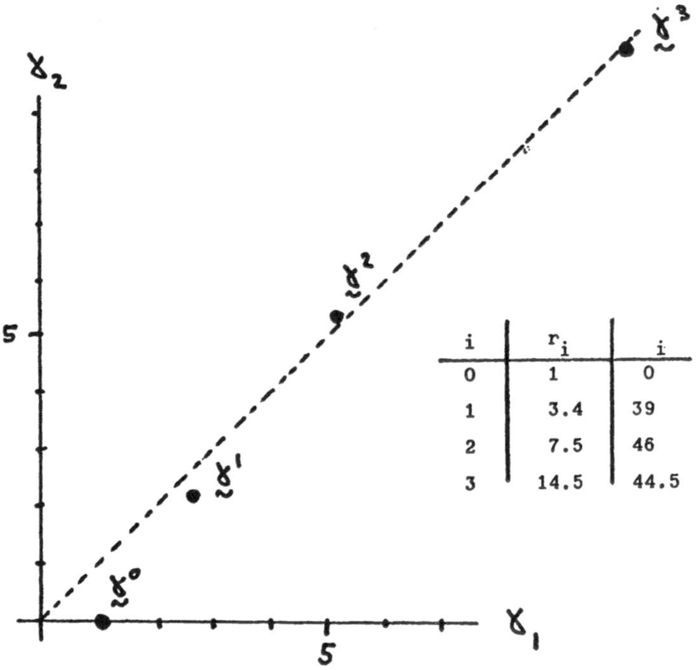

i	r_i	i
0	1	0
1	3.4	39
2	7.5	46
3	14.5	44.5

Figure 1. Evolution of A-posteriori Parameter

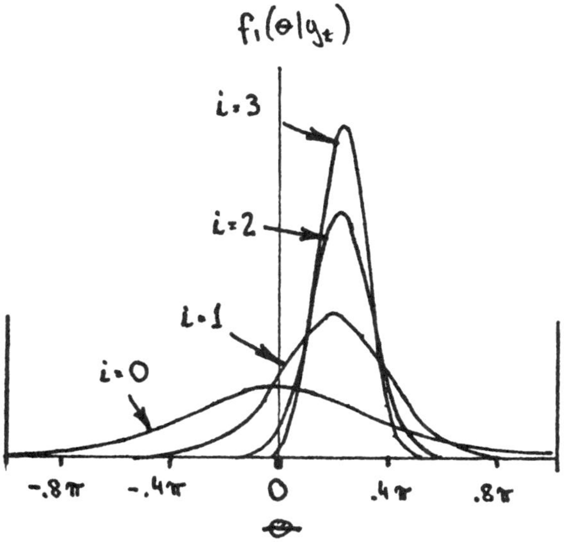

Figure 2. A-posteriori Densities

When these three expressions are used in (18) and suitable cancellation is performed, there remains the detection statistic

$$\Lambda(y_t) = \frac{K(\underset{\sim}{\gamma}^0)}{K(\underset{\sim}{\gamma}^1)} \exp\left(-\frac{E}{N_0}\right) \tag{20}$$

or, using (16),

$$\Lambda(y_t) = \frac{I_o[r(\underset{\sim}{\gamma}^1)]}{I_o[r(\underset{\sim}{\gamma}^0)]} \exp\left(-\frac{E}{N_0}\right) \tag{21}$$

where E is the signal energy. This is, of course, precisely the same statistic obtained if one uses the conventional "generalized L. R." approach

$$\Lambda(y_t) = \int_0^{2\pi} \Lambda(y_t|\theta)\, f_1(\theta)d\theta \tag{22}$$

The only significant difference is that, in the present procedure, we have explicitly obtained a Bayesian estimate during computation of the detection statistic. A pictorial summary of the required processing is shown in Figure 3.

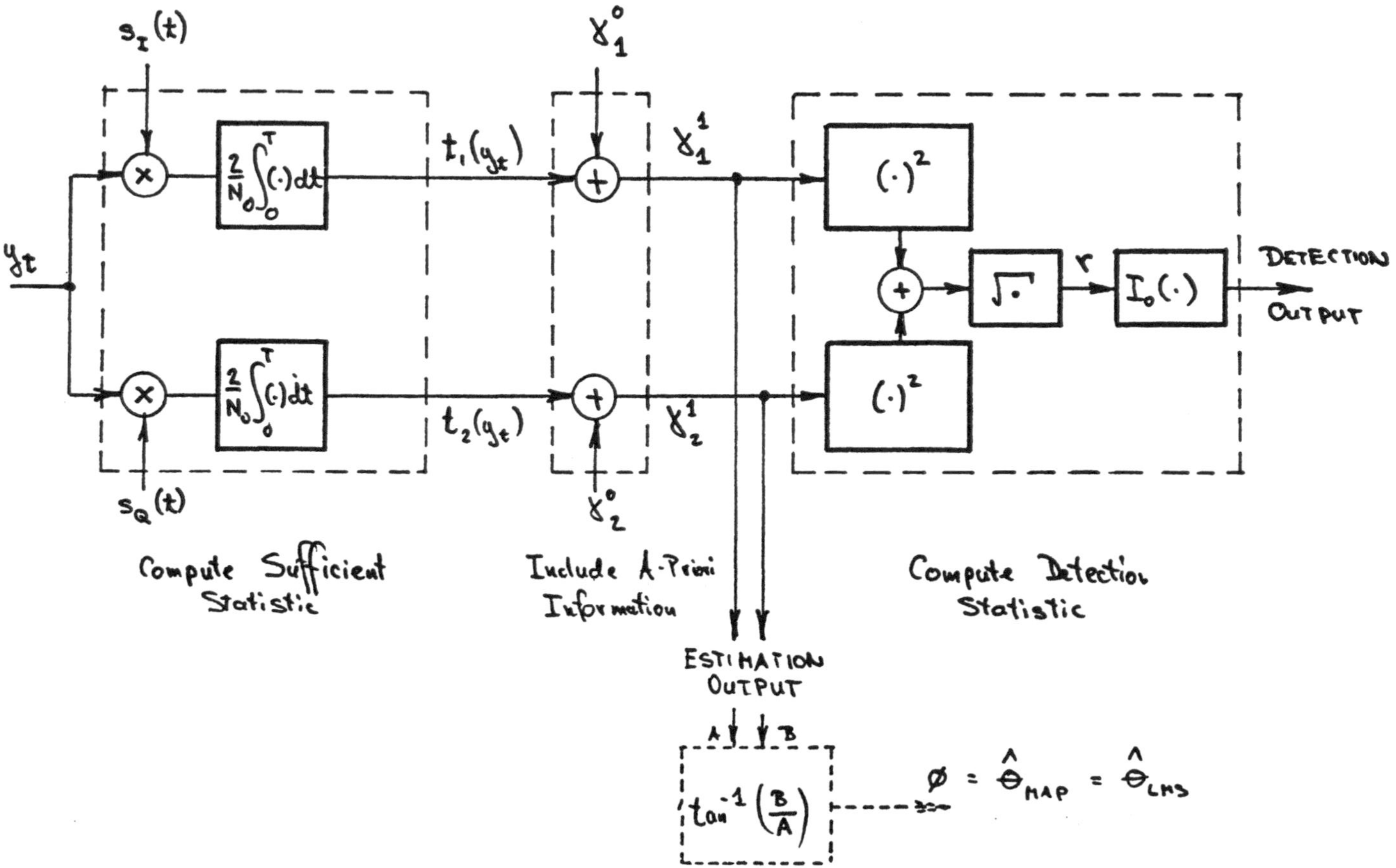

Figure 3. Simultaneous Phase Estimator/Detector

EXAMPLE 2. (Discrete Observations)

In this and the following example, the signal will be exactly known but the noise contains uncertain parameters; this results in both hypotheses being composite. Specifically, we will solve

$$H_0 : y_t = n_t \qquad\qquad 0 \le t \le T \qquad (23)$$
$$H_1 : y_t = s(t) + n_t$$

where s(t) is a known, twice differentiable, finite-energy signal and n_t is stationary, zero-mean, first order Gauss-Markov noise whose autocorrelation and power spectral density functions are, respectively,

$$R_n(\tau) = \frac{a^2}{2q} e^{-q|\tau|} \quad , q > 0 \qquad (24)$$

$$S_n(f) = \frac{a^2}{(2\pi f)^2 + q^2} \qquad (25)$$

In the finite-dimensional (time-sampled) solution, both a^2 and q will be considered uncertain. In the continuous-time solution of Example 3, estimation of a^2 would be mathematically singular and hence only q will be considered uncertain. All results presented in this note are discussed in more detail in [3].

Suppose k evenly spaced time-samples from $\{y_t, 0 \le t \le T\}$ are arranged into a vector $\underset{\sim}{Y}_k$; then

$$H_0 : \underset{\sim}{Y}_k = \underset{\sim}{n}_k$$
$$H_1 : \underset{\sim}{Y}_k = \underset{\sim}{s}_k + \underset{\sim}{n}_k \qquad (26)$$

and components of the noise vector $\underset{\sim}{n}_k$ satisfy the autoregression

$$n_i + \beta n_{i-1} = \rho^{-1} e_i \qquad i = 1,2,\ldots k \qquad (27)$$

where

$$\beta = -\exp(-q\Delta) \tag{28}$$

$$\rho^{-1} = \frac{a^2}{2q}\,[1 - \exp(-2q\Delta)] \tag{29}$$

$$\Delta = T/k \tag{30}$$

The e_i are a sequence of independent unit Gaussian random variables. Note that $\rho > 0$, $-1 < \beta < 0$, and these uncertain noise parameters are unambiguously related to a^2 and q. They will be considered as random variables and are to be estimated while $\underset{\sim}{s}_k$ is being detected. The Markovian nature of the noise is best exploited by temporarily conditioning all expressions on an "initial value" n_0; for ease of notation, this will not be explicitly indicated in the equations.

<u>Estimation</u>. Conditioned on H_0, the observation is noise. Using (27) the joint conditional pdf of the components of $\underset{\sim}{Y}_k$ (conditioned on β, ρ, and y_o) can be written as a product of transition densities

$$f_0(\underset{\sim}{Y}_k|\beta,\rho) = \left[\frac{\rho}{2\pi}\right]^{\frac{k}{2}} \exp\left\{-\frac{\rho}{2}\left[\beta^2\sum_{i=1}^{k} y_{i-1}^2 + 2\beta\sum_{i=1}^{k} y_i y_{i-1} + \sum_{i=1}^{k} y_i^2\right]\right\} \tag{31}$$

Factorization of (31) reveals a four-dimensional sufficient statistic

$$\underset{\sim}{t}_0(\underset{\sim}{y}_k) = \left[\sum_{i=1}^{k} y_{i-1}^2,\ \sum_{i=1}^{k} y_i y_{i-1},\ \sum_{i=1}^{k} y_i^2,\ k/2\right] \tag{32}$$

and shows the natural conjugate class of densities for β and ρ to be

$$f_0^c(\beta,\rho;\underset{\sim}{\psi}) = K(\underset{\sim}{\psi})\ \rho^{\psi_4} \exp\left\{-\frac{\rho}{2}\left[\beta^2\psi_1 + 2\beta\psi_2 + \psi_3\right]\right\} \tag{33}$$

Parameter $\underset{\sim}{\psi}$ indexes these densities just as $\underset{\sim}{\gamma}$ does under H_1, and $K(\underset{\sim}{\psi})$ is a complicated normalizing constant written in the Appendix. Recall that $\rho > 0$ and $-1 < \beta < 0$. Equation (33) has the properties of a pdf if $\underset{\sim}{\psi}$ satisfies $\psi_2^2 - \psi_1\psi_3 < 0$ and $\psi_1, \psi_3, \psi_4 > 0$.

Suppose $\underset{\sim}{\psi}_0$ satisfies these conditions and one uses $f_0^c(\beta,\rho;\ \underset{\sim}{\psi}_0)$ as a prior and (31) as the conditional pdf in Bayes' rule. Then the a-posteriori pdf is $f_0(\beta,\ \rho;\ \underset{\sim}{\psi}_k)$, where

$$\underset{\sim}{\psi}_k = \underset{\sim}{\psi}_0 + \underset{\sim}{t}_0(\underset{\sim}{Y}_k) \tag{34}$$

The sufficient statistic (33), and hence $\underset{\sim}{\psi}_k$ and the a-posteriori pdf, can be computed recursively if components of $\underset{\sim}{Y}_k$ arrive sequentially. This solves the H_0 estimation problem if the a-priori density is natural conjugate. If not, the technique can be modified as shown in [2].

The a-posteriori pdf just found is conditioned on y_0. For large k this may be of little significance. If desired, the conditioning can be eliminated through multiplication by the conditional pdf of y_0; this density is Gaussian with zero mean and variance $[\rho(1-\beta^2)]^{-1}$. As will shortly be illustrated, it is also possible to choose the a-priori pdf from a class which will <u>become</u> natural conjugate after processing y_0.

Under H_1, the noise samples can be reconstructed by subtracting $\underset{\sim}{s}_k$ from $\underset{\sim}{y}_k$. It follows that $f_1(\underset{\sim}{Y}_k|\beta,\rho)$ is found by replacing y_i with $y_i - s_i$ in (31); hence $\underset{\sim}{t}_1(\underset{\sim}{Y}_k)$ is found by making the same substitution in (32). The natural conjugate densities $f_1^c(\beta,\ \rho;\ \underset{\sim}{\gamma})$ have the same form as (33), and their indexing parameter $\underset{\sim}{\gamma}$ is updated using $\underset{\sim}{t}_1(\underset{\sim}{Y}_k)$ in a manner similar to (34). Conditioning on y_0 is accounted for as already discussed.

<u>Detection</u>. The detection statistic is found by substituting both sets of apriori and aposteriori densities, along with the parameter-dependent L.R. given by the ratio of (31) to its analog under H_1, into (8). After considerable cancellation there remains

$$\Lambda(\underset{\sim}{Y}_k) = \frac{K(\underset{\sim}{\gamma}_0)\ K(\underset{\sim}{\psi}_k)}{K(\underset{\sim}{\psi}_0)\ K(\underset{\sim}{\gamma}_k)} \tag{35}$$

where K(.) is the normalizing constant of (33) as given in the Appendix. Note that (35) has a complicated nonlinear dependence on $\underset{\sim}{Y}_k$ and that, in this particular case, the statistics $\underset{\sim}{t}_0(\underset{\sim}{y})$ and $\underset{\sim}{t}_1(\underset{\sim}{y})$ suffice for both estimation and detection.

EXAMPLE 3. (Continuous Observations)

The continuous-time solution is best approached by time-sampling as above, then letting $k\to\infty$. Use of the Karhunen-Loève Theorem is difficult because eigenfunctions and eigenvalues of the noise kernel depend upon the random parameters.

If H_0 or H_1 is assumed true, the remaining estimation problem is straightforward to solve. Consider H_1; Striebel [6] has shown that the a-posteriori pdf converges if $(\mathcal{Y}, \mathcal{A}, P_{\underset{\sim}{\theta}})$ is complete for all $\underset{\sim}{\theta} \in \Theta$, $\{Y_t, 0 \leq t \leq T\}$ is separable and continuous in probability, and $\{P_{\underset{\sim}{\theta}}\}$ is dominated by P_*. The limit of the a-posteriori pdf is found by using the Radon-Nikodym (R-N) derivative $d P_{\underset{\sim}{\theta}}/d P_*$ in place of the conditional pdf in Bayes' rule, or by letting $k\to\infty$ in the sampled solution. Even if both estimation problems are solved in this fashion, convergence of (8) is in general formidable to investigate.

In the example treated here, all measures are Gaussian and hence either equivalent (mutually absolutely continuous) or singular. Necessary and sufficient conditions for equivalence (as well as, in some cases, closed-form expressions for the Radon-Nikodym derivations) are known [7]. Equivalence within the classes $\{P_{\underset{\sim}{\theta}}\}$ and $\{P_{\underset{\sim}{\eta}}\}$ guarantees convergence of the estimators, and equivalence between the classes is sufficient for convergence of (8). Measures on the observation space defined by (33) thru (35) are equivalent if the parameter a^2 is fixed, i.e., if the ratio of spectral density functions approaches unity for large f. Thus a^2 cannot be considered uncertain in the continuous time solution.

Estimation. Assume H_0 true, a^2 known, and q random. The required R-N derivative is (see [7] p. 433)

$$\frac{dP_q}{dP_*} = \left[\frac{2q}{a^2}\right]^{1/2} \exp\left\{-\frac{1}{2a^2}\left[q^2\int_0^T n_t^2\,dt + q\,(n_T^2 + n_0^2 - a^2T)\right]\right\} \tag{36}$$

Equation (36) can be directly employed in Bayes' rule; however, the resulting estimator "batch" processes the observation. A more useful form of the processor is recursive in the sence that it can be terminated at any time instant in [0,T] without that instant being a-priori specified. If s(t) is periodic, the estimator can run indefinitely; when terminated it will yield the a-posteriori pdf. To this end, partition the observation interval into k subintervals

$$[0,T] = \{0\} \cup (0,t_1] \cup \cdots \cup (t_{k-1}, t_k] \tag{37}$$

Now write (36) for the interval $0 < t \leq t_i$, conditioned on y_0; this can be accomplished by dividing by the pdf of y_0, and yields

$$\frac{dP_q}{dP_*}[y_t, 0<t\leq t_i | y_0] = \sqrt{2\pi}\exp\left\{-\frac{1}{2a^2}\left[q^2\int_0^{t_i} y_t^2\,dt - q(a^2 t_i - a^2 0 + y_0^2 - y_{t_i}^2)\right]\right\} \tag{38}$$

An analogous expression is obtained for the interval $t_{i-1} < t \leq t_i$ and we observe that, except for a normalizing constant

$$\frac{dP_q}{dP_*}\left[y_t, 0<t\leq t_k | y_0\right] = \prod_{i-1}^{k} \frac{dP_q}{dP_*}\left[y_t, t_{i-1}<t\leq t_i | y_{i-1}\right] \tag{39}$$

This multiplicative form is precisely analogous to the use of transition densities in the discrete case, and permits recursive processing of observed sample functions on contiguous half-open intervals. It is employed in Bayes' rule

$$f_0(q|y_t, 0<t\leq t_k) = K_0 \frac{dP_q}{dP_*}[y_t, 0<t\leq t_k]\, f_0(q) \tag{40}$$

where the conditioning on y_0 is again omitted for the sake of clarity. Thus, (38) and (39) can be factored to find sufficient statistics and recognize the natural conjugate prior density. The former is

$$\underset{\sim}{t}_0(y_t, t_{i-1}<t\leq t_i) = \begin{bmatrix} \int_{t_{i-1}}^{t_i} y_t^2\,dt \\ a^2(t_i - t_{i-1}) + y_{t_{i-1}}^2 - y_{t_i}^2 \end{bmatrix} \tag{41}$$

and, for sequential observations, is updated additively

$$\ell_0(y_t, 0<t\leq t_k) = \sum_{i=1}^{k} \ell_0(y_t, t_{i-1}<t\leq t_i) \tag{42}$$

The natural conjugate class of densities for q is truncated Gaussian,

$$f_0^c(q;\underset{\sim}{\psi}) = K(\underset{\sim}{\psi}) \exp\left\{-\frac{1}{2a^2}\left[q^2\psi_1 - q\psi_2\right]\right\} \quad q>0,\ \psi_1>0 \tag{43}$$

Its normalizing constant is

$$K(\psi) = \left[\frac{\psi_1}{a^2}\right]^{1/2} \Omega\left[\frac{\psi_2}{2a\sqrt{\psi_1}}\right] \tag{44}$$

where $\Omega(.)$ is the logarithmic derivative of the unit Gaussian distribution,

$$\Omega(x) = \frac{\exp\left(-\frac{1}{2}x^2\right)}{\int_{-\infty}^{x} \exp\left(-\frac{1}{2}\lambda^2\right)d\lambda} \tag{45}$$

Suppose (for simplicity only) that the a-priori pdf (given y_0) of q is natural conjugate with parameter $\underset{\sim}{\psi}_0$. Using (43) and (38) in Bayes' rule (40), the a-posteriori pdf at time t_i is found to be natural conjugate with

$$\underset{\sim}{\psi}_i = \underset{\sim}{\psi}_0 + t_0(y_{[o,t_i]}) \tag{46}$$

Knowledge of the a-posteriori density is sufficient to make any desired point estimate of q under H_0.

Estimation under H_1 is, as before, analogous to the above. One merely replaces H_0 with H_1, $y_t - s(t)$, and $\underset{\sim}{\psi}$ with $\underset{\sim}{\gamma}$ throughout.

Detection. The a-posteriori pdf's are conditioned on y_0. This must be accounted for prior to the use of (2). One convenient way of doing so is to process a point observation y_0. Suppose that the a-priori pdf of q is the same under both hypotheses and has the form

$$f_Q(q) = K\, q^{-1/2} \exp\left\{-\frac{1}{2a^2}\left[Aq^2 - Bq\right]\right\} \tag{47}$$

where A and B are constants. After y_0 is processed, the a-posteriori pdf's will be natural conjugate with parameters

$$\underset{\sim}{\psi} = \left[A\ ,\ B - 2y_0^2\right]^T \tag{48}$$

$$\underset{\sim}{\gamma} = \left[A\ ,\ B - 2(y_0 - s_0)^2\right]^T \tag{49}$$

but will not be conditioned on y_0. Thus the effect of y_o may be "washed out" by choosing a prior in the class (47); when the recursive processing of the estimators begins, the densities on q will have the more tractable natural conjugate form. Since the ratio of prior densities is unity and the a-posteriori densities are natural conjugate, the continuous-time version of (8) may in this case be written

$$\Lambda(y_t) = \frac{f_0^c(q;\underset{\sim}{\psi}_i)}{f_1^c(q|\underset{\sim}{\gamma}_i)}\ \Lambda(y_t|q) \tag{50}$$

The last term is the simple-hypothesis L.R. for arbitrary fixed q. This is well known; appropriate expressions are given as (54) and (55) in the Appendix. The form of the conjugate pdf's is given by (43). Using (46) through (49), $\underset{\sim}{\psi}_i$ and $\underset{\sim}{\gamma}_i$ may be written in terms of the parameters A and B of the a-priori density and the sufficient statistics $\underset{\sim}{t}_0(y_t)$ and $\underset{\sim}{t}_1(y_t)$.

When the indicated substitutions are made the argument q does (as it must) cancel and one is left with the detection statistic

$$\Lambda(y_t) = \left[\frac{A + \int_0^T y_t^2\,dt}{A + \int_0^T [y_t - s(t)]^2 dt}\right]^{1/2} \Omega\left[\frac{B + a^2T - y_0^2 - y_T^2}{2a\left(A + \int_0^T y_t^2\,dt\right)}\right]$$

$$\Omega^{-1}\left[\frac{B + a^2T - [y_0 - s(0)]^2 - [y_T - s(T)]^2}{2a\left\{A + \int_0^T [y_t - s(t)]^2 dt\right\}}\right] \tag{51}$$

$$\exp\left\{a^{-2}\left[s'(0)y_0 + s'(T)y_T - \int_0^T s''(t)y_t dt - \frac{1}{2}\int_0^T [s'(t)]^2 dt\right]\right\}$$

This is compared to a threshold in the usual way. Though complicated, it is much easier to compute than is the discrete result.

CONCLUSIONS

The general estimator/detector structure is appealing because, in addition to explicitly relating estimation and detection, it is to a large extent free of assumptions about optimality criteria. Even the existence of sufficient statistics is not necessary in (8); they are needed only to obtain recursive and continuous-time solutions. The technique presented here has been used to solve "conventional" composite signal-hypothesis problems. In each case, the detector realization was identical to that obtained using the more common "generalized likelihood ratio" approach while an estimate of the uncertain parameter was obtained as a fringe benefit.

For first order Gauss-Markov noise, the quadratic content of the observed sample function and the difference of the squares of its end-point values are sufficient for estimating the spectral pole of the noise under the "noise alone" hypothesis. Under the "signal plus noise" hypothesis, sufficient estimation statistics are the same but the signal must first be subtracted from the observation. Additional statistics are necessary for detection; these appear in the last factor of (56). Similar results have been obtained for stationary M-th order Gauss Markov noise with all M spectral poles unknown. These may be found in [3]: The quadratic content of the first M-1 derivatives of the observation (or the observation minus the signal) are sufficient for estimation, but the detection statistics are again much more complicated. No attempt has been made to evaluate the performance of these detectors.

APPENDIX

If ψ_4 is an integer multiple of 1/2, the normalizing constant of (33) is given by the following expressions. Though not impressive for the insight they provide, they are certainly suitable for machine evaluation. Let,

$$n = \psi_4$$

$$m = \psi_4 - \frac{1}{2}$$

$$D = \psi_1\psi_3 - \psi_2^2$$

$$P = \psi_1 - 2\psi_2 + \psi_3$$

$$N_i = \frac{(i-1)!\,i!}{(2i)!}$$

$$M_i = \frac{(2i)!}{(i!)^2}$$

Then

$$K^{-1}(\psi) = \frac{2(2n)!}{n!}\left(\frac{\psi_1}{2D}\right)^n \left\{\frac{1}{2D}\left[(\psi_1 - \psi_2)\sum_{i=1}^{n} N_i\left(\frac{4D}{\psi_1 P}\right)^i + \psi_2 \sum_{i=1}^{n} N_i \left(\frac{4D}{\psi_1\psi_3}\right)^i\right] + \frac{1}{\sqrt{D}}\left[\tan^{-1}\left(\frac{\psi_1-\psi_2}{\sqrt{D}}\right) + \tan^{-1}\left(\frac{\psi_2}{\sqrt{D}}\right)\right]\right\} \tag{52}$$

provided that n is an integer; or, if m is an interger, then

$$K^{-1}(\psi) = \frac{(m+1)!m!2^{3(m+1)}\Gamma(m+3/2)\psi_1{}^m}{\sqrt{2}D^{m+1}(2m+2)!}$$

$$\left[\frac{(\psi_1-\psi_2)}{\sqrt{P}} \sum_{i=0}^{m} M_i \left(\frac{D}{4\psi_1 P}\right)^i\right.$$

$$\left. + \frac{\psi_2}{\sqrt{\psi_3}} \sum_{i=0}^{m} M_i \left(\frac{D}{4\psi_1\psi_3}\right)^i \right] \qquad (53)$$

The L.R. required in (31) may be found in any text on signal detection. Care must be exercises to retain all terms of the likelihood ratio, including those which are usually lumped into the detection threshold:

$$\ln \Lambda(y_t|q) = a^{-2}\left\{\int_0^T [q^2 s(t) - s''(t)]y_t dt\right.$$

$$\left. + [qs(0) - s'(0)]y_0 + [qs(T) + s'(T)]y_T\right\} - \frac{d}{2} \qquad (54)$$

where d is the "detectability index"

$$d = a^{-2}\left\{q^2 \int_0^T s^2(t)dt + q\,[s^2(T) + s^2(0)] + \int_0^{T_1} [s'(t)]^2 dt\right\} \qquad (55)$$

and prime denotes differentiation. When these expressions and the estimation solutions are used in (31), considerable simplification occurs. Using (23) for the normalizing constants of the a-posteriori pdf's, one obtains the detection statistic of (51).

REFERENCES

1. H. Raiffa and R. Schlaiffer, Applied Statistical Decision Theory, MIT Press, Cambridge, 1961.

2. T. G. Birdsall and J. O. Gobien, "Sufficient Statistics and Reproducing Densities in Simultaneous Sequential Detection and Estimation," IEEE Trans. Information Theory, IT-19, No. 6, pp 760-768, Nov 1973.

3. J. O. Gobien, "Simultaneous Detection and Estimation: The Use of Sufficient Statistics and Reproducing Probability Densities," PhD Dissertation, The University of Michigan, Ann Arbor, 1973.

4. H. L. Van Trees, Detection, Estimation, and Modulation Theory Part I, Wiley, New York, 1968.

5. A. Viterbi, Principles of Coherent Communication, McGraw-Hill, New York, 1966.

6. C. T. Stribel, "Densities for Stochastic Processes," Annals of Math Statistics, vol 30, pp 559-567, 1959.

7. J. Hajek, "On Linear Statistical Problems in Stochastic Processes," Czeck Math Journal, vol 12, pp 404-443, 1962.

PHASE COMPENSATION RECEIVERS FOR OPTICAL COMMUNICATION

STANLEY R. ROBINSON

Air Force Institute of Technology
School of Engineering
Department of Electrical Engineering
Wright-Patterson AFB, Ohio 45433

I. Introduction

The characteristics of optical communication systems, coupled with the increased demand of urban population centers will bring about an increased use of optical systems in short haul, high rate, line of sight communication links, especially in situations where other alternatives, such as microwave or "wired" communication systems, are not practical. The magnitude of this use will, to a large extent, depend on the outage time that such systems suffer when weather conditions are sufficiently severe, e.g. the reduced visibility conditions caused by clouds, fog or haze. This paper considers phase compensation receivers as a means of improving the performance of optical communication systems which may be operated in a low visibility environment. Although a great deal has been written on the use of phase compensation for such applications as imaging through the earth's atmosphere and spatially concentrating an optical beam in the far field, relatively little has been written on the application of improving the reliability of optical communication. The reader is referred to the literature for other applications as well as discussions of implementation issues for the hardware [1-5].

The general breakdown of the receiver structure for optical communication applications is illustrated in Figure 1. We are free to choose the receiver optics (and associated measurements), the phase estimator that processes the detector outputs and the processor for the communication function. Of these, our primary concern will be with the receiver optics and the associated processing for phase control.

In the following sections, the effects of low visibility channels on the received fields are first presented. We then discuss the phase control needed to improve communication performance. Finally, we present the communication performance of such an adaptive receiver. Our results illustrate that for certain applications of interest, an adaptive receiver would allow reliable communication in a low visibility environment while a nonadaptive receiver would be inoperable.

II. Effects of Low Visibility Channels on Received Fields

The effect of a low visibility channel, composed of clouds, fog, haze, etc., on a signal field propagating through it is to corrupt and attenuate the signal field in space, frequency and time. Generally, a plane wave signal field transmitted into such a channel will lose its spatial coherence by the time it reaches the receiver; this change of coherence and the possibility of restoring it through phase compensation are our major concern.

The Doppler frequency spread of the input signal due to the relative motion of the scattering particles is also important. The reciprocal of this spread is a measure of channel coherence time [6, 7]. It provides an estimate of the time over which the channel characteristics do not change significantly and hence of the time over which an adaptive receiver may reasonably make measurements.

Time spreading or frequency-selective fading of the transmitted signal is also encountered in low visibility channels. For simplicity, we will assume that the signal bandwidth is much less than the channel coherence bandwidth [6, 8]. Hence time spreading will not be of direct interest to us.

There is however a relationship between channel coherence time and time spreading effects which is quite important if phase compensation is to be feasible. First, if such an adaptive receiver is to be useful in improving communication performance, the bit interval τ should be much less than the channel coherence time T_c. To see this, assume the bit interval is on the order of the coherence time. Presumably, if there is enough signal to track phase over a coherence time, then there is enough signal to transmit that bit reliably and the added complexity of a phase compensator is not really justified. At the same time, the bit interval τ should be larger than the time spreading effect in order to avoid intersymbol interference (or frequency selective fading) [6-10]. Channels for which both conditions are met are known as underspread channels [6, 7]. Note however that the underspread requirement need not apply to the entire aperture field, but rather to the smallest portion of the field that is resolvable by the receiver. If the resolvable portions of the received field are locally underspread, then presumably with added receiver complexity, the individual portions could be phase compensated to yield an overall improvement in communication performance.

To this point, only crude parameters such as coherence time and coherence bandwidth have been used to describe the effects of the channel on the received field. In addition, we will assume that the complex envelope of the received field is a zero mean, Complex Gaussian Random Process in space and time [10-13]. This model is suggested by the large number of independent particles that scatter the light which contributes to the signal field in a low visibility channel and can be partially justified by an application of the central limit theorem. For simplicity, we assume that the real and imaginary parts of the complex envelope are independent and identically distributed; the complete specification of the statistics of the signal field is then given by:

$$\begin{aligned} &E[U_s(\bar{r},t)] = 0 \\ &E[U_s(\bar{r},t)U_s(\bar{r}',t')] = 0 \qquad (1) \\ &E[U_s(\bar{r},t)U_s^*(\bar{r}',t')] = R_s(\bar{r},t,\bar{r}'t') \end{aligned}$$

where $\bar{r}$ is a two-dimensional vector in the receiver's aperture plane. The set of conditions in (1) imply that $R_s(\bar{r},t,\bar{r}',t')$ is a real function and is twice the correlation function of either the real or imaginary part of the field.

It is useful to have a discrete coherence cell representation for the field in a measurement plane. Such a description is motivated by the following. Although equations (1) specify the field statistics completely, the determination of $R_s(\overline{r},t,\overline{r}',t')$ in detail is quite difficult, so that practically it can only be described (crudely) by simple parameters. One such measure, known as the coherence distance D_c, is defined by:

$$R_s(\overline{r},t,\overline{r}',t') = 0 \quad ; \quad |\overline{r}-\overline{r}'| > D_c, \text{ all } t,t' \tag{2}$$

It is the minimum distance beyond which samples of the field are statistically independent. The coherence cell model for the field is a simple approximation in which it is assumed that the field is constant over an area (known as a coherence area) and is statistically independent from the field at other coherence cells. For simplicity, we use (2) as an indication of the coherence distance D_c.

Conceptually the aperture plane can be broken into areas D_c on a side and the field over each cell can be assumed to be spatially constant and independent from the field in other cells. The cells are labeled with the index i and the field at the ith cell is denoted by $A_i(t)\exp[+j\phi_i(t)]$ where it is explicitly indicated that the field at a cell is still a temporal random process. Hereafter, we shall suppress the (t) notation i the cell fields; it is to be understood that the temporal dependence is still present.

It is instructive to see how coherence distance is related to other obvious but crude properties of the channel. Consider the signal field illustrated in Figure 2. Note that because of scattering the signal energy incident on the receiver aperture has a wide angular distribution. For that case, it is reasonable to model the aperture plane field by a plane wave decomposition

$$U_s(\overline{r},t) = \sum_i a_i \exp[+j\overline{k}_i\overline{r}] \tag{3}$$

where the a_i are statistically independent, identically distributed complex Gaussian random variables with

$$\begin{aligned} E[a_i] &= 0 \\ E[|a_i|^2] &= \sigma_a^2 \end{aligned} \tag{4}$$

and the sum is over all directions of propagation (indicated by the $\bar{k}$ vectors) from which we expect signal energy. To relate the problem to coherence distance, we compute the correlation function of the field

$$E[U_s(\bar{r},t)U_s^*(\bar{r}',t)] = \sigma_a^2 \sum_i \exp[+j\bar{k}_i\cdot(\bar{r}-\bar{r}')] = R_s(\bar{r},t,\bar{r}',t) \qquad (5)$$

where we have exploited the properties of the a_i's in the calculation. We further assume that the energy is incident from all directions within a cone with half-angle Θ_0 so that the sum may be approximated by an integral [14-15] to yield the correlation function:

$$R_s(\bar{r},t,\bar{r}',t) = \sigma_a^2 \int_{|\bar{k}|<k\sin\Theta_0} \exp[+j\bar{k}\cdot(\bar{r}-\bar{r}')]d\bar{k} = \pi(k\sin\Theta_0)^2\, \sigma_a^2 \left[\frac{J_1(k|\bar{r}-\bar{r}'|\sin\Theta_0)}{k|\bar{r}-\bar{r}'|\sin\Theta_0}\right] \qquad (6)$$

If we define D_c as the width of the main lobe of $R_s(\bar{r},t,\bar{r}',t)$, then to a good approximation:

$$D_c \approx \frac{0.610\lambda}{\sin\Theta_o} \qquad (7)$$

Equation (7) gives us a rough measure of coherence cell sizes in the receiver aperture as related to the crude angular spectrum of the signal propagating from the channel.

III. Phase Control for Improved Communications Performance

Consider a direct detection receiver designed for digital communication through a "clear" atmospheric channel, i.e. clear weather and little turbulence. As shown in Figure 3, it consists of an optical filter, lens and a photodector [13, 16]. The detector will respond to the background noise field incident on the aperture as well as the signal field, so the optical filter is used to limit the effects of the noise field. Since background noise has an extremely large angular distribution, it will be approximately uniformly spread over the focal plane, as shown in Figure 3. If the expected signal field is a plane-wave, the field stop pinhole should be roughly the size of a diffraction limited spot, so that the detector intercepts the signal field while still discriminating spatially against background noise. Here, the receiver exploits the extreme spatial coherence of the signal field to discriminate against background noise, i.e. the receiver responds to the only spatial mode in which a signal was expected.

Suppose meteorological conditions change and the increased particle content in the atmosphere (fog, haze, etc.) causes the signal to be corrupted in space, time and frequency. Ignoring issues of distortion in time and frequency, the result would be that the power in the "unscattered" plane wave component (which the system in Figure 3 was designed to receive) is attenuated by $\exp[-\alpha L]$, where α is the extinction coefficient and L is the path length [17-21]. A crude rule of thumb at visible wavelengths is that the path loss is roughly 10^{-2} when the path length equals the visibility [20]. Thus, there may be little energy in the unscattered field so that the system in Figure 3 is doomed to failure in such low visibility conditions, even though there may be significant signal energy in the scattered component of the received field [16, 22-32].

If we suppose the signal field has a conical angular distribution, it seems reasonable that the field stop should be opened to allow the detector to intercept all the signal power; however, the price paid to collect all the signal power is the increase in background noise which is also collected. This suggests that if a method could be used to improve the spatial coherence of the signal field, then an improvement in performance could be realized by a background noise limited direct-detection receiver that exploits the spatial coherence of the signal field.

Thus, a receiver which uses direct detection in the communication measurement portion of the receiver should use a phase control that maximizes the signal power in an Airy-disc-sized detector. For such a performance measure, the phase control can be shown to be the minimum-mean-square-error (MMSE) estimate of the aperture field phase, when the estimation error is "small enough" [33]. Such a phase control can be viewed as a mode compressor which transforms the spatially corrupt signal into essentially one mode and thus allows the receiver to again spatially discriminate against the background noise fields. Although not discussed here, similar comments on phase compensation as a mode compressor apply to improve the communication performance of homodyne or heterodyne systems, which respond to only one signal mode [11, 42].

Thus, we conclude that closed-loop MMSE phase estimators provide the phase control required to improve communication performance. The structure and performance of such estimators have been presented in an earlier paper [34]. In order to use these results, we will assume that a coherence cell model is used to represent the aperture field. In addition, we suppose that the cell amplitudes $\{A_i\}$ and any Doppler frequency offset are known exactly at the receiver. This implies that the channel coherence time discussed earlier is now due just to the phase random processes of each coherence cell; we hereafter denote it as phase coherence time T_ϕ.

IV. Low Visibility Communication Performance

This section develops the relation between phase compensated signal detector power and communication performance. Both direct detection and homodyne communication receivers are presented. The signaling format considered will be binary pulse position modulation with the performance measure the probability of a bit error*.

A. Binary Receiver and Performance

Direct detection optimum receiver structures and performance are available when the detector output, conditioned on the message sent, is a Poisson process.+ The Poisson process model is valid in a limiting case commonly known as weakly coherent quantum channels, where the maximum average number of detected signal photons (and noise photons) per space-time mode is small [11]. This assumption is quite reasonable for a wide field of view receiver operating in a low visibility environment. At the other extreme, the Poisson process model is valid when the detector field is known exactly; this corresponds nicely to a phase compensated signal where the field amplitudes are presumed known. Therefore, the model is valid for the uncompensated/compensated comparisons which are of interest.

A reasonable receiver for the Poisson case is a photon counting receiver; its performance is given approximately by

$$P_e \approx \tfrac{1}{2} \exp[-\varepsilon_d] \qquad (8)$$

The error exponent is a nonlinear function of several parameters to be described subsequently; it is discussed in detail in [11, 13, 35]. For our purposes, limiting forms of ε_d are of interest.

$$\varepsilon_d = \begin{cases} \frac{\alpha\gamma}{4} \; ; & \gamma < 1.0 \quad (9a) \\ \alpha\zeta \; ; & \gamma > 1.0 \quad (9b) \end{cases}$$

where

ζ is a constant ranging from 0.1 to 1, depending weakly on γ for $1 < \gamma < 10$.

α is the average number of signal photons per bit = $n_s'[\tau]$.

γ is the ratio of average number of detected signal photons to detect background noise photons plus dark counts = $\frac{n_s'}{n_b'+n_d}$.

τ is the bit interval length.

When (9a) is valid, the communication receiver is said to be background noise-dark current limited, while for (9b) to be valid, the receiver is signal shot noise limited [11, 13, 35].

When a homodyne receiver is used to communicate, the optimum receiver and its performance is also easily obtained from classical results [13, 36, 37],

$$P_e \approx \tfrac{1}{2} \exp[-\varepsilon_h] \tag{10}$$

$$\varepsilon_h = \frac{P_s[\tau]\eta}{hf_o} \tag{11}$$

where: P_s is the signal power in the single spatial mode detected by the homodyne receiver.

τ is the bit interval length.

η is the detector quantum efficiency.

hf_o is the energy in a photon.

---

*Throughout this paper, it is assumed that there is no intersymbol interference so that one shot probability is a useful performance measure.

+We assume thermal noise in the detector is negligible.

The reader should be aware that "known signal" or conditional error expressions have been used, even though the field in the aperture is a Gaussian process. The conditional statements, which are quite different from the Gaussian signal detection results, are used to be consistent with the known amplitude assumption made for the phase estimator.

B. Compensated/Uncompensated Communication

Consider the idealized communication receiver structure shown in Figure 4. The communication receiver is either homodyne, in which case the hardware in the dotted box is included, or it is direct detection. The optical filter has a bandwidth of W_0 Hertz; both the local oscillator complex envelope and the size of the detector field stop can be specified.

In order to proceed, the average detector power must be related to the aperture field exiting from the phase plate. Suppose there are M coherence cells in the aperture, each with area A_c, so that the total aperture area A_a is easily expressed as,

$$A_a = MA_c \tag{12}$$

We denote the variance of the phase estimate of the ith coherence cell by σ_i^2,

$$\sigma_i^2 = E[(\phi_i - \hat{\phi}_i)^2] \tag{13}$$

To further simplify the discussion, suppose "most" of the A_i's and σ_i^2's are equal and that M is large enough so that terms independent of M are small relative to terms proportional to M. In that case, the total average power intercepted by the receiver becomes

$$\overline{P_T} = MA_cA^2 \tag{14}$$

Except for diffraction effects, the average signal power collected by an Airy-disc-sized detector can be expressed as [33],

$$E\,[P_d]\Big|_{\substack{\text{no}\\ \text{compensation}}} = A_c A^2 = \frac{\overline{P_T}}{M} \tag{15}$$

$$E\,[P_d]\Big|_{\substack{\text{exact}\\ \text{compensation}}} = M A_c A^2 = \overline{P_T} \tag{16}$$

$$E\,[P_d]\Big|_{\substack{\text{all cells}\\ \text{track well}}} = M A_c A^2 (1 - \sigma^2/2)^2 = (1 - \sigma^2/2)^2\, \overline{P}_T \tag{17}$$

where Parsevals theorem has been used to relate the focal plane power to the power collected in the receiver aperture.

In subsequent discussions, error exponents will be compared for detector powers that range over several orders of magnitudes, so that the difference between phase tracking well (17), i.e. $\sin(\phi_i - \hat{\phi}_i) \simeq \phi_i - \hat{\phi}_i$, and phase tracking exactly (16) will be unimportant.* We henceforth assume that phase compensation yields a detector power as though the compensation were exact.

Consider the direct detection receiver, as shown in Figure 4. If the signal power were spread over a large angular spectrum, the detector for an uncompensated system would be chosen to be roughly the size of the blur circle so as to intercept all the signal power. That is,

$$n_s' = \frac{\eta}{hf_o}\,\overline{P}_T \tag{18}$$

where η is the detector quantum efficiency and hf_o is the energy of a photon. The background noise field would cause noise counts,

$$n_b' = \frac{\eta}{hf_o}\,M(N_{ob}W_o) \tag{19}$$

where $N_{ob}W_o$ is the background noise power intercepted by a diffraction limited field of view antenna [13]. Thus, when thermal noise is negligible,

*For example, with $\sigma^2 < \frac{1}{4}$, $(1 - \sigma^2/2)^2 > 0.765$.

the uncompensated direct detection communication receiver performance is determined by (8, 9) with

$$\text{Uncompensated} \left\{ \quad \gamma = \frac{\overline{P}_T}{M(N_{ob}W_0)} \quad , \ \alpha = \frac{\overline{P}_T\eta}{hf_o}[\tau] \right. \tag{20}$$

Note we have omitted n_d in equation (20). We chose to include dark current by modeling it as due to an equivalent background noise field. For a specified $N_{ob}W_o$, this implies that dark current can be included in subsequent expressions by an appropriately increased equivalent noise power $(N_{ob}W_o)'$.

At the other extreme, if phase compensation were used, the direct detection system would use a diffraction-limited-spot-size detector, which would result in

$$n_s' = \frac{\overline{P}_T\eta}{hf_o} \qquad , \ n_b' = \frac{N_{ob}W_o\eta}{hf_o} \tag{21}$$

so that compensated communication performance is governed by*:

$$\begin{matrix}\text{Phase} \\ \text{Compensated}\end{matrix} \left\{ \ \gamma = \frac{\overline{P}_T}{N_{ob}W_o} \quad ; \quad \alpha = \frac{\overline{P}_T\eta}{hf_o}[\tau] \right. \tag{22}$$

Thus, the effect of phase compensation on direct detect (negligible thermal noise) communication has been to effectively improve γ by a factor of M. It is also easily seen that phase compensation will significantly improve communication performance only when the receiver is background noise-dark current limited, say $\gamma < 0.1$ and M is large enough to insure $M\gamma > 1.0$.

Now consider a homodyne system used in Figure 4. The error exponent for the uncompensated system is given by

*This statement is true when dark current is included only if the dark current scales with the detector area. Although not true in general, it is considered a good approximation for our discussion [38, 39].

$$\text{Uncompensated} \left\{ \varepsilon_h \approx \frac{\overline{P}_T}{M} \frac{\eta}{hf_o} [\tau] \right. \tag{23}$$

The compensated system should use a small detector (narrow field of view), resulting in an error exponent

$$\text{Compensated:} \left\{ \varepsilon_h \approx \frac{\overline{P}_T[\tau]\eta}{hf_o} \right. \tag{24}$$

The improvement in performance is due to the fact that phase compensation has concentrated all the signal power into the one mode that the homodyne system is designed to receive.

Note that when homodyne (or heterodyne) receivers are used in communication, phase compensation is useful only when the received field is divided into many coherence cells (or modes); this conclusion is independent of background noise. The noise that phase compensation reduces when the homodyne communication receiver is properly designed is the inherent quantum noise in the measurement.

In contrast, the direct detection communications receiver can easily respond to all the signal modes, but does so at the expense of background noise-dark current, so phase compensation is useful primarily when these can be reduced by using a smaller detector. Such a conclusion is not a property of the measurement (as it was in homodyning). For example, if W_o were small enough*, the simple direct detection (negligible thermal noise) receiver, without compensation, would be extremely useful in the low visibility environment.

C. Regimes of Improved Performance

The performance of the communication system as related to their uncompensated counterparts, are now examined in detail. The issue of power split-

*The narrow optical bandwith would have to be achieved for signals with a large angular spread. This angle requirement severely limits the usefulness of currently available narrowband optical filters.

ting at beam combiners will introduce relatively small changes in the subsequent expressions and hence will be ignored. Several approaches to the comparisons will be taken, depending on what parameter is assumed fixed.

We first seek operating conditions for which phase tracking will significantly improve performance. To do so, suppose that sufficient signal power per coherence cell is received so that phase compensation is possible. That is, in terms of average detected photons per coherence cell, the requirement for $\sigma^2 < \frac{1}{4}$ becomes [34],

Homodyne: Phase Track $\quad n_s T_\phi \geq 4 \quad$ (25a)

Direct Detect: Phase Track $\quad \frac{(n_s)^2}{n_b} T_\phi \geq 4 \quad$ (25b)

where n_s and n_b are the average number of detected signal and background photons per coherence cell, respectively.

Note that (25b) is written for the case when the detector is "background noise" limited.* In fact, we will assume that the direct detection system are always background limited, and will require (somewhat arbitrarily):

$$\gamma = \frac{n_s}{n_b} < 0.1 \tag{26}$$

Otherwise (crudely), the system will not be significantly restricted by background noise and a direct detection, wide field of view receiver with negligible thermal noise will perform nearly as well as the more complex compensated receiver. Equations (25) and (26) establish a range of signal power (per cell) that will be considered interesting. Assume the minimum signal power per cell (25) is achieved with near equality, then (26) becomes

Homodyne: $\quad T_\phi > \frac{40}{n_b} \quad$ (27a)

Direct Detect: $\quad T_\phi > \frac{400}{n_b} \quad$ (27b)

Equation (27) presents a restriction on T_ϕ so that when the minimum power for phase tracking is received, the resulting system is still background noise

*We will no longer explicitly indicate the possibility of dark current. It is to be included through the background term as described earlier.

limited. The value of n_b is determined by the optical filter bandwidth and N_{ob}. For example, at visible wavelengths during daytime [11, 13, 40],

$$N_{ob} \approx 10^{-25} \text{ W/Hz} \tag{28}$$

while a good filter bandwidth is 10A° [11, 13]. Suppose $\lambda = 1$ μm, so that [41],

$$W_o \approx 10^{12} \text{ Hz} \tag{29}$$

In addition, with $hf_o = 2 \times 10^{-19}$ J at $\lambda = 1\mu m$ and letting $\eta = 1$, n_b becomes,

$$n_b = \frac{\eta}{hf_o} N_{ob} W_o \approx \tfrac{1}{2} \times 10^6 \quad \text{events/sec} \tag{30}$$

Equation (30) will be used as a typical number for background noise counts using a 10A° filter at visible wavelengths. To complete the background dominated noise discussion, substitute (30) into (27) to obtain,

$$\text{Homodyne:} \qquad T_\phi > 80 \times 10^{-6} = 80 \text{ μsec.} \tag{31a}$$

$$\text{Direct Detect:} \qquad T_\phi > 0.8 \times 10^{-3} = 0.8 \text{ msec} \tag{31b}$$

We conclude that if sufficient power is available to phase track, then for typical background noise at visible wavelengths, a direct detection communication system would be background limited for phase coherence time greater than 0.8 m sec. Stated alternatively, if $n_s/n_b \approx 0.1$, then a direct detection communication system is background noise limited and for typical values of n_b direct detection phase tracking would be possible only if $T_\phi > 0.8$ ms. This restriction may realisitically be met by a low visibility channel. Note that if a very good optical filter, with bandwidth 0.1A°, even for signals with a large angular spectrum, was available, the direct detection restriction would be greater than 80 ms, a coherence time that may not be realistic. Thus, 0.1A° is roughly the filter bandwidth required at visible wavelengths to make the phase compensation receiver unappealing.

Suppose a homodyne receiver structure is used for both communication and phase tracking. Furthermore only a 10A° filter is available and equations (25a) and (31a) are satisfied so that phase compensation seems sensible. Using (25a) as an approximate equality, the error exponent for the uncompensated receiver, equation (23) becomes:

$$\varepsilon_h = n_s[\tau] = \frac{4\tau}{T_\phi} \tag{32}$$

We define communication to be unreliable when $\varepsilon_h \leq 1$ so that (32) provides a relation between phase coherence time and the bit interval for which uncompensated communication would be unreliable. The compensated system would increase the error exponent (32) by M, which in many instances would make the system reliable. If the signaling rate is defined as

$$R = \frac{1}{2\tau} \qquad \text{bits}/_{\text{sec}} \tag{33}$$

the dividing line between reliable and unreliable uncompensated performance can be plotted as a function of rate and coherence time. The result is presented in Figure 5, which is plotted for $n_s = 4/T_\phi$, i.e. just sufficient signal power per cell is available to homodyne phase track. Thus for example if T_ϕ is 1 msec, then reliable uncompensated communication is possible up to roughly 2 Kbits/$_{\text{sec}}$; while rates above 2 Kbits/$_{\text{sec}}$ will yield unacceptable performance unless the receiver phase compensates. Note the result here is independent of the number of coherence cells (except for the improvement in ε_h).

A simple variation of the previous example is to change to direct detect communication while homodyne phase tracking. The error exponent for the uncompensated receiver, at the required power level per cell, is given by

$$\varepsilon_d = \frac{M(n_s)^2\tau}{4n_b} = \frac{4M\tau}{(T_\phi)^2 n_b} = \frac{2M}{(T_\phi)^2 n_b R} \tag{34}$$

Using equation (30) in (34) the dividing lines between reliable/unreliable communication for various value of M are plotted in Figure 6. Note the much higher signaling rate required to make compensation attractive, and that the required rate depends on the number of modes. The higher signaling rate reflects the fact that direct detection performance varies as $(n_s)^2$ and that it is proportional to M. The M dependence arises because the uncompensated receiver is designed to respond to all the modes. Recall that a 10A° filter was assumed in computing n_b; if a 100A° (wide angle) filter is used, the curves are the same as Figure 6 except number of modes M will be increased by 10, as indicated by the bracketed labels.

Now continue the assumption of having minimum power for tracking and being background noise limited, but use direct detection for phase tracking and homodyne for communication. The uncompensated error exponent is given by (using 25b),

$$\epsilon_h = n_s[\tau] = 2\sqrt{\frac{n_b}{T_\phi}}\,[\tau] = \sqrt{\frac{n_b}{T_\phi}}\,\frac{1}{R} \tag{35}$$

On substituting (30), (35) is plotted in Figure 7. Observe the required rate is quite small as compared to the rate in Figure 6. This simply indicates that the direct detection phase tracker requires more power than the homodyne method and the implicit increase in power allows the homodyne communication receiver to communicate at a higher rate before becoming unreliable.

Finally, suppose direct detection is used for both purposes. With the ususal assumptions, the uncompensated error exponent becomes,

$$\epsilon_d = \frac{(n_s)^2}{n_b}\quad \frac{\tau M}{4} = \frac{\tau M}{T_\phi} = \frac{M}{2RT_\phi} \tag{36}$$

Note the result is independent of n_b. (The only requirement is that $\gamma < 0.1$, which determines a minimim T_ϕ as discussed earlier.) Equation (36) is plotted in Figure 8. Although the rate required is larger than all the other systems considered, it is not outside the region of possibility. For example, if the 10A° filter is used in the wide field of view system, T_ϕ is 10ms and M is 10^4, phase compensation is attractive for rates greater than 0.5 Mbits/$_{sec}$. This result is encouraging when in addition we consider the structural simplicity offered by the direct detection system.

Summarizing the results of the past few paragraphs, it was shown that <u>if</u> there was just enough signal power available per coherence cell to track phase well, then for phase coherence times in tens of milliseconds, phase compensation appeared to be attractive relative to nonadaptive receivers for signaling rates from kilobits (all homodyne system) to megabits (all direct detection) per second. However, the underlying assumptions for this analysis should be emphasized. The comparisons were made holding the signal power per cell fixed, while varying the other parameters. This meant, for example, that as M increased, so did the required total signal power intercepted by the receiver. Therefore, it is important to also examine the problem in terms of other system parameters.

Suppose $\overline{P_T}$ is uniformly spread over an angular spectrum with Θ_o radians half angle. An indication of the size of M as related to other variables is easily computed. As an example, assume the receiver has a square aperture D_o on a side, and that the coherence distance D_c, is as defined in equation (7), then

$$M = \frac{A_a}{A_c} = \frac{(D_o)^2}{(D_c)^2} = \left(\frac{D_o}{\lambda}\right)^2 (\sin\Theta_o)^2 \left(\frac{\pi}{3.83}\right)^2 \tag{37}$$

To illustrate the large number of cells that are possible, consider that a typical receiver operated at near visible wavelengths (say $\lambda = 1\mu m$) may have $D_o = 10$ cm and expect $\Theta_o \geq 1.0$ mr so that M would be about 10^4.

It was shown that M is the improvement in the error exponent of an uncompensated homodyne communication receiver, while in direct detection receivers (with negligible thermal noise) the signal to noise count ratio γ is increased by M. Thus the large magnitude of M indicates the very significant improvement in the error exponent that is possible if phase compensation can be implemented. The improvement in the error exponent of a compensated homodyne or direct detection (with $M\gamma \leq 1.0$) communication receiver as compared to the corresponding uncompensated system is presented in Figure 9.

Since the improvement is plotted in terms of optical power, there is a natural tendency to interpret Figure 9 as the additional signal powe· required to make the uncompensated error exponent as large as the compensated one. Such a conclusion is certainly true, but may be misleading unless interpreted with care; since depending on the signal power and communication rate, the compensated system may be operating with a ridiculously small error probability. For example, assume a homodyne system is used for communication as well as phase tracking, and that just enough signal power is available to phase track. Suppose $\frac{D_o}{\lambda} = 10^5$, $T_\phi = 1$ ms, and $R = 100$ Kbits/sec so that

$$P_e \Big|_{\text{uncompensated}} \approx \tfrac{1}{2} \tag{38}$$

and

$$P_e \Big|_{\text{compensated}} \approx \tfrac{1}{2}\exp[-200] \approx 6.91 \times 10^{-88} \tag{39}$$

Clearly if we increased the power of the uncompensated system by 40dB as indicated in Figure 9 to achieve the same performance as the compensated system, the system would be overdesigned. However, if the desired error probability was

$$P_e = \tfrac{1}{2}\exp[-15] = 1.5 \times 10^{-7} \tag{40}$$

then using (32), we find the uncompensated system would require 28.7 dB more power. Thus Figure 9 should be interpreted as an upper bound on the power increase needed to make the uncompensated system reliable.

We are unable to make further statements about the feasibility of phase compensation for low visibility communication without further knowledge of actual channel characteristics. The major requirements are: (1) The aperture must collect enough signal power to make phase tracking possible. (2) Th. phase coherence time must be long enough relative to the bit interval length that can be supported by the channel without suffering intersymbol interference, so that a high enough signaling rate can be used to make compensation attractive.

V. Conclusions

This paper investigated the use of phase compensation receivers to improve the reliability of a line of sight optical communication system. The optimum phase control, which concentrates spatially the signal power in the focal plane, is the MMSE estimate of the phase of the aperture field, when the estimation error is "small enough". Using currently available optical filters and maintaining the minimum signal power per cell required to track phase, the direct detection (uncompensated) system is background noise limited for $T_\phi > 0.8$ms. For channels whose coherence time is at least that long, the region of significant communication improvement due to phase compensation depends on the signaling rate, number of coherence cells in the aperture and the precise phase coherence time. For example, if T_ϕ is 10ms, M is 10^4 and there is just enough power to track phase, an all homodyne system would require compensation above 0.2 Kbits/sec while the "all direct detect" system (no thermal noise) would require compensation above 0.5 Mbits/sec. Required rates for hybrid systems, which fall in between the two extremes are also presented.

We conclude that the phase compensation receivers presented show promise for improving optical communication through low visibility channels.

REFERENCES

1. See for example the Special Issue on Adaptive Optics, Journal of the Optical Society of America, Vol. 67, No. 3, March 1977.

2. R. A. Muller and A. Buffington, J. Opt. Soc. Am., Vol. 63, p. 647, 1973.

3. J. W. Hardy, J. Feinlab and J. C. Wyant, "Real-Time Phase Correction of Optical Imaging Systems", Presented at the Opt. Soc. Am. Topical Meeting on Optical Propagation Through Turbulence, 1974.

4. W. B. Bridges, P. T. Brunner, S. P. Lazzara, T. A. Nussmeier, T. R. O'Meara, J. A. Sanguinet and W. P. Brown, Applied Optics, Vol. 13, p. 291, 1974.

5. J. E. Pearson, W. B. Bridges, L. W. Horwitz, T. J. Walsh, and R. F. Ogrodnik, "Atmospheric Turbulence Compensation Using Coherent Optical Adaptive Techniques", Presented at the Opt. Soc. Am. Topical Meeting on Optical Propagation Through Turbulence, 1974.

6. R. S. Kennedy, Fading and Dispersive Communication Channels, Wiley, New York, 1969.

7. H. L. Van Trees, Detection, Estimation and Modulation Theory: Part III, Wiley, New York, 1972.

8. P. A. Bello, "Characterization of Randomly Time-Invariant Linear Channels," IEEE Transactions Communication Systems, pp. 360-393, December 1963.

9. R. S. Kennedy and I. L. Lebow, "Signal Design for Dispersive Channels," IEEE Spectrum, pp. 231-237, March 1964.

10. T. P. McGarty, "On the Structure of Random Fields Generated by Multiple Scatter Media," Ph.D. Thesis, Department of Electrical Engineering, M.I.T., June 1971

11. R. S. Kennedy, "Communication Through Optical Scattering Channels: An Introduction," Proc. IEEE, Vol. 58, No. 10, pp. 1651-1665, October 1970.

12. N. S. Myung, "Propagation of Random Fields in Moving Scatter Media," S.M. Thesis, Department of Electrical Engineering, M.I.T., September, 1974.

13. E. V. Hoversten, "Optical Communication Theory", Laser Handbook, F. T. Arecchi and E. D. Schulz-DuBois, Eds, Amsterdam: North Holland, pp. 1626-1650, 1972.

14. J. W. Goodman, Introduction to Fourier Optics, McGraw-Hill, New York 1968.

15. A Papoulis, Systems and Transforms with Applications in Optics, McGraw-Hill, New York, 1968.

16. R. S. Kennedy and S. Karp, Editors, "Optical Space Communication," NASA SP-217, 1969.

17. H. C. VanDehulst, Light Scattering by Small Particles, Wiley, New York, 1957.

18. M. Kerker and Milton, The Scattering of Light and Other Electromagnetic Radiation, Academic Press, New York, 1969.

19. D. Deirmendjian, Electromagnetic Scattering on Spherical Polydispersions, American Elsevier, New York, 1969.

20. W. E. K. Middleton, Vision Through the Atmosphere, University of Toronto Press, Toronto, Canada, 1958.

21. D. E. Setzer, "Influence of Scattered Radiation on Narrow Field of View Communication Links," Applied Optics, Vol. 10, No. 1, 109-113, January 1971.

22. T. S. Chu and H. C. Hogg, "Effects of Precipitation on Propagation at 0.63, 3.5, and 10.6 Microns," Bell System Tech. J., Vol. 47, pp 723-761, June 1968.

23. H. M. Heggestad, "Optical Communication through Multiple Scattering Media," Res. Lab. Electron., M.I.T., Tech. Rep. 474, November 22, 1968.

24. G. W. Kattawar and G. N. Plass, "Influence of Particle Size Distribution on Reflected and Transmitted Light from Clouds," Applied Optics, Vol. 7, pp. 869-878, May 1968.

25. R. M. Lerner and A. E. Holland, "The Optical Scatter Channel," Proc. IEEE, Vol. 58, No. 10, pp. 1547-1563, October 1970.

26. E. A. Bucher, R. M. Lerner, and C. W. Niesson, "Some Experiments on the Propagation of Light Pulses Through Clouds," Proc. IEEE, Vol. 58, No. 10, pp. 1564-1568, October 1970.

27. E. A. Bucher, "Computer Simulation of Light Pulse Propagation for Communication through Thick Clouds," Applied Optics, Vol. 12, pp. 2391-2400, October 1973.

28. E. A. Bucher and R. M. Lerner, "Experiments on Light Pulse Communication and Propagation through Atmospheric Clouds," Applied Optics, Vol. 12, pp. 2401-2414, October 1973.

29. R. A. Dell Imagine, "A Study of Multiple Scattering of Optical Radiation with Application to Laser Communication," Advances in Communication Systems, Vol. II, Academic Press, 1966, pp. 5-50.

30. R. S. Kennedy, "Low Visibility Short-Haul Communication in the Atmosphere," Report of the NSF Grantee-User Semi-Annual Meeting, Boulder, Colorado, May 29-30, 1974.

31. J. Baird, J. Clark, "Communications under Poor Visibility Conditions with Emphasis on Direct Detection Techniques," Report of the NSF Grantee-User Semi-Annual Meeting, Washington University, St. Louis, MO., November 14-15, 1973.

32. J. R. Clark, J. R. Baird and R. S. Rearden, Jr., "Low Visibility Optical Communications: Received Signal Level as a Function of Receiver Field of View," Applied Optics, Vol. 15, No. 2, pp. 314-316, February 1976.

33. S. R. Robinson, "Spatial Phase Compensation Receivers for Optical Communication," Ph.D. dissertation, Department of Electrical Engineering and Computer Science, Massachusetts Institute of Technology, Cambridge, May 1975.

34. S. R. Robinson, "Phase Estimators with Predetection Feedback for Optical Communication," IEEE Trans. on Comm., Vol. COM-24, No. 11, November 1976, pp. 1231-1238.

35. E. A. Bucher, "Error Performance Bounds for Two Receivers for Optical Communication and Detection," Applied Optics, Vol. 11, No. 4, pp. 884-889, April 1972.

36. H. L. Van Trees, Detection, Estimation and Modulation Theory: Part I, Wiley, New York, 1968.

37. J. M. Wozencraft and I. M. Jacobs, Principles of Communication Engineering, Wiley, New York, 1965.

38. H. Melchior, et al., "Photodetectors for Optical Communication Systems," Proc. IEEE, Vol. 58, No. 10, pp. 1466-1486, October 1970.

39. See for example, "Silicon Photodiode Application Notes," Application Notes D3000B-1, printed by the Electro-Optics Division of EG&G, Inc., Salem, Massachusetts.

40. N. S. Kopeika and J. Bordogna, "Background Noise in Optical Communication Systems," Proc. IEEE, Vol. 58, No. 10, pp. 1571-1577, October 1970.

41. W. K. Pratt, Laser Communication Systems, Wiley, New York, 1969.

42. G. Q. McDowell, "Pre-distortion of Local Oscillator Wavefront for Improved Optical Heterodyne Detection through a Turbulent Atmosphere," Sc.D. Thesis, Department of Electrical Engineering, M.I.T.. April 1971.

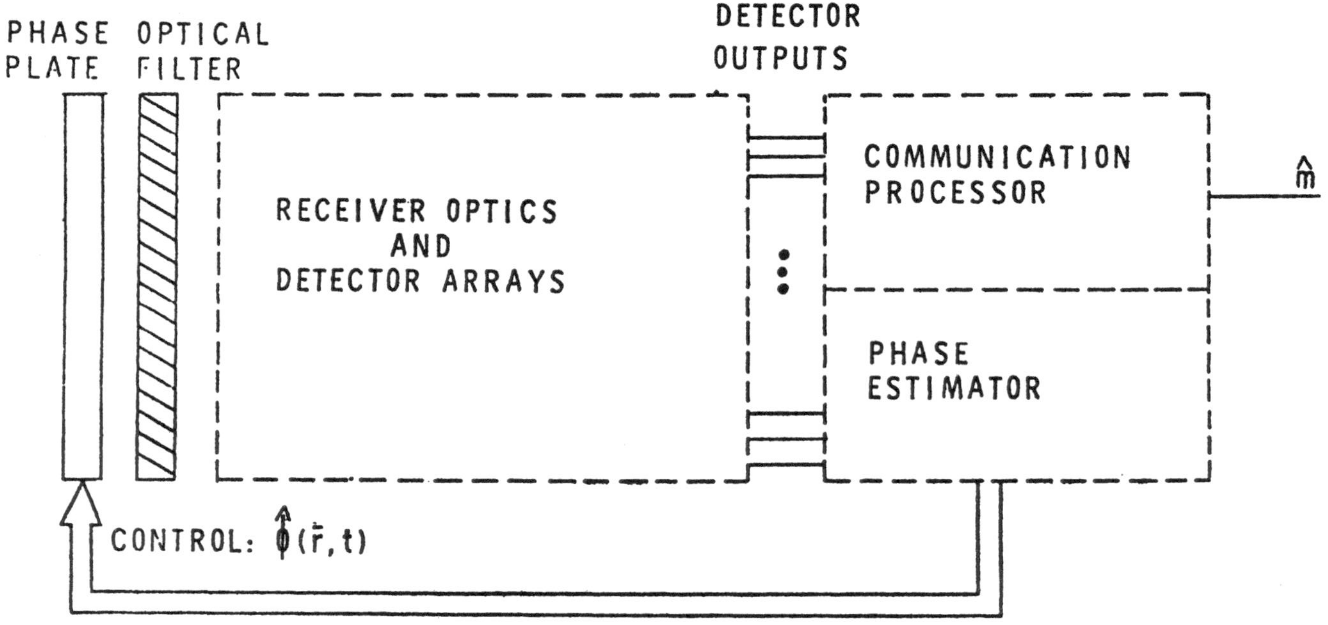

FIG. 1: GENERAL PHASE COMPENSATION RECEIVER STRUCTURE

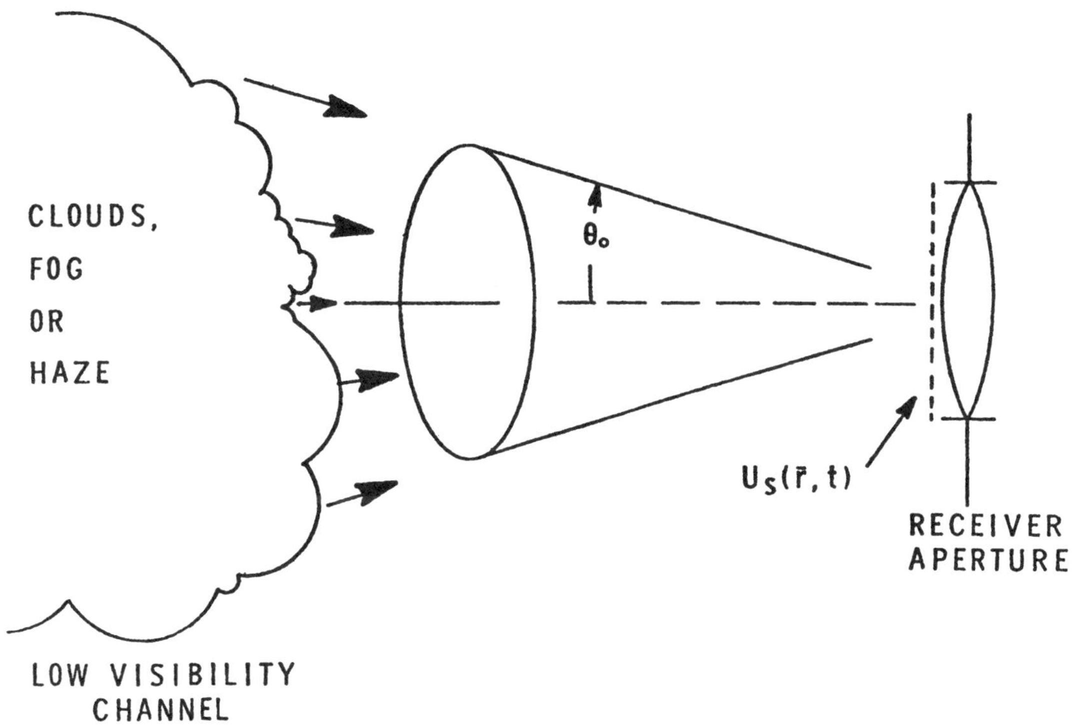

FIG. 2: ANGULAR DISTRIBUTION OF SIGNAL FIELD

FIG. 3: DIRECT DETECTION OPTICAL COMMUNICATION SYSTEM FOR CLEAR WEATHER CHANNEL WITH LITTLE TURBULENCE

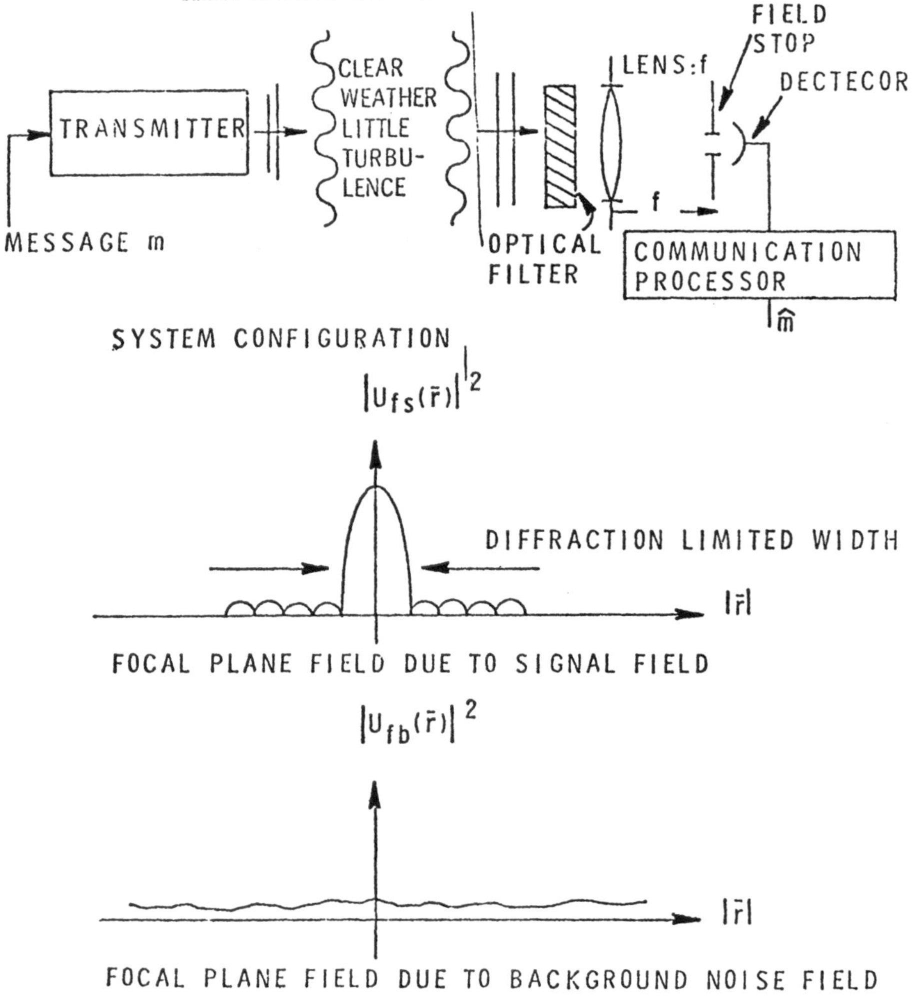

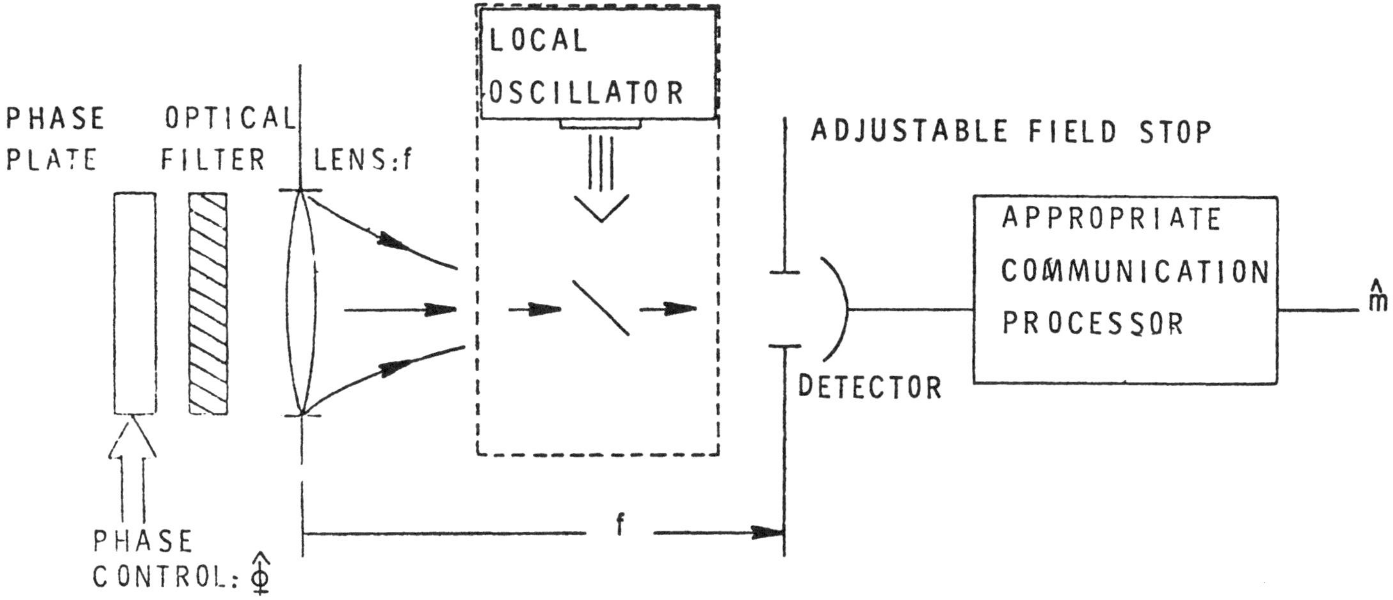

FIG. 4: IDEALIZED PHASE COMPENSATED COMMUNICATION RECEIVER

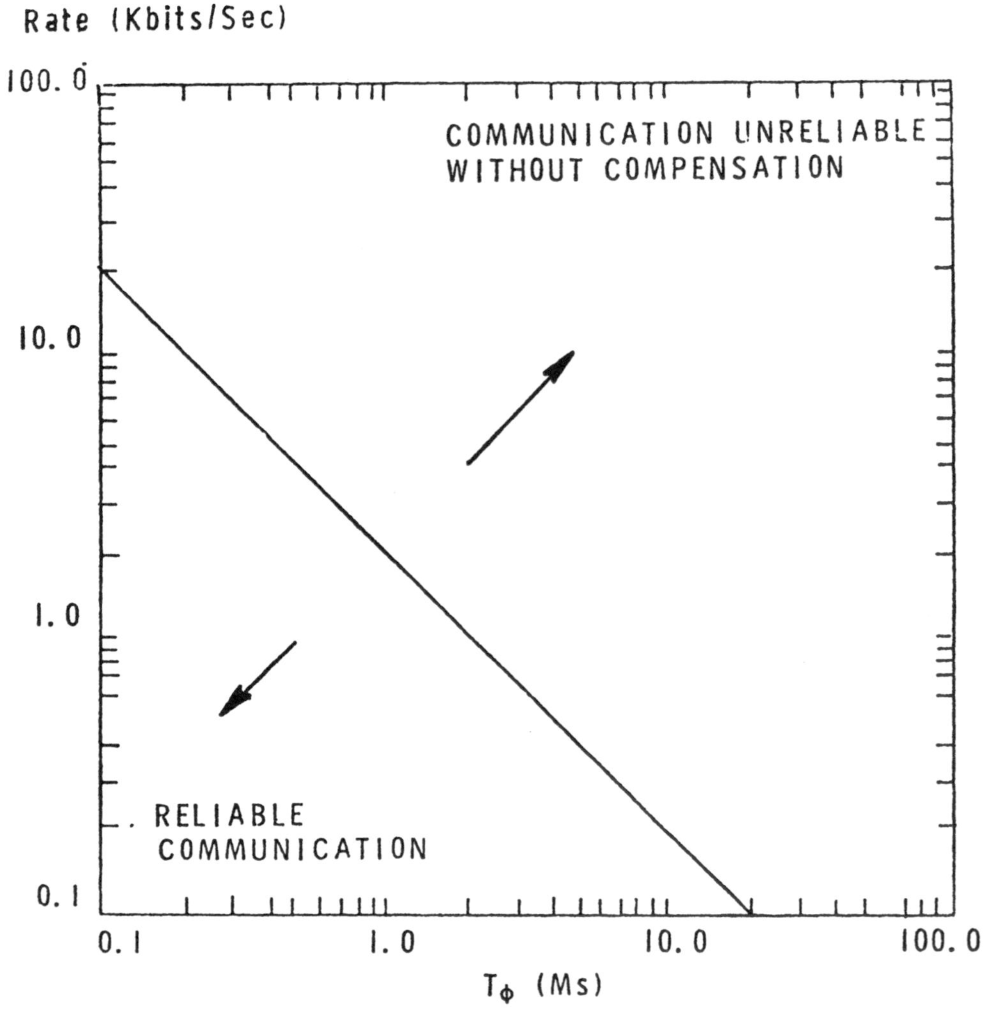

FIG. 5: UNCOMPENSATED HOMODYNE COMMUNICATION WITH JUST SUFFICIENT POWER TO HOMODYNE PHASE TRACK

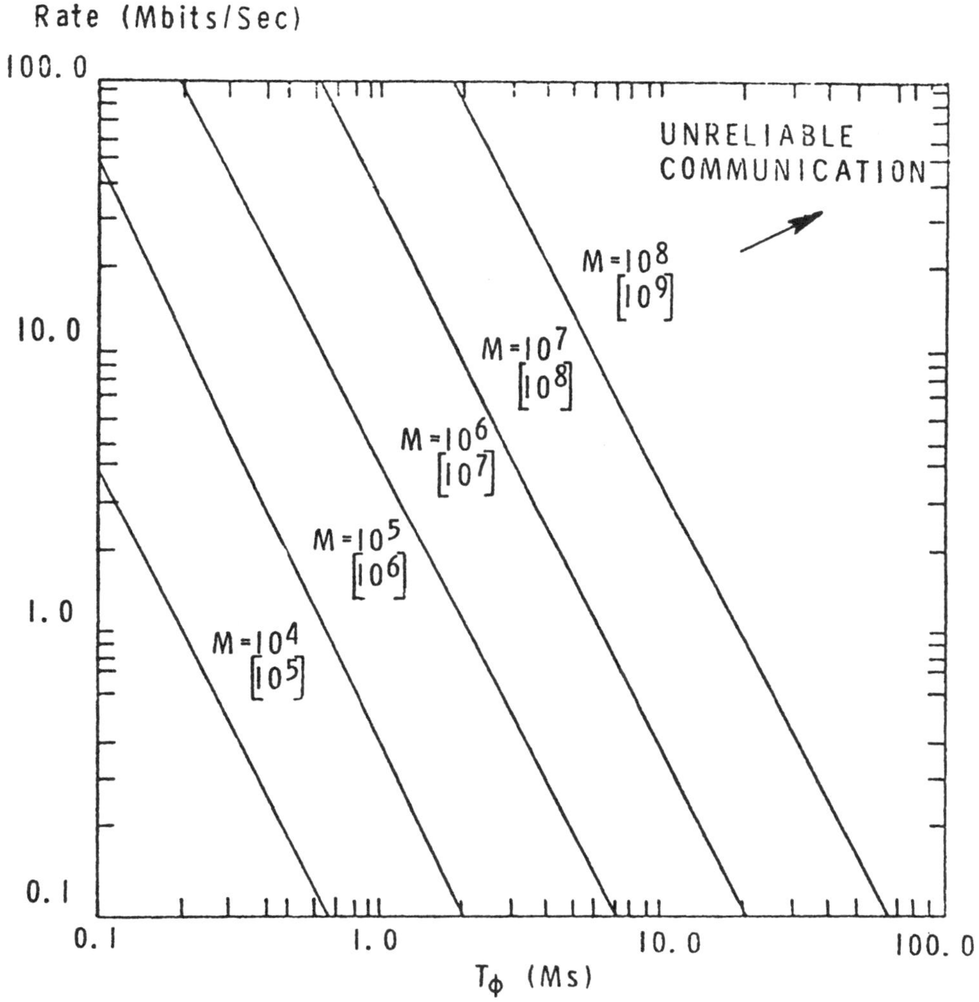

FIG. 6: UNCOMPENSATED DIRECT DETECTION (NO THERMAL NOISE) COMMUNICATION WITH JUST SUFFICIENT POWER TO HOMODYNE PHASE TRACK; M COHERENCE CELLS IN THE RECEIVER APERTURE

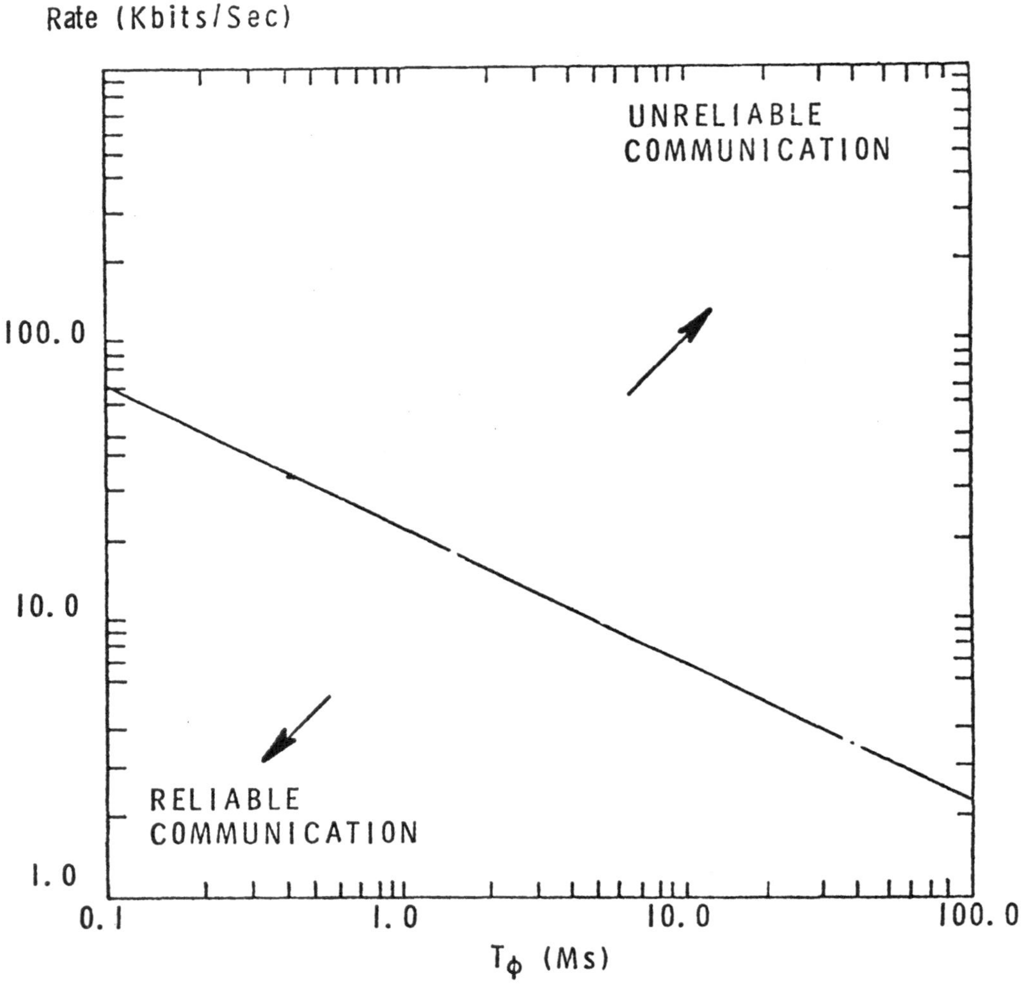

FIG. 7: UNCOMPENSATED HOMODYNE COMMUNICATION WITH JUST SUFFICIENT POWER TO DIRECT DETECT PHASE TRACK

FIG. 8: UNCOMPENSATED DIRECT DETECTION (NO THERMAL NOISE) COMMUNICATION WITH JUST SUFFICIENT POWER TO DIRECT DETECT PHASE TRACK; M COHERENCE CELLS IN THE RECEIVER APERTURE

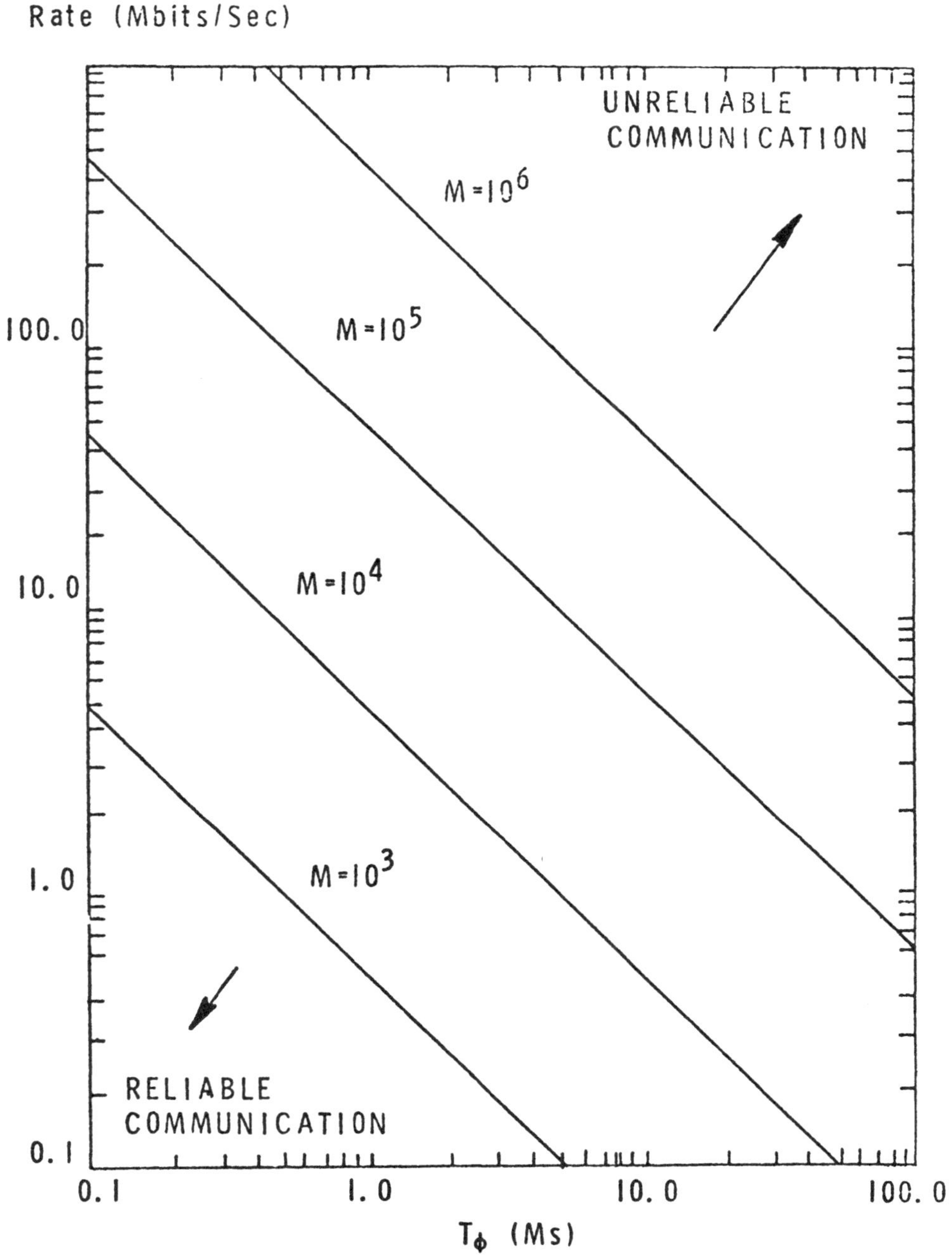

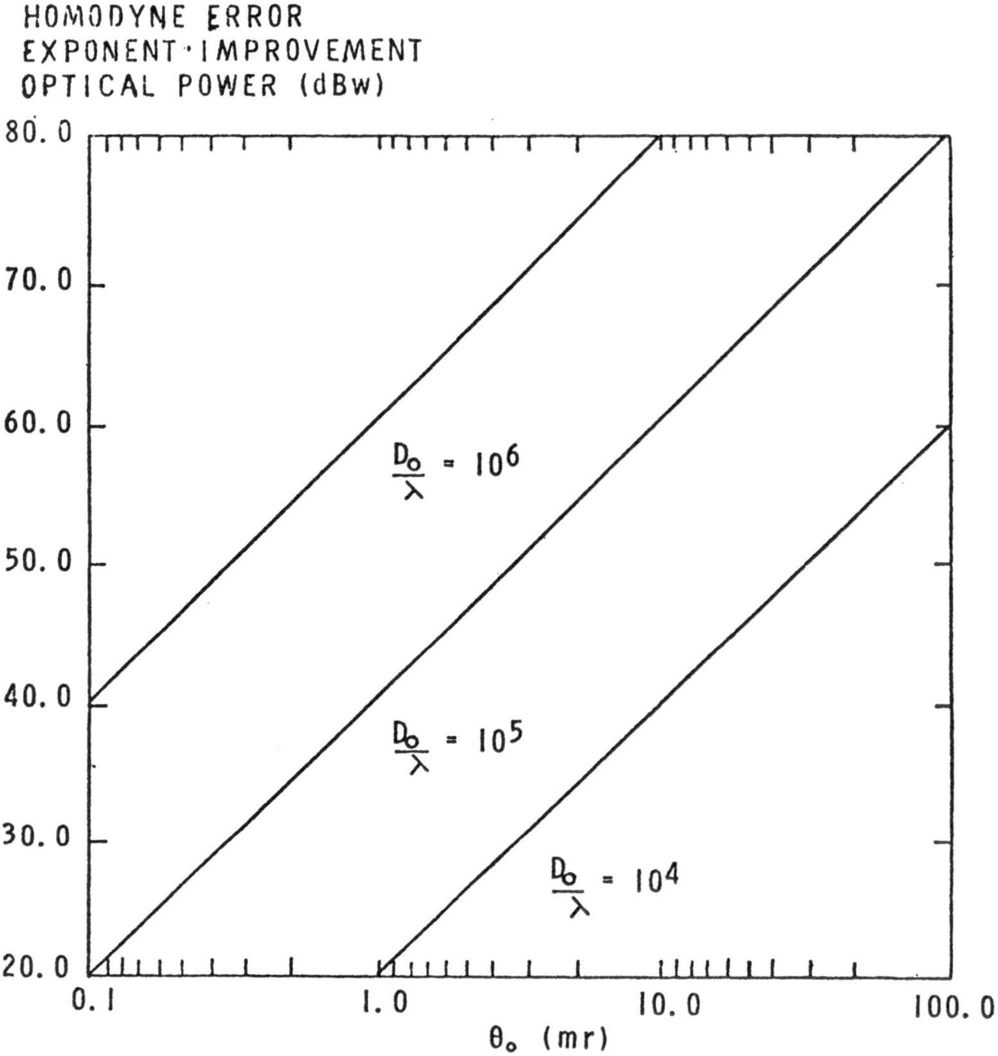

FIG. 9: IMPROVEMENT OF ERROR EXPONENT OF A COMPENSATED HOMODYNE COMMUNICATION SYSTEM OVER AN UNCOMPENSATED SYSTEM

MODELLING OF THE SEA SURFACE SCATTERING CHANNEL AND UNDERSEA COMMUNICATIONS*

A.N. Venetsanopoulos

Department of Electrical Engineering, University of Toronto, Toronto, Ontario, Canada

ABSTRACT. This paper deals with the sea surface scattering channel. The channel consisting of a set of correlated scatterers, distributed over a time-varying, random surface is described and modelled as a random, time-varying, linear filter. A brief description of the pertinent channel-signal characteristics is given and the ideas of quasistationarity and channel system autocorrelation function are introduced. The problem of near optimum detection and signal design for binary communication over purely random, linear, time-varying undersea acoustic channels is then formulated and the results are applied to the surface scattering channel.

1. INTRODUCTION

This paper studies the problem of modelling of the sea surface scattering channel. The implications of the results of this study to the problem of signal design for binary communication are also considered. Situations where the channel can be modelled as a linear time-varying stochastic filter, corrupted by independent, additive, zero-mean, Gaussian noise at the channel output are dealt with. The solution of this problem is of interest in those cases where the transmitted signal is corrupted by surface scattering and contaminated by additive noise.

While the paper mainly deals with the problem of communication over the undersea scattering channel, the approach is also useful

* This work has been supported by the National Research Council of Canada under Grant No. A-7397.

in other situations. Problems where communication over a random, linear, time-varying channel is of interest, subject to an average power constraint of the transmitted signals (e.g. communication over tropospheric links, orbiting dipole links, planet surfaces etc.), can be formulated in a similar way, by introducing the appropriate microscopic channel models, thus relating their various system function statistics to the physical parameters involved. In addition, the problem of active detection of a target with direct and indirect returns, through a random dispersive medium, as well as the problem of passive detection through a scattering channel, can benefit from similar studies.

To formulate and solve the problem of interest, it is necessary to combine and unify the theories of stochastic linear filtering, surface scattering and signal detection and design. The next section of this paper is consequently devoted to procedures for modelling the general underwater channel as a linear, time-varying, stochastic filter and gives a brief review of the macroscopic channel-signal characteristics. In addition, the concepts of channel system autocorrelation function and quasistationarity are introduced.

2. CHARACTERIZATION OF LINEAR TIME-VARYING STOCHASTIC CHANNELS

Throughout the paper it will be assumed that the input signals are narrowband described by

$$s(t) = \text{Re}[z(t) \exp(j\ 2\pi\ f_o t)] \tag{1}$$

where $z(t)$ is a low pass complex function (the envelope) and f_o is the carrier frequency.

The effect of the medium will be represented by a linear, time-varying, stochastic filter with impulse response $h_1(t,\tau)$, interpreted as the response of time t, due to an impulse of time $t-\tau$, where τ denotes the "age" of the impulse. Moreover using complex representation for the channel we set

$$h_1(t,\tau) = \text{Re}[\tilde{h}(t,\tau)] = 2\text{Re}[h(t,\tau)\exp(j\ 2\pi\ f_o\tau)] \tag{2}$$

where $h(t,\tau)$ can be interpreted as the complex envelope of the channel impulse response and is a low pass complex random function centered at the frequency of the carrier of the narrowband input signal.

Since the channel impulse response $h(t,\tau)$ is a stochastic process in two variables two single Fourier transforms $m(f,\tau)$ and $n(t,\phi)$ can be defined, depending on whether the t or the τ variable in $h(t,\tau)$ is Fourier transformed. $m(f,\tau)$ is referred to as the scattering function, while $n(t,\nu)$ is referred to as the time-varying system function. A second Fourier transform with respect to the remaining time variable then yields the bifrequency function $H(f,\nu)$.

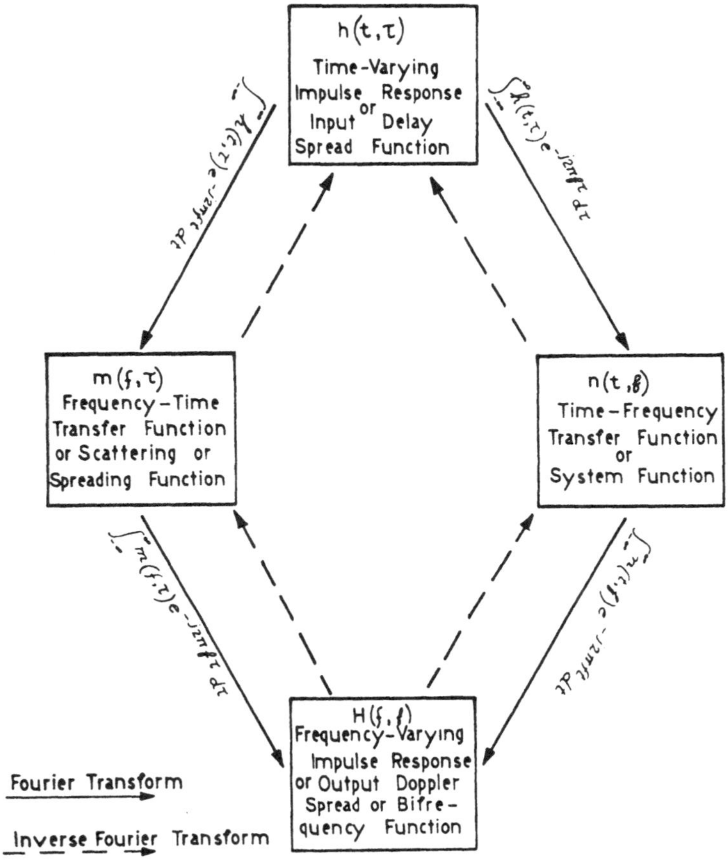

Fig. 1. Channel system functions.

Due to linearity and the narrowband assumption we can use the complex envelope $z(t)$ and the complex impulse response $h(t,\tau)$, to approximate the complex envelope of the channel output $p(t)$, in the following way

$$p(t) = \int_{-\infty}^{\infty} h(t,\tau) z(t-\tau) d\tau \tag{3}$$

Other relations between the Fourier transforms of these quantities are possible and are demonstrated in figure 2 [1-2].

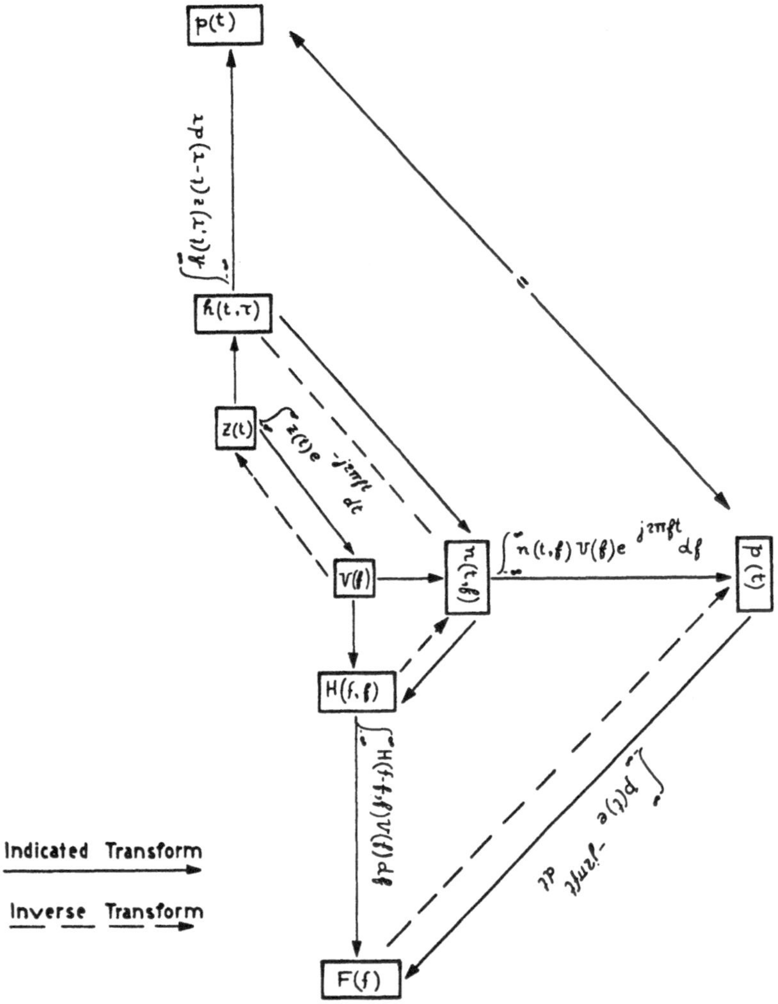

Fig. 2. Input-output relations.

Since the channel functions are complex quantities, for each function, two second-order moments of the form $P = \langle \overset{*}{h}h \rangle$ and $P' = \langle hh \rangle$ can be defined, where the pointed brackets symbolize ensemble average, the star means complex conjugate, and where h is used as a general symbol denoting any of the four channel functions previously defined. For narrowband signals and purely (zero mean) stochastic channels it can be shown that the primed second moment functions (i.e. where the complex conjugate is not used) are very much smaller than the unprimed functions, and in many analyses they are neglected [3]. The second-moment function found most convenient is the system autocorrelation function

$$P_n(t_1,t_2;\phi_1,\phi_2) = \langle n^*(t_1,\phi_1)\ n(t_2,\phi_2)\rangle \tag{4}$$

A quasistationary channel can be defined by making the change of variables

$$t_1 = t' - \tfrac{1}{2}t'' \qquad \phi_1 = \phi' - \tfrac{1}{2}\phi''$$

$$t_2 = t' + \tfrac{1}{2}t'' \qquad \phi_2 = \phi' + \tfrac{1}{2}\phi''$$

The function $P_n(t_1,t_2;\phi_1,\phi_2)$ can then be rewritten in terms of the new variables in the form

$$P_{n_{t',\phi'}}(t'',\phi'') = \langle n^*\ (t' - \tfrac{1}{2}t'',\phi' - \tfrac{1}{2}\phi'')\ n(t' + \tfrac{1}{2}t'',\phi' + \tfrac{1}{2}\phi'')\rangle \tag{5}$$

If θ_{max}[hertz] and γ_{max}[seconds] are respectively the maximum rate at which this function varies in the t' and ϕ' direction, and if W is the signal bandwidth and L+M is the signal time duration and channel memory, then the channel is quasistationary if [4]

$$W << \frac{1}{\gamma_{max}} \quad \text{and} \quad L+M << \frac{1}{\theta_{max}} \tag{6}$$

The usual ocean channel as well as the channel investigated in the experimental work reported in section 4 appear to be quasistationary according to this definition [5]. Moreover for a quasistationary channel it is easy to obtain the following relation [1-2], where the indices t' and ϕ' have been dropped from the system autocorrelation function notation for simplicity.

$$\theta_g(t'',\phi'') = P_n(t'',\phi'')\theta_s(t'',\phi'') \tag{7}$$

where

$$\theta_g(t'',\phi'') = \int_{-\infty}^{\infty} \langle p^*(t'-\tfrac{1}{2}t'')p(t'+\tfrac{1}{2}t'')\rangle \exp[-j\ 2\pi\phi''t']dt'$$

$$\theta_s(t'',\phi'') = \int_{-\infty}^{\infty} \langle z^*(t'-\tfrac{1}{2}t'')z(t'+\tfrac{1}{2}t'')\rangle \exp[-j\ 2\pi\phi''t']dt'$$

relating in a simple way the θ (ambiguity [6-7]) functions of the input and output complex envelopes to the channel system autocorrelation function.

The brief description given here is adequate for a microscopic analysis of the scattering channel, summarized in Section 3 of this paper. State variable models for random linear time-varying channels have been also considered in the literature [8-9].

3. THEORETICAL ANALYSIS OF THE SURFACE SCATTERING CHANNEL

The problem of diffraction of waves at uneven surfaces has received increasing attention in the past 20 years. This has resulted in a large number of reports and papers in the open literature. A good survey of the literature up to the year 1969 is reported in [10]. Attempts to describe the undersea acoustic channel as a random filter are reported in [11-14] under various assumptions.

The determination of the statistics of $\tilde{h}(t,\tau)$ or $\tilde{n}(t,f)$ its Fourier transform with respect to the τ variable, in terms of the parameters characterizing a wind-driven sea surface consisting of correlated scatterers is developed in [12-13], for an Eckart model [15], using the Fraunhofer approximation of geometrical optics, which presupposes a very directional transmitter. A summary of the results obtained is presented here.

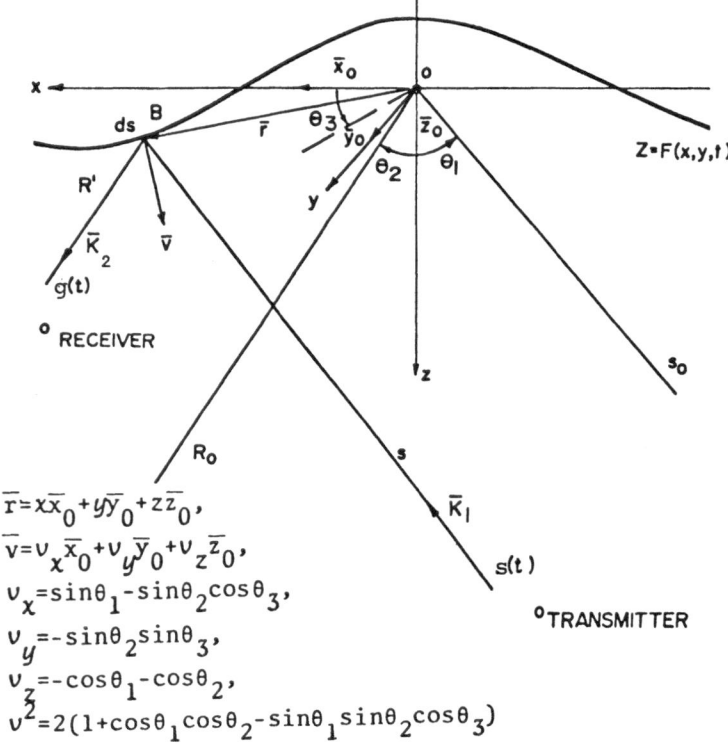

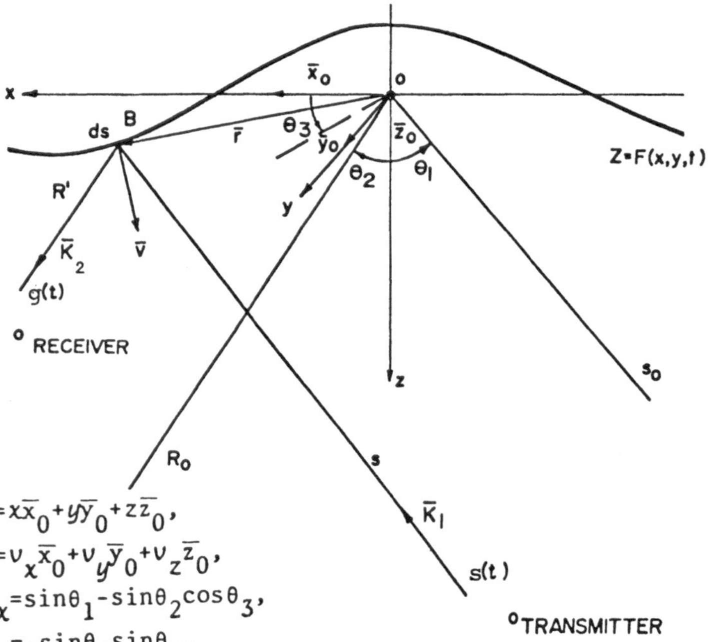

$\bar{r}=x\bar{x}_0+y\bar{y}_0+z\bar{z}_0,$

$\bar{v}=v_x\bar{x}_0+v_y\bar{y}_0+v_z\bar{z}_0,$

$v_x=\sin\theta_1-\sin\theta_2\cos\theta_3,$

$v_y=-\sin\theta_2\sin\theta_3,$

$v_z=-\cos\theta_1-\cos\theta_2,$

$v^2=2(1+\cos\theta_1\cos\theta_2-\sin\theta_1\sin\theta_2\cos\theta_3)$

Fig. 3. The surface scattering channel.

Additional approximations and assumptions are:

1. The "pressure release" boundary condition and the "far field" approximation [16].
2. The characteristic period of surface time variations is assumed to be large compared to differences in retardation times within the active scattering area and the period of acoustical frequencies of interest.
3. Shadowing and multiple reflections are neglected.
4. The source and receiver are directional, the beam pattern being such that a square area of length L is illuminated on the surface.
5. The dimensions of the illuminated area are much larger than the wavelength of the incident acoustic radiation.
6. The surface is assumed to have a Gaussian wave height distribution.
7. The autocorrelation function of the surface has the form

$$\rho(\tau) = e^{-\tau^2/T^2} \tag{8}$$

where τ is the space-time distance between two points on the surface with coordinates x_1, y_1, t_1, and x_2, y_2, t_2 such that

$$\tau^2 = (x_2 - x_1)^2 + (y_2 - y_1)^2 + K^2(t_2 - t_1)^2, \tag{9}$$

T is a space-time correlation distance, and K is a parameter describing the surface correlation velocity.

With these approximations the following results are obtained [12-13].

1. The specular part of the system function is given by

$$\tilde{n}_d(t,f) = \langle \tilde{n}(t,f) \rangle = 2jf(L^2v^2/cR_oS_ov_z)\exp[-\tfrac{1}{2}V^2f^2 - j(2\pi f/c)(R_o+S_o)]Q(f), \tag{10}$$

where R_o and S_o are the distances between the specular point and the transmitter and receiver respectively, c is the velocity of sound v_x, v_y, v_z, and v are combinations of direction cosines defining the ray directions, $Vf = 2\pi v_z\sigma/\lambda$ is the Rayleigh parameter, and

$$Q(f) = \operatorname{sinc}\left(\frac{2\pi f v_x L}{c}\right) \operatorname{sinc}\left(\frac{2\pi f v_y L}{c}\right). \tag{11}$$

An equivalent expression is

$$\tilde{n}_d(t,\delta) = \tilde{n}_{ds}(t,\delta)\ \exp(-\frac{V^2\delta^2}{2}) \tag{12}$$

where $\tilde{n}_{ds}(t,\delta)$ is the system function of a mirror surface. Note that the precise form of the Q function in terms of two sinc functions arises from the assumption that the illuminated area on the surface is a square of length L. More realistic illumination functions result in a $Q(\delta)$ that differs in detail, but not in any qualitative way, from the one given here.

The autocorrelation of the channel system function is defined by

$$P_r(t_1,t_2;\delta_1,\delta_2)=\langle\tilde{n}^*(t_1,\delta_1)\tilde{n}(t_2,\delta_2)\rangle-\tilde{n}_d^*(t_1,\delta_1)\tilde{n}_d(t_2,\delta_2) \tag{13}$$

The general form of this expression is fairly complex, however by assuming the surface to be either only slightly rough or extremely rough simplified expressions can be obtained. For the slightly rough case $(V^2\delta_o^2 \ll 1)$

$$P_r(t_1,t_2;\delta_1,\delta_2)=P_{rsp}(t,t;\delta_o,\delta_o)\cdot\exp[-\frac{K^2}{T^2}(t_2-t_1)^2 -j\frac{2\pi(R_o+S_o)}{c}(\delta_2-\delta_1)]\cdot\ell(\theta_1,\theta_2,\theta_3)\cdot Q(\delta_2-\delta_1) \tag{14}$$

where $P_{rsp}(t,t;\delta_o,\delta_o)$ is the autocorrelation function evaluated for the specular direction, and at a "center frequency" δ_o, given by

$$P_{rsp}(t,t;\delta_o,\delta_o)=4\pi\cos^2\theta_1\ \frac{T^2}{R_o^2S_o^2}\ \frac{L^2}{\lambda_o^2}\ (V\delta_o)^2\ \exp(-V^2\delta_o^2) \tag{15}$$

Also

$$\ell(\theta_1,\theta_2,\theta_3) = \frac{v^4}{4v_z^2\cos^2\theta_1}\exp\left[-\frac{\pi^2(v_x^2+v_y^2)T^2}{\lambda_o^2}\right] \tag{16}$$

is a scattering function depending only on the scattering angles $(\theta_1,\theta_2,\theta_3)$, $c = \lambda_o\delta_o$.

If the surface is very rough the result in the specular direction is

$$P_{rsp}(t_1,t_2;\delta_1,\delta_2)=P_{rsp}(t,t;\delta_o,\delta_o)\exp\left[-\tfrac{1}{2}V^2(\delta_2-\delta_1)^2-\frac{V^2\delta_o^2K^2(t_2-t_1)^2}{T^2}\right]\cdot\exp\left[-j\frac{2\pi(R_o+S_o)}{c}(\delta_2-\delta_1)\right] \quad (17)$$

where now

$$P_{rsp}(t,t;\delta_o,\delta_o) = 4\pi\cos^2\theta_1 \frac{T^2}{R_o^2S_o^2}\frac{L^2}{\lambda_o^2}\frac{1}{V^2\delta_o^2} \quad (18)$$

Further, for directions away from the specular, the expression is again modified by a scattering function which in the rough surface case takes the form

$$\ell(\theta_1,\theta_2,\theta_3)=\frac{v^4}{4v_z^2\cos^2\theta_1}\exp[-(\frac{\pi T}{Vc})^2(v_x^2+v_y^2)] \quad (19)$$

Plots of the scattering function $\ell(\theta_1,\theta_2,\theta_3)$ for the two cases show that for a slightly rough surface most of the energy is scattered in the neighborhood of the specular direction, while for very rough surface scattering in the backward direction becomes more important and eventually even dominates.

It is also noted that in other respects the main difference between the slightly rough surface approximation and the very rough surface approximation is that in the former the magnitude of $P_r(t_1,t_2;\delta_1,\delta_2)$ increases with Rayleigh parameter while in the latter it decreases. In fact, using the exponential integral function a closed-form expression for the normalized autocorrelation function in the specular direction can be derived, which is zero for $V\delta_o = 0$ and $V\delta_o = \infty$ and it peaks in the vicinity of $V\delta_o = 1$ [5].

The previously described Eckart-type approximation is concerned with the so-called Fraunhofer limits and presupposes a very directional transmitter. Theoretical analyses based on a Gulin-type model [17], which retain some second order terms and obtain the so-called Freshnel limit were recently developed. These analyses, though more complex mathematically, result in a better channel model in those cases where the transmitter directionality constraints are not satisfied. An investigation of the effect of the beam pattern of a source exciting an acoustic scattering channel using the Freshnel corrected physical optics approximation is reported in [18]. Gulin-type approximations for the statistical description of a sea surface scattering channel are presented in [19-23] and a theoretical analysis for a surface scattering channel with several bounces can be found in [24].

4. EXPERIMENTAL RESULTS

In an effort to verify the results of the various theoretical calculations which have been performed, a great variety of experimentation has been reported, primarily in the last 15 years. The general problem of measurement of linear, time-varying stochastic channels was discussed in [25-26]. The experimental methods employed for the measurement of sound reflection from the sea surface can be classified into field experimentation in the ocean and physical scale modelling.

Ocean studies offer the advantage of experimentation with the real channel, but suffer from a number of severe drawbacks and have produced results of only limited usefulness in the confirmation of basic theory [27]. The most severe of these disadvantages are the problem of accurate calibration, the problem of measurement and the inability to isolate the scattered wave from multiple propagation modes, extraneous sources of sound, and the presence of several scattering and reflecting mechanisms.

The physical scale modelling method was adopted by a number of investigators (see Fortuin [10] for a relatively complete list). The general approach is to construct a water tank in which the sound source and receiver can be precisely controlled. The location and structure of the scattering surface is also precisely controlled and is usually either an artificially constructed random rough but fixed surface or the upper surface of the water excited by a constant velocity air stream. The scaling ratios vary from 1:100 to 1:1000 and a great variety of geometries have been employed. Medwin and Clay [28] reported such experimental results, while Gazahnes and Leandre [29] confirmed the validity of the Eckart-type model of [12-13] by experimental measurements. McDonald, Schultheiss and Spindel [30-32], obtained a number of results of surface statistics of a wind-driven surface by direct measurement. Measurements were made of one and two-dimensional probability densities of the surface heights, correlation functions, and spectra. The characteristics of the communication channel utilizing reflections from the rough surface were also obtained by the use of impulse response measurements. Direct measurements and first and second order moments were obtained. An improved physical scale model facility was used to conduct experiments in order to verify theoretical predictions in the scattering of sound from wind driven water surfaces [33-35]. The statistics of the surface were measured for the conditions of the scattering measurements. Theoretical predictions for each statistic and parameter set measured were also computed and were found to be in good agreement with the experimental results.

5. DETECTION AND SIGNAL DESIGN FOR COMMUNICATION OVER THE SURFACE SCATTERING CHANNEL

The problem of detection and signal design over purely random, linear, time-varying channels has received considerable attention in the literature [36-43]. The application of the theory to the kind of channel described in this paper, for realistic channel models, has not received sufficient attention. In this section two different approaches are taken, depending on the output signal-to-noise ratio. For very small signal-to-noise ratio it is well known that the optimum detector for a stochastic signal with additive Gaussian noise is an estimator-correlator. This detection is optimum in the sense that it minimizes the total error probability. It is also well known that the estimator-correlator should be designed to maximize the output signal-to-noise ratio coefficient C defined by

$$C = \frac{[\langle S/H_1\rangle - \langle S/H_o\rangle]^2}{\mathrm{Var}(S/H_o)} \tag{20}$$

where S is the output of the estimator-correlator at the end of the observation period T, $\langle S/H_o\rangle$ and $\langle S/H_1\rangle$ are the expectations of S under the two hypotheses H_o: signal is absent and H_1: signal is present respectively, and $\mathrm{Var}(S/H_o)$ is the variance of S under the hypothesis H_o. The estimator part of the estimator-correlator is then a linear filter with impulse response given by

$$K(t,\tau) = \begin{cases} 4R(t-\tau,t) & 0<\tau<t \\ 0 & \text{otherwise} \end{cases} \tag{21}$$

where $R(t_1,t_2) = \langle p^*(t_1)p(t_2)\rangle$ is the autocorrelation function of the signal at the channel output. (The factor of 4 is arbitrary and is used to simplify subsequent expressions.)

It is now possible to choose signal waveforms at the channel input that maximize C. The choice of signals can be guided by the following theorem:

Theorem [44]

For deterministic input signals to a purely random, linear, time-varying, very noisy, quasistationary channel the coefficient C satisfies the upper bound

$$\overline{C} = \left(\frac{E_{in}}{N_o}\right)^2 \frac{|P_n(0,0)|^2}{2} \tag{20}$$

Iff

$$|\theta_s(t'',\delta'')| = E_{in} \; |P_n(0,0)|^{-1} \; |P_n(t'',\delta'')| \tag{21}$$

and the input waveform has the form

$$z(t) = \exp(-\, at^2 + bt + c) \tag{22}$$

where a,b,c are arbitrary complex constants with Re(a) > 0.

Note that the previous theorem states that the θ function of the transmitted signal should be identical in form with the system autocorrelation function of the channel. It also states that the signal should be a Gaussian pulse, possibly with linear frequency modulation; however if the signal is to be matched to the channel this implies that the channel impulse response should have this Gaussian pulse form also. It is interesting to note that the simplifying results used to obtain equations (17) and (19) of this paper result in this kind of channel; also a commonly used model for reverberation has this form [3]. Thus although the upper bound is not realized in general, those two channel models would, in fact result in realization of the bound.

If the signal-to-noise ratio is large the simple output signal-to-noise ratio C of equation (20) is no longer adequate since it does not take into account the change in variance resulting from the signal. Hence a new criterion C_λ is defined by [45-47]

$$C_\lambda = \frac{[\langle S/H_1\rangle - \langle S/H_o\rangle]^2}{\lambda \mathrm{Var}(S/H_1) + (1-\lambda)\, \mathrm{Var}(S/H_o)} \tag{23}$$

For λ = 0 this criterion is the same as C used above, but it is shown that depending on the range of signal-to-noise ratio expected λ should range in the vicinity of 0.25[47-48].

Let the output signal be denoted by

$$g(t) = \mathrm{Re}[p(t)\, \exp(j\; 2\pi f_o t)] \tag{24}$$

where p(t) is the complex envelope. The second and fourth order moments of p(t) are defined respectively by

$$R(t_1,t_2) = \langle p^*(t_1)\ p(t_2)\rangle \tag{25}$$

$$P(t_1,t_2,t_3,t_4)\rangle \langle p^*(t_1)\ p(t_2)\ p(t_3)\ p^*(t_4)\rangle \tag{26}$$

It is then possible to expand the envelope function in a Karhunen-Loève series of the form

$$p(t) = \sum_n p_n\ f_n(t) \tag{27}$$

where the $f_n(t)$ are the eigenfunctions corresponding to $R(t_1,t_2)$ satisfying

$$\int_0^T R(t_1,t_2)\ f_n(t_1)dt_1 = \lambda_n f_n(t_2) \tag{28}$$

so that the λ_n are the corresponding eigenvalues.

A general requirement for "C_λ-optimality" of the channel input signal under the assumption of white noise at the output is then that it should produce at the channel output a stochastic waveform with second and fourth-order statistics satisfying for all m and n

$$\sum_r \lambda_r \quad \langle p_n{}^* p_m p_r p_r{}^*\rangle + [2N_o\lambda_n + \frac{N_o^2 - k}{\lambda}]\lambda_n\delta_{nm} = 0 \tag{29}$$

Where N_o is the noise spectral level for the assumed white noise at the channel output and where k is an arbitrary constant [47-49]. A somewhat more complicated condition can also be derived for colored noise [50-51].

The requirement simplifies in the case of a Gaussian channel where it takes the form

$$\sum_r \lambda_r{}^2 + \lambda_n{}^2 + 2N_o\lambda_n + \frac{N_o^2 - k}{\lambda} = 0 \tag{30}$$

This result implies further that in the Gaussian channel the input signal should be designed to produce a specified number of equal eigenvalues at the channel output depending on the signal-to-noise ratio. The number of eigenvalues is given by the largest interger less than $\mu=\lambda^{\frac{1}{2}}\ (E_o/N_o)$; the eigenvalues are $\lambda^{-\frac{1}{2}}\ N_o$ or zero, and the maximum value of the C_λ achievable is $(E_o/N_o)/2(\lambda^{\frac{1}{2}} + \lambda)$ where E_o is the output signal energy. Such a result indicates that in the case of a Gaussian channel, binary on-off signaling, additive white Gaussian noise and adequately large signal-to-noise ratios, only transmitter optimisation is needed, while a simple form of the estimator-correlator receiver, namely that of a power detector, should be retained.

A related result, valid for the Gaussian channel with white noise at the output is that the signal should be designed with an ambiguity function $\theta_s(t'',f'')$ satisfying

$$\int_{-\infty}^{\infty}\int_{-\infty}^{\infty} |\theta_s(t'',\delta'')|^2 |P_n(t'',\delta'')|^2 dt''d\delta'' = \frac{E_{in}^2 \; P_n^2(0,0)}{\mu} \qquad (31)$$

where μ is the above mentioned number of non-zero eigenvalues and $P_n(t'',\delta'')$ was defined in (5).

Since the equality in (31) cannot be achieved for $\mu<2$, (31) implies that for low signal-to-noise ratio the signal ambiguity function should match the channel system autocorrelation function so as to maximize the left side of (31). For large signal-to-noise ratio on the other hand, the signal ambiguity function should be chosen to differ from the channel system autocorrelation function.

Example

To demonstrate the previous principles, for the slightly rough surface scattering model described by (14), three plots of the channel system autocorrelation function were obtained. In plotting the curves, the angles θ_1, θ_2, and θ_3 are taken to be 45°, 60° and 0° respectively. Since the sea surface is slightly rough, σ is taken to be 0.5 feet and a reasonable value for L is 10 feet (the illuminated area of radiation is then well within the first Fresnel zone). Since T << L, T is chosen to be 1. For smooth surfaces, a reasonable value for K is 0.1. The velocity of sound c is 4,950 feet per second. The plot of $|P_r(0,\delta'')|$ can be seen in figure 4, δ_o= 5KHz. All the plots are normalized so that the maximum value at $t''=\delta''=0$ is 1. The δ''-axis plot resembles a sinc function with its first zero at 1.5 KHz and the side lobes die away very quickly. The plot is found to approach quite closely to a Gaussian curve $\exp(-0.8\delta''^2)$ which is shown in dotted line in the same diagram.

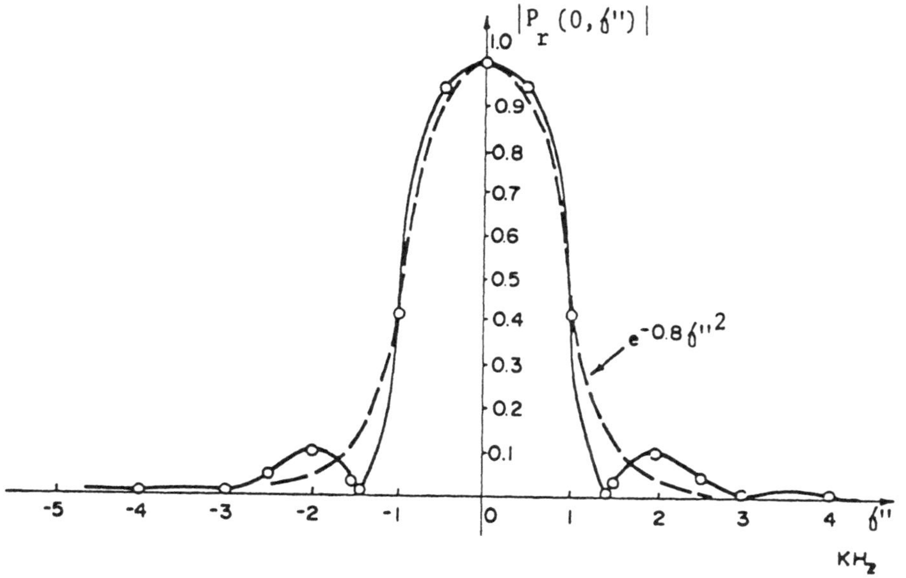

Fig. 4. Plot of normalized $|P_r(0,\delta'')|$ vs. δ'' for slightly rough surface.

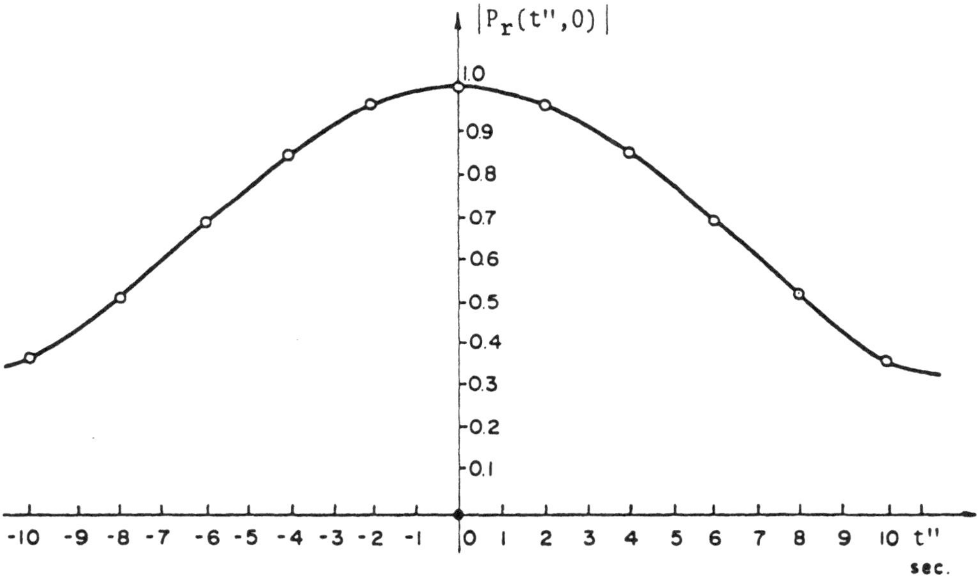

Fig. 5. Plot of normalized $|P_r(t'',0)|$ vs. t'' for slightly rough surface.

Figure 5 shows the t" axis plot which is clearly a Gaussian shape. Figure 6 shows the $|P_r|$ function plotted over a quadrant of the t"-δ" plane. Because of the symmetry about the origin, the shape in the other three quadrants can be pictured easily. Using the approximation that along the δ"-axis, the curve is $\exp(-0.8\delta''^2)$ and also since the plot is Gaussian along the t"-axis, it can be concluded that this ambiguity function approaches quite closely that of a Gaussian monotone pulse. From figure 6, the level contour at the magnitude 0.5 will give an ellipse with the major axis aligned with the t"-axis. The signal that matches the channel is therefore of the form

$$z(t) = \left(\frac{2K_o^2}{\pi}\right)^{1/4} \exp(-K_o^2 t^2)$$

where K_o is a function of the sea surface statistics.

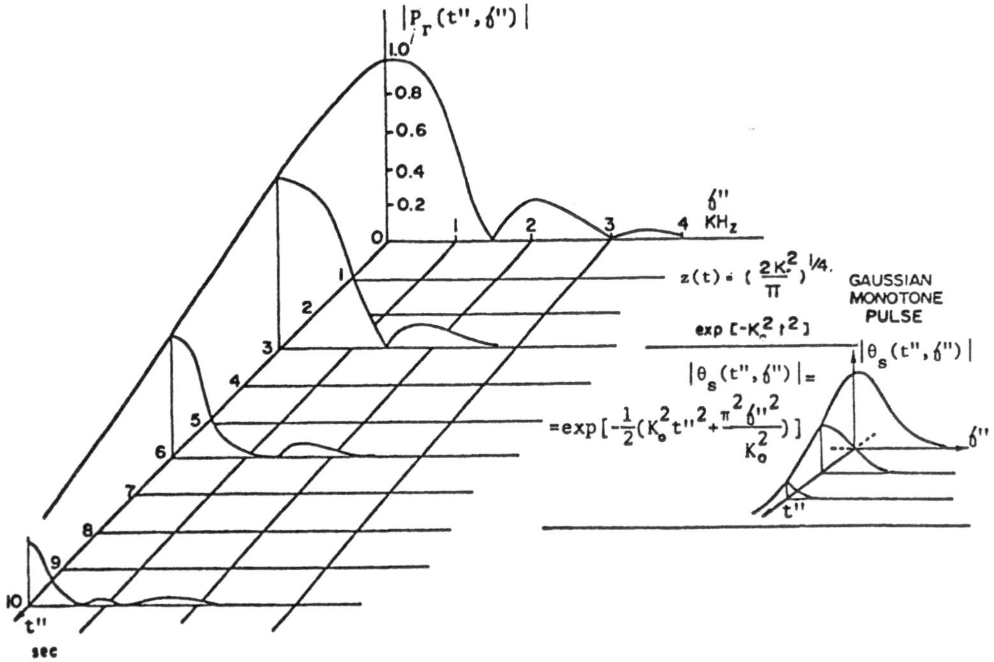

Fig. 6. t"-δ" plot of normalized $|P_r(t'',\delta'')|$ for slightly rough surface.

For the case of an extremely rough surface, as described in (17) and (19), examples of optimum transmitted waveforms have been reported in the literature [48].

Finally [52] employed computer simulation techniques to simulate an Eckart-type model and thus evaluate fixed-time and sequential detection schemes under small signal-to-noise ratio conditions and for a "very rough" surface channel.

6. CONCLUSIONS

This paper has presented a survey of the literature dealing with the formulation and solution of the binary communication problem over the sea surface scattering channel. The results presented in section 5, based on the Eckart-type channel model of section 3, indicate the form of signal waveshapes needed for optimum or suboptimum transmission of information under various signal-to-noise ratios and surface conditions. Research is under way to determine the desirable transmitter and receiver structures for improved theoretical models, recently reported in the literature, and for cases where the specular component at the channel output is significant.

REFERENCES

1. A.N. Venetsanopoulos and F.B. Tuteur, "Mathematical Modelling of the Undersea Acoustic Channel as a Linear, Time-Varying Stochastic Filter", Technical Report no. 1, U.S. Navy Underwater Sound Laboratory Research Contract N00-140-69-C-0045, Yale University, May 1969.
2. A.N. Venetsanopoulos, "Channel Modelling", Communications Technical Report no. 71-1, Department of Electrical Engineering, University of Toronto, Oct. 1971.
3. A.N. Venetsanopoulos, "Detection and Signal Design for Undersea Communications", Doctoral Thesis, available from University Microfilms Inc., Order No. 70-16, 573, Yale University, August 1969.
4. P.A. Bello, "Characterization of Randomly Time-Variant Linear Channels", IEEE Trans. on Com. Syst., vol. CS-11, No. 4, pp. 360-383, 1963.
5. R.C. Spindel, A.N. Venetsanopoulos, J.F. McDonald and F.B. Tuteur, "Propagation and Communications Through Underwater Acoustic Channels", Final Technical Report, U.S. Navy Underwater Sound Laboratory Research Contract N00-140-69-C-0045, Filed with the National Technical Information Service under AD 717 678, Yale University, Aug. 1970.
6. P.M. Woodward, "Probability and Information Theory with Applications in Radar", Pergamon Press, Oxford, 1953.

7. C.E. Cook and M. Bernfeld, "Radar Signals", Academic Press, 1967.
8. R.R. Kurth, "Distributed Parameter State Variable Techniques Applied to Communication over Dispersive Channels", Sc.D. Thesis, Department of Electrical Engineering, M.I.T., June 1969.
9. R.R. Kurth, "A Distributed Parameter State Variable Model for Time-Variant Channels", Sperry Rand Co., Aug. 1970.
10. L. Fortuin, "Survey of Literature on Reflection and Scattering of Sound Waves at the Sea Surface", J. Acoust. Soc. Amer., vol. 47, No. 5 (Part 2), pp. 1209-1228, 1970.
11. A.W. Ellinthorpe and A.H. Nuttall, "Theoretical and Empirical Results on the Characterization of Undersea Acoustic Channels", First IEEE Annual Communications Convention, Boulder, Colorado, June 1965.
12. A.N. Venetsanopoulos and F.B. Tuteur, "Physical Modelling of the Undersea Acoustic Channel", Technical Report no. 2, U.S. Navy Underwater Sound Laboratory Research Contract N00-140-69-C-0045, Yale University, June 1969.
13. A.N. Venetsanopoulos and F.B. Tuteur, "Stochastic Filter Modelling for the Sea-Surface Scattering Channel", J. Acoust. Soc. Amer., vol. 49, No. 4 (Part 1), pp. 1100-1107, April 1971.
14. L. Fortuin, "The Sea Surface as a Random Filter for Underwater Sound Waves", J. Acoust. Soc. Amer., vol. 52, No. 1 (Part 2), pp. 302-315, 1972.
15. C. Eckart, "Scattering of Sound from the Sea Surface", J. Acoust. Soc. Amer., vol. 25, pp. 566-570, 1953.
16. P. Beckmann and A. Spizzichino, "Scattering of Electromagnetic Waves from Rough Surfaces", Pergamon Press, New York, 1963.
17. E.P. Gulin, "Amplitude and Phase Fluctuation of a Sound Wave Reflected from a Sinusoidal Surface", Soviet Physics-Acoustics, vol. 8, No. 3, pp. 223-227, January-March, 1963.
18. J.F. McDonald and R.C. Spindel, "Implications of Fresnel Corrections in a Non-Gaussian Surface Scatter Channel", J. Acoust. Soc. Amer., vol. 50, No. 3 (Part 1), pp. 746-757, 1971.
19. J.F. McDonald, "Fresnel Corrected Second Order Interfrequency Correlations for a Surface Scatter Channel", IEEE Transactions on Communications, vol. COM-22, No. 2, pp. 138-145, Feb. 1974.
20. J.F. McDonald and P.M. Schultheiss, "Asymptotic Frequency Spread in Surface Scatter Channels at Large Rayleigh Numbers", J. Acoust. Soc. Amer., vol. 57, No. 1, pp. 160-164, Jan. 1975.
21. J.F. McDonald and F.B. Tuteur, "Calculation of the Range-Doppler Plot for a Doubly Spread Surface Scatter Channel at High Rayleigh Parameters", J. Acoust. Soc. Amer., vol. 57, No. 5, pp. 1025-1029, May 1975.
22. J.F. McDonald, F.B. Tuteur and J.G. Zornig, "Spatial Interfrequency Correlation Effects in a Surface Scatter Channel", J. Acoust. Soc. Amer., vol. 59, No. 6, pp. 1284-1293, June 1976.

23. F.B. Tuteur, J.F. McDonald and H. Tung, "Second-Order Statistical Moments of a Surface Scatter Channel with Multiple Wave Direction and Dispersion", IEEE Transactions on Communications, vol. COM-24, No. 8, pp. 820-831, Aug. 1976.
24. F.B. Tuteur, "Underwater Acoustic Scatter Channels with Several Bounces", J. Acoust. Soc. Amer., vol. 60, No. 4, pp. 840-842, Oct. 1976.
25. P.A. Bello, "On the Measurement of a Channel Correlation Function", IEEE Transactions on Info. Theory, vol. IT-10, No. 4, pp. 381-383, Oct. 1964.
26. P.A. Bello, "Measurement of Random Time-Variant Linear Channels", IEEE Transactions on Info. Theory, vol. IT-15, No. 4, pp. 469-475, July 1969.
27. H.W. Marsh, "Sound Reflection and Scattering from the Sea Surface", J. Acoust. Soc. Amer., vol. 35, No. 2, pp. 240-244, Feb. 1963.
28. H. Medwin and C.S. Clay, "Dependence of Spatial and Temporal Correlation of Forward-Scattered Underwater Sound on the Surface Statistics I. Experiment", J. Acoust. Soc. Amer., vol. 47, No. 5 (Part 2), pp. 1419-1429, May 1970.
29. C. Gazahnes and J. Leandre, "Notes des Membres et Correspondants et Notes Présentées ou Transmises par Leurs Soins", Academie Des Sciences, Paris, vol. 276, No. 15, pp. 635-638, 1973.
30. J.F. McDonald and R.C. Spindel, "A Computer Aided Experimental Characterization of an Acoustic Scattering Channel", The Symposium on Computer Processing in Communications, Polytechnic Institute of Brooklyn, April 8-10, 1969.
31. R.C. Spindel, "An Experimental Investigation of a Statistically Distributed Scattering Surface and Acoustic Scattering", Ph.D. Thesis, Yale University, Feb. 1971.
32. R.C. Spindel and P.M. Schultheiss, "Acoustic Surface-Reflection Channel Characterization through Impulse Response Measurements", J. Acoust. Soc. Amer., vol. 51, No. 6 (Part 1), pp. 1812-1824, 1972.
33. J.G. Zornig and J.F. McDonald, "A High Speed Microprogrammed System for Generation and Acquisition of Signals", Rev. Sci. Instrum., vol. 44, No. 9, pp. 1217-1222, Sept. 1973.
34. J.G. Zornig and F.B. Tuteur, "Measurement of the Statistics of a Surface Scatter Channel", Technical Report CS-1, Yale University, April 1974.
35. J.G. Zornig and J.F. McDonald, "Experimental Measurement of the Second-Order Interfrequency Correlation Function of the Random Surface Scatter Channel", IEEE Transactions on Communications, vol. COM-23, No. 3, pp. 341-347, March 1975.
36. T. Kailath, "Optimum Receivers for Randomly Varying Channels", Information Theory, Fourth London Symposium, Ed. Colin Cherry, Wash., pp. 109-122, 1961.
37. R.S. Kennedy and I.L. Lebow, "Signal Design for Dispersive Channels", IEEE Spectrum, pp. 231-237, March 1964.
38. T. Kailath, "A Note on Least-Squares Estimates from Likelihood Ratios", Info. and Control, vol. 13, pp. 534-540, 1968.

39. T. Kailath, "An Innovations Approach to Least-Squares Estimation, Part I: Linear Filtering in Additive White Noise" IEEE Transactions in Aut. Control, vol. AC-13, No. 6, pp. 646-655. Dec. 1968.
40. T. Kailath and P. Frost, "An Innovations Approach to Least Squares Estimation, Part II: Linear Smoothing and Additive White Noise", Transactions on Aut. Control, vol. AC-13, No. 6, pp. 655-660, Dec. 1968.
41. R.F. Daly, "Signal Design for Efficient Detection in Dispersive Channels", IEEE Transactions on Info. Theory, vol. IT-16, No. 2, pp. 206-213, March 1970.
42. S.K. Chow and A.N. Venetsanopoulos, "Optimal On-Off Signaling Over Linear Time-Varying Stochastic Channels", IEEE Transactions on Info. Theory, vol. IT-20, No. 5, pp. 602-609, Sept. 1974.
43. E. Conte, M. Longo and E. Mosca, "Signal Design for Low-Error Probability in Fading Dispersive Channels", IEEE Transactions on Communications, vol. COM-24(5), pp. 567-575, May 1976.
44. A.N. Venetsanopoulos and F.B. Tuteur, "Upper Bound of the Signal-to-Noise Ratio at the Output of a Threshold Receiver", IEEE Transactions on Info. Theory, vol. IT-17, No. 6, pp. 753-755, Nov. 1971.
45. P. Rudnick, "A Signal-to-Noise Property of Binary Decisions", Nature, vol. 193, No. 4815, pp. 604-605, Feb. 10, 1962.
46. K. Fukunaga and T.F. Krile, "A Minimum Distance Effectiveness Criterion", IEEE Transactions on Info. Theory, vol. IT-14, No. 5, pp. 780-782, Sept. 1968.
47. A.N. Venetsanopoulos and F.B. Tuteur, "Near Optimum Detection and Signal Design for Communications over Purely Random, Very Noisy Undersea Acoustic Channels", Technical Report no. 3, U.S. Navy Underwater Sound Laboratory Research Contract N00-140-69-C-0045, Yale University, pp. 67, July 1969.
48. A.N. Venetsanopoulos and F.B Tuteur, "Transmitter-Receiver Optimization for Active Signaling over Undersea Acoustic Channels", IEEE Transactions on Communication Technology, vol. COM-19, No. 5, pp. 649-659, Oct. 1971.
49. F.B. Tuteur and A.N. Venetsanopoulos, "A Suboptimal Approach to Transmitter-Receiver Design for Communication over Random Channels", Proceedings of the 14th Midwest Symposium on Circuit Theory, University of Denver, pp. 15.6-1 - 15.6-7, May 6-7, 1971.
50. A.N. Venetsanopoulos and F.B. Tuteur, "Suboptimum Detection and Signal Design for Communications over Purely Random, Undersea Acoustic Channels", Technical Report no. 5, U.S. Navy Underwater Sound Laboratory Research Contract N00-140-69-C-0045, Yale University, pp. 79, Aug. 1969.
51. A.N. Venetsanopoulos and F.B. Tuteur, "Signal Design for Stochastic Linear Channels with Additive Colored Noise", Proceedings of the IEEE National Telecommunications Conference, Houston, Texas, IC-1-IC-6, Dec. 4-6, 1972.

52. C.J.M. Turnbull and A.N. Venetsanopoulos, "Simulation Techniques in Undersea Communications", Proceedings of the 1971 IEEE International Conference on Engineering in the Ocean Environment, San Diego, California, pp. 369-372, Sept. 21-23, 1971.

Panel Discussion

on

CHANNEL STATISTICS AND MODELS *

Chairman:	Professor J.L. Lacoume	CEPHAG, Grenoble Cedex, France
Panel:	Professor G. Tacconi	Istituto de Elettrotecnica, Genova, Italy
	Dr. G. Jourdain	CEPHAG, Grenoble Cedex, France
	Dr. P.T. Nielsen	SHAPE Technical Centre, The Hague, Holland
	Professor J.G.Proakis	North Eastern University, Boston, U.S.A.
Contributors:	Mr. B.G. West	GEC-Marconi Electronics, Great Baddow, U.K.
	Dr. G. Dieterich	Standard-Elektrik Lorenz, Stuttgart,Germany
	Prof.A.N.Venetsanopoulos	Univ. of Toronto,Canada
	Prof. S.R.Robinson	Wright-Patterson AFB, U.S.A.

The panel discussion was divided into four topics, each one introduced by a member of the panel.

Channels with Spatially Extended Transmitter and Receiver

This topic was introduced by Prof. Lacoume, as a special case of the analysis for point source and receiver presented previously at the Institute [Review of Methods for Description of Random Transmission Channels - Lacoume and Jourdain]. He considered only the time-stationary case, and defined the "mutual coherency" function of a signal propagating between two points as the cross-spectral-density between the two points

* The editor wishes to express his thanks to Dr. Farrell for the preparation of this report.

normalised by the product of the power-spectral densities at each point. This coherency function is a very important parameter in the design of the receiver signal processing array. The modulus of the coherency always lies between zero and unity; when unity, the propagation is totally coherent; when zero, totally incoherent. A deterministic channel (e.g., as in the optical case) with a totally coherent input clearly has a totally coherent output, and a coherent array could be designed. When the input to the deterministic channel is totally incoherent, then the coherency function between two points in the output reproduces the diffraction pattern of the source, focussed on one of the points. Thus an important result emerges: even with a totally incoherent source, some coherency is found at the receiver (i.e. a coherence length is involved). Also, there is some kind of equivalence between time and frequency; that is, by using a large bandwidth signal and only two sensors, a family of sensors can be simulated.

Prof. Tacconi continued the discussion by considering some of the basic physical properties of underwater transmission channels. Previous models of the channel had taken into account the direct path, bottom bounce, and surface scattering, but had neglected volume scattering. This last factor turns out to be more important than expected, however, and so the model had to be modified to take account of it. The new model of the ocean which results can be thought of as a quantity of domains (blobs) with differing refractive index. The size, lifetime and velocity of the blobs are "statistically determinate" processes, and measurements have been carried out to characterise their parameters.

The relevance of studies on propagation through random media (i.e. high ionospheric investigations) to the above problems was pointed out.

Measurement of Scattering Functions

This topic was introduced by Madame Jourdain, who began by making two points about scattering functions. First, the scattering function is mathematically simple; and second, it has physical meaning. Some problems were then mentioned. The wide-sense-stationary uncorrelated scatterers (WSSUS) hypothesis is not always valid, or it may only be valid for a small range of frequency and time. Simple test signals (e.g. monochromatic waves) may be used to test for the WSSUS condition, but the results need to be interpreted carefully since the WSSUS channel may be randomly time-varying, or reverberating. A further problem is that the scattering function may have a mean value, or specular component, which must be removed before further processing or interpretation. Another difficulty concerns the choice of signals for testing the channel. A good choice is to use signals with an ambiguity function peak close to the origin of the time-frequency domains - the "performant" signals mentioned in the previously referenced presentation by Prof. Lacoume and Madame Jourdain. If the medium is underspread (product of the frequency spread by the time spread less than unity), however, then simpler signals can be used (e.g. monochromatic waves or impulse signals). A final problem concerns the duration over which a statistical average is valid; that is, to determine the scattering function, which is a statistical property, the medium must be statistically characterised.

Dr. Nielsen continued with a few words relating the general topic to the problem of measuring the statistical properties of the troposcatter medium. He thought that the WSSUS model for the troposcatter case was confidently established experimentally. The medium could be assumed, on average, to be stationary for a few tens of minutes. Concerning the assumption of uncorrelated scattering, Dr. Nielsen pointed out that a channel which may be uncorrelated for one signal may not be for another. It depends on the bandwidth of the transmission; this must be much less than the velocity of light divided by the scale of the turbulence, if the uncorrelated condition is to hold. For a 10 MHz signal, for example, the critical scale of turbulence becomes something like 30 metres, so that the uncorrelated assumption is quite good for most cases. In the troposcatter case, the multipath delay spread, or dispersion, tends to be of more interest than the more general scattering function. Measurements of dispersion of course involve a transmitter and a receiver as well as the medium, so that the results can be misleading, particularly for short paths with small dispersions. Deconvolution of the transmitter and receiver responses is required in order to get the true response of the medium.

Mr. West showed by means of a simple argument that the normalised channel capacity versus signal-to-noise ratio curve for a Rayleigh fading channel is only about 2 dB worse than the corresponding curve for the non-fading case. Various aspects of the problems of determining and realising channel capacity on fading channels were then discussed.

Signal and Receiver Design in Fading and Dispersive Media

Prof. Proakis then introduced some signal design techniques for fading channels. In the underspread channel case (e.g. HF and troposcatter) perhaps the oldest technique is FDM, or parallel sub-channel transmission. Here intersymbol interference (ISI) is minimised by making the pulse (bit) duration much greater than the multipath spread, and simultaneously much less than the channel coherence time. Inter-Channel Interference (ICI) is minimised by having a frequency guard-band between sub-channels at least equal to the Doppler spread. Thus roughly 30 sub-channel 100 Hz bandwidth each can be packed into a 3 KHz HF bandwidth, bit duration on each sub-channel being about 13 msec (e.g. KINEPLEX). If the signalling (bit)rate can be lower, then a spread-spectrum diversity (redundancy) technique, as in the RAKE receiver, offers a performance advantage over FDM. The signalling bandwidth is much greater than the coherence bandwidth, and orthogonal signalling is used. The receiver consists of an automatic (set by channel measurements) transversal filter with tap spacings equal to the reciprocal of the signal bandwidth. More recently, equalisation methods have come into vogue, applied to channels where the entire bandwidth available is used for high speed serial data transmission. In this case there is ISI, and equalisation is needed to remove it. Linear adaptive equalisation does not work, because it cannot compensate for spectral nulls. Decision feedback equalisers have been shown to work, however (e.g. a 10 Kbit/sec tropo system build by Signatron), and the Viterbi algorithm can also be used; the Viterbi algorithm is several dB better (depending on the channel characteristics) than decision feedback equalisation.

In the overspread channel case (e.g. the underwater acoustic channel) again there are a number of solutions to the transmission problem. An FDM parallel sub-channel scheme can be used, but as the channel is overspread each sub-channel bit cannot be coherently integrated over the full bit period; instead, the bit period must be broken up into sub-periods, each of which may be coherently integrated. A RAKE type technique is possible, but channel measurements cannot be used to set the equaliser, only energy detection at each tap. As a last technique, Prof. Proakis described an EDC coding method, useful for both underspread and overspread channels, but particularly for the overspread case. A set of N frequency tones is selected, covering the available channel bandwidth in such a way that adjacent tones are separated by more than the coherence bandwidth of the channel. Thus during transmission the tones will fade independently. A binary fixed-weight code is used, with block length N, and M code words, where $M = 2^K$ and K is the number of information bits in each code word. Each digit position in a code word maps into one of the N tones; if the digit is a ONE, that tone is sent, if a ZERO, it is not. This is multi-tone on/off keying. The optimum maximum likelihood detector for this scheme is a soft-decision detector which consists of a set of N filters followed by square-law devices (energy detection because the channel is overspread), from which M decision variables may be found (corresponding to the M possible transmitted code words), and the largest selected. The reason for using a fixed weight code is that the maximum likelihood detector is relatively simple, since no bias terms are needed. A suitable code might be a Hadamard code with $N = 20$ (i.e. 20 tones needed), $M = 32$ (i.e. $K = 5$), and code word weight 10 (i.e.,10 tones keyed for each transmission). The minimum distance of the code is 10, and the "effective diversity" is $10/2 = 5$. This coding scheme thus requires less bandwidth than an equivalent scheme with no coding having the same effective diversity (160 tones are required to get diversity 5 with 32 possible waveforms). Concatenated block codes, or convolutional codes, will give even better performance than the simple coding scheme described.

Prof. Venetsanopoulos then spoke on the difficult problem of describing the performance of a system without having to prespecify the structure of the transmitter and receiver; that is, without initially specifying the input signal of its optimum receiver. If a system could be so described,then hopefully it could be globally optimised with respect to some performance criterion (e.g. error rate). Most work to date has either specified the signal, and then optimised the receiver structure, or vice-versa. Various signal and receiver design aspects were mentioned to give a deeper understanding of the problem, and the relationship between signal design and receiver design in certain idealised circumstances was used to give an insight into their joint optimisation in more typical situations with real channels (e.g. the underwater channel).

Finally, Dr. Dieterich spoke on the advantages of using the vector space approach to analyse and design synchronous digital data transmission systems.

Optical Communication

Prof. Robinson described some of the difficulties of modelling optical propagation channels. Obtaining simple input-output relations for an optical system is difficult, for example, if the propagation medium is other than free space. This is particularly true if the medium is turbulent, but quite a lot of work has been done on turbulent atmospheric and underwater (green light) optical propagation. Another difficult aspect is that of lining up an optical link between an aircraft and a ground site because it involves statistical modelling of reflected optical fields subject to clutter (rather like the radar problem). A final related optical problem is that of "speckle" on coherently recovered images, due to scattering, etc., as the statistical properties of speckle are not well understood.

Part VII

NON-STATIONARY AND NON-GAUSSIAN SIGNAL/NOISE ANALYSIS

organised by Dr. G. A. Richards

REPRESENTATION OF BIVARIATE DISTRIBUTIONS WITH APPLICATIONS

John B. Thomas

Department of Electrical Engineering,
Princeton University, Princeton, N.J. 08540

ABSTRACT. In the general area of communication theory, some of the most difficult problems involve non-Gaussian processes and/or nonlinear systems. Such problems arise particularly in modulation, detection, and estimation systems. Bivariate distributions are required in the analysis of the second-order properties of nonlinear systems, in the construction of certain classes of Markov processes, in the synthesis of optimal detectors when the input observations are not independent, and in many other problems in communications. This paper considers the series representation of bivariate distributions in the Barrett-Lampard manner by orthonormal polynomials. Attention is centered on the structure of the bivariate distribution when the marginals are known. Some unifying results are given as well as some extensions of the results of Barrett and Lampard and of Brown. Using this representation a class of Markov processes is constructed, the bandwidth of memoryless systems is investigated, and applications to optimal detectors are discussed.

INTRODUCTION

We begin with a bivariate density function $h(x,y)$ with marginal densities $f(x)$ and $d(y)$. Let ν be a measure on the x-y plane and let $L^2(d\nu)$ denote the space of all real Baire functions defined on the x-y plane that are square integrable with respect to the measure ν. If we define the inner product $(u,v) = \int uv\, d\nu$ on this space, then it becomes a Hilbert space. Let the measure ν be defined by

$$\nu(E) = \iint_E f(x)d(y)dxdy.$$

If the bivariate density function satisfies the condition that

$$\iint \frac{[h(x,y)]^2}{f(x)d(y)} dxdy < \infty , \qquad (1)$$

then the function $h(x,y)/f(x)d(y)$ belongs to the Hilbert space.

It now follows from the properties of Hilbert space [1] that there exist series expansions of the form

$$h(x,y) = f(x)d(y) \sum_{n=0}^{\infty} \sum_{m=0}^{\infty} c_{nm}\theta_n(x)\phi_m(y) , \qquad (2)$$

where $\theta_n(x)$ and $\phi_m(y)$ are sequences of functions that are orthonormal with respect to the weighting functions $f(x)$ and $d(y)$, respectively; the coefficients are given by

$$c_{nm} = \iint \theta_n(x)\phi_m(y)h(x,y)dxdy = E\{\theta_n(X)\phi_m(Y)\} \qquad (3)$$

and the convergence is in the sense that

$$\lim_{N,M\to\infty} \iint \left[\frac{h(x,y)}{f(x)d(y)} - \sum_{n=0}^{N} \sum_{m=0}^{M} c_{nm}\theta_n(x)\phi_m(y)\right]^2 f(x)d(y)dxdy=0 \qquad (4)$$

Parseval's relation is

$$\iint \frac{[h(x,y)]^2}{f(x)d(y)} dxdy = \sum_{n=0}^{\infty} \sum_{m=0}^{\infty} (c_{nm})^2 \qquad (5)$$

For any two sequences of orthonormal functions, if the coefficients are defined by Eq. (3), then Eq. (5) is a necessary and sufficient condition for the series expansion to represent the bivariate density function in the way described by Eq. (4). In general, the orthonormal functions need not be complete. However, if both sequences are complete, then the series expansion does represent the bivariate density function in the way described by Eq. (4).

There are obviously many sequences of orthonormal functions that may be used in Eq. (2), and the resulting two-dimensional expansions may become quite complicated. One way to treat the problem is to look at the coefficient matrix $\underline{C}$ where

$$\underline{C} = [c_{nm}] = \begin{bmatrix} c_{00} & c_{01} & c_{02} \cdots \\ c_{10} & c_{11} & c_{12} \cdots \\ c_{20} & c_{21} & c_{22} \cdots \\ \vdots & \vdots & \vdots \end{bmatrix} \tag{6}$$

and to consider ways to simplify this infinite matrix. Fortunately an important and useful special case occurs when $c_{nm} = 0$ for $n \neq m$; that is, when the matrix $\underline{C}$ is diagonal. Now the series expansion of the bivariate density function can be written as

$$h(x,y) = f(x)d(y) \sum_{n=0}^{\infty} a_n \theta_n(x) \phi_n(y) \tag{7}$$

where a_n has been written for c_{nn}. The following theorem shows that the sequences of orthonormal functions can always be chosen to result in a diagonal series expansion.

Theorem 1 (Lancaster [2,3]): Consider the bivariate density function $h(x,y)$ with marginals $f(x)$ and $d(y)$. If Eq. (1) holds, then $h(x,y)$ possesses the diagonal series expansion of Eq. (7), where $\theta_n(x)$ and $\phi_n(y)$ are sequences of functions that are orthonormal with respect to the weight functions $f(x)$ and $d(y)$, respectively; $\theta_0(x) \equiv \phi_0(y) \equiv 1$; the coefficients are given by

$$a_n = \iint \theta_n(x) \phi_n(y) h(x,y) dxdy \ ;$$

and the convergence is in the sense that

$$\lim_{N\to\infty} \iint \left[\frac{h(x,y)}{f(x)d(y)} - \sum_{n=0}^{N} a_n \theta_n(x) \phi_n(y)\right]^2 f(x)d(y)dxdy = 0$$

Throughout this paper we will assume that all bivariate density functions satisfy the integrability condition of Eq. (1). This integrability condition is fairly weak and, loosely speaking, most bivariate density functions do satisfy it. Thus, on the strength of Theorem 1, any bivariate density function that we consider will possess a diagonal series expansion. It follows from the Schwarz inequality that the form of convergence in Theorem 1 implies the weaker form of convergence

$$\lim_{N\to\infty} \iint \left| h(x,y) - f(x)d(y) \sum_{n=0}^{N} a_n\theta_n(x)\phi_n(y)\right| dxdy = 0$$

Barrett and Lampard [4] seem to have been the first to recognize the engineering significance of using diagonal series expansions to represent bivariate density functions. They considered the class Λ of bivariate density functions that possess a diagonal series expansion where each n-th orthonormal function is given by an n-th degree polynomial. The orthonormal polynomials can be obtained by using the Gram-Schmidt procedure to orthonormalize the sequences $\{1, x, x^2, x^3, \ldots\}$ and $\{1, y, y^2, y^3, \ldots\}$, using $f(x)$ and $d(y)$ as the respective weighting functions. In particular, they noted that the Gaussian bivariate density function belongs to this class. Brown [5] gave necessary and sufficient conditions for a bivariate density function to belong to the class Λ.

The following theorem on conditional expectations is easily proved [6]:

Theorem 2: Let $h(x,y)$ be a bivariate density function having the diagonal series expansion of Eq. (7). Let $u(x)$ be a Baire function such that $\int [u(x)]^2 f(x)dx < \infty$, and let $b_n = \int \theta_n(x)u(x)f(x)dx$. Then

$$E\{u(X)/y\} = \underset{N\to\infty}{\text{l.i.m.}} \sum_{n=0}^{N} a_n b_n \phi_n(y). \tag{8}$$

Notice that the function $u(x)$ is not required to possess a convergent expansion in terms of the orthonormal functions. Also, notice that, if only a finite number of the b_n are nonzero, then the expression for the conditional expectation holds with probability one. In particular, the following equation holds with probability one:

$$E\{\theta_n(X)/y\} = a_n\phi_n(y). \tag{9}$$

Recall that the minimum mean squared error estimate of one random variable in terms of another is given by the conditional expectation of the first random variable, conditioned on the second random variable [7]. Thus Theorem 2 relates diagonal series expansions of bivariate densities to minimum mean squared error estimation. In particular, we see that, for class Λ bivariate density functions, the minimum mean squared error estimate of one random variable in terms of the other is linear.

NON-DIAGONAL EXPANSIONS

The concept of the diagonalness of the coefficient matrix $[c_{ij}]$ can be generalized by considering matrices which are zero except for the main diagonal and for the first i diagonals above and the first j diagonals below the main diagonal.

Definition 1. If a bivariate density h(x,y) is expressible as in Eq. (2) where $\{\theta_m(x)\}$ and $\{\phi_n(y)\}$ are orthonormal polynomials, then h(x,y) is in the class Γ_{ij} of bivariate densities if and only if $c_{mn} = 0$ for $n - m > i$ and for $m - n > j$ where $i,j = 0,1,2,\ldots$. Note that the class Γ_{00} is the same as the class Λ of Barrett and Lampard [4]. The class Γ_{ij} can be characterized by the conditional expectations. The following theorem gives this characterization. The special case where $i = 0$ and $j = 0$ is due to Brown [5].

Theorem 3: If the bivariate density h(x,y) is expressible as in Eq. (2) where $\theta_m(x)$ and $\phi_n(y)$ are orthonormal polynomials, of order m and n respectively, then h(x,y) is in the class Γ_{ij} if and only if

(i) $E\{X^k | Y = y\}$ is a polynomial of order $\leq k+i, \forall k \geq 0$,

and (10)

(ii) $E\{Y^k | X = x\}$ is a polynomial of order $\leq k+j, \forall k \geq 0$,

for $i,j = 0,1,2,\ldots$.

Proof: Assume h(x,y) is in class Γ_{ij}. Then the conditional expectation $E\{X^k|Y = y\}$ is

$$E\{X^k|Y = y\} = \int_{-\infty}^{\infty} x^k \frac{h(x, y)}{d(y)} dx \tag{11}$$

On substituting for h(x,y) from Eq. (2) and interchanging the order of integration and summation, we have

$$E\{X^k|Y=y\} = \sum_{m=0}^{\infty} \sum_{\substack{n=m-j \\ n\geq 0}}^{m+i} c_{mn}\phi_n(y) \int_{-\infty}^{\infty} \theta_m(x)x^k f(x)dx \tag{12}$$

Since x^k is representable as a linear combination of $\theta_0(x)$, $\theta_1(x)$,...,$\theta_k(x)$, the integral in Eq. (12) is zero for $m > k$; therefore $E\{X^k|Y = y\}$ is a polynomial of order less than or equal to $k + i$, $\forall k \geq 0$. Similarly, $E\{Y^k|X = x\}$ is a polynomial of order less than or equal to $k + j$, $\forall k \geq 0$. This implies that $E\{\theta_m(X)|Y = y\}$ is a polynomial of order less than or equal to $m + i$; therefore, by Eq. (3), $c_{mn} = 0$ for $n > m + i$. In the same way, if $E\{Y^k|X = x\}$ is a polynomial of order less than or equal to $k + j$, $\forall k \geq 0$, then $c_{mn} = 0$ for $m-n>j$. In general, the following two remarks hold for the matrix $[c_{ij}]$:

Remark 1. Since $\theta_0(x)$ and $\phi_0(y)$ are each unity, it follows from Eq. (3) that $c_{i0} = c_{0i} = 0$, $\forall i > 0$, and $c_{00} = 1$.

Remark 2. $E\{\theta_i(X)|Y = y\} = \sum_{\ell=0}^{\infty} k_{i\ell}\phi_\ell(y)$

if and only if $c_{ij} = k_{ij}$, $\forall j \geq 0$; $\forall i \geq 0$.

A similar remark can be made by interchanging $\theta_i(x)$ and $\phi_j(y)$ and interchanging rows and columns. Note that the rows and columns of $[c_{ij}]$ are determined by the conditions given in Remark 2. Also it is clear that either the equation

$$E\{\theta_i(X)|Y = y\} = c_{ii}\phi_i(y) \ , \ \forall i \geq 0 \tag{13}$$

or the equation

$$E\{\phi_j(Y) | X = x\} = c_{jj}\theta_j(x) \quad , \forall j \geq 0 \tag{14}$$

constitutes a set of necessary and sufficient conditions for $[c_{ij}]$ to be diagonal.

Despite the attractiveness of Eq. (10) and its relation to Balakrishnan's interesting work [8], it is not clear how useful is the class Γ_{ij} for $i,j \neq 0$. The only case for which a number of meaningful results have been obtained is the $\Gamma_{00} = \Lambda$ class of Barrett and Lampard. We proceed now to discuss some applications of this Λ class.

BANDWIDTH EFFECTS OF MEMORYLESS SYSTEMS

In this section, we consider the effect of a memoryless nonlinearity $g(\cdot)$ on the spectrum of a random process. The process $\eta(t)$ will be a second-order random process that is mean-square continuous and second-order stationary. Let $L_2(f)$ denote the class of all real Baire functions [9] that are square integrable with respect to $f(\cdot)$, the common univariate density of $\eta(t)$. With the inner product defined as $(u,v) = \int u(x)v(x)f(x)dx$, $L_2(f)$ becomes a Hilbert space. Throughout, we will assume that the nonlinearity $g(\cdot)$ belongs to $L_2(f)$ and also that it is not a constant.

We shall use the notation $X = \eta(t)$ and $Y = \eta(t + \tau)$. We assume that the bivariate density $h(x,y;\tau)$ of X and **Y** can be represented as a diagonal expansion as in Eq. (7) so that

$$h(x,y;\tau) = f(x)f(y) \; [1 + \sum_{n=1}^{\infty} a_n(\tau)\theta_n(x)\theta_n(y)] \tag{15}$$

where $\{\theta_n(\cdot)\}$ is a complete sequence of orthonormal functions in $L_2(f)$ with $\theta_0(x) \equiv 1$. The coefficients $a_n(\tau)$ are given by Eq. (3); that is,

$$a_n(\tau) = E\{\theta_n[\eta(t)]\theta_n[\eta(t + \tau)]\} \;, \tag{16}$$

and the convergence is in the mean-square sense of Eq. (4).

As previously stated, this diagonal-series expansion has been used successfully by many authors to analyze the effect of a non-linearity on a random process. For example, Barrett and Lampard [4] investigated the "cross-covariance property," while Leipnik [10] gave a more general treatment of such expansions. Also, Blachman [11] treated the uncorrelated output components of a memoryless non-linearity, and Brown [12,13] considered the reconstruction of the spectrum after the nonlinearity.

We now consider the sequence of functions $a_n(\cdot)$. Since the functions $\theta_n(\cdot)$ are normalized, it follows that $a_n(0) = 1$.

Lemma 1: The functions $a_n(\cdot)$ are nonnegative definite since $a_n(\tau)$ is the autocorrelation function of the random process $\theta_n[\eta(t)]$.

Lemma 2: If $g(\cdot)\epsilon L_2(f)$, then $g[\eta(t)]$ is mean-square continuous, as proved in [14].

Therefore, the functions $a_n(\tau)$ are continuous, nonnegative-definite functions. It follows from Bochner's theorem [15,p.126] that $a_n(\tau)$ is of the form

$$a_n(\tau) = \int_{-\infty}^{\infty} \cos\omega\tau dA_n(\omega) \tag{17}$$

where $A_n(\omega)$ is a bounded nondecreasing real function, the spectral distribution function of the process $\theta_n[\eta(t)]$.

We assume now that the process $\eta(t)$ is passed through the nonlinearity $g(\cdot)$, resulting in the output $\mu(t) = g[\eta(t)]$. Since $g(\cdot)\epsilon L_2(f)$, it admits the representation [11,16]

$$g(x) = \sum_{n=0}^{\infty} b_n\theta_n(x) \tag{18}$$

where b_n is given by

$$b_n = \int_{-\infty}^{\infty} g(x)\theta_n(x)f(x)dx \tag{19}$$

and where the convergence is in $L_2(f)$. Let $R(\tau)$ denote the autocovariance function $E\{[\mu(t) - E\mu(t)][\mu(t + \tau) - E\mu(t + \tau)]\}$ of the output $\mu(t)$. Then it is easily seen that

$$R(\tau) = \sum_{n=1}^{\infty} b_n^2 a_n(\tau) \tag{20}$$

since $b_0 = E\{g[\eta(t)]\} = E\{\mu(t)\}$ and $a_n(0) = 1$. The sequence $\{b_n\}$ is square summable since $g(\cdot)\epsilon L_2(f)$, and $|a_n(\tau)| \leq 1$; thus, the series in (20) converges absolutely and uniformly. We now transform both sides of (20) to obtain $D(\omega)$, the spectral distribution function of the output. Because of uniform convergence, the series can be transformed term by term to yield

$$D(\omega) = \sum_{n=1}^{\infty} b_n^2 A_n(\omega) \quad . \tag{21)}$$

The spectral distribution of each random process $b_n\theta_n[\eta(t)]$ contributes additively to the output spectral distribution. Since the spectral distributions $A_n(\omega)$ depend in a complicated manner on the second-order properties of the input process, it is difficult to draw general conclusions about the shape of the output spectrum. However, specific results can be obtained by considering special classes of random processes.

Spherically invariant processes

A centered spherically invariant random process (SIRP) is a random process [17] whose n-th order characteristic function is a function of an n-th order non-negative definite quadratic form. For our purposes, we need consider only positive definite quadratic forms so that the n-th order characteristic function of the SIRP denoted by $\eta_1(t)$ is $\psi_n(u_n R_n u_n')$ where u_n is an n-dimensional row vector and R_n is an n x n positive definite matrix, the covariance matrix. These processes are a generalization of the Gaussian process and are the most general class for which minimum mean-squared-error estimates admit linear solutions. However, as we shall see, the class is not very large.

Let $\eta_1(t)$ be a spherically invariant random process with zero mean and normalized autocovariance function $\rho_1(\tau)$. It follows from [18] that $h_1(x,y;\tau)$ can be represented as

$$h_1(x,y;\tau) = \int_0^\infty h_G(x,y;\tau)dF(v) \tag{22}$$

where $h_G(x,y;\tau)$ is the bivariate density function of a Gaussian process with mean zero, variance v^2, and normalized autocovariance function $\rho_1(\tau)$; here $F(\cdot)$ is a probability distribution function concentrated on the positive half line.

It follows from Eq. (22) that $\eta_1(t)$ is a SIRP if and only if it is equivalent to a random process of the form [19]

$$\eta_1(t) = A\eta_2(t)$$

where A is an arbitrary random variable and $\eta_2(t)$ is a zero mean Gaussian process independent of A. As shown in [19], the random processes $A\eta_2(t)$ and $|A|\eta_2(t)$ have the same family of finite dimensional distributions and, hence, are undistinguishable for our purposes. Since $\eta_2(t)$ is a zero-mean Gaussian process with variance v^2 and normalized autocovariance function $\rho_1(\tau)$, it is well-known [20] that the bivariate density function $h_G(x,y;\tau)$ admits a Mehler expansion of the form of Eq. (15) with $a_n(\tau) = \rho_1^n(\tau)$ and with

$$\theta_n(x) = H_n(x|v)/(n!)^{1/2} \tag{23}$$

where $H_n(\cdot)$ is the n-th Hermite polynomial.

When $\eta_1(t)$ is passed through the nonlinearity $g(\cdot)$ with mean

$$\overline{g} = \int_{-\infty}^{\infty} g(x)f(x)dx \tag{24}$$

and

$$b_n(v) = (2\pi n!)^{-1/2} v^{-1} \int_{-\infty}^{\infty} [g(x) - \bar{g}] H_n(x/v) \exp(-x^2/2v^2) dx,$$

the autocovariance function of the output $\mu_1(t)=g[\eta_1(t)]$ is given by

$$R_1(\tau) = \int_{-\infty}^{\infty} \int_{-\infty}^{\infty} [g(x) - \bar{g}][g(y) - \bar{g}] h_1(x,y;\tau) dxdy \qquad (25)$$

On substituting for h_1 and interchanging the order of integration, we have

$$R_1(\tau) = \int_0^{\infty} \left[\sum_{n=0}^{\infty} b_n^2(v) \rho_1^n(\tau) \right] dF(v) = \sum_{n=0}^{\infty} c_n \rho_1^n(\tau) \qquad (26)$$

The interchange of the order of integration is justified by Fubini's theorem [21,p.269], and the expectation can be brought through the infinite series by the Lebesgue convergence theorem [21,p.229]. It can be shown in a straightforward fashion that $b_0(v) = 0$ almost everywhere with respect to the measure induced by F. Hence

$$R_1(\tau) = \sum_{n=1}^{\infty} c_n \rho_1^n(\tau). \qquad (27)$$

We see that $c_n \geq 0$ since it is given by

$$c_n = \int_0^{\omega} b_n^2(v) dF(v) \qquad (28)$$

Let $\eta_1(t)$ have spectral density $S_1(\omega)$, the Fourier transform of $v^2EA^2\rho_1(\tau)$. Then $S_{o1}(\omega)$, the spectral density of the ac output of the nonlinearity, is obtained as the Fourier transform of (27) and becomes

$$S_{o1}(\omega) = \sum_{n=1}^{\infty} \frac{c_n}{\sigma^{2n}} [S_1(\omega)]^{*n} \qquad (29)$$

where $\sigma^2 = v^2EA^2$ and where $[S_1(\omega)]^{*n} = (S*\dots*S)(\omega)$ is the (n - 1)-fold convolution of $S_1(\omega)$ with itself. It is well known [22] that convolution is a smoothness-increasing operation; that is, $(S*S)(\omega)$ is smoother than $S(\omega)$ by almost any test (modulus of continuity, moduli of smoothness of high order, number of derivatives, total variation, etc.). Thus, the output of a memoryless nonlinearity with a SIRP input has a spectrum that is smoother than the input spectrum and tends to be more evenly distributed over the ω-axis.

We now suppose that $S_1(\omega) = 0$ for $|\omega| > B$ so that $\eta_1(t)$ is a strictly bandlimited input process. Let Ω be the support of $S_1(\omega)$. Then the support of $[S_1(\omega)]^{*n}$ is

$$\mathrm{supp}\{[S_1(\omega)]^{*n}\} = \underbrace{\Omega\oplus\ldots\oplus\Omega}_{n \text{ times}} = \oplus^n\Omega \tag{30}$$

where the set operation $\oplus$ is defined for two sets A and B as

$$A \oplus B = \{c | c = a + b, a\in A, b\in B\}.$$

It is not difficult to show that

$$\infty > m(\oplus^n\Omega) \geq nm(\Omega) \tag{31}$$

where m is Lebesgue measure. We consider the following measure of bandwidth of $\eta_1(t)$:

$$B_1[\eta_1(t)] = m(\mathrm{supp}\{S_1(\omega)\}) = m(\Omega) \tag{32}$$

Now $B_1[\mu_1(t)]$ is the bandwidth of the output. Thus (29) implies that $B_1[\mu_1(t)] < \infty$ if and only if there exists an N such that $c_n = 0$, for all $n > N$. Notice that this condition is equivalent to the nonlinearity being a polynomial, establishing the following.

<u>Theorem 4</u>: The output process $\mu_1(t)$ is strictly bandlimited if and only if a) the SIRP input process is strictly bandlimited and b) the nonlinearity $g(\cdot)$ is a polynomial.

We now consider the case where the input spectrum need not be strictly bandlimited but has a finite second moment. In this case, we will take as the bandwidth measure the root-mean-square bandwidth, given by the positive square root of the second moment of the normalized ac spectral mass; that is,

$$B_2[\eta_1(t)] = \left[\sigma^{-2} \int_{-\infty}^{\infty} \omega^2 S_1(\omega)d\omega \right]^{1/2} \tag{33}$$

Notice that $\sigma^{-2n}[S_1(\omega)]^{*n}$ can be interpreted as the density function of a sum of n independent identically distributed random variables with mean zero and variance $B_2^2[\eta_1(t)]$. It follows that

$$\int_{-\infty}^{\infty} \omega^2 \sigma^{-2n}[S_1(\omega)]^{*n} d\omega = nB_2^2[\eta_1(t)] \tag{34}$$

Thus (29) yields

$$\int_{-\infty}^{\infty} \omega^2 S_{o1}(\omega)d\omega = B_2^2[\eta_1(t)] \sum_{n=1}^{\infty} nc_n \tag{35}$$

Since the squared output bandwidth $B_2^2[\mu_1(t)]$ is

$$B_2^2[\mu_1(t)] = R_1^{-1}(0) \int_{-\infty}^{\infty} \omega^2 S_{o1}(\omega)d\omega \tag{36}$$

and

$$R_1(0) = \sum_{n=1}^{\infty} c_n \tag{37}$$

it is seen that the output bandwidth is

$$B_2[\mu_1(t)] = B_2[\eta_1(t)] \left[\sum_{n=1}^{\infty} nc_n \right]^{1/2} \Big/ \left[\sum_{n=1}^{\infty} c_n \right]^{1/2} \tag{38}$$

Thus $B_2[\mu_1(t)] \geq B_2[\eta_1(t)]$. Here equality holds if and only if $c_n = 0$, for all $n \geq 2$, i.e., if and only if $g(\cdot)$ is affine. This can be summarized in the following theorem.

<u>Theorem 5</u>: If $\eta_1(t)$ is an SIRP and has a finite root-

mean-square bandwidth $B_2[\eta_1(t)]$, the root-mean-square bandwidth of $\mu_1(t) = g[\eta_1(t)]$ is greater than or equal to $B_2[\eta_1(t)]$. Equality holds if and only if $g(\cdot)$ is affine.

It follows from Theorem 5 that an SIRP, say $A\eta_2(t)$, is no longer an SIRP after passage through a memoryless nonlinearity $g(\cdot)$ except in the degenerate case where $g(\cdot)$ is affine. This change is always accompanied by a spreading of the spectrum. This conclusion holds for <u>any</u> nonlinearity (see [23] and [24] for a treatment of bandwidth relationships for differentiable detectors with particular emphasis on envelope detectors). Also it follows from Theorem 4 that an SIRP passed through any type of limiter is never strictly band limited.

Let us consider the output $\mu_1(t)$ of an invertible nonlinearity $g(\cdot)$ whose input $\eta_1(t)$ is an SIRP process with finite root-mean-square bandwidth. We suppose that $\mu_1(t) = g[\eta_1(t)]$ is passed through another nonlinearity $h(\cdot)$ resulting in the output $\nu_1(t) = h[\mu_1(t)] = h[g[\eta_1(t)]]$. It is clear that there exists a class C of nonlinearities $h(\cdot)$ that result in the minimum bandwidth for $\nu_1(t)$; that is, C is the class of all affine functions of $g^{-1}(\cdot)$. If $h(x) = ag^{-1}(x) + b$, then $\nu_1(t) = a\eta_1(t) + b$ has the same root-mean-square bandwidth as $\eta_1(t)$. However, if $H(\cdot)\notin C$, then $\nu_1(t) = h[g[\eta_1(t)]] = (h\cdot g)[\eta_1(t)]$. In this case, $(h\cdot g)(\cdot)$ is not affine, and by Theorem 5, $\nu_1(t)$ will have a larger bandwidth than $\eta_1(t)$. Notice that minimization of the bandwidth coincides with restoration of the normality of $\eta_2(t)$ in the SIRP $A\eta_2(t)$. Thus, in the class of invertibly transformed SIRP's, an SIRP process is characterized by having minimum bandwidth.

Gaussian processes and others

The Gaussian process in a special case of the SIRP obtained when the random variable A is degenerate; for example, with distribution function $F(v) = u(v-1)$, where $u(v)$ is the unit step function. Now $EA^2 = 1$ and $\sigma^2 = v^2$, the variance of the Gaussian random process. With these changes in notation, Theorems 4 and 5 hold as do the comments after Theorem 5.

As discussed in [25], other processes can be found for which Theorems 4 and 5, or their equivalents, can be shown to apply. One class, which we have called Gaussian-related, is that where the bivariate density can be expressed is in Eq. (15) with $a_n(\tau) = \rho_2^n(\tau)$ where $\rho_2(\tau)$ is the normalized autocorrelation function of the process. This is the class of bivariate densities having the same coefficients $a_n(\tau)$ as in the Gaussian case but with different orthonormal functions $\theta_n(\cdot)$. When such a process is passed through a memoryless nonlinearity $g(\cdot)$, results can be derived which are similar to those for the SIRP or Gaussian case. The reason, of course, is that the output spectral density can be written in the form of Eq. (29).

In the same way, similar remarks can be made about other special classes of random processes; for example [25], certain Markov processes and mixture processes. However, lack of space precludes further discussion here.

A CLASS OF MARKOV SEQUENCES

We consider a general symmetric bivariate density function satisfying Eq. (1), whose diagonal series expansion is given by

$$h(x,y) = f(x)f(y) \sum_{m=0}^{\infty} a_m \theta_m(x)\theta_m(y) \tag{39}$$

For notational convenience, define the kernel

$$K(x,y) = \frac{h(x,y)}{\sqrt{[f(x)f(y)]}} \tag{40}$$

where $K(x,y) = 0$ if either $f(x)$ or $f(y)$ is zero. Let A denote the operator induced by the kernel $K(x,y)$; that is, $v = Au$ is defined on the space of measurable square integrable functions by

$$v(x) = \int K(x,y)u(y)dy \tag{41}$$

It can be shown [26] that A is a self-adjoint completely continuous operator. Let $\phi_m(x) = \sqrt{[f(x)]}\theta_m(x)$. It follows that

$$A\phi_m = a_m\phi_m. \tag{42}$$

Thus $\{\phi_m\}$ and $\{a_m\}$ are the (orthonormal) eigenfunctions and eigenvalues of the operator A. Notice that

$$A(A\phi_m) = A(a_m\phi_m) = a_m^2\phi_m \tag{43}$$

In general, for any positive integer k, we have

$$A^k\phi_m = a_m^k\phi_m \tag{44}$$

where the operator A^k and the kernel $K^k(x,y)$ are defined by

$$A^k u = A(A^{k-1}u) \tag{45}$$

and

$$K^k(x,y) = \int K(x,z)K^{k-1}(z,y)dz \tag{46}$$

and where $A^1 = A$. The kernel $K^k(x,y)$ is simple the kth iterate of the kernel $K(x,y)$. It follows [27] that

$$K^k(x,y) = \sum_{m=0}^{\infty} a_m^k\phi_m(x)\phi_m(y) \tag{47}$$

where the convergence is in the mean square sense as in Eq. (4).

Let a function $h(x,y,k)$ be defined by

$$h(x,y,k) = \sqrt{[f(x)f(y)]}K^k(x,y) \tag{48}$$

Then we have

$$h(x,y,k) = \int \frac{h(x,z)h(y,z,k-1)}{f(z)} dz \tag{49}$$

where $h(x,y,1) = h(x,y)$. From this we see that $h(x,y,k)$ is a bivariate density function. Also, the diagonal series expansion is given by

$$h(x,y,k) = f(x)f(y) \sum_{m=0}^{\infty} a_m^k \theta_m(x)\theta_m(y) \tag{50}$$

Therefore, we conclude that, in the diagonal series expansion of a symmetric bivariate density, the coefficient sequence can be raised to any positive integral power and the resulting diagonal series expansion will also be that of a bivariate density function. Rewriting Eq. (49), we get

$$\frac{h(x,y,k)}{f(x)} = \int \frac{h(x,z)}{f(x)} \frac{h(y,z,k-1)}{f(z)} dz \tag{51}$$

It can be shown [Ref. [28], p.89] that this last expression is a form of the Chapman-Kolmogorov equation, yielding the following theorem:

Theorem 6: Let $\{X_n\}$ be a second-order stationary, discrete-time random process. Assume that the bivariate density of X_n and X_{n+1} is given by Eq. (39). Then $\{X_n\}$ is a Markov process if and only if, for any positive integer k, the bivariate density of X_n and X_{n+k} is given by Eq. (50).

Many bivariate densities possess a diagonal series expansion where the coefficients are of the form $a_m = \lambda^m$. For example, if the orthonormal functions $\theta_m(x)$ corresponds to the orthonormalized Hermite polynomials, generalized Laguerre polynomials or Jacobi polynomials, and if the univariate densities are chosen appropriately [29], then a bivariate density in the form of Eq. (39) exists where the coefficients are given by $a_m = \lambda^m$, where λ is any number in [0,1). In the case of orthonormalized Hermite polynomials or Gegenbauer polynomials [29], the parameter λ can be any number in (-1,1). Recall that Legendre polynomials and both kinds of Chebychev polynomials belong to the class of Gegenbauer polynomials. Also, it is not hard to show that, if the orthonormal functions $\theta_m(x)$ are uniformly bounded (in both x and m), then a bivariate density will always exist where the coefficients are of the form $a_m = \lambda^m$. Naturally, any strict monotonic transformation on the random variables will not change the coefficient sequence. The class of Markov sequences constructed from bivariate densities of this type is such that the bivariate density for any two random variables in a sequence will have the same functional form as that for two adjacent random variables in the sequence. For example, consider a Gaussian-Markov sequence. Then we have

$$f(x,y;k) = \frac{1}{2\pi\sigma^2(1-\rho^{2k})^{\frac{1}{2}}} \exp\left[\frac{-1}{2\sigma^2(1-\rho^{2k})}(x^2-2\rho^k xy+y^2)\right],$$

where k is any positive integer.

It is possible to derive a number of second-order properties for these random sequences. For example, let g(x) be a memoryless nonlinearity such that

$$\int_{-\infty}^{\infty} g^2(x)f(x)dx < \infty \tag{52}$$

Also let X and Y be random variables with bivariate density given by Eq. (39). It follows that

$$E\{g(X)g(Y)\} = \sum_{m=0}^{\infty} b_m^2 a_m \tag{53}$$

where

$$b_m = \int g(x)\theta_m(x)f(x)dx \tag{54}$$

Now suppose R(k) is the autocovariance function of the Markov sequence $\{X_n\}$. On setting $g(x) = x-\mu$ in Eq. (53), we have

$$R(2k) = \sum_{m=0}^{\infty} b_m^2 a_m^{2|k|} \geq 0 \tag{55}$$

for all positive integer k. Thus
Theorem 7: Either R(k) = 0 for all non-zero integer k or R(2k) > 0 for all integer k.

See [6] for a further discussion of second-order properties.

ACKNOWLEDGEMENTS

This research is supported by the National Science Foundation under Grant-24187 and by the U.S. Army Research Office under Contract DAAG29-75-G-0192.

REFERENCES

1. P.R. Halmos, Introduction to Hilbert Space and the Theory of Spectral Multiplicity, Chelsa, N.Y., 1957.
2. H.O. Lancaster, "The Structure of Bivariate Distributions", Ann. Math. Stat., Vol. 29, pp.719-736, 1958.
3. H.O. Lancaster, "Correlations and Canonical Forms of Bivariate Distributions", Ann. Math. Stat., Vol. 34, pp.532-538, 1963.
4. J.F. Barrett and D.G. Lampard, "An Expansion for Some Second-Order Probability Distributions and Its Application to Noise Problems", IRE Trans. Information Theory, Vol. IT-1, pp.10-15, March 1955.
5. J.L. Brown, "A Criterion for the Diagonal Expansion of a Second-Order Probability Distribution in Orthogonal Polynomials", IRE Trans. Information Theory, Vol. IT-4, p.172, December 1958.
6. G.L. Wise and J.B. Thomas, "A Characterization of Markov Sequences", Journal of The Franklin Institute, Vol. 299, No. 4, pp.269-278, April 1975.
7. J.B. Thomas, An Introduction to Applied Probability and Random Processes, p.105, Wiley, New York, 1971.
8. A.V. Balakrishnan, "On a Characterization of Processes For Which Optimal Mean-Square Systems are of Specified Form", IRE Trans. Information Theory, Vol. IT-6, pp.490-500, September 1960.
9. W. Feller, An Introduction to Probability Theory and Applications, Vol. II, pp.104-106, Wiley, N.Y., 1971.
10. R. Leipnik, "Integral Equations, Biorthonormal Expansions, and Noise", SIAM J. Appl. Math., Vol. 7, pp.6-30, March 1959.
11. N.M. Blachman, "The Uncorrelated Output Components of a Nonlinearity", IEEE Trans. Inform. Theory, Vol. IT-14, pp.250-255, March 1968.
12. J.L. Brown, Jr., "Recovery of Bandlimited Gaussian Noise Spectra After Nonlinear Transformation and Bandlimiting", IEEE Int'l. Conf. on Communications, Conf. Record, Vol. 7, pp.43-1 to 43-4, 1971.
13. J.L. Brown, Jr., "Some Results on Nonlinear Transformation of Bandlimited Stochastic Signals", Proc. of the Eighth Annual Princeton Conference on Information Sciences and Systems, pp.53-56, March 1974.
14. G.L. Wise and J.B. Thomas, "On a Completeness Property of Series Expansions of Bivariate Densities", J. Multivariate Analysis, Vol. 5, pp.243-247, 1975.
15. H. Cramer and M.R. Leadbetter, Stationary and Related Stochastic Processes, New York: Wiley, 1967.
16. See Reference 1, p.27.

17. A.M. Vershik, "Some Characteristic Properties of Gaussian Stochastic Processes", Theory of Prob. and Its Appl., Vol. 9, pp.353-356, 1964.
18. K. Yao, "A Representation Theorem and Its Applications to Spherically-Invariant Random Processes", IEEE Trans. Information Theory, Vol. IT-19, pp.600-608, September 1973.
19. G.L. Wise and N.C. Gallagher, Jr., "A Representation for Spherically Invariant Random Processes", Proceedings of the Fourteenth Annual Allerton Conference on Circuit and System Theory, pp.460-469, September-October, 1976.
20. See Reference 4, for example.
21. H.L. Royden, Real Analysis, London, Macmillan, 1968.
22. H.S. Shapiro, Smoothing and Approximation of Functions, New York: Van Nostrand Reinhold, pp.8-9, 1969.
23. P.A. Bello, "On the RMS Bandwidth of Nonlinearly Envelope Detected Narrowband Gaussian Noise," IEEE Trans. Inform. Theory., Vol. IT-11, pp.236-239, April 1965.
24. N. Abramson, "Nonlinear Transformations of Random Processes," IEEE Trans. Inform. Theory, Vol. IT-13, pp.502-505, July 1967.
25. G.L. Wise, A.P. Traganitis, and J.B. Thomas, "The Effect of a Memoryless Nonlinearity on the Spectrum of a Random Process," IEEE Trans. on Information Theory, Vol. IT-23, No. 1, pp.84-89, January 1977.
26. A.N. Kolmogorov and S.V. Fomin, Elements of the Theory of Functions and Functional Analysis, Vol.2, p. 120, Graylock Press, Balt., Md., 1961.
27. F. Riesz and B. Sz.-Nagy, Functional Analysis, p.243, Ungar, New York, 1955.
28. J.L. Doob, Stochastic Processes, pp.80-91, Wiley, New York, 1953.
29. P. Beckmann, Orthogonal Polynomials for Engineers and Physicists, Appendix I, The Golem Press, Boulder, Colorado, 1973.

SIMULATION OF SIGNAL AND NOISE IN A NON-LINEAR CHANNEL

G.A. Richards

Marconi Research Laboratories, GEC-Marconi Electronics Limited, Great Baddow, Chelmsford, Essex, U.K.

1. INTRODUCTION

Several general purpose programs have been developed to simulate the effect of a channel on a 'signal'. The signal can be described as a modulated carrier where the modulation is specified by its statistical parameters or by a sequence of sample values of a representative signal. The component parts of the channel are described by transfer characteristics, the overall channel being composed of a cascade of these transfer characteristics. When the channel is linear the simulation can be carried out entirely in the time or frequency domain. For a non-linear channel some of the actions of the system are most conveniently described in the time domain and others in the frequency domain, e.g. a non-linear amplifier will be described in the time domain and filters in the frequency domain. A useful tool for transferring between the two domains is the Fast Fourier Transform [1].

An approach as outlined above would be sufficient to evaluate the distortion introduced into the signal by the channel through intermodulation or cross-modulation in multi-carrier situations. In some applications the behaviour of the signal is required in a noisy channel. If the channel is linear the effect of the channel on the signal and noise can be evaluated separately and there is no major problem. For a non-linear channel this separation is not possible and the complexity of the simulation is increased enormously. It is possible to use standard simulation programs by adding noise samples to the signal and performing many runs. Averaging of the results gives the required answers. This approach has two main disadvantages,

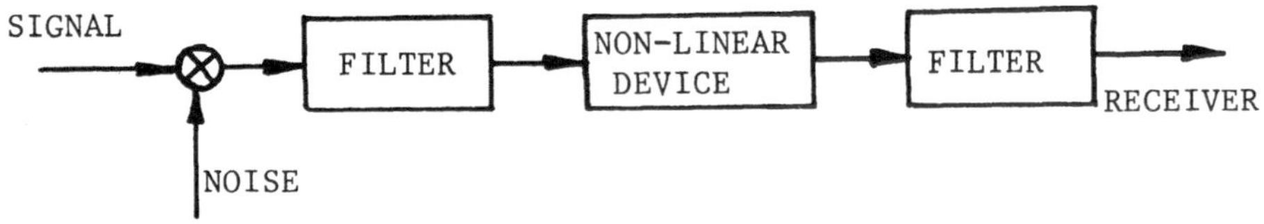

Fig. 1 Type of Channel to be Considered.

viz. the cost of the multiple runs (possibly several thousand runs) and the difficulty of assigning confidence limits to the results. The purpose of this paper is to explore possible methods of analytically treating the noise in a simulation environment.

The type of channel considered will be as shown in fig. 1. Saturating types of non-linear amplifiers, e.g. class-C or TWT amplifiers, which introduce AM-PM distortion, will be used for the non-linear component in the channel throughout this paper. The approach considered is based on a simulation program which uses a deterministically defined signal sequence.

The paper is structured as follows. Section 2 outlines the possible ways of tackling the problem. In Section 3 a representation of the non-linear device is developed which gives a characterization of the signal after the non-linearity. Two approaches to the representation of the noise are given in sections 4 and 5.

2. POSSIBLE APPROACHES TO THE PROBLEM

Characterization of the noise at the output of the non-linear device requires a knowledge of the structure of the input and output signals. The form of characterization will depend on the processing to be carried out after the non-linear device. For the type of channel shown in fig. 1 the noise would need to be specified at a number of time samples corresponding to the order of the filter after the non-linearity. Complete specification of the noise would require the joint probability density function at a sufficient number of time samples. The evolution of this density function with time would be described by the Fokker-Planck equation, a partial differential equation. The coefficients of the equation would depend on the non-linear function and the input and output filter characteristics. In addition the coefficients will be time dependent due to the signal. The numerical solution of this equation is not a feasible proposition.

Since a complete specification of the noise is not feasible approximate approaches have been considered. One possible approach is to specify the noise power levels as a function of time. To propagate the noise power through the filter requires the autocorrelation function which will be a function of two times instead of the time difference due to the dependence on the signal. Methods developed by Shimbo [2] and Kelly [3] could be used to generate the autocorrelation function at the output of the filter. An approach using a polynomial representation is described in sections 3 and 4.

An approach as described above is essentially the same as assuming that the noise after the non-linearity is gaussian since only the second moments of the noise are used. An alternative approach is to approximate the noise probability density function at the output of the non-linear device. In section 5 the form of this density function is considered and method of generating it via moments is discussed. An approximation to the probability density function is made using a sum of gaussian distributions.

3. ANALYSIS OF A NON-LINEAR DEVICE

The types of non-linear device which are encountered in satellite channels are TWT and class-C amplifiers and limiters. In each case the non-linearity is a function of the amplitude of the input signal. For amplifiers there may also be AM-PM conversion, resulting in a complex transfer characteristic. The following development will concentrate on the case where the non-linearity is a saturating amplifier.

The approach adopted for the analysis of a non-linear device consists of the following three stages:

(i) Representation of the device characteristics by a power series.

(ii) Definition of an equivalent characteristic of the device for processing the signal, which is obtained by averaging over the noise.

(iii) The 'output noise' is obtained by taking all those terms in the output which do not contribute to the output signal.

3.1 REPRESENTATION OF THE DEVICE CHARACTERISTICS

Let the input to the non-linear device be of the form

$$e(t) = A(t) \exp\{ j \left[\omega_o t + \phi(t)\right]\} \qquad \ldots(3.1.1)$$

The resulting output will be of the form, see fig. 2,

$$z(t) \quad = \quad g\{A(t)\} \; \exp\{j[\omega_o t + \phi(t)]\} \qquad \ldots(3.1.2)$$

where g{.} is the non-linear transfer characteristic. Note that if there is AM-PM conversion then g{.} will be a complex function of the input amplitude.

Re-write eqn. (3.1.2) as

$$z(t) \quad = \quad h\{A(t)\} \;\; e(t) \qquad \ldots(3.1.3)$$

where

$$h\{A\} \quad = \quad \frac{g\{A\}}{A} \qquad \ldots(3.1.4)$$

is the transfer gain of the device.

In general the gain of the device will be measured as a function of the input power, $A^2 = ee^*$. [e^* is the complex conjugate of e].

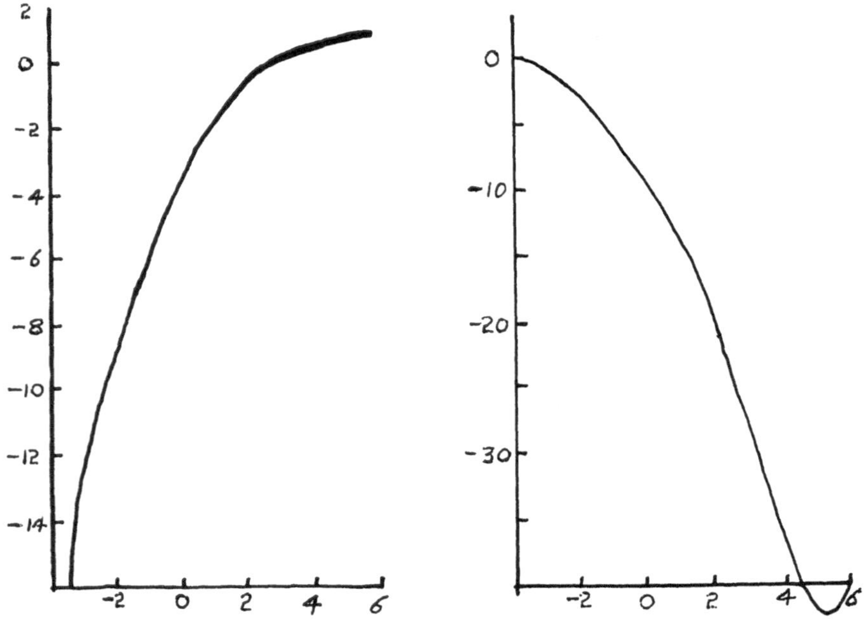

(a) Amplitude (dB) (b) Phase (degrees)

Fig. 2 Typical Device Characteristics

The gain function h{.} will be represented by the finite power series

$$h\{A\} = \sum_{n=o}^{L} b_n A^{2n} \qquad ...(3.1.5)$$

where the coefficients b_n are possibly complex.

Then, from eqn. (3.1.3)

$$z = \sum_{n=o}^{L} b_n e^{n+1} e^{*n} \qquad ...(3.1.6)$$

This representation of the device characteristics will be used throughout the paper.

3.2 EXPANSION FOR SIGNAL PLUS NOISE

The input to the non-linearity will consist of a signal $S(t)\exp\{j\omega_o t\}$ and a bandlimited gaussian noise $V(t)\exp\{j\omega_o t\}$.

To simplify the notation let

$$z(t) = Z(t)\exp\{j\omega_o t\} \qquad ...(3.2.1)$$

In the following development the dependence on time is not explicitly written for notational convenience.

$$Z = \sum_{n=o}^{L} b_n (S+V)^{n+1} (S^*+V^*)^n \qquad ...(3.2.2)$$

The output can now be expanded into signal and noise terms as

$$Z = \sum_{n=o}^{L} \sum_{p=o}^{n} \sum_{q=o}^{n+1} \binom{n}{p} \binom{n+1}{q} b_n S^{n+1-q} S^{*n-p} V^q V^{*p} \qquad ...(3.2.3)$$

3.3 'OUTPUT SIGNAL'

The output from the non-linear device consists of a mixture of 'signal x signal', 'signal x noise', and 'noise x noise' terms. The 'pure' output components 'signal x signal' and 'noise x noise' terms can be directly identified in the output as signal and noise components. The problem is how to interprete the 'signal x noise' terms. The effect of the noise is to jitter the signal about the operating point, fig. 3. The output will also jitter about the noiseless output level. Because

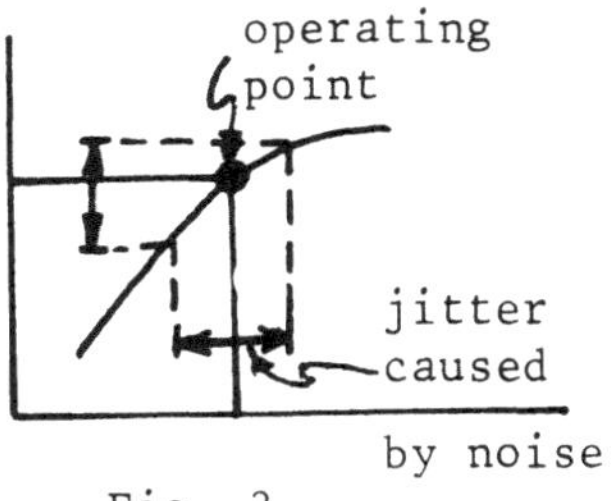

Fig. 3

of the non-linearity this jitter produces a bias in the output which cannot be identified as noise. The mean value of the 'signal x noise' terms are therefore included with the signal terms to produce a 'mean output signal'. This mean output is obtained below by averaging eqn. (3.2.3) over the noise.

Assume that the input noise has independent real and imaginary components which are zero-mean gaussian noise processes, each with a noise power σ^2. The noise amplitude A_v will be Rayleigh distributed and the phase θ_v uniformly distributed with A_v and θ_v independent.

Then

$$E\{V^q V^{*p}\} = E\{A_v^{p+q}\} \; E\{\exp j(q-p)\theta_v\} \qquad \ldots(3.3.1)$$

where E{.} denotes the expectation operator.

But

$$E\{\exp j(q-p)\theta_v\} = \begin{cases} 0 \text{ for } p \neq q \\ 1 \text{ for } p = q \end{cases} \qquad \ldots(3.3.2)$$

and

$$E\{A_v^{2p}\} = p!\,(2\sigma^2)^p \qquad \ldots(3.3.3)$$

(Note that $2\sigma^2$ is the total input noise power)

The mean output signal is therefore given by

$$\overline{S} = \sum_{n=o}^{L} \sum_{p=o}^{n} \binom{n}{p} \binom{n+1}{p} b_n \, p!\,(2\sigma^2)^p \, S^{n+1-p} \, S^{*n-p}$$

$$= \sum_{n=o}^{L} \sum_{p=o}^{n} \binom{n}{p} \binom{n+1}{n-p} (n-p)!\, b_n (2\sigma^2)^{n-p} \, S^{p+1} \, S^{*p}$$

$$= \sum_{p=o}^{L} \sum_{n=p}^{L} \binom{n}{p} \binom{n+1}{n-p} (n-p)!\, b_n (2\sigma^2)^{n-p} \, S^{p+1} \, S^{*p} \qquad \ldots(3.3.4)$$

on interchanging the order of the summations.

Define the new coefficients

$$B_p = \sum_{n=p}^{L} \binom{n}{p} \binom{n+1}{n-p} (n-p)!\, b_n \, (2\sigma^2)^{n-p} \qquad \ldots(3.3.5)$$

Then the output signal is given by

$$\overline{S} = \sum_{p=o}^{L} B_p \, S^{p+1} \, S^{*p} \qquad \ldots(3.3.6)$$

The coefficients B_p define an equivalent device characteristic for processing the signal analytically for a specified input noise power. An equivalent channel, fig. 4 can be defined for the signal for which standard programs can be used to simulate the effect of the channel on the signal.

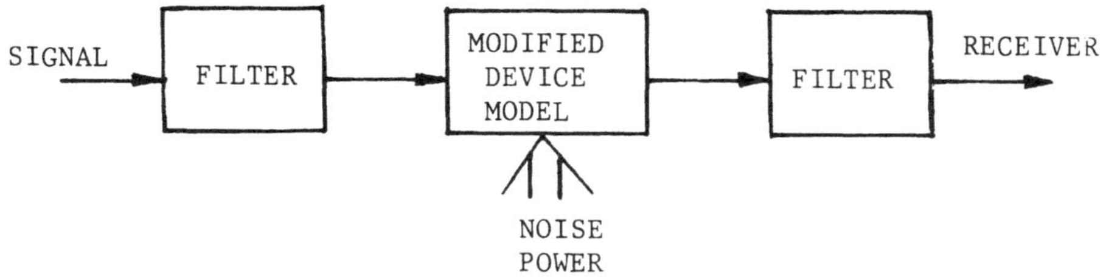

Fig. 4 Equivalent Signal Channel

For a class-C amplifier several orders of polynomial were used to fit the characteristics. For two levels of input noise power the resulting equivalent signal characteristics are shown in fig. 5 and fig. 6 for a 9-th degree polynomial fit. In the low noise case the equivalent characteristic is similar to the device characteristic but as the noise power is increased, the level of the jitter is greater and there is a considerable change in the equivalent characteristic.

3.4 'OUTPUT NOISE'

The components in the 'signal x noise' terms which do not contribute to the mean signal are grouped with the 'noise x noise' terms and the total considered as the output noise from the non-linear device.

From eqn. (3.2.3) and (3.3.4) this noise is given by

$$N = \sum_{n=o}^{L} \sum_{p=o}^{n} \sum_{\substack{q=o \\ \neq p}}^{n+1} \binom{n}{p} \binom{n+1}{q} b_n \, S^{n+1-q} \, S^{*\,n-p} \, V^q \, V^{*\,p} \quad \ldots(3.4.1)$$

An equivalent characteristic for processing the noise can be derived as follows by interchanging the summation over n with those over p and q :

$$N = \sum_{p=o}^{L} \sum_{\substack{q=o \\ \neq p}}^{L+1} \sum_{n=n_\ell}^{L} \binom{n}{p} \binom{n+1}{q} b_n \, S^{n+1-q} \, S^{*\,n-p} \, V^q \, V^{*\,p} \quad \ldots(3.4.2)$$

where $n_\ell = \max(p, k-1)$

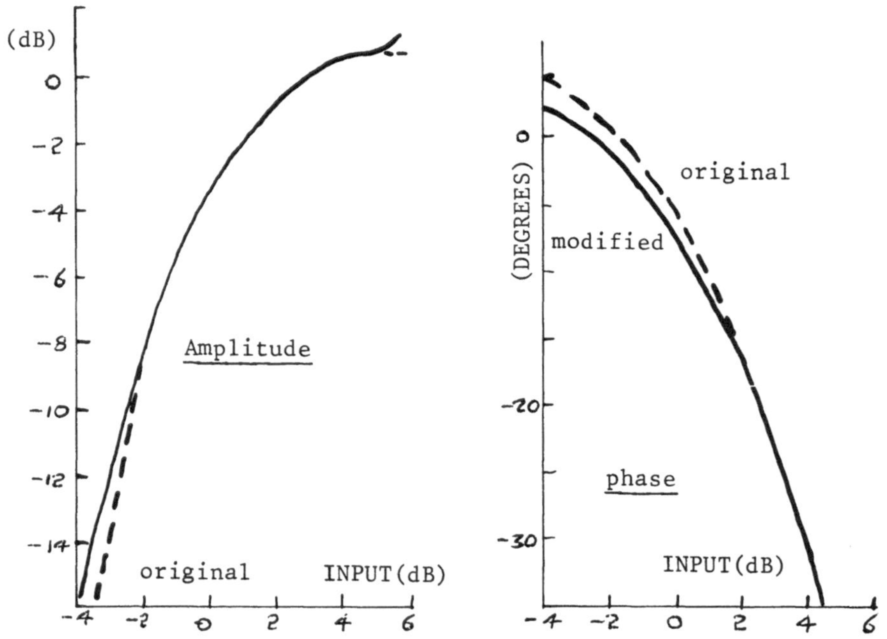

Fig. 5 MODIFIED SIGNAL CHARACTERISTICS
(-20 dB NOISE)

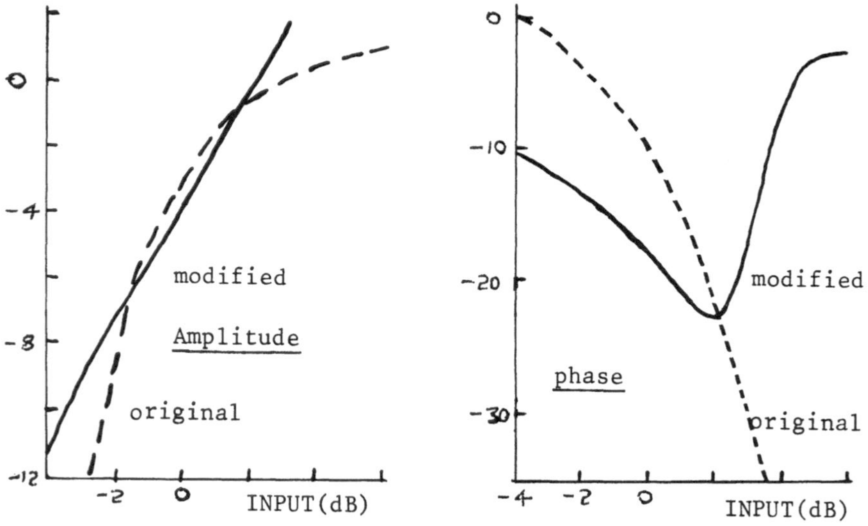

Fig. 6 MODIFIED SIGNAL CHARACTERISTICS
(-10 dB NOISE)

Define the 'equivalent coefficients'

$$C_{pq} \triangleq \sum_{n=n_\ell}^{L} \binom{n}{p}\binom{n+1}{q} b_n S^{n+1-q} S^{*n-p} \qquad \ldots(3.4.3)$$

Then eqn. (3.4.2) can be written as

$$N = \sum_{p=o}^{L} \sum_{q=o}^{L+1} C_{pq} V^q V^{*p} \qquad \ldots(3.4.4)$$

where $C_{pp} \triangleq 0$

The problem now remains of how to characterise this output noise. The characteristic given by (3.4.4) is non-linear and is also time-dependent due to the dependence of the coefficients C_{pq} on the signal level. The nature of this output noise will depend on the operating point on the device characteristics (signal level) and the swing over the characteristics (noise power). If the device is operated over the central linear region and with a low noise power the output noise will remain approximately gaussian, see fig.9. For a higher noise level or for high or low signal levels the noise is non-gaussian. This situation is considered in more detail later in section 5 when the distribution of the output noise is considered. The next section characterises the noise by its time varying autocorrelation function.

4. CHARACTERIZATION OF THE NOISE BY ITS AUTOCORRELATION FUNCTION

The simplest characterization of the noise following a non-linear device is to use the noise power. This will be time-dependent due to the signal x noise components. Due to filtering of the noise the autocorrelation function is required to obtain the noise power at the receiver. To retain phase information on the noise two complex autocorrelation functions are used.

Let suffices 1 and 2 denote evaluation of the process at times t_1 and t_2, respectively. Define the two autocorrelation functions

$$R = E\{N_1 N_2\} \qquad \ldots(4.1)$$

and

$$R_* = E\{N_1 N_2^*\} \qquad \ldots(4.2)$$

If $N = x+jy$, then the autocorrelation and crosscorrelation functions of the real and imaginary parts are as follows

$$E\{x_1 x_2\} = \tfrac{1}{2}\mathcal{R}\{R_* + R\} \quad \dots(4.3)$$

$$E\{y_1 y_2\} = \tfrac{1}{2}\mathcal{R}\{R_* - R\} \quad \dots(4.4)$$

$$E\{x_1 y_2\} = \tfrac{1}{2}\mathcal{I}\{R - R_*\} \quad \dots(4.5)$$

4.1 DERIVATION OF THE AUTOCORRELATION FUNCTIONS

From eqn. (3.4.4) the autocorrelation functions are obtained as

$$R = \sum_{p=o}^{L} \sum_{q=o}^{L+1} \sum_{r=o}^{L} \sum_{s=o}^{L+1} C_{pq1}\, C_{rs2}\, E\{V_1^q\, V_1^{*p}\, V_2^s\, V_2^{*r}\} \dots(4.1.2)$$

and

$$R_* = \sum_{p=o}^{L} \sum_{q=o}^{L+1} \sum_{r=o}^{L} \sum_{s=o}^{L+1} C_{pq1}\, C^*_{rs2}\, E\{V_1^q\, V_1^{*p}\, V_2^r\, V_2^{*s}\} \dots(4.1.2)$$

where C_{pqk} is C_{pq} evaluated at t_k.

It can be shown that for circular gaussian noise

$$E\{V_1^q\, V_1^{*p}\, V_2^s\, V_2^{*r}\} = M(q, p, s)\, \delta_{q+s-p-r} \quad \dots(4.1.3)$$

where δ_k is the Kroneker delta function and

$$M(q, p, r) = (2\sigma^2)^{q+r} \sum_{s=s_\ell}^{s_u} \frac{q!p!r!(q+r-p)!}{s!(q-s)!(p-s)!(s-p+r)!}\, \rho^{p-s}\, \rho^{*q-s}$$

$$\text{where } s_\ell = \max(o, p-r) \quad \dots(4.1.4)$$

$$s_u = \min(k, p)$$

$$\rho = \frac{1}{\sigma^2}\left[E\{\alpha_1 \alpha_2\} + jE\{\alpha_1 \beta_2\}\right] \quad \dots(4.1.5)$$

is the normalized correlation function of the circular input noise $V = \alpha + j\beta$.

Using the result (4.1.3), the two autocorrelation functions are

$$R = \sum_{p=o}^{L} \sum_{q=o}^{L+1} \sum_{r=r_{\ell 1}}^{r_{u1}} C_{pq1}\, C_{r(q+r-p)2}\, M(q, p, r) \quad \dots(4.1.6)$$

$$R_* = \sum_{p=o}^{L} \sum_{q=o}^{L+1} \sum_{r=r_{\ell 2}}^{r_{u2}} C_{pq1} C^*_{r(p+r-q)2} M(q, p, q+r-p) \quad \ldots(4.1.7)$$

where
$$r_{\ell 1} = \max(o, p-k)$$
$$r_{u1} = \min(L, L+1+p-k)$$
$$r_{\ell 2} = \max(o, k-p)$$
$$r_{u2} = \min(L, L+1+k-p)$$

A subroutine has been written, for inclusion in an in-house simulation program, to generate these autocorrelation functions. The program uses a predetermined sampling rate and these functions are stored at these sampling intervals in matrix form. The diagonal components of the matrix give the noise power at each time sample; Fig. 7 and 8 show the variation of the output noise power as a function of the input signal level for two different input noise levels. Also shown on these figures are the results obtained from one thousand runs of the in-house simulation program. These results are obtained by rotating the noise reference axes such that the noise components are independent. The angle of rotation is given by

$$\psi = \tfrac{1}{2} \arctan \left[\frac{2\sigma_x \sigma_y \rho}{\sigma_x^2 - \sigma_y^2} \right]$$

where σ_x^2 and σ_y^2 are the variances of the two components and ρ the correlation coefficient.

The corresponding noise powers are

$$\sigma_\alpha^2 = \tfrac{1}{2} (\sigma_x^2 + \sigma_y^2) - \gamma$$

$$\sigma_\beta^2 = \tfrac{1}{2} (\sigma_x^2 + \sigma_y^2) + \gamma$$

$$\gamma = \frac{\sigma_x \sigma_y \rho}{\sin 2\psi}$$

where α and β are the rotated noise components.

Under some circumstances when the signal level is relatively constant the autocorrelation functions can be averaged over time by averaging the coefficients C_{pq} over the signal levels.

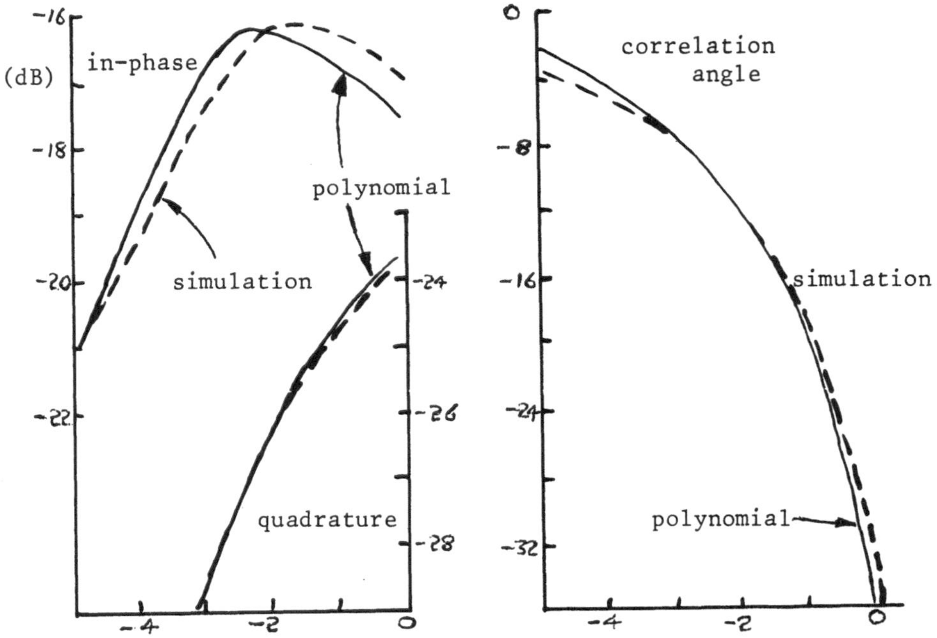

Fig. 7 OUTPUT NOISE POWER (-20 dB)

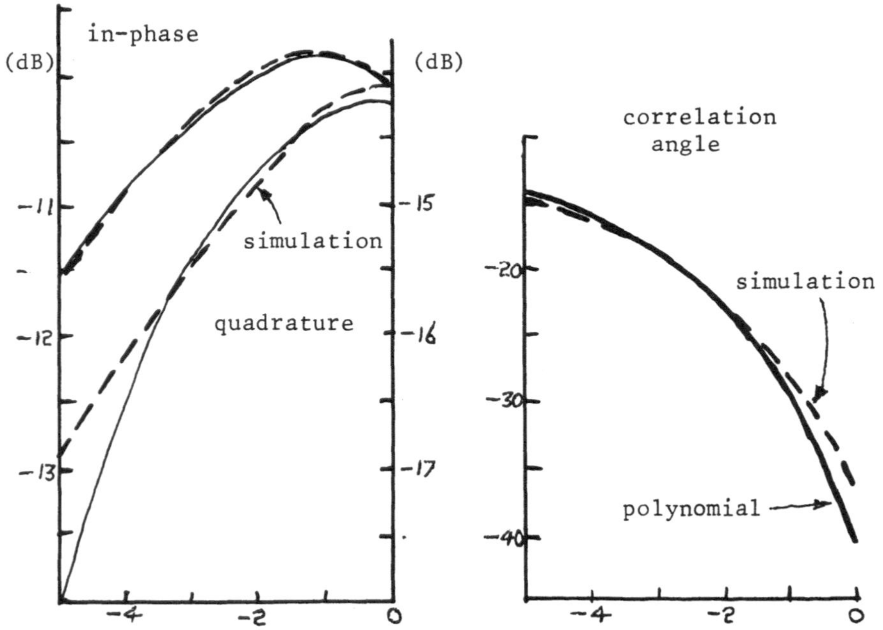

Fig. 8 OUTPUT NOISE POWER (-10 dB)

4.2 FILTERING OF THE NOISE

The Wiener-Klintchine theory cannot be used to obtain the output power spectral density due to the time-varying nature of the noise. To obtain the autocorrelation function of the noise after the filter proceed formally as follows:

The output of a filter with impulse response h(t) for input N(t) is

$$Z(t) = \int h(t-\tau) \; N(\tau) \; d\tau \qquad \ldots(4.2.1)$$

Let $R_Z = E\{Z_1 Z_2\}$

and $R_{Z*} = E\{Z_1 Z_2^*\}$

Then

$$R_Z = \iint h(t_1-\alpha) \; h(t_2-\beta) \; R_N(\alpha,\beta) \; d\alpha \; d\beta \qquad \ldots(4.2.2)$$

where $R_N(\alpha,\beta) = E\{N(\alpha) \; N(\beta)\}$

Define the double Fourier transform

$$S_N(u, v) \underline{\Delta} \iint R_N(\alpha,\beta) \exp \{-j \; (\alpha u+\beta v)\} \; d\alpha \; d\beta$$

Using this inverse transform in eqn. (4.2.2) and the inverse transform of the filter transfer function H(w), gives,

$$R_Z = \frac{1}{(2\pi)^2} \iint H(u)H(v)S_N(u,v)\exp \{j(ut_1+vt_2\} \; du \; dv$$

Similarly $\qquad \ldots(4.2.3)$

$$R_{Z*} = \frac{1}{(2\pi)^2} \iint H(u)H^*(-v)S_{N*}(u,v)\exp \{j(ut_1+vt_2)\} \; du \; dv \qquad \ldots(4.2.4)$$

These expressions have been evaluated in the simulation program using the double Fourier Transform procedure contained in the Singleton FFT algorithm [1] .

5. CHARACTERIZATION OF NON-GAUSSIAN NOISE

In the previous section the second moment was used to characterize the noise. This section will consider the distribution of the noise. To give an indication of the type of distribution that results from a saturating type of device an arctan non-linearity was considered. For this non-linearity the output distribution functions can be found analytically. The effect of a filter on the density function can also be illustrated. An approximation to the density function is considered which is a sum of gaussian functions with different means and variances. Following this discussion a method of generating this approximation to the output distribution is considered.

5.1 ARCTAN NON-LINEARITY

To generate the type of probability density function that results from a saturating non-linearity, consider the transfer characteristic

$$y = \tan^{-1} x \qquad \text{...(5.1.1)}$$

For x a gaussian random process with mean m and variance σ^2, the density function of y is given by [4]

$$f(y) = (1+\tan^2 y) . \frac{1}{\sqrt{2\pi\sigma}} \exp\{ \frac{-1}{2\sigma^2} (\tan y - m)^2 \} \qquad \text{...(5.1.2)}$$

Fig. 9 shows the variation of this function with σ for a zero-mean input process. For a small σ the output distribution is essentially still of gaussian shape. As σ is increased the distribution changes to a double peaked function. These distributions are symmetric because of the zero-mean input process. By including a mean value to simulate the effect of a signal the output distribution becomes non-symmetric as shown in fig. 10. The distribution still remains essentially gaussian for small noise levels.

If the input noise has a correlation coefficient ρ the joint density function of y is given by

$$f(y_1, y_2) = (1+t_1^2)\ (1+t_2^2)\ \frac{1}{2\pi C}$$

$$\exp\{- \frac{1}{2C^2} \left[(t_1-m)^2 - 2\rho(t_1-m)(t_2-m) + (t_2-m)^2\right]\} \qquad \text{...(5.1.3)}$$

where $C = \sigma^2 \sqrt{1 - \rho^2}$

and $t_k \equiv \tan y_k$

This function is shown in fig. 11 for three values of the correlation coefficient ρ and a zero-mean input noise.

In the previous sections of the paper the input to the non-linear device was a complex noise. Now consider the non-

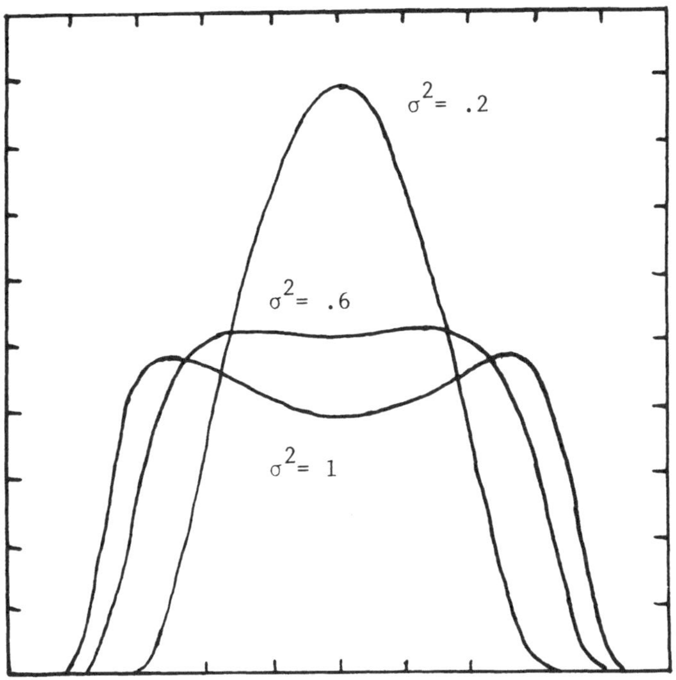

Fig. 9 OUTPUT PDF FROM REAL ARCTAN DEVICE (ZERO MEAN)

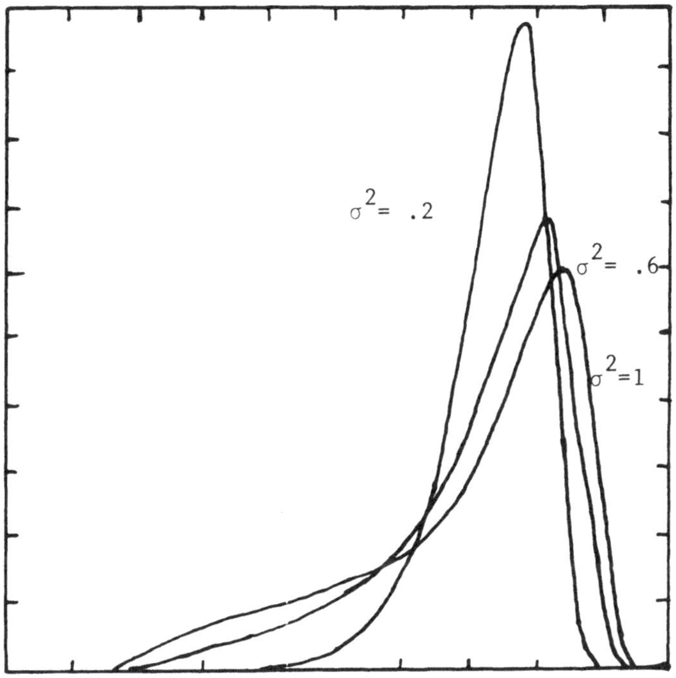

Fig. 10 OUTPUT PDF FROM REAL ARCTAN DEVICE (MEAN=1)

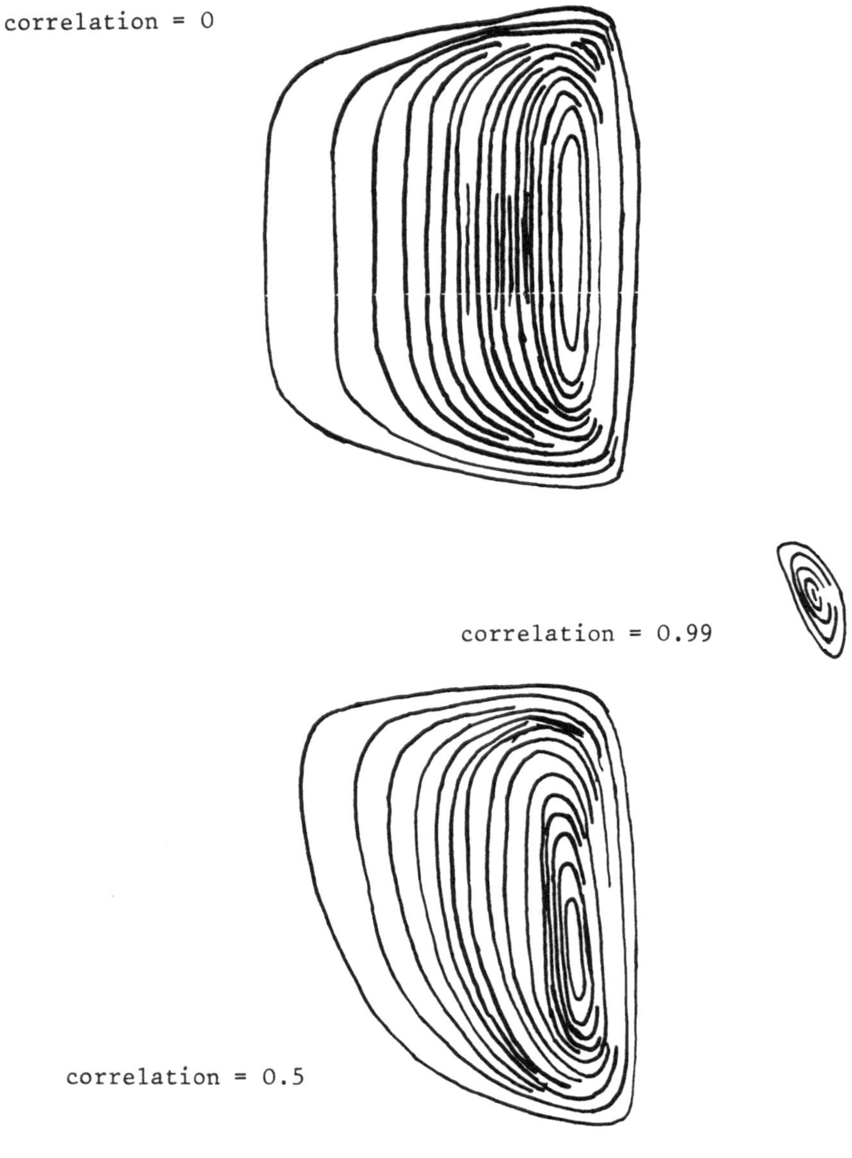

Fig. 11 EFFECT OF INPUT CORRELATION ON OUTPUT JOINT DENSITY FUNCTION.

linear transformation of the noise

$$V = A\,e^{j\phi} = x+jy \qquad \ldots(5.1.4)$$

to

$$N = \tan^{-1}A\,e^{j\phi} = \alpha + j\beta \qquad \ldots(5.1.5)$$

The joint density function of α and β can be derived as

$$f(\alpha,\beta) = \frac{t}{\mu}\,(1+t^2)\,p(t\cos\theta,\ t\sin\theta) \qquad \ldots(5.1.6)$$

where $p(x,y)$ is the joint density function of the input process

$$t = \tan\mu$$
$$\mu = \sqrt{\alpha^2+\beta^2}$$
$$\theta = \tan^{-1}[\beta/\alpha]$$

For a gaussian input process, fig. 12 shows plots of this output density function for two values of the input noise correlation coefficient.

5.2 APPROXIMATION OF OUTPUT NOISE DENSITY FUNCTION

A method of approximating density functions of the form found in the previous section is to use an expansion in terms of gaussian density functions [5] as follows:

For a real noise let the approximation be

$$\overline{f}(y) = \sum_k c_k\, g(y-b_k,\sigma_k) \qquad \ldots(5.2.1)$$

where

$$g(x,\sigma) = \frac{1}{\sqrt{2\pi}\,\sigma}\exp\left\{-\frac{x^2}{2\sigma^2}\right\} \qquad \ldots(5.2.2)$$

Fig. 13 shows a five term fit to a density function obtained from an arctan non-linearity. The fit was obtained by using a damped least squares optimisation procedure[6]. This method of approximation is readily extended to the case of complex input noise and a characteristic with AM-PM conversion.

In practice it is not feasible to form the output density function directly. It is possible however to generate the moments of the noise at the output of the non-linear device. The polynomial representation derived in section 3.4 can be used to generate these moments. Direct multiplication of the poly-

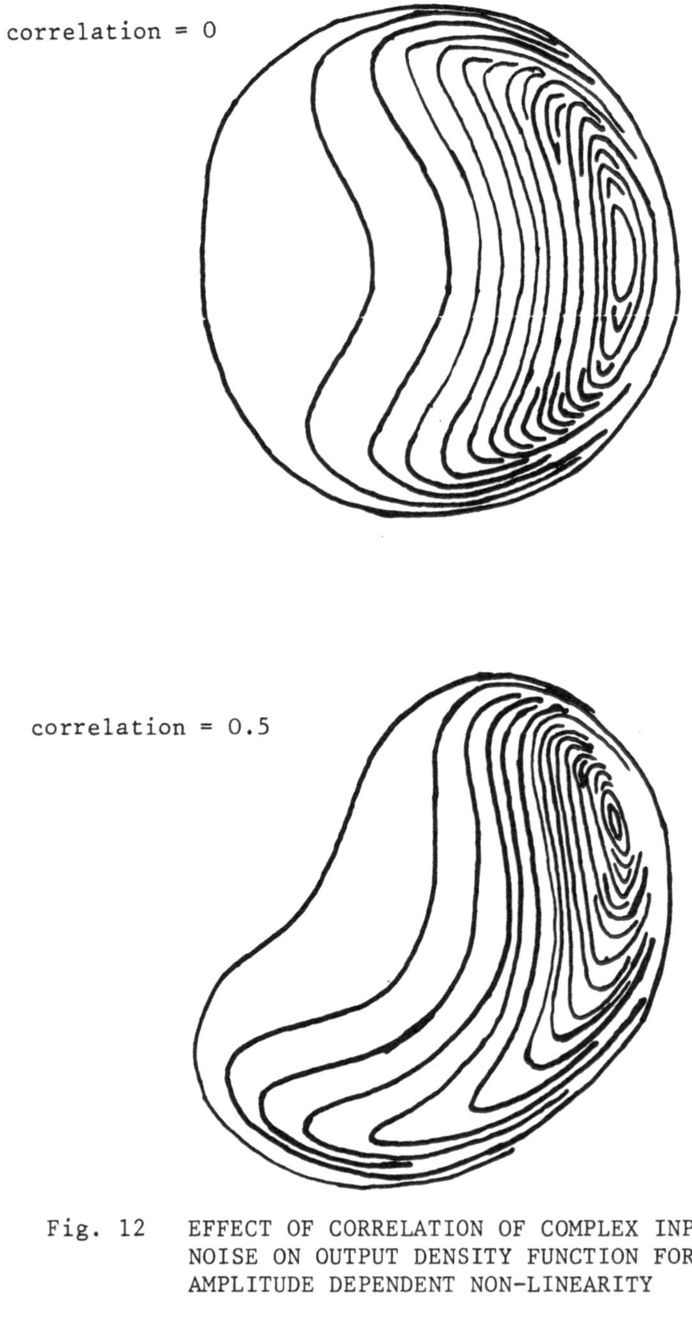

Fig. 12 EFFECT OF CORRELATION OF COMPLEX INPUT NOISE ON OUTPUT DENSITY FUNCTION FOR AMPLITUDE DEPENDENT NON-LINEARITY

nomials and then averaging over the input noise is not computationally feasible due to the extremely large number of terms involved. A recursive procedure for generating the moments has been used as follows for the complex noise case.

Let the input noise $V = x+jy$ and the output noise $N = \alpha+j\beta$. From eqn. (3.4.4), α and β can be written as polynomials in x and y from

$$\alpha+j\beta = \sum_{n=o}^{2L+1} \sum_{m=o}^{n} (a_{nm} + jb_{nm}) x^{n-m} y^{m} \qquad \ldots(5.2.3)$$

where

$$a_{nm} + jb_{nm} = \sum_{k=k_1}^{k_2} \sum_{r=r_1}^{r_2} (-1)^{m-r} j^{m} \binom{k}{r} \binom{n-k}{m-r} C_{n-k,k} \qquad \ldots(5.2.4)$$

$$k_1 = \max(0, n-L)$$
$$k_2 = \min(n, L+1)$$
$$r_1 = \max(0, m-n+k)$$
$$r_2 = \min(k,m)$$

Define the generalized moments

$$\mu_{nm}^{rs} = E\{x^r y^s \alpha^n \beta^m\} \qquad \ldots(5.2.5)$$

Now

$$\mu_{n+1m}^{rs} = E\{x^r y^s \alpha^{n+1} \beta^m\}$$

$$= \sum_k \sum_p a_{kp} E\{x^{r+k} y^{s+p} \alpha^n \beta^m\}$$

$$\therefore \mu_{n+1m}^{rs} = \sum_k \sum_p a_{kp} \mu_{nm}^{(r+k)(s+p)} \qquad \ldots(5.2.6)$$

Corresponding expressions hold for increases in m by replacing a_{kp} by b_{kp}. Higher generalized moments are required for lower order moments in order to generate the higher moments. Since only the μ_{nm}^{oo} are required not all of the generalized moments need be generated. Having generated these moments it is possible to fit the moments obtained from the approximation (5.2.1) by a non-linear equation solving routine or by optimization.

5.3 FILTERING OF THE NON-GAUSSIAN NOISE

In this section only a real noise and non-linear device is considered but the results can be straightforwardly extended to the complex case.

To filter the noise after the non-linear device requires the joint density function at several time samples. The generation of this joint density function requires additional information unless the noise at the input of the non-linearity is uncorrelated. The generation of the joint density function when the input noise is correlated is considered in the next section. For uncorrelated input noise the joint density function at two time samples is obtained, using eqn. (5.2.1) as

$$\bar{f}(y_1,y_2) = \bar{f}(y_1)\,\bar{f}(y_2)$$

$$= \sum_k \sum_n c_k\, c_n\, g(y_1-b_k,\sigma_k)\, g(y_2-b_n,\sigma_n) \quad \ldots(5.3.1)$$

and for m time samples as

$$\bar{f}(y_1,\ldots y_m) = \prod_{k=1}^{m}\{\sum_n c_n\, g(y_k-b_n,\sigma_n)\} \quad \ldots(5.3.2)$$

Let the output of the filter be z, which is given in terms of the input at several time samples as

$$z_n = \sum_{k=o}^{m} \lambda_k\, y_{n-k} \quad \ldots(5.3.3)$$

where for convenience λ_o is defined as unity.

By a direct transformation of probabilities the density function of z is given by

$$p(z_n) = \int\ldots\int f(y_{n-m},\ldots y_{n-1},z_n - \sum_{k=1}^{m} \lambda_k y_{n-k})dy_{n-m}\ldots dy_{n-1} \quad \ldots(5.3.4)$$

If $C(u_o,\ldots u_m)$ is the joint characteristic function of y and $M(\omega)$ is the characteristic function of z, then

$$M(\omega) = \frac{1}{(2\pi)^{m+1}} \int \ldots \int C(u_o, \ldots u_m)$$

$$\exp \{-j \sum_{k=1}^{m} u_k \, y_{n-k} - j \, (z_n - \sum_{k=1}^{m} \lambda_k y_{n-k}) u_o + j\omega z_n \}.$$

$$dy_{n-m} \ldots dy_{n-1} \; du_o \ldots du_m \; dz_n$$

$$= \frac{1}{(2\pi)^{m+1}} \int \ldots \int C(u_o, \ldots u_m)$$

$$\exp\{ j(\omega - u_o) z_n + j \sum_{k=1}^{m} (u_o \lambda_k - u_k) y_{n-k} \}$$

$$dy_{n-m} \ldots dy_{n-1} \; du_o \ldots du_m \; dz_n$$

$$= \int \ldots \int C(u_o, \ldots u_m) \delta(\omega - u_o) \prod_{k=1}^{m} \delta(u_k - \lambda_k u_o) du_o \ldots du_m$$

$$\therefore \quad M(\omega) = C(\omega, \lambda_1 w, \ldots \lambda_m \omega) \qquad \ldots(5.3.5)$$

For the case when the noise at the input to the non-linearity is uncorrelated this equation reduces to

$$M(\omega) = \prod_{k=o}^{m} C(\lambda_k \omega) \qquad \ldots(5.3.6)$$

This result provides a simple method of filtering the non-gaussian noise using the approximate density function given by eqn. (5.2.1). The characteristic function of the approximation is easily formulated since the characteristic function of each gaussian component is known.

To illustrate the effect of a simple filter on the shape of the density function this method was implemented for two filters, viz. (i) $\lambda_o = \lambda_1 = \frac{1}{2}$

(ii) $\lambda_o = .529, \lambda_1 = .0625, \; \lambda_2 = .096, \; \lambda_3 = .3125$

The input density functions considered were derived from an arctan non-linearity.

The characteristic function was evaluated and inverted using the FFT to give the filtered density functions shown in fig. 13 and 14.

In the earlier sections the noise was time-varying. This would result in the approximate density functions given by

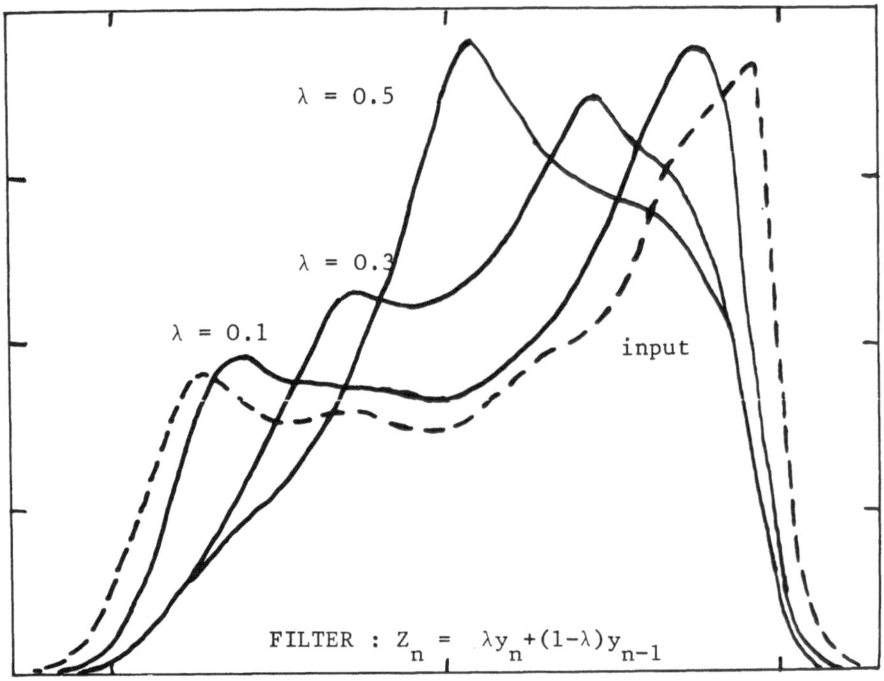

Fig. 13 FIVE TERM GAUSSIAN APPROXIMATION (MEAN = .4)

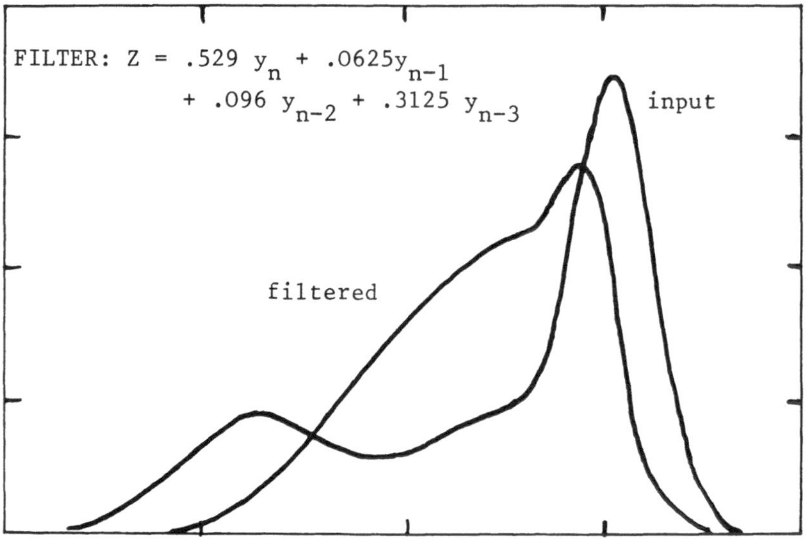

Fig. 14 FIVE TERM GAUSSIAN APPROXIMATION (MEAN = .4)

eqn. (5.2.1) having different parameter sets at each time sample. This only affects the filtering operation in that the characteristic function at each time sample of the input to the filter being different.

5.4 JOINT PROBABILITY DENSITY FUNCTION

The joint density function in time for the non-gaussian noise after the non-linear device could be approximated by

$$\overline{f}(y) = \sum_k c_k \, g(y-b_k, \Lambda_k) \qquad \ldots(5.4.1)$$

where y is the vector of the noise values at each time sample, b_k is the vector of mean values and Λ_k the correlation matrix.

Given the joint moments of the noise after the non-linearity at all time samples, the approximation to the density function given by eqn. (5.4.1) could be formulated. However it is not computationally feasible to generate these moments. An alternative approach is suggested as follows:

For two time samples eqn. (5.3.1) could be generalized as

$$\overline{f}(y_1, y_2) = \sum_k \sum_n c_k \, c_n \, g(y_1-b_k, \sigma_k, \, y_2-b_n, \sigma_n, \lambda_{kn}) \qquad \ldots(5.4.2)$$

where g is the bivariate gaussian density function with variances σ_k^2, σ_n^2 and correlation λ_{kn}.

By assuming a constant correlation coefficient $\lambda_{kn}/\sigma_k \, \sigma_n$ for all k and n only one correlation parameter is required. This parameter can be matched to the output correlation function obtained in section 4. This process has yet to be examined in any detail.

6. SUMMARY

This paper has presented some techniques suitable for simulating signal and noise in non-linear communication channels. The methods in section 4 have been programmed into the in-house simulation program. Some of the techniques in section 5 have been programmed and others are still being examined. Due to the action of the filter the assumption of gaussian noise at the receiver may be adequate for many applications. In the remainder of cases the sum of gaussian approximation should provide a suitable tool.

ACKNOWLEDGEMENT

The work recorded in this paper was developed under contract to ESTEC, Noordwijk, Holland. The author would like to thank the Directors of ESTEC and the Marconi Research Laboratories for permission to publish this paper. I should also like to thank my colleagues who have been involved in this work.

REFERENCES

1. R.C. Singleton. An algorithm for computing the mixed radix FFT. IEEE Trans. on Audio and Electroacoustics, Vol. AU-17, No. 2, June 1969 pp.93-103.

2. J.C. Fuenzalida, O. Shimbo, W.L. Cook. Time domain analysis of intermodulation effects caused by non-linear amplifiers. COMSAT Technical Review, Vol.3, No. 1, Spring 1973, pp. 89-140.

3. R.W. Kelly & P.R. Hariharan. Ideal limiting of periodic signals in random noise. IEEE Trans. on Aerospace and Electronic Systems, Vol. RES-7, No. 4, July 1971, pp.644-651

4. W.B. Davenport & W.L. Root. Random Signals and Noise McGraw Hill, 1958.

5. D.L. Alspach & H.W. Sorenson. Approximation of density functions by a sum of Gaussians for non-linear Bayesian estimation. Proc. of a symposium on non-linear estimation theory and its applications, 1970, San Diego, pp. 19-31.

6. E.M.L. Beale Numerical Methods. Nonlinear Programming (Ed Abadie) North Holland Publishing Company 1967.

DIGITAL TRANSMISSION OVER NONLINEAR CHANNELS - A VOLTERRA-SERIES ANALYSIS

Ezio Biglieri

Istituto di Elettrotecnica dell'Università, Napoli, Italy

ABSTRACT. *This paper examines the transmission of digital signals over nonlinear channels with memory. The approach taken uses a Volterra series expansion of the channel characteristics, and provides a general analytical tool to evaluate the performance of communication systems operating on those channels.*

Volterra series for bandpass channels are also introduced. As an application, a nonlinear digital satellite channel is modeled.

1. INTRODUCTION

The analysis of nonlinear channels with memory and the evaluation of digital transmission schemes operating on them are important practical problems. For example, in telephone lines used for data transmission, the advent of equalization, and hence of new precision in transmission, revealed that nonlinear distortion -- arising principally from inaccuracies in companding -- is a serious source of performance impairment. It has been conjectured that, for data transmission systems operating at rates higher than 4800 bits/s, the error rate is almost entirely determined by nonlinear distortion[1].

Another important example of nonlinear channel arises from digital satellite communications, where the on-board amplifiers, operated at or near saturation for better efficiency, exhibit strongly nonlinear characteristics.

In this paper, Volterra series will be used to analyze digital communication systems -- both baseband and passband -- operating over nonlinear channels with memory. The Volterra series approach has been taken because it provides a general analytical tool to deal with these channels. Although it may suffer some drawbacks in certain situations, its generality makes it attractive in several instances.

We shall first examine baseband transmission through a channel described by a Volterra series. The case of more than one signal, and of a signal with added noise entering the channel are considered.

We shall then turn our attention to bandpass signals and systems. Bandpass Volterra series are derived; as an example of application, a digital satellite channel is modeled using this theory.

2. CHARACTERIZING A NONLINEAR CHANNEL

For linear channels, the input-output relationship is fully described by their impulse response; on the other hand, if the channel is nonlinear but memoryless, and sufficiently well behaved, an input-output relationship can be obtained by expanding the nonlinear characteristics in a power series.

More generally, for a nonlinear channel with memory that satisfies certain regularity conditions, Volterra series provide a generalization of these two output representations. Actually, they turn out to be limiting cases of channels that can be described using Volterra series.

To motivate the general expression of the input-output relationship for a nonlinear system, let us first consider a simple example. Assume that the system is created by cascading a linear, time-invariant system with impulse response $h(t)$ and a nonlinear, memoryless system with an analytic input-output relationship $z(t)=f[v(t)]$ (fig.1).

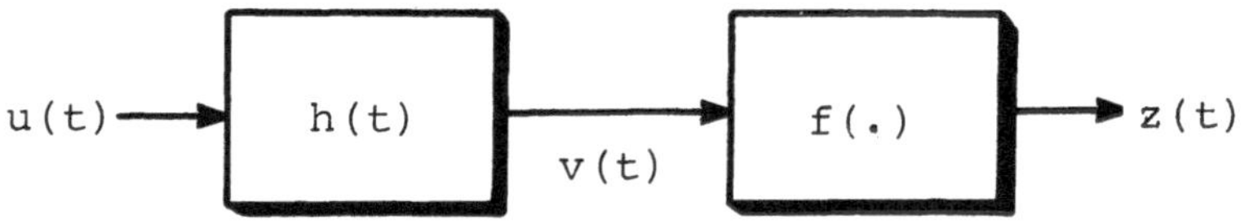

Fig.1

Let

$$f(\cdot) = \sum_{k=1}^{\infty} \gamma_k \frac{(\cdot)^k}{k!}$$

be the Taylor series expansion of $f(\cdot)$, where we assume that $f(0)=0$.

Denoting by $z(t)$ the system output when the input is $u(t)$, we get

$$z(t) = f\left[\int_{-\infty}^{\infty} h(\tau)u(t-\tau)d\tau \right]$$

$$= \sum_{k=1}^{\infty} \frac{\gamma_k}{k!}\left[\int_{-\infty}^{\infty} h(\tau)u(t-\tau)d\tau \right]^k$$

$$= \sum_{k=1}^{\infty} \frac{\gamma_k}{k!} \int_{-\infty}^{\infty} d\tau_1 \cdots \int_{-\infty}^{\infty} d\tau_k \prod_{r=1}^{k} h(\tau_r) \prod_{r=1}^{k} u(t-\tau_r)$$

Letting

$$h_k(\tau_1,\tau_2,\ldots,\tau_k) \triangleq \frac{\gamma_k}{k!} \prod_{r=1}^{k} h(\tau_r) \tag{2.1}$$

the input-output relationship takes the following form:

$$z(t) = \sum_{k=1}^{\infty} \int_{-\infty}^{\infty} d\tau_1 \cdots \int_{-\infty}^{\infty} d\tau_k \; h_k(\tau_1,\ldots,\tau_k) \prod_{r=1}^{k} u(t-\tau_r) \tag{2.2}$$

Without position (2.1), eq. (2.2) is the most general form of input-output relationship for a time-invariant nonlinear system with memory that meets some regularity conditions. It can be seen that the "Volterra kernels" $h_k(\ldots)$, a generalization of linear system impulse response, completely describe the system behavior. Thus, the problem of characterizing a nonlinear system with memory reduces to the problem of computing its Volterra kernels.

We have seen that, for the simple channel of fig.1, the Volterra kernels are given by eq. (2.1). In the more general situation of fig.2 -- a system which can be modeled as a memoryless nonlinearity $f(.)$ preceded and followed by two linear systems -- we get

$$(2.3) \qquad h_k(\tau_1,\ldots,\tau_k) = \frac{\gamma_k}{k!}\int_{-\infty}^{\infty} d\tau\, h''(\tau) \prod_{r=1}^{k} h'(\tau_r-\tau)$$

where γ_k are the power series coefficients of $f(\cdot)$,and $h'(\cdot)$, $h''(\cdot)$ are the impulse responses of the linear systems preceding and following the nonlinearity.

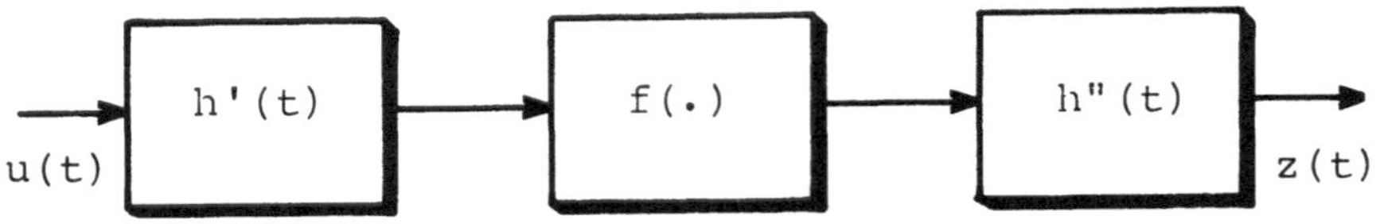

Fig.2

It can be observed that, in both equations (2.1) and (2.3) specifying Volterra kernels, these kernels turn out to be symmetric functions of their arguments, i.e., their values do not change when their arguments are permuted. The assumption of symmetry for the kernels does not entail any loss of generality [2] ; so, throughout this paper all Volterra kernels will be assumed to be symmetric, unless otherwise specified.

The most relevant feature of the Volterra series approach is that it offers a general, explicit input-output relationship. On the other side, it suffers some drawbacks that may limit its applicability.

First of all, for a Volterra series representation to be applicable to a nonlinear system, this system is required to satisfy some regularity conditions. For example, hysteresis and switching sytems cannot be described using this approach.

Furthermore Volterra kernels, which are needed to characterize a nonlinear system, cannot be parametrized in the general case, not unlike the impulse response for linear systems. Finally, for certain nonlinearities a huge number of terms in (2.2) must be retained for an accurate description,, thus reducing the applicability of this technique for practical computations.

As an example, to get a finite number of terms in (2.2) to describe the system of fig.1, the function $f(\cdot)$ must be approximated by a polynomial. Thus, the computational effort needed to evaluate (2.2) will be proportional to the degree of this polynomial, which in some cases may be very large.

3. SOME APPLICATIONS

The expression of the output from a channel described by a Volterra series will now be specialized to some cases of practical interest.

Consider first a baseband, linearly modulated digital signal

$$(3.1) \qquad x(t) = \sum_{n=-\infty}^{\infty} c_n \, q(t-nT)$$

where $(c_n)_{n=-\infty}^{\infty}$ is a real, discrete-time stationary random process and q(t) is a deterministic waveform. T denotes the signalling period.

Suppose that this signal is transmitted over a nonlinear channel. To derive an expression for the channel output using Volterra series, we can assume that (3.1) is obtained by passing the signal

$$(3.2) \qquad u(t) = \sum_{n=-\infty}^{\infty} c_n \, \delta(t-nT)$$

through a linear, time-invariant system whose impulse response is q(t). This linear system will be included in the channel structure, and the Volterra kernels of the channel will be modified accordingly.

If the Volterra kernels of the channel are $h_k(\ldots)$, we get at its output

$$(3.3) \qquad v(t) = \sum_{k=1}^{\infty} \sum_{n_1} \cdots \sum_{n_k} c_{n_1} \cdots c_{n_k} h_k(t-n_1T, \ldots, t-n_kT)$$

where the indices $n_1, \ldots, n_k$ run from $-\infty$ to ∞. In particular, if the signal v(t) is sampled at $t=t_o$, we get

$$(3.4) \qquad v(t_o) = \sum_{k=1}^{\infty} \sum_{n_1} \cdots \sum_{n_k} c_{n_1} \cdots c_{n_k} H_k(n_1, \ldots, n_k)$$

where

$$(3.5) \qquad H_k(n_1, \ldots, n_k) \triangleq h_k(t_o-n_1T, \ldots, t_o-n_kT).$$

Eq. (3.4) can also be rewritten as follows:

$$(3.6)\quad v(t_o) = c_o H_1(0) + \sum_{n_1 \neq 0} c_{n_1} H_1(n_1) + \sum_{k=2}^{\infty} \sum_{n_1} \cdots \sum_{n_k} c_{n_1} \cdots c_{n_k} H_k(n_1, \dots, n_k)$$

where one can easily recognize the various output contributions: useful signal, intersymbol interference produced by the linear distortion, nonlinear interference.

Assume now that the channel input is a real random process $u(t)$. Under a mild regularity assumption (weak continuity), $u(t)$ can be given the following series representation, valid for $-\infty < t < \infty$ [3]:

$$(3.7)\quad u(t) = \sum_n \xi_n s_n(t)$$

where $(\xi_n)_{n=-\infty}^{\infty}$ is a real, discrete-time random process such that

$$(3.8)\quad E\xi_n \xi_m = \delta_{mn}$$

At the channel output we get

$$(3.9)\quad z(t) = \sum_{k=1}^{\infty} \sum_{n_1} \cdots \sum_{n_k} \xi_{n_1} \cdots \xi_{n_k} \sigma_k(t; n_1, \dots, n_k)$$

where

$$(3.10)\quad \sigma_k(t; n_1, \dots, n_k) =$$

$$= \int_{-\infty}^{\infty} d\tau_1 \dots \int_{-\infty}^{\infty} d\tau_k \; h_k(\tau_1, \dots, \tau_k) \prod_{r=1}^{k} s_{n_r}(t - \tau_r)$$

Consider next the sum of two random processes, say

$$(3.11)\quad u(t) = x(t) + y(t)$$

and suppose that this signal is entering a nonlinear channel. This situation arises for instance when two different signals share the same channel, or when the input signal is corrupted by additive noise.

Assuming that both x(t) and y(t) can be expanded in the following form

$$x(t) = \sum_n \xi_n s_n(t)$$

$$y(t) = \sum_n \eta_n r_n(t)$$

and defining

$$\rho_{k-i,i}(t;n_1,\ldots,n_k) \triangleq \binom{k}{i} \int_{-\infty}^{\infty} d\tau_1 \cdots \int_{-\infty}^{\infty} d\tau_k \cdot$$

$$h_k(t-\tau_1,\ldots,t-\tau_k)\cdot s_{n_1}(\tau_1)\cdots s_{n_{k-i}}(\tau_{k-i})\cdot$$

$$r_{n_{k-i+1}}(\tau_{k-i+1})\cdots r_{n_k}(\tau_k)$$

we can write the channel output v(t) as

$$v(t) = \sum_{k=1}^{\infty} \sum_{i=0}^{k} \sum_{n_1} \cdots \sum_{n_k} \xi_{n_1} \cdots \xi_{n_{k-i}} \eta_{n_{k-i+1}} \cdots \eta_{n_k}$$

$$\rho_{k-i,i}(t;n_1,\ldots,n_k)$$

Notice that the terms $\rho_{k-i,i}(\ldots)$ account for the interaction between the two processes x(t) and y(t). In particular, $\rho_{k,0}(\ldots)$, $1<k<\infty$, give the output corresponding to the input x(t) alone, whereas $\rho_{0,k}(\ldots)$, $1<k<\infty$, do the same for the input y(t) alone.

Consider finally the problem of deriving first-order statistics of the signal at the output of a nonlinear channel. In particular, we are interested in the computation of the quantities

$$E\{\Omega(X)\}$$

and

$$\Pr\{X \leq \lambda\}$$

where Ω is a known function, λ a given real quantity and X a random variable representing the channel output at a given instant of time. Quantities like these are met, for instance, when one wants to compute the error probability for a digital transmission scheme [10] or the probability that the channel output exceeds a given threshold.

Exact evaluation of these quantities is generally a very hard task. However, one can approximate them, or get tight upper and lower bounds starting from the knowledge of a few moments of the random variable X (see, e.g., [6-9]). To obtain such moments, simple computational algorithms are available (see [10] , where Volterra series are applied to the computation of error probability for multilevel PAM transmission over nonlinear channels with memory).

4. BANDPASS NONLINEAR CHANNELS

In this section, the results presented previously -- an input-output relationship valid for nonlinear systems with memory -- are specialized to bandpass nonlinear systems.

Consider these systems, and a bandpass input. The analytic signal associated with the input can be expressed as

$$x(t) = A(t)e^{j\{\omega_0 t + \theta(t)\}} \tag{4.1}$$

where $A(t)$ and $\theta(t)$ are baseband signals, and ω_0 is the center frequency of the power spectrum of $x(t)$.

Let $\tilde{x}(t)$, $\tilde{y}(t)$ denote the complex envelope of the input and the output of the channel; if the channel is linear we have the known result

$$\tilde{y}(t) = \frac{1}{2}\int_{-\infty}^{\infty} \tilde{h}(\tau)\tilde{x}(t-\tau)d\tau \tag{4.2}$$

where $\tilde{h}(t)$ is the equivalent low-pass impulse response of the channel.

When the channel is nonlinear but memoryless, the channel output will generally include several spectral zones, centered around multiples of the frequency ω_0.

Assuming that the nonlinearity is followed by a zonal filter that stops all the spectral components other than that centered at ω_0, the complex envelope of the output will be given by

$$(4.3) \qquad \tilde{y}(t) = F[A(t)]\, e^{j\{\theta(t) + \phi[A(t)]\}}$$

where F(.) and ϕ(.) are two real functions that characterize the bandpass nonlinearity.

Clearly, (4.2) and (4.3) describe two limiting cases of nonlinear systems with memory. In the general case, the complex envelope of the first spectral zone output signal is given by [11] :

$$(4.4) \qquad y(t) = \sum_{k=0}^{\infty} L_k \int_{-\infty}^{\infty} d\tau_1 \cdots \int_{-\infty}^{\infty} d\tau_{2k+1}$$

$$\cdot \tilde{h}_{2k+1}(\tau_1, \ldots, \tau_{2k+1})$$

$$\cdot \prod_{r=1}^{k} \tilde{x}^*(t-\tau_r) \prod_{s=k+1}^{2k+1} \tilde{x}(t-\tau_s)$$

where

$$(4.5) \qquad L_k \triangleq \binom{2k+1}{k} 2^{-2k-1}$$

and $\tilde{h}_{2k+1}(\ldots)$ are the "equivalent low-pass Volterra kernels" that characterize the bandpass channel. These kernels can be derived from the real kernels as follows:

(i) Take the Fourier transform of $h_k(\tau_1, \ldots, \tau_k)$, say $H_k(\omega_1, \ldots, \omega_k)$. Due to the bandpass hypothesis, this function differs significantly from zero only in small neighborhoods of the points $(\pm\omega_0, \ldots, \pm\omega_0)$.

(ii) Write $H_k(\omega_1, \ldots, \omega_k)$ as a sum of 2^k functions with arguments $(\omega_1 \pm \omega_0, \ldots, \omega_k \pm \omega_0)$, each one being significantly different from zero only in the neighborhood of the origin.

(iii) Retain only the function with arguments

$$(\underbrace{\omega_1+\omega_0,\ldots,\omega_{k'}+\omega_0}_{k'},\underbrace{\omega_{k'+1}-\omega_0,\ldots,\omega_{2k'+1}-\omega_0}_{k'+1})$$

($k=2k'+1$). The inverse Fourier transform of this function is the equivalent low-pass Volterra kernel $\tilde{h}_k(\ldots)$.

Example 1

As a simple example, take a sinusoidal input with complex envelope

$$\tilde{x}(t) = Ae^{j\theta}$$

Then, the complex envelope of the output signal is given by

$$\tilde{y}(t) = e^{j\theta}\sum_{k=0}^{\infty} L_k A^{2k+1}\beta_{2k+1}$$

where

$$\beta_{2k+1} = 2H_{2k+1}(\underbrace{\omega_0,\ldots,\omega_0}_{k+1},\underbrace{-\omega_0,\ldots,-\omega_0}_{k})$$

in accordance with [2] .

Example 2

Consider now the transmission of a digital signal over a bandpass nonlinear channel. Assume that the channel has been modeled in such a way that the complex envelope of the input signal takes the form

$$\tilde{x}(t) = \sum_{n=-\infty}^{\infty} c_n\delta(t-nT)$$

where $(c_n)_{n=-\infty}^{\infty}$ is a complex, discrete-time stationary random process. Using (4.4), we get

$$\tilde{y}(t) = \sum_{k=0}^{\infty} L_k \sum_{n_1}\cdots\sum_{n_{2k+1}} c_{n_1}\cdots c_{n_{k+1}} c^*_{n_{k+2}}\cdots c^*_{n_{2k+1}} \cdot \tilde{h}_{2k+1}(t-n_1T,\ldots,t-n_{2k+1}T)$$

5.MODELING A DIGITAL SATELLITE CHANNEL

As an example of actual computation of a bandpass Volterra series, let us consider the model represented in fig.3, which is usually assumed for digital satellite channels. Here a nonlinear memoryless part, representing the on-board traveling-wave tube amplifier, is preceded and followed by two bandpass linear systems. The first represents the cascade of earth station transmitting filter and satellite input filter; the other represents the cascade of satellite output filter and earth station receiving filter.

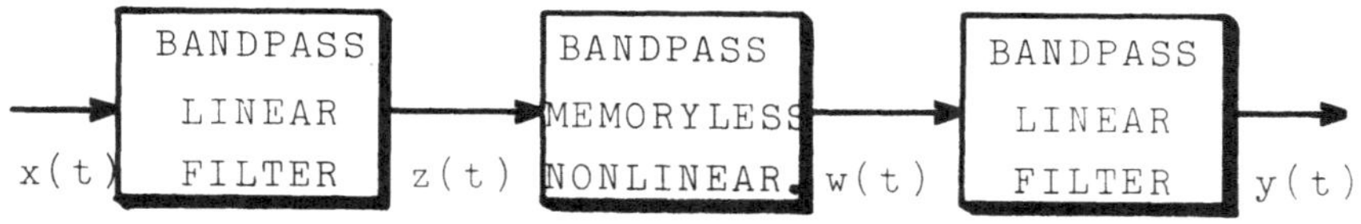

Fig.3

Assume that h'(.),h"(.) are the impulse responses of the two linear systems, both of whose transfer functions H'(.),H"(.) are centered around the center frequency ω_0; assume also that the memoryless nonlinear device has an analytic input-output relationship

$$(5.1) \qquad w(t) = \sum_{n=1}^{\infty} \gamma_n \frac{z^n(t)}{n!}$$

The Volterra series representation for this channel has kernels given by (2.2). Consider now the complex envelope of the first spectral zone of the output signal. Computing the Fourier transform of the kernels (2.3) and using the procedure described in Section 4, we obtain

$$(5.2) \qquad \tilde{h}_{2k+1}(\tau_1,\ldots,\tau_{2k+1}) =$$

$$= \frac{\gamma_{2k+1}}{(2k+1)!} \int_{-\infty}^{\infty} \tilde{h}''(\tau) \prod_{r=1}^{k} \tilde{h}'^{*}(\tau_r-\tau) \prod_{s=k+1}^{2k+1} \tilde{h}'(\tau_s-\tau)\,d\tau$$

where $\tilde{h}'(.)$, $\tilde{h}''(.)$ are the equivalent low-pass impulse responses of the linear filters (see [11] , where this theory is also applied to the computation of error probability for a PSK digital satellite system).

Let us now turn our attention to the characterization of the nonlinear device of fig.3. Usually, band-pass nonlinear devices are described using the functions $F(.)$ and $\phi(.)$ of eq. (4.3). So, to characterize completely the satellite channel we should be able to derive the coefficients γ_n of (5.1) from these two functions.

To do this, observe that letting $\tilde{h}'(.)=\tilde{h}''(.)=\delta(.)$, i.e., removing the two linear filters from the scheme of fig.3, we get

$$\tilde{y}(t) = \sum_{k=0}^{\infty} \frac{\gamma_{2k+1}}{k!(k+1)!2^{2k+1}} [\tilde{x}^*(t)]^k [\tilde{x}(t)]^{k+1} \tag{5.3}$$

$$= e^{j\theta(t)} \sum_{k=0}^{\infty} \frac{\gamma_{2k+1}}{k!(k+1)!2^{2k+1}} [A(t)]^{2k+1}$$

Thus, comparing (5.3) with (4.3),

$$F(A)e^{j\phi(A)} = \sum_{k=0}^{\infty} \frac{\gamma_{2k+1}}{k!(k+1)!2^{2k+1}} A^{2k+1} \tag{5.4}$$

Eq.(5.4) relates the functions $F(.)$ and $\phi(.)$ with the coefficients γ_{2k+1} (which may become complex if the nonlinearity involves a nonzero $\phi(.)$, i.e., a phase shift). As an example, expand the left-hand side of (5.4) in the form

$$F(A)e^{j\phi(A)} = \sum_{\ell=1}^{L} b_\ell J_1(\delta_\ell \alpha A)$$

where α is a real constant, $b_1,\dots,b_L$ are complex numbers, and the δ_ℓ's can take the value $\delta_\ell=\ell$ [4] or are zeros of $J_1(.)$ [5] . Then, recalling the Taylor series expansion of the Bessel function $J_1(.)$, we get

$$\gamma_{2k+1} = \alpha^{2k+1} \sum_{\ell=1}^{L} b_\ell \delta_\ell^{2k+1} \quad .$$

REFERENCES

[1] R.W.Lucky,"Modulation and detection for data transmission on the telephone channel", in J.K.Skwirzynski,ed., New Directions in Signal Processing in Communication and Control, Noordhoff, Leiden (Holland),1975

[2] E.Bedrosian and S.O.Rice,"The output properties of Volterra systems (nonlinear systems with memory) driven by harmonic and Gaussian inputs",IEEE Proc., vol.59,p.1699 ff,December 1971

[3] S.Cambanis and E.Masry,"On the representation of weakly continuous stochastic processes", Information Sciences,vol.3, p.277 ff,1971

[4] O.Shimbo and P.J.Pontano,"A general theory for intelligible crosstalk between frequency-division multiplexed angle-modulated carriers",IEEE Trans.Commun.,vol.COM-24,p.999ff,1976

[5] W.C.Lindsey,J.K.Omura,K.T.Woo,T.C.Huang,L.Biederman, "Investigation of modulation/coding tradeoff for military satellite communications.Vol.2: System modeling and analysis", LINCOM Technical Report, Jan.1977

[6] M.G.Krein,"The ideas of P.L.Čebyšev and A.A.Markov in the theory of limiting values of integrals and their further developments",Am.Math.Soc.Transl., Ser.2,vol.12,1951

[7] K.Yao and R.M.Tobin,"Moment space upper and lower bounds for digital systems with intersymbol interference", IEEE Trans. Inform.Theory, vol.IT-22, p.65 ff,Jan.1976

[8] K.Yao and E.Biglieri,"Moment inequalities for error probabilities in digital communication systems", submitted to NTC'77

[9] S.Benedetto and E.Biglieri, "A computational method for solving noise problems", in J.K.Skwirzynski,ed.,New Directions in Signal Processing in Communication and Control, Noordhoff, Leiden (Holland),1975

[10] S.Benedetto,E.Biglieri and R.Daffara,"Performance of multilevel baseband digital systems in a nonlinear environment", IEEE Trans.Commun., vol.COM-24, p.1166 ff, October 1976

[11] M.Ajmone Marsan,S.Benedetto,E.Biglieri and R.Daffara,"Analytical evaluation of the performance of digital satellite links",Convegno Internazionale delle Comunicazioni, Genova,Italy, October 1977

The Output Signal and Noise from a Bandpass Nonlinearity Involving AM-to-PM Conversion

Nelson M. Blachman

Office of Naval Research, US Navy
223 Old Marylebone Rd., London NW1 5TH

The input to any bandpass nonlinearity can be expressed as $A(t)\cos\phi(t)$, where $\phi(t)$ includes a term of the form $2\pi Ft$ and a more slowly varying term; it can be represented by $A(t)e^{j\phi(t)}$, of which it's the real part. We now suppose that this input is the sum of a signal $a(t)\cos\alpha(t)$ plus interference $b(t)\cos\beta(t)$. Hence,

$$Ae^{j\phi}=ae^{j\alpha}+be^{j\beta}$$

with

$$A=|ae^{j\alpha}+be^{j\beta}| \text{ and } e^{j\phi}=\frac{ae^{j\alpha}+be^{j\beta}}{A} \qquad (1)$$

The output of the bandpass nonlinearity takes the form

$$g(A)\cos[\phi+\gamma(A)]=Re\{g(A)e^{j\phi+j\gamma(A)}\}=Re\{g(A)\frac{ae^{j\alpha}+be^{j\beta}}{A}e^{j\gamma(A)}\}, \qquad (2)$$

where $g(A)$ expresses the amplitude nonlinearity and $\gamma(A)$ the AM-to-PM conversion.

We suppose that the interference is statistically independent of the signal and that its amplitude and phase have joint probability density function $p(b)/2\pi$. The output signal, as explained in reference [1], is therefore the real part of

$$\int_0^\infty\int_0^{2\pi}\frac{p(b)}{2\pi}\,g(A)\,\frac{ae^{j\alpha}+be^{j\beta}}{A}\,e^{j\gamma(A)}d\beta\;db=h(a)e^{j\alpha+j\theta(a)}, \qquad (3)$$

where A is given by (1). Because this average of the output (2) over the statistics of the interference takes the form appearing on the right in (3), we see that the output signal generally suffers both amplitude distortion and AM-to-PM conversion.

The probability density function $p(b)$ may take the form of a Rayleigh distribution, and the integration in (3) can be carried out for any given $g(\ )$ and $\gamma(\ )$, although numerical methods may be required. When the input signal consitutes only a small fraction of the total input power, however, a slight generalization of the approach presented in reference [2] yields a considerable simplification. Replacing the $g(A)$ there by $g(A)e^{j\gamma(A)}$, we find that the output signal is the real part of

$$\frac{1}{2}ae^{j\alpha}\int_0^\infty p(A)\ \{\frac{1}{A}\, g(A)\, e^{j\gamma(A)} + \frac{d}{dA}\,[g(A)e^{j\gamma(A)}]\}\ dA. \qquad (4)$$

When the interference is gaussian noise or the sum of many signals, each contributing only s amall part of their total power, plus any amount of additive gaussian noise, the input amplitude A has a Rayleigh distribution with density $p(A)=(A/\sigma^2)e^{-A^2/2\sigma^2}$. Substituting it into (4) and integrating by parts [2], we find that the output signal is the real part of

$$\frac{1}{2}\, ae^{j\alpha}\int_0^\infty Ag(A)e^{j\gamma(A)}p(A)dA/\sigma^2\,. \qquad (5)$$

Because the input signal was assumed weak in (4) and (5) [and the $p(b)$ in (3) was therefore essentially identical with $p(A)$ $(b\approx A)$], the nonlinearity and AM-to-PM conversion seen in (3) do not appear in [(4) or] (5). The input signal simply suffers an attenuation by a factor equal to half the absolute value of the integral and a constant phase shift equal to its argument. The evaluation of (5) can be very easy; if, for example, $g(A)e^{j\gamma(A)}$ can be expressed as an odd polynomial with complex coefficients $\Sigma_k c_k A^{2k+1}$ (which, with enough terms, should always provide an adequate approximation), the output signal is the real part of

$$\Sigma_k\ (k+1)!(2\sigma^2)^k c_k\ a(t)e^{j\alpha(t)} \qquad (6)$$

In addition to this weak output signal, the output of the nonlinearity contains output noise and intermodulation which, in total, is essentially the same as the output that would be obtained if the input were just gaussian noise with the power spectrum of the total input. To study this output we can use a slight generalization of the approach presented in reference [3], which applied diagonal expansions of bivariate probability density functions as treated by Barrett and Lampard [4]. By making use of the work of Reed [5] and Campbell [6], we can obtain a diagonal expansion of the quadrivariate distribution of

$A = A(t)$, $\phi = \phi(t)$, $A' = A(t+\tau)$, and $\phi' = \phi(t+\tau)$, which has density [5], [6]

$$p(A,\phi,A',\phi') = \frac{AA'}{4\pi^2\sigma^4(1-r^2)}\exp\left[-\frac{A^2+A'^2-2rAA'\cos(\phi'-\phi-\Psi)}{2\sigma^2(1-r^2)}\right], \qquad (7)$$

where $\sigma^2 r(\tau)\cos\Psi(\tau)$ is the correlation function of the total input $A(t)\cos\phi(t)$ and is the Fourier transform of its power spectral density $\frac{1}{2}\sigma^2[S(f)+S(-f)]$. Here $\Psi(\tau)$ includes a term of the form $2\pi F\tau$ and a more slowly varying term; it is an odd function of τ while $r(\tau)$ is an even function with $r(0)=1$. $S(f)$, the Fourier transform of $r(\tau)e^{j\Psi(\tau)}$, with unit area, describes that portion of the narrowband input spectrum in the neighborhood of frequency F; if it is symmetric about F, $\Psi(\tau)$ is simply $2\pi F\tau$.

Expanding (7) as a Fourier series in terms of its argument $\phi'-\phi-\Psi$, we find that it becomes

$$p(A,\phi,A',\phi') = \frac{AA'}{4\pi^2\sigma^4(1-r^2)} \exp\left[\frac{-A^2-A'^2}{2\sigma^2(1-r^2)}\right] \cdot \sum_{m=-\infty}^{\infty} I_m\left(\frac{rAA'/\sigma^2}{1-r^2}\right) e^{mj(\phi-\phi'+\Psi)}.$$

Making use of the Hille-Hardy formula [5], [6] now, we get

$$p(A,\phi,A',\phi')=p(A,\phi)p(A',\phi')\sum_{m=-\infty}^{\infty}\sum_{k=0}^{\infty} F_{mk}^{-1}a_{mk}(\tau)f_{mk}(A,\phi)f_{mk}^{*}(A',\phi')$$

with

$$F_{mk}=\frac{(|m|+k)!}{k!},\quad f_{mk}(A,\phi)=\left(\frac{A}{\sigma\sqrt{2}}\right)^{|m|} L_k^{(|m|)}\left(\frac{A^2}{2\sigma^2}\right)e^{mj\phi}, \tag{8}$$

$$p(A,\phi)=\frac{A}{2\pi\sigma^2}e^{-A^2/2\sigma^2},\quad a_{mk}(\tau)=r^{|m|+2k}(\tau)\,e^{mj\Psi(\tau)}$$

where $L_k^{(m)}(x)=x^{-m}e^{x}(d/dx)^k(x^{m+k}e^{-x})/k!$ is the generalized Laguerre polynomial, the asterisk indicates the complex conjugate, F_{mk} is the mean-squared absolute value of $f_{mk}(A,\phi)$ when averaged over $p(A,\phi)$, and $F_{mk}a_{mk}(\tau)$ is the correlation function of $f_{mk}(A,\phi)$, i.e., the mean value of $f_{mk}^{*}(A,\phi)f_{mk}(A',\phi')$ when averaged over $p(A,\phi,A',\phi')$.

The functions $f_{mk}(A,\phi)$ form a complete orthogonal set under the weighting $p(A,\phi)$, and the expansion (8) of $p(A,\phi,A',\phi')$ is exactly of the diagonal form discussed in [4] and [3] except for the doubling of the number of independent variables and the appearance of complex functions. [This expansion is not analogous to the double diagonal series for the joint bivariate density function of a signal and noise because it cannot be factored as the product of a density $q(A,A')$ times another density $q'(\phi,\phi')$.] From the analysis in [3] it thus follows that the output $G(A,\phi)$ of the nonlinearity can be expressed as a summation of uncorrelated terms

$$G(A,\phi)=\sum_{m=-\infty}^{\infty}\sum_{k=0}^{\infty}F_{mk}^{-1}g_{mk}f_{mk}(A,\phi)\ \text{with}\ g_{mk}=E\{G(A,\phi)f_{mk}^{*}(A,\phi)\}, \tag{9}$$

where $E\{\ \}$ here denotes the average over the distribution $p(A,\phi)$.

In particular, when $G(A,\phi) = g(A) \cos [\phi+\gamma(A)]$, the integration with respect to ϕ yields $g_{mk}=0$ for $m = 1$, and it leaves the result

$$g_{1k} = g_{-1k}^{*} = \frac{1}{2} \int_0^\infty g(A) e^{j\gamma(A)} \frac{A}{\sigma\sqrt{2}} L_k^{(1)}\left(\frac{A^2}{2\sigma^2}\right) p(A)\, dA, \qquad (10)$$

where $p(A)=(A/\sigma^2)\exp(-A^2/2\sigma^2)$ is the Rayleigh probability density of A. Since $L_0^{(m)}(x)$ is identically 1 for every m, it is easily seen that the $k=0$ terms of (9) correspond to the output signal (5) in the preceding case. These terms represent the undistorted component of the output and here include undistorted versions of all input signals (and of the input noise as well), although they will in general suffer a constant phase shift.

Because the terms of the series (9) are uncorrelated with one another (for all time shifts τ), the correlation function of the output $G(A,\phi)$ is simply the sum of the correlation functions of the separate terms of (9), viz.,

$$\sum_{m=-\infty}^{\infty} \sum_{k=0}^{\infty} F_{mk}^{-1} |g_{mk}|^2 a_{mk}(\tau). \qquad (11)$$

The power spectral density of the output is the Fourier transform of (11), which is easy to determine because $a_{mk}(\tau)$ is evidently the product of a power of the Fourier transform of $S(f)$ times a power of the Fourier transform of $S(-f)$, the exponents of the two powers being k and $m + k$ or vice versa, accordingly as m is negative or positive, with $S(f)$ the normalized one-sided input spectral density. The output spectral density can thus be expressed as a summation of convolutions of this input spectrum, viz.,

$$\sum_{m=-\infty}^{\infty} \sum_{k=0}^{\infty} F_{mk}^{-1} |g_{mk}|^2 S(f \operatorname{sgn} m)^{*|m|} * [S(-f)*S(f)]^{*k} \qquad (12)$$

The asterisks here denote convolution; the superior asterisks indicate $(|m| - 1)$-fold and $(k-1)$-fold autoconvolution. When the number following the asterisk is zero, it yields a delta function $\delta(f)$. Thus, for $k=0$ and $|m|=1$ the corresponding terms of (12) combine to give just a constant times the input spectrum, which represents the undistorted output component.

Returning to the expression (10) for g_{1k}, we can determine its value when $g(A)e^{j\gamma(A)}$ is approximated by a polynomial $\Sigma_h c_h A^{2h+1}$. For this purpose we substitute both this polynomial and the Rodrigues formula $L_k^{(1)}(x) = x^{-1} e^x (d/dx)^k (x^{k+1} e^{-x})/k!$ into (10), and we integrate k times by parts, getting

$$g_{1k} = g_{-1k}^{*} = \frac{1}{2} (-1)^k \Sigma_h (h+1)! \binom{h}{k} (\sqrt{2}\,\sigma)^{2h+1} c_h . \qquad (13)$$

(This expression can also be obtained by making use of the result derived in the appendix of reference [5], but the latter does not make clear that the terms vanish for which $h<k$.) Equations (5) and (10), however, shed more light upon the effect of the AM-to-PM conversion than do (6) and (13).

Equation (5) gives us the output signal of the nonlinearity when the input signal represents only a small proportion of the total input power, and (12) gives us the power spectrum of the output noise (and intermodulation). For the case of a bandpass nonlinearity with no d.c. or harmonics in its output, all of the coefficients $|g_{mk}|^2$ needed in (12) can be obtained from (10) [or (13)].

REFERENCES

1. N.M. Blachman, "The signal × signal, noise × noise, and signal × noise output of a nonlinearity," IEEE Trans. Information Theory, vol. IT-14, pp. 21-27, January, 1968.

2. N.M. Blachman, "Band-pass nonlinearities," IEEE Trans. Information Theory, vol. IT-10, pp. 162-164, April, 1964.

3. N.M. Blachman, "The uncorrelated output components of a nonlinearity," IEEE Trans. Information Theory, vol. IT-14, pp. 250-258, March, 1968.

4. J.F. Barrett and D.G. Lampard, "An expansion of some second-order probability distributions and its application to noise problems," IRE Trans. Information Theory, vol. IT-1, pp. 10-15, March, 1955.

5. I.S. Reed, "On the use of Laguerre polynomials in treating the envelope and phase components of narrow-band gaussian noise," IRE Trans. Information Theory, vol. IT-5, pp. 102-105, September, 1959.

6. L.L. Campbell, "A general analysis of post-detection correlation," IEEE Trans. Information Theory, vol. IT-11, pp. 409-415, July, 1965.

MOMENT SPACE ERROR BOUNDS IN DIGITAL COMMUNICATION SYSTEMS*

K. Yao

Department of System Science, University of California
Los Angeles, California, USA

ABSTRACT. In many digital communication systems, the error probability can be expressed as the statistical expectation of a function of the random interference variable. We consider a bounding technique based on an isomorphism theorem from the theory of moment spaces to evaluate upper and lower bounds to the error probability. In the isomorphism theorem, an equivalence is established between the convex hull of a curve generated by an arbitrary set of continuous kernel functions and the convex body generated by the generalized moments of these kernels. By proper selection of the kernels, the error probability can be identified as one of the moments, while other moments of the interference, which depend upon the system parameters, can be chosen such that they can be readily evaluated. Thus bounds to error probability can be obtained from the envelopes of the convex body. Various forms of error bounds are presented. Specific examples of error bounds for linear intersymbol interference channels, co-channel interferences, and spread-spectrum multiple-access interference channels are shown to be tight and computationally tractable.

1. INTRODUCTION

In many practical digital communication systems, we encounter various random interferences besides the additive Gaussian noise. In high data-rate digital systems, there are intersymbol interfer-

* This work has been supported by the Electronics Program of the US Office of Naval Research.

ences from adjacent random data. In satellite and microwave systems, there may be co-channel interferences from other users in the same passband of the receiver. In code division spread-spectrum multiple-access communication systems, other coded signals appear as random interferences to each user. In all the above described and in many other communication systems, the error probability can be expressed as the statistical expectation of a function of the random interference variable Z. Since the set of possible realizations of Z is usually very large, it is generally not possible to perform this expectation directly. We consider a new bounding technique based on an isomorphism theorem from the theory of moment spaces to evaluate upper and lower bounds to the error probability. In the isomorphism theorem, an equivalence is established between the convex hull of a curve generated by an arbitrary set of continuous kernel functions defined over a compact interval and the compact convex body generated by the generalized moments of these kernels with respect to all probability distributions defined on the compact interval. By proper selection of the kernels, the error probability can be identified as one of the moments, while other moments of the interference, which depend upon the system parameters, can be chosen such that they can be readily evaluated. Then the upper and lower bounds to error probability can immediately be obtained from the upper and lower envelopes of the compact convex body.

Consider a quite general linear binary digital communication system where the decision random variable at a given sampling instant is given by

$$y = ah + Z + n \,, \tag{1.1}$$

where a is the input data taking values ± 1 with equal probability, h is the desired system response,

$$Z = \sum_{i=1}^{M} X_i \tag{1.2}$$

is the interference random variable with independent known symmetric discrete random variables X_i, and n is a zero-mean Gaussian random variable of variance σ^2. All the random variables a, X_i and n are assumed to be independent. The maximum interference distortion

$$D = \sum_{i=1}^{M} \text{Max}|X_i| \tag{1.3}$$

is assumed to be finite. Additional details are given for each specific problem of interest in Section 4.

Then the error probability P_e can be expressed as

$$P_e = E_Z\{Q((h+Z)/\sigma)\} \qquad (1.4a)$$

$$= E_Z\{[Q((h+|Z|)/\sigma)) + Q((h-|Z|)/\sigma)]/2\} \quad , \qquad (1.4b)$$

where $Q(x)$ is the complementary probability distribution function of a zero-mean and unit variance Gaussian random variable. In general, the direct evaluation of P_e by (1.2a) or (1.2b) can be formidable when M, the number of terms in Z, is large.

2. THEORY OF MOMENT SPACES

A basic result from the theory of moment spaces ([1-3]) will now be used to obtain upper and lower bounds to the error probability given by (1.4).

Theorem Let Z be a random variable with a probability distribution function $G_Z(z)$ defined over a finite closed interval $I = [a,b]$. Let $k_1(z), k_2(z), \ldots, k_N(z)$ be a set of N continuous functions defined on I. The generalized moment of the random variable Z induced by the function $k_i(z)$ is

$$m_i = \int_I k_i(z)dG_Z(z) = E_Z\{k_i(z)\} \quad , \quad i=1,\ldots,N \qquad (2.1)$$

We denote the N-th moment space $\mathcal{M}$ by

$$\mathcal{M} = \{\underline{m} = (m_1,\ldots,m_N) \in \mathbb{R}^N | m_i = \int_a^b k_i(z)dG_Z(z), \ i=1,\ldots,N\}$$

where G_Z ranges over the set of probability distributions defined in $I = [a,b]$. Then $\mathcal{M}$ is a closed, bounded, and convex set. Now let $\mathcal{C}$ be the curve $\underline{r} = (r_1,\ldots,r_N)$ traced out in $\mathbb{R}^N$ by $r_i = k_i(z)$ for z in I. Let $\mathcal{H}$ be the convex hull of $\mathcal{C}$. Then $\mathcal{H} = \mathcal{M}$.

Since the expression for the error probability given in (1.4a) or (1.4b) can be represented as a moment of a continuous function, then the error probability can be identified as one of the coordinates in $\mathcal{M}$. The remaining coordinates can then be chosen according to the manner in which one wishes to utilize the moments of the intersymbol interference statistics. Then, noting that $\mathcal{M} = \mathcal{H}$, we can find upper and lower bounds to the error probability in terms of the remaining moments by direct evaluation of the upper and lower envelopes of $\mathcal{H}$. As an example, let N=2. By selecting $k_1(z) = Q((h+z)/\sigma)$, then $m_1 = P_e$. Choose $k_2(z)$ such that its generalized moment m_2 is easily evaluable. In Fig. 1, we plot $k_1(z)$ versus $k_2(z)$ and obtain the bold curve $\mathcal{C}$. Its convex $\mathcal{H}$, which is equal to $\mathcal{M}$, is given by the shaded body. Then the vertical line at m_2 piercing $\mathcal{M}$ yields a lower envelope value ℓ and an upper envelope value u. Thus,

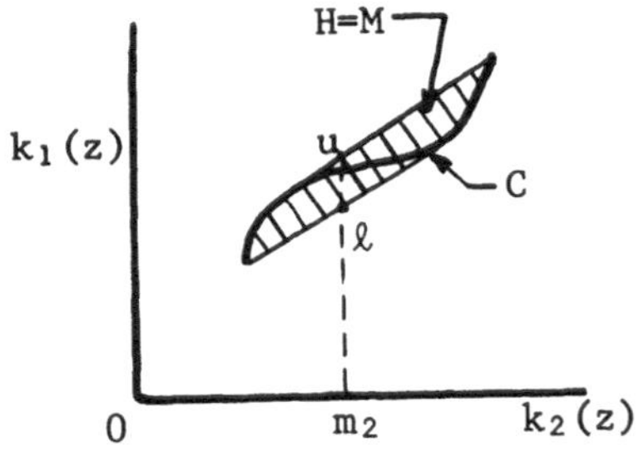

Fig. 1. Generation of Compact Convex Hull.

the bounds on P_e are given by $\ell \leq P_e \leq u$.

From Fig. 1, it is clear that in order to obtain tight error bounds, we want the convex body $\mathcal{H} = \mathcal{M}$ to be "thin". Of course we want to select $k_2(z)$ such that not only is the convex body thin, but its generalized moment is easily evaluable. Otherwise, we can select $k_2(z) = k_1(z)$ and obtain a convex body consisting of a finite straight line segment. Then $m_2 = m_1 = P_e$. But $m_1 = P_e$ was originally postulated to be computationally intractable and thus m_2 is not available. Thus, in the application of the Isomorphism Theorem, there are conflicting requirements. On the one hand, we want $k_2(z)$ to be similar to $k_1(z)$ such that the resulting body is thin. On the other hand, we require $k_2(z)$ such that its generalized moment is readily evaluable.

3. VARIOUS FORMS OF MOMENT SPACE BOUNDS

In Section 3, we shall consider various forms of moment space error bounds.

3.1 Absolute moment bounds

Let $k_1(z)$ be the function given by the expression inside the bracket of the r.h.s. of (1.4b) and let $k_2(z) = |z|$ with $I = [0,D]$ where D is given by (1.3). Direct evaluation shows $\mathcal{C}$ is always a convex $\cup$ function. Then the upper envelope of $\mathcal{H}$ is a straight line ([1]). Thus,

$$P_e \leq [(Q((h-D)/\sigma) + Q((h+D)/\sigma))/2 - Q(h/\sigma)](m_2/D) + Q(h/\sigma) \tag{3.1a}$$

$$[Q((h-m_2)/\sigma) + Q((h+m_2)/\sigma)]/2 \leq P_e \tag{3.1b}$$

The lower bound of (3.1b) is equivalent to that in [4] obtained by using Jensen's inequality.

3.2 Variance bounds

Let $k_1(z)$ be the function given by the expression inside the bracket of the r.h.s. of (1.4b) and let $k_2(z) = z^2/D$ with $I=[0,D]$. The shape of the resulting curve $\mathscr{C}$ depends on h, D, and the SNR and is generally not a convex function. Thus, the analytical descriptions of the upper and lower envelopes, while fundamentally not complicated, are quite detailed. As discussed in [1], these variance bounds are equivalent to those obtained by [5] and [6] and are summarized in Table I of [6].

3.3 Even order moment bounds

Let $k_1(z)$ be the function given by the expression inside the brackets of the r.h.s. of (1.4b), let $k_2(z) = dz^m$, where d is a positive number and m is an even positive integer, and let $I=[0,D]$. The case of $m=2$ yields the variance bounds. In [7], relevant equations for obtaining the error bounds were discussed. In particular, detailed results for $m=4$ were presented.

3.4 Linear exponential bounds

Let $k_1(z)$ be the function given by the expression inside the brackets of the r.h.s. of (1.4a), let $k_2(z) = \exp(c(h+z))$, where c is an arbitrary real-valued number, and $I=[0,D]$. In [1], it is shown that for $c \le c_0 = -(h+D)/\sigma^2$, $\mathscr{C}$ is a convex $\cap$ function while, for $c_1 = -(h-D)/\sigma^2 \le c$, $\mathscr{C}$ is a convex $\cup$ function. For c inside $[c_0,c_1]$, $\mathscr{C}$ has only one point of inflection. Thus, the upper (lower) envelope of $\mathscr{H}$ (and thus the bound on P_e) is given by the convex $\cap(\cup)$ part of $\mathscr{H}$ or a straight line from the lower (upper) end point tangent to $\mathscr{C}$. (See Fig. 1 in [1].) Detailed derivation of bounds is given in [1]. The complexity of the solutions of these bounds is comparable to that of variance bounds, except now the user has the freedom to use any c to obtain an upper and a lower bound. For any h, D, and SNR, there is an optimum c for the upper bound and an optimum c for the lower bound. Of course, even if we do not use the optimum c, we can still obtain a bound. Often such bounds can still be quite tight.

3.5 Quadratic exponential bounds

Let $k_1(z)$ be the function given by the expression inside

the brackets of the r.h.s. of (1.4b), let $k_2(z) = \exp(cz^2)$, where c is an arbitrary real-valued function, and $I-[0,D]$. The geometry of the curve $\mathscr{C}$ here is quite similar to that in the linear exponential case. However, since the quadratic exponential function with the appropriate c can approximate the complementary Gaussian probability distribution function more closely than that of the linear exponential function, the bounds obtained in Section 3.5 are usually tighter than those of Section 3.4.

3.6 Linear combination exponential type bounds

In the linear combination linear exponential bounding approach, let $k_1(z)$ be that kernel function considered in Section 3.4 and let

$$k_2(z) = \sum_{j=1}^{J} d_j \exp(c_j(h+z)) , \tag{3.6.1}$$

with $I=[-D,D]$. In the linear combination quadratic exponential bounding approach, let $k_1(z)$ be that kernel function considered in Section 3.5 and let

$$k_2(z) = \sum_{j=1}^{J} d_j \exp(c_j z^2) , \tag{3.6.2}$$

with $I=[0,D]$. There are no known optimum c_j's and d_j's in the Chebychev norm sense [1]. However, even with non-optimum c_j's and d_j's , we can obtain very tight bounds.

4. RANDOM INTERFERENCE COMMUNICATION SYSTEM PROBLEMS

In this section, we shall consider three commonly encountered random interference communication system problems and show the techniques discussed in Section 3 can yield tight and computationally tractable error bounds for these problems.

4.1 Intersymbol interference problem

We consider a standard model of a linear PAM system with independent and equiprobable binary (±1) random input data $\{a_i\}$. The system, with an overall impulse response $h(t)$, is perturbed with an additive zero-mean Gaussian noise $n(t)$, independent of $\{a_i\}$, of variance σ^2 . Sampling the received signal in multiples of T seconds, the decision random variable at $t=0$ is given by (1.1), where $h=h(0)$ is assumed to be positive, $h_i = h(-iT)$, $n=n(0)$, and the intersymbol interference $Z=\Sigma_i' a_i h_i$ is assumed to exist only for $-M \le i \le M, i \ne 0$. The distortion $D=\Sigma_i' |h_i|$ is assumed to be finite.

In order to study the usefulness of the error bounds for the intersymbol interference problem, consider the commonly used overall system impulse response specified by a Chebychev pulse given by

$$h(t) = 0.4023 \cos\left[\frac{2.839|t|}{T} - 0.7553\right] \exp(-0.4587|t|/T) + 0.7162 \cos\left[\frac{1.176|t|}{T} - 0.1602\right] \exp(-1.107|t|/T)$$

for $-\infty < t < \infty$ such that $h = h_0 = 1$, M is taken as 40, and $D = 0.28$.

In Table 1, upper and lower error bounds for $SNR(dB) = 10 \log_{10}(h^2/\sigma^2)$ of 4dB, 8dB, and 12dB, using the absolute moment bounds (Section 3.1), the variance bounds (Section 3.2), and the fourth-order moment bounds (Section 3.3), are given. In order to obtain these error bounds, we need to evaluate the appropriate moments of z. From [4], it is known that the absolute moment $E\{|x|\}$, is not computationally tractable. However, the absolute moment can be upper bounded by the standard deviation σ_Z and lower bounded by the cubic power of σ_Z divided by the square root of the fourth-order moment of Z [4]. The variance of Z and the fourth-order moment of Z are given by

$$E[Z^2] = \sigma_Z^2 = \sum_{i=-M}^{M}{}' |h_i|^2$$

$$E[Z^4] = \left[\sum_{i=-M}^{M}{}' h_i^2\right]^2 + 4 \sum_{i=-M}^{M-1}{}' h_i^2 \sum_{j=i+1}^{M}{}' h_j^2 .$$

In Table 2, upper and lower bounds are evaluated for the same three SNR cases, using the linear exponential bounding techniques (Section 3.4). The linear exponential moment is easily evaluated by using

$$E[\exp(c(h+Z))] = [\exp(ch)] \prod_{i=-M}^{M}{}' \cosh ch_i .$$

The detailed definitions of the different c's and the corresponding error bounds are given in [1]. However, we note c_0 and c_1 are given explicitly in Section 3.4. Each value of c_T, c_B, c_1^*, and c_2^* need to be evaluated from a relatively simple non-linear equation which is a function of h, D, and σ^2. $\hat{c}_U$ and $\hat{c}_L$ are the optimum values of c's for the upper and lower bounds.

In Table 3, corresponding error bounds are evaluated using the quadratic exponential bounding technique (Section 3.5). The quadratic exponential moment is obtained from

SNR		Absolute Moment	Variance Moment	4th Power Moment
4dB	upper bound	$6.12E^{-2}$	$5.77E^{-2}$	$5.86E^{-2}$
	lower bound	$5.46E^{-2}$	$5.77E^{-2}$	$5.67E^{-2}$
8dB	upper bound	$9.20E^{-3}$	$6.84E^{-2}$	$7.25E^{-3}$
	lower bound	$5.83E^{-3}$	$6.75E^{-3}$	$6.16E^{-3}$
12dB	upper bound	$3.07E^{-4}$	1.05^{-4}	$8.25E^{-5}$
	lower bound	$4.01E^{-5}$	$6.14E^{-5}$	$4.78E^{-5}$

Table 1. Various Power Moment Bounds for Chebychev Pulse Channel

	c_0	c_1	c_T	c_B	$c_2{}^*$	$c_1{}^*$	$\hat{c}_U$	$\hat{c}_L$
(1)	$5.81E^{-2}$	$6.20E^{-2}$	$5.93E^{-2}$		$5.78E^{-2}$		$5.78E^{-2}$	
	$5.31E^{-2}$	$5.74E^{-2}$		$5.63E^{-2}$		$5.76E^{-2}$		$5.76E^{-2}$
(2)	$7.01E^{-3}$	$8.91E^{-3}$	$7.63E^{-3}$		$6.83E^{-3}$		$6.82E^{-3}$	
	$4.56E^{-3}$	$6.55E^{-3}$		$6.01E^{-3}$		$6.68E^{-3}$		$6.69E^{-3}$
(3)	$8.22E^{-5}$	$1.19E^{-4}$	$9.33E^{-5}$		$6.92E^{-5}$		$6.84E^{-5}$	
	$2.03E^{-5}$	$5.68E^{-5}$		$4.33E^{-5}$		$6.28E^{-5}$		$6.29E^{-5}$

upper and lower bounds for SNR: (1) 4dB, (2) 8dB, (3) 12dB.

Table 2. Linear Exponential Moment Bounds for Chebychev Pulse Channel

$$E[\exp(cZ^2)] = \left[\exp\left(2c\sum_{i=1}^{M} h_i{}^2\right)\right] \times \left[\prod_{i=1}^{M} \cosh 2ch_i{}^2\right] \times \left[\prod_{i=1}^{M-1}\prod_{j=i+1}^{M} \cosh 2ch_ih_j\right]$$

with $h_i = h_{-i}$, $i=1,\ldots,M$ and $h_i = 0$, $|i| > M$.

	c_0	c_1	c_T	c_B	c_2^*	c_1^*	$\hat{c}_U$	$\hat{c}_L$
(1)	$5.77E^{-1}$	$5.77E^{-1}$	$5.77E^{-1}$		$5.77E^{-1}$		$5.77E^{-1}$	
	$5.77E^{-1}$	$5.77E^{-1}$		$5.77E^{-1}$		$5.77E^{-1}$		$5.77E^{-1}$
(2)	$6.77E^{-3}$	$6.76E^{-3}$	$6.77E^{-3}$		$6.76E^{-3}$		$6.76E^{-3}$	
	$6.75E^{-3}$	$6.75E^{-3}$		$6.76E^{-3}$		$6.76E^{-3}$		$6.76E^{-3}$
(3)	$7.74E^{-5}$	$6.64E^{-5}$	$7.23E^{-5}$		$6.60E^{-5}$		$6.58E^{-5}$	
	$6.35E^{-5}$	$5.20E^{-5}$		$6.39E^{-5}$		$6.46E^{-5}$		$6.46E^{-5}$

upper and lower bounds for SNR: (1) 4dB, (2) 8dB, (3) 12dB .

Table 3. Quadratic-Exponential Moment Bounds for Chebychev Pulse Channel

Detailed results on the different c's and the corresponding error bounds are given in [8]. While the evaluation of the quadratic exponential moment is more involved than that of the linear exponential moment, the evaluation of the c's and error bounds is similar. As can be seen, the error bounds in Table 3 are tighter than those in Table 2.

4.2 Co-channel interference problem

We consider a standard model for an *m*-ary CPSK system in the presence of Gaussian noise and cochannel interference. We assume that this interference is composed of the sum of K unequal amplitude PSK signals which lie within the passband of the receiver. We also assume that no intersymbol interference effects are present.

In the Nth time interval, let the received waveform y(t) be given by

$$y(t) = \sqrt{2S}\cos(\omega_0 t + \theta) + \sum_{j=1}^{K} \sqrt{2I_j}\cos(\omega_j t + \theta_j + \mu_j) + n(t)$$

where S is the signal power, ω_0 is the angular frequency of the signal, and $\theta = 2\pi k/m$, $k \in \{0,1,2,\ldots,m-1\}$ is the phase angle associated with the kth message. Here I_j is the (known) power of the jth interferer, ω_j is the angular frequency of the jth interferer, and θ_j is the phase modulation of this interferer; μ_j is assumed to be a uniform random variable over $[0,2\pi)$, with μ_j independent of $\mu_i (i \neq j)$ and also independent of the wide-sense stationary zero-mean Gaussian noise $n(t)$ of variance σ^2. The receiver is assumed to contain an ideal limiter followed by an ideal phase detector. For simplicity, we shall only consider the binary CPSK problem. The general m-ary CPSK problem is discussed in [9].

We now consider the error probability P_e of the system described above. We define the relative phase angles λ_j by $\lambda_j = (\omega_j - \omega_0)t + \theta_j + \theta + \mu_j$, $j=1,2,\ldots,K$. The $\{\lambda_j\}$ form a set of independent uniformly distributed random variables over $[0,\pi)$ for each t. For a fixed relative phase vector $\lambda = (\lambda_1, \lambda_2, \ldots, \lambda_K)$, the conditional error probability $P_e(\lambda)$ for equiprobable binary data can be expressed as $P_c(Z) = Q(\rho\sqrt{2}(1+Z))$, where $\rho = \sqrt{S}/\sigma$, $R_j = (I_j/S)^{1/2}$, and $Z = \Sigma_{j=1}^{K} R_j \cos \lambda_j$. Then the error probability $P_e = E[P_c(Z)]$.

Thus, we can use the linear exponential error bounding approach (Section 3.4) by selecting $k_1(z) = Q(\rho\sqrt{2}(1+z)$ and $k_2(z) = \exp[c(1+z)]$. In particular, the linear exponential moment is now given by

$$E[\exp(c(1+z))] = \exp[c] \prod_{j=1}^{K} I_0(cR_j) \quad .$$

In Fig. 2, we consider an equipower binary CPSK example with $R_j = R_0$ for all j, the number of interferers K taken to be 3, the carrier-to-interference ratio $\mathrm{CIR(dB)} = 10 \log_{10} 1/KR_0^2$ taken to be 12dB, and the carrier-to-noise ratio defined by $\mathrm{CNR(dB)} = 10 \log_{10} \rho^2$. We see in Fig. 2 that even the simple "(c_0,c_1)" upper bound and the $c_1{}^*$ lower bound provide very tight bounds on the computationally more involved exact error probability [12]. The lower ([11]) bound is McLane's fourth-order moment bound [11] and the upper ([10]) bound is the peak-limited bound of Rosenbaum and Glave [10].

4.3 Spread spectrum multiple access problem

We consider a direct code modulation asynchronous spread spectrum multiple access (SSMA) communication system.

The time delays τ_i and phase angles θ_i of different users are r.v.'s. The input to each receiver consists of the sum of all K users' signals and additive white Gaussian noise. Each receiver

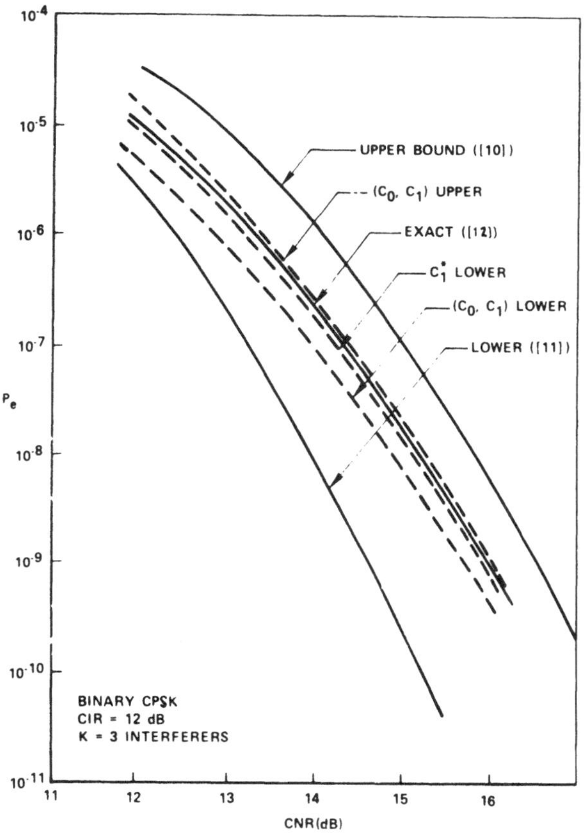

Fig. 2. Error Bounds for Equipower Cochannel Interference Channel.

consists of a matched filter matched to its corresponding code word. Without loss of generality, we consider the first receiver. Then we assume it is completely synchronized to its own code word. Thus, $\theta_1=\tau_1=0$ But $\theta_2,\ldots,\theta_K$ and $\tau_2,\ldots,\tau_K$ are independent and uniformly distributed r.v.'s. Thus, the output of the matched filter of the first receiver is given by

$$y= \frac{P}{2}\int_0^T a_1^2(t)b_1(t)dt + \frac{P}{2}\sum_{i=2}^{K}\int_0^T a_i(t-\tau_i)b_i(t-\tau_i)a_1(t)dt \cos\theta_i$$

$$+ \int_0^T n(t)a_1(t)\cos w_c t dt = hb_{1,0} + Z + n \, .$$

The first term of y represents the desired signal, the second term, Z, represents the interference from the other (K-1) users, and the last term, n, is a Gaussian r.v. of zero mean and variance σ^2. All users are assumed to have equal power P. Here, the information data of the ith user is defined by

$b_i(t) = \Sigma_{n=-\infty}^{\infty} b_{i,n} P_T(t-nT)$, where the $b_{i,n}$'s are i.i.d r.v.'s taking values +1 and -1 with equal probability, and $P_T(\cdot)$ is a unit height rectangular window from 0 to T. The code waveform is defined by $a_i(t) = \Sigma_{j=-\infty}^{\infty} a_j^{(i)} P_{T_c}(t-jT_c)$, where $a_j^{(i)}$ is the code sequence of the ith user and consists of +1 and -1 which is periodic with period p and of chip length T_c .

Now the error probability of the first user, assuming all K users' code words have been specified, is again expressible in the probability form of (1.4a) or (1.4b). The expectation operation E_Z in (1.4) is taken over all the r.v.'s τ_i, $b_{i,-1}$, $b_{i,0}$, and θ_i in Z. In [13], the second-order moment, the fourth-order moment, and the linear exponential moment of the SSMA random variable Z were obtained. The details of these evaluations are quite involved and will not be repeated here. Thus, the error bounds discussed in Sections 3.2, 3.3, and 3.4 are applicable. It turns out in SSMA problems that, for small length codes and (or) large number of users, the normalized distortion can be large. Thus, there is a need for making $k_2(z)$ as close to $k_1(z)$ as possible. In [13], we considered an n=3 Gold code with K=2 users, and a normalized distortion of D'=0.7143. We used the linear combination exponential type bounds of Section 3.6. The results in Table 4 show that, even for a system with such large distortion, our error bounds are still tight. Furthermore, the usual Gaussian noise approximation of $P_e^G = Q(h/(\sigma^2 + \sigma_Z^2)^{1/2})$ often used in SSMA applications does not yield good results in this situation.

dB	P_e^U	P_e^L	P_e^G
8	$1.46E^{-2}$	$1.46E^{-2}$	$1.46E^{-2}$
12	$4.06E^{-4}$	$4.05E^{-4}$	$4.68E^{-4}$
16	$9.69E^{-5}$	$9.13E^{-5}$	$15.2\ E^{-5}$
20	$1.74E^{-6}$	$1.66E^{-6}$	$27.6\ E^{-6}$

Table 4. Linear Combination Linear Exponential Bounds for SSMA

5. CONCLUSIONS AND EXTENSIONS

We have considered three communication systems with random interferences. The error probabilities of these systems were bounded by utilizing a result from the Isomorphism Theorem in the theory of moment spaces. These bounds are optimum in the sense that no other bounds using that given generalized moment can be tighter.

Many generalizations and extensions are possible. In the application direction, any communication system performance analysis problem that involves some random parameters can probably use the moment space bounding technique. One such example, dealing with the effect of quantization noise in a finite-precision digitally-implemented receiver's performance, is considered in [14].

In this paper, we have only considered two-dimensional moment space error bounds. Recently, various forms of multi-dimensional moment space error bounds have also been found. In [15], error bounds based on Krein's principal representation are given. In [16], error bounds based on geometrical arguments, as well as on algorithms for the evaluation of boundary hyperplanes of polyhedrons, have been obtained. These multi-dimensional moment bounds generally provide very tight bounds with varying degrees of additional computational complexity.

References

1. K. Yao and R.M. Tobin, "Moment space upper and lower error bounds for digital systems with intersymbol interference," *IEEE Trans.Inform.Theory*, vol.IT-22, Jan. 1976, pp. 65-74.
2. M. Dresher, S. Karlin, and L.S. Shapley, "Polynomial Games," in *Contributions to the Theory of Games*, Annals of Mathematics Studies, No. 24, Princeton, NJ: Princeton Univ. Press, 1950, pp. 161-180.
3. M. Dresher, "Moment spaces and inequalities," *Duke Math.J.*, vol. 20, June 1953, pp. 261-271.
4. P.J. McLane, "Lower bounds for finite intersymbol interference error rates," *IEEE Trans.Commun.Technol.*, vol. COM-22, June 1974, pp. 853-857.
5. F.E. Glave, "An upper bound on the probability of error due to intersymbol interference for correlated digital signals," *IEEE Trans.Inform.Theory*, vol. IT-18, May 1972, pp. 356-363.
6. J.W. Matthews, "Sharp error bounds for intersymbol interference," *IEEE Trans.Inform.Theory*, vol. IT-19, July 1973, pp.440-447.
7. T.Y. Yan, *Moment Space Error Bounds for Digital Communication Systems with Intersymbol Interference Based on N^{th} Moment*, UCLA M.S. Thesis, Fall 1975. Also UCLA Tech.Rpt. Eng-7608, Jan. 1976.
8. K. Yao, "Quadratic-exponential moment error bounds for digital communication systems," *Conf.Rec.Tenth Ann. Asilomar Conf. on Circuits, Systems, and Computers*, Nov. 1976, pp. 99-103.

9. R.M. Tobin and K. Yao, "Upper and lower error bounds for coherent phase-shift-keyed (CPSK) systems with cochannel interference," *IEEE Trans.Commun.*, vol.COM-25, Feb. 1977, pp. 281-287.

10. A.S. Rosenbaum and F.E. Glave, "An error probability upper bound for coherent phase-shift-keying with peak-limited interference," *IEEE Trans.Commun.*, vol.COM-22, Jan. 1974, pp. 6-16.

11. P.J. McLane, "Error rate lower bounds for digital communication with multiple interference," *IEEE Trans.Commun.*, vol.COM-23, May 1975, pp. 539-543.

12. A.S. Rosenbaum, "Binary PSK error probabilities with multiple cochannel interferences," *IEEE Trans.Commun.Technol.*, vol.COM-18, June 1970, pp. 241-253.

13. K. Yao, "Error probability of asynchronous spread spectrum multiple access communication systems," *IEEE Trans.Commun.*, vol.COM-25, Aug. 1977.

14. S. Reisenfeld and K. Yao, "On upper and lower bounds of error probability of a finite precision digitally implemented receiver," *Proc.Fifteenth Ann. Allerton Conf. on Circuit and Sys.*, Oct. 1977.

15. K. Yao and E. Biglieri, "Moment inequalities for error probabilities in digital communication systems," *Proc.Nat.Telecommun. Conf.*, Dec. 1977.

16. M.A. King, *Multi-Dimensional Moment Bounds with Applications to Communication Theory*, UCLA Ph.D. Thesis, 1977.

MODELLING AND ANALYSIS OF IMPULSIVE NOISE

J. W. Modestino, B. Sankur

Rensselaer Polytechnic Institute,
Troy, New York, U.S.A.

University of Boğaziçi, İstanbul, Turkey

ABSTRACT. A model for impulsive noise is developed and its statistical properties are described. The model consists of the linear combination of a low-density shot noise process and white Gaussian noise. Locally optimum receiver structures for baseband and narrowband impulsive noise are derived and their dependence upon the noise model parameters is analyzed.

1. MODELLING OF IMPULSIVE NOISE

In an increasing number of important applications the prevailing noise environment can be realistically described by an impulsive noise process. A physical-statistical model of impulsive noise phenomena is thus potentially quite rich in its applications. Analytically tractable statistical representations of such non-Gaussian processes are necessary for signal processing as well as experimentation. It is well known that algorithms for optimum signal detection and estimation depend critically on the adequate description of channel statistics. Some examples of communication channels where the prevailing noise is impulsive are a) ELF (3-300 Hz) and VLF (3-30 KHz) bands where the primary source of disturbance can be attributed to atmospheric discharge phenomena [1]. b) The radio frequency range comprising HF, VHF, UHF where the impulsive noise originates from various natural sources such as galactic and solar noise or it could be man-made type in metropolitan areas [2]. c) Wideband FM systems become susceptible to click noise. which is impulsive in nature, near the threshold range [3]. d) In wire communication impulsive noise may arise from lightning as well as from switching and accidental hits during maintenance and repair work [4]. e) Electronic surveillance and reconnaissance systems subject to wideband jamming are very susceptible to impulsive noise disturbances [5]. In addition, the proposed model can adequately describe noise phenomena in such diverse phenomena as: f) "Popcorn Noise" encountered in certain semiconductor devices due to contaminated and imperfect junction surfaces. g) EEG wave recordings where the impulsive bursts are due to muscle contractions by nerve impulses [6].

The main concern in proposing the existing models of impulsive noise has been to provide good fit to the first order statistics of the measured noise data [4]. These models prove useful only locally but are otherwise unsatisfactory for a variety of reasons. In particular, they fail to reconcile themselves with the underlying physics of the phenomena and they cannot be generalized to higher order statistics. The model to be presented has a proper physical basis and is analytically tractable. Furthermore, with the dynamic nature inherent in the model it has been possible to use modern filtering, estimation and detection techniques [7].

1.1. Representation of baseband impulsive noise:

An intuitively satisfying and analytically tractable model for impulsive noise processes in the ELF band results from the consideration of a low density shot noise process additively combined with Gaussian noise [1]. The Gaussian component represents the low level and homogeneous background noise, while the shot noise accounts for the large dynamic range.

The shot noise $y(t)$ is assumed to be generated by a random impulse train exciting a finite state linear time invariant system. In state variable form the shot noise can be expressed as

$$\dot{\underline{x}}(t) = \underline{\underline{F}}\,\underline{x}(t) + \underline{b}\,u(t) \qquad (1)$$

$$y(t) = \langle \underline{c}, \underline{x}(t) \rangle \qquad \underline{x}(t_o) = \underline{x}_o$$

where $\underline{\underline{F}}$ $\underline{b}$, $\underline{c}$ are constant nxn, nxl, nxl matrices respectively; $\underline{x}(t)$ denotes the state vector and $\langle \; , \; \rangle$ stands for the inner product operation.

The driving process is a scalar of the form

$$u(t) = \sum_{i=0}^{N(t)} u_i \; \delta(t-t_i) \tag{2}$$

which is a point process with $\delta(\cdot)$ denoting the Dirac-delta function and $\{N(t), \; t \geq t_o\}$ is a general counting process with arrival times $\{t_i\}$ and $\{u_i\}$ is a sequence of independent identically distributed (i.i.d.) random variables. In (2) we have assumed for convenience $N(t_o) = 0$. The ELF noise has therefore the representation

$$n(t) = y(t) + w(t) \tag{3}$$

$\{w(t)\}$ being a zero-mean white Gaussian noise (WGN) with double sided spectral density $\sigma_0^2 = N_o/2$ watts/Hz and is also independent of the driving point process $\{u(t)\}$. The solution to (1) is given by

$$x(t) = \underline{\underline{\Phi}}(t-t_o) \; \underline{x}(t_o) + \int_{t_o}^{t} \Phi(t-\tau) \; \underline{b} \; u(\tau) d\tau, \tag{4}$$

where $\underline{\underline{\Phi}}(t) \triangleq \exp \{\underline{\underline{F}}t\}$ is the state transition matrix satisfying the equation

$$\dot{\underline{\underline{\Phi}}}(t) = \underline{\underline{F}} \; \underline{\underline{\Phi}}(t) \; ; \quad \underline{\underline{\Phi}}(t_o) = \underline{\underline{I}}.$$

Defining the impulse response function $h(t) = \langle \underline{c}, \; \underline{\underline{\Phi}}(t)\underline{b} \rangle$, (3) can be expressed as

$$n(t) = w(t) + \langle \underline{\underline{c}}, \; \underline{\underline{\Phi}}(t-t_o)\underline{x}(t_o) \rangle + \sum_{i=0}^{N(t)} u_i h(t-t_i). \tag{5}$$

1.2. Representation of narrowband impulsive noise:

The narrowband impulsive noise can be represented as an additive combination of a bandpass shot noise and a bandpass Gaussian noise. Both noise processes are assumed to be narrowband about some center frequency ω_0 rad/sec so that

$$n(t) = \sqrt{2}\,\{n_c(t) \quad \cos \quad \omega_0 t + n_s(t) \quad \sin \omega_0 t\} \;, \tag{6}$$

where $n_c(t)$ and $n_s(t)$ are the inphase and quadrature components of the noise, respectively.

In terms of the envelope $A(t)$ and phase $\Theta(t)$ the expression in (6) can also be written as

$$n(t) = \sqrt{2}\, A(t) \quad \cos\,(\omega_0 t - \Theta(t)). \tag{7}$$

It will be convenient to define a complex envelope process $\tilde{n}(t)$ (tilda above denotes a complex quantity), such that

$$\tilde{n}(t) = n_c(t) - jn_s(t). \tag{8}$$

In (8) $\tilde{n}(t)$ is a lowpass quantity and it completely determines the bandpass process through the relation

$$\tilde{n}(t) = \sqrt{2}\ \mathrm{Re}\ \{\tilde{n}(t)\ e^{+j\omega_0 t}\}\ . \tag{9}$$

The complex envelope $\tilde{n}(t)$ in terms of its components can be written as

$$\tilde{n}(t) = \tilde{y}(t) + \tilde{w}(t), \tag{10}$$

where $\tilde{w}(t) \triangleq w_c(t) - jw_s(t)$ and $\{w_c(t)\}$, $\{w_s(t)\}$ are both zero mean Gaussian processes with variances σ_0^2.

The complex envelope of a shot process $y(t)$ can be described by the complex state equations

$$\dot{\tilde{\underline{x}}}(t) = \tilde{\underline{\underline{F}}}\ \tilde{x}(t) + \tilde{\underline{b}}\ \tilde{u}(t) \tag{11}$$

$$\tilde{y}(t) = \langle\tilde{\underline{c}},\ \tilde{\underline{x}}(t)\rangle \qquad \tilde{\underline{x}}(t_o) = \tilde{\underline{x}}_o.$$

The input $\tilde{u}(t)$ is a complex point process of the form

$$\tilde{u}(t) = \sum_{i=0}^{N(t)} \tilde{u}_i\ \delta(t-t_i) \tag{12}$$

and

$$\tilde{u}_i = u_{c_i} - j\ u_{s_i}\ . \tag{13}$$

The inphase and quadrature components of the ith impulse amplitude are defined as

$$u_{c_i} = u_i \cos \theta_i \tag{14a}$$

$$u_{s_i} = u_i \sin \theta_i, \tag{14b}$$

with $\{u_i\}$ being an i.i.d. sequence of positive random variables and $\{\theta_i\}$ an i.i.d. sequence of random variables uniformly distributed on $[-\pi,\pi]$ and is assumed independent of $\{u_i\}$.

Using the complex state transition matrix $\underline{\underline{\Phi}}(t)$, the inphase component can be written as

$$y_c(t) = \mathrm{Re}\{<\underline{\tilde{c}}, \tilde{\underline{\underline{\Phi}}}(t-t_o)\underline{\tilde{x}}(t_o)>\} - \mathrm{Re}\{\int_{t_o}^{t} <\underline{\tilde{c}}, \tilde{\underline{\underline{\Phi}}}(t-s)\underline{\tilde{b}}> \tilde{u}(s)ds\}. \tag{15}$$

The quadrature component $y_s(t)$ can be similarly expressed. One can identify the inphase and quadrature components of the bandpass filter as

$$h_c(t) = \mathrm{Re}\{<\underline{\tilde{c}}, \tilde{\underline{\underline{\Phi}}}(t)\underline{\tilde{b}}>\} \tag{16a}$$

$$h_s(t) = \mathrm{Im}\{<\underline{\tilde{c}}, \tilde{\underline{\underline{\Phi}}}(t)\underline{\tilde{b}}>\} \tag{16b}$$

$$\text{and} \quad h_{LP}(t) = (h_c^2(t) + h_s^2(t))^{1/2}, \tag{16c}$$

where the subscript LP emphasizes the lowpass nature of these functions. Using (16a, 16b) in (15) and neglecting the initial conditions, the components of the narrowband impulsive noise can be expressed as:

$$n_c(t) = w_c(t) + \sum_{i=0}^{N(t)} u_i\, h_c(t-t_i) \cos \theta_i - u_i\, h_s(t-t_i) \sin \theta_i \tag{17}$$

and

$$n_s(t) = w_s(t) + \sum_{i=0}^{N(t)} u_i\, h_s(t-t_i) \cos \theta_i + u_i\, h_c(t-t_i) \sin \theta_i. \tag{18}$$

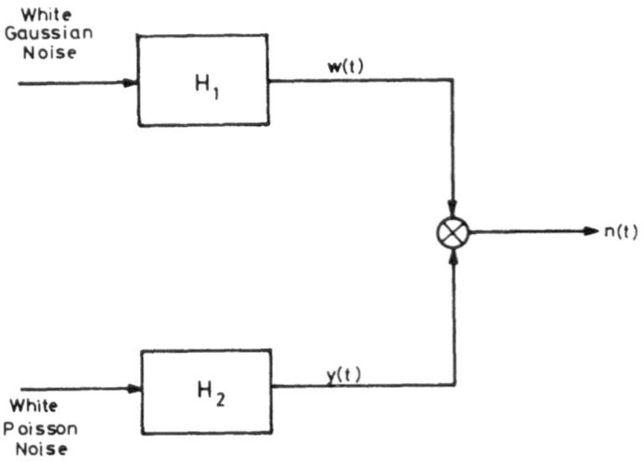

Fig. 1. Generation Scheme for Impulsive Noise.

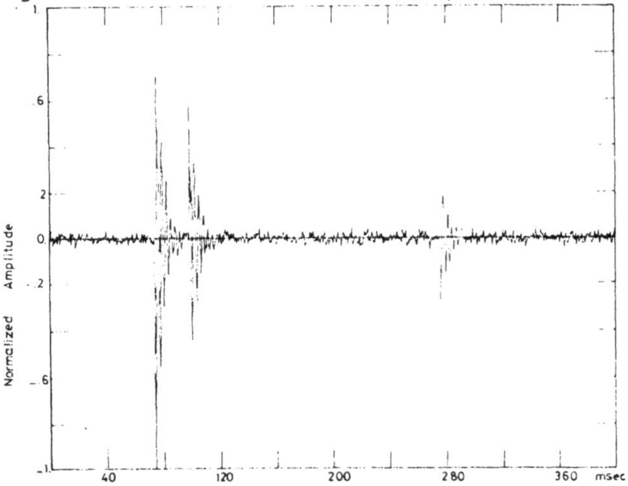

Fig. 2. A typical ELF Noise Sample Function.

A block diagram for the generation scheme of impulsive noise is shown in Fig. 1. where H_1 and H_2 represent two independent linear dynamical systems. In Fig. 1 we have considered an arbitrary system, H_1, thus allowing for Gaussian component to be colored as well.

1.3. Choice of model parameters

The first quantity to be specified is the probability density function (p.d.f.) of the impulse amplitudes. The two p.d.f.'s that fit the observed ELF noise statistics are:

(i) The power Rayleigh distribution

$$f_u(u) = \frac{\alpha}{R_0^\alpha} |u|^{\alpha-1} \exp\{ -(\frac{|u|}{R_0})^\alpha\} \qquad 0 < \alpha \leq 2 \qquad (19)$$

$$-\infty < u < \infty ,$$

which has zero mean and variance $R_0^2\, \Gamma(1-2/\alpha)$, where $\Gamma(\cdot)$ is the Gamma function. The power Rayleigh distribution is completely defined by the scale parameter R_0 and the exponent α.

(ii) Log-normal distribution

$$f_u(u) = \frac{1}{\sqrt{2\pi}\, \sigma_u |u|} \exp\left[- \frac{(\ln|u|-m_u)^2}{2\sigma_u^2} \right] \qquad -\infty < u < \infty, \qquad (20)$$

with mean m_u and variance $e^{2\sigma_u^2}$. It is rather useful to characterize log-normal distribution by the parameter

$$V_d = 10 \log_{10} \frac{E\{u^2\}}{E^2\{|u|\}} = 4.34\, \sigma_u^2. \qquad (21)$$

For the case of VLF noise it suffices to consider the one-sided version of these distributions.

The counting process $\{N(t),\ t \geq t_o\}$ is generally assumed to be a renewal process. A Poisson-Poisson model for the counting process has been suggested, since some impulsive noise phenomena in the atmosphere tend to occur in packets of activity.

Accordingly, the packets of activity arrive in time with rate λ' and within each packet the impulses occur according to another independent Poisson distribution with rate λ'' $(\lambda'' > \lambda')$. Since mathematical complications preclude this approach, in this work a constant rate Poisson model has been chosen so that

$$P\ \{N(t)-N(t_o) = k\} = e^{-\lambda(t-t_o)} \frac{\lambda(t-t_o)^k}{k!} \qquad k=0,1,\ldots. \qquad (22)$$

Finally, typical choices for system dynamics are shown in Table I, [1]. The appropriate scale constant c in Table I is easily determined as

$$\gamma^2 \triangleq \frac{Var\{y(t)\}}{\sigma_0^2}. \qquad (23)$$

It can be observed that this parameter controls the impulsive nature of the noise; in fact as $\gamma \to 0$ the impulsive noise tends toward a pure Gaussian noise, whereas it tends to a shot noise process for $\gamma \to \infty$. A procedure has been developed [9] to calibrate the model parameters to fit the noise statistics derived from the model to those obtained from the actual observed data. Typical results of this empirical adjustment procedure are summarized in Table II for ELF/VLF noise recorded at selected geographical sites. Figure 2 illustrates a simulated noise waveform with parameter values derived from entries of Table II.

Table I

Some Useful Low Order Linear System Models

H(s)	h(t)	Var{y(t)}
$\frac{c}{s+a}$	$ce^{-at}u_{-1}(t)$	$\frac{\lambda c^2 \sigma_u^2}{2a}$
$\frac{c}{(s+a)^2+\omega_c^2}$	$\frac{c}{\omega_c} e^{-at} \sin \omega_c t \, u_{-1}(t)$	$\frac{\lambda c^2 \sigma_u^2}{4a(a^2+\omega_c^2)}$
$\frac{c(s+a)}{(s+a)^2+\omega_c^2}$	$c\, e^{-at} \cos \omega_c t \, u_{-1}(t)$	$\frac{\lambda c^2 \omega_c^2 \sigma_u^2}{4a(a^2+\omega_c^2)}$

Table II

Typical Results of the Empirical Adjustment Procedure

	Noise Type	λ(pulses/sec)	a(sec^{-1})	ω_c(rad/sec)	Power Rayleigh Distributed Pulse Amplitudes α	γ	Lognormal Distributed Pulse Amplitudes v_d	γ
ELF Noise*	Saipan-High Level	12.8	250	2000	0.25	4.4	15	4.4
	Malta-Moderate Level	11.3	250	2000	0.25	2.6	15	2.0
	Norway-Low Level	9.3	333	2000	0.50	1.2	13	1.0
VLF Noise^x	Singapore in Spring	15.0	200	-	.55	11.2	-	-
	England in Winter	15.0	200	-	.60	8.9	10.9	7.9

* Data taken from [1]

x Date taken from [[illegible]

2. STATISTICAL CHARACTERISTICS

The amplitude probability distribution (APD) and characteristic function (CF) of the impulsive noise process can be calculated using (5) and (17,18).

2.1. Characteristic functions

The CF of baseband impulsive noise is given by

$$\Psi_n(v) = E\{\exp(jvn(t))\} \tag{24}$$

It is shown that [1] for times sufficiently far removed from t_o, so that initial transient disturbances have been dissipated, the first order ch.f has the form:

$$\Psi_n(v) = \exp\{\lambda\int_0^\infty\left[\psi_u(vh(\xi))-1\right]d\xi - \frac{\sigma_0^2}{2}\} \tag{25}$$

and the second order ch.f is given by

$$\begin{aligned}\psi_n(v_1,v_2;\tau) &= \exp\{\lambda\int_0^\infty \left[\psi_u\{v_1h(\xi)-v_2h(\xi+\tau)\}-1\right]d\xi\} \\ &\cdot\exp\{\lambda\int_0^\infty\left[\Psi_u(v_2h(\xi))-1\right]d\xi-\frac{\sigma_0^2}{2}(v_1^2+v_2^2)\}\end{aligned} \tag{26}$$

where $\Psi_u(\cdot)$ is the univariate ch.f of the random amplitude $\{u_i\}$ possessing common p.d.f. $f(u)$.

Similarly the joint ch.f. of the inphase and quadrature components of the narrowband impulsive noise can be obtained starting from

$$\Psi_{n_c,n_s}(v_1,v_2) \triangleq E\{\exp jv_1n_c(t) + jv_2n_s(t)\}\ . \tag{27}$$

Substitution of (17,18) in (27) gives [9]

$$\Psi_{n_c,n_s}(v_1,v_2) = \exp\{\int_0^\infty E\left[J_0((v_1^2+v_2^2)^{\frac{1}{2}}\, uh_{LP}(\tau))-1\right]d\tau - \frac{\sigma_0^2}{2}(v_1^2+v_2^2)\}, \tag{28}$$

where $J_0(\cdot)$ denotes the ordinary Bessel function of order zero and the expectation is over the impulse amplitudes u.

For $\lambda\to 0$ (28) reduces to a bivariate Gaussian density. Furthermore, one has in the limit

$$\lim_{\lambda\to\infty} \psi_{n_c,n_s}(v_1,v_2) = \exp\{ -\frac{\sigma_0^2}{2}(v_1^2+v_2^2)(1+\gamma^2)\} \tag{29}$$

The last expression is a bivariate Gaussian density and this result follows intuitively from the central limit theorem. Similarly, the quadrivariate p.d.f. of the inphase and quadrature components of VLF noise as well as the higher order statistics have been derived in [9] .

2.2. Amplitude probability distributions of baseband and narrow-band impulsive noise processes

The p.d.f.'s of impulsive noise can be obtained by inverse Fourier transforming the ch.f.'s in (25) and (28), respectively. In particular, observing that the ch.f. of the narrowband impulsive noise is circularly symmetric and using Bochner's theorem [8] results in the following expression for the p.d.f. of the envelope A(t):

$$f_A(A) = \int_0^{\infty}\beta\, J_0(\beta A)\, \exp\{-\sigma_0^2 \frac{\beta^2}{2} + \lambda \int_0^{\infty}[J_0(\beta u h_{LP}(\tau))-1]\,d\tau\} \tag{30}$$

with $\beta = (v_1^2+v_2^2)^{\frac{1}{2}}$.

Similarly, the exceedance probability is given by

$$Q_A(A) = \int_A^{\infty} f_A(x)\,dx =$$

$$\int_0^{\infty}\{1-J_0(\beta A)- \frac{\pi\beta A}{2}\,[J_0(\beta A)H_1(\beta A)-J_1(\beta A)H_0(\beta A)]\}\; \psi_{n_c,n_s}(\beta)\,d\beta \tag{31}$$

where $H_n(\cdot)$ is the Struve function of order n, [10] . It is very tedious to compute numerically the expressions in (30) and (31), mainly because the integrands are highly oscillatory functions.

A computationally efficient alternate expression for the p.d.f. and the exceedance probability can be obtained by considering a series expression in ascending powers of λ. For low density shot processes such a series converges very rapidly and one need to consider only the first few terms. Thus the exceedance probability of the VLF noise envelope is given approximately by [9] .

$$Q_A(A) \simeq (1-\lambda T)\ \exp\{-\frac{A^2}{2}\} + \lambda T E\left\{M(quh_{LP}(\tau),A)\right\} \qquad (32)$$

where the expectation is over the impulse amplitudes $\{u\}$ and epochs $\{\tau\}$ uniformly distributed in $[0,T]$. $M(\cdot,\cdot)$ denotes the Marcum-Q function defined as

$$M(a,b) = \int_b^\infty x\ \exp\{-\frac{a^2+x^2}{2}\}\ I_0(ax)dx \qquad (33)$$

I_0 being the modified Bessel function of order zero and where

$$q = \gamma/(\lambda\ \mathrm{Var}\{u\} \int_0^\infty h_{LP}^2(\tau)d\tau)^{\frac{1}{2}}, \quad T \doteq 4/a. \qquad (34)$$

APD plots of VLF noise envelope calculated from (32) are shown in Figs. 3-4. These curves are plotted on a Rayleigh graph (where a Rayleigh distribution plots as a straight line with a slope of $-\frac{1}{2}$).

Observe that APD's in the low amplitude region plot as straight lines parallel to the Rayleigh density function. This is undoubtedly due to the presence of a Gaussian component. At larger amplitudes APD curves show radical departures from Rayleigh density behaviour indicating the broad tailed nature of VLF noise APD's. The latter effect is due to the presence of the shot noise. The APD's are plotted for various values of γ in Fig. 3. One can observe that the APD's are depressed with respect to the Rayleigh density by approximately $-20 \log\gamma$ db, which reflects the fact that the contribution of the Gaussian noise component to the total .m.s. value of the noise is $1/(1+\gamma^2) \simeq 1/\gamma^2$ for $\gamma \gg 1$. The dependence of the impulsive noise APD'S upon other model parameters is illustrated in [9]. Very good agreement has been obtained between VLF noise data measured at different seasons and geographical locations [1] and the APD curves of the model using (31). In Fig. 4 the theoretical curves are made to fit measured data through a calibration procedure described in [12].

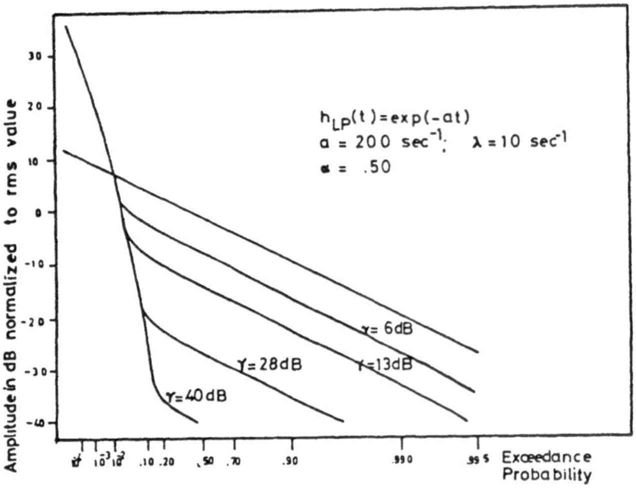

Fig. 3: APD of VLF Noise for Various Values of

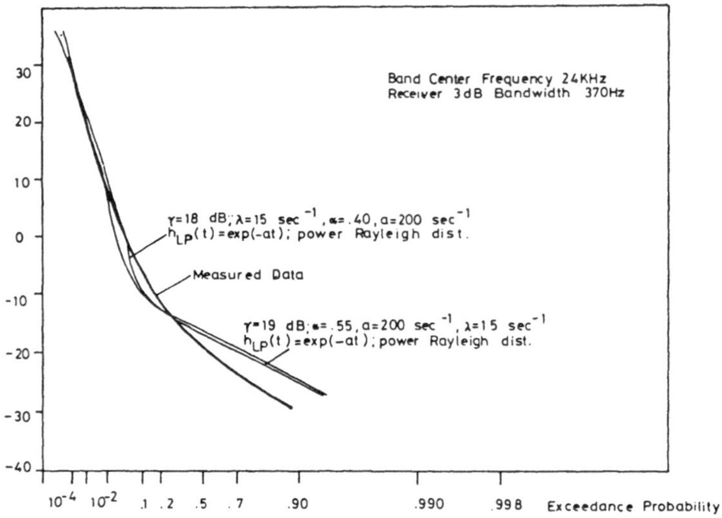

Fig. 4: APD of VLF Noise Envelope Comparison with Actual Data Taken at Slough in England in Winter. (Clarke et al. 11).

3. LOCALLY OPTIMUM NONLINEAR RECEIVERS

The optimum receiver for a signaling scheme additively combined with noise can be obtained by implementing the likelihood ratio. The resulting signal processing operations, however, depend explicitly upon the signal strength at the receiver. Often, however, signal-to-noise like quantities are unknown or very tedious to measure. One can resort than to locally most powerful receiver structures. This is the receiver structure that maximizes the derivative of the power function for vanishingly small signal amplitudes, subject to a fixed alarm probability. Thus, by optimizing the performance in the critical small signal region, adequate performance can be obtained in the less critical moderate to large signal-to-noise regions.

3.1. Baseband processes

The locally optimum receiver structure for baseband processes is shown in Fig. 5. Under the assumptions of small signal levels, independence of samples and stationarity of the noise process the optimum zero-memory nonlinear (ZMNL) characteristics are given by:

$$g(r) = - \frac{d\ell n f_n(r)}{dr} \qquad (35)$$

where $f_n(r)$ denotes the univariate p.d.f. of the ELF noise samples. Typical examples of ZMNL's for ELF noise are shown in Fig. 6. It is interesting to note that those ZMNL's process small amplitude input signals linearly. This is intuitively correct since at low levels the amplitude statistics are predominantly Gaussian and the optimum processor in additive Gaussian noise is a linear device. With increasing λ, large amplitudes are severely suppressed with more emphasis being placed on the small amplitude samples which are potentially information bearing signal components.

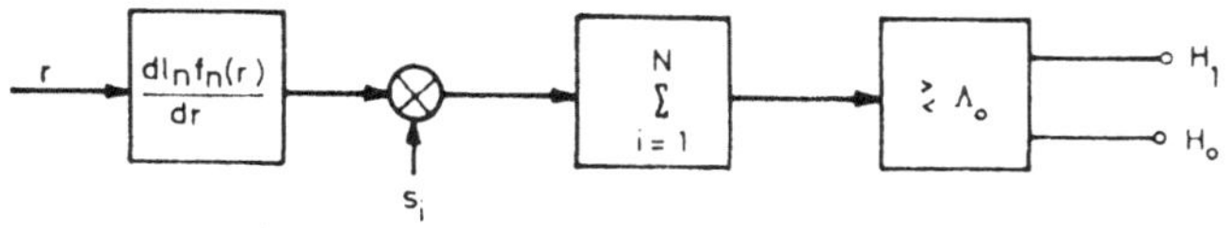

Fig. 5: Locally Optimum Receiver for a Baseband Binary Communication Systems.

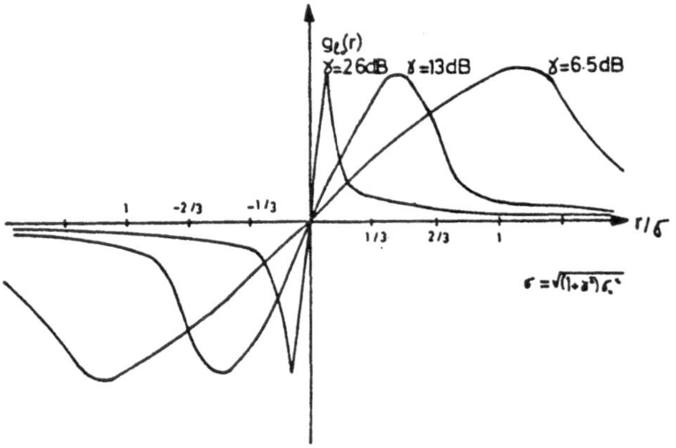

Fig. 6: Optimum ZMNL Function in ELF Noise; α=.35
λ=10 sec^{-1}; a =200 sec^{-1}; h(t) = exp(-at).

3.2. Narrowband processes [14]

The hypothesis testing problem for narrowband processes can be posed as:

$$H_0 : \underline{\tilde{r}} = \underline{\tilde{n}} \tag{36}$$

$$H_1 : \underline{\tilde{r}} = \nu\underline{\tilde{s}} - \underline{\tilde{n}}$$

The observables consist of a sequence of complex amplitudes $\tilde{r}_1, \tilde{r}_2, \ldots, \tilde{r}_N$, whose real and imaginary part are obtained by sampling respectively the outputs of appropriate inphase and quadrature channel filters. Similarly the complex vectors $\underline{\tilde{n}}$, $\underline{\tilde{s}}$ represent the M-samples of the inphase and quadrature components of the noise and signal processes, respectively. In particular the complex amplitudes of the signal and/or noise are measured with respect to a known carrier frequency and a reference phase. The Neyman-Pearson test for independent sampling becomes

$$\Lambda_N(\underline{r};\Theta) = \prod_{i=1}^{N} \frac{P_{n_c,n_s}(r_{c_i}-\nu s_{c_i}, r_{s_i}-\nu r_{s_i})}{P_{n_c,n_s}(r_{c_i}, r_{s_i})} \gtrless \Lambda_o \tag{37}$$

The locally optimum receiver structure for narrowband processes can be obtained using an extension of Neyman-Pearson lemma [13]

$$\frac{\partial \Lambda_N(\tilde{\underline{r}};\nu)}{\partial \nu}\bigg|_{\nu=0} \gtrless \Lambda_o \tag{38}$$

where

$$\frac{\partial \Lambda(\tilde{\underline{r}};\nu)}{\partial \nu}\bigg|_{\nu=0} = -\sum_{i=0}^{N} \frac{S_{c_i}\dfrac{\partial p_{n_c,n_s}(r_{c_i},r_{s_i})}{\partial r_{c_i}} - S_{s_i}\dfrac{\partial p_{n_c,n_s}(r_{c_i},r_{s_i})}{\partial r_{s_i}}}{p_{n_c,n_s}(r_{c_i},r_{s_i})} \tag{39}$$

Due to the circularly symmetric nature of the p.d.f. (39) reduces to

$$T_N(\tilde{\underline{r}}) = \sum_{i=1}^{N} \mathrm{Re}\{r_i\ s_i^*\}\ g(R_i) \tag{40}$$

where $g(R) = -\frac{1}{R}\frac{d\ell n f(R)}{dR}$ and $R = (r_c^2+r_s^2)$.

Note that even in the absence of a quadrature signal component ($S_{q_i}=0$, $i=1,\ldots,N$) the test statistics depend upon both the inphase and quadrature components of the noise through the envelope term. A computationally efficient approximation to the locally optimum ZMNL characteristics $g_{\ell_o}(R)$ is given by [9] :

$$g_{\ell_o}(R) = 1 - \frac{\lambda T\ E\{b\ \exp(-\frac{R^2+b^2}{2})\ I_1(Rb)\}}{(1-\lambda T)R\ \exp(-\frac{R^2}{2}) - \lambda T\ E\{\exp(1-\frac{R^2+b^2}{2})\ I_0(Rb)\}} \tag{41}$$

with $b=quh_{LP}(\tau)$ q and T as in (34).

The block diagram of locally optimum receiver structure in narrowband impulsive noise is shown in Fig. 9; Fig. 10 illustrates typical bandpass locally optimum ZMNL characteristics. The signal processing properties of these ZMNL's are similar to the observations made for the ELF case.

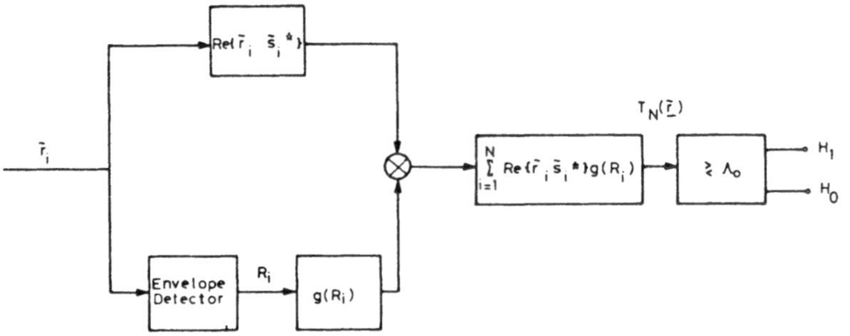

Fig. 7. Locally optimum receiver structures when inphase and quadrature noise components possess circularly symmetric joint distribution.

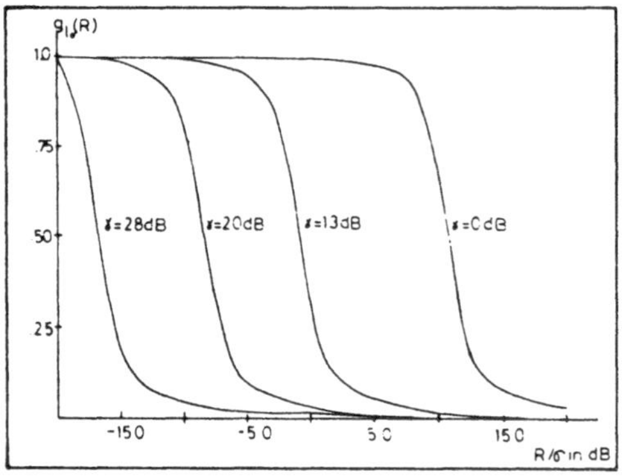

Fig. 8. Locally optimum ZMNL characteristics in VLF noise; $a = 200\ sec^{-1}$, $\alpha = .5$; $\lambda = 10\ sec^{-1}$; $h_{LP}(t) = \exp(-at)$.

5. CONCLUSIONS

A model for impulsive noise has been developed and analyzed consisting of an additive combination of Gaussian and shot noise processes. The model is intuitively satisfying the underlying physics of the phenomena, it has been characterized by a few significant parameters and it can be easily generalized and calibrated to adequately represent impulsive noise phenomena in a variety of frequency bands and physical environments.

This noise model is quite in its potential applications. It has been applied to evaluate the performance of various digital modulation/demodulation strategies operating in impulsive noise channels. Future work is directed toward developing optimum realizable estimation and filtering algorithms for signal and noise processes modelled as an additive combination of filtered Poisson and Gaussian noise processes.

REFERENCES

1. J.W. Modestino, "Modelling of ELF Noise", Technical Report 493, Lincoln Labaratory, M.I.T., December 1971.
2. E.N. Skomal, "Distribution and Frequency of Unintentionally Generated Man-Made VHF-UHF Noise in Metropolitan Areas", IEEE Trans. Electromagnetic Comp., Vol. EMC-7, pp. 263-278, Sept. 1968.
3. S.O. Rice, "Noise in FM Receivers", Ch. 25 in Time Series Analysis, (Ed. M. Resonblatt). New York: Wiley 1963.
4. B.W. Stuck and B. Kleiner, "A Statistical Analysis of Telephone Noise", Bell Sys. Tech. J., Vol. 53, pp. 1263-1320, Sept. 1974.
5. D.L. Synder, Random Point Processes, J. Wiley, Toronto, 1975.
6. A.Segall and T.Kailath, "Orthogonal Functions of Independent-Increment Processes", IEEE Trans.Info. Th., Vol. IT-22, pp. 287-298, May 1976.
7. J.W.Modestino, "Optimum Detection of M-ary Orthogonal Signals in ELF Noise Environments", Technical Report 494, Lincoln Laboratory, M.I.T., April 1972.
8. S.Bochner, "Lectures on Fourier Integrals", Annals of Math. Studies, No: 42, Princeton Univ. Press.
9. B.Sankur, "Modelling of Impulsive Noise and Analysis of Linear and Nonlinear Reception in ELF/VLF Noise", Ph.D Dissertation, Dept. Elec. Syst. Eng., Rensselaer Polytechnic Institute, 1976.

10. J.S. Gradstheyn, I.M.Rhyzik, Tables of Integrals, Series and Products, Acamedic Press 1965.

11. C. Clarke, "Atmospheric Radio Noise Studies Based on Amplitude Probability Measurements at Slough, England, during International Geophysical Year", Proc. IEEE, Vol. 109B, pp. 394-404, Sept. 1962.

12. J.W. Modestino and B. Sankur, "Modelling and Simulation of ELF/VLF Noise", presented at the 7th Annual Pittsburgh Conf. on Modelling and Simulation, Pittsburgh, PA, USA, 1976.

13. F.L. Lehmann, Testing Statistical Hypothesis, New York, J.Wiley, 1959.

14. J.W. Modestino, "Locally Optimum Receiver Structures for Known Signals in NonGaussian Narrowband Noise", Allerton Conference, Oct. 1975.

Part VIII

ADAPTIVE SIGNAL PROCESSING

ADAPTIVE SIGNAL PROCESSING FOR DETECTION AND COMMUNICATION

B. Picinbono

Laboratoire des Signaux et Systèmes †
ESE - Plateau du Moulon - 91190 - GIF - France

1. INTRODUCTION

The classical theories of detection and estimation are usually presented in a non adaptive context |1||2|. In such a context it is assumed that some statistical properties of the signals and the noise are known beforehand. As an example it is well known that the Wiener theory of estimation needs the knowledge of the second order properties of the signals. Similarly, in detection theory, the optimal receiver makes use of a likelihood ratio which requires the knoledge of the signal and noise probability distributions.

On the other hand virtually every practical system of detection or estimation is adaptive (this expression is now very general and will be made more precise in the following). Indeed it is in practice almost always necessary to take into account, at least approximately, the fluctuations of the signal and noise parameters which cannot be described by a given a priori probability distribution.

Unfortunately the concept of optimality becomes fuzzy in adaptive situations. Let us illustrate this point by discussing two simple examples.

First consider the mean-square linear estimate (M.S.L.E.) of a random variable (r.v.) in term of another vectorial r.v. $\vec{x}$. It is well known that the estimate $\hat{y}$ of y can be found from the sole knowledge of the correlation matrix of $\vec{x}$ and the inter - correlation between y and $\vec{x}$. When these two correlations are not known it may be possible to estimate them. If we replace

† Laboratoire associé à l'Université de Paris Sud

the unknown correlations by their estimates we obtain an adaptive solution for the initial problem. But no optimal criterion is used in this replacement, and we cannot assert that this procedure is in a precise sense optimal.

The same situation appears in the detection problem of a deterministic signal in an additive Gaussian noise. When the correlation function of the noise is known, the optimal receiver is the matched filter. If this correlation is unknown but can be estimated, we can use this estimate to construct an adaptive matched filter. But this solution is not necessarily optimal and a search for truly optimal solutions has been conducted in some particular cases, but complete solutions have not yet been obtained [3][4].

The general problem is more complex because one needs to introduce the concepts of complexity (or simplicity) and recursivity in order to select among the many possible adaptive solutions the most interesting one. Indeed, some adaptive solutions are too complicated to be practical and it is usually difficult to evaluate the relative costs of the systems using a simpler solution. Nevertheless recursive algorithms are usually needed as far as computer implementation is concerned.

Because it is evidently not possible in a single lecture to present all aspects of the subject matter, we selected a few topics which appear to have a tutorial value but also can introduce in current research. Some other material can be found for example in Ref. [5].

A. ESTIMATION

2. OPTIMAL LINEAR ESTIMATION

In this section we briefly recall classical results concerning the statistical and the least square linear estimation.

2.1. Statistical estimation.

We wish to estimate a r.v. y in terms of the observation vector $\vec{X}$, which is also random. We suppose that the ensemble $(y, \vec{X})$ is a zero-mean second-order variable. Its second order properties are defined by the covariances

$$\Gamma_X \triangleq E[\vec{X}\vec{X}^T] \; ; \; \Gamma_{xy} \triangleq E[y\vec{X}] = \vec{c}. \qquad (2\text{-}1)$$

Because the estimation is taken as linear we can write the estimate $\hat{y}$ as

$$\hat{y} = \vec{a}^T \vec{X} = \vec{X}^T \vec{a}. \qquad (2\text{-}2)$$

The vector $\vec{a}$ in Eq.(2-2)will be obtained after selection of a criterion of optimality.

When only the second-order properties are known, the simplest possible criterion is the minimization of mean-square error

$$\varepsilon^2 \triangleq E\left[(y - \hat{y})^2\right]. \qquad (2\text{-}3)$$

In the Hilbert space of random variables in which the scalar product is defined by

$$< u, v > \; = \; E\left[u\,v\right], \qquad (2\text{-}4)$$

ε^2 is the square of the distance between y and $\hat{y}$. The <u>projection theorem</u> follows, which states that $\hat{y}$ is the projection of y on the supspace $H(\vec{X})$ spanned by $\vec{X}$. Thus $\hat{y}$ is defined by the equation

$$y - \hat{y} \perp H(\vec{X}) \qquad (2\text{-}5)$$

which can be written

$$E\left[\hat{y}\,\vec{X}\right] = E\left[y\,\vec{X}\right] \qquad (2\text{-}6)$$

or, from Eqs (2-2) and (2-1)

$$\Gamma_X\,\vec{a} = \Gamma_{xy} = \vec{c}. \qquad (2\text{-}7)$$

We suppose that Γ_X is strictly positive. It thus follows that

$$\vec{a} = \Gamma_X^{-1}\,\vec{c}. \qquad (2\text{-}8)$$

Using this result in (2-2) the optimal estimate $\hat{y}$ can be written

$$\hat{y} = \vec{c}^{\,T}\,\Gamma_X^{-1}\,\vec{X}. \qquad (2\text{-}9)$$

Let us now apply to the present situation some of the concepts introduced in the introduction. The estimator in (2-9) is evidently <u>optimal</u>, in a statistical sense. However it is not <u>adaptive</u>, because the covariances are known beforehand. Furthermore, it is not <u>simple</u>, because of the inversion of the correlation matrix Γ. Finally, it is evidently not <u>recursive</u>.

2.2. Adaptive estimation.

We no longer assume, in the present section, that the covariances defined in Eq. (2-1) are known. We assume instead that a sequence of observations $(y_i, \vec{X}_i)$ in time is known.

Let us introduce a linear estimate

$$\hat{y}_n = \vec{a}_n^T \vec{X}_n = \vec{X}_n^T \vec{a}_n \qquad (2\text{-}10)$$

defined as in Eq. (2-2), and use the least-square criterion such that $\vec{a}_n$ minimizes

$$\eta_n^2 = \sum_{1}^{n}{}_i (y_i - \vec{a}_n^T \vec{X}_i)^2. \qquad (2\text{-}11)$$

It is worthwile to explain more precisely the meaning of Eq.(2-11). The observations are $(y_i\ \vec{X}_i)$ for $1 \leq i \leq n$ and the aim is to obtain an expression of y_n only in terms of $\vec{X}_n$ which minimizes the least-square criterion η_n^2 .

The best example of such a situation appears in the prediction problem for a stationary time series. We observe the process from $t = -r + 1$ till $t = n$ and X_i is the vector with components $x_{i-1}, x_{i-2}, \ldots, x_{i-r}$ while y_i is x_i. Thus at $t = n$ we obtain the best linear estimate of x_n which is observed in terms of $x_{n-1}, x_{n-2}, \ldots, x_{n-r}$. If there is a local stationarity this linear estimate can also applied for x_{n+1}, which is not observed, and that is the basic prediction problem.

After calculations similar to those given in the previous section, we find that the vector $\vec{a}_n$ is

$$\vec{a}_n = \Gamma_n^{-1} \vec{c}_n \qquad (2\text{-}12)$$

where

$$\Gamma_n \triangleq \sum_{1}^{n}{}_i \vec{X}_i \vec{X}_i^T \qquad (2\text{-}13)$$

$$c_n \triangleq \sum_{1}^{n}{}_i y_i \vec{X}_i \qquad (2\text{-}14)$$

This vector $\vec{a}_n$ defines (through Eq. 2-10) an estimate $\hat{y}_n$ which is evidently both optimal and adaptive.

Indeed no a priori knowledge of the covariances (2-1) has been introduced in (2-11). The estimate, however, is neither simple nor recursive. Note incidentally that it is the one which follows the intuitive ideas on adaptation outlined in the introduction. Indeed Eq. (2-12) can be considered as a particular case of Eq. (2-8) in which the unknown covariances are replaced by their estimates (2-13) and (2-14) obtained only from the data. The true estimates are evidently $n^{-1}\ \Gamma_n$ and $n^{-1}\ c_n$, but the factor n^{-1} disappears in (2-12).

3. ADAPTIVE AND RECURSIVE ESTIMATION

To compute the $\vec{a}_n$ defined in Eq. (2-12) it is interesting to use a recursive form of the estimate.

For this purpose we shall make use of the well known matrix inversion formula

$$[A + b\, b^T]^{-1} = A^{-1} - \gamma\, A^{-1}\, b\, b^T\, A^{-1} \tag{3-1}$$

where

$$\gamma = (1 + b^T A^{-1} b)^{-1}. \tag{3-2}$$

In this expression A is assumed to be positive definite, which secures the existence of A^{-1}.

We will use this expression to calculate recursively the matrix Γ_n^{-1}, written as

$$K_n \triangleq \Gamma_n^{-1}. \tag{3-3}$$

From Eq. (2-13), we obtain a recurrence on Γ_n which is

$$\Gamma_n = \Gamma_{n-1} + \vec{X}_n\, \vec{X}_n^T, \tag{3-4}$$

and we deduce from the previous equations (3-1) and (3-2) that

$$\boxed{K_n = K_{n-1} - \gamma_n\, K_{n-1}\, \vec{X}_n\, \vec{X}_n^T\, K_{n-1}} \tag{3-5}$$

with

$$\boxed{\gamma_n = (1 + \vec{X}_n^T\, K_{n-1}\, \vec{X}_n)^{-1}} \tag{3-6}$$

Thus we have obtained a recursive expression for the matrix Γ_n^{-1} in term of the data $\vec{X}_n$. Let us notice that γ_n in Eq. (3-6) is a scalar.

There is evidently a problem of initial values of the algorithm, because the rank of $X_n\, X_n^T$ is one and there is no inverse matrix. The same problem appears obviously on (2-12) and it becomes possible to introduce the inverse matrix Γ_n^{-1} only after some observations $\vec{X}_n$.

It is interesting for the following to deduce the quantity $K_n\, \vec{X}_n$ from Eq. (3-5). We obtain directly

$$K_n\, \vec{X}_n = K_{n-1}\, \vec{X}_n (1 - \gamma_n\, \vec{X}_n^T\, K_{n-1}\, \vec{X}_n) \tag{3-7}$$

and if we replace γ_n by the value obtained from Eq. (3-6) we deduce

$$K_n \vec{X}_n = \gamma_n K_{n-1} \vec{X}_n. \tag{3-8}$$

Now we will calculate recursively the vector $\vec{a}_n$ defined by Eq. (2-12) which can be written as

$$\vec{a}_n = K_n \vec{c}_n. \tag{3-9}$$

The vector $\vec{c}_n$ is also obtained recursively from Eq. (2-14) by

$$\vec{c}_n = \vec{c}_{n-1} + y_n \vec{X}_n, \tag{3-10}$$

and by using Eqs. (3-5) and (3-8) we obtain

$$\vec{a}_n = (K_{n-1} - K_n X_n X_n^T K_{n-1})(\vec{c}_{n-1} + y_n \vec{X}_n) \tag{3-11}$$

which gives after simplification

$$\boxed{\vec{a}_n = \vec{a}_{n-1} + K_n \vec{X}_n (y_n - \vec{X}_n^T \vec{a}_{n-1}).} \tag{3-12}$$

In conclusion from the Eqs. (3-5), (3-6) and (3-12) we have obtained a recursive expression of the vector $\vec{a}_n$ which is stricly equivalent to Eqs. (2-12), (2-13) and (2-14), but avoids the matrix inversion appearing in (2-12).

From $\vec{a}_n$ we can deduce with (2-10) an estimate which is adaptive, optimal, recursive and relatively simple. The only numerical problem is to compute the matrix K_n from the data with Eqs. (3-5) and (3-6).

4. CONVERGENCE AND ADAPTIVITY.

For any recursive algorithm it is important to study convergence problems. But the solutions are very different in the stationary case from the non-stationary case.

4.1. Stationary Adaptation.

If the data $(y_n, \vec{X}_n)$ are the values of a stationary stochastic process (s.p.), it is clear from Eqs. (2-13) and (2-14) that $n^{-1} \Gamma_n$ and $n^{-1} \vec{c}_n$ converge respectively to Γ_X and $\vec{c}$ defined by Eq. (2-1). As the term n^{-1} diappears in Eq. (2-12), we deduce that the vector $\vec{a}$ introduced in the statistical theory is the limit of $\vec{a}_n$. This limit can be obtained either with (2-12) or with (3-12) which are strictly identical.

But it is interesting to notice that there are many other algorithms like (3-12) which can define a sequence of $\vec{a}_n$ converging to $\vec{a}$. Among them it is convenient to choose the simplest possible. As the major problem in Eq. (3-12) is to calculate the matrix K_n, it is natural to try the replacement of this matrix by a scalar μ_n, which gives the recursive algorithm

$$\vec{a}_n = \vec{a}_{n-1} + \mu_n \vec{X}_n(y_n - \vec{X}_n^T \vec{a}_{n-1}). \qquad (4\text{-}1)$$

This algorithm is used in the method of stochastic approximation |6| and it is possible to prove that if the coefficients μ_n are positive and satisfy

$$\sum_{1}^{\infty} \mu_n = \infty \qquad \sum_{1}^{\infty} \mu_n^2 < \infty, \qquad (4\text{-}2)$$

then a_n converges almost surely and in quadratic mean to a |7| |8|

Thus we have now obtained a very simple method which gives with Eq. (2-10) an asymptotically optimal adaptive estimate. Indeed we notice the $\vec{a}_n$ in Eqs. (3-12) and (4-1) are quite different, but only converging to $\vec{a}$.

4.2. Non-stationary Adaptation.

The stationary adaptation is rarely interesting for practical uses. Indeed the main interest of adaptive methods is to follow the fluctuations of the system. If, for example, we consider Eqs. (2-13) or (2-14) we see that there is no difference between the contributions in the past. But it is clear that if the correlation has changed at any time instant n, it is necessary to suppress the contribution at the previous time instants.

Thus it is convenient to introduce an adaptation time, or a finite memory, in the algorithm.

Many ways are possible for this purpose. We can for example modify Eqs. (3-4) and (3-10) in such a way that

$$\Gamma_n = \alpha\, \Gamma_{n-1} + \vec{X}_n \vec{X}_n^T \qquad (4\text{-}3)$$

$$\vec{c}_n = \alpha\, \vec{c}_{n-1} + y_n \vec{X}_n. \qquad (4\text{-}4)$$

After very similar calculation we obtain $\vec{a}_n$ with exactly the same algorithm (3-12), while there is a modification in the calculation of K_n. The constant α can be considered as defining the memory of the adaptation.

But if we use an algorithm of adaptation with a finite memory it becomes impossible to obtain a convergence to a, even in the stationary case. There is always a residual noise, and in general faster adaptation, or shorter memory, leads to more noisy adaptive processes. We will discuss this point in the case of the last and most used algorithm of adaptive systems.

4.3. The Gradient Algorithm.

The last simplification which can be introduced in Eq. (3-12) is to replace in (4-1) μ_n by a constant μ, which gives the algorithm commonly used, particular by Widrow |9||10|,

$$\vec{a}_n = \vec{a}_{n-1} + \mu \vec{X}_n (y_n - \vec{X}_n^T \vec{a}_n). \qquad (4\text{-}5)$$

This algorithm can be considered as a particular cas of gradient technique algorithm. Moreover it is also possible to see that it has a finite memory directly connected to the value of μ|10||11|. If μ is very small we obtain a long memory and a small residual error. But in general it is preferable to realize a good adaptation, even with a residual noise.

B. DETECTION.

5. ADAPTIVE DETECTION IN SPHERICALLY INVARIANT NOISE.

We consider in this section the detection problem of a deterministic signal in an additive noise. In all the following the observation is considered as a vector $\vec{x}$ in a finite or infinite dimension space $\mathcal{X}$. This vector can be obtained by many means, and we recall briefly the most used. Firstly the observation can be a real time series x_n observed during a finite interval [1 N], and in this case $\mathcal{X} = R^N$.

When the observation is a continuous signal observed in a finite time interval [OT] two methods can be used. Firstly we can simply sample the observation, and if sampling interval is very small, we obtain N samples which are a good approximation of the continuous signal. On the other hand it is possible to expand the observation in the basis of a set of orthonormal functions $u_n(t)$ which gives

$$x(t) = \sum_{1}^{\infty} {}_n\, x_n\, u_n(t), \quad 0 \leqslant t \leqslant T. \qquad (5\text{-}1)$$

Thus we obtain a vector $\vec{x}$ with components x_n. Moreover we have a large choice of basis $u_n(t)$, and we will use this freedom in the following.

Now the fundamental problem of detection is to decide from the observation $\vec{x}$ if we have a signal present, or only the noise. If the density probability $p(\vec{x})$ of the observation without signal is known, the optimal receiver is achieved by computing the like lihood ratio

$$L(\vec{x}) = \frac{p(\vec{x} - \vec{s})}{p(\vec{x})} \quad (5\text{-}2)$$

which is compared to a threshold [1][2]. In Eq. (5-2) the vector $\vec{s}$ is evidently the signal vector.

If the noise is Gaussian we have

$$p(\vec{x}) = \alpha \exp - \frac{1}{2} \vec{x}^T \Gamma^{-1} \vec{x}, \quad (5\text{-}3)$$

where Γ is the correlation matrix of the noise, defined as in Eq. (2-1). Thus we deduce immediatly that the optimal receiver has to calculate the test function

$$T(\vec{x}) = \vec{s}^T \Gamma^{-1} \vec{x} \quad (5\text{-}4)$$

which is linear in $\vec{x}$. The system which calculates $T(\vec{x})$ from the observation $\vec{x}$ is called the matched filter.

The assumption of Gaussian noise is not very convenient for many applications and a lot of work was presented to obtain more general results. They are unfortunately in general very complicated.

However it is still possible to obtain interesting results in the case of spherically invariant (S.I.) noise |12||13|. It is not our purpose in this lesture to describe in details the properties of such processes, and we can only recall that the class of all the S.I. processes is identical to the class of all the Gaussian processes multiplied by an independent random variable, such that a S.I. process can be written

$$x(t) = \sqrt{w}\, y(t). \quad (5\text{-}5)$$

In this expression $y(t)$ is a zero-mean stationary Gaussian process with unit power, $E[y^2(t)] = 1$, and a given normalized correlation function $\tilde{\Gamma}(\tau)$; moreover w is a random variable, independent of $y(t)$, which represents the power of $x(t)$.

The physical interest of such a model in that it can represente a very common phenomenon in noise problems which is the slow fluctuations of the power written as

$$x(t) = \sqrt{w(t)}\; y(t). \qquad (5\text{-}6)$$

As w(t) has very slow variations, it can be assumed that it is constant in an observation interval T for the detection. Thus we obtain the model (5-5).

It is clear that the probability distribution of x(t) depends on p(w), probability density of w. Thus from (5-2) the structure of the optimal receiver is dependent on p(w) and many examples have been studied |14||15|.

But it appears that the S.I. process like (5-5) have a very nice property which allows to obtain a structure of the optimal receiver independent of p(w) |14|. Indeed if we observe such a process on a finite time interval OT it is possible to estimate without error the value of w. For example if we sample x(t) by dividing the time interval OT in n intervals we obtain

$$\lim_{n\to\infty} \frac{1}{n} \vec{x}_n^T \tilde{\Gamma}_n^{-1} \vec{x}_n = w. \qquad (5\text{-}7)$$

where $\tilde{\Gamma}_n$ is deduced from the correlation function $\tilde{\Gamma}(\tau)$ of y(t).

Moreover this property is independent of the presence of the signal, and Eq. (5-7) is valid for $\vec{x}_n = \vec{b}_n + \vec{s}_n$ or $\vec{x}_n = \vec{b}_n$ at least in the case of non singular detection of s.

This very important property is indicated by saying that the left term of (5-7) is a reference noise alone, R.N.A.

More generally a function of the observation $f(\vec{x})$ is a R.N.A., if it is independent of the signal, which means that

$$f(\vec{b}) = f(\vec{s} + \vec{b}), \qquad (5\text{-}8)$$

Moreover if there is a subspace of the observation space such that all elements of this subspace are independent of the signal, this subspace is also called R.N.A. space.

If now we come back to our detection problem it is possible to show |3| that the optimal receiver combines the use of the matched filter (5-4) and an adjustable threshold which is driven by the power measurement given by (5-7). We can also obtain a similar structure by using a perfect automatic gain control (A.G.C.) at the imput of the matched filter. This A.G.C. is evidently performed by using Eq. (5-7). Thus we have a good example of adaptive detection.

Unfortunately the perfect A.G.C. is not simple to implement, and we will present some approximative solutions of this problem.

6. AUTOMATIC GAIN CONTROL IN DETECTION

We have seen in the previous section the interest of using a perfect A.G.C. system before the matched filter in the case of the detection of a known signal in a S.I. process.

The other interest of the perfect A.G.C. is its R.N.A. property. Indeed the detection system are generally used for the detection at very low signal to noise ratio, but it is important for such systems to have a good behaviour for strong signals. If we are using a classical A.G.C., the strong signals are controlled as a fluctuation of the noise power and there is a diminution of their amplitude. Thus we obtain a loss of performance. Conversely in the case of a perfect A.G.C., there is no diminution of the amplitude of the signal because its R.N.A. property. It is therefore very interesting to study in more details its structure and implementation.

Thus let us consider the estimate (5-7) obtained for finite n

$$\hat{w}_o = \frac{1}{n} \vec{x}_n^T \tilde{\Gamma}_n^{-1} \vec{x}_n . \qquad (6\text{-}1)$$

It is an estimate of the power $E[x_n^2]$ of the r.v. $\vec{x}_n$ obtained by sampling the signal x(t). It is important to note that $\tilde{\Gamma}_n$ is the covariance matrix of the vector $\vec{y}_n$ obtained by sampling y(t) which has a unit power. Thus the diagonal elements of $\tilde{\Gamma}_n$ are equal to 1, and $\tilde{\Gamma}_n$ is sometimes called the normalized covariance of $\vec{x}$.

This estimate $\hat{w}_o$ is unbiased and is optimal in the class of quadratic unbiased estimates of the power, which means that its variance is minimal.

It is very easy to show that $\hat{w}_o$ can be written as

$$\hat{w}_o = \frac{1}{n} \vec{\xi}_n^T \vec{\xi}_n , \qquad (6\text{-}2)$$

where $\vec{\xi}_n$ is deduced from $\vec{x}$ by using $\tilde{\Gamma}_n$ in order that its components are uncorrelated and with the same power.

Thus the physical interpretation of (6-1) is that, because of our knowledge of $\tilde{\Gamma}_n$, it is possible to use a "whitening" of the vector $\vec{x}$ into the vector $\vec{\xi}$.

Without knowledge of $\tilde{\Gamma}_n$, the power estimate is evidently given by the standard equation

$$\hat{w}_s = \frac{1}{n} \vec{x}_n^T \vec{x}_n . \qquad (6\text{-}3)$$

Now we will use the ideas presented previously to construct an adaptive power estimate, which will be used for the adaptive A.G.C.

Let us suppose that the input of the detection system is a time series x_n, for example obtained by sampling a continuous signal. The problem is to estimate rapidly the power in order to control its fluctuation.

To simplify this problem and to use the results of the first sections of this lecture, we will suppose that the observation x_n in the case where there is no signal can be modeled by a stationary autoregressive (A.R.) time series, such that

$$x_n = \sum_{1}^{r} a_i x_{n-i} + u_n. \qquad (6\text{-}4)$$

In this equation r is the order of the regression and u_n is a discrete stationary white noise with power $\sigma_u^2 = E[u_n^2]$. If we introduce the random vector $\vec{X}_n$ defined by

$$\vec{X}_n^T = [x_{n-1}, x_{n-2}, \ldots, x_{n-r}] \qquad (6\text{-}5)$$

we can write x_n in a more convenient from

$$x_n = \vec{a}^T \vec{X}_n + u_n. \qquad (6\text{-}6)$$

The properties of A.R. models have been extensively studied |16| and one of their most interesting properties is the relation between the regression vector $\vec{a}$ and the normalized correlation matrix, generally called the Yule-Walker equation. To establish this equation we calculate the correlation vector $E[x_n \vec{X}_n]$. From Eq. (6-6) we obtain

$$E[x_n \vec{X}_n] = E[\vec{X}_n \vec{X}_n^T] \vec{a} + E[u_n \vec{X}_n]. \qquad (6\text{-}7)$$

The last term of this equation is zero because u_n is white and $\vec{X}_n$ is a linear combination of $\{u_i\}$ for $i < n$. Thus we can write

$$\vec{c} = \Gamma \vec{a} \qquad (6\text{-}8)$$

which is evidently the same equation as (2-8) It is clear that because of the stationarity the vector $\vec{c}$ and the matrix Γ are independent of n. Moreover if we divide $\vec{c}$ and Γ by the power σ_x^2 of x_n we obtain the relation

$$\tilde{c} = \tilde{\Gamma} \vec{a} \qquad (6\text{-}9)$$

where the symbol $\sim$ means that we use normalized covariances. Thus the vector of the regression $\vec{a}$ is tha same if we consider x_n or $\sqrt{w}\, x_n$, i.e. if there is a modulation of the power.

In conclusion if the correlation of the A.R. noise x_n is known, we can calculate $\vec{a}$ by

$$\vec{a} = \Gamma^{-1} \vec{c}. \qquad (6\text{-}10)$$

If that is not the case we can operate as in Sec 3 and for example calculate recursively $\vec{a}_n$ by using the simplest algorithm (4-5) which is in our case

$$\vec{a}_n = \vec{a}_{n-1} + \mu \vec{X}_n (x_n - \vec{X}_n^T \vec{a}_{n-1}). \qquad (6\text{-}11)$$

An A.G.C. acting on x_n needs an estimation of its power on a finite interval. For this purpose we can first use the standard method of (6-2) which is

$$\hat{w}_{s,n} = \frac{1}{n} \sum_{1}^{n} {}_i \, x_i^2 \qquad (6\text{-}12)$$

But the whitening method used in Eq. (6-1) can also be used. Indeed it is clear (6-6) that $x_n - \vec{a}^T \vec{X}_n$ is white, and its power is σ_u^2. To obtain σ_u^2 we can start from Eq. (6-6) and write

$$\sigma_x^2 \triangleq E(x_n^2) = \vec{a}^T \, \Gamma \, \vec{a} + \sigma_u^2 \qquad (6\text{-}13)$$

because the independence between u_n and $\vec{X}_n$.

But we have seen that $\Gamma = \sigma_x^2 \tilde{\Gamma}$, which gives

$$\sigma_x^2 = (1 - \vec{a}^T \tilde{\Gamma} \vec{a})^{-1} \sigma_u^2 \qquad (6\text{-}14)$$

$$= (1 - \vec{a}^T \vec{\tilde{c}})^{-1} \sigma_u^2. \qquad (6\text{-}15)$$

From all the discussion we can introduce the power estimate similar to (6-1) which in our model can be vritten

$$\hat{w}_n(\vec{b}) = \frac{1}{n} (1 - \vec{b}^T \vec{\tilde{c}})^{-1} \sum_{1}^{n} {}_i (x_i - \vec{b}^T \vec{X}_i)^2. \qquad (6\text{-}16)$$

If $\vec{b} = 0$ we obtain the standard estimate $\hat{w}_s$. If $\vec{b} = \vec{a}$ we obtain the optimal estimate $\hat{w}_{on}$. If now $\vec{b}$ is also an estimate of the regression vector $\vec{a}$, as for example in (6-11), we obtain an adaptive estimate which can be used to construct an adaptive A.G.C.

Let us now discuss some general features concerning this A.G.C. which are illustred by the figures presented in the lecture and obtained by computer simulation.

The first point which can be noticed is that the stochastic algorithm (6-11) is sufficient to have in power measurement almost the same results as when a is perfectly known.

The second point is that the influence of a strong modulation of thepower is very weak in the measurement of $\vec{a}$. This confirms the fact that the regression vector is only depending on the normalized covariance.

Finally if we are only interested by the control of the power, it is not necessary to take into account the term $(1 - \vec{b}^T \tilde{c})^{-1}$, which gives significant simplifications for the problem.

A more complete discussion of this A.G.C. system and its application to detection will be presented in a future paper.

7. DETECTION AND R.N.A. SPACES

7.1. Signals and noise.

In this section we will discuss with more details the concept of reference noise alone (R.N.A.) and its application to detection problems. Some preliminary materials were already presented |17| |18|.

At first we recall the principal class of signals appearing in detection problems. The first and also simplest class, corresponds to the case of the detection of one deterministic signal. In communication context this situation appears for example in the case of two antipodal signals. The second one appears particularly in communication problems when we have to select one among M possible signals in noise |19|. This problem belong to the class of multiple hypothesis testing |20|. In the observation space $\mathcal{X}$, there is a subspace spanned by all the linear combination of the M signals which called the signal subspace S. The subspace of $\mathcal{X}$ orthogonal to S is evidently also a R.N.A. subspace. The third important case appears when the signal is random. Even in this case it is sometimes possible to introduce a R.N.A. subspace. That is particularly the case in spatial detection when S sensors receive the same random signal which the M noises are only correlated. By choosing correctly the observation space, it is clear that because of the condition on the signals, there is also a signal subspace, and then a R.N.A. subspace.

Let us now consider the noise. In general it is assumed to be Gaussian, and we use also here this assumption. As in this case the correlation means independence, it is natural to choose the observation space in order to obtain independent noises in the two subspaces previously introduced, which allows to neglect completely the noise space for the detection. This method is particularly used if we introduce properly a Reproducing Kernel Hilbert Space (R.K.H.S.), and we will first shortly discuss this method to introduce another one |21||22|.

7.2. Hilbert Spaces H(A).

Let us suppose, for simplification, that the observation space $\mathfrak{X}$ is R^N, i.e. that it has a finite dimension. The extension to continuous signal for which infinite dimension is required follows straightforward. Thus the vectors signal and noise $\vec{s}$ and $\vec{n}$ are elements of R^N.

Now we can give to $\mathfrak{X}$ the structure of an Hilbert space by introducing a convenient scalar product $<u, v>$.

Consider a symmetric positive matrix A. It is easy to show that we can define a scalar product by

$$<u, v> \triangleq \vec{u}^T A \vec{v} \qquad (7\text{-}1)$$

and therefore associate to A the Hilbert space H(A) whose elements are vectors of R^N and with the scalar product defined by (7-1). If $A = I$ we obtain evidently the usual scalar product.

An orthonormal basis of H(A) is a set of vectors $\{\vec{u}_i\}$ such that

$$\vec{u}_i^T A \vec{u}_j = \delta_{ij} \qquad (7\text{-}2)$$

and we have a large degree of freedom to choose such a basis.

The components of vector $\vec{n}$ in this basis are evidently

$$n_i = \vec{u}_i^T A \vec{n} \qquad (7\text{-}3)$$

which allows the expansion

$$\vec{n} = \sum_{1}^{N}{}_i \; n_i \vec{u}_i . \qquad (7\text{-}4)$$

Let us now suppose that $\vec{n}$ is a random vector, as for example the observation $\vec{x}$ with noise only. If $\vec{n}$ is zero-mean, that is the same for n_i, from Eq. (7-3). On the other hand the correlation between the components of $\vec{n}$ is, from (7-3)

$$E[n_i \; n_j] = \vec{u}_i^T A \Gamma A \vec{u}_j \qquad (7\text{-}5)$$

where Γ is the covariance matrix of $\vec{n}$.

It is interesting to specify the class of Hilbert space defined by A and basis which give uncorrelated components. From Eq. (7-5) we see that it is the case if

$$\vec{u}_i^T \, A \, \Gamma \, A \, \vec{u}_j \;=\; \lambda_i \, \delta_{ij}. \qquad (7\text{-}6)$$

Two natural solutions appear.

a) R K H S.

If we suppose that $A = \Gamma^{-1}$ we see that Eq. (7-6) is satisfied for any orthonormal basis with $\lambda_i = 1$. Indeed such a basis is defined by Eq. (7-2) where $A = \Gamma^{-1}$ which is equivalent to Eq. (7-6) with $\lambda_i = 1$.

In conclusion in the Hilbert space $H(\Gamma^{-1})$ all the orthonormal basis give a decomposition of the random vector with components which are uncorrelated and have the same variance equal to 1.

It is interesting to explain the reproducing property in $H(\Gamma^{-1})$. For this purpose let us introduce the vector $\vec{\gamma}$ defined by

$$\vec{\gamma}_j^T \;=\; [\Gamma_{1j}, \Gamma_{2j}, \ldots, \Gamma_{Nj}] \qquad (7\text{-}7)$$

which is the jth column of the covariance matrix.

We find easily by multiplication that the components of the vector $\Gamma^{-1} \vec{\gamma}_j$ are

$$(\Gamma^{-1} \vec{\gamma}_j) \;=\; \delta_{ij} \qquad (7\text{-}8)$$

and we can deduce that

$$\langle \vec{u}, \vec{\gamma}_j \rangle \;=\; u_j \qquad (7\text{-}9)$$

which is the reproducing property of the scalar product introduced.

b) Karhunen-Loève expansion.

Let us suppose now that $A = I$. In this case we must have simultaneously

$$\vec{u}_i^T \, \vec{u}_j \;=\; \delta_{ij} \qquad (7\text{-}2')$$

$$\vec{u}_i^T \, \Gamma \vec{u}_j \;=\; \lambda_i \, \delta_{ij}. \qquad (7\text{-}6')$$

We deduce that the $\{u_i\}$ are the eigenvector of the matrix Γ with eigenvalue λ_i

$$\Gamma \vec{u}_i = \lambda_i \vec{u}_i. \qquad (7\text{-}10)$$

In this Hilbert space H(I) there is only one basis for which we obtain a decomposition with uncorrelated components. The expansion of n in this basis is the well known Karhunen Loève expansion.

7.3. Detection in $H(\Gamma^{-1})$ and H(I).

We will here only consider the detection of one deterministic signal, because that is sufficient to give the principles of the methods and introduce to another one.

First we recall that if we consider only the observation space without introducing a structure of Hilbert Space, the detection problem is solved by calculating the test function

$$T(\vec{x}) = \vec{s}^T \Gamma^{-1} \vec{x} \qquad (7\text{-}11)$$

as indicated in Eqs. (5-2), (5-4).

Let us now consider the same problem in the R K H S $H(\Gamma^{-1})$. As we are completely free to choose an orthonormal basis in this space, we can take

$$\vec{u}_1 = (\vec{s}^T \Gamma^{-1} \vec{s})^{-1/2} \vec{s} \qquad (7\text{-}12)$$

which defines the signal subspace S and we call $S_\perp$ the R.N.A. subspace, which is orthogonal to S.

Clearly Eq. (7-12) can be valid only if

$$\vec{s}^T \Gamma^{-1} \vec{s} < +\infty \qquad (7\text{-}13)$$

which is the condition of regular detection of $\vec{s}$ in the noise defined by Γ. Thus this condition is equivalent to indicate that $\vec{s} \in H(\Gamma^{-1})$.

Now we can use the projection theorem and decompose the observation vector as

$$\vec{x} = x_s \vec{u}_1 + \vec{x}_\perp \qquad (7\text{-}14)$$

with

$$x_s = \vec{u}_1^T \Gamma^{-1} \vec{x}$$

$$x_s = (\vec{s}^T \Gamma^{-1} \vec{s})^{-1/2} \cdot \vec{s}^T \Gamma^{-1} \vec{x} \qquad (7\text{-}15)$$

$$= \alpha \; T(\vec{x})$$

Moreover $\vec{x}$ is a R.N.A. statiscally independent of x_S. This fact is due to the choice of $H(\Gamma^{-1})$ and the Gaussian assumption.

In conclusion we have by a "projection method" reduced the detection problem in $\mathfrak{X}$ to another one in S, subspace of dimension 1.

That is the principal advantage of this method. But it is clear that in S we have always to calculate the same test function $T(\vec{x})$, and therefore there is no physical interpretation of this expression.

In H(I) the solution is not so simple. Indeed we have seen that the basis is not arbitrary, but given by Eq. (7-10) and if $\vec{s}$ is not an eigenvector of Γ, it is not possible to introduce a subspace S of dimension one. Therefore $\vec{s}$ must be expanded in the basis $\vec{u}_i$ defined by Eq. (7-10) and

$$\vec{s} = \sum_{1}^{N}{}_i \; s_i \; \vec{u}_i . \qquad (7\text{-}16)$$

As a matter of fact the principal difficulty in the use of Karhunen-Loève expansion is the calculation of the $\vec{u}_i$ and λ_i of Eq. (7-10). Therefore it is interesting to study the detection problem in H(I) without introducing the statistical independence of the coordinates which leads to the K.L. expansion.

7.4. Detection in H(I) with a R.N.A.

Let us suppose that the observation space has the structure of H(I), with scalar product defined by (7-2'), and that the signal subspace S has the dimension d. In the case of communications with M signals we have

$$1 \leqslant d \leqslant M. \qquad (7\text{-}17)$$

The case of only one signal gives evidently $d = 1$.

The R.N.A. space, or the noise space is clearly orthogonal to $S(S_\perp)$, and any vector $\vec{x}$ of can be decomposed by projection in

$$\vec{x} = \vec{x}_1 + \vec{x}_2 \qquad (7\text{-}18)$$

where $\vec{x}_1$ belongs to S and $\vec{x}_2$ to $S_\perp$.

Evidently $\vec{x}_2$ is a R.N.A. because there is never signal in $S_\perp$.

When we have M possible signals $\vec{s}_i$ to detect, the theory is a direct extension of the case of only one signal. In Gaussian noise the optimal receiver must compute the M functions

$$T_i(\vec{x}) = \vec{s}_i^{\,T} \, \Gamma^{-1} \, \vec{x} \qquad (7\text{-}19)$$

and select the greatest one. It is clear that $T_i(\vec{x})$ has exactly the same structure as $T(\vec{x})$ in Eq. (7-11). In Eq. (7-19), Γ^{-1} is the inverse of the correlation matrix of the noise in the observation space,

$$\Gamma \triangleq E[\underline{n}\ \underline{n}^T]. \qquad (7\text{-}20)$$

By using the decomposition of $\vec{n}$ in two orthogonal subspaces we can also decompose the matrix Γ with four matrices

$$\Gamma_{ij} \triangleq E[\vec{n}_i \ \vec{n}_j^{\,T}] \qquad (7\text{-}21)$$

and it is interesting to calculate the corresponding decomposition of Γ^{-1} written as

$$\Gamma^{-1} = \begin{pmatrix} A & C \\ C^T & B \end{pmatrix} \qquad (7\text{-}22)$$

By assuming that $\Gamma\Gamma^{-1} = 1$, we obtain after simple calculations

$$A = (\Gamma_{11} - \Gamma_{12} \ \Gamma_{22}^{-1} \ \Gamma_{21})^{-1} \qquad (7\text{-}23)$$

$$B = (\Gamma_{22} - \Gamma_{21} \ \Gamma_{11}^{-1} \ \Gamma_{12})^{-1} \qquad (7\text{-}24)$$

$$C = - A \, \Gamma_{12} \ \Gamma_{22}^{-1}.$$

The matrix A and B are positive and it is interesting to give an interpretation of their structure in term of estimation.

For this purpose let us introduce the best mean-square linear estimate $\hat{n}_1 = \phi \, \vec{n}_2$ in term of $\vec{n}_2$. This estimate is obtained as in section 2 by assuming that $\vec{n}_1 - \hat{n}_1$ is uncorrelated with $\vec{n}_2$, or

$$E[(\vec{n}_1 - \phi \, \vec{n}_2) \ \vec{n}_2^{\,T}] = 0 \qquad (7\text{-}26)$$

which given

$$\hat{n}_1 = \Gamma_{12} \; \Gamma_{22}^{-1} \vec{n}_2. \tag{7-27}$$

Now the covariance matrix of the error of estimation

$$\vec{z} = \vec{n}_1 - \hat{n}_1 \tag{7-28}$$

is

$$\Gamma_z \triangleq E[\vec{z} \; \vec{z}^T] = \Gamma_{11} - \Gamma_{12} \; \Gamma_{22}^{-1} \; \Gamma_{21} = A^{-1}. \tag{7-29}$$

With this decomposition it is now possible to give an interesting interpretation of the test function $T_i(\vec{x})$ given by Eq.(7-19). Indeed as $\vec{s}_i$ is in S, its decomposition like as (7-18) is

$$\vec{s}_i = \vec{s}_i + 0 \tag{7-30}$$

which gives

$$T_i(\vec{x}) = \vec{s}_i^T A \vec{x}_1 + \vec{s}_i^T C \vec{x}_2. \tag{7-31}$$

With the Eqs. (7-23), (7-25) and (7-27) and after noticing that $\vec{x}_2 = \vec{n}_2$, because of the R.N.A. property, we obtain finally

$$T_i(\vec{x}) = \vec{s}_i^T \Gamma_z^{-1}(\vec{x}_1 - \hat{n}_1) \tag{7-32}$$

Let us now discuss some interesting features abtained from this expression.

1. We see that the R.N.A. subspace $S_\perp$ is only used to calculate $\hat{n}_1$, which is the best mean-square linear estimate of the noise in S.
2. The test function $T_i(\vec{x})$ is calculated in the signal subspace S only.
3. In this subspace the structure of $T_i(\vec{x})$ is exactly the same as in Eq. (7-19), but the covariance matrix is now Γ_z defined by (7-29).
4. In the case of the detection of only one signal, Γ_z is a scalar, and $T(\vec{x})$ can also be written

$$T(\vec{x}) = \alpha \; \vec{s}^T(\vec{x}_1 - \hat{n}). \tag{7-33}$$

This structure shows that even in colored noise we can use to compute $T(\vec{x})$ the structure of an optimal receiver for white noise (correlator receiver) but we must use in this case the estimation of the noise in the direction of the signal.

Now let us briefly discuss the use of R.N.A. spaces for adaptive detection. If the statistical properties of the noise are not very well known we can perform some measurements in the subspace $S_{\perp}$ orthogonal to the signal. In this subspace the results cannot be perturbed by the presence of the signal,and therefore are exactly the same under the two hypothesis H_0 and H_1. A more precise discussion of this point will be discussed elsewhere..

REFERENCES

[1] C.W. HELSTROM, Statistical Theory of Signal Detection Pergamon, New York, 1968.

[2] H.L. VAN TREES, Detection, Estimation and Modulation Theory Wiley, New York, 1968.

[3] G. VEZZOSI and B. PICINBONO, Détection d'un signal certain dans un bruit sphériquement invariant. Structure et comparaison des différents récepteurs Ann. Télécomm. 27, 95, 1972.

[4] G. VEZZOSI, What is optimality for an adaptive detection system, Nato A.S.I. on Signal Processing Academic Press, New York, 1973.

[5] C. TSYPKIN, Adaptation and learning in Automatic systems, Academic Press, New York, 1971.

[6] D. SAKRISON, Stochastic approximation, a recursive method for solving regression problems, Advances in communication systems, Av. Balakrishnan Ed, Academic Press, New York, 1966.

[7] C. MACCHI, Itération stochastique et traitements numériques adaptatifs, thèse de Doctorat d'Etat, Paris 1972.

[8] C. and O. MACCHI, Un théorème d'itération stochastique multidimensionnelle, Ann. Inst. Henri Poincaré, 7-193, 1971.

[9] B. WIDROW and al., Adaptive noise cancelling : principles and applications, Proc. of I.E.E.E., 63, 1962,1975.

[10] B. WIDROW and al., Stationary and non stationary learning charasteristics of the L M S adaptive filter, Nato A.S.I. Aspects of signal processing, G. Tacconi Ed., Part 1, p. 335, Reidel Publishing Dordrecht 1976.

[11] O. MACCHI, Résolution adaptative de l'équation de Wiener Hopf, to be published in Ann. Inst. Henri Poincaré.

[12] B. PICINBONO, Spherically invariant and compound gaussian stochastic processes, I.E.E.E. Trans. Inf. Theor. I.T. 16, 77, 1970.

[13] K. YAO, A representation theorem and its application to spherically invariant random processes, I.E.E.E. Trans. Inf. Theor. I.T. 19, 600, 1973.

[14] B.PICINBONO and G. VEZZOSI, Détection d'un signal certain dans un bruit non stationnaire et non gaussien, 1 calcul du récepteur optimal, Ann. Télécom. 25, 433, 1970.

[15] J. GOLDMAN, Detection in the presence of spherically symetric random vectors, I.E.E.E. Trans. Inf. Theor. I.T. 22, 52, 1976.

[16] J. MAKHOUL, Linear prediction, a tutorial review, Proc. I.E. E.E., 63, 561, 1975.

[17] B. PICINBONO, Estimation et détection, revue de Cethedec 50, 59, 1977.

[18] B. PICINBONO, Introduction to detection and estimation, the same reference as 10, p. 163.

[19] A.J. VITERBI, Principles of Coherent Communication, Mc Graw Hill, New York, p. 216, 1966.

[20] Sec ref 2, p. 46.

[21] T. KAILATH, A projection method for signal detection in colored gaussian noise, I.E.E.E. Trans. Inf. Theor. I.T. 13, 441, 1967.

[22] T. KAILATH, Approach to detection and estimation problems 1, deterministic signal in gaussian noise, I.E.E.E. Trans. Inf. Theor. I.T., 17, 530, 1971.

ADAPTIVE FILTERS*

John G. Proakis

Department of Electrical Engineering, Northeastern University, Boston, Mass. 02115, U.S.A.°

ABSTRACT. A brief review is presented of the fundamentals of adaptive filtering using a transversal filter with coefficients that are recursively adjusted to minimize the mean square error at the output of the filter. The convergence properties of the recursive algorithm are discussed and the variance of the self-noise generated at the filter output is derived. Some applications of adaptive filters are also given.

1. INTRODUCTION

During the last ten to fifteen years adaptive or self-optimizing filters have been widely used in communication and control systems in which the statistics of the signals to be filtered are either unknown a priori or, in some cases, slowly time-variant (nonstationary signals) [1]-[10]. Although several types of adaptive filter structures have been described in the literature, by far the most practical and, probably, the simplest structure is the linear transversal (nonrecursive) filter shown in Figure 1. This is an all-zero filter with adjustable coefficients, denoted by c_k, $k=0, 1, \ldots, N-1$. It may be viewed as a discrete-time filter with transfer function

$$C(z) = \sum_{k=0}^{N-1} c_k z^{-k} \tag{1}$$

* This work has been supported in part by the National Science Foundation under Grant ENG 76-00848.

° Present address.

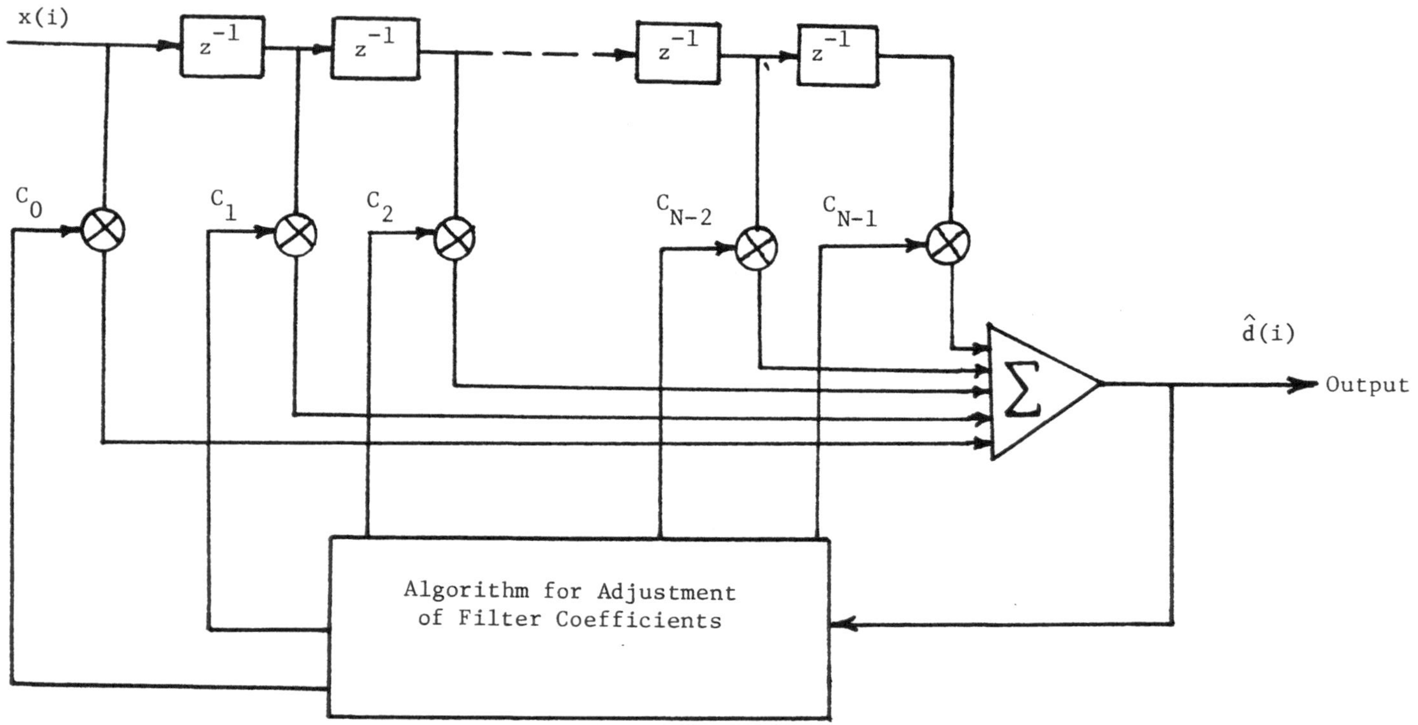

Figure 1: Adaptive Linear Transversal Filter

where z^{-1} is interpreted as a unit of delay.

Since the transversal filter contains no poles, other than the ones at $z=0$, it is free of stability problems associated with the more general adaptive linear filter structure that has adjustable poles as well as zeros.

In passing, we mention an alternative and completely equivalent realization of the all-zero filter shown in Figure 1. It consists of a comb filter in cascade with a parallel bank of resonators [10] and has been termed the "frequency-sampling realization". Each resonator has an adjustable gain and phase shift. This filter structure has been used in speech signal processing and has also been considered for equalization of telephone channels.

An important consideration in the use of an adaptive filter is the criterion for optimizing the filter coefficients $\{c_k\}$. For example, a desirable performance index in a digital communication system is the average probability of error. Consequently, one might consider the implementation of an adaptive filter having coefficients that are adjusted to minimize the average probability of error. In this situation, however, one is faced with the problem that the performance index (average probability of error) is a highly nonlinear function of the filter coefficients. Although a number of algorithms are known for finding the minimum (or maximum) of a highly nonlinear function of several variables, they are usually unsuitable for adaptive filtering primarily because the signal statistics are unknown and, possibly, time-variant. Also, in some cases, the performance index possesses many relative minima (or maxima), so that one is not certain whether the adaptive filter has converged to the optimum solution or to one of the relative minima (or maxima). For these reasons, many desirable performance indeces such as the average probability of error are impractical to implement.

One criterion that does not suffer from such practical difficulties when used in conjunction with an all-zero filter is the mean-square-error (MSE) criterion. This criterion leads to a quadratic performance index as a function of the filter coefficients $\{c_k\}$ and possesses a single minimum. Hence, the MSE criterion is chosen primarily for its mathematical convenience and ease in implementation, even though it may not be the ultimate index of performance.

A number of tutorial treatments of adaptive filters based on the MSE criterion have been published in the literature [2], [5], [7], [8]. These papers deal with a recursive algorithm for adjusting the filter coefficients, originally proposed by Widrow and Hoff [1]. The algorithm employs estimates of the gradients to update the coefficients. A generalization of the Widrow and Hoff algorithm is obtained by using linearly filtered estimates of the gradients in the adjustment algorithm [8].

In this survey paper, our intent is to briefly review the fundamentals of adaptive filtering based on the MSE criterion and to present some applications of adaptive filters in spectral estimation, in noise whitening and speech signal compression.

2. ADAPTIVE TRANSVERSAL FILTER BASED ON MSE CRITERION

Let $x(k)$, $k=\ldots, -2, -1, 0, 1, 2, \ldots$ denote the input signal sequence to the transversal filter shown in Figure 1 and let its output be denoted by

$$\hat{d}(i) = \sum_{k=0}^{N-1} c_k \, x(i-k) \quad , \quad k=\ldots, -2, -1, 0, 1, 2, \ldots \tag{2}$$

The sequence $\hat{d}(i)$ is called the estimate of the desired output. Let us assume that there exists a desired signal sequence $d(i)$ to which the estimate $\hat{d}(i)$ can be compared. An error signal sequence is defined as

$$\varepsilon(i) \equiv d(i) - \hat{d}(i) \tag{3}$$

The performance index for optimizing the filter, based on the MSE criterion, is

$$\xi(\underline{C}) \equiv E|\varepsilon(i)|^2 \tag{4}$$

where $E(u)$ denotes the expected value of the random variable u.

Under the assumption that the input sequence is statistically stationary, the performance index $\xi(\underline{C})$ is a quadratic function of the filter coefficients. It can be expressed in the form

$$\xi(\underline{C}) = E|d(i)|^2 - 2\mathrm{Re}(\underline{C}^*, \underline{b}) + (\underline{C}^*, \underline{\psi}\, \underline{C}) \tag{5}$$

where the notation (u,v) denotes the inner product of two vectors $\underline{u}$ and $\underline{v}$, $\underline{C}$ is an $(N\times 1)$ column vector representing the filter coefficients, $\underline{b}$ is an $(N\times 1)$ column vector of crosscorrelation coefficients

$$b(k) = E[d(i)x^*(i-k)] \quad , \quad k=0, \ldots, N-1 \tag{6}$$

and $\underline{\psi}$ is an (NxN) correlation matrix with elements

$$\psi(k-j) = E[x(i-j)x^*(i-k)] \quad , \quad k,j=0, 1, \ldots, N-1 \tag{7}$$

An asterisk denotes the complex conjugate of a quantity and Re(u) denotes the real part of the complex variable u.

When $\underline{b}$ and $\underline{\psi}$ are known, the minimization of $\xi(\underline{C})$ is easily carried out to yield the result

$$\underline{C}_{opt} = \underline{\psi}^{-1}\,\underline{b} \tag{8}$$

where $\underline{C}_{opt}$ denotes the optimum filter coefficients. (In the case of nonstationary signals the matrix $\underline{\psi}$ and the vector $\underline{b}$ vary with time and, hence, $\underline{C}_{opt}$ is also a function of time.) The MSE corresponding to $\underline{C}_{opt}$ is

$$\begin{aligned} \xi_{min} \equiv \xi(\underline{C}_{opt}) &= E\{[d(i) - \hat{d}(i)]d^*(i)\} \\ &= E|d(i)|^2 - (\underline{b}^*, \underline{C}_{opt}) \end{aligned} \tag{9}$$

When $\underline{b}$ and $\underline{\psi}$ are known, it is also possible to obtain $\underline{C}_{opt}$ recursively by any one of a number of "hill-climbing" methods such as the method of steepest descent, the conjugate gradient method [11] and the Fletcher-Powell method [12]. These recursive methods for obtaining $\underline{C}_{opt}$ may be described by the algorithm

$$\underline{C}(i+1) = \underline{C}(i) + \Delta(i)\,\underline{S}(i) \quad , \quad i=0, 1, \ldots \tag{10}$$

where $\underline{C}(i)$ is the vector of filter coefficients at the ith iteration, $\Delta(i)$ is the step size parameter at the ith iteration and $\underline{S}(i)$ is the direction vector at the ith iteration. The initial vector $\underline{C}(0)$ is selected arbitrarily.

The method for choosing $\Delta(i)$ and $\underline{S}(i)$ depends on the particular algorithm that is used. In general, the direction vector $\underline{S}(i)$ is some linear function of the gradient vector

$$\underline{g}(i) = \frac{d\xi[\underline{C}(i)]}{d[\underline{C}(i)]}$$

$$= \psi\ \underline{C}(i) - \underline{b}$$

$$= -E[\varepsilon(i)\ \underline{X}^*(i)] \tag{11}$$

where $\underline{X}(i)$ denotes the vector of input samples that are used in the estimate $\hat{d}(i)$ of the desired signal. That is $\underline{X}^T(i) = [x(i), x(i-1), \ldots, x(i-N+1)]$. For example, in the method of steepest descent, $\underline{S}(i) = -\underline{g}(i)$, while in the conjugate gradient method the direction vector is given by the expression $\underline{S}(i) = \beta(i-1)\ \underline{S}(i-1) - \underline{g}(i)$ where $\beta(i-1)$ is a scale factor that is selected to satisfy the generalized orthogonality condition $[\underline{S}(i), \psi\ \underline{S}(i-1)] = 0$.

In the filtering problems under consideration neither the autocorrelation matrix ψ nor the vector of crosscorrelations $\underline{b}$ is known. However, since we have access to the received signal sequence $\{x(i)\}$ and the desired signal sequence $\{d(i)\}$, both ψ and $\underline{b}$ can be estimated from these sequences and the estimates can be used in (8) to yield an estimate of $\underline{C}_{opt}$, i.e.,

$$\hat{\underline{C}}_{opt} = \hat{\psi}^{-1}\ \hat{\underline{b}} \tag{12}$$

where $\hat{\psi}$, $\hat{\underline{b}}$ and $\hat{\underline{C}}_{opt}$ are estimates of ψ, b, and $\underline{C}_{opt}$, respectively. Alternatively, we may use a recursive algorithm similar in form to the one given in (10). The first approach is predominately used in encoding (compression) of speech based on linear prediction. The second method is used extensively in equalization algorithms for digital communication systems [8], in adaptive antenna systems [3] and in identification of dynamical systems [2].

We focus our attention on the recursive algorithm approach. In particular, we observe that an unbiased estimate of the gradient $\underline{g}(i)$ at the ith iteration is simply

$$\hat{\underline{g}}(i) = -\varepsilon(i)\ \underline{X}^*(i) \tag{13}$$

where $\underline{X}(i)$ is the vector of the input signal used in the estimate $\hat{d}(i)$, as defined previously. Now, if the estimate $\hat{\underline{g}}(i)$ is used in place of $\underline{g}(i)$ in the method of steepest descent, the resulting recursive algorithm is

$$\underline{\hat{C}}(i+1) = \underline{\hat{C}}(i) - \Delta(i)\, \underline{\hat{g}}(i) \quad , \quad i=0,\ 1,\ \ldots \tag{14}$$

with $\underline{\hat{C}}(0)$ selected arbitrarily.

The selection of the step size $\Delta(i)$ at each iteration poses somewhat of a problem when $\underline{\psi}$ and $\underline{b}$ are unknown. In practice, a fixed step size Δ is used instead of a variable step size. One advantage to using a fixed step size is the simplicity of the resulting algorithm. That is, the computational burden involved in selecting the step size parameter for each iteration is eliminated. However, the use of a fixed step size in the presence of noise results in a self-noise at the filter output even when the signal is stationary. The variance of the self-noise and its dependence on Δ is given in the next section.

With a fixed step size, the recursive algorithm for adjusting the coefficients of the adaptive filter is simply

$$\begin{aligned} \underline{\hat{C}}(i+1) &= \underline{\hat{C}}(i) - \Delta \underline{\hat{g}}(i) \\ &= \underline{\hat{C}}(i) + \Delta \varepsilon(i)\, \underline{X}^*(i) \end{aligned} \tag{15}$$

This algorithm was originally proposed by Widrow and Hoff [1].

A generalization of the steepest descent algorithm given above is [8]

$$\underline{\hat{C}}(i+1) = \underline{\hat{C}}(i) + \Delta \underline{\hat{S}}(i) \tag{16}$$

where $\underline{\hat{S}}(i)$ is some linear function of $\underline{\hat{g}}(k)$, k=i, i-1, ... For example, we consider linear filtering of the estimated gradient vectors to obtain $\underline{\hat{S}}(i)$. In particular, the use of a single pole lowpass filter with transfer function

$$H(z) = \frac{1}{1 - \beta z^{-1}} \tag{17}$$

and an input sequence $-\underline{\hat{g}}(i)$ yields the output sequence $\underline{\hat{S}}(i)$ where

$$\underline{\hat{S}}(i) = \beta \underline{\hat{S}}(i-1) - \underline{\hat{g}}(i) \quad , \quad \underline{\hat{S}}(0) = -\underline{\hat{g}}(0) \tag{18}$$

The expression in (18) is similar in form to the formula for computing the direction vectors in the conjugate gradient method, except that the parameter β in (18) is fixed. The selection of β is discussed in the next section. In general, more complex filters (with more poles and zeros) can be used in filtering the estimates of the gradients $\hat{\underline{g}}(i)$ to yield $\hat{\underline{S}}(i)$.

3. STABILITY AND SELF-NOISE OF THE RECURSIVE ALGORITHM

In this section we consider the problems of stability and self-noise of the recursive algorithm given in (16). For illustrative purposes the single-pole lowpass filter given in (17) is used for filtering the estimates of the gradients. Thus, the direction vectors $\hat{\underline{S}}(i)$ are computed according to (18).

The conditions for stability of the algorithm are obtained by examining the mean values of (16) and (18). Since $\hat{\underline{g}}(i)$ is an unbiased estimate of $\underline{g}(i)$, we have

$$E[\hat{\underline{C}}(i+1)] = \underline{C}(i+1) = \underline{C}(i) + \Delta\underline{S}(i)$$

$$E[\hat{\underline{S}}(i)] = \underline{S}(i) = \beta\underline{S}(i-1) - \underline{g}(i) \tag{19}$$

The mean value relations given by (19) represent a discrete-time feedback control system as shown in Figure 2. The plant to be controlled can be represented by N one-dimensional discrete-time filters each having a z-transform

$$G(z) = \frac{\Delta}{z - 1} \tag{20}$$

The output of the plant is delayed by one unit in time and fed back ghrough the matrix transformation $\underline{\psi}$. In addition, there is a filter with transfer function H(z), which in this case is specified by (17), but which in general can be any arbitrary rational function of z, designed to yield a stable closed loop system with a good transient response.

Since H(z) and G(z) are in cascade, the two filters can be combined into a single filter with transfer function

$$H(z)\,G(z) = \frac{\Delta}{z - (1+\beta) + \beta z^{-1}} \quad , \tag{21}$$

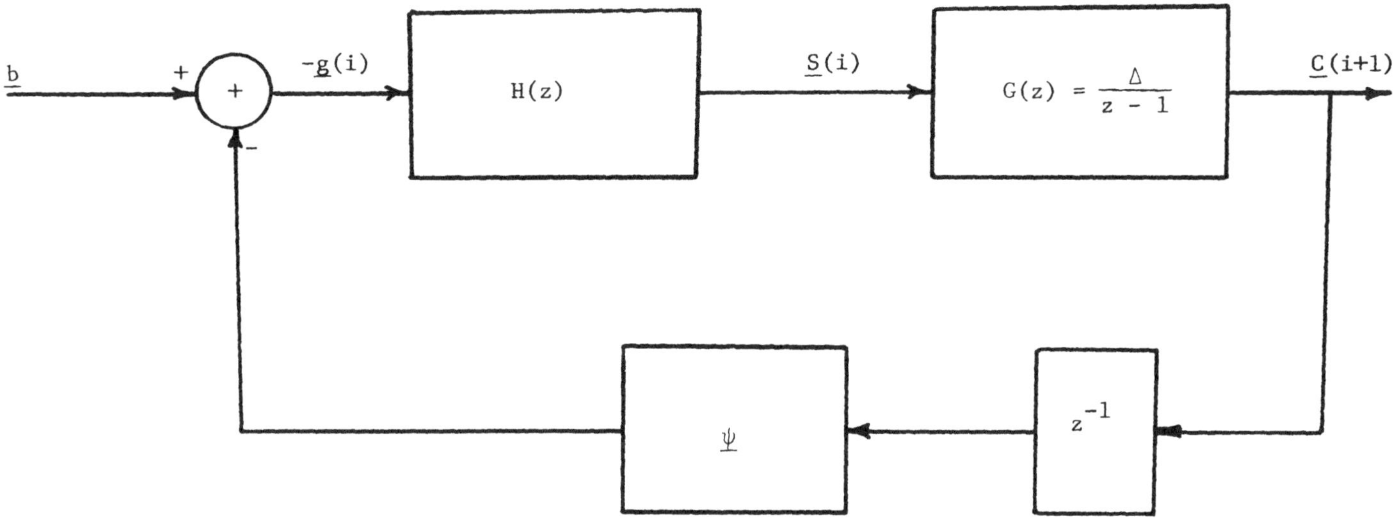

Figure 2: Closed-loop Control System Representation of the Recursive Algorithm

which corresponds to the input-output difference equation

$$\underline{C}(i+1) = (1+\beta)\ \underline{C}(i) - \beta\underline{C}(i-1) - \Delta\underline{g}(i) \tag{22}$$

Equation (22) represents a set of N difference equations. Since $\underline{g}(i) = \underline{\psi}\ \underline{C}(i) - \underline{b}$, each component in the gradient vector $\underline{g}(i)$ is coupled to the N filter coefficients $\underline{C}(i)$ through the matrix $\underline{\psi}$. In order to determine the conditions for stability of the algorithm it is mathematically convenient to decouple the two vectors $\underline{g}(i)$ and $\underline{C}(i)$ by performing a linear transformation. The appropriate transformation is obtained by noting that $\underline{\psi}$ is Hermitian and, hence, it can be represented as

$$\underline{\psi} = \underline{U}\ \underline{\Lambda}\ \underline{U}^{*T} = \underline{U}\ \underline{\Lambda}\ \underline{U}^{-1} \tag{23}$$

where $\underline{U}$ is the normalized modal matrix of $\underline{\psi}$ and $\underline{\Lambda}$ is a diagonal matrix with diagonal elements equal to the eigenvalues of $\underline{\psi}$, all of which are real and positive. Now, if one defines the transformed vectors $\underline{g}'(i) = \underline{U}^{-1}\ \underline{g}(i)$, $\underline{b}' = \underline{U}^{-1}\ \underline{b}$ and $\underline{C}'(i) = \underline{U}^{-1}\underline{C}(i)$,

$$\underline{g}'(i) = \underline{\Lambda}\ \underline{C}'(i) - \underline{b}' \tag{24}$$

which, when combined with the transformed version of (22) yields

$$\underline{C}'(i+1) = [(1+\beta)\ \underline{I} - \Delta\underline{\Lambda}]\ \underline{C}'(i) - \beta\underline{C}'(i-1) + \Delta\underline{b}' \tag{25}$$

where $\underline{I}$ denotes the identity matrix.

A difference equation of a similar form is obtained by premultiplying (25) by $\underline{\Lambda}$ and substituting $\underline{g}'(i) + \underline{b}'$ for $\underline{\Lambda C}'(i)$. The result is

$$\underline{g}'(i+1) - [(1+\beta)\ \underline{I} - \Delta\underline{\Lambda}]\ \underline{g}'(i) + \beta\underline{g}'(i-1) = \underline{0} \tag{26}$$

The N second-order homogeneous difference equations in (26), which are identical to the homogeneous difference equations in (25) obtained by setting $\underline{b}'=\underline{0}$, have a solution (transient response) of the form

$$g_k(i) = A_1 p_{1k}^i + A_2 p_{2k}^i \quad , \quad \begin{array}{l} i=0, 1, \ldots \\ k=0, 1, 2, \ldots, N-1 \end{array} \tag{27}$$

where A_1 and A_2 are arbitrary constants and

$$\left.\begin{array}{l} p_{1k} \\ p_{2k} \end{array}\right\} = \frac{1+\beta-\Delta\lambda_k}{2} \pm \sqrt{\left(\frac{1+\beta-\Delta\lambda_k}{2}\right)^2 - \beta} \tag{28}$$

The set $\{\lambda_k\}$ denote the N (possibly nondistinct) eigenvalues of $\underline{\psi}$. Stability of the algorithm is assured provided that $g_k(i)\to 0$ as $i\to\infty$ for all k or, equivalently, $|p_{1k}|<1$ and $|p_{2k}|<1$ for all k. These conditions are satisfied if

$$0<\beta<1 \quad , \quad 0<\Delta< \frac{2(1+\beta)}{\lambda_{max}} \tag{29}$$

where λ_{max} is the largest eigenvalue of $\underline{\psi}$. Under these conditions $\underline{g}(i)\to\underline{0}$ and

$$\underline{C}'(i) \to \underline{C}'_{opt} = \underline{\Lambda}^{-1}\, \underline{b}' \tag{30}$$

as $i\to\infty$. Thus, the mean value relations for $\underline{C}(i)$ and $\underline{g}(i)$ converge when the conditions in (29) are satisfied.

The rate of convergence increases as $\beta\to 1$ and $\Delta\to 2(1+\beta)/\lambda_{max}$. Ideally, it is desirable to design the adaptive filter so that its transient response decays to zero as rapidly as possible. This implies that Δ and β should be selected as large as possible within the limits imposed for stability. On the other hand, when Δ and β are large, the self-noise generated by the recursive algorithm has a large variance, as discussed below. Since a large self-noise is undesirable in steady-state operation of the adaptive filter, a two-step procedure is sometimes used in practice. That is, during initial adaptation large values of Δ and β are selected, and these are reduced appropriately to yield an acceptable self-noise variance in steady-state operation of the filter.

Let us now turn our attention to the self-noise generated by the recursive algorithm as a result of using noisy estimates of the gradients and a fixed step size Δ. This noise appears at the output of the adaptive filter as an additive noise term and serves to increase ξ_{min}, given in (9), by an amount which is denoted as ξ_Δ. It is easily shown [5] that the variance of the self-noise can be expressed in terms of the "transformed" filter coefficients as

$$\xi_\Delta = \lim_{i=\infty} \sum_{k=0}^{N-1} \lambda_k E|\hat{c}'_k(i) - c'_{k\,opt}|^2$$

$$= \sum_{k=0}^{N-1} \lambda_k E|\hat{c}'_k - c'_{k\,opt}|^2 \quad , \tag{31}$$

where (31) assumes steady-state operation.

The problem is to relate the term $E|\hat{c}'_k - c'_{k\,opt}|^2$ to the parameters Δ and β. This may be done as follows. In the absence of measurement noise the "transformed" coefficients of the filter are given by the set of decoupled difference equations in (25). The use of noisy estimates of the gradient vector $\underline{g}'(i)$ in the recursive algorithm can be modeled by the addition of a zero-mean noise vector $\underline{n}'(i)$ to the right-hand side of (25). The noise causes the tap coefficients $\hat{\underline{C}}'(i)$ to fluctuate about the optimum value $\underline{C}'_{opt}$. Let us assume that the noise $\underline{n}'(i)$ is white†, i.e.,

$$E[n'_k(i)\, n^{*\prime}_k(j)] = E|n'_k(i)|^2 \delta_{ij} \tag{32}$$

Then, the steady-state mean square value of the fluctuation of $\hat{c}'_k$ about its optimum value is

$$E|\hat{c}'_k - c'_{k\,opt}|^2 = \frac{E|n'_k(i)|^2(1+\beta)}{(1-\beta)(\Delta\lambda_k)(2+2\beta-\Delta\lambda_k)} \tag{33}$$

It has been shown in [5] that the noise variance in (33) can be expressed as

$$E|n'_k(i)|^2 = \Delta^2\, \xi_{min}\, \lambda_k \tag{34}$$

By combining (31), (33) and (34) we obtain

† In steady-state operation the noise $\underline{n}'(i)$ is stationary although, in general, it is not strictly white. The assumption of noise whiteness is met approximately because of the variability in the input samples $\{x(k)\}$ and the fluctuation of the filter coefficients due to the adaptation process.

$$\xi_\Delta = \frac{\Delta\xi_{min}}{2(1-\beta)} \sum_{k=0}^{N-1} \lambda_k [1 - \frac{\Delta\lambda_k}{2(1+\beta)}]^{-1} \tag{35}$$

It is interesting to compare the variance of the self-noise obtained with and without filtering of the estimated gradient vectors. In order for this comparison to be meaningful, one must normalize the dc gain of the filter H(z) to unity, as indicated in [13]. This normalization is easily accomplished as shown in [14] by replacing Δ in (35) by $\Delta(1-\beta)$. Thus, the normalized variance is

$$\xi_\Delta^n = \frac{\Delta\xi_{min}}{2} \sum_{k=0}^{N-1} \lambda_k [1 - \frac{\Delta(1-\beta)\lambda_k}{2(1+\beta)}]^{-1} \tag{36}$$

The variance of the self-noise obtained without filtering of the noisy gradient vectors is simply obtained from (35) or (36) by setting $\beta=0$. Thus,

$$\xi_\Delta = \frac{\Delta\xi_{min}}{2} \sum_{k=0}^{N-1} \lambda_k [1 - \frac{\Delta\lambda_k}{2}]^{-1} \tag{37}$$

Comparison of (36) with (37) reveals that each term in the sum of (36) is smaller than the corresponding term in (37). Therefore, filtering the noisy gradients reduces the variance of the self-noise generated in the recursive algorithm.

In many cases of practical interest Δ is chosen so that $\Delta\lambda_k \ll 1$. Under this condition, (36) and (37) reduce to

$$\begin{aligned} \xi_\Delta &\simeq \frac{\Delta\xi_{min}}{2} \sum_{k=0}^{N-1} \lambda_k \\ &\simeq \frac{\Delta\xi_{min}}{2} \text{trace } \underline{\psi} \\ &\simeq \frac{\Delta\xi_{min}}{2} N\,\psi(0) \end{aligned} \tag{38}$$

where $\psi(0) = E|x(k)|^2$ is the element along the diagonal of $\underline{\psi}$. This expression shows that the variance of the self-noise increases linearly with Δ and the number N of taps used in the filter. Consequently, an increase in the filter size, which has the effect of decreasing ξ_{min} tends to increase the self-noise ξ_Δ. As N is increased, a point is reached where the decrease in

ξ_{min} is completely offset by the increase in ξ_Δ. Thereafter, any further increase in N actually leads to a deterioration in the performance of the adaptive filter. Finally, the result in (38) illustrates that, when $\Delta\lambda_k << 1$, there is no advantage to filtering the noisy gradient vectors.

4. SOME APPLICATIONS OF ADAPTIVE FILTERS

Numerous applications of adaptive filters have been described in the literature. Some of the more noteworthy applications include: (1) adaptive antenna systems in which adaptive filters are used for beam steering and for providing nulls in the beam pattern to remove undesired interference [3]; (2) digital communication receivers in which adaptive filters are used to provide equalization of intersymbol interference and for channel identification [4]-[8], [15]-[17]; (3) adaptive noise cancelling techniques in which an adaptive filter utilizes a reference noise waveform to estimate and, thus, eliminate a noise component in some desired signal [9] (several interesting applications of adaptive filters in medicine are discussed in this reference); (4) system modeling in which an adaptive filter is used as a model for an unknown dynamical system [2]; (5) linear filtering and prediction of signals [2].

A detailed survey of all these applications is too ambitious a task. Instead, we shall briefly discuss the application of adaptive filtering in spectral estimation, in noise whitening and in speech signal compression.

Let us assume that we have a random sequence $\{x(n)\}$ which we model as being generated by passing a white noise sequence through an all-pole filter having a transfer function

$$V(z) = \frac{G}{1 - \sum_{k=1}^{p} a_k z^{-k}} \tag{39}$$

where the $\{a_k\}$ and G are the filter parameters. In other words, the sequence x(n) is an autoregressive process of order p:

$$x(i) = a_1 x(i-1) + a_2 x(i-2) + \dots + a_p x(i-p) + Gu(i) \tag{40}$$

where u(i) is a unit variance white noise sequence. The filter parameters are unknown and are to be identified from observation of the sequence $\{x(i)\}$.

This problem is common to several applications. For example, in speech compression based on linear prediction [18], the filter in (39) represents the model for the vocal tract that produces the sampled signal sequence $\{x(i)\}$. Usually, p=10 is adequate for this application. Thus, a segment of about 100 to 200 samples taken at the Nyquist rate is represented by about ten coefficients that determine the poles plus the gain parameter G. (An additional parameter of the speech signal that must be estimated from $\{x(i)\}$ is the pitch period.) It is interesting to note that in this application a short segment (100 to 200 samples) of the sequence $\{x(i)\}$ is used to estimate the matrix $\underline{\psi}$ and an efficient algorithm due to Levinson [19] is commonly employed in inverting $\underline{\psi}$. An alternative appears to be the recursive algorithm described above in which the segment of the sequence $\{x(i)\}$ is recycled several times until the estimates of $\{a_k\}$ converge to within some specified tolerance.

Another application of this signal model is in parametric spectral analysis. Suppose it is desired to obtain the spectral density of $\{x(i)\}$. If $\{x(i)\}$ is well-modeled as an autoregressive process, then the spectral density is simply

$$|V(e^{j\omega})|^2 = V(z)V(z^{-1})\Big|_{z=e^{j\omega}} \tag{41}$$

Thus, estimation of the filter parameters in V(z) leads to an estimate of the spectral density. The advantage of this procedure over the conventional approach based on the computation of the discrete Fourier transform of the sequence $\{x(i)\}$ is increased frequency resolution.

A third application is in adaptive noise whitening. That is, given $\{x(i)\}$ we wish to filter the sequence to produce a white noise output. Obviously, if $\{x(i)\}$ is an autoregressive process the desired filter is an all-zero filter with zeros coinciding with the poles of V(z). Thus, the problem reduces to estimating the parameters $\{a_k\}$ in V(z).

A method for estimating the $\{a_k\}$ parameters in these problems is to pass the observed random sequence through an adaptive all-zero filter with transfer function

$$A(z) = 1 - \sum_{k=1}^{p} \hat{a}_k z^{-k}$$

$$= \sum_{k=0}^{p} c_k z^{-k} \tag{42}$$

where $\{\hat{a}_k\}$ denotes the estimates of $\{a_k\}$, $c_o=1$ and $c_k=-\hat{a}_k$, $1\leq k\leq p$. Thus, A(z) may be viewed as a transversal filter with the coefficient c_o normalized to unity and the remaining coefficients variable. Alternatively, the adaptive filter may be viewed as a single-step linear prediction filter, as illustrated in Figure 3.

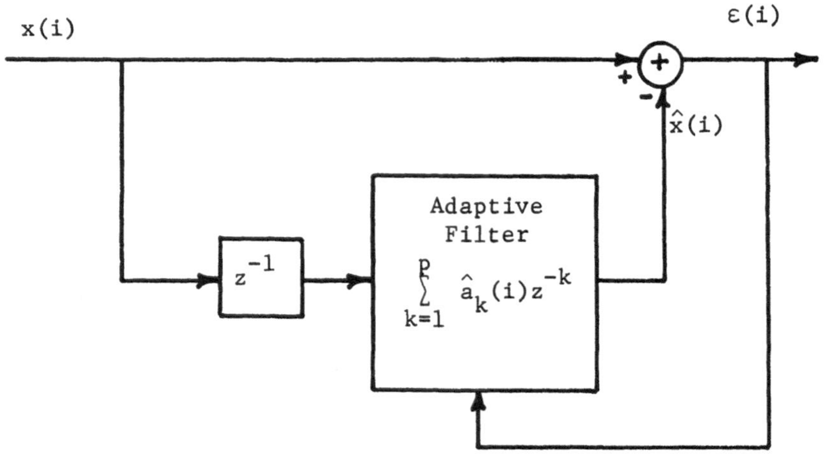

Figure 3: Adaptive Linear Prediction

That is, the filter output sequence is simply the error sequence

$$\varepsilon(i) = x(i) - \hat{x}(i) \tag{43}$$

where $\hat{x}(i)$ denotes the predicted value of x(i), i.e.,

$$\hat{x}(i) = \sum_{k=1}^{p} \hat{a}_k(i)x(i-k) \tag{44}$$

The $\{\hat{a}_k(i)\}$ may be recursively estimated, as for example by means of the (steepest descent) algorithm

$$\underline{\hat{a}}(i+1) = \underline{\hat{a}}(i) + \Delta\varepsilon(i)\underline{X}^*(i) \tag{45}$$

Clearly, if $\{x(i)\}$ is actually an autoregressive process of order p and if $\hat{a}_k=a_k$ for $1\leq k\leq p$, then the sequence $\varepsilon(i)$ is a white noise sequence with variance G^2. Otherwise, $\varepsilon(i)$ is not a white noise sequence. The degree to which $\varepsilon(i)$ resembles a white noise process depends on how well the sequence $\{x(i)\}$ is modeled by an autoregressive process. Similar statements can be made with respect to how well the estimated spectrum resembles the true spectrum of the sequence $\{x(i)\}$ and to how well the speech signal synthesized from the estimates of the filter parameters resembles the original speech signal.

We conclude this discussion of adaptive filtering by noting that the problem of tracking nonstationary signals [20] appears to be a fruitful area for research as well as the possibility of employing more general adaptive filtering algorithms [21].

REFERENCES

1. B. Widrow and M. Hoff, Jr., "Adaptive switching circuits", IRE WESCON Conv. Rec., pt. 4, pp. 96-104, 1960.
2. B. Widrow, "Adaptive filters", in Aspects of Network and System Theory, R. E. Kalman and N. DeClaris, Ed., New York: Holt, Rinehart and Winston, 1971, pp. 563-587.
3. B. Widrow, P. E. Mantey, L. J. Griffiths and B. B. Goode, "Adaptive antenna systems", Proc. IEEE, vol. 55, pp. 2143-2159, December 1967.
4. R. W. Lucky, "Automatic equalization for digital communication", Bell Syst. Tech. J., vol. 44, pp. 547-588, April, 1965.
5. J. G. Proakis and J. H. Miller, "An adaptive receiver for digital signaling through channels with intersymbol interference", IEEE Trans. Inform. Theory, vol. IT-15, pp. 484-497, July 1969.
6. A. Gersho, "Adaptive equalization of highly dispersive channels for data transmission", Bell Syst. Tech. J., vol. 48, pp. 55-70, January 1969.
7. D. A. George, R. R. Bowen, and J. R. Storey, "An adaptive decision-feedback equalizer", IEEE Trans. Commun. Technol., vol. COM-19, pp. 281-293, June 1971.
8. J. G. Proakis, "Advances in equalization for intersymbol interference", in Advances in Communication Systems, vol. 4, (Ed. A. J. Viterbi), New York: Academic Press, 1975, pp. 123-198.
9. B. Widrow, J. R. Glover, Jr., J. M. McCool, J. Kaunitz, C. S. Williams, R. H. Hearn, J. R. Zeidler, E. Dong, Jr., and R. C. Goodlin, "Adaptive Noise Cancelling: Principles and Applications", Proc. IEEE, vol. 63, pp. 1692-1716, December 1975.
10. J. G. Proakis, "Adaptive digital filters for equalization of telephone channels", IEEE Trans. Audio and Electroacoustics,

vol. AU-18, pp. 195-200, June 1970.

11. F. S. Beckman, "The solution of linear equations by the conjugate gradient method", in Mathematical Methods for Digital Computers, A. Ralston and H. S. Wilf, Eds., New York: Wiley, 1960.
12. R. Fletcher and M. J. D. Powell, "A rapidly convergent descent method for minimization", Comput. J., vol. 6, pp. 163-168, 1963.
13. J. R. Glover, Jr., "Comments on channel identification for high speed digital communications", IEEE Trans. Automat. Contr., vol. AC-20, p. 823, December 1975.
14. J. G. Proakis, "Authors Reply", IEEE Trans. Automat. Contr., vol. AC-20, pp. 823-824, December 1975.
15. F. R. Magee, Jr., and J. G. Proakis, "Adaptive maximum-likelihood sequence estimation for digital signaling in the presence of intersymbol interference", IEEE Trans. Inform. Theory (Corresp.), vol. IT-19, pp. 120-124, January 1973.
16. R. W. Lucky, J. Salz, and E. J. Weldon, Jr., Principles of Data Communication, New York: McGraw-Hill, 1968.
17. H. E. Nichols, A. A. Giordano and J. G. Proakis, "MLD and mse algorithms for adaptive detection of digital signals in the presence of interchannel interference", IEEE Trans. Inform. Theory, vol. IT-23, pp. 563-575, September 1977.
18. J. Makhoul, "Linear prediction: A tutorial review", Proc. IEEE, vol. 63, pp. 561-580, April 1975.
19. N. Levinson, "The Wiener RMS error criterion in filter design and prediction", J. Math. Phys., vol. 25, pp. 261-278, 1947.
20. B. Widrow, J. M. McCool, M. G. Larimore and C. R. Johnson, Jr., "Stationary and nonstationary learning characteristics of the LMS adaptive filter", Proc. IEEE, vol. 64, pp. 1151-1162, August 1976.
21. B. Picinbono, "Adaptive signal processing for detection and communication", in Proceedings of NATO Advanced Study Institute on Communication Systems and Random Process Theory (this volume), Darlington, England, August 8-20, 1977.

Panel Discussion

on

ADAPTIVE FILTERING FOR HIGH-SPEED MODEMS *

The Chairman, Professor F.L. Stumpers (Philips Laboratories, Eindhoven, Netherlands), introduced the other panel members, Professor L.E. Franks (University of Massachusetts, Amherst, U.S.A.) and Professor J.G. Proakis (North Eastern University, Boston, U.S.A.).

L.E. Franks: The problem I am going to describe is quite old but has had little recent discussion. It concerns the structure of adjustable filters. Is the tapped delay line, which was described by Professor Proakis in his lecture, the only or the best structure? Are there other structures with fewer adjustable parameters which could be used?

The structures I shall consider are:

(a) a parallel filter, realising a transfer function of the form $\sum_i a_i \phi_i$ where the a_i are adjustable gains and the ϕ_i are linear time-invariant filter functions

(b) the transversal filter, a special case in which $\phi_m = \gamma^m$; here γ may be a simple delay as in the usual tapped delay line equaliser, or a first-order allpass network as in the Laguerre-function equalisers of Wiener and Lee (circa 1931). Very little has been done with more general transfer functions.

---

* This text was prepared by Professor K.W. Cattermole, University of Essex, Colchester, U.K.

(c) the cascade filter, realising a transfer function which is the product of terms due to individual sections. The logarithm of the transfer function is the sum of the logarithmic terms, and this is a natural structure for compensation of gain and phase. Bode devised adjustable sections, in the form of a bridged-T with two ganged resistors, which had constant impedance but adjustable gain in a specific fairly narrow frequency band: these were called 'bump equalisers', and could be designed to give positive or negative bumps with assigned centre frequency and bandwidth. Other cascade structures used elements realising a family of polynomials. Up to 5 or 6 sections were commonly used in carrier system equalisers.

Carrier systems on open-wire or coaxial cable in the beginning used polynomial or bump sections, not in adaptive, but in manually adjustable, equalisers; these are probably the oldest equaliser structures. The second-generation coaxial systems used 'cosine equalisers' which were in fact transversal structures; later systems reverted to bumps and polynomials. There seems to be no settled opinion on the 'best' form of equaliser. Clearly it will depend to some extent on the type of channel; for instance, bumps are good at correcting irregularities near the band edge, while polynomials are more suitable for correcting the broad slope changes which are often due to temperature sensitivity, and cosine equalisers have yet other roles.

Now let me turn to the problem of the adjustment algorithm. One can adopt a minimum mean-squared error criterion, and minimise either total error in continuous time, as in Wiener filters, or the error at uniformly spaced points in time, which is more appropriate for digital systems. To adjust the gain coefficients a, one estimates the gradients from the cross-correlation of the error with a set of suitable derived signals. Now an important property of the structure is: how easy is it to derive these auxiliary signals? In the parallel structure, they are simply the outputs of the several parallel filters. In the cascade structure, they are obtained by passing the output of the equaliser through a further bank of filters whose transfer functions are the logarithms of the respective section characteristics, which for polynomial sections are quite simple in form.

Let us consider the overall problem. The signal x is fed into a channel which is characterised by its dispersion and by an additive noise process. It is followed by an equaliser. The output $\tilde{x}$ of the equaliser must be, in some sense, close to the signal x. If the dispersion were fixed, it could be permanently equalised; but it is random, not in the sense of time-varying but in being chosen randomly within a sample space ω. It can be represented by a frequency function $G(f,\omega)$. The equaliser, having a frequency function H dependent on parameters $a_1 \ldots a_n$ is in $L^2 \times R^n$. It is effectively confined to some predetermined bandwidth B.

To design an optimum parallel equaliser, we must consider two minimisations:

(i) for a given channel, the mean squared error must be minimised by choice of the a_i

(ii) the ϕ_i must be chosen so that, over the ensemble of channel dispersions, the mean value of this minimum is minimised. This decomposes the problem into the adjustment algorithm and the structure of the filters. The adjustment of the a_i amounts to minimising a distance, expressed as a weighted norm: the expressions depend on the particular channel realisation. This topic was covered in Professor Proakis' lecture. The second minimisation problem, which requires us to form the expectation over the sample space ω and minimise over the ϕ_i, is much more difficult to formulate except under a few special circumstances. The topic has been studied by Mauren and myself at Bell Laboratories, by P. McLane at Queens, by Dr. W. Steenart (University of Ottawa, Canada) and Dr. W.A. Gardner (University of California, Davis, U.S.A.) who are here today. I will briefly describe the results so far.

One special case has deterministic gain but random phase. This can be solved by means of an integral equation; the solution resembles a Karhunen-Loève expansion, which is intuitively satisfying. An even more specific case is random delay with constant gain. With delay spread τ, the solution depends on the product $B\tau$; circumstances can be defined in which the tapped-delay-line filter is optimal.

The wider problem remains largely unsolved; but I think it is an important one, and would like to have your reactions to it.

Discussion elicited little further on the problem of choosing a filter structure, which does not seem to have been studied so intensively as the adjustment algorithm for a given structure. Professor Franks agreed with a questioner that the important property was good performance with a given, and not too large, number of adjustable elements; he suggested that the Karhunen-Loève expansion, which minimises mean squared error for a given number of basis functions, is useful. Professor Proakis pointed out that one of the structures described in his lecture used non-overlapping resonators to generate basis functions; his treatment was equivalent to optimising a bump equaliser.

F.L. Stumpers: My colleagues at Philips Laboratories have used a filter structure which is dual to the usual one. An elementary filter is designed to have a 'one' and a series of zeros in the frequency domain. Given a design for a single filter of this type, an array may be derived by modulation processes. Multiplication by cos $2\pi f_r t$ or sin $2\pi f_r t$ translates the frequency response by f_r. An array of elements translated by frequencies f_r can be combined with weights a_r. We can construct a pair of filter arrays designated as 'sine' and 'cosine' filters; this combination can do anything which could be done with tapped-delay-line filters. The filters can be incorporated in an adaptive receiver which compensate for multipath and other transmission distortions. The apparatus can be built from filters and modulators as I have stated, but it can also be realised by means of a microprocessor. *

When two signals are modulated on to quadrature carriers, four equalisers are needed: one to correct each of the desired signal paths, and one to annul each of the crosstalk paths. This is another possibility for the adaptive filter. Such a structure has been made in the MBLE Laboratories in Brussels, and works well.

J.G. Proakis: I will try to summarise some recent work on high-speed adaptive modems in the U.S.. My impression is that the problem of equalising the voice-band channel is now regarded as solved, for rates up to 9600 bit/s on a telephone line. The channel is well-behaved in that it has no spectral nulls; linear equalisers can deal with intersymbol interference quite well, and are commonly used.

There is a lot of interest, especially military, in equalising the tropo-scatter channel and the h.f. channel. The Sylvania-Signatron team has developed an equaliser for the tropo-scatter channel using decision-feedback techniques which can deal with spectral nulls; this apparently works well. Raytheon, with a

---

* There are recent papers by P.J. van Gerwen, N.A.M. Verhoeckx, H.A. van Essen and F.A.M. Snijders, IEEE Transactions on Communications, COM-25, 238-250 (February 1977): and by F. de Jager and M. Christiaens, Philips Technical Review, vol. 37, 1977, No. 1, pp.10-24.

different technical approach which I do not recall in detail, have gained a major contract. The C.N.R. company, run by P. Bello and D. Chase who have done a lot of related work, has built an equipment using a Viterbi type decoder to deal with intersymbol interference, also with decoding for a convolutional code of rate one-half; but I have not seen the results of any tests over a tropo channel.

The problem of the h.f. channel is, I think, more difficult; it varies faster than the tropo channel, and the bandwidth is less, so the number of bits between major changes of channel properties is much smaller.

Discussion touched on the problem of telephone lines connected via acoustic couplers; bandwidth limitation and non-linearity reduce the achievable rate below those normal on good voice-band lines. Mr. P. Cochrane (Post Office Research Centre, Ipswich, U.K.) pointed out that a British tropo-scatter link combats fading and interference problems by using a return signal to control transmitter power; could the equalisation problem be eased at the receiver by similar control of the transmitted spectrum? Professor Proakis was aware of schemes for adapting the transmitted data rate but not of other adaptations. Professor Franks quoted as another application of decision-directed equalisation the work of Bell Northern Research on 'digit-under-voice' transmission over microwave links. Dr. P.T. Nielson (SHAPE Technical Centre, The Hague, Netherlands) pointed out the timing problems which could arise with decision-directed feedback. If a conventional timing circuit is placed after a decision-feedback equaliser, there may be enough interaction to make its design and operation rather difficult. The Sylvania tropo-scatter modem avoided this by a special timing extractor, controlled by a comparison of the average weighting on the first and last tape on the forward part of the equaliser which, as Professor Proakis pointed out, used analogue circuits with half-bit-interval spacing.

Dr.E. Olcayto (University of Strathclyde, Glasgow, Scotland) referred to experiments on h.f. channel equalisation. Their adaptive equaliser had used decision feedback. Data transmitted at 2.4 kbit/s over a 700 km. route in the U.K. had been recorded and subsequently processed by computer to implement the desired algorithm. The results were, on the whole, not very promising. The average error rate was reduced, but on several episodes with poor error rate the processing had little effect and a few were actually made worse. He attributed the difficulties to the rapidity of changes, as remarked upon by Professor Proakis. The multipath structure was very difficult; a reversal of 'stronger' and 'weaker' roles between two paths was not uncommon, and this causes obvious synchronising problems. Professor Proakis thought it likely that failure of synchronism accounted for the poorer episodes. In general, the decision-feedback equaliser is capable of using the inherent diversity of resolvable multipath propagation, almost to the extent of the Viterbi algorithm; on h.f. paths the dispersion is sufficiently large that the Viterbi scheme is not very practical, whereas the decision-feedback method is.

There was a short further discussion on the cost of adaptive modems, which for tropo-scatter channels of high rate (10 Mbit/s and upwards) and good performance remains very expensive; then the meeting closed at the appointed hour.

Part IX

STOCHASTIC CALCULUS

organised by Dr. M. H. A. Davis

MARTINGALE INTEGRALS AND STOCHASTIC CALCULUS

M.H.A. Davis

Department of Computing and Control,
Imperial College, London SW7 2BZ

INTRODUCTION

Brownian motion has played a major role in the mathematical formulation of many problems in filtering, communication and stochastic control theory. This is partly due to its immediate use as a representation for "white noise", but mainly due to the availability of the powerful Ito stochastic calculus which enables explicit answers to be given to questions which, if not formulated in terms of Brownian motion, would be quite intractable.

Within the last ten years or so the scope of stochastic calculus has been broadened to encompass general martingales, and the purpose of these notes is to outline some of these developments. Martingales occur naturally in almost any information processing problem involving sequential acquisition of data: for example the sequence of estimates of a random variable based on increasing observation record, and the sequence of likelihood ratios in a sequential hypothesis test, are martingales.

As is well known, the characteristic features of Ito calculus arise from the fact that sample paths of Brownian motion have unbounded variation, so that terms in a Taylor series expansion which would ordinarily be of second order get promoted to first order. It turns out that similar properties occur with any continuous martingale because any such process must, like Brownian motion, have non-zero quadratic variation. This fact is a consequence of perhaps the central result of modern martingale theory, the Doob-Meyer decomposition of supermartingales. With the quadratic variation properties in hand, stochastic integrals and stochastic calculus can be constructed for continuous martingales in much the same way as for Brownian motion. The following notes give an outline of these results. For a full account the reader is referred to Kunita and Watanabe [5], where the key ideas were originally introduced, or to Meyer's Course on Stochastic Integrals [9].

Terminology, etc., for stochastic processes

Recall that a stochastic process is a family of random variables $\{x_t, t \geq 0\}$ all defined on a fixed probability space (Ω, F, P). Associated with such a process is an increasing family of σ-fields $X_t = \sigma\{x_s, s \leq t\}$, i.e. X_t is the smallest σ-field of subsets of Ω containing all events of the form $\{\omega : x_s(\omega) < a\}$ for $a \in R$ and $s \leq t$. Often it is convenient to consider bigger σ-fields than X_t, for example $F_t = \sigma\{x_s, y_s, s \leq t\}$ where $\{y_t\}$ is another process. In this case $\{x_t\}$ does not generate $\{F_t\}$, but x_t is F_t-measurable for each t, and we say $\{x_t\}$ is *adapted* to F_t. Martingale properties are defined for a process $\{x_t\}$ relative to a given fixed family of σ-fields F_t. (x_t, F_t) is a martingale if x_t is adapted to F_t, $E|x_t| < \infty$, and $E[x_t|F_s] = x_s$ for $t \geq s$. (x_t, F_t) is a supermartingale if $E[x_t|F_s] \leq x_s$ and a submartingale if $E[x_t|F_s] \geq x_s$. Super- and sub-martingales can be visualized as a player's fortune in a game where the odds are unfavourable and favourable, respectively.

BROWNIAN MOTION AND ITO CALCULUS

Brownian motion

A *Brownian motion* $\{b_t, t \geq 0\}$ is a stochastic process with stationary, normal, independent increments; it is *standard* if $b_0 = 0$ with probability 1 and $\text{var}(b_1) = 1$. Thus a standard BM is defined by

(i) $(b_t - b_s)$ and $b_{t'} - b_{s'})$ are independent for non-overlapping $(s,t), (s',t')$

(ii) $(b_t - b_s) \sim N(0, t-s)$; $b_0 = 0$ a.s.

It is easy to show that the covariance function of $\{b_t\}$ is $r(t,s) = t \wedge s$ $(= \min(t,s))$ and in particular $\text{var}(b_t) = t$. There is a simple construction for BM: take a sequence $X_1, X_2 ..$ of independent $N(0,1)$ random variables and an orthonormal basis $\{\phi_i\}$ for $L_2[0,1]$ and for $t \in [0,1]$ define

$$b_t^n = \sum_{i=1}^{n} X_i \int_0^t \phi_i(s)\,ds. \tag{1}$$

Then it is easily shown, using the Parseval equality, that b_t^n is a Cauchy sequence in quadratic mean (i.e. $E(b_t^n - b_t^m)^2 \to 0$ as $n, m \to \infty$) and the limit process b_t has the right covariance function, $t \wedge s$. What is also true but more difficult to show (see [1],[7]) is that

b_t^n converges uniformly in $t \in [0,1]$ to b_t, a.s. (= almost surely, i.e. with probability 1). This means that $\{b_t\}$ has continuous sample paths (except for a set of measure zero, which can be discarded), which is the key property both in the mathematical development of stochastic calculus and in the use of Brownian motion as a model for "noise". Some of the difficulties inherent in the idea of continuous-time white noise can be seen immediately from (1). Recall that, formally, white noise is a normal process $\{\xi_t\}$ with covariance function $r(t,s) = \delta(t-s)$. If we take the indefinite integral $w_t = \int_0^t \xi_u du$ then for $s < t$

$$\operatorname{cov}(w_t, w_s) = E\int_0^t \xi_u du \int_0^s \xi_v dv$$

$$= \int_0^s \int_0^t \delta(u-v)\, du dv = s.$$

Thus $\operatorname{cov}(w_t, w_s) = t \wedge s$, so that w_t is a Brownian motion, i.e. ξ_t can be regarded as the derivative of BM, dw_t/dt. Now the approximating processes b_t^n in (1) are differentiable and

$$\dot{b}_t^n = \sum_{i=1}^n X_i \phi_i(t),$$

but it is easy to see that $\dot{b}_t^n$ will not in general converge in any sense as $n \to \infty$. In fact b_t is differentiable nowhere, and white noise must be interpreted either, as here, in terms of its integral or else in terms of random generalized functions [4].

The feature of BM underlying the Ito calculus is its quadratic variation property. For $t \in [0,1]$ fix an integer n and define

$$Q_t^n = \sum_{0 \le k < 2^n t} (b((k+1)/2^n) - b(k/2^n))^2$$

and let $t_n = \max\{k/2^n : k/2^n \le t\}$. Denote $Y_k = b((k+1)/2^n) - b(k/2^n)$; then the Y_k are independent and $N(0, 1/2^n)$-distributed, so that $EY_k^2 = 1/2^n$, $EY_k^4 = 3/2^{2n}$. Thus

$$EQ_t^n = \sum_k EY_k^2 = t_n$$

and $$\operatorname{var}(Q_t^n) = E(Q_t^n - t_n)^2 = 2^n t_n (3/2^{2n} - 1/2^{2n})$$

$$= 2t_n/2^n.$$

Applying the Chebyshev inequality

$$\sum_n P[|Q_t^n - t_n| > \varepsilon] < \frac{1}{\varepsilon^2}\sum_n \operatorname{var}(Q_t^n) \le \frac{1}{\varepsilon^2} t$$

so that according to the Borel-Cantelli lemma $|Q_t^n - t_n| < \varepsilon$ eventually, a.s., i.e. $Q_t^n \to t$ a.s. This result means that the sample paths of $\{b_t\}$ are of unbounded variation (have infinite length), since

$$Q_t^n = \sum_k Y_k^2 \le \max_k |Y_k| \cdot \sum_k |Y_k| .$$

Let $m_n = \max_k |Y_k|$. Then $m_n \to 0$ as $n \to \infty$, a.s., since the paths of b_t are continuous. But $\sum_k |Y_k| \ge Q_t^n/m_n$ and hence the total variation is unbounded.

Martingale properties and Ito calculus

Let $\mathcal{B}_t = \sigma\{b_s, s \le t\}$. Then $(b_t, \mathcal{B}_t)$ is a martingale since, for $t > s$, $b_t - b_s$ is independent of $\mathcal{B}_s$, so that

$$E[b_t | \mathcal{B}_s] = E[b_s + (b_t - b_s) | \mathcal{B}_s] = b_s. \tag{2}$$

Also

$$E[b_t^2 - b_s^2 | \mathcal{B}_s] = E[(b_t - b_s)^2 | \mathcal{B}_s] = t - s$$

(using (2) and the independent increments property). This shows that b_t^2 is a submartingale and $b_t^2 - t$ is a martingale. It will be seen below that this latter property characterizes Brownian motion.

Suppose F_t is an increasing family of σ-fields such that $\mathcal{B}_t \subset F_t$ and $(b_{t'} - b_t)$ is independent of F_t for $t' > t$. Let $\{\phi_t\}$ be a process adapted to F_t and satisfying

$$E\int_0^t \phi_s^2 ds < \infty, \qquad t > 0. \tag{3}$$

Then the Ito integral

$$X_t = \int_0^t \phi_s db_s$$

is defined as the quadratic mean limit of a sequence of approximating sums of the form $\Sigma\, c_i(b_{t_{i+1}} - b_{t_i})$. (Note that, even when ϕ_t is deterministic, X_t cannot be defined as an ordinary Stieltjes integral since the sample paths of b_t do not have bounded variation.) X_t is a martingale of F_t and has a version with continuous sample paths. Now suppose ϕ_t satisfies only the weaker condition

$$\int_0^t \phi_s^2 ds < \infty, \quad t > 0 \quad \text{a.s.} \tag{4}$$

Define the following sequence of stopping times

$$T_n = \inf\{t : \int_0^t \phi_s^2 ds > n\}.$$

Then $T_n \uparrow \infty$ a.s. (this follows directly from (4)) and $\phi_t^n = \phi_t I_{(t \le T_n)}$ is an adapted process satisfying (3), so that the process

$$X_t^n = \int_0^t \phi_s^n db_s$$

is well defined.

It is possible to show that X_t^n converges in probability to a random variable X_t which we denote $\int_0^t \phi_s db_s$, and clearly $X_t^n = X_{t\wedge T_n}$. This shows that X_t can be taken to have continuous sample paths, but the *martingale property is lost*. A *local martingale* is a process M_t such that $M_{t\wedge S_n}$ is a uniformly integrable martingale for some sequence of stopping times $S_n \uparrow \infty$ a.s. Thus the Ito integral of a process satisfying (4) is in general a local martingale and (3) is a sufficient but not necessary condition for the integral to be a martingale.

Suppose α_t is an adapted process with integrable sample functions, ϕ_t satisfies (3), and Y_0 is an F_0-measurable random variable. Define

$$Y_t = Y_0 + \int_0^t \alpha_s ds + \int_0^t \phi_s db_s .$$

This is an example of a *local semimartingale* which in general is a process which is the sum of a process of bounded variation and a local martingale. If $F: R \to R$ is a C^2 function with first and second derivatives F', F'' then the *Ito differential rule* states that

$$F(Y_t) = F(Y_0) + \int_0^t F'(Y_s) dY_s + \tfrac{1}{2}\int_0^t F''(Y_s)\phi_s^2 ds$$

$$= F(Y_0) + \int_0^t \{F'(Y_s)\alpha_s + \tfrac{1}{2}F''(Y_s)\phi_s^2\} ds + \int_0^t F'(Y_s)\phi_s db_s \quad (5)$$

Thus $F(Y_t)$ is also a local semimartingale. Note that even if ϕ_s satisfies (3) there is no guarantee that $F'(Y_s)\phi_s$ will, though it satisfies (4) since $F'(Y_t)$ is continuous in t. This indicates that local martingales arise naturally in the context of Ito stochastic calculus.

The appearance of the "second-order" term in the formula (5) is connected with the quadratic variation property of the Brownian path. This can be seen quite explicitly by calculating from scratch the simplest stochastic integral, $\int_0^t b_s db_s$. Since the integrand b_s is continuous, step function approximations to it are obtained simply by sampling, and we get

$$\int_0^t b_s db_s = \lim_{n\to\infty} \sum_k b(k/2^n)[b((k+1)/2^n) - b(k/2^n)].$$

But, denoting $b_k = b(k/2^n)$, we have

$$2b_k(b_{k+1} - b_k) = b_{k+1}^2 - b_k^2 - (b_{k+1} - b_k)^2$$

so that

$$\int_0^t b_s db_s = \lim_{n\to\infty} \tfrac{1}{2}(b_{t_n}^2 - Q_t^n) = \tfrac{1}{2}(b_t^2 - t)$$

which we can write in the form

$$d(b_t^2) = 2b_t db_t + dt,$$

exemplifying (5).

The programme now is to extend the scope of the stochastic calculus by considering integrals with respect to general continuous martingales, not just Brownian motion. In view of the above, it seems that the key factor in this will be the quadratic variation properties of continuous martingales. To study this we first have to introduce the ideas of supermartingale decomposition.

DECOMPOSITION OF SUPERMARTINGALES

Discrete-time case

Suppose $\{F_k, k=1,2..\}$ is an increasing family of σ-fields and $\{X_k\}$ a process adapted to F_k such that $E|X_k| < \infty$ for all k. Define $D_k = X_k - X_{k-1}$ and write

$$X_k = \{X_0 + \sum_{i=1}^{k} (D_i - E[D_i|F_{i-1}])\} + \sum_{i=1}^{k} E[D_i|F_{i-1}]$$

$$= M_k - A_k.$$

Then M_k is a *martingale* and A_k is *adapted to* F_{k-1} with $A_0 = 0$. Suppose $X_k = M_k' - A_k'$ is another similar decomposition of X_k, and let

$$Z_k = M_k' - M_k = A_k' - A_k.$$

Then $Z_k = M_k' - M_k$ is a martingale, so that $E[Z_{k+1}|F_k] = Z_k$, and $Z_k = A_k' - A_k$ is adapted to F_{k-1}, so that $E[Z_{k+1}|F_k] = Z_{k+1}$. Thus $Z_k = Z_0 + A_0' - A_0 = 0$ and the decomposition is unique. If X_k is a supermartingale, i.e. $X_k \geq E[X_{k+1}|F_k]$ then $E[D_i|F_{i-1}] \leq 0$ and A_k is an *increasing process*. This is the *Doob decomposition* of discrete-time supermartingales.

Now suppose X_k is a positive supermartingale; then since $A_k = M_k - X_k$ we have $EA_k \leq EM_k = EM_0$, so that $A_k \uparrow A_\infty$ where A_∞ is a random variable with $EA_\infty < \infty$. If $EX_k \to 0$ then X_k is called a *potential*. In this case it is easily seen that M_k is uniformly integrable (since both X_k and A_k are) and from the martingale convergence theorem [8, VI T6], $M_k \to M_\infty$ and $M_k = E[M_\infty|F_k]$. But $X_k \to 0$, so that $M_\infty = A_\infty$. This shows that every potential X_k is of the form

$$X_k = E[A_\infty|F_k] - A_k$$

for some increasing process A_k which is adapted to F_{k-1}. A_k is constructed from X_k using the facts that $A_0 = 0$ and $A_k - A_{k-1} = E[X_{k-1} - X_k | F_{k-1}]$.

Continuous-time case

In continuous time it is clear that there can be no *unique* decomposition of a supermartingale into the difference of a martingale and an increasing process. Take for example a standard Poisson process $N_t = \sum_i I_{(t \geq T_i)}$ where $T_0 = 0$ and $S_i = T_i - T_{i-1}$ are independent exponentially distributed (with parameter 1) random variables. Since N_t is an increasing process, $-N_t$ is automatically a supermartingale, so one possible decomposition is: $-N_t = 0 - N_t$. But N_t is also an independent increments process, so that $N_t - EN_t = N_t - t$ is a martingale, and another decomposition is therefore $-N_t = (t-N_t) - t$. To pick out a unique decomposition we have to formulate the continuous-time analogue of the discrete-time property that A_k is F_{k-1}-measurable, i.e. A_k is known a little bit "in advance". The appropriate concept turns out to be *predictability.* The reader is referred to [3] for a full discussion of this, but a sketch is as follows.

Recall that a process $\{X_t, t \geq 0\}$ defined on a probability space (Ω, F, P) can be thought of as a function $X: \Omega \times R^+ \to R$. $\{X_t\}$ is a *measurable* process if this function is measurable with respect to the product σ-field $F*S$ in $\Omega \times R^+$ (S = Borel sets of R^+) and this is necessary if we want to define sample function integrals such as $\int_0^t X_s(\omega)ds$. However in considering stochastic integrals we are often specifically interested in processes which are adapted to some fixed increasing family F_t of sub-σ-fields of F, and it is then appropriate to consider other σ-fields in $\Omega \times R^+$ which are connected with adapted processes. Let SP_r denote the family of all processes (i.e. real-valued functions on $\Omega \times R^+$) which are adapted to F_t and have right-continuous sample paths, i.e. for each (t,ω), $X_{t+}(\omega) = X_t(\omega)$ where $X_{t+} = \lim_{s \downarrow t} X_s$. The *optional* σ-field O in $\Omega \times R^+$ is the smallest σ-field with respect to which all $X \in SP_r$ are measurable. Similarly P, the *predictable* σ-field in $\Omega \times R^+$ is the σ-field generated by SP_ℓ, the *left*-continuous adapted processes. Note that $P \subset O$; indeed if $X \in SP_\ell$ and $Y_t := X_{t+}$ then $X_t^\delta := Y_{t-\delta}$ is in SP_r and $X_t^\delta \to X_t$ as $\delta \downarrow 0$, so that left-continuous processes can be expressed as limits of right continuous ones. This does not work the other way round since $X_{t+\delta}$ may not be adapted to F_t for $\delta > 0$.

Another way in which the predictable σ-field can be generated is the following. Let H be the class of subsets of $\Omega \times R^+$ consisting of finite unions of rectangles of the form $A \times]u,v]$ where $A \in F_u$, or $A \times \{0\}$ where $A \in F_0$. Note that each $H \in H$ corresponds to a left-continuous process $I_H(\omega,t) = 1$ if $(\omega,t) \in H$, $= 0$ otherwise. Thus $\sigma(H) \subset P$ and in fact $\sigma(H) = P$, since if $X \in SP_\ell$ then $X_t = \lim_{n\to\infty} X_t^n$

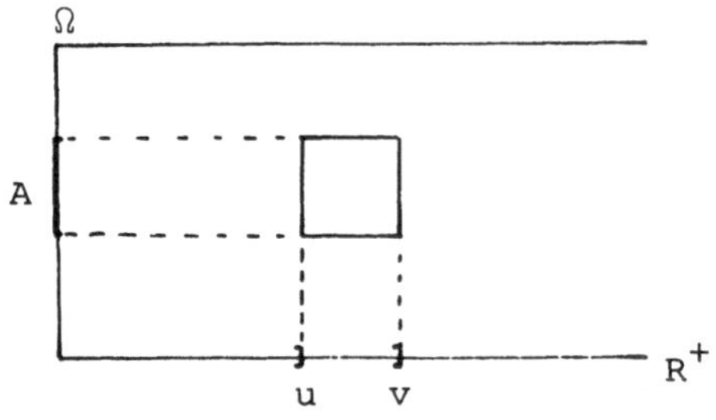

Fig. 1. Generating set for $\mathcal{H}$. $A \in \mathcal{F}_u$

where

$$X^n(\omega,t) = \begin{cases} X_{k/2^n}(\omega)\, I_{]k/2^n,(k+1)/2^n]}(t) & , k \le 2^{2^n} \\ 0 & k > 2^{2^n}. \end{cases}$$

Now $X_{k/2^n}$ can be expressed as a limit of $\mathcal{F}_{k/2^n}$-measurable step functions, and hence X_t can be approximated by sums of indicator functions of sets in $\mathcal{H}$.

There is an important tie-up between the above ideas and stopping times. Recall that a stopping time of $\mathcal{F}_t$ is a positive real valued random variable T such that $\{\omega : T(\omega) \le t\} \in \mathcal{F}_t$ for each t. If S and T are two stopping times such that $S \le T$ a.s., the *stochastic interval* $[S,T]$ is the subset $\{(\omega,t) : S(\omega) \le t \le T(\omega)\}$ of $\Omega \times R^+$ (with analogous definitions for $]S,T[$, etc.). In particular, the *graph* of T is $[T] := [T,T]$. T is *predictable* if there is an increasing sequence of stopping times T_n which "announces the arrival of T", i.e. such that $T_n \le T$, $T_n(\omega) < T(\omega)$ if $T(\omega) > 0$ and $T_n \uparrow T$ a.s. The main result is the following: *A positive random variable* T *is a stopping time if and only if its graph* [T] *is an optional set and* T *is predictable iff* [T] *is a predictable set.* It is easy to see that the graph of a predictable stopping time is predictable: if T_n announces T then $[T,\infty[\; = \bigcap_n]T_n,\infty[$ and $]T_n,\infty[$ is predictable since $I_{]T_n,\infty[}$ is left-continuous. But then $[T] = [T,\infty[\setminus]T,\infty[$ is predictable. However, to produce an announcing sequence T_n, given that T has a predictable graph, is considerably more tricky.

As an example of a predictable stopping time, consider a standard Brownian motion b_t and let $\mathcal{F}_t = \sigma\{b_s, s \le t\}$. Then $S_a = \inf\{t : |b_t| \ge a\}$ is predictable since it is obviously announced by $S_{a-1/n}$. For a time which is *not* predictable, consider the Poisson process N_t. If T_1 is the first jump time of N_t then it is fairly easy to see that any stopping time T of $\mathcal{F}_t = \sigma\{N_s, s \le t\}$ which is $\le T_1$ must be of the form $T = t_0 \wedge T_1$ for some constant t_0 (since "nothing has happened" before T_1). Thus the only stopping time strictly less than T_1 is 0, so that T_1 cannot be announced.

Returning now to supermartingale decomposition, suppose X_t is a supermartingale with respect to a given family F_t of sub-σ-fields of F on a probability space (Ω, F, P). The fundamental result of Meyer states that: X_t *has a unique decomposion,* $X_t = M_t - A_t$, *where* M_t *is a martingale and* A_t *a predictable increasing process, if and only if* X_t *belongs to class (DL).* The latter condition, a technical one almost always satisfied in applications, requires that the family of random variables $\{X_{S\wedge a} : S \text{ is a stopping time of } F_t\}$ be uniformly integrable (u.i.) for each $a > 0$.

There are several different proofs of the Meyer decomposition, the most elementary being due to Murali Rao [10] (this proof is also given in §3.3 of [6]) who constructs the increasing process as a limit of discrete-time approximations. The idea is as follows. First the general case is easily reduced (see [8, VII T31]) to the case where X_t is u.i. Then the martingale convergence theorem asserts that $X_t \to X_\infty$ a.s. and in L_1 and that $X_t \geq E[X_\infty | F_t]$. We can then write

$$X_t = E[X_\infty | F_t] + (X_t - E[X_\infty | F_t]).$$

The first term on the right is a martingale and the second a potential (a positive supermartingale whose expectation coverges to 0). Thus it suffices to consider decomposion of potentials. If X_t is a potential then for each n the discrete-time process $\{X_{k/2^n}\ k = 0,1,2..\}$ is a potential with respect to the σ-fields $F_{k/2^n}$. Applying the discrete-time decomposition, we see that there is an F_∞-measurable random variable A^n such that

$$X_{k/2^n} = E[A^n_\infty | F_{k/2^n}] - A^n_{k/2^n}.$$

It follows from the uniform integrability of X_t that the random variables $\{A^n_\infty, n = 1,2..\}$ are u.i. Now u.i. sets of random variables are weakly compact [8, II T23], so there exists a subsequence, also denoted A^n_∞ and a random variable A_∞ such that $A^n_\infty \overset{w}{\to} A_\infty$ (i.e. $E\theta A^n_\infty \to E\theta A_\infty$ for any bounded r.v. θ). Define $M_t = E[A_\infty | F_t]$. From the discrete-time decomposition we know that for dyadic rational t,s, with $s \leq t$,

$$E[A^n_\infty | F_s] - X_s \leq E[A^n_\infty | F_t] - X_t$$

and hence, taking the limit, $M_s - X_s \leq M_t - X_t$. Thus $A_s := M_s - X_s$ is an increasing process and it can be shown that A_t is predictable (roughly, this is because $A_t = \lim A^n_{t_n}$ where $t_n = \max\{k/2^n : k/2^n \leq t\}$). This expresses X_t, as required, in the form $M_t - A_t$ where A_t is a predictable increasing process.

STOCHASTIC INTEGRALS FOR CONTINUOUS MARTINGALES

Throughout this section F_t will denote a fixed increasing family of σ-fields, and X_t a martingale of F_t which has continuous sample paths. Two observations are immediate: (i) X_t is *locally bounded,* i.e there is a sequence T_n of stopping times such that $|X_{t\wedge T_n}| \le n$ and $T_n \uparrow \infty$ a.s. (take $T_n = \inf\{t : |X_t| \ge n\}$; this works for any continuous process); (ii) X_t cannot have bounded variation sample paths, because if it did then we could write $X_t = A_t^1 - A_t^2$, where A_t^1, A_t^2 are continuous (and *a fortiori* predictable) increasing processes. But then $A_t^1 = 0 + A_t^1 = X_t + A_t^2$ are two alternative decompositions of the submartingale A_t^1 (any increasing process is automatically a submartingale) and this contradicts the uniqueness of the Meyer decomposition.

Since X_t is a martingale, $E[X_t^2 - X_s^2|F_s] = E[(X_t - X_s)^2|F_s] \ge 0$ and this shows that X_t^2 is a submartingale. Invoking the Meyer decomposition we can write

$$X_t^2 = M_t + \langle X\rangle_t$$

where M_t is a martingale and $\langle X\rangle_t$ a continuous increasing process (the reason for the curious notation will appear). We have already noted that for a BM b_t, $b_t^2 - t$ is a martingale, so that $\langle b\rangle_t = t$. Now t is the quadratic variation of the Brownian path, and it turns out this property holds in general, i.e. $\langle X\rangle_t$ *is the quadratic variation of the martingale* X_t. To show this, first observe that it is no restriction to assume that X_t and $\langle X\rangle_t$ are bounded, since they are always locally bounded, as indicated above. We want to show that there is a sequence of partitions $0 = t_0^n < t_1^n .. < t_{k(n)}^n = t$ such that $\max_k (t_{k+1}^n - t_k^n) \to 0$ as $n \to \infty$ and

$$\sum_{k=0}^{k(n)} (X_{t_{k+1}^n} - X_{t_k^n})^2 \to \langle X\rangle_t \quad \text{as } n \to \infty. \tag{6}$$

Temporarily dropping the superscript n we have

$$\begin{aligned}
&E(\sum_k (X_{t_{k+1}} - X_{t_k})^2 - \langle X\rangle_t)^2 \\
&= E(\sum_k \{(X_{t_{k+1}} - X_{t_k})^2 - \langle X\rangle_{t_{k+1}} - \langle X\rangle_{t_k}\})^2 \\
&= \sum_k E\{(X_{t_{k+1}} - X_{t_k})^2 - \langle X\rangle_{t_{k+1}} - \langle X\rangle_{t_k}\}^2 \\
&\le 2\sum_k E(X_{t_{k+1}} - X_{t_k})^4 + 2\sum_k E(\langle X\rangle_{t_{k+1}} - \langle X\rangle_{t_k})^2
\end{aligned} \tag{7}$$

where the second equality follows from the fact that

$$E[(X_{t_{k+1}} - X_{t_k})^2 | F_{t_k}] = E[\langle X\rangle_{t_{k+1}} - \langle X\rangle_{t_k} | F_{t_k}]$$

and the inequality just uses $(A+B)^2 \leq 2(A^2+B^2)$. Now

$$E\sum_k (\langle X\rangle_{t_{k+1}} - \langle X\rangle_{t_k})^2 \leq E\{\max_k(\langle X\rangle_{t_{k+1}} - \langle X\rangle_{t_k}).\langle X\rangle_t\}$$

and the right-hand side converges to zero as $n \to \infty$ since $\langle X\rangle_t$ is bounded and continuous. A similar argument using also the Hölder inequality also applies to the first term in (7). Thus convergence takes place in quadratic mean in (6), so that a subsequence converges a.s. as required.

This result shows again that a continuous martingale X_t cannot have bounded variation sample paths unless it is constant because if it did the quadratic variation $\langle X\rangle_t$ would be zero; but then $E\langle X\rangle_t = E(X_t - X_0)^2 = 0$ so that $X_t = X_0$ a.s.

For an example where $\langle Y\rangle_t$ is easily calculated, consider the Ito integral $Y_t = \int_0^t \phi_s db_s$. Applying the Ito rule we have

$$Y_t^2 - Y_0^2 = 2\int_0^t \phi_s Y_s db_s + \int_0^t \phi_s^2 ds$$

and hence

$$\langle Y\rangle_t = \int_0^t \phi_s^2 ds. \tag{8}$$

That this is the quadratic variation of Y_t follows also from the fact that Y_t can be regarded as a Brownian motion with random time change [7].

Martingale integrals

Denote by CM the set of continuous martingales with respect to F_t. Take $X_t \in$ CM and let ϕ_t be a process of the form

$$\phi_t(\omega) = \begin{cases} c_i(\omega) & t \in]t_i, t_{i+1}] \\ 0 & t > t_n \end{cases}$$

where c_i is F_{t_i}-measurable for each i. (Note that ϕ_t is left-continuous). Now define

$$Y_t = \int_0^t \phi_s dx_s = \sum_i c_i (X_{t_{i+1}\wedge t} - X_{t_i \wedge t}). \tag{9}$$

Then $Y_t \in$ CM, and

$$E\,Y_t^2 = E\int_0^t \phi_s^2 d\langle X\rangle_s. \qquad (10)$$

Note that this is similar to (8) except that "$d\langle X\rangle_s$" replaces "ds". (10) is true because putting $\Delta X_i = X_{t_{i+1}\wedge t} - X_{t_i\wedge t}$ we have for $i < j$

$$E[\phi_i\phi_j\Delta X_i\Delta X_j] = E[\phi_i\phi_j\Delta X_i E(\Delta X_j|F_{t_j})] = 0$$

and hence

$$\begin{aligned} EY_t^2 &= \sum E(\phi_i\Delta X_i)^2 \\ &= \sum E[\phi_i^2 E(\Delta X_i^2|F_{t_i})] \\ &= \sum E[\phi_i E(\langle X\rangle_{t_{i+1}\wedge t} - \langle X\rangle_{t_i\wedge t}|F_{t_i})] \\ &= \sum E[\phi_i^2(\langle X\rangle_{t_{i+1}\wedge t} - \langle X\rangle_{t_i\wedge t})] \\ &= E\int_0^t \phi_s^2 d\langle X\rangle_s. \end{aligned}$$

For any $U,V \in CM$, define the process $\langle U,V\rangle_t$ by

$$\langle U,V\rangle_t = \tfrac{1}{4}\{\langle U+V\rangle_t - \langle U-V\rangle_t\}.$$

Then $\langle U,V\rangle_t$ is a continuous process of bounded variation, and it is easily checked that it has the following property:

$$E[\langle U,V\rangle_t - \langle U,V\rangle_s|F_s] = E[(U_t-U_s)(V_t-V_s)|F_s] = E[U_tV_t-U_sV_s|F_s]$$

i.e. $U_tV_t - \langle U\ V\rangle_t$ is a martingale. (In particular, $\langle U,U\rangle_t = \langle U\rangle_t$.) Now if Y is as in (9) and U is any continuous martingale, a calculation similar to the above shows that

$$\langle Y,U\rangle_t = \int_0^t \phi_s d\langle X,U\rangle_s \qquad (11)$$

and Y is the only martingale satisfying this for all U, since if Y' also satisfied it we would have $0 = \langle Y,U\rangle_t - \langle Y',U\rangle_t = \langle Y-Y',U\rangle_t$ for all U. Taking $U = Y - Y'$ gives $\langle Y-Y'\rangle_t = 0$ so that $E\langle Y-Y'\rangle_t = E(Y_t-Y_t')^2 = 0$. Thus property (11) characterizes Y.

In view of the linearity of the integral and property (10), we can extend the definition to any integrand ϕ which is the limit of a Cauchy sequence of step functions ϕ^n, in the sense that $E\int_0^t(\phi_s^n - \phi_s^m)^2 d\langle X\rangle_s \to 0$ as $n,m\to\infty$. It turns out that all predictable processes such that

$$E\int_0^t \phi_s^2 d\langle X\rangle_s < \infty \quad \text{for each } t \qquad (12)$$

are included. We thus have the following result: *for any* $X \in CM$ *and predictable proces* ϕ *satisfying* (12) *there is a unique* $Y \in CM$ *such that* (11) *is satisfied for all* $U \in CM$. This is the stochastic integral $\int_0^t \phi_s dX_s$. This formulation of the integral is due to Kunita and Watanabe [5].

STOCHASTIC CALCULUS

Having defined stochastic integrals for martingales, the next step is to formulate a generalization of the Ito differential rule. As before, this is most naturally formulated in terms of local semi-martingales. A continuous local semimartingale is a process Z_t that can be written in the form

$$Z_t = A_t + X_t \tag{13}$$

where A_t and X_t are continuous processes, A_t has sample functions of bounded variation, and X_t is a local martingale. Suppose $T_n \uparrow \infty$ is a sequence of stopping times such that $X_{t\wedge T_n}$ is a martingale for each n, and let ϕ_t be a predictable process such that

$$\int_0^t \phi_s^2 d\langle X\rangle_s < \infty \quad \text{a.s.}$$

Now define

$$S_n = \inf\{t : \int_0^t \phi_s^2 d\langle X\rangle_s \geq n\}.$$

Then $U_n = S_n \wedge T_n \uparrow \infty$ a.s. and the integral

$$Y_t^n = \int_0^t \phi_s I_{(s\leq U_n)} dX_s$$

is well defined for each n. Thus there is a well defined local martingale, denoted $Y_s = \int_0^t \phi_s dX_s$ such that $Y_s = Y_s^n$ for $s \leq U_n$.

Let $Z_t = (Z_t^1, Z_t^2, ..Z_t^\ell)$ where each Z_t^i is a continuous local semimartingale as in (13), and let $F : R^\ell \to R$ be twice continuously differentiable. Then the Kunita-Watanabe differential rules states that

$$F(Z_t) - F(Z_0) = \sum_{i=1}^{\ell} \int_0^t \frac{\partial F}{\partial x_i}(Z_s) dX_s^i + \sum_{i=1}^{\ell} \int_0^t \frac{\partial F}{\partial x_i}(Z_s) dA_s^i + \tfrac{1}{2} \sum_{i,j}^{\ell} \int_0^t \frac{\partial^2 F}{\partial x_i \partial x_j}(Z_s) d\langle X^i, X^j\rangle_s. \tag{14}$$

The proof of this is along the same lines as that of the Ito formula: one takes the Taylor series expansion and finds that some second-order terms must be retained because of the quadratic variation property of X_t^i. In the scalar case (14) becomes simply

$$F(Z_t) - F(Z\ ) = \int_0^t F(Z_s)sX_s + \int_0^t F(Z_s)dA_s + \tfrac{1}{2}\int_0^t F\ (Z_s)d\langle X\rangle_s .$$

Note that this agrees with the Ito rule when $X_t = \int_0^t \phi_s db_s$ since then, as mentioned above, $d\langle X\rangle_t = \phi_t^2 dt$. Another special case worthy of note is the product formula for $X, Y \in CM$:

$$X_tY_t - X_0Y_0 = \int_0^t X_s dY_s + \int_0^t Y_s dX_s + \langle X,Y\rangle_t .$$

As a first application of the differential formula (14), let us show that the stochastic differential equation

$$dM = MdX \ , \quad M_0 = 1$$

(where $X \in CM$ is given) has a unique solution, which is

$$M_t = \exp(X_t - \tfrac{1}{2}\langle X\rangle_t) .$$

Indeed, applying (14) to the function $F(x) = e^x$ and semimartingale $X_t - \frac{1}{2}\langle X\rangle_t$ shows that

$$dM_t = M_t dX_t - \tfrac{1}{2}M_t d\langle X\rangle_t + \tfrac{1}{2}M_t d\langle X\rangle_t$$

so that M_t is a solution. If M_t' is another, we can apply (14) again to calculate M_t'/M_t (note that $M_t > 0$ always). This gives

$$d\left(\frac{M'}{M}\right) = \frac{1}{M}\, dM' - \frac{1}{M^2}dM - \frac{1}{M^2}d\langle M',M\rangle + \frac{M'}{M^3}d\langle M,M\rangle .$$

Now $dM' = M'dX$ and $dM = MdX$ so that from (11)

$$d\langle M,M'\rangle = Md\langle X,M'\rangle = MM'd\langle X\rangle ,$$

and $d\langle M,M\rangle = M^2 d\langle X\rangle$. Using these facts we see that $d(M'/M) = 0$, i.e. $M_t = M_t'$ a.s. M_t is a generalization of the "Girsanov exponential formula"

$$\exp\left(\int_0^t \phi_s^2 db_s - \tfrac{1}{2}\int_0^t \phi_s^2 dt\right)$$

which plays a big role in the representation of likelihood ratios in problems of detecting signals in additive white noise.

A characterization of Brownian motion

Brownian motion b_t is a particularly simple martingale in that its quadratic variation $\langle b\rangle_t = t$ is deterministic. There are other processes X_t such that $\langle X\rangle_t = t$: for example the centralized standard Poisson process $X_t = N_t - t$. This has independent increments and $\mathrm{var}(X_t - X_s) = t - s$, so that $E[X_t^2 - X_s^2 | F_s] = t - s$ (of course, in this case $\langle X \rangle_t$ is not the sample quadratic variation since X_t is not continuous). Nevertheless *Brownian motion is the only continuous martingale* X_t *such that* $\langle X \rangle_t = t$. To see this, suppose (X_t, F_t) is such a martingale, fix $\theta \in R$ and apply (14) to the function $F(X) = e^{i\theta X}$. Since $d\langle X \rangle_t = dt$ we get

$$d(e^{i\theta X_t}) = i\theta e^{i\theta X_t} dX_t - \tfrac{1}{2}\theta^2 e^{i\theta X_t} dt.$$

Now $|i\theta e^{i\theta X_t}| = |\theta|$ so that the stochastic integral is a martingale; hence

$$E[e^{i\theta X_t} - e^{i\theta X_s} | F_s] = -\tfrac{1}{2}\theta^2 \, E[\int_s^t e^{i\theta X_u} du | F_s].$$

Now multiplying through by $e^{-i\theta X_s}$ and integrating over any set $A \in F_s$ gives

$$\int_A E[e^{i\theta(X_t - X_s)} - 1 | F_s] dP = -\tfrac{1}{2}\theta^2 \int_A E[\int_s^t e^{i\theta(X_u - X_s)} du | F_s] dP$$

$$= -\tfrac{1}{2}\theta^2 \int_s^t \int_A e^{i\theta(X_u - X_s)} dP du$$

Define

$$y(t) = \int_A e^{i\theta(X_t - X_s)} dP = \int_A E[e^{i\theta(X_t - X_s)} | F_s] dP.$$

Then the above equation says

$$y(t) - PA = -\tfrac{1}{2}\theta^2 \int_s^t y(u) du$$

i.e. $\qquad \dot{y} = -\tfrac{1}{2}\theta^2 y \;, \quad y(s) = PA.$

This has the unique solution $y(t) = PA\, e^{-\frac{1}{2}\theta^2 (t-s)}$, which, bearing in mind the definition of $y(t)$, we can write

$$\int_A \{E[e^{i\theta(X_t - X_s)} | F_s] - e^{-\frac{1}{2}\theta^2(t-s)}\} dP = 0.$$

Since the integrand is F_s-measurable and the integral is zero for all $A \in F_s$, it follows that

$$E[e^{i\theta(X_t - X_s)} | F_s] = e^{-\frac{1}{2}\theta^2 (t-s)}. \tag{15}$$

Now $e^{-\frac{1}{2}\theta^2(t-s)}$ is the characteristic function of the $N(0,t-s)$ distribution. Thus $(X_t - X_s) \sim N(0,t-s)$ and it is easily seen from (15) that $(X_t - X_s)$ is independent of F_s, in particular of $(X_{t'} - X_{s'})$ for t', s' s. So X_t is a Brownian motion.

Finally, let us mention the immediate application of this result to the innovations process, which plays a major role in filtering theory (see for example [2]). Suppose b_t is a BM and Z_t another process such that $E|Z_t| < \infty$ for all t and $(b_t - b_s)$ is independent of $F_s = \sigma\{b_u, Z_u, u \leq s\}$. Let

$$dy_t = Z_t dt + db_t$$

(this is a model for a signal Z_t observed in additive white noise). Let $\mathcal{Y}_t = \sigma\{y_s, s \leq t\}$ and denote $\hat{Z}_t = E[\mathcal{Y}_t| \ _t]$, $\tilde{Z}_t = Z_t - \hat{Z}_t$. The *innovations process* is then

$$\nu_t = y_t - \int_0^t \hat{Z}_s ds$$

$$= \int_0^t \tilde{Z}_s ds + b_t. \tag{16}$$

It is easy to check that $(\nu_t, \mathcal{Y}_t)$ is a martingale (this just depends on the properties of conditional expectation) and it clearly has continuous sample paths. Now, from (16), ν_t is of the form $a_t + b_t$ where a_t has bounded variation. Therefore the quadratic variation of ν_t is the same as that of b_t, i.e.$\langle\nu\rangle_t = t$, since a_t has zero quadratic variation. So ν_t is a Brownian motion. This is a very striking result since it implies in particular that ν_t is a normal process, without any restriction on the distributions of Z_t

REFERENCES

1. J.M.C. Clark, An introduction to stochastic differential equations on manifolds, in: Geometric Methods in System Theory, ed. by D.Q. Mayne and R.W. Brockett, Reidel, Dordrecht, 1973.
2. M.H.A. Davis, Linear Estimation and Stochastic Control, Chapman and Hall, London, 1977.
3. C. Dellacherie and P.A. Meyer, Probabilités et Potentiel (2ème ed.), Hermann, Paris, 1975.
4. I.N. Gelfand and N.Ya. Valenkin, Generalized Functions, Vol. IV, Academic Press, New York, 1964.
5. H. Kunita and S. Watanabe, On square-integrable martingales, Nagoya Math. J., 30 (1967), 209-245.
6. R.S. Liptser and A.N. Shiryaev, Statistics of Stochastic Processes, Nauka, Moscow, 1974.
7. H.P. McKean, Stochastic Integrals, Academic Press, New York, 1969.
8. P.A. Meyer, Probability and Potentials, Blaisdell, Waltham, Mass., 1966.
9. P.A. Meyer, Un cours sur les intégrales stochastiques, Seminaire de Probabilités X, Lecture Notes in Mathematics 511, Springer-Verlag, Berlin, 1976.
10. C.M. Rao, On decomposition theorems of Meyer, Math. Scand. 24 (1969), 66-78.

APPENDIX: THE GENERALIZED STRATONOVICH INTEGRAL

In many cases it is possible to define the stochastic integral differently, so that it obeys the 'ordinary' rules of calculus. Consider again the stochastic integral $\int_0^t b_s db_s$ for Brownian motion b_t. Suppose we define

$$\oint_0^t b_s db_s = \lim_{n\to\infty} \sum_k b((k+\tfrac{1}{2})/2^n)[b((k+1)/2^n) - b(k/2^n)]$$

(i.e. the integrand is sampled at the mid-point of the interval $[k/2^n,(k+1)/2^n]$ instead of at the left-hand end as for the Ito integral.) Denoting by I_n the sum on the right, we have, with the same notation as before

$$I_{n-1} = \sum_{k\text{ odd}} b_k(b_{k+1} - b_{k-1})$$

$$= \sum_k b_k(b_{k+1} - b_k) + \sum_{k\text{ odd}} (b_{k+1} - b_k)^2$$

The first term on the right is the approximation to the Ito integral and it is easy to see that the second converges to $\tfrac{1}{2}t$ as $n\to\infty$. Thus

$$\oint_0^t b_s db_s = \tfrac{1}{2}b_t^2.$$

This is the *Stratonovich integral*, and has the same value as if calculated by the ordinary rules of calculus. Of course, it is *not a martingale* (in fact,as noted before, b_t^2 is a submartingale).

The general formulation is as follows. Suppose X_t, Y_t are two continuous local semimartingales. We define

$$\oint_0^t Y_s dX_s = \int_0^t Y_s dX_s + \tfrac{1}{2}\langle X,Y\rangle_t,$$

where the integral on the right is the stochastic integral as defined previously and $\langle X,Y\rangle_t$ is short for $\langle M^X,M^Y\rangle_t$, M^X and M^Y being the martingale components of X,Y respectively. Denoting this integral by Z_t, take a C^3 function F and calculate $F(Z_t)$ using (14). This gives

$$dF(Z_t) = F'(Z_t)dZ_t + \tfrac{1}{2}F''(Z_t)Y_t^2 d\langle X\rangle_t \tag{18}$$

Now $F'(Z_t)$ is itself a semimartingale: indeed, replacing F by F' in (18) we get

$$dF'(Z_t) = F''(Z_t)Y_t dX_t + \tfrac{1}{2}F''(Z_t)Y_t d\langle X,Y\rangle_t + \tfrac{1}{2}F'''(Z_t)Y_t^2 d\langle X\rangle_t$$

which shows that the local martingale component of $F'(Z_t)$ is $F''(Z_t)Y_t dX_t$. Hence, using property (11) of the stochastic integral,

$$\langle F'(Z), Z\rangle_t = \langle \int F''(Z)YdX, \int YdX\rangle_t$$
$$= \int_0^t F''(Z_s)Y_s^2 d\langle X\rangle_s$$

But, recalling the definition (17), this means that (18) can be written simply as

$$F(Z_t) - F(Z_0) = \displaystyle{\not}\int_0^t F'(Z_s)dZ_s$$

which shows that the generalized Stratonovich integral $\not\int$ obeys the ordinary rules of calculus. But again, the martingale properties are lost, and the integral is only defined for continuous local semimartingale integrands. It is however useful in doing calculations since the 'extra terms' in (14) do not appear. The Ito version of any expression involving Stratonovich integrals can always be re-derived using the correspondence (17). The main use of the Stratonovich integral, though, is in stochastic modelling it appears naturally when considering the relation between idealized 'white noise' and band-limited approximations to it. The reader is referred to [1] for further details.

DETECTION, MUTUAL INFORMATION AND FEEDBACK ENCODING: APPLICATIONS OF STOCHASTIC CALCULUS

M. H. A. Davis

Department of Computing and Control, Imperial College, London SW7 2BZ, England.

INTRODUCTION

In this paper, the stochastic calculus developed in the companion paper [1] is applied to calculate likelihood ratios and mutual information for signals in additive noise. These quantities are essentially connected with absolute continuity of measures and Radon-Nikodym derivatives, which are indeed simply the function-space equivalents of the likelihood ratios of elementary hypothesis-testing theory. The general formulation, and the connection with martingale theory, is as follows.

Let (Ω, F_T, P_o) be a probability space and P_1 another measure on (Ω, F_T) which is absolutely continuous with respect to P_o (i.e. $P_oA = 0 \Rightarrow P_1A = 0$; we write $P_1 << P_o$). Then the Radon-Nikodym Theorem asserts the existence of a non-negative random variable L such that

$$P_1A = \int_A L(\omega)dP_o(\omega) \; , \quad A \in F_T \quad . \tag{1}$$

Taking $A = \Omega$ we see in particular that $E_oL = P_1\Omega$ (E_o and E_1 will denote expectation with respect to P_o, P_1 respectively) so that P_1 is a probability measure if and only if $E_oL = 1$. If $\{F_t, 0 \leq t \leq T\}$ is an increasing family of σ-fields, for example those generated by some observed process X_t over the interval $[0,T]$, then $L_t := E_o[L|F_t]$ is a non-negative martingale with expectation $E_o L_t = 1$. Conversely if L_t is any such martingale then the final value L_T can be used to define a new measure P_1 via (1); thus there is a one-to-one correspondence between such martingales and probability measures absolutely continuous with respect to P_o. The so-called Girsanov Theorem, in its generalized form due to Wong [10], characterizes these martingales and describes what happens to processes which are F_t-martingales under P_o when the measure is changed to P_1. Specialized to Girsanov's original form [4] this gives the likelihood ratio formula for detection of signals in additive white noise and shows that every measure absolutely continuous with respect to Wiener measure essentially corresponds to a process in this form. These results are outlined in the next two sections. We then consider some problems in communication theory relating to the additive white gaussian channel. Since mutual information is expressed in terms of Radon-Nikodym derivatives the likelihood ratio formulas can be used to calculate this. We obtain as an almost immediate corollary the fact that the channel capacity cannot be increased by feedback. However, noise-free feedback often results in simplified source encoding. We give the optimal coding for transmitting a gaussian random variable through an additive white noise channel with feedback.

ABSOLUTE CONTINUITY AND THE GIRSANOV THEOREM

Consider a probability space (Ω, F_T, P_o) as above and a probability measure P_1 *mutually* absolutely continuous with respect to P_o (i.e. $P_1 << P_o$ and $P_o << P_1$). Define L and L_t as before and suppose that L_t is continuous; then L_t and $1/L_t$ are locally bounded in the following sense. Doob's submartingale inequality states that for any n,

$$P_o[\sup_{t \le T} L_t > n] \le \frac{1}{n} E_o L_T = \frac{1}{n} \quad . \tag{2}$$

Now let R_n be a time which is equal to T if $L_t < n$ for all $t \in [0,T]$ and equal to the first time at which $L_t \ge n$ otherwise. Define S_n similarly with L_t replaced by $1/L_t$ $(= dP_o/dP_1)$. Then $1/n \le L_{t \wedge R_n \wedge S_n} \le n$ and (2) implies that $R_n \wedge S_n \uparrow T$ a.s. This means that the differential formula (1.14)[‡] can be applied

‡ This refers to equation (14) of the companion paper [1].

to calculate $ln\ L_t$. This gives

$$ln\ L_t = \int_0^t \frac{1}{L_s}\, dL_s - \frac{1}{2}\int_0^t \frac{1}{L_s^2}\, d\langle L\rangle_s \quad . \tag{3}$$

Define $M_t := \int_0^t (1/L_s)dL_s$. By the above argument this is a continuous local martingale, and the second term on the right of (3) is just $\frac{1}{2}\langle M\rangle_t$. Thus L_t can be written in the form

$$L_t = \exp(M_t - \tfrac{1}{2}\langle M\rangle_t) \tag{4}$$

and this associates another local martingale M_t with the new measure P_1 . M_t is useful in elucidating what happens to processes X_t which are martingales under the original measure P_o when the measure is changed to P_1 . First, the formula for conditional expectation under P_1 is

$$E_1[X|F_t] = \frac{E_o[XL|F_t]}{E_o[L|F_t]} \tag{5}$$

(Here X is any - say bounded - F_t-measurable random variable).

Now suppose W_t is an F_t-adapted process and $t > s$; then (5) becomes

$$E_1[W_t|F_s] = \frac{E_o[W_tL_t|F_s]}{L_s}$$

and it follows that W_t *is a martingale under measure* P_1 *if and only if the process* W_tL_t *is a martingale under* P_o .

Now suppose X_t is a P_o-martingale and define

$$W_t = X_t - \langle X,M\rangle_t$$

where M_t is as in (3) . Then Y_t is a P_1-local martingale. To check this it is only necessary to apply the product formula

$$W_tL_t - W_oL_o = \int_0^t W_s dL_s + \int_0^t L_s dW_s + \langle X,L\rangle_t \tag{6}$$

which follows from the general differential formula (1.14). Recall from [1] that L_t satisfies

$$L_t = 1 + \int_0^t L_s dM_s$$

so that by the Kunita-Watanabe characterization (1.11) of the stochastic integral we have

$$\langle L,X\rangle_t = \int_0^t L_s d\langle M,X\rangle_t \quad .$$

Substituting this in (6) gives

$$W_t L_t = W_o + \int_0^t W_s dL_s + \int_s^t L_s dX_s \quad .$$

Thus $W_t L_t$ is a P_o-local martingale, being a sum of stochastic integrals, and hence W_t is a P_1-local martingale. Thus the effect of changing the measure to P_1 is that a martingale X_t remains a local martingale if we subtract from it the bounded variation term $\langle X,M\rangle_t$, where M_t is the local martingale associated with P_1 via the exponential formula (4). Another crucial observation is that the *quadratic variations of* X_t *and* W_t *are the same*, since these processes only differ by a bounded variation term which has no effect on the quadratic variation. Thus we can write

$$\langle W\rangle_t^{(P_1)} = \langle X\rangle_t^{(P_o)} \ .$$

The reason for formulating the above results is that we often want to consider processes of the form: bounded variation process + martingale, which are in applications models of signals in additive noise. Since the bounded variation term above is given in terms of M_t and not of L directly, it is natural to ask whether we can start with M_t as the primary entity, i.e. take an arbitrary continuous local martingale M_t and *define* L_t by (4). Unfortunately this does not always work. We saw in [1] that for any M_t, the corresponding L_t satisfies

$$L_t = 1 + \int_0^t L_s dM_s \quad .$$

Now if L_t is to correspond to a change of measure then we must have $E_o L_t = 1$ for all t. But

$$\int_0^t L_s dM_s$$

is in general only a *local* martingale and there is no guarantee that its expectation is zero. We can, however, say a bit more; in fact L_t is always a *supermartingale*. Indeed, suppose T_n is an increasing sequence of stopping times such that $T_n \uparrow T$ a.s. and $L_{t\wedge T_n}$ is a martingale. Then for $t > s$

$$\begin{aligned} E_o[L_t|F_s] &= E[\lim_n L_{t\wedge T_n}|F_s] \\ &\le \lim_n E[L_{t\wedge T_n}|F_s] \qquad \text{(Fatou's Lemma)} \\ &= \lim_n L_{s\wedge T_n} = L_s \quad . \end{aligned}$$

Now any supermartingale has monotonically decreasing expectation and since here $L_o = 1$ a.s. we have

$$1 = E_oL_o \ge E_oL_t \ge E_oL_T$$

for any $t \in [0,T]$. Thus $E_oL_t = 1$ for all t if and only if $E_oL_T = 1$ and thus we only have one expectation to check in order to assure that L_t is a *bone fide* martingale.

Let us now specialize the above results to the Brownian case, i.e. we suppose X_t is a Brownian motion and define

$$M_t = \int_o^t \phi_s dX_s$$

where ϕ_t is an adapted process satisfying

$$\int_o^t \phi_s^2 ds < \infty \qquad \text{a.s.}$$

Then (4) becomes

$$L_t = \exp\left(\int_o^t \phi_s dX_s - \tfrac{1}{2}\int_o^t \phi_s^2 ds\right) \quad . \tag{7}$$

Assuming that $E_oL_T = 1$ we can define a measure P_1 as before by taking $dP_1/dP_o = L_T$. Now from (1.11) we have

$$\langle M,X\rangle_t = \int_o^t \phi_s d\langle X,X\rangle_s = \int_o^t \phi_s ds \quad .$$

By the preceding argument, $W_t := X_t - \int_0^t \phi_s ds$ is a P_1-local martingale whose quadratic variation is equal to that of X_t, i.e. $\langle W\rangle_t^{(P_1)} = t$. But according to the characterization of Brownian motion given in [1] this means that *W_t is a Brownian motion under P_1*. Writing

$$X_t = \int_0^t \phi_s ds + W_t \tag{8}$$

we now see the effect of the change of measure: the process X_t which was (integrated) white noise under measure P_0 gets transformed into (integrated) signal + white noise under measure P_1. This is the result as originally presented by Girsanov [4]; the generalized version was discovered by Wong [10]. The beauty of the result is that there are so few restrictions on the form of the process ϕ_t. In applications ϕ_t normally represents a transmitted signal which depends in some way on a message process Z_t. If there is feedback from receiver to transmitter then ϕ_t could also depend on the past $X_0^t = \{X_s, 0 \le s \le t\}$ of the received signal X_t. Then (8) will be of the form

$$dX_t = \phi(t, X_0^t, Z_t)dt + dW_t \tag{9}$$

which represents a very complicated functional relationship between X_t, Z_t and W_t. However if, for example, Z_t and W_t are independent, we have only to start with a probability space (Ω, F_T, P_0) carrying independent processes Z_t and X_t, with Z_t having the desired message distributions and X_t being Brownian motion, and to change the measure to P_1 using Girsanov's exponential formula (7). Then the same processes X_t and Z_t will be in the signal + noise form (9), as desired. We have only to be assured that the measure transformation "works", i.e. that $E_0 L_T = 1$. Needless to say, finding conditions under which this holds has been an area of much research. The simplest condition, already given in [4], is that ϕ_t be bounded: $|\phi_t(\omega)| \le K$ for all (t,ω). There are a variety of weaker conditions, for details of which the reader is referred to [9].

DETECTION OF SIGNALS IN ADDITIVE GAUSSIAN WHITE NOISE

Suppose we have a probability space (Ω, F_T, P) and know that P must be one of two given probability measures, P_0 or P_1. We have to decide which it is, having observed a sample function $\{X_t(\omega), 0 \leq s \leq T\}$ of a continuous process X_t defined on (Ω, F_T, P). This problem is the same as the classical Neyman-Pearson case if we agree to regard X_t as a function-space-valued random variable. Let C be the space of all real-valued continuous functions on $[0,T]$ and S be the Borel σ-field of C (i.e. the σ-field generated by the open sets of C in the topology of uniform convergence). If $\{X_t\}$ is a measurable process then the mapping

$$\omega \mapsto \{X(s,\omega), 0 \leq s \leq T\}$$

defines a measurable function $X : (\Omega, F_T) \to (C,S)$, so that X_t is actually a C-valued random variable (C is called the *sample space*). The sample space measures of the process corresponding to the measures P_0, P_1 on (Ω, F_T) are μ_0, μ_1 defined by

$$\mu_i(S) = P_i(X^{-1}(S)) \qquad S \in S, \; i = 0,1 .$$

Suppose P_0, P_1 are mutually absolutely continuous and let $L = dP_1/dP_0$. Then clearly μ_0, μ_1 are also mutually absolutely continuous, and it is not hard to see that

$$E_0[L|X](\omega) = \frac{d\mu_1}{d\mu_0}(X(\omega)) \tag{10}$$

where $X = X^{-1}(S) = \sigma\{X_s, 0 \leq s \leq T\}$. Now, just as in the finite-dimensional case, the best test of P_0 versus P_1 (in either the Neyman-Pearson or the Bayesian sense) is "choose P_1 if $d\mu_1/d\mu_0\,(X(\omega)) > \varkappa$" where $\varkappa$ is a constant depending on significance levels or prior probabilities and error costs.

Let us return to the case, considered in the previous section, where X_t is (under measure P_1) given by (9), and the message process Z_t and Brownian motion noise W_t are assumed to be independent. For the sake of simplicity we assume that ϕ_t is *bounded*; then the Girsanov exponential *always* has expectation 1 so that the measure transformation always "works". Under more natural conditions such as (15) below the following results are still true, with minor modifications, but much more care is needed in getting them; see [7,8,9]. First define

$$L(\omega) = \exp(-\int_0^T \phi(t,X_0^t, Z_t)dW_t - \tfrac{1}{2}\int_0^T \phi^2(t,X_0^t, Z_t)dt) \tag{11}$$

and let P_0 be the probability measure defined by $dP_0/dP_1 = L$. Then according to the Girsanov theorem, the process

$$dW_t + \phi_t dt = dX_t$$

is Brownian under P_0. Further, under P_0 the distributions of $\{Z_t\}$ are the same as under P_1 and $\{X_t\}, \{Z_t\}$ are independent. The reason for these latter facts is roughly as follows.

Let $Z = \sigma\{Z_s, 0 \le s \le T\}$ be the σ-field generated by the process $\{Z_t\}$. Then since $\{W_t\}$ and $\{Z_t\}$ are independent, the conditional distributions of $\{W_t\}$ given Z are still those of Brownian motion. Now take a bounded function $f : R^n \to R$ and n times $t_1 \ldots t_n \in [0,T]$ and calculate

$$E_0[f(Z_{t_1} \ldots Z_{t_n})] = E_1[f(Z_{t_1} \ldots Z_{t_n})E_1(L|Z)] .$$

Let $L_z(\omega)$ be $L(\omega)$ conditioned on a particular trajectory of $\{Z_t\}$ say $Z_t(\omega) = z_t$, $0 \le t \le T$. Then since W_t is still Brownian, L_z is in the form of a Girsanov exponential, i.e.

$$L_z(\omega) = \exp(-\int_0^T \phi_z^1(t,X_0^t)dW_t - \tfrac{1}{2}\int_0^T \phi_z^2(t,X_0^t)dt)$$

where $\phi_z(t,X_0^t) = \phi(t,X_0^t, z_t)$. Now because ϕ_z is bounded,

$E_1 L_z(\omega) = 1$ and since this applies for all z, we see that $E_1[L|Z] = 1$ a.s. Thus $E_0[f(Z_{t_1} \ldots Z_{t_n})] = E_1[f(Z_{t_1} \ldots Z_{t_n})]$

which shows that $\{Z_t\}$ has the same distributions under either measure. In a similar manner, take bounded $g : R^n \to R$ and calculate

$$E_o[f(Z_{t_1}..Z_{t_n})g(X_{t_1}..X_{t_n})] = E_1[f(Z_{t_1}..Z_{t_n})E_1(g(X_{t_1}..X_{t_n})L|Z)].$$

By the same argument as above, for each z the process $dX^z = \phi_z(t)dt + dW_t$ is Brownian under the measure whose density is L_z. Thus, since X_t itself is Brownian under P_o we have

$$E_1[g(X_{t_1}..\ X_{t_n})L|Z] = E_o[g(X_{t_1}\ ..\ X_{t_n})]$$

and hence

$$E_o[f(Z_{t_1}..Z_{t_n})g(X_{t_1}..X_{t_n})] = E_1[f(Z_{t_1}..Z_{t_n})]E_o[g(X_{t_1}..X_{t_n})] .$$

But P_o and P_1 coincide on Z so that $E_1 f = E_o f$ and the the above equation shows that $\{X_t\}$ and $\{Z_t\}$ are independent under P_o.

Let $(D,\mathcal{D})$ be the sample space of the message process $\{Z_t\}$ (this could be another copy of (C,S) if $\{Z_t\}$ has continuous sample functions, or, more generally, a space of right-continuous functions). Let $\Psi : \Omega \to D \times C$ and $Z : \Omega \to D$ be the maps that take ω into the sample functions $\{(Z_s(\omega),X_s(\omega)), 0 \le s \le T\}$ and $\{Z_s(\omega), 0 \le s \le T\}$ respectively. Then the joint sample space measure of $\{Z_t, X_t\}$ is given by

$$\mu_{ZX}(A) = P_1(\Psi^{-1}(A)) \qquad A \in \mathcal{D}*S$$

and the sample space measure of $\{Z_t\}$ by

$$\mu_Z(B) = P_1(Z^{-1}(B)) \qquad B \in \mathcal{D} \quad .$$

Now according to the above argument $\{X_t\}$ and $\{Z_t\}$ are independent under measure P_o, and hence

$$\mu_Z * \mu_W(A) = P_o(\Psi^{-1}(A)) \qquad A \in \mathcal{D}*S$$

where μ_W is Wiener measure on (C,S) and $\mu_Z * \mu_W$ denotes the product measure. Now, from (11), $L = dP_o/dP_1 > 0$ a.s. and hence $P_1 \ll P_o$ with Radon-Nikodym derivative

$$\frac{dP_1}{dP_o}(\omega) = \frac{1}{L(\omega)} = \exp(\int_0^T \phi_t dW_t + \tfrac{1}{2}\int_0^T \phi_t^2\, dt)$$

$$= \exp(\int_0^T \phi_t dX_t - \tfrac{1}{2}\int_0^T \phi_t^2\, dt) \quad . \tag{12}$$

Let $\Lambda : D\times C \to R$ be the function defined by

$$\Lambda(z,x) = \exp(\int_0^T \phi(t,x_o^t,z_t)dx_t - \tfrac{1}{2}\int_0^T \phi^2(t,x_o^t,z_t)dt)$$

Then (12) simply says that

$$\frac{dP_1}{dP_o}(\omega) = \Lambda(\Psi(\omega))$$

and using this together with (10) shows that

$$\frac{d\mu_{ZX}}{d(\mu_Z * \mu_W)}\ (z,x) = \Lambda(z,x) \tag{13}$$

this gives us the density of the *joint* measure μ_{ZX} with respect to $\mu_Z * \mu_X$. In hypothesis testing, where the process $\{X_t\}$ only is observed, we are interested in the *marginal* measure μ_X , which is, in view of (13), certainly mutually absolutely continuous with respect to Wiener measure μ_W . To calculate the derivative it is quickest to use the innovations representation

$$dX_t = \hat{\phi}_t dt + d\nu_t \quad .$$

(Recall that $\hat{\phi}_t := E_1[\phi(t,X_o^t, Z_t)|X_o^t]$ and that this equation *defines* ν_t , which is then a P_1-Brownian motion). Now define a measure $\tilde{P}_o$ by the derivative

$$\frac{d\tilde{P}_o}{dP_1} = \exp(-\int_0^T \hat{\phi}_t d\nu_t - \tfrac{1}{2}\int_0^T \phi_t^2 dt)$$

where again $\tilde{P}_o\Omega = 1$ since $\hat{\phi}_t$ is bounded. Then applying the Girsanov theorem once more we see that $dX_t = d\nu_t + \hat{\phi}_t dt$ is a $\tilde{P}_o$-Brownian motion. Now $\hat{\phi}_t$ is a function of X_o^t, and applying exactly the same reasoning as above we see that

$$\frac{d\mu_X}{d\mu_W}(x) = \exp\left(\int_o^T \hat{\phi}(t,x_o^t)dx_t - \tfrac{1}{2}\int_o^t \hat{\phi}^2(t,x_o^t)dt\right) \quad . \tag{14}$$

This is Kailath's famous Likelihood Ratio formula for detecting the signal ϕ_t in additive white noise [7,8]. Measure μ_X corresponds to "signal present" as in (9) whereas the "signal absent" case corresponds to $\{X_t\}$ being pure noise, i.e. a Brownian motion with measure μ_W. One interesting aspect of the LR formula is the connection it establishes between filtering and detection: the statistic to be computed for detection involves the estimate $\hat{\phi}_t$ of the signal (under measure μ_X). Finding this is a non-linear filtering problem the solution to which is known in principle but in general prohibitively complex to implement. Kailath has suggested [7] that the LR formula should be regarded as providing "structural information", and that a good test statistic will be obtained if an implementable estimate, say $\bar{\phi}_t$, is used in place of $\hat{\phi}_t$ in (14). It has yet to be established in any generality how good such a test would be.

MUTUAL INFORMATION AND FEEDBACK ENCODING [2,5,6]

In communication theory parlance, the equation

$$dX_t = \phi(t,X_o^t, Z_t)dt + dW_t$$

represents the additive white gaussian channel with noise-free feedback, the presence of the feedback channel being indicated by the dependence of the transmitted signal ϕ_t on previous received signals X_o^t. Physically it is reasonable to assume that ϕ_t is subject to an average power constraint:

$$E_1\phi_t^2 < p_o \qquad\qquad 0 \leq t \leq T \quad . \tag{15}$$

In calculating the channel capacity, the central quantity required is the *mutual information* $I(X,Z)$ between the message $\{Z_t\}$ and received signal $\{X_t\}$. This is easily evaluated since its definition is in terms of Radon-Nikodym derivatives which are already to hand. (These formulae are still valid under condition (15)). The mutual information is given by:

$$I(X,Z) = \int_{D\times C} ln \frac{d\mu_{ZX}}{d(\mu_Z*\mu_X)}(z,x) \quad d\mu_{ZX}(z,x)$$

$$= E_1 \; ln \frac{d\mu_{ZX}}{d(\mu_Z*\mu_X)}(Z,X) \quad .$$

Now an application of the Fubini theorem shows that

$$\frac{d\mu_{ZX}}{d(\mu_Z*\mu_X)} = \frac{d\mu_{ZX}}{d(\mu_Z*\mu_W)} \frac{d\mu_W}{d\mu_X} \quad .$$

But the derivatives on the right are given by (13) and (14) respectively. Thus

$$ln \frac{d\mu_{ZX}}{d(\mu_Z*\mu_X)} = \int_0^T (\phi_t - \hat{\phi}_t)dX_t - \tfrac{1}{2}\int_0^T (\phi_t^2 - \hat{\phi}_t^2)dt$$

$$= \int_0^T (\phi_t - \hat{\phi}_t)(\phi_t dt + dW_t) - \tfrac{1}{2}\int_0^T (\phi_t^2 - \hat{\phi}_t^2)dt$$

$$= \tfrac{1}{2}\int_0^T (\phi_t - \hat{\phi}_t)^2 dt + \int_0^T (\phi_t - \hat{\phi}_t)dW_t \quad .$$

Under P_1 and assuming (15) the last integral is a stochastic integral with zero expectation. Thus the mutual information is

$$I(X,Z) = \tfrac{1}{2} E_1 \int_0^T (\phi_t - \hat{\phi}_t)^2 dt \; .$$

A couple of observations follow immediately from this result. First, if the noise power is σ^2 instead of 1 , i.e. the signal equation is

$$dX_t = \phi_t dt + \sigma dW$$

then it is easy to see that

$$I(X,Z) = \frac{1}{2\sigma^2} E_1 \int_0^T (\phi_t - \hat{\phi}_t)^2 dt$$

and thus, in view of (15),

$$I(X,Z) \leq \frac{p_o}{2\sigma^2} T \quad . \tag{16}$$

But $p_o/2\sigma^2$ is the capacity of the white gaussian channel *without* feedback [3]. This gives us the well-known result that *the channel capacity cannot be increased by feedback.*
Now consider how equality might be attained in the bound (16). This would happen in particular if $E_1\phi_t^2 = p_o$ and $\hat{\phi}_t = 0$, and this can be arranged for example in a signal of the form

$$\phi(t, X_o^t, Z_t) = A(t, X_o^t)(g(Z_t) - \hat{g}_t) \tag{17}$$

where $\hat{g}_t = E_1[g(Z_t)|X_o^t]$. (Note that such a signal always requires feedback). For this signal $\hat{\phi}_t = 0$ is automatic and $A(t,X_o^t)$ can be chosen so that the power constraint is satisfied with equality. This exemplifies the point that, even though it does not increase capacity, a feedback channel may substantially simplify the task of source encoding.

Finally, let us mention one case, due to Shiryaev [9,§16.4] and Ihara [5] where the optimal encoding can be explicitly worked out. This is where the message is a single random variable Z with distribution $N(0,\gamma)$ and the objective is to choose ϕ_t so as to minimize the mean square error $E_1[Z - \hat{Z}_t]^2$, where $\hat{Z}_t = E_1[Z|X_o^t]$. The optimal signal for this purpose is similar in form to (17) above; in fact it is

$$\phi(t, X_o^t, Z) = A(t)(Z - \hat{Z}_t)$$

where $A(\cdot)$ is a *deterministic* function to be determined below.

Let us calculate the performance of this scheme for any given function $A(\cdot)$. Defining

$$d\tilde{X}_t = dX_t + A(t)\hat{Z}_t dt$$

we have

$$d\widetilde{X}_t = A(t)Z + \sigma dW$$

and clearly $E_1[Z|\widetilde{X}_0^t] = E_1[Z|X_0^t]$. But calculating $E_1[Z|\widetilde{X}_0^t]$ is a standard Kalman filtering problem and the error variance $e(t) := E_1[Z - \hat{Z}_t]^2$ is given by [9, Theorem 10.1]

$$\dot{e}(t) = -\frac{1}{\sigma^2} A^2(t)e^2(t) , \quad e(0) = \gamma$$

which has the solution

$$e(t) = \gamma(1 + \frac{\gamma}{\sigma^2} \int_0^t A^2(s)ds)^{-1} .$$

Since $\phi_t = A(Z - \hat{Z}_t)$ the power constraint (15) becomes

$$A^2(t)e(t) \leq p_o .$$

Taking equality in this and solving for $A(t)$ gives

$$A(t) = \sqrt{p_o\gamma} \exp(p_o t/2\sigma^2)$$

$$e(t) = \gamma \exp(- p_o t/\sigma^2) .$$

We now want to show that no other admissible signal of whatever form can do better than this. This involves *rate distortion theory*. The rate distortion function for the $N(0,\gamma)$ random variable Z with mean-square error distortion function is [3]

$$R(\varepsilon) = \tfrac{1}{2} \ln \max(\gamma/\varepsilon^2, 1)$$

and this is the minimum mutual information between Z and any other random variable U such that $E[U - Z]^2 \leq \varepsilon$. Thus for any r.v. U ,

$$I(Z,U) \geq R(E[U - Z]^2) \geq \tfrac{1}{2} \ln(\gamma/E[U - Z]^2) .$$

Using this, (16), and a standard inequality of mutual information, shows that for any signal ϕ_t ,

$$p_o T/2\sigma^2 \geq I(Z,X_o^t) \geq I(Z,\hat{Z}_T^\phi) \geq \tfrac{1}{2} \ln(\gamma/E(Z - \hat{Z}_T^\phi]^2) \tag{18}$$

where $\hat{Z}^{\phi}_t = E_1[Z|X^t_o]$ with X_t given by (9). But

$$\tfrac{1}{2}\, ln(\gamma/e(t)) = p_o T/2\sigma^2$$

so that taking $\phi_t = A(t)(Z - \hat{Z}_t)$ we have equality throughout in (18), which shows that this signal is optimal both in the sense of minimizing the mean-square error $E[Z - \hat{Z}_T]^2$ and of maximizing the mutual information $I(Z,X)$.

Note that with the optimal signal the innovations process is

$$d\nu_t = dy_t - \hat{\phi}_t dt = dy_t$$

so that the *received signal is Brownian motion*. This is intuitively reasonable: under the optimal scheme, only *new information* is transmitted, and the transmitter does not waste energy by transmitting signals which are already predictable from past observations. However it should be pointed out that the above derivation depends explicitly on the normal distribution of Z , and the type of optimal signal could be entirely different if Z had some other distribution.

ACKNOWLEDGEMENT

Some of this material was originally organized for presentation in a series of lectures at the Institute for Optimization and System Theory, Royal Institute of Technology, Stockholm. I am grateful to Professor L. E. Zachrisson for providing this opportunity and for his hospitality.

REFERENCES

1. M. H. A. Davis, Martingale Integrals and Stochastic Calculus, this volume.
2. T. E. Duncan, On the calculation of mutual information, SIAM J. Applied Math. 19 (1970), 215-220.
3. R. G. Gallager, Information Theory and reliable communication, Wiley, New York 1968.
4. I. V. Girsanov, On transforming a certain class of stochastic processes by absolutely continuous substitution of measures, Theory of Probability and its Applications V (1960), 314-330.
5. S. Ihara, Optimal coding in white gaussian channel with feedback, in 2nd USSR-Japan Symposium on Probability Theory,

Lecture Notes in Mathematics vol. 330, Springer-Verlag, Berlin, 1973.

6. T. T. Kadota, M. Zakai and J. Ziv, Mutual information of the white gaussian channel with and without feedback, IEEE Trans. Information Theory vol 11-17 (1971), pp 368-371.
7. T. Kailath, The strucure of Radon-Nikodym derivatives with respect to Wiener and related measures, Annals of Math. Stat. 42 (1971), 1054-1067.
8. T. Kailath and M. Zakai, Absolute continuity and Radon-Nikodym derivatives for certain measures relative to Wiener measure, Annals of Math. Stat. 42 (1971), 130-140.
9. R. S. Liptser and A. N. Shiryaev, Statistics of Stochastic processes, Nauka, Moscow 1974 (in Russian; English translation forthcoming from Springer-Verlag).
10. E. Wong, Recent progress in stochastic processes, IEEE Trans. Information Theory, 11-19 (1973), 262-275.

THE DESIGN OF ROBUST APPROXIMATIONS TO THE STOCHASTIC DIFFERENTIAL EQUATIONS OF NONLINEAR FILTERING

J.M.C. Clark

Department of Computing and Control
Imperial College, London SW7 2BZ

INTRODUCTION

The Ito calculus is undoubtedly a natural vehicle for the expression of the underlying ideas of the theory of nonlinear filtering. But for the formulation of models on which practical designs are to be based it has its limitations. For this purpose it is reasonable to subject any filter model to the following tests. It should be statistically robust (in a sense that will be defined later); this is especially important for models based on the idealization of white noise by Brownian motion. It should also prescribe an "output", a conditional mean or probability, for all data inputs that are likely to be met with in practice. Interpreted literally, the stochastic integral representations of filtering theory fail these tests.

The purpose of this paper is to present reformulations of the main results of filtering theory that do pass the tests, and to show how they give rise to approximation algorithms with similar properties. Three basic results will be considered; they will be presented in the next three sections in "before and after" pairs: the classical result followed by its reformulation. The first result, the so-called "Bayes" formula for conditional expectations, is included to demonstrate the key idea; namely that conditional expectations have natural versions that depend continuously on the observation data. The second and third sections consider the equations of evolution of conditional probabilities and densities for signals that are Markov chains and diffusion processes; here the classical stochastic-differential-equation models transform into ordinary and partial differential equations.

The final section is mainly concerned with the "sure" convergence and statistical robustness of discrete approximations to the model of the Markov chain filter, though approximations to the partial differential equations of the diffusion filter are also briefly discussed.

The paper is written very much in the spirit of McShane's study [1] of stochastic differential equations; but it has not been necessary to introduce his general apparatus of first- and second-order belated integrals and canonical extensions since, for this problem, ordinary calculus suffices.

1. A FORMULA FOR CONDITIONAL EXPECTATIONS

In the remainder of the paper, unless otherwise stated, all random variables will be assumed to be defined on an underlying probability space $(\Omega,\mathcal{F},P)$. Upper-case letters will be used to denote random variables (and also, unfortunately, matrices) and lower-case letters to denote functions on other spaces.

Suppose, for $0\le t\le T$, (S_t) is a stochastic process representing a signal, (W_t) is an independent Brownian motion representing noise and (Y_t) is an observation process of the "signal plus white noise" variety

$$Y_t = \int_0^t S_u du + W_t. \tag{1.1}$$

Let $\mathcal{Y}_t$ denote the σ-field generated by $(Y_s;0\le s\le t)$. The following representation of a conditional expectation is due, in its full generality, to Kallianpur and Striebel [2] though special versions occur in the papers of Bucy [3], Wonham [4] and Zakai [5].

Theorem 1 Suppose that

$$\int_0^T S_u^2 du < \infty \text{ a.s.} \tag{1.2}$$

Let F be a random variable with finite mean. Then for $0\le t\le T$,

$$E[F|\mathcal{Y}_t](\omega) = \frac{\int F(\omega')\exp[\int_0^t S_u(\omega')dY_u(\omega) - \frac{1}{2}\int_0^t S_u^2(\omega')du]P(d\omega')}{\int \exp[\dots]P(d\omega')} \tag{1.3}$$

a.s. [P]

where the stochastic integrals in this expression are defined over $(\Omega\times\Omega',\mathcal{F}\times\mathcal{F},P\times P)$.

If the signal belongs to a certain class of semimartingales that includes most cases of interest, the formula can be modified as follows.

Theorem 2 Suppose S_t is a semimartingale of the form

$$S_t = B_t + M_t \tag{1.4}$$

where B_t is a process with paths of bounded variation, and M_t is a martingale with continuous sample paths, and where for all positive k

$$E[e^{k\int_0^t d<M>_t}] < \infty, \quad E[e^{k\int_0^t |dB_t|}] < \infty. \tag{1.5}$$

Here $<M>_t$ is the quadratic variation of M_t. Suppose F is independent of (W_t) and has finite expectation. Let $\hat{f}_t$ be defined on C[O,T] by

$$\hat{f}_t(y) = \frac{E[F \exp(y(t) - \int_0^t y(u)\,dS_u - \tfrac{1}{2}\int_0^t S_u^2 du)]}{E[\exp(y(t) - \ldots)]} \tag{1.6}$$

Then $\hat{f}_t$ is locally Lipshitz continuous with respect to the uniform norm on C[O,T] and $\hat{f}_t(Y)$ is a version of $E[F|\mathcal{Y}_t]$.

The full proof of this theorem is full of technical detail and will be given elsewhere [6], but the underlying argument is elementary enough. The terms in the exponents of formula (1.3) are integrated "by parts" and Y_t is replaced by y(t) to yield (1.6).

The theorem shows that $\hat{f}_t$, regarded as an estimate of F, is statistically robust in the following sense. Suppose the signal S_t and the variable F are functions of some underlying process (X_t). Let the probability space Ω be the product of the space Ω_X of the sample paths of (X_t), which we shall assume is complete separable metric, and the space C[O,T] of sample paths of (Y_t). The distributions on Ω can then be endowed with the usual topology of weak convergence. Suppose F is a bounded continuous functional f(X). If we regard $\hat{f}(Y)$ as a least-squares estimate of f(X), then Theorem 2 tells us that the mean square error $E_{\tilde{P}}[(f(X)-\hat{f}(Y))^2]$ will vary continuously as the distribution $\tilde{P}$ varies. Consequently if a "true" distribution P is close in a weak sense to the idealized distribution P determined by (1.1) then the "true" mean square error will not be significantly different from the idealized one. These questions of robustness are considered in more detail in [6].

2. FILTERS FOR MARKOV CHAINS

Suppose in a filtering problem the process (X_t) to be estimated is a Markov chain with a finite number of states (which we shall identify with the integers 1,2,...,n). Suppose the (column) vector of probabilities $p_t^i = P(X_t=i)$ satisfy $\dot{p} = A_t p$ where A_t is a Markov intensity matrix continuous in t. Suppose the observations (Y_t) form an m-vector process satisfying for $0 \le t \le T$,

$$dY_t = g(X_t)dt + dW_t , \quad Y_0 = 0 \tag{2.1}$$

where (W_t) is an m-vector of Brownian motions independent of each other and (X_t). The following filter representation is essentially due to Wonham [4], though we shall present the result in terms of unnormalized conditional probabilities, a form proposed by Zakai [5]. If in Theorem 1 F is taken to be the indicator functions of the states of X, then these unnormalized probabilities correspond to the numerator of the ratio in (1.3).

Theorem 3 Let $\mathcal{Y}_t$ be the σ-field generated by $(Y_s : s\leq t)$ and let G^k be the diagonal matrix with $(i,i)^{th}$ element $g^k(i)$. If (X_t) and (Y_t) are defined as above then

$$P(X_t=i\,|\,\mathcal{Y}_t) = \frac{Q_t^i}{\Sigma_j\, Q_t^j} \tag{2.2}$$

where the process $Q_t \equiv (Q_t^1,\ldots,Q_t^n)$ satisfies the stochastic differential equation

$$dQ_t = A_t Q_t dt + (\Sigma_k G^k dY_t^k)Q_t, \quad Q_0 = p_0. \tag{2.3}$$

The following is a sharpening of this result.

Theorem 4 Let y be an m-vector of continuous functions on [0,T]. Let $L_t(y)$ denote the diagonal matrix exponential

$$L_t(y) \equiv \exp[\Sigma_k (G^k y^k(t) - \tfrac{1}{2}(G^k)^2 t)]. \tag{2.4}$$

Suppose $r_t(y)$ is the solution of the ordinary linear differential equation

$$\frac{dr}{dt} = L_t^{-1}(y) A_t L_t(y) r, \quad r_0(y) = p_0 \tag{2.5}$$

If

$$\hat{p}_t^i(y) \equiv \frac{L_t(y)_{i,i} r_t^i}{\Sigma_i L_t(y)_{i,i} r_t^i} \tag{2.6}$$

then $\hat{p}_t^i$ is locally Lipschitz in y (for the uniform Euclidean norm) and $\hat{p}_t^i(Y)$ is a version of $P(X_t= i\,|\,\mathcal{Y}_t)$.

Here the "uniform Euclidean norm" refers to

$$\|y\| \equiv \sup(|y(t)| : 0 \leq t \leq T),$$

where $|\cdot|$ is the Euclidean norm of a vector. We shall denote the Euclidean norm of a matrix by $|\cdot|$, as well.

Proof We first establish $r_t(y)$ is locally Lipschitz in y uniformly in t. It is easy to verify this property holds for $B_t(y) \equiv L_t^{-1}(y)A_tL_t(y)$. Furthermore a standard theorem in the theory of ordinary differential equations tells us that (2.6) has a unique solution with bounds depending continuously on $\|y\|$.

Let z be a second n-vector continuous function. Then it follows from (2.5) that

$$|r_t(y) - r_t(z)| \le \int_0^t B_s(y)\,|r_s(y) - r_s(z)|\,ds$$

$$+ \int_0^t |B_s(y) - B_s(z)|\,|r_s(z)|\,ds$$

$$\le \int^t B_s(y)\,|r_s(y) - r_s(z)|\,ds$$

$$+ k(\|y\|, \|z\|)\,\|y-z\|$$

where k(.,.) is some continuous function. Gronwall's inequality then gives the uniform local Lipschitz condition. For the same to be true of $\hat{p}_t(y)$ we require that $\Sigma_i L_{ii} r^i$, the denominator in (2.6), be bounded away from zero for all t locally in y. Since A is a Markov intensity matrix its off-diagonal terms are all non-negative. The matrix $L^{-1}AL$ also has this property. Consequently $\dot{r}_t^i \ge a_{ii}(t) r_t^i$ for all r_t in the non-negative orthant of R^n. Since p_0 lies in this set it follows that $r_t^i \ge (\exp\int_0^t a_{ii} ds) p_0^i$. The positivity of at least one p_0^i and the continuity of the $a_{ii}(t)$ then give the result required.

It remains to verify that $\hat{p}_t(Y)$ is the conditional probability vector of X_t. Let

$$q_t(y) = L_t(y) r_t(y). \qquad (2.7)$$

If $q_t(Y)$ is then expanded by Ito's rule we find it satisfies (2.3). Such an equation has an a.s. unique solution and so $q_t(Y) = Q_t$ a.s. The result then follows from (2.2). This completes the proof.

It should be emphasized that there is nothing "canonical" about the pair of equations (2.5) and (2.6) representing the filter. It is left to the reader to verify that if ρ_t satisfies

$$\frac{d\rho}{dt} = (e^{-\Sigma_k G^k y^k} A\, e^{\Sigma_k G^k y^k} - \tfrac{1}{2}\Sigma_k (G^k)^2)\rho \qquad \rho_0 = p_0 \qquad (2.8)$$

and $\hat{p}_t^i$ is set to be the normalized form of $(\exp\Sigma_k g^k(i) y^k)\rho_t^i$, this is an equivalent definition of $\hat{p}_t$.

A consequence of Theorem 4 is that the solution Q_t of the stochastic differential equation (2.3) can be expressed as a continuous functional of several driving processes Y_t^k. In the context of the theory of stochastic differential equations this result is unusual and it is worth studying from that point of view. Consider the vector differential equation

$$dx = a(x)dt + b(x)dy \tag{2.7}$$

where y(t) is a scalar differentiable function. If the vector field b is constant, y(t) enters additively and it is a simple matter to verify that the solution x is a continuous functional of y; if b is variable it is possible, under mild conditions, to make it constant by continuously transforming coordinates so that one axis is aligned along an integral curve of b. x is therefore continuous in y for general b. This argument was used, in essence, by Wong and Zakai [7] to establish that the solution of ordinary differential equations forced by noise converges, as the noise tends to Brownian motion, to the solution of a stochastic differential equation (of Stratonovich type). McShane [1, p.215] and Sussman [8] make the point that the continuous extension of x(y) to all continuous functions y defines a generalized solution of (2.7) which provides, when y(t) is endowed with an appropriate distribution, a natural alternative definition for the solution of stochastic differential equations of McShane's canonical-extension type [1] or of Stratonovich type [8]. Suppose now a second driving term c(x)dz is added to (2.7), where z(t) is differentiable. Though a solution still exists, it is no longer in general a continuous functional of its forcing terms. The one case when continuity is preserved is where the vectors b and c "commute":

$$\Sigma_j\left(\frac{\partial b^i}{\partial x^j}c^j - \frac{\partial c^i}{\partial x^j}b^j\right) = 0.$$

Then it is possible under coordinate transformation to align axes along integral curves of both b and c, and the previous argument can be applied. It is this property that is possessed by the filtering equations; commutivity of the vector fields translates into the commutivity of the diagonal matrices G^k; $k=1,\ldots,m$.

McShane [1, p.226] has also considered the natural representation of Wonham's filter. It seems to me that the representation of the filter by the ordinary differential equation (2.5) is simpler and more flexible in application than its formulation in terms of canonical extensions, but this is a question of individual preference.

Both Theorems 3 and 4 can be generated to cover the practically important case where the observation function g is a differentiable function of time t. If $L_t(y)$ is modified to be

$$L_t(y) = \exp[\Sigma_k (G_t^k y^k(t) - \int_0^t \dot{G}_s^k y^k(s)ds - \tfrac{1}{2}\int_0^t (G_s^k)^2 ds)] \tag{2.8}$$

the representation given by (2.5) and (2.6) is still valid.

3. FILTERS FOR DIFFUSION PROCESSES

Since the early papers of Bucy [3], Kushner [9] and Stratonovich [10] on the subject, an extensive literature has developed on the theory of filtering for diffusion processes. A recent addition is the paper [11] by Levieux; this contains a theorem on the existence of unnormalized conditional probability densities for a diffusion process. A version of this result is presented in Theorem 5, and its "robust" modification is given in Theorem 6. For notational simplicity, only equations with time-invariant coefficients are considered.

Let (X_t) be a diffusion on R^n with a probability density $p_t(x)$ governed by the Fokker-Planck equation

$$\frac{\partial}{\partial t}p_t(x) = Ap_t(x) \equiv -\Sigma_i \frac{\partial}{\partial x^i}(f_i(x)p_t(x)) + \tfrac{1}{2}\Sigma_{ij}\frac{\partial}{\partial x^i \partial x^j}(\sigma_{ij}(x)p_t(x)). \tag{3.1}$$

Let (Y_t) be a scalar observation process of the form (2.1) but where g is now a function on R^n. The following conditions will be required.

(C1) A is uniformly elliptic; that is, for some $\lambda > 0$ for all $z_1,\ldots,z_n$,

$$\Sigma_{ij}\ \sigma_{ij}(x)z_i z_j \geq \lambda\Sigma_i z_i^2.$$

(C2) The coefficients f_i and σ_{ij} are elements of $C^{n_0}(R^n)$ where $n_0 = \text{int}[\frac{n}{2}+1]$ and they and their first-order derivatives are bounded.

(C3) g is bounded and differentiable and its first-order derivatives belong to $L^2(R^n)$.

(C4) $p_0(x)$, the density of X_0, belongs to the Sobolev space $H^{n_0}(R^n)$ and, for some positive k and t, $p_0(x) < k\rho_t(x)$ where ρ_t is the solution of $\dot{\rho} = A\rho$ with ρ_0 as the Dirac delta function.

(C5) p_0, g, f_i and σ_{ij} are bounded and have continuous bounded derivatives up to fifth order.

Theorem 5 If conditions (C1) to (C4) are satisfied the equation

$$dQ_t(x) = AQ_t(x)dt + g(x)Q_t(x)dY_t, \; Q_0(x) = p_0(x) \qquad (3.2)$$

has an a.s. unique solution and is almost everywhere positive on $[0,T] \times R^n \times \Omega$. Furthermore

$$\hat{P}_t(x) = \frac{Q_t(x)}{\int_{R^n} Q_t(x)dx} \qquad (3.3)$$

is the conditional probability density of X_t given $\mathcal{Y}_t$.

This theorem is a special case of Theorem III.3 in [11].

Theorem 6 If the conditions (C1) and (C5) are satisfied, for any $y \in C[0,T]$ there exists a unique solution $r_t(x,y)$, $0 \le t \le T$, to the partial differential equation

$$\frac{\partial}{\partial t} r_t(x) = e^{-(g(x)y(t) - \frac{1}{2}g^2(x)t)} A[e^{(g(\cdot)y(t) - \frac{1}{2}g^2(\cdot)t)} r_t(\cdot)](x)$$

$$r_0(x) = p_0(x). \qquad (3.4)$$

The solution is locally Lipschitz in y uniformly in t. If in addition (C2), (C3) and (C4) are satisfied then the conditional probability density of X_t

$$\hat{P}_t(x) = \frac{e^{g(x)Y_t - \frac{1}{2}g^2(x)t} r_t(x,Y)}{\int_{R_n} e^{g(x)\ldots t} r_t(x,Y)dx} \quad \text{a.s.} \qquad (3.5)$$

Proof The steps are identical in pattern to those in the proof of Theorem 4, though, of course, different existence and estimate theorems have to be invoked. Existence and uniqueness of the solution to (3.4) follows from Theorems 12 and 16 in Friedman [12], Chapter 1. With the condition (C5) it follows from an inequality on fundamental solutions ((6,12) Ch.1 [12]) that $r_t(x,y)$ and its first- and second-order derivatives with respect to x share a common bound depending continuously on $\|y\|$. Let B_t^y denote the differential operator defined by the right-hand side of (3.4). Then from (3.4)

$$r_t(x,y) - r_t(x,z) = \int_0^t B_s^y(r_s(x,y) - r_s(x,z))ds + \int_0^t (B_s^y - B_s^z) r_s(x,z)ds.$$

Let $h_s(x)$ denote the integrand of the second integral. All the derivatives with respect to x and coefficients contained in this expression are locally bounded in y and z and differentiable in the arguments y(s), z(s). Hence it is easy to verify that for each x $|h_s(x)| \le k(\|y\|, \|z\|)\,\|y-z\|$ for some continuous $k(\cdot,\cdot)$. Since $r_t(\cdot,y) - r_t(\cdot,z)$ is the solution of

$$\frac{\partial\rho}{\partial t} = B_t^y \rho + n_t. \quad \rho_0 = 0$$

it also follows from the inequality on fundamental solutions that r_t is locally Lipschitz in y uniformly in t.

To prove the second part, let

$$R_t(x) = e^{-(g(x)Y_t - \frac{1}{2}g^2(x)t)} Q_t(x)$$

where Q_t is defined as in (3.2). Now expand $R_t(x)$ by Ito's rule for each x. The resulting integral expression, which holds a.s., coincides with the integral in t of (3.4) with r_t replaced by R_t and y(t) replaced by Y_t. But we have established this equation has a unique solution, for each sample path of Y_t, which is $r_t(x,Y)$. So $r_t(x,Y) = R_t(x)$ a.s. and the conclusion of the theorem then follows from Theorem 5.

With suitable smoothness conditions versions of both Theorems 5 and 6 exist for time-varying coefficients. The form of the equations in Theorem 5 are unchanged; in Theorem 6 the exponential coefficients are changed in a fashion analogous to the modification of L_t in (2.8).

4. DISCRETE APPROXIMATIONS

If a Markov-chain filter is to be made out of analogue devices, the equations (2.5) and (2.6) and their variations such as (2.8) provide natural models for its construction. If, however, the device that operates on the observation process is digital, the filter model must first be reduced to a difference method that generates a discrete approximation. This section is concerned with ways of deriving such difference methods. The Markov-chain case will be treated almost exclusively; approximations for diffusion-process filters will be touched upon only briefly.

We shall adopt the definitions and notation of Section 2. To simplify the presentation it will be assumed that A is constant, and the input is a single process y(t). Also we shall define approximations to the unnormalized conditional probability vector $q_t(y)$ as defined in (2.7) rather than to the normalized conditional probability; an approximation to the latter is obtained by simple normalization. Though $q_t(y)$ does not satisfy an ordinary differential equation (we are not assuming that y(t) is differentiable) a difference method for approximating it can be constructed in the following way: select a convergent difference method for the ordinary differential equation for r_t and, if the approximant at t = nh is taken to be r_n^h, set

$$q_n^h \equiv L_{nh} r_n^h . \tag{4.1}$$

q_n^h will then approximate $q_t(y)$ and can be expressed in difference form. The derivation of the following methods, which are all of practical interest, will illustrate this procedure. Since sample paths y(t) will generally be highly oscillatory there is little point in employing a high order method; so the methods are all of first order.

First consider the Euler method applied to (2.5). The step-length is h.

$$r_{n+1}^h = r_n^h + h\,L_{nh}^{-1} A L_{nh} r_n^h \qquad r_0^h = p_0 . \tag{4.2}$$

Note that $L_{nh} = e^{Gy(nh) - \frac{1}{2}G^2 nh}$ and depends y only through y(nh). The corresponding difference method for q_n^h, obtained by substitution in (4.2), is

$$q_{n+1}^h = e^{G\Delta y - \frac{1}{2}G^2 h}(1 + Ah) q_n \qquad q_0 = p_0 . \tag{4.3}$$

where Δy is short for y((n+1)h) - y(nh). In practical applications measurements are not made of the "integrated" observation y(t) but of some smooth approximation to its "derivative" such as $\Delta y/h$. Consequently, though the two methods (4.2) and (4.3) are mathematically equivalent it will often be simpler to perform calculations in form (4.3) rather than (4.2).

For Markov chains that arise as approximations to diffusion processes, explicit methods, such as the Euler method, may suffer problems of numerical instability. To improve stability we might use an implicit Euler method:

$$(1 - hL_{(n+1)h}^{-1} A\, L_{(n+1)h}) r_{n+1}^h = r_n^h , \; r_n^h = p_0 \tag{4.4}$$

which translates into

$$(1 - hA) q_{n+1}^h = e^{G\Delta y - \frac{1}{2}G^2 h} q_n^h , \; q_0^h = p_0 . \tag{4.5}$$

Again, probably (4.5) is computationally the simpler of the two forms.

The following method, called by McShane [1, p.205] the Cauchy-Maruyama approximation, is the natural discretization of what would be McShane's canonical-extension equation for the process Q_t given by (2.3). It is the simplification of (4.3) obtained by expanding the exponential and discarding $h^2, h\Delta Y, \Delta Y^3$ and terms of higher order.

$$q^h_{n+1} = (I + (A-\tfrac{1}{2}G^2)h + G\Delta y + \tfrac{1}{2}G^2\Delta y^2)q^h_n. \tag{4.6}$$

The following theorem summarizes the convergence properties of these methods. Let $w_y(h)$ denote the modulus of continuity of y:

$$w_g(h) = \max\{|y(t)-y(s)| : 0 \le s,t \le T, |s-t| \le h\}.$$

Theorem 7 If q^h_n is generated by either (4.3) or (4.5) then for some continuous $k(\cdot)$ and for all n,h with $0 \le nh \le T$,

$$|q_{nh}(y) - q^h_n(y)| \le k(\|y\|)(h + w_y(h)). \tag{4.7}$$

If q^h_n is given by (4.6) then for some continuous k'

$$|q_{nh}(y) - q^h_n(y)| \le k'(\|y\|)(h + w_y(h) + \frac{w_y^3(h)}{h}). \tag{4.8}$$

Proof The argument is an elementary one of numerical analysis. Suppose, for instance, q^h_n is defined by (4.3) and the corresponding r^h_n by (4.2). It is clearly sufficient to establish a similar bound for $|r_{nh}-r^h_n|$. Let $e_n = r_{nh}-r^h_n$. Then it follows from (2.5) and (4.2) that

$$e_{n+1} = e_n + hB_{nh}e_n + \int_{nh}^{(n+1)h} (B_t-B_{nh})r_t dt + \int_{nh}^{(n+1)h} B_{nh}(r_t-r_{nh})dt$$

where B_t is short for $L_t^{-1}AL_t$. But for some k_1,k_2 depending continuously on $\|y\|$ r_t and B_t are bounded on $[0,T]$ by k_1 and for $nh \le t \le (n+1)h$,

$$|B_t-B_{nh}| \le k_2 w_y(h).$$

Hence

$$|e_{n+1}| \le |1 + hB_{nh}|\,|e_n| + h(k_1k_2w_y(h) + k_1^3h)$$

and the result then follows from the usual reduction of recursive inequalities. The proof for the second case is almost identical. For the third case we find, on substitution, that

$$r^h_{n+1} = hL^{-1}_{(n+1)h}AL_{nh}r^h_n + [e^{-(G\Delta y-\frac{1}{2}G^2h)}(I+G\Delta y+\tfrac{1}{2}G^2(\Delta y^2-h))]r^h_n.$$

The expression in [.] differs from I by a term of order Δy^3. This contributes the extra term in the bound.

Theorem 7 shows that methods (4.3) and (4.5) converge for all sample paths y(t) of the observations and that the Cauchy-Maruyama approximation also converges for all sample paths of practical significance ($w_y(h) = o(h^{1/3})$). Other convergent methods, tailored to a particular problem, can be designed in a similar fashion. It also follows from the theorem that the convergence is uniform over families of sample paths that satisfy a uniform Hölder condition:

$$\{y : w_y(h) \leq kh^{\alpha}\}.$$

This allows us to define a concept of "robust" rate of convergence in the following way.

The idealised distribution of (Y_t) that is defined by (2.1) is absolutely continuous with respect to that of Brownian motion. Consequently, if P denotes the idealised distribution, by Lévy's modulus theorem for Brownian motion, for all $\alpha < \frac{1}{2}$

$$P(\lim_{h \to 0} \frac{w_Y(h)}{h^{\alpha}} = 0) = 1. \tag{4.7}$$

The following assertion then follows immediately from Theorem 7.

Corollary 8 If q_n^h is defined by either (4.3) or (4.5) or (4.6) and P is any member of the family of distributions P of the process (Y_t) satisfying (4.7) (which contains the idealised distribution), then for all $\alpha < \frac{1}{2}$

$$P(\lim_{h \to 0} \sup_t \frac{|q_t(Y) - q_{n(t)}^h(Y)|}{h^{\alpha}} = 0) = 1$$

where $n(t) = \mathrm{int}[\frac{t}{h}]$.

Since the family of distribution in the corollary certainly contains all practical distributions of interest, we can say that the methods of this section have a robust rate of convergence of order (almost) $h^{\frac{1}{2}}$. In a sense, however, this invokes a weak concept of robust rate since it is only defined in terms of the order of the error as $h \to 0$. A stronger concept of rate "h^{α}" would be that $P(\sup_t |q_t(Y) - q_{n(t)}^h(Y)| < Kh^{\alpha}) = 1$, (where K is random) should hold over the class of distributions P of interest. Other statistical measures of error may also be used, such as the root-mean-square error at T. McShane [1] has shown that the root-mean-square error of the Cauchy-Maruyama approximation is, under the idealised distribution, of order h. It would be interesting to determine whether this can be interpreted for a suitable class of distributions as a robust rate in the stronger sense above.

The desirability of the property of convergence in a difference method should not be over-emphasized. A basic method for approximating the stochastic filter equation (2.3) is the explicit method

$$q_{n+1}^h = q_n^h + Aq_n^h h + Gq_n^h \Delta y \tag{4.8}$$

where, as before, $\Delta y = y((n+1)h) - y(nh)$. It is well known that $q^h_{n(t)}(Y)$ will converge to $q_t(Y)$ almost surely under the idealised distribution. However for any differentiable y it will converge to the solution q'_t of $\dot{q} = (A+G\dot{y})q$ which, as a simple analysis shows, is not the same as $q_t(y)$. In practice, however, the method will produce satisfactory approximations, for a given step length h, as long as h is greater than the "dependence time" of the noise W_t. Loosely speaking, one might define the dependence time to be the shortest step length such that the increment ΔW is independent of the previous noise. An apparent advantage of the three methods considered earlier is that they seem to be insensitive to this parameter.

Kushner [13] and Levieux [11] have both developed numerical methods for the approximation of the conditional probability densities of filtered diffusion processes. Both methods are based on the stochastic differential equations of the unnormalised form of the conditional probability density. If we adopt the notation and definitions of Section 3 Levieux's method is based on the "semi-discretization" (discrete in t, but not in x)

$$(I-Ah)q^h_{n+1}(x) = Gq^h_n(x)\Delta Y$$

where A is the elliptic operator in the Fokker-Planck equation of X_t, $q^h_n(x)$ is the approximant at $t = nh$ of Q_t and $\Delta Y = Y((n+1)h) - Y(nh)$. It is shown [11] that $q^h_{n(t)}$ converges almost surely to Q_t in a strong L^2-sense. Kushner's method is more complicated to describe, but it is closely related to the full discretization of the formula

$$q^h_{n+1}(x) = e^{G\Delta Y - \frac{1}{2}G^2 h} P^h_{nh} q^h_n(x)$$

where P^h_t is an approximation (as is $I + Ah$) to the transition probability density $P(X_{t+n} = x'/X_t = x)$. He proves his method converges almost surely in a weak L^2-sense. It is reasonable to conjecture on the basis of our previous analysis that Levieux's method would not have the "sure" convergence properties discussed at the beginning, and that Kushner's method would have these properties. Levieux, however, has emphasized the importance for numerical stability of an implicit form of difference equation. A suitable compromise might be the analogue in function space of the difference method (4.5):

$$(I-hA)q^h_{n+1}(x) = e^{G\Delta y - \frac{1}{2}G^2 h} q^h_n(x),$$

but a formal analysis is required before this can be confirmed.

REFERENCES

1. E.J. McShane, Stochastic Calculus and Stochastic Models, Academic Press, New York, 1974.

2. G. Kallianpur, C. Striebel, Estimation of stochastic systems: artibrary system process with additive white noise observation errors, Ann. Math. Stat. 39, 1968, pp.785-801.

3. R.C. Bucy, Nonlinear filtering theory, IEEE Trans. Automatic Control 10, 1965, pp.198-199.

4. W.M. Wonham, Some applications of stochastic differential equations to optimal nonlinear filtering, J. SIAM Control, Ser. A, 2, 1965, pp.347-369.

5. M. Zakai, On the optimal filtering of diffusion processes, Z. Wahrscheinlichkeitstheorie verw. Geb. 11, 1969, pp.230-243.

6. J.M.C. Clark, The local robustness of Bayes estimates with respect to weak perturbations of distribution (in preparation).

7. E. Wong, M. Zakai, On the convergence of ordinary integrals to stochastic integrals, Ann. Math. Stat. 36, 1965, pp.1560-1564.

8. H. Sussman, An interpretation of stochastic differential equations as ordinary differential equations which depend on a sample point, Bull. Am. Maths. Soc. 83, 1977, pp.296-298.

9. H.J. Kushner, On the differential equations satisfied by conditional probability densities of Markov processes, with applications, J. SIAM Control, 2, 1964, pp.106-119.

10. R.L. Stratonovich, Conditional Markov processes, Theor. Prob. Appl. 5, 1960, pp.156-178.

11. F. Levieux, Conception d'algorithmes parallélisables et convergents de filtrage récursif non-linéaire, Rapport Laboria No. 235, IRIA, 1977.

12. A. Friedman, Partial Differential Equations of Parabolic Type, Prentice-Hall, New Jersey, 1964.

13. H.J. Kushner, Probability Methods for Approximations in Stochastic Control and for Elliptic Equations, Academic Press, 1977.

A VIEW AT STOCHASTIC CALCULUS FOR POINT PROCESSES (tutorial)

by P. BREMAUD *

IRIA/LABORIA, B.P. 105, 78150 Rocquencourt, FRANCE

1. Martingales in stochastic systems theory.

In this tutorial, we will attempt to demonstrate the relevance of martingale concepts to the theory of dynamical point process systems. In stochastic systems theory, the dynamics of interest are the state dynamics and the information dynamics. When using the phrase "information dynamics" we have in mind those situations in which the available observations vary with time, and for which the decisions to be taken at each time t are based solely on the information collected at instant t. This is typically the case in recursive estimation, sequential analysis and optimal control. In the martingale approach the information dynamics play a central part and this approach is therefore particularly well suited to the analysis of the above problems. It is to be opposed to (or rather complemented by) the measure theoretic approach in which point processes are viewed as discrete random measures. In the latter approach, realizations are given at once in their totality, whereas in the martingale approach a realization is progressively unverled as time flows. Thus, the counting process N_t (N_t is the number of points in the interval $[o,t]$) is a natural description, and hence stochastic process theory is the adequate tool (rather than random measure theory).

We first give a short review of vocabulary and notations relative to stochastic processes and their associated information patterns.

Histories (information patterns)

Let X_t be some process (say, taking its values in R^m), and let, at each time $t > o$, $\mathcal{F}_t^x$ be the smallest σ-field that contains all events of the form :

$$\{\omega \mid X_{t_i}(\omega) \in C_i \ , \ 1 \leqslant i \leqslant n\} \ ,$$

where n is a strictly positive integer, $t_i \leqslant t$ and $C_i \in B^m$ – the borelian σ-field of R^m – for all i, $1 \leqslant i \leqslant n$. The family $\mathcal{F}_t^x$ is called the internal history of X_t. In general, a family $\mathcal{F}_t$ of increasing σ-fields is called a history, and if $\mathcal{F}_t \supset \mathcal{F}_t^x$, $\forall t \geqslant o$, it is called a history of X_t. Clearly $\mathcal{F}_t^x$ is the smallest history of X_t.

When $\mathcal{F}_t$ is a history of X_t, one also says that X_t is adapted to $\mathcal{F}_t$. Let x_t be some process (say, R^n – valued) adapted to $\mathcal{F}_t^x$, the internal history of X_t. Then, for each t :

$$(*) \quad x_t(\omega) = \varphi(t, \mathcal{X}_t(\omega))$$

where $\mathcal{X}_t(\omega) = \{X_s(\omega), s \in [o,t]\}$. More precisely, (*) is a short way for writing :

$$x_t(\omega) = \psi_t(X_{s_{t,i}}(\omega), \ i \geqslant 1)$$

where the $s_{t,i}$ are times in $[o,t]$, and ψ_t is a measurable mapping from $(R^m)^\infty$ into R^n (this is a theorem, not just an observation!)

If $X_t = (Y_t, Z_t)$ where Y_t and Z_t are stochastic processes respectively R^k and R^l valued with $k + l = m$, we denote $\mathcal{F}_t^x$ by $\mathcal{F}_t^Y \vee \mathcal{F}_t^Z$.

Martingales and noise

Let $\mathcal{F}_t$ be a history and let M_t be a R-valued process adapted to $\mathcal{F}_t$ and such that, for all $o \leqslant s \leqslant t$, $E[|M_t|] < \infty$ and :

$$(1.1) \qquad E[M_t \mid \mathcal{F}_s] = M_s \ ; \quad P \text{ a.s.}$$

Then M_t is called a $(P, \mathcal{F}_t)$-martingale.

If M_t is a martingale, then its increments are uncorrelated, i.e., if $o \leqslant t_1 \leqslant t_2 \leqslant t_3 \leqslant t_4$, then :

$$E[(M_{t_4} - M_{t_3})(M_{t_2} - M_{t_1})] = o$$

Example:

Let X_t be an integrable R-valued process with independant increments, o mean $(E[X_t] = o, \ \forall t \geqslant o)$. Then X_t is a $(P, \mathcal{F}_t^X)$-martingale. Also if X_t^2 is integrable, $X_t^2 - E[X_t^2]$ is a $(P, \mathcal{F}_t^X)$-martinga-

le. A classical illustration is the Wiener process W_t, a process with independent increments such that for all $o \leq s \leq t$, $W_t - W_s$ is a Gaussian random variable mean o, variance t-s.
Here W_t as well as $W_t^2 - t$ are $(P, \mathfrak{F}_t^W)$-martingales.

In the case of the Wiener martingale W_t, the increments are independent and it is this property that has made it a popular model for integrated white noise. A martingale M_t can be thought of as a form of integrated noise, but not white since the independence property for the increments is replaced by uncorrelation. Another illustration along these lines is the Poisson process (to be featured in paragraph 2) which plays a central role in models of shot noise.

Martingales and information patterns

Let U be a square integrable, random variable and let X_t be some process. If U and the process X_t are not independent, then X_t carries some information about U that is not trivial. Suppose that at time t, only $\mathcal{X}_t = \{X_s, s \in [o,t]\}$ is observed and suppose that we want to find a functional of $\mathcal{X}_t$ that "approximates" U, i.e. we want to find, among all square integrable random variables of the form :

$$\varphi(t, \mathcal{X}_t)$$

one that is a close as possible to U. If the distance between U and the estimate $\varphi(t, \mathcal{X}_t)$ is given by

$$E[|U - \varphi(t, \mathcal{X}_t)|^2]^{\frac{1}{2}} ,$$

then the best approximation is the conditional expectation

$$\hat{U}_t = E[U / \mathfrak{F}_t^X]$$

Clearly $\hat{U}_t$ is a $(P, \mathfrak{F}_t^X)$-martingale. The above example of martingale clearly shows that martingales are subjacent to all problems in which increasing information patterns play a role. Moreover, this example carries some generality : indeed, it can be shown that any uniformly integrable $(P, \mathfrak{F}_t)$-martingale has the form :

$$M_t = E[U / \mathfrak{F}_t] ,$$

where $U = M_\infty = \lim_{t \to \infty} M_t$, where the lim is P a.s. and L_1 (see any textbook on martingale theory).

Martingales and state dynamics.

Martingales appear quite naturally in the modeling of systems which are differential systems driven by noise. A differential system of the form

$$\dot{X}_t = f(t,X_t)$$

when corrupted by white noise, takes the form :

$$(1.2) \qquad \dot{X}_t = f(t,X_t) + \beta_t$$

where β_t is the white noise. The correct interpretation of (1.2) is :

$$(1.2.') \qquad X_t = X_o + \int_o^t f(s,X_s)\, ds + W_t$$

where W_t is a Wiener process (see the tutorials of E. Wong and M.H.A. Davis in this volume). Equations such as (1.2.') are known under the name of Ito differential equations. More generally, one can consider states X_t that satisfy semi-martingale equations.

$$(1.3) \qquad X_t = X_o + \int_o^t f_s\, ds + m_t$$

where m_t is a $(P, \ {}_t)$-martingale. Such semi-martingale equations occur in a natural way in queueing theory (see [1] and the tutorial of R. Boel) and more generally in point process theory. A few examples will be given later on.

2. Doubly stochastic Poisson processes and stochastic intensities.

Point processes.

First, we give a few definitions concerning point processes. A point process (over $[o,\infty]$) is a sequence T_n of $[o,\infty]$ - valued random variables such that $T_o = o$ (convention!) and :

$$T_n < \infty \implies T_n < T_{n+1}$$

Put $T_\infty = \lim \uparrow T_n$, and say that the point process is non-explosive if $T_\infty = \infty$ almost surely.

Define the counting process N_t by :

$$(2.1)\qquad N_t = \begin{cases} n & \text{if } t \in [T_n, T_{n+1}) \\ \infty & \text{if } t \in [T_\infty, \infty) \end{cases}$$

By extension, we call N_t a point process. It is a right continuous process, non-negative, integer valued, and for each t, N_t counts the T'_ns in $(o,t]$ (T_o is excluded from this enumeration, i.e. $N_o \equiv o$).

Doubly stochastic point processes

Let $\mathcal{F}_t$ be a history of the form :

$$(2.2)\qquad \mathcal{F}_t = \mathcal{F}_\infty^Y \vee \mathcal{F}_t^N$$

that is to say, for each $t > o$, $\mathcal{F}_t$ is the smallest σ-field that contains all the events of $\mathcal{F}_t^N$ and the events of the form :

$$\{\omega \mid Y_{t_i}(\omega) \in C_i,\ 1 \leqslant i \leqslant n\} ,$$

where Y_t is a R^1-valued process, n is a strictly positive integer, $t_i \in [o,\infty)$ and $C_i \in \mathcal{B}^1$, $\forall\, i$, $1 \leqslant i \leqslant n$. We are going to define a doubly stochastic Poisson process (doubly stochastic with respect to the process Y_t) as a point process N_t such that :

$$(*)\ E[e^{iu(N_t-N_s)} / \mathcal{F}_s] = \exp\{(e^{iu}-1) \int_s^t \lambda_v\, dv\}$$

where λ_t is a non negative measurable stochastic process such that for each t, λ_t is $\mathcal{F}_\infty^Y$ - measurable and $\int_o^t \lambda_s\, ds < \infty$, a.s.

A typical case would be : λ_t is adapted to $\mathcal{F}_\infty^Y$, i.e. :

$$\lambda_t = \lambda(t, \mathcal{Y}_t) ,$$

Another particular case is :

$$Y_t = \Lambda$$

$$\lambda_t = \Lambda$$

where Λ is a non-negative random variable (this is the classical example of a doubly stochastic Poisson process).

Let as examine a few consequences of (*). First, since

$\mathcal{F}_s = \mathcal{F}_\infty^Y \vee \mathcal{F}_s^N$ and the R.H.S. of (*) is $\mathcal{F}_\infty^Y$-measurable, we obtain by conditioning both sides of (*) by $\mathcal{F}_s^N \vee \mathcal{F}_\infty^Y$:

$$(*1) \quad E[e^{iu(N_t-N_s)} / \mathcal{F}_\infty^Y] = \exp\{(e^{iu}-1) \int_0^t \lambda_v \, dv\}$$

and therefore :

$$(*2) \quad E[e^{iu(N_t-N_s)} \mid \mathcal{F}_s^N \vee \mathcal{F}_\infty^Y] = E[e^{iu(N_t-N_s)} \mid \mathcal{F}_\infty^Y]$$

The above equality being true for all real u, N_t-N_s is thus independent of $\mathcal{F}_s^N$ conditionally to $\mathcal{F}_\infty^Y$ (this is a well known result on conditional independence :

$$E[e^{iu X} / \mathcal{F} \vee \mathcal{G}] = E[e^{iu X} / \mathcal{F}], \quad \forall U \in (-\infty;+\infty)$$

is equivalent to the conditional independence of X and $\mathcal{G}$ given $\mathcal{F}$). In other words :

(i) <u>Conditionally to $\{Y_t, t \in [0,\infty]\}$, N_t has independent increments</u>.
Also, by (*1) :

$$(2.3) \qquad P[N_t-N_s = k / \mathcal{F}_\infty^Y] = e^{-\int_s^t \lambda_u \, du} \frac{(\int_s^t \lambda_u \, du)^k}{k\,!}$$

In other words :

(ii) <u>Conditionally to $\{Y_t, t \in [o,\infty)\}$, for all $o \leqslant s \leqslant t$, N_t-N_s is distributed according to a Poisson law of parameter</u> $\int_s^t \lambda_u du$.

<u>Stochastic intensity</u>.

The process λ_t is called the $\mathcal{F}_t$-<u>intensity</u> of the doubly stochastic point process N_t. Note that λ_t is in general a <u>stochastic</u> process ! If it is a deterministic process, then N_t is a Poisson process in the classical sense.

Suppose, for the sake of simplicity, that λ_t is a <u>bounded</u> process.

Then :

$$M_t = N_t - \int_0^t \lambda_s \, ds$$

is integrable (since, by (2.3), $E[N_t] = E[\int_0^t \lambda_s \, ds]$, and λ_t is bounded). Moreover, it is not difficult to show, using (i) that M_t is a $(P, \mathfrak{F}_t)$ martingale. We call it the fundamental $\mathfrak{F}_t$-martingale associated to N_t.

Writing down the martingale property for M_t, we obtain

$$E[N_t - N_s \mid \mathfrak{F}_s] = E[\int_s^t \lambda_u \, du \mid \mathfrak{F}_s]$$

If in addition to being bounded, λ_t is right-continuous, then using Lebesgue averaging theorem and Lebesgue dominated convergence theorem, we obtain :

(2.4) $\quad$ a.s. $\lim_{t \downarrow s} \frac{1}{t-s} E[N_t - N_s \mid \mathfrak{F}_s] = \lambda_s$

Equality (2.4) reminds of the definitions of stochastic intensity found in the technical litterature and aptly express the idea that λ_t is a "<u>local characteristic</u>" of N_t, or, in some sense, the "<u>derivative</u>" $\frac{dN_t}{dt}$.

However relying on (2.4) for a definition of stochastic intensity is not safe from a purely mathematical stand point. Also, a definition such in the form of (2.4) certainly appeals to intuition but its operational value is very small (*). In extending the notion of stochastic intensity to point processes that are not doubly stochastic Poisson processes we will choose

(!) $\quad M_t \stackrel{d}{=} N_t - \int_0^t \lambda_s \, ds = (P, \mathfrak{F}_t)\text{-martingale}$

as our starting point. More precisely, let $\mathfrak{F}_t$ be a history of N_t and let λ_t be a (bounded for the sake of simplicity) measurable process adapted to $\mathfrak{F}_t$. The, if (!) holds, we say that λ_t is the $\mathfrak{F}_t$-intensity of N_t.

(*) Of course, at this point, the above statements seem gratuitous. As to the first one, it is generally recognized that seldom, if ever, correct mathematical analysis has emerged from definition (2.4) unless drastic unpractical conditions were imposed upon λ_t concerning the second statement, a look at the achievements of both approaches to the notion of stochastic intensity should convince the reader. The martingale approach includes all the results of the approach (2.4), and much more !

Does intensity characterizes the probability measure ?

We have seen that in the case of doubly stochastic Poisson processes with a bounded intensity, (!) was a consequence of (*). A question arises naturally, if we want to take (!) as a general definition of intensity : does (!) characterizes P ? More precisely : if N_t is a point process such that (!) holds for a bounded process λ_t adapted to $\mathcal{F}_t = \mathcal{F}_\infty^Y \vee \mathcal{F}_t^N$, is it a doubly stochastic point process with the $\mathcal{F}_t$-intensity λ_t, i.e. is (*) (or equivalently (i) and (ii)) verified ? The answer "yes" was given by Watanabe in the case of Poisson processes (non doubly stochastic in [2]. The proof of Watanabe's result given by the present author in [3] applies to doubly stochastic point processes with minor modifications and yields the same result.

What now if we take (!) as a general definition of the $\mathcal{F}_t$-intensity λ_t (bounded) of a point process N_t ? Jacod in [4] has answered the question in the case where :

(2.5) $\qquad \mathcal{F}_t = \mathcal{F}_o \vee \mathcal{F}_t^N ,$

is the following way : if λ_t is a non negative mesurable process adapted to $_t$, and if P and P' are two probabilities such that :

$$N_t - \int_o^t \lambda_s \, ds = (P, \mathcal{F}_t)\text{-martingale},$$

and

$$N_t - \int_o^t \lambda_s \, ds = (P', \mathcal{F}_t)\text{-martingale},$$

and if more over P and P' coïncide on $\mathcal{F}_o$ ($P(A) = P'(A)$, $\forall A \in \mathcal{F}_o$), then :

$$P \equiv P' \quad \text{on} \quad \mathcal{F}_\infty$$

This result is an <u>uniqueness</u> result. It admits a converse (still due to Jacod [4]) : If $\mathcal{F}_t$ has the form (2.5), if λ_t is a non negative measurable process adapted to $\mathcal{F}_t$, and if P_o is a probability on $\mathcal{F}_o$, then there exists one (and only one in view of the above uniqueness result) probability P such that :

$$P \equiv P_o \quad \text{on} \quad \mathcal{F}_o ,$$

and

$$N_t - \int_o^t \lambda_s \, ds = (P, \mathcal{F}_t)\text{-martingale}.$$

This is the <u>existence</u> result.

We insist that the existence and uniqueness result is not restricted to doubly stochastic Poisson processes.

The existence and uniqueness result gives a full status of definition to (!), and shows that (!) is the right formalism for (2.4).

3. Stochastic integrals, martingales and Stieltjes calculus.

First encounter with stochastic calculus and stochastic integrals

Let N_t be a doubly stochastic Poisson process with the $\mathfrak{F}_t$-intensity λ_t where $\mathfrak{F}_t = \mathfrak{F}_\infty^Y \vee \mathfrak{F}_t^N$, and suppose that λ_t is bounded (this assumption only serves technical purposes and is not essential). We have already seen that :

$$M_t \overset{d}{=} N_t - \int_0^t \lambda_s \, ds$$

is a $(P, {}_t)$-martingale. It is not difficult to show that :

$$(3.1) \qquad m_t \overset{d}{=} M_t^2 - \int_0^t \lambda_s \, ds = (N_t - \int_0^t \lambda_s \, ds)^2 - \int_0^t \lambda_s \, ds$$

is also a $(P, {}_t)$-martingale. As we will shortly see :

$$(3.2) \qquad m_t = \int_0^t C_s \, dM_s \quad ,$$

where C_t is a left continuous process adapted to $\mathfrak{F}_t$. It turns out that this is not a coincidence, and that all $(P, \mathfrak{F}_t)$-martingales m_t have the form (3.2), where C_t is not quite a left continuous process adapted to $\mathfrak{F}_t$, but "almost" : in fact $C_t(\omega)$ is a (t,ω) point wise limit of processes adapted to $\mathfrak{F}_t$ and left continuous (it is called a $\mathfrak{F}_t$-predictable process). This is due to Jacod [4]. Note : in (3.2), the integral is defined for each ω as a Stieltjes-Lebesgue integral ; more precisely

$$\int_0^t C_s \, dM_s = \sum_{n \geq 1} C_{T_n} 1(T_n \leq t) - \int_0^t C_s \lambda_s \, ds \quad .$$

Let us show that m_t defined by (3.1) has indeed the form (3.2) ; to do so we need the following rule of Stieltjes calculus

$$(3.3) \quad f(t)\, g(t) = f(o)\, g(o) + \int_0^t f(s)\, dg(s) + \int_0^t g(s-)\, df(s)$$

where f and g are functions of bounded variation on finite intervals, right continuous, with left hand limits.
Applying the above formula of integration by parts to

$M_t = N_t - \int_0^t \lambda_s \, ds$, we obtain

$$M_t^2 = \int_0^t M_s \, dM_s + \int_0^t M_{s-} \, dM_s = 2 \int_0^t M_{s-} dM_s + \int_0^t (M_s - M_{s-}) dM_s$$

that is to say, since $M_t - M_{t-} = N_t - N_{t-}$.

$$M_t^2 = 2 \int_0^t M_{s-} dM_s + N_t$$

Subtracting $\int_0^t \lambda_s \, ds$ to both sides of the above equality :

$$m_t = M_t^2 - \int_0^t \lambda_s \, ds = \int_0^t (2\, M_{s-} + 1)\, dM_s$$

In other words, (3.2) is verified with $C_t = 2\, M_{t-} + 1$. As we have already mentioned, the integration of left continuous process with respect to the fundamental martingales yields martingales. More precisely, the following is a weak form of the <u>direct integration theorem</u> of [5] and [6] :

<u>Let</u> N_t <u>be a doubly stochastic point process with the bounded</u> $\mathfrak{F}_t$-<u>intensity</u> λ_t $(\mathfrak{F}_t = \mathfrak{F}_t^Y \vee \mathfrak{F}_t^N)$. <u>Let</u> C_t <u>be any process adapted to</u> $_t$, <u>left continuous and such that</u> $E[\int_0^t |C_s| \lambda_s \, ds] < \infty$ <u>for all</u> $t > o$. <u>Then</u> $m_t = \int_0^t C_s \, dN_s$ is a $(P, \mathfrak{F}_t)$-<u>martingale</u>.

This result is also valid when N_t is not a doubly stochastic Poisson process, but more generally a point process with the bounded $\mathfrak{F}_t$-intensity λ_t, the $\mathfrak{F}_t$-intensity being defined by the martingale property (!), and $\mathfrak{F}_t$ being an arbitrary history of N_t.

Let us see how this can be used to obtain a result quite analogous to Girsanov's theorem (see E. Wong and M.H.R. Davis tutorials in this volume).

<u>Likelihood ratio for point processes</u>

In this subsection, we will change probabilities and therefore we must be more precise as to which probability is in force in such expressions as "almost surely". Also, as we will see, a <u>change in probability entails a change in intensity</u>, and therefore we must give additional information about P, writing "$(P, \mathfrak{F}_t)$-intensity", rather than the short form " $\mathfrak{F}_t$-intensity".

Let us start with a point process N_t defined on some probability space $(\Omega, \mathfrak{F}, P)$ and let $\mathfrak{F}_t$ be a history of N_t. Suppose that N_t admits the bounded $(P, \mathfrak{F}_t)$-intensity λ_t, that is to say

$$(3.4) \qquad M_t \stackrel{d}{=} N_t - \int_0^t \lambda_s \, ds = (P, \mathfrak{F}_t)\text{-martingale.}$$

Our goal is to construct a probability $\tilde{P}$ on $(\Omega, \quad)$ such that N_t admits the $(\tilde{P}, \mathfrak{F}_t)$-intensity $\tilde{\lambda}_t$, where $\tilde{\lambda}_t = \mu_t \lambda_t$, μ_t being some bounded left continuous process adapted to $\mathfrak{F}_t$. In other words we want $\tilde{P}$ to be such that :

$$(3.5) \qquad \tilde{M}_t \stackrel{d}{=} N_t - \int_0^t \mu_s \lambda_s \, ds = (\tilde{P}, \mathfrak{F}_t)\text{-martingale.}$$

In order to avoid some technical complications that are not essential to our purpose, we will suppose that we work on the interval $[0,1]$ rather than on $[0,\infty)$. We look for a $\tilde{P}$ to be absolutely continuous with respect to P, and therefore it suffices to define it by its Radon-Nikodym derivative $d\tilde{P}/dP$ which we take equal to L_1, where :

$$(3.6)\quad L_t = L_0 \left(\prod_{0<T_n\leq t} \mu_{T_n} \right) \exp\left\{ \int_0^t (1-\mu_s)\, \lambda_s \, ds \right\} , \quad t \in [0,1]$$

where $L_0 \equiv 1$.

Clearly, L_1 is non negative. In order that $d\tilde{P}/dP = L_1$ be a valid definition of $\tilde{P}$ we must have $E[L_1] = 1$. More can be proven, namely L_t is a $(P, \mathfrak{F}_t)$-martingale, mean 1 (the mean of a martingale is a constant. This is done by rewriting L_t as a "differential equation" :

$$(3.7)\qquad L_t = L_0 + \int_0^t L_{s-}\, (\mu_s - 1)\, d(N_s - \int_0^s \lambda_u \, du)$$

Proving (3.7) is not difficult ; indeed in between jumps we see from (3.6) that :

$$dL_s = L_s\, (\mu_s - 1) = L_{s-}(M_s - 1)$$

and at a jump $(s = T_n)$.

$$dL_s = \Delta L_s = L_s - L_{s-} = L_{s-}(\mu_s - 1)$$

combining the two last equalities, we obtain (3.7).

Now, using the integration theorem (see [5] or [6] for details), we obtain from (3.7) the martingale property for L_t, and therefore $E[L_1] = E[L_t] = E[L_0] = 1$, which validates the definition of $\tilde{P}$ by :

$$d\tilde{P}/dP = L_1$$

Now it remains to prove (3.5). As is explained in M.H.A. Davis' tutorial (present volume), it is sufficient to show that :

$$L_t\, \tilde{M}_t = (P, \mathfrak{F}_t)\text{-martingale}$$

Details can be found in [6]. Here we will be content with giving an outline of the proof that shows the use of stochastic calculus, and in particular of formula (3.3). Indeed :

$$L_t\, \tilde{M}_t = \int_0^t L_{s-}\, d\tilde{M}_s + \int_0^t \tilde{M}_s\, dL_s$$

now :

$$\int_0^t L_{s-} d\tilde{M}_s = \int_0^t L_{s-}\, dM_s - \int_0^t L_{s-}(\mu_s - 1)\, \lambda_s \, ds$$

and :

$$\int_0^t \tilde{M}_s \, dL_s = \int_0^t \tilde{M}_s L_{s-}(\mu_s - 1) dM_s = \int_0^t \tilde{M}_{s-} L_{s-}(\mu_s - 1) dM_s + \int_0^t L_{s-}(\mu_s - 1) dN_s$$

(where we have used : $\tilde{M}_s - \tilde{M}_{s-} = N_s - N_{s-}$). Combining the two above equalities, we obtain :

$$L_t \tilde{M}_t = \int_0^t L_{s-}(\tilde{M}_{s-}(M_s - 1) + \mu_s) \, dM_s$$

We see that $L_t \tilde{M}_t$ is of the form (3.2) with a C_t that is left-continuous, adapted to $\mathfrak{F}_t$. The integrability condition in the direct integration theorem is easy to verify in view of the boundedness of μ_t and λ_t, and therefore $L_t \tilde{M}_t$ is a $(P, \mathfrak{F}_t)$-martingale, and $\tilde{M}_t$ is therefore a $(\tilde{P}, \mathfrak{F}_t)$-martingale, q.e.d.

<u>Remark</u> : the Radon Nikodym derivative is better known in the statistical or engineering litterature under the name "likelihood ratio" and is used in estimation or detection problems. For instance, if one wishes to discriminate between two Poisson hypotheses with different intensities λ and $\mu\lambda$, one gets for the likelihood ratio over the interval $[0,1]$

$$l_1 = \mu^{N_1} \exp\{(1-\mu)\lambda\}$$

Now l_1 is nothing but L_1 of (3.6) with

$$\lambda_t \equiv \lambda \ , \ \mu_t \equiv \mu$$

4. Applications of the martingale approach.

So far the reader has only had a glimpse of stochastic calculus for point processes. A complete survey of the achievement of martingale methods should include filtering (recursive estimation) and optimal control which are two of the nicest theoretical contributions of the martingale approach. Fortunately, the tutorial of R. Boel in this volume adresses the filtering problem. As to the control problem, we refer to the article Boel and Varaiya [7], and also to references [8] and [9].

Although there is no space in this tutorial for a detailed inventory of potential applications of the martingale approach to point processes, we shall mention two areas of engineering where this point of view might be useful.

Communications.

(i) optical communications : we refer to D. Snyder's tutorial in this volume where a nice example (optimal tracking of a laser source) is treated.

(ii) cryptography : the reference here is our article [10]. A brief summary of its content follows. We want to send information from a stochastic source to a friendly receiver. The stochastic source produces a sequence of o's and 1's, which is an equiprobable i.i.d. sequence. It is possible to send this information under the form of a point process (or a sequence of numbers, the interarrival times) using such a coding procedure that the point process in question is a standard Poisson process (i.e. Poisson, intensity 1). The receiver (friend) will be able to decipler this "white" sequence and all that is needed for this purpose is a triple (a, b,T) of non negative numbers, the key of the encryption procedure.

Because whatever key is used the transmitted point process is Poisson, intensity 1, the procedure is 100 $^o/_o$ security level is never attained, and therefore one must expect some "catch" in the method of [10]. There is indeed one : the transmission is not perfect and there is a probability of error $P_E > o$. However, by an appropriate choice of the key, P_E can be made as close to o as desired, at least in principle. We believe that there is a kind of "Heisenberg principle" relating the security of an encryption procedure and the probability of error P_E, but this is only a very vague conjecture (one would have to find an adequate measure of "security").

We are not going to describe the encryption procedure and its associated decryption procedure in detail, refering to [10] for this ; however, here is a brief description.

When a digit is received, a sequence of i.i.d. exponential variables $S_1, S_2, \ldots$ is generated. If the digit is 1, the common distribution has mean a, if the digit is o, it has mean b. Also a number N is selected. N is the first n for which $S_1 + \ldots S_n > T$. Now the point process T_n defined by $T_o = o$, $T_1 = S_1, \ldots, T_n = S_1 + \ldots + S_n, \ldots$, is a doubly stochastic Poisson process with the intensity Λ where

$$P[\Lambda = a] = P[\Lambda = b] = \frac{1}{2}$$

In other words, if N_t be the associated counting process, then

$$N_t - \int_o^t \Lambda \, ds = (P, \mathfrak{F}_t)\text{-martingale}$$

where $\mathfrak{F}_t = \sigma(\Lambda) \vee \mathfrak{F}_t^N$, $\sigma(\Lambda)$ being the σ-field generated by Λ. It is easy to check that the $\mathfrak{F}_t^N$-intensity of N_t is $\hat{\lambda}_t = E[\Lambda / \mathfrak{F}_t^N]$, i.e. :

$$N_t - \int_o^t E[\Lambda / \mathfrak{F}_s^N] \, ds = (P, \mathfrak{F}_t^N)\text{-martingale}$$

If we perform the change of time

$$t \to \tau(t)$$

defined by

$$\int_o^{\tau(t)} \hat{\lambda}_s \, ds = t$$

then N_t is transformed into

$$N'_t = N_{\tau(t)}$$

and it follows by Doob's optimal sampling theorem that :

$$N_{\tau(t)} - \int_o^{\tau(t)} \hat{\lambda}_s \, ds = N'_t - t = (P, \mathfrak{F}_t^{N'})\text{-martingale}$$

Hence, by Watanabé's theorem N'_t is Poisson standard. The corresponding interarrival times being called S'_n, the sender transmits $S'_1, \ldots, S'_N$, which are i.i.d., exponential, mean 1. This is done for all digits in succession, and since these digits are independent and equiprobable, the resulting sequence is also i.i.d, exponential, mean 1, or, to put it otherwise, a Poisson process, intensity 1.

The nice thing about the change of time $t \to \tau(t)$ is that it is invertible, and the blocks $S_1, \ldots, S_N$ can be recovered. It then remains to perform on each block $S_1, \ldots, S_N$ a bayesian test to decide whether the mean is a or b (it is at this point that the probability of error P_E is introduced).

Computer networks.

(i) Jacksonian networks.
The Jacksonian networks of queues serve as a fairly accurate model for computer communications networks (see Kleinrock [11])and any theoretical insight concerning their behavior is of conceptual interest to the design and control of such networks. In particular, it would be interesting to validate the routing models in A. Segall's tutorial (this volume) at least in the simplest analytical case, that of Jacksonian networks. A first step in this direction is made in [12] and [13] using martingale methods.

(ii) Satellite communications
Here we refer to an article by A. Segall [14] where a potential application of point process theory to the ALOHA-type communications link is described. One should also mention a recent attempt (Baras and Levine, unknown reference) to use martingale methods for solving estimation problems in transportation networks (the problem there is set up as a queueing problem, quite analogously to satellite communications).

(iii) Dynamic assignment in queues.
A simple example is given in [15] for dynamic file assignment in a computer network. The method of [15] can also be applied to the problems of priority assignment or server assignment. Even if the martingale methods do not provide a tractable way of computing the optimal solutions to dynamic assignment, they at least provide qualitative answers that will save non negligible amount of simulation time !

5. Conclusion

We will end this tutorial with a reading list for the beginner. It is a very short one.

First, for additional motivation we propose the survey [22] (at the time when [22] was written, the field was somewhat new, and the review given there does not give an account of the recent developments in control and of the very important queueing applications ; however it is a very clear exposition of the basic concepts of the martingale approach ; it deals with jump processes, also called marked point processes).

Reference [6] contains elementary proofs of the results stated in this tutorial.

Reference [16] is a very thorough review of all the important theoretical results (predictable projection, Radon Nikodym derivative) and of some applications (filtering, queues). It also gives an extensive bibliography on the subject.

Reference [4] gives a definitive form to the theoretical basis of the martingale approach to point processes. The required mathematical background is somewhat important.

References

[1] P. Brémaud (1975) Estimation de l'état d'une file d'attente.. Adv. Appl. Proba. 7, 4, pp. 845-863.

[2] S. Watanabe (1964) On discontinuous additive functionals and Levy measures of a Markov process, Jap. J. Math. 34, pp. 53-70.

[3] P. Brémaud (1975) An extension of Watanabe's theorem, J. Appl. Proba ; 12, 2, pp. 396-399.

[4] J. Jacod (1975) Multivariate point processes : predictable projection, Radon-Nikodym derivatives, representation of martingales, Z. für Wahrsch. 31, 3, pp. 235-253.

[5] P. Brémaud (1972) A martingale approach to point processes Ph.D. Thesis, U. of Cal. Bkley, Memo ERL. M 345.

[6] P. Brémaud (1974) The martingale theory of point processes over the real half line admitting an intensity, in Control Theory, Numerical Methods and Computer Syst. Modelling, Lect. Notes in Econ. and Math. Syst., 107, Springer-Verlag, pp. 519-542.

[7] R. Boel, P. Varaiya (1977) Optimal Control of Jump processes, SIAM J. Control.

[8] P. Brémaud (1977) Optimal cancellation of arrivals, preprint

[9] P. Brémaud, J.M. Pietri (1977) The role of martingale theory in Dynamic Programming for continuous time stochastic control Preprint.

[10] P. Brémaud (1975) On the information carried by a stochastic point process, Cahiers du CETHEDEC, 43.

[11] L. Kleinrock Queueing Systems.

[12] P. Brémaud (1977) On the output theorem of Queueing Theory, via filtering, Preprint.

[13] P. Brémaud (1977) Streams of a M/M/1 feedback stationary queue, preprint.

[14] A. Segall (1976) Recursive estimation for discrete time point processes with application to ALOHA-type computer systems IEEE Transactions, IT-22, 4.

[15] A. Segall (1976) Dynamic file assignment in a computer Network, IEEE Transactions, AC-21, 2, pp. 161. 173.

[16] P. Brémaud, J. Jacod (1977) Processus ponctuels et martingales : une revue des résultats récents sur la modélisation et le filtrage. Adv. Appl. Proba. 9, 3.

[17] R. Boel, tutorial in this volume

[18] E. Wong, idem.

[19] M.H.A. Davis, idem.

[20] D. Snyder, idem.

[21] A. Segall, idem.

[22] P. Varaiya (1975) The martingale theory of jump processes, IEEE Transactions AC-20, 1, pp. 34-42.

Panel Discussion

on

'ARE ITO CALCULUS AND MARTINGALE THEORY USEFUL IN PRACTICE?' *

Chairman: Professor W.L. Root — University of Michigan, U.S.A.

Panel:
Dr. J.M.C. Clark — Imperial College, London, UK
Professor B. Picinbono — Centre National de la Recherche Scientifique, France.
Dr. M.H.A. Davis — Imperial College, London, UK
Professor E. Wong — University of California, U.S.A.
Dr. P. Brémaud — IRIA/LABORIA, France
Professor D.L. Snyder — Washington University, U.S.A.
Professor A. Segall — Technion IIT, Haifa, Israel

Opening the discussion the Chairman, Professor W.L. Root, suggested that the panel members might start by discussing what distinguishes a stochastic calculus from any other. There are many sorts of integrals: Riemann, Riemann-Stieltjes, Lebesgue, Ito and so on. An integral is a linear functional with certain properties**. Why isolate a collection of these and call them stochastic integrals and how are they different from other integrals? Turning to their usefulness in practice, Professor Root pointed out that this is a broad question. A mathematician might say that a certain branch of algebra is

---

* The editor wishes to thank Dr. M.H.A. Davis for the preparation of this report.

** See, for example, S.J. Taylor, 'Introduction to Measure and Integration' Cambridge Univ. Press 1973 for further information. Good introductions to Ito integrals and stochastic calculus are to be found in E. Wong, 'Stochastic Processes in Information and Dynamical Systems', McGraw Hill (1971) and L. Arnold, 'Stochastic Differential Equations', Wiley (1974).

extremely useful in its application to topology. Is that practice or not? It certainly is for him, but of course what is at issue here is 'practice' in a sense closer to 'realization of equipment'. Nonetheless one does not want to enforce too narrow a view.

Professor E. Wong said that the need for a new calculus arose because Ito was faced with trying to deal with integrals with respect to a function, namely a sample function of Brownian motion, which was neither differentiable nor of bounded variation. As such, serving as an integrator it did not fall in the realm of classical calculus, so something new had to be invented, and he chose one way of defining it, so that it gave rise to the Martingale property. Many years later the question arose as to whether it was important to preserve the rules of ordinary calculus at least as far as the physical interpretation of the integrals is concerned. For example, if we replace the Brownian motion by a smooth process, approximating it and taking the limit, this limit is not the Ito integral but something else related to the so-called Stratonovich integral*. This fact has implications regarding how the integrals are interpreted, but not in how they are manipulated.

Clarifying the distinction between Ito and Stratonovich integrals, Dr. Clark gave the definition of the Ito integral $\int x_t \, dy_t$ as the limit of Riemann sums

$$\sum_i x(t_i)\,(y(t_{i+1}) - y(t_i))$$

so that x_t is evaluated at the left-hand end of the interval $|t_i, t_{i+1}|$. Stratonovich was concerned more with the ordinary calculus and defined a 'centred' integral where the sums are of the form

$$\sum_i x\left(\frac{t_i + t_{i+1}}{2}\right)\,(y(t_{i+1}) - y(t_i))$$

Both of these converge under nice conditions and the relation between them is precisely

$$\int_0^t x_t \, \overline{d}y_t = \int_0^t x_t \, dy_t + \tfrac{1}{2} \langle x,y \rangle_t \qquad (1)$$

---

* E. Wong and M. Zakai, 'On the relation between ordinary and stochastic differential equations', Int. J. Engng. Sci 3 (1965) 213-229.

where "$\bar{d}$" denotes the Stratonovich integral and $\langle x,y\rangle_t$ is the triangular brackets process (quadratic variation) introduced by Dr. Davis in his talk*. The Ito integral can be defined for a very large class of integrands x_t whereas the Stratonovich integral is essentially limited to semimartingale integrands for which $\langle x,y\rangle_t$ is defined. But this is usually the situation in practice and one then has the choice of either using Ito calculus or of converting everything to Stratonovich form via (1) and then using ordinary calculus; the two formulations are interchangeable.

Summarizing the difference between stochastic integrals and 'plain old integrals' Professor Root stressed that stochastic integrals were an attempt to deal with a class of integrands (unbounded) for which none of the classical integration theories apply, and not an attempt to get a different answer to the same problem.

Professor Picinbono said that Brownian motion was first studied as a physical phenomenon and Einstein stated that to explain diffusion the variance had to be proportional to time t. The mathematical formulation as developed for example by Paul Lévy, came later, as an abstraction and idealization of the physical problem. Clarifying the relation between the two is the subject of discussion today. By way of illustration Professor Picinbono considered the classical problem of detecting a signal in white gaussian noise**. For a deterministic signal s(t), in (mathematical) white noise y(t), it is extremely simple to compute the likelihood ratio, which is

$$\exp\{\int_0^T s(t)\,y(t)\,dt - \tfrac{1}{2}\int_0^T s^2(t)dt\}.$$

The last term is a constant, so the test is to compute

$$T(y) = \int_0^T s(t)\,y(t)\,dt$$

* In this volume. See also J.M.C. Clark's article in "Geometric Methods in System theory" ed. D.Q. Mayne and R.W. Brockett, Reidel. 1973.

** H.L. van Trees, 'Detection Estimation and Modulation theory, Part I', Wiley 1968.

and decide "signal present" if T(y) exceeds some threshold. Of course, T(y) is the output of the matched filter. Now if you consider random signals you have exactly the same expression for the likelihood ratio, but the first integral must be interpreted in the Ito sense. And we no longer have precisely the matched filter because the second term is now also random*.

Now consider the same problem with physical white gaussian noise, interpreted as a process with a flat spectrum in some bandwidth containing the signal. (There is no information about the noise outside this bandwidth). If you consider the detection problem with a deterministic signal and assume that the observation time is much larger than the inverse of the bandwidth then again you get the same expression for the likelihood ratio. Therefore you obtain again the matched filter, and that is the reason this is so often used in practice. However for the case of random signals, the situation is not so satisfactory. It is necessary to assume that the signal is gaussian and then one finds that there is an extra term in the likelihood ratio formula. From the practical standpoint it is important to investigate this term and how it depends on the signal and noise bandwidths.

Professor Wong said that he believed the last expression mentioned by Professor Picinbono (with the extra term) was true even for non gaussian signals. This leads to the work of Professor Balakrishnan** who has shown that if you adopt a stochastic calculus which is consistent with ordinary calculus then the likelihood ratio expression is similar to that given by Professor Picinbono but with the integral being interpreted á la Balakrishnan.

Dr. Davis made two comments about the application of Martingale theory. The first was that in any textbook on the subject one finds Martingales defined in terms of some specified increasing family of sigma-fields. In practical terms these correspond to an increasing record of observations of some process. Thus Martingale theory is likely to be relevant in situations where one is concerned specifically with the temporal evolution of the processes rather than steady-state or asymptotic analysis.

---

* T. Kailath 'A general likelihood ratio formula for random signals in gaussian noise', IEEE Trans. Inf. theory IT-15 (1969) 350-361.

** A.V. Balakrishnan, 'Radon-Nikodym Derivatives of a class of weak distributions on Hilbert space', Applied Math and Opt. 3 (1977) 209-226.

Perhaps this is why it has been of more immediate application in control theory where one is dealing with feedback control and real-time processing. The second comment was to point out that for the Kalman-Bucy filter one does not need Martingale theory and Ito calculus at all. The whole theory can be couched in terms of Hilbert spaces and Wiener integrals*. It is only when one comes to consider non-linear operations that the need for stochastic calculus appears.

Dr. Segall represented the status of Martingale vis-a-vis applications as follows:

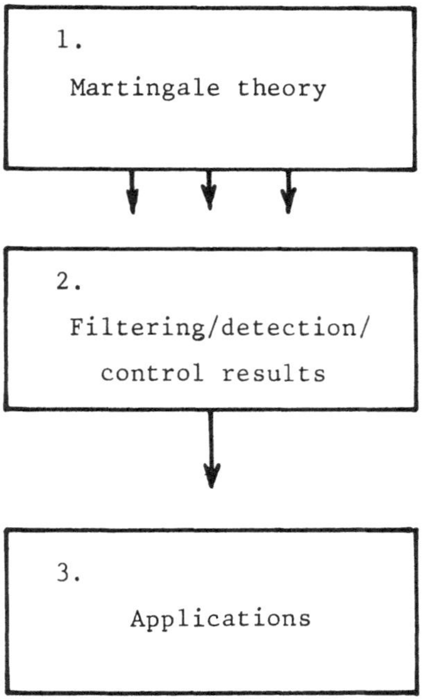

---

* M.H.A. Davis, 'Linear Estimation and Stochastic control', Chapman and Hall 1977.

There is an extensive theory surrounding the Martingale concept, represented by Box 1, and using this a variety of results in communication and control problems have been obtained, as in Box 2. However these occupy an intermediate position between theory and application in that many of them - for example the likelihood ratio formula for signal detection - are just representations for the solutions. One still has a nonlinear estimation problem to solve, and the question in moving from Box 2 to Box 3 is how you actually estimate the signal. Thus the situation as of today is that although there are plenty of representational results (Box 2), few connections with real applications have been made, partly because the solutions are often extremely complex. The advantage of Martingale methods in solving the problem of Box 2 is that greater generality is achieved. For example the likelihood ratio formula given by Professor Picinbono can be derived in a variety of ways if the signal and noise are independent, but the fact that the same formula holds when there is feedback, so that the signal depends in part on the past of the noise, would be very hard to show without Martingale methods.

Dr. Bremaud pointed out that other mathematical theories have been in the state represented by Dr. Segall's diagram. For example you could put "spectral representation of stationary processes" in Box 1 and "Wiener-Kolmogorov filtering" in Box 2. Again it took some time for real applications to emerge, and perhaps this augurs well for Martingale Theory. Turning to stochastic integrals, Dr. Brémaud remarked that the panellists had only been concerned with continuous-Martingale integrals where there were some difficulties of interpretation, but there are other stochastic integrals with respect to bounded-variation Martingales like the Poisson process, which are just Stieltjes integrals, for which no difficulties of interpretation occur. What remains is just the spirit of Martingale calculus. There are some applications - in the sense of Box 2 - of these Poisson-like Martingales, but these have so far been mainly in areas like queuing theory rather than communications*. However Dr. Brémaud predicted that 'real' applications would appear by the time of the next Darlington meeting.

Prof. Snyder drew up a "profit and loss" statement about the use of Martingale methods. On the profit side were :

---

* See for example P. Brémaud, 'Estimation of queues and of machine disorder by the semimartingale method', Advances in Applied Prob. 7 (1975) 845-863.

(i) Martingale theory exists in a highly developed state and is applicable to the kind of problems discussed at this Institute; so it cannot be ignored.

(ii) It is a unifying theory, in that it enables problems involving white gaussian noise on the one hand, and point process models on the other, to be treated by a common systematic method.

(iii) It is broad in scope, providing a framework which accommodates non-gaussian processes, non-linear transformations and feedback; all important questions in communications theory.

To set against this, the loss side of the statement was :

(i) In a different sense from (iii) above, it is narrow in scope. Martingales seem appropriate in situations where there is increasing information, or a dynamic information pattern of some sort. There are definitely applications where that is not the case - for example in image transmission - and Martingales are then no longer an appropriate tool.

(ii) Modelling difficulties. There is a potential difficulty in the use of Martingale mathematics to develop models for physical situations, arising from the fact that the theory is a very general one and the definitions are abstract. To be specific, suppose we are formulating a model for some point-process physical data N_t. Then we want to talk about its rate λ_t. Now in Martingale terms λ_t is the unique process that satisfies

$$E\left| \int_o^T c_t \, dN_t \right| = E\left| \int_o^T c_t \, \lambda_t \, dt \right|$$

for all signals c_t in some class, and this is a 'global' definition. On the other hand the physicist's or engineer's approach would be to look at the fine structure associated with the point process, using the idea that the intensity represents the local behaviour of the process (the probability that there is going to be a point in some small interval of time), and to build up a model from physical considerations of how the point process was generated in the first place. This has to be reconciled with the Martingale formulation*.

* A discussion of this point is given in A. Segall and T. Kailath, 'The modelling of randomly modulated jump processes', IEEE Trans. Inf. Theory IT-21 (1975) 135-143.

(iii) The long "learning period": there is a fairly large amount of superstructure associated with Martingales that has to be learnt, and from the engineer's point of view this means not learning something else. A decision has to be made as to whether it is worth it.

Dr. Clark thought the panellists had been skirting around the 'question behind the question' to which the discussion is addressed. This is not "Are Martingales useful in some theoretical way?" but "Do they give good stochastic models for modelling detection devices and filters and so on?". The simple answer is "No". It is misleading to talk about stochastic integrals isolated from their role as components of stochastic models of devices. A good model is one that has certain smoothness properties. In general you have some input process to your device and an output which is a likelihood ratio or conditional probability, etc. We can represent the device by a mapping Q which takes inputs y into outputs :

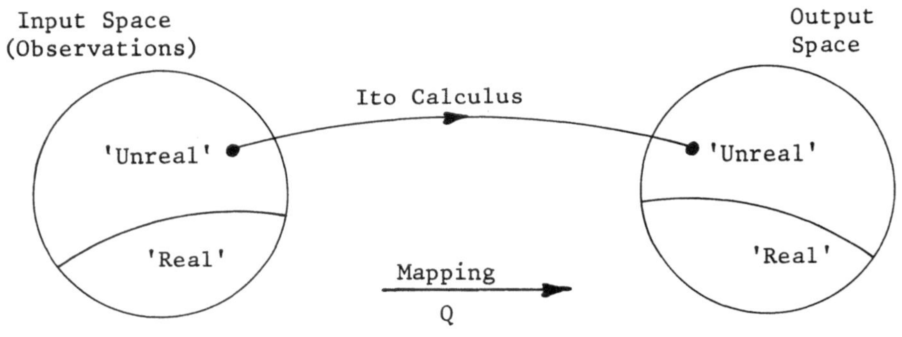

The models we get cover a very large class of inputs but some will be realistic - say, differentiable - inputs, while others are 'unreal' (unbounded variation). We divide up the input space in this way and the mapping Q gives a similar dichotomy of types of output. It is important for a good model that Q should be continuous in y, and this is where Ito calculus falls down, because basically it defines a map from unreal inputs to unreal outputs and says nothing about continuity. One has to go further to get models which describe the map on the whole input space, and this is important since of course one is really interested in its operation on 'realistic' inputs*. Now there are examples

* This question is developed in E.J. McShane 'Stochastic calculus and stochastic models'. Academic Press, 1974.

like Balakrishnan's Detection example, Wong's and Zakai's approximation of stochastic differential equations by ordinary differential equations, or Stratonovich differential equations, where things seem to work nicely. The reason for this is that underlying them all is a continuous map*, and this is true of most practical situations of interest though not of stochastic differential equations in general. If there is not a continuous map present then the whole question is totally ill-posed and one will be designing a vacuous theory which cannot be applied in practice. If you want a "Which?"** recommendation on the 'best buy' in Stochastic Calculus then it is clearly Stratonovich. It doesn't give you the same theoretical generalizations, but for practical purposes it is clearly the best, or rather, almost the best: one might consider McShane's formulation*** an even better buy.

---

* See J.M.C. Clark's article, this volume.

** The British equivalence of "Consumer Report". It publishes the results of objective tests of consumer products.

*** E.J. McShane, op. cit.

Part X

SIGNAL PROCESSING APPLICATIONS OF STOCHASTIC CALCULUS

organised by Dr. M. H. A. Davis

A MARTINGALE APPROACH TO RANDOM FIELDS

E. Wong

University of California, Berkeley
and Harvard University

1. INTRODUCTION

A random field $\{X_t, t \in T\}$ is a stochastic process with a multi-dimensional parameter, i.e. $T \subset R^n$. As in the one dimensional case, one is often interested in signal processing problems involving an observation equation of the form

$$(1.1) \qquad \eta_t = X_t + \zeta_t , \qquad t \in T$$

where η is the observed process, X is the signal, and ζ represents a corrupting noise. In one dimension the noise process is often associated with the superposition of a large number of pulses of comparable amplitudes and of extremely short durations. Such situations (e.g. shot noise) give rise to a process ζ_t which is Gaussian and white. By a white noise we mean a process with a correlation function given by

$$(1.2) \qquad E\zeta_t\zeta_s = N_o\delta(t-s)$$

when δ is the Dirac delta function. Of course, Gaussian white noise is an abstraction and to deal with it effectively has required the development of a stochastic calculus, which began in the form of Wiener integrals and acquired its full generalization and power in recent years as a calculus of martingales [1]. The critical step was taken by Ito, who established the connection between "white noise integrals" and martingales, and derived the differentiation rule which is basic to the stochastic calculus.

Since Gaussian white noise is but an abstraction and since to handle it requires a rather elaborate machinery, the question arises as to why it is so widely used. The answer lies in the fact that with appropriate interpretations it can be said that the values of a Gaussian white noise at different times are statistically independent, and this independence gives rise to a major simplification to the analysis of those signal processing problems where the noise is assumed to be white and Gaussian. Results which owe their existence to the simplification include some of the best known formulas of filtering and detection theory.

For processes with a multidimensional parameter, i.e., for random fields, a similar motivation for using Gaussian white noise as a model exists. There is no difficulty in extending the definition of white noise (1.2) to the multidimensional case. The difficulty lies in develiping a stochastic calculus to deal with it. To generalize martingales and their calculus to multi-parameter processes turned out to be far from straightforward. However, for the two-dimensional case, the essential elements of the stochastic calculus are now known, and some preliminary results on applying the calculus to problems in filtering and detection are also in hand. While the full extent of its usefulness remains to be seen, a martingale approach to random fields can now be said to exist, at least for the two-parameter case.

2. WIENER PROCESS AND MARTINGALES

Our first task is to make precise the idea of a white Gaussian noise. In one dimension this is done by viewing it as the formal derivative of a Wiener process or Brownian motion. We shall employ the same device for the multidimensional case. Let R^2_+ denote the positive quadrant of the plane R^2, and let $\{W_t, t \in R^2_+\}$ be a two-parameter Gaussian process with zero mean and a covariance function given by

$$EW_{(t_1,t_2)}W_{(s_1,s_2)} = N_o \min(t_1,s_1)\min(t_2,s_2) \tag{2.1}$$

We shall call N a Wiener process. If we denote by A_t the rectangle to the left and below t then we can view W_t formally as

$$W_t = \int_{A_t} \zeta_s ds \tag{2.2}$$

where ζ_t is a Gaussian white noise, alternatively

$$\zeta_t = \frac{\partial^2}{\partial t_1 \partial t_2} W_t \qquad t = (t_1,t_2) \tag{2.3}$$

It is useful to define a process $\{W(A)\}$ parameterized by Borel sets A in the plane as a Gaussian process with zero mean and covariance property

(2.4) $$EW(A)W(B) = N_o \text{Area}(A \cap B)$$

If we set $W_t = W(A_t)$ then W_t is a Wiener process. Alternatively, given a Wiener process W_t, a set-parametered process W(A) can be defined in terms of it. Hence, {W(A)} and $\{W_t\}$ are equivalent and we shall refer to W(A) also as a Wiener process. Formally, we have

(2.5) $$W(A) = \int_A \frac{\partial^2 W_t}{\partial t_1 \partial t_2} dt_1 dt_2 = \int_A \zeta_t dt$$

The values of W(A) for nonoverlapping areas are independent and this captures the independence property of white Gaussian noise.

For two points t and s in the plane, denote

(2.6) $$t \succ s \quad \text{if } t_1 \geq s_1 \text{ and } t_2 \geq s_2$$

and

(2.7) $$t \curlywedge s \quad \text{if } t_1 \leq s_1 \text{ and } t_2 \geq s_2$$

The relation $\succ$ is a partial ordering for points in R_+^2 with respect to which martingales can now be defined. Let T be a rectangle in R_+^2 of the form $T = \{s: 0 \leq s_1 \leq a,\ 0 \leq s_2 \leq b\}$. We say a random field $\{W_t,\ t \in T\}$ is a martingale if

(2.8) $$t' \succ t \text{ implies that } E\{M_{t'} | M_s,\ s \prec t\} = M_t \quad \text{w.p.1}$$

We can generalize the concept of a martingale by introducing a family of σ-fields. We shall say that $\{\mathscr{F}_t,\ t \in T\}$ is an increasing family of σ-fields if

$$t' \succ t \text{ implies } \quad \mathscr{F}_{t'} \supset \mathscr{F}_t$$

A random field $\{M_t,\ t \in T\}$ is said to be adapted to $\{\mathscr{F}_t, t \in T\}$ if for each t M_t is $\mathscr{F}_t$-measurable and it is said to be a martingale with respect to $\{\mathscr{F}_t,\ t \in T\}$ of it is adapted and if

(2.8') $$t' \succ t \text{ implies that } E\{M_{t'} | \mathscr{F}_s,\ s < t\} = M_t \quad \text{w.p.1}$$

A Wiener process $\{W_t,\ t \in R_+^2\}$ is a martingale under definition (2.8). Further if $\{\mathscr{F}_t,\ t \in R_+^2\}$ is an increasing family of σ-fields such that W(A) is $\mathscr{F}_t$-independent whenever A does not intersect A_t, then $\{W_t,\ t \in R_+^2\}$ is a martingale with respect to $\{\mathscr{F}_t,\ t \in R_+^2\}$ and we shall say for such a case that $\{W_t,\ \mathscr{F}_t,\ t \in R_+^2\}$ is a Wiener process.

3. STOCHASTIC INTEGRATION

Let $\{W_t,\ \mathscr{F}_t,\ t \in T\}$ be a Wiener process. The first integrals defined with respect to W were of the form

$$\int_T f(s)W(ds) \tag{3.1}$$

where f is a deterministic function which satisfies

$$\int_T f^2(s)ds < \infty \tag{3.2}$$

Integrals of the form (3.1) are known as Wiener integrals and have long been known.

The first generalization to (3.1) was to replace the deterministic integrand by a random integrand. Let $\{\phi_t, t \in T\}$ be an $\mathscr{F}_t$-adapted random field which satisfies

$$\int_T E\phi_s^2 ds < \infty \tag{3.3}$$

Then the integral

$$M = \int_T \phi_s W(ds) \tag{3.4}$$

can be defined as a straightforward generalization of the Ito integral. The integral M is defined as the quadratic-mean limit of a sequence of approximating sums, i.e.,

$$M = \lim_{n \to \infty} \text{in q.m.} \sum_{i,j} \phi(t_{ij}^{(n)})W(\Delta_{ij}^{(n)}) \tag{3.5}$$

where $\{t_{ij}^{(n)}\}$ is a rectangular partition of T for each n, $\Delta_{ij}^{(n)}$ denotes the increment

$$\Delta_{ij}^{(n)} = t_{i+1,j+1}^{(n)} - t_{i+1,j}^{(n)} - t_{i,j+1}^{(n)} + t_{ij}^{(n)} \tag{3.6}$$

and

$$\max_{i,j} \Delta_{ij}^{(n)} \xrightarrow[n \to \infty]{} 0 \tag{3.7}$$

One of the most important properties of the Ito integral preserved in the generalization (3.4) is its martingale property, which can be stated as

$$E(M|\mathscr{F}_t) = \int_{A_t} \phi_s W(ds) \tag{3.8}$$

Equation (3.5) implies that if we define

$$M_t = \int_{A_t} \phi_s W(ds) \tag{3.9}$$

then $\{M_t, \mathscr{F}_t, t \in T\}$ is a martingale.

The generalization of the Ito integral to the two-parameter case (indeed to the n-parameter case) is routine. By itself, it is hardly sufficient to yield a usable two-parameter stochastic calculus. For that we need certain completeness results and a differentiation formula. These proved to be far more elusive, and a great deal more of the structure underlying two-parameter martingales had to be uncovered before the desired results were finally derived.

The first completeness results were obtained by Wong and Zakai, who posed and ansered the following question in [1]:

Q_1 Suppose that $\{W_t, t \in T\}$ is a Wiener process, and $\mathscr{F}_{Wt}$ denotes the σ-field generated by $\{W_s, s \in A_t\}$. Let $\{M_t, t \in T\}$ be an $\mathscr{F}_{Wt}$-martingale such that $EM_T^2 < \infty$. Question: is M_t necessarily representable as a stochastic integral of the form (3.9)?

In one dimension the answer is "yes", and this completeness result represents an important fundamental property of the Ito integral. For the two-parameter case the answer turned out to be "no". To represent every $\mathscr{F}_{Wt}$-martingale, we need not only stochastic integrals of the form (3.4) but stochastic integrals of a second type, which we shall write in the form

$$\int_{T\times T} \psi_{s,s'} W(ds)W(ds') \tag{3.10}$$

Observe that it is an integral on T×T, hence a four-fold integral. Clearly, (3.10) needs to be defined, but we shall postpone doing so for a moment. Henceforth, we shall refer to (3.4) as a type-I stochastic integral and (3.10) as a type-II integral. Wong and Zakai showed that every square-integrable $\mathscr{F}_{Wt}$-martingale can be represented in the form

$$M_t = M_0 + \int_{A_t} \psi_s W(ds) + \int_{A_t\times A_t} \psi_{s,s'} W(ds)W(ds') \tag{3.11}$$

where 0 denotes the origin. (3.11) provides a full answer to Q_1.

We now return to the problem of defining (3.10). We shall define a general multiple integral of the form

$$I(\psi;\mu,\nu) = \int_{T\times T} \psi_{s,s'} \mu(ds)\nu(ds') \tag{3.12}$$

where μ and ν can each be either the Lebesgue measure or a Wiener process, $\psi_{s,s'}$ is $\mathscr{F}_{s\vee s'}$-measurable for each (s,s'), and

$$\int_{s\curlywedge s'} E\psi^2_{s,s'} ds ds' < \infty \tag{3.13}$$

where we recall that $s\curlywedge s'$ means ($s_1 \leqslant s_1'$ and $s_2 \geqslant s_2'$). $I(\psi;\mu,\nu)$ is defined by

$$(3.14) \qquad I(\psi;\mu,\nu) = \lim_{n\to\infty} \text{ in q.m.} \sum_{\substack{i<k \\ j>\ell}} \psi(t_{ij}^{(n)}, t_{k\ell}^{(n)})\mu(\Delta_{ij}^{(n)})\nu(\Delta_{k\ell}^{(n)})$$

where $t_{ij}^{(n)}$ and $\Delta_{ij}^{(n)}$ are defined as in (3.5). Observe that the summation condition ($i<k$, $j > \ell$) implies that only the values of $\psi_{s,s'}$ on the set $s\wedge s'$ affect the integral.

By taking $\mu(ds)$ and $\nu(ds)$ alternatively as ds and $W(ds)$ we get four different types of integrals for (3.12). The case of

$$\int \psi_{s,s'} ds ds'$$

is uninteresting because it is representable as an ordinary Lebesgue integral. The case of

$$\int \psi_{s,s'} W(ds)W(ds')$$

gives us the type-II stochastic integral. The remaining two possibilities

$$\int \psi_{s,s'} W(ds)ds', \qquad \int \psi_{s,s'} dsW(ds')$$

will be called mixed integrals and they play an essential role in the stochastic calculus of two-parameter martingales.

4. DIFFERENTIATION FORMULAS AND WEAK MARTINGALES

The next natural question that arises concerns transformation rule for stochastic integrals, i.e., generalization of the Ito differentiation formula. This turned out to be rather difficult, but a restricted version of the differentiation formula was obtained in [1]. Suppose that M_t is an $\mathscr{F}_{Wt}$-martingale which is also a type-I stochastic integral, i.e., of the form (3.9). Let $F(u,t)$, $u \in R$, $t \in R_+^2$, be a suitably differentiable function such that

$$(4.1) \qquad X_t = F(M_t, t)$$

is again an $\mathscr{F}_{Wt}$-martingale. Then the representation of X in terms of stochastic integrals is given by

$$(4.2) \qquad X_t = X_0 + \int_{A_t} F'(M_s, s)M(ds) + \int_{A_t \times A_t} F''(M_{s\vee s'}, s\vee s')M(ds)M(ds')$$

where F' and F'' denote differentiations of F with respect to the first variable, $s\vee s' = (\max(s_1,s_1'), \max(s_2,s_2'))$, and $M(ds) = \phi_s W(ds)$.

As a transformation formula (4.2) is severely limited. It deals only with transformations of type-I stochastic integrals which result in martingales. The general question is the following:

Suppose that X_t is of the form

$$X_t = \int_{A_t} \theta_s ds + \int_{A_t} \phi_s W(ds) + \int_{A_t \times A_t} \psi_{s,s'} W(ds)W(ds') \tag{4.3}$$

and $F(x,t)$ is a suitable differentiable function. Can $F(X_t,t)$ again be expressed as the sum of three such integrals?

The answer is once again "no". This means that the two types of stochastic integrals together with the Lebesgue integral are still not enough to give us a complete stochastic calculus.

To get a complete stochastic calculus we need the mixed integrals defined at the end of the last section. The substance of the differentiation formula is that if X_t is of the form

$$X_t = \int_{A_t} \theta_s ds + \int_{A_t} \phi_s W(ds) + \int_{A_t \times A_t} \psi_{s,s'} W(ds)W(ds') + \int_{A_t \times A_t} f_{s,s'} W(ds)ds' + \int_{A_t \times A_t} g_{s,s'} ds W(ds') \tag{4.4}$$

then $F(X_t,t)$ is again the sum of five such integrals. The full differentiation formula is rather complicated and will not be given here. (See [2])

Stochastic integrals of the two types are martingales. The mixed integrals clearly cannot be martingales, because there would be no need to introduce them otherwise. However, they are weak martingales. [3] A process $\{X_t, \mathscr{F}_t, t \in T\}$ is said to be a weak martingale if X is $\mathscr{F}$-adapted and

$$E[X(\Delta_t) | \mathscr{F}_t] = 0 \tag{4.5}$$

where

$$X(\Delta_t) = X_{(t_1+\Delta_1, t_2+\Delta_2)} - X_{(t_1+\Delta_1, t_2)} - X_{(t_1, t_2+\Delta_2)} - X_t \tag{4.6}$$

All martingales are also weak martingales, but not conversely. We note that (4.5) is a natural generalization of the alternative definition for martingales in one dimension given by

$$E[X(t+\Delta) - X(t) | \mathscr{F}_t] = 0$$

The substance of the general differentiation formula can now be taken to mean that a suitably differentiable function of a weak martingale plus a Lebesgue integrable is again a weak martingale plus a Lebesgue integral.

5. LIKELIHOOD RATIO FORMULAS

Let us return to the observation equation (1.1) with which we began our discussion. Integrating both sides of (1.1) we get a new observation equation

$$Y_t = \int_{A_t} X_s ds + W_t, \qquad t \in T \tag{5.1}$$

where we now consider Y as the observation and W the noise process. Let $\mathscr{F}_t$ denote the σ-field generated by $\{X_s, W_s, s \in A_t\}$ and let $\mathscr{F}_{yt}$ denote the sub-σ-field of $\mathscr{F}_t$ generated by $\{Y_s, s \in A_t\}$. If $\mathscr{P}$ denotes the probability measure, then our earlier assumption that the noise is white and Gaussian is equivalent to say that W_t is a Wiener process under $\mathscr{P}$. Let us also assume that the noise is independent of the signal then W is a Wiener process with respect to $(\mathscr{P}, \{\mathscr{F}_t\})$, and it was shown in [4] that there exists a new probability measure $\mathscr{P}_0$ such that Y_t is a Wiener process with respect to $(\mathscr{P}_0, \{\mathscr{F}_t\})$. The density (Radon-Nikodym derivative) of $\mathscr{P}$ with respect to $\mathscr{P}_0$ is given by

$$\frac{d\mathscr{P}}{d\mathscr{P}_0} = \exp\{\int_T X_s Y(ds) - \frac{1}{2}\int_T X_s^2 ds\} \tag{5.2}$$

Observe that since Y is a Wiener process, under $\mathscr{P}_0$ the first integral is a type-I stochastic integral.

The likelihood ratio L_t defined by

$$L_t = E_0\{\frac{d\mathscr{P}}{d\mathscr{P}_0} \Big| \mathscr{F}_{yt}\} \tag{5.3}$$

plays a prominent role in both hypothesis testing (detection) and estimation problems, and it is important to find explicit expressions for it. First, it can be observed immediately that by virtue of its definition L_t must be a $(\mathscr{P}_0, \{\mathscr{F}_{yt}\})$ martingale. Since Y_t is a $(\mathscr{P}_0, \{\mathscr{F}_{yt}\})$ Wiener process, this means that we can write L_t in the form

$$L_t = 1 + \int_{A_t} \phi_s Y(ds) + \int_{A_t \times A_t} \psi_{s,s'} Y(ds) Y(ds') \tag{5.4}$$

The integrands ϕ and ψ were identified in [4], and (5.4) can be rewritten as

$$L_t = 1 + \int_{A_t} \hat{X}(s|s) L_s Y(ds) + \int_{A_t \times A_t} [\rho(s,s'|s\vee s') + \hat{X}(s|s\vee s')\hat{X}(s'|s\vee s')] L_{s\vee s'} Y(ds) Y(ds') \tag{5.5}$$

where

$$(5.6) \qquad \hat{X}(s|s) = E[X_s | \mathscr{F}_{ys}]$$

and

$$(5.7) \qquad \rho(s,s'|s\vee s') = \mathrm{Cov}(X_s X_{s'} | \mathscr{F}_{ys\vee s'})$$

Equation (5.4) shows that L_t is expressible in terms of (5.6) and (5.7) which are conditional moments of X given the observation. Unlike the one-dimensional case, the likelihood ratio now depends on the second as well as the first moment. The next obvious step was to treat (5.5) as an integral equation in $\underset{\sim}{L}$ and solve it to obtain an explicit expression for L_t in terms of $\hat{X}$ and ρ. This proved to be rather difficult and had to await further developments in the stochastic calculus. The desired expression is derived in [5] and has the form

$$(5.8) \quad L_t = \exp\{\int_{A_t} \hat{X}(s|s)Y(ds) - \frac{1}{2}\int_{A_t} \hat{X}^2(s|s)ds + \int_{A_t\times A_t} \rho(s,s'|s\vee s')[Y(ds)-\hat{X}(s|s\vee s')ds][Y(ds')-\hat{X}(s'|s\vee s')ds'] - \frac{1}{2}\int_{A_t\times A_t} \rho^2(s,s'|s\vee s')ds\,ds'\}$$

Equation (5.8) is an exceedingly interesting formula, and its form could not have been predicted from its one-dimensional counterpart.

6. INNOVATIONS AND RECURSIVE FILTERING

Consider the observation equation (5.1) once again, and pose the following question:

Suppose that Z_t is a martingale with respect to $(\mathscr{P},\{\mathscr{F}_{yt}\})$. What is the general form for Z_t?

This turns out to be a difficult question and a satisfactory answer is not yet known. Although this question may appear to be the most natural generalization of the innovations representation problem in one dimension, it is actually not. In terms of the form of the answer and the usefulness of the answer, the most natural generalization of the innovation problem is the following.

What is the general form of a <u>weak</u> martingale with respect to $(\mathscr{P},\{\mathscr{F}_{yt}\})$?

The answer is that if Z_t is a $(\mathscr{P},\{\mathscr{F}_{yt}\})$ weak martingale then it must be of the form

$$(6.1) \quad Z_t = Z_0 + \int_{A_t} \phi_s \hat{Y}(ds|s) + \int_{A_t\times A_t} f_{s,s'}\hat{Y}(ds|s\vee s')ds' + \int_{A_t\times A_t} g_{s,s'}ds\hat{Y}(ds'|s\vee s')$$

$$+\int_{A_t \times A_t} \psi_{s,s'}[\hat{Y}(ds|s\vee s')\hat{Y}(ds'|s\vee s')-\rho(s,s'|s\vee s')ds\,ds']$$

where $\hat{Y}$ is defined by

$$(6.2) \qquad \hat{Y}(ds|t) = Y(ds)-\hat{X}(s|t)ds$$

Equation (6.1) leads rather quickly to a recursive formula for computing $\hat{X}$, when the signal X satisfies a modelling equation of an appropriate type. Without attempting to get the most general result possible, assume that X is a Gaussian process which satisfies the differential equation

$$(6.3) \qquad \frac{\partial^2}{\partial t_1 \partial t_2} X_t = \alpha(t)\frac{\partial}{\partial t_1} X_t+\beta(t) \frac{\partial}{\partial t_2} X_t+\gamma(t)X_t+\zeta_t$$

where ξ_t is again a white Gaussian noise. We can rewrite it as

$$(6.4) \qquad d_{t_1}d_{t_2}X_t-\alpha(t)d_{t_1}X_t dt_2-\beta(t)dt_1 d_{t_2}X_t-\gamma(t)X_t dt_1 dt_2 = V(dt)$$

where V is a Wiener process. Now, it is easily shown that if we define a process M by

$$(6.5) \qquad d_{t_1}d_{t_2}\hat{X}(t|t)-\alpha(t)d_{t_1}\hat{X}(t|t)dt_2-\beta(t)dt_1 d_{t_2}\hat{X}(t|t)$$
$$=\gamma(t)\hat{X}(t|t)dt_1 dt_2 = M(dt)$$

then M must be a weak martingale with respect to $(\mathscr{P}, \mathscr{F}_{Y_t})$. This means that M_t must be of the form given by (6.1). Furthermore, X is Gaussian and the observation equation is linear. It follows that $\hat{X}(t|t)$ must be linear in Y, and for M the last term in (6.1) must vanish. It also means that the integrands in the first three integrals of (6.1) must be deterministic functions for M. Indeed, these integrands can all be shown to be expressible in terms of the covariance function ρ, and M(dt) can be expressed as

$$(6.6) \qquad M(dt) = \frac{1}{N_0}\rho(t,t|t)\hat{Y}(dt|t)$$
$$+\frac{1}{N_0}\int_0^{t_2} d_{t_2}\rho(t_1,t_2;t_1,s_2|t)\hat{Y}(dt_1 ds_2|t)$$
$$+\frac{1}{N_0}\int_0^{t_1} d_{t_1}\rho(t_1,t_2;s_1,t_2|t)Y(ds_1 dt_2|t)$$

Equations (6.5) and (6.6) together with the definition (6.2) for Y reveal the nature of recursion for $\hat{X}(t|t)$. They show that $d_{t_1}d_{t_2}X(t|t)$ depends not only on $\hat{X}(t|t)$ but also on $\hat{X}(s|t)$ for s on the boundary ∂A_t of the rectangle A_t. Instead of (6.5), a more interesting recursion is in terms of the boundary data

$$\hat{X}(\partial t) = \{\hat{X}(s|t), \quad s \in \partial A_t\}$$

If $t' \succ t$, then $\hat{X}(\partial t')$ can be computed from $\hat{X}(\partial t)$ and the observation in the area between A_t and $A_{t'}$. The details can be found in [6].

REFERENCES

1. E. Wong and M. Zakai, "Martingales and stochastic integrals for processes with a multidimensional parameter," Zeit. Wahrscheinlichkeitstheorie, Vol. 29, pp. 109-122, 1974.
2. E. Wong and M. Zakai, "Differentiation formulas for stochastic integrals in the plane," Stochastic Processes and Their Applications, In press.
3. R. Cairoli and J.B. Walsh, "Stochastic integrals in the plane," Acta Math., Vol. 134, pp. 111-183, 1975.
4. E. Wong, "A likelihood ratio formula for two-dimensional random fields," IEEE Trans. Information Theory, Vol. IT-20, pp. 418-422.
5. E. Wong and M. Zakai, "Likelihood ratios and transformation of probability associated with two-parameter Wiener processes," Zeit. Wahrscheinlichkeitstheorie. In press.
6. E. Wong, "Recursive causal linear filtering for two-dimensional random fields," IEEE Trans. Information Theory. In press.

JUMP PROCESSES IN FILTERING AND DETECTION

R. Boel*

Engineering School, State University Ghent
Gent, Belgium.

ABSTRACT. The use of martingale theory for stochastic dynamical systems is illustrated by means of the filtering problem. Both the innovations approach and the measure transformation techniques are discussed. Also detection, prediction and smoothing problems are related to a corresponding filtering problem. Finally, the need for sufficient statistics is stressed, in order to get explicit recursive algorithms.

1. MARTINGALES AND JUMP PROCESSES

We begin by defining the local description of a jump process, i.e. its decomposition into a martingale and an increasing, predictable process. A very good introduction, explaining the intuitive meaning of the mathematical concepts, is given by Varaiya [17]. More details can be found in [2], while Jacod [9,10] gives a thorough mathematical discussion.
On a measure space $(\Omega,\mathcal{F})$, we are given the increasing family $(\mathcal{F}_t)$ of sub-σ-algebras of $\mathcal{F}$, $t \in R_+$ and $(Y,\mathcal{Y})$ a Borel subset of R^n. The basic jump process

$$Y_t(\omega) = Y_n(\omega),\ T_n(\omega) \leqslant t < T_{n+1}(\omega)$$

defined by the strictly increasing sequence of $\mathcal{F}_t$-stopping times (T_n), is $\mathcal{F}_t$-adapted (i.e. Y_t is $\mathcal{F}_t$-measurable), with values in Y. This can also be represented by the counting processes $(A \in \mathcal{Y})$:

*The author is with NFWO (Belgian Foundation for Scientific Research).

$$P^Y(A,t) = \sum_{s\leq t} I_{\Delta Y_s \in A}$$

(a completely analogous theory can be stated for $P^Y(A,t) = \sum_n I_{Y_n \in A} \cdot I_{T_n \leq t}$). With the notation $\mathcal{Y}_t = \sigma(Y_s, s \leq t)(\subset \mathcal{F}_t)$, $(\mathcal{Y}_t)$ is the smallest increasing family of σ-algebras with respect to which (Y_t) is adapted. This family of σ-algebras is also generated by all $P^Y(A,t)$, and the jump times T_n are $\mathcal{Y}_t$-stopping times.

On $(\Omega,\mathcal{F})$ we now define a probability measure $\mathcal{P}$, such that the jump times are "totally unexpected". Then, to each $P^Y(A,t)$, there corresponds a continuous, increasing, $\mathcal{F}_t$-adapted process $\tilde{P}(A,t)$, and a similar $\mathcal{Y}_t$-adapted $\tilde{P}^Y(A,t)$ such that

$$Q(A,t) = P^Y(A,t) - \tilde{P}(A,t),$$

resp. $Q^Y(A,t) = P^Y(A,t) - \tilde{P}^Y(A,t)$

is an $(\mathcal{F}_t,\mathcal{P})$-martingale, resp. $(\mathcal{Y}_t,\mathcal{P})$-martingale (more accurately: locally square integrable martingales).

Continuous, adapted processes, such as the extrinsic (resp. intrinsic) local descriptions $\tilde{P}(A,t)$ (resp. $\tilde{P}^Y(A,t)$) are also predictable. Roughly speaking, an $\mathcal{F}_t$-predictable process is the limit of a sequence of left-continuous $\mathcal{F}_t$-adapted processes. When we need the predictable version of an adapted process (X_t), we will simply take (X_{t-}). For a more rigorous treatment, see [3,6,11]. A process $f(t,y,\omega)$, indexed by $y \in Y$, will be called predictable if it is predictable for each fixed y, and is jointly measurable as a mapping from $(R_+ \times Y \times \Omega)$ to R^n.

Note also that if

$$\tilde{P}(A,t) = \int_0^t \int_A g(y,s)\,n(dy,ds)$$

with $n(A,t)$ $\mathcal{Y}_t$-predictable, $g(y,t)$ only $\mathcal{F}_t$-adapted, then the intrinsic local description is

$$\tilde{P}^Y(A,t) = \int_0^t \int_A \hat{g}(y,s)\,n(dy,ds)$$

with the notation $\hat{g}(y,t,\omega) = E(g(y,t,\omega)|\mathcal{Y}_t)$.

To prove this, note that

$$E(g(y,s)|\mathcal{Y}_t) - E(g(y,s)|\mathcal{Y}_s)$$

is a $\mathcal{Y}_t$-martingale.

We can now state the following martingale representation theorem which is very important for filtering applications.

Theorem 1 : To each $(\mathcal{Y}_t,\mathcal{P})$-square integrable martingale (M_t), there corresponds a unique $\mathcal{Y}_t$-predictable process $k_t(y,\omega)$ such that :

$$E \int_0^\infty \int_Y k_s^2(y,\omega)\tilde{P}^Y(dy,ds) < \infty$$

and

$$M_t = M_o + \int_o^t \int_Y k_s(y,\omega) Q^Y(dy,ds,\omega) \tag{1}$$

The most intuitive proof is given by Davis [7]. Other proofs can be found in [2,9,10]. It shows that all $\mathcal{Y}_t$-martingales only jump at the jump times T_n of the basic process, and behave continuously (predictable) in between.

<u>Examples</u>

1. <u>Point process</u> : $Y = \{1\}$, $Y_t = \sum_n I_{T_n \leq t} = N_t$

   with interarrival distribution

   $$\mathcal{P}(T_n - T_{n-1} | \mathcal{Y}_{T_{n-1}}) = F_n(t,\omega)$$

   Then : $Q^Y(\{1\},t) = Q_t^Y = N_t - \sum_n \int_{T_{n-1}\wedge t}^{T_n \wedge t} \dfrac{F(d(t-T_{n-1}),\omega)}{1-F(t-T_{n-1},\omega)}$

Special cases are

a. Poisson process with rate λ : $F_n(t) = 1-e^{-\lambda t}$

$$Q_t^Y = N_t - \lambda . t$$

b. Markovian point process : $F_n(t) = 1-e^{-\lambda_n t}$,

$$Q_t^Y = N_t - \sum_n \lambda_n (T_n \wedge t - T_{n-1} \wedge t).$$

2. <u>Queueing process</u> : $Y = \{+1,-1\}$, $Q_t = A_t - D_t$, the number of customers in the system, is the difference of the number of customers who arrived up to time t, A_t, and those who have already left the system, D_t. When the interarrival times and service times are exponentially distributed, with rates λ_n and μ_n, depending on $Q_t = n$, the local description is :

   $\tilde{P}^Y(\{+1\},t) = \int_o^t \lambda_{Q_{s-}} .ds$; $\tilde{P}^Y(\{-1\},t) = \int_o^t \mu_{Q_{s-}} .1_{Q_{s-}>0} .ds$

   $$\tilde{P}^Y(Y,t) = \int_o^t [\lambda_{Q_{s-}} + \mu_{Q_{s-}} .1_{Q_{s-}>0}] ds$$

3. <u>Gamma-Poisson process</u> : $Y = [0,\infty)$, $Y_t = \sum_n Y_n . I_{T_n \leq t}$

   distributed (rate λ), and Y_n independent, gamma distributed.

   The local description is :

   $$\tilde{P}^Y(A,t) = \lambda . t . \int_A \frac{\beta^r}{\Gamma(r)} . v^{r-1} e^{-\beta v} dv.$$

   Many other examples can be found in Snyders book [13].

Jacod [9] has shown that the family $(\tilde{P}^Y(A,t),\ A \in \mathcal{Y})$ completely defines $\mathcal{P}$ on $(\Omega, \mathcal{Y}_\infty)$, where Ω is the space of all sample paths Y_t. In fact, for absolutely continuous changes of probality measure the <u>translation theorem</u> gives a formula relating the change of local description and the change of measure (even on $\mathcal{F}_t$).

Theorem 2 : Given a jump process (Y_t), $t \in [0,T]$ on $(\Omega, \ , \ )$, with local description $\tilde{P}(A,t)$. Let $\phi(y,t)$ be an $\mathcal{F}_t$-predictable process, such that $1 + \phi(y,t) \geq 0$ and

$$
\begin{aligned}
L_t(\phi) &= 1 + \int_0^t L_{s-}(\phi) \, . \int_Y \phi(y,s) Q(dy,ds) \\
&= \prod_{\substack{s \leq t \\ Y_{s-} \neq Y_s}} [1 + \phi(Y_s, s)] \, . \exp\left(-\int_0^t \int_Y \phi(y,s) \tilde{P}(dy,ds)\right)
\end{aligned}
\tag{2}
$$

exists, and satisfies $E \, L_T = 1$. Then $\frac{d\mathcal{P}_1}{d\mathcal{P}} = L_T(\phi)$ defines a new probability measure on $(\Omega, \mathcal{F}_T)$ with respect to which the local description is :

$$
\tilde{P}_1(A,t) = \int_0^t \int_A [1 + \phi(y,s)] \, \tilde{P}(dy,ds) \tag{3}
$$

Conversely, if $\mathcal{P}_1$ is absolutely continuous with respect to $\mathcal{P}$ on $(\Omega, \mathcal{F}_T)$, then there exists an $\mathcal{F}_t$-predictable process $\phi(y,t)$ such that (2) and (3) hold, and $L_T(\phi) = \frac{d\mathcal{P}_1}{d\mathcal{P}}$. If the local description $\tilde{P}^Y(A,t)$ is used in (2) and (3) and the probability measures are restricted to $\mathcal{Y}_T$, $\phi(y,t)$ has to be $\mathcal{Y}_t$-predictable.
A proof can be found in [2,9], and is an application of a general theorem of van Schuppen and Wong [16]. The above theorem is most useful to transform a complicated probability measure $\mathcal{P}_1$, into an "easy" probability measure $\mathcal{P}$ (see chapters 3 and 4 for applications). As an example, let N_t be a self-exciting point process with local description (under $\mathcal{P}_1$) :

$$
\tilde{P}_1^Y(t) = \int_0^t \lambda_s(\omega) ds
$$

where (λ_t) is $\mathcal{Y}_t$-adapted. Then under the probability measure defined by

$$
\frac{d\mathcal{P}}{d\mathcal{P}_1} = L_T\left(\frac{1-\lambda_{s-}}{\lambda_{s-}}\right) = \prod_{\substack{s<T \\ N_{s-} \neq N_s}} \left(\frac{1}{\lambda_{s-}}\right) \, . \exp\left(-\int_0^t (1-\lambda_s) ds\right)
$$

N_t will be a Poisson process with rate 1. ($\tilde{P}_t^Y = t$). Note also :
$\frac{d\mathcal{P}_1}{d\mathcal{P}} = L_T(\lambda_{s-}) = \frac{1}{L_T\left(\frac{1-\lambda_{s-}}{\lambda_{1-}}\right)}$. A rather restrictive condition for the above method to work is : $+\varepsilon < \lambda_s < K$, $\varepsilon > 0$; $K < \infty$.

2. INNOVATIONS APPROACH TO FILTERING

This section is based on work by van Schuppen [14], Segall, Davis and Kailath [13]. For a mathematical treatment, see Brémaud [3]. The jump process of chapter 1 is now considered as the observation, while on $(\Omega, \mathfrak{F}, \mathcal{P})$ there also is given a semi-martingale (X_t), the unobserved state. An $(\mathfrak{F}_t, \mathcal{P})$-semi-martingale, is a sum of an $\mathfrak{F}_t$-predictable process and an $(\mathfrak{F}_t, \mathcal{P})$-martingale, roughly the solution of a differential equation perturbed by additive white noise. It has been used as the basic process to model stochastic dynamical systems [1,10,13,14]. We will assume here that it has the form

$$X_t = X_o + \int_o^t f_s \, ds + M_t \tag{4}$$

where (f_t) is $\mathfrak{F}_t$-adapted, (M_t) is an $(\mathfrak{F}_t, \mathcal{P})$-martingale. The observed counting processes $P^Y(A,t)$ are both $(\mathfrak{F}_t, \mathcal{P})$ and $(\mathcal{Y}_t, \mathcal{P})$-semi-martingales, with the decompositions :

$$\begin{aligned} P^Y(A,t) &= \int_o^t \int_A g(y,s)\, n(dy,ds) + Q(A,t) \\ &= \int_o^t \int_A \hat{g}(y,s_-)\, n(dy,ds) + Q^Y(A,t) \end{aligned} \tag{4'}$$

where $g(y,t)$ is $\mathfrak{F}_t$-adapted, $n(A,t)$ $\mathcal{Y}_t$-predictable. The martingales $(Q^Y(A,t), A \in \mathcal{Y})$ are called the innovations. For simplicity we assume $\mathfrak{F}_t = \sigma(X_s, Y_s, s \leqslant t)$.
The filtering problem then tries to find

$$\hat{X}_t = E(X_t | \mathcal{Y}_t)$$

Using (4) it is easily verified that the $(\mathcal{Y}_t, \mathcal{P})$-semi-martingale $(\hat{X}_t)$ has the unique decomposition

$$\hat{X}_t = E(X_o) + \int_o^t \hat{f}_s \, ds + m_t$$

where m_t is a $(\mathcal{Y}_t, \mathcal{P})$-martingale. Applying the martingale representation theorem, leads to

<u>Theorem 3</u> : The estimate (X_t) of the state (X_t) given by (4), satisfies the equation :

$$\hat{X}_t = E(X_o) + \int_o^t \hat{f}_s \, ds + \int_o^t \int_Y k(y,s)\, Q^Y(dy,ds) \tag{5}$$

$k(y,t)$ = predictable version of

$$\frac{E[X_{t-}(g(y,t) - \hat{g}(y,t)) | \mathcal{Y}_t]}{\hat{g}(y,t)} \tag{6}$$

proof : Only the form of $k(y,t)$ has to be verified. It suffices to check that with (X_t) defined by (4) and $(\hat{X}_t)$ by (5-6), for every $\mathcal{Y}_t$-martingale N_t (i.e. for all $\mathcal{Y}_t$-predictable $\phi(y,t)$ such that $N_t = \int_0^t \int_Y \phi(y,s) Q^{Y^t}(dy,ds))$ we have

$$E(X_t . N_t) = E(\hat{X}_t . N_t) \tag{7}$$

Then $E[(X_t - \hat{X}_t) N_t] = 0$ for all N_t, and the filtering error is orthogonal to "the information $\mathcal{Y}_t$".

Using the Ito-Meyer differentiation rule [11] and the fact that stochastic integrals are martingales, hence have zero expectation, (7) can be worked out to give :

$$E[\int_0^t \int_Y X_{s-} \phi(y,s)[g(y,s) - \hat{g}(y,s)] n(dy,ds) + \int_0^t N_s . f_s ds + \sum_{s \leq t} \phi(Y_s,s) \Delta Y_s . \Delta M_s] = E[\int_0^t N_s . \hat{f}_s ds + \sum_{s \leq t} \phi(Y_s,s) k(Y_s,s) \Delta Y_s]$$

Assuming $E\, \Delta Y_s . \Delta M_s . \phi(Y_s,s) = 0$ for all ϕ (roughly saying that (M_t) and (Y_t) have no common jumps, with probability 1), this reduces to :

$$E[\int_0^t \int_Y \phi(y,s) X_{s-} [g(y,s) - \hat{g}(y,s)] n(dy,ds)] = E[\int_0^t \int_Y \phi(y,s) k(y,s) \hat{g}(y,s) n(dy,ds)]$$

The $\mathcal{Y}_t$-predictable process $k(y,t)$ defined by (6) does indeed satisfy this equation.

Example [12] : (Z_t) is a finite state Markov process with transition probabilities

$$\mathbb{P}(Z_{t+dt} = j | Z_t = i) = P_{ij}(t) dt \qquad i \neq j$$

$$\mathbb{P}(Z_{t+dt} = i | Z_t = i) = 1 + P_{ii}(t) dt, \quad P_{ii} = - \sum_{j \neq i} P_{ij}$$

For our purposes this is best described by

$$X_t = \begin{pmatrix} I_{Z_t = 1} \\ \vdots \\ I_{Z_t = m} \end{pmatrix}$$

which has the semi-martingale representation :

$$X_t = X_0 + \int_0^t P(s) X_s ds + M_t$$

where $(P(s))_{ij} = P_{ij}(s)$. This Markov process cannot be observed, but it influences the local description of a jump process (Y_t), i.e. if $Z_t = i$, then

$$\tilde{P}^Y(dy,ds) = g_i(y,s)n(dy,ds) = g(y,s)^T X_s n(dy,ds)$$

with $g(y,s) = (g_1(y,s) \ldots g_m(y,s))^T$. Here $g_i(y,t)$ and $n(A,t)$ are assumed $\mathcal{Y}_t$-predictable. The innovations are then :

$$Q^Y(A,t) = P^Y(A,t) - \int_0^t \int_A g(y,s)^T \hat{X}_{s-} n(dy,ds) \qquad (9)$$

Equations (5) and (6) give, using $X_t X_t^T = \text{diag } X_t$:

$$\hat{X}_t = E(X_o) + \int_o^t P(s).\hat{X}_s.ds + \int_o^t \int_Y k(y,s-) Q^Y(dy,ds) \qquad (10)$$

$$k(y,t) = \frac{[\text{diag } \hat{X}_t - \hat{X}_t \hat{X}_t^T]\, g(y,t)}{g(y,t)^T \hat{X}_t} \qquad (11)$$

Notice that equation (4'), (5) and (6) do not in general form a recursive filter, because $\hat{g}_t$, $\hat{f}_t$ and k_t cannot be calculated unless the conditional distribution of X_t given $\mathcal{Y}_t$, is known. In the example above a recursive filter is obtained because $\hat{X}_t$ is actually a sufficient statistic. Its i-th component is the conditional probability of Z_t being in state i. Moreover the state was a Markov process, and the equations were linear. Compare this with the Kalman filter. Brémaud [4] has solved the problem of estimating Q_t, the queneing process of example 2, chapter 1, given D_t, the observation of the departures. There the characteristic function (conditioned on $\mathcal{Y}_t$)

$$E(e^{iuX_t} | \mathcal{Y}_t)$$

is calculated. It turns out that $E(I_{Q_t=n} | \mathcal{Y}_t)$ is an infinite dimensional sufficient statistic, for which recursive innovations equations can be found.

3. DETECTION,PREDICTION AND SMOOTHING

Suppose $\mathcal{P}_1$ and ϕ in theorem 2, are replaced by $\mathcal{P}_\alpha$ and ϕ_α indexed by an unknown parameter α. Then the Radon-Nikodym derivative $L_t(\phi_\alpha)$ calculated in (2) can be used for hypothesis testing and maximum likelihood estimation. When only the observation σ-algebra $\mathcal{Y}_t$ is available, one should use $E(L_t(\phi_\alpha) | \mathcal{Y}_t)$. Fortunately, this is easy to calculate when a filtering problem has first been solved. Let the extrinsic local description of Y be

$$\tilde{P}(A,t) = \int_o^t \int_A g_\alpha(y,s) n(dy,ds)$$

and let $\hat{g}_\alpha(y,t) = E_\alpha(g_\alpha(y,t)|\mathcal{Y}_t)$ (E_α is expectation under measure $\mathcal{P}_\alpha$). Then the innovations are

$$Q^Y_\alpha(A,t) = P^Y(A,t) - \int_0^t \int_A \hat{g}_\alpha(y,s)n(dy,ds)$$

It has been shown in [2] that

$$E_{\alpha_o}\left(\frac{d\mathcal{P}_\alpha}{d\mathcal{P}_{\alpha_o}}\Big|\mathcal{Y}_t\right) = \prod_{\substack{s\leq t\\ Y_{s-}\neq Y_s}} \left[\frac{\hat{g}_\alpha(Y_{s-},s)}{\hat{g}_{\alpha_o}(Y_{s-},s)}\right]\cdot \exp -\int_0^t \int_Y \left[\frac{\hat{g}_\alpha(y,s)}{\hat{g}_{\alpha_o}(y,s)}-1\right]\hat{g}_{\alpha_o}(y,s)n(dy,ds) \tag{12}$$

Assuming $g_\alpha(y,s) = 1$, this can be written as :

$$E_{\alpha_o}(L_t(\phi_\alpha)|\mathcal{Y}_t) = L_t(E_{\alpha_o}(\phi_\alpha(y,t)|\mathcal{Y}_t))$$

with

$\phi_\alpha(y,t) = g_\alpha(y,t)-1$. Brémaud and Yor [6] call this the separation of detection and filtering.

Example : Suppose the transition matrix $P_\alpha(s)$ and the jump rates $g_\alpha(y,s)$ depend on an unknown parameter α. The conditional likelihood ratio is then given by

$$E_{\alpha_O}\left(\frac{d\mathcal{P}_\alpha}{d\mathcal{P}_{\alpha_o}}\Big|\mathcal{Y}_t\right) = \prod_{\substack{s\leq t\\ Y_{s-}\neq Y_s}} \left[\frac{g_\alpha(Y_{s-},s)^T\hat{X}^\alpha_s}{g_{\alpha_o}(Y_{s-},s)^T\hat{X}^{\alpha_o}_s}\right]\cdot \exp -\int_0^t \int_Y \left[g_\alpha(y,s)^T\hat{X}^\alpha_{s-} - g_{\alpha_o}(y,s)^T\hat{X}^{\alpha_o}_s\right]n(dy,ds)$$

where X^α_t satisfies the recursive equations (9), (10), (11) with P,g replaced by P_α, g_α.

The prediction problem can also be reduced to a related filtering problem. With the model of chapter 2, let $t_1>t$. Then :

$$E(X_{t_1}|\mathcal{Y}_t) = E[\,E(X_{t_1}|\mathcal{F}_t)\,|\mathcal{Y}_t]$$

and we have to filter the $\mathcal{F}_t$-adapted process $E(X_{t_1}|\mathcal{F}_t)$. When $(X_t,\mathcal{F}_t)$ is a Markov process we have to filter $P_{t_1-t}(X_t)$.

For the smoothing problem, take $t_1<t$ and note that

$$M_\tau = E(X_{t_1}|\mathcal{Y}_{\tau\vee t_1})-E(X_{t_1}|\mathcal{Y}_{t_1})$$

is a $(\mathcal{Y}_t,\mathcal{P})$-martingale. Hence :

$$E(X_{t_1}|\mathcal{Y}_t) = \hat{X}_{t_1} + \int_{t_1}^{t}\int_{Y} k^s(y,s)Q^Y(dy,ds)$$

for some predictable process $k^s(y,t)$. The derivation of $k^s(y,t)$ is conceptually similar to the proof of theorem 3, but more messy.

4. MEASURE TRANSFORMATION APPROACH

Consider the following well known formula : let $L_t(\phi)=E(\frac{d\mathcal{P}_1}{d\mathcal{P}}|\mathcal{Y}_t)$, then

$$E_1(X_{t_1}|\mathcal{Y}_t) = E(X_{t_1}L_{t_1}|\mathcal{Y}_t)/L_t$$

This suggest that one replaces the complicated filter problem on $\mathcal{P}_1$, by a filter problem for $\mathcal{P}$. When one can find $\mathcal{P}$ such that state and observation process are independent, the observation process has independent increments and the state (X_t) is Markovian, one obtains the following linear equations (see [2,5]):

$$E(L_t(\phi)g(X_t)|\mathcal{Y}_t) = \pi_t(g) = E(g(X_t))+\int_o^t \pi_{s-}(\phi_s H_{t,s}(g))Q^Y(dy,ds)$$

where $H_{t,s}(g) = E(g(X_t)|\mathcal{X}_s)$.

Then $E_1(g(X_t)|\mathcal{Y}_t) = \pi_t(g)/\pi_t(1)$. See [2,5] for examples where this leads to a closed form solution for Markovian birth-death processses.

Van Schuppen [15] has used this method, in discrete time, with $g(X_t) = \exp(iu X_t)$. Then $\pi_t(g)$ is the conditional characteristic function. In [15] this is combined with the concept of conjugate pairs of distributions: the prior distribution of X_o is chosen such that $E(e^{iuX_t}|\mathcal{Y}_t)$ always remains of the same form. A simple example is the Poisson process with unknown rate λ, a priori gamma-distributed, with parameters β,r. ($E \exp i u \lambda = (1-\frac{iu}{\beta})^{r}$). After observing N_t jumps up to time t, λ is still conditionally gamma distributed.

$$E(\exp i u \lambda|\mathcal{Y}_t) = (1 - \frac{iu}{\beta+t})^{r+Y_t}.$$

In continuous time the author knows of only 2 examples where the unobserved state changes dynamically : Kalman filter (Gaussian--Gaussian is a conjugate pair), and the above Poisson process with λ replaced by a gamma-process (see Frost [8]). Further research is necessary in this area. However it does provide an operational definition of a sufficient statistic. A stochastic process (Z_t) is a sufficient statistic for the filtering model of chapter 2, if

i) $Z_t = Z_o + \int_o^t f(s,Z_s)ds + \int_o^t\int_Y k(y,s-,Z_{s-})Q^Y(dy,ds)$

with f(z), k(y,t,z) deterministic functions.

ii) $E(\exp(i\, u\, X_t)|\mathcal{Y}_t) = E(\exp(i\, u\, X_t)|\, Z_t)$.

In the discrete time case this definition is equivalent to the classical definition from statistics.

Acknowledgment : The author is very grateful, for help in various forms, to dr. P. Brémaud, prof. J. van Schuppen, prof. P. Varaiya and prof. E. Wong.

References

1. R. Boel : Some examples of semi-martingale models in filtering and stochastic control, 1977 Johns Hopkins Conf. on Info. and Systems.
2. R. Boel, P. Varaiya and E. Wong : Martingales on jump processes, I and II, SIAM J. Control, 13,5 (1975).
3. P. Brémaud : La méthode des semi-martingales en filtrage quand l'observation est un processus ponctuel marqué, Séminaire de Probabilités X, Springer, 1976.
4. P. Brémaud : Estimation de l'état d'une file d'attente et du temps de panne d'une machine par la méthode de semi-martingales, Adv. Appl. Prob., 7, 845-863 (1975).
5. P. Brémaud : Filtrage récursif pour une observation mixte par la méthode de la probabilité de référence, preprint, 1976.
6. P. Brémaud and M. Yor : Changes of filtration and of probability measures, IRIA report, 1977.
7. M.H.A. Davis : The representation of martingales of jump processes, SIAM J. Control and Optimization, 14, 4(1976).
8. P. Frost : Examples of linear solutions to non-linear estimation problems, 5th Princeton Conf. on Info. Sci. and Systems, 1971.
9. J. Jacod : Multivariate point processes : predictable projection, Radon-Nikodym derivatives, representation of martingales, Zeitschrift f. Wahrscheinlichkeitsth., 31, 3 (1975).
10. J. Jacod and J. Memin : Caractéristiques locales et conditions de continuité absolue pour les semi-martingales, Zeitschr. f. Wahrscheinlichkeitsth., to appear.
11. P.A. Meyer : Un cours sur les intégrales stochastiques, Séminaire de Probabilités X, Springer, 1976.
12. A. Segall : Recursive estimation from discrete-time point processes, IEEE-T-Info.Th., 22, 4(1976).
13. A. Segall, M. Davis and T. Kailath : Nonlinear filtering with counting observations.

13'. D. Snyder : Random Point Processes, Wiley, New York (1977).

14. J. van Schuppen : Filtering, prediction and smoothing for counting process observations, a martingale approach, SIAM J. Appl. Math., to appear.

15. J. van Schuppen : Representations and filtering problems for discrete time processes, Proc. 1977 JACC.
16. J. van Schuppen and E. Wong : Translation of local martingales under a change of law, Ann. of Prob., 2, 5 (1974).
17. P. Varaiya : The martingale theory of jump processes, IEEE-T-AC, 20, 1(1975).

APPLICATIONS OF STOCHASTIC CALCULUS FOR POINT PROCESS MODELS ARISING IN OPTICAL COMMUNICATION†

Donald L. Snyder

Washington University
St. Louis, Missouri 63130 U.S.A.

INTRODUCTION

Point process models arise in optical communication systems that employ direct detection to convert optical energy into electrical energy. This is a consequence of the fundamental quantum-mechanical consideration of the energy-conversion process [1, and references therein]. An implication of this quantum-mechanical consideration that is important for our discussion may be stated in terms of the following notation. Let $I(t,\vec{r})$ denote the intensity of a coherent light-field incident at time t and position $\vec{r}$ on an "ideal" direct-detection device; by "ideal" we mean that the device has infinite bandwidth, a fixed gain, and no internal noise. These conditions are met to a reasonable approximation under some conditions by a photomultiplier and an avalanche photodiode. The incident light results in photoelectron conversions in the detector. Let $N(T \times A)$ denote the number of these that occur in a time interval $T \in [0,\infty)$ and a spatial region $A \in \mathcal{A} \subset R^2$, where $\mathcal{A}$ denotes the two-dimensional region occupied by the detector. Also, let $\mathcal{N}_t$ denote the history of conversions up to time t; if $(t_i,\vec{r}_i)$ is the time of occurrence and location of the i*th* photoconversion, then $\mathcal{N}_t = \{(t_1,\vec{r}_1), (t_2,\vec{r}_2), \ldots, (t_{N(t)},\vec{r}_{N(t)}); N(t)\}$, where $N(t) = N([0,t) \times \mathcal{A})$. Here, N(t) is the total number of photoconversions up to time t regardless of their location in the detector surface. Finally, let $c(\vec{r},\rho)$ denote a square subregion of $\mathcal{A}$ containing $\vec{r}$ and having

†This work was supported by the National Science Foundation under Grant ENG76-11565 and by the National Institutes of Health under Grant RR00396 from the Division of Research Resources.

sides of length ρ. Then, quantum-mechanical considerations imply [1]:

$$\Pr(N([t,t+\tau) \times c(\vec{r},\rho)) = 1 | N_t)$$
$$= \Pr(N([t,t+\tau) \times c(\vec{r},\rho)) \geq 1 | N_t)$$
$$= \alpha\ I(t,\vec{r})\tau\rho^2 + o(\tau\rho^2), \qquad (1)$$

where α is a constant. A consequence of these infinitesimal properties of N is that N is a time-space Poisson process on $[0,\infty) \times \mathcal{A}$ with mean measure [2]

$$W(T \times A) = E[N(T \times A)] = \int_T \int_A \alpha\ I(t,\vec{r}) d^2\vec{r}\ dt. \qquad (2)$$

Here, $\alpha^{-1} W(T \times A)$ is the energy incident on the region A of the detector during T. Thus,

$$\Pr[N(T \times A) = n] = \frac{W^n(T \times A)}{n!} \exp[-W(T \times A)],$$

and $N(T_1 \times A_1)$ is statistically independent of $N(T_2 \times A_2)$ if $T_1 \times A_1$ and $T_2 \times A_2$ are disjoint sets in $[0,\infty) \times \mathcal{A}$. Also if $\lambda(t,\vec{r})$ denotes the instantaneous average rate at time t at which photoelectron conversions occur at position $\vec{r}$ (that is, $\lambda(t,\vec{r})$ is the limit of $(\tau\rho^2)^{-1} E[N([t,t+\tau) \times c(\vec{r},\rho))]$ as $\tau \to 0$ and $\rho \to 0$), then

$$\lambda(t,\vec{r}) = \alpha\ I(t,\vec{r}) \qquad (3)$$

is proportional to the incident light intensity at $(t,\vec{r})$.

In optical communication, the optical field is not coherent even for coherent sources such as a laser; rather, it is incoherent or stochastic. The randomness can occur intentionally due to modulation of the optical source by an information bearing signal, and it can also occur unintentionally due to propagation of the optical field in a randomly dispersive medium such as the clear turbulent atmosphere, haze, or a cloud [3,4]. In this case, $I(t,\vec{r})$ is a nonnegative random process on $[0,\infty) \times \mathcal{A}$, and N is not a Poisson process. However, N is conditionally a Poisson process with rate λ given $\{I(t,\vec{r}); (t,\vec{r}) \in [0,\infty) \times \mathcal{A}\}$; processes of this type are called time-space doubly-stochastic Poisson processes [2, Chapters 6,7; 5].

More complex interactions can occur in optical-communication systems with the result that the photoelectron conversion process cannot be adequately modeled as a doubly-stochastic Poisson process. These interactions can arise, for example, due to feedback intentionally designed into the communication system. The photoelectron conversion rate at $(t,\vec{r})$ then becomes dependent upon the history N_t of conversions up to time t. The point processes that must be used to model these effects are termed self-exciting [2, Chapter 5; 5].

The natural occurrence of point process models in direct-detection optical-communication systems has been one of the substantial motivations for the investigation in recent years of signal processing, communication, and control issues for point process observations. These investigations have relied upon the heavy, if not exclusive, use of the stochastic calculus for discontinuous processes. In order to be more specific about the kinds of problems encountered and the approaches that have been used for their solution, we shall outline two practical issues that have recently been addressed; the first is optical tracking and the second is sequential decision making. Detailed derivations for these problems are presented elsewhere [6-8] and, therefore, will not be repeated here.

OPTICAL TRACKING

The work we now outline was performed in collaboration with I. B. Rhodes of Washington University and E. V. Hoversten of the COMSAT Corporation [6,7].

When a narrow beam of light used as a carrier in an optical communication system propagates through clear-air turbulence, its angle of arrival at the receiver and its intensity fluctuate randomly [3,9]. As a result, there is a requirement for active tracking in order to maintain optical alignment between the source and the receiver. The manner in which this is accomplished in practice may be described in terms of the block diagram shown in Figure 1. This summarizes the major operations performed in an optical communication receiver; it is abstracted from a one gigabit-per-second optical-communication system that has been "brass boarded" and which achieves a design goal of less than one microradian angular tracking error. The various subsystems perform the following functions.

1. optical preprocessing. This accounts for initial operations performed on the incident optical field including the effects of the collection optics, such as a telescope, and any spatial or temporal filtering used to reduce the effects of background radiation, such as with a field stop, a bandpass interference filter, and temporal gating. Also included in this portion of the system, and important for our consideration, are the elements used to effect tracking to maintain optical alignment. These might include servo-driven gimbal arrangements, bender bimorphs, and Risley prisms, all of which are electromechanical devices. It will be noted later in our development that these devices are assumed to be described by linear, stochastic

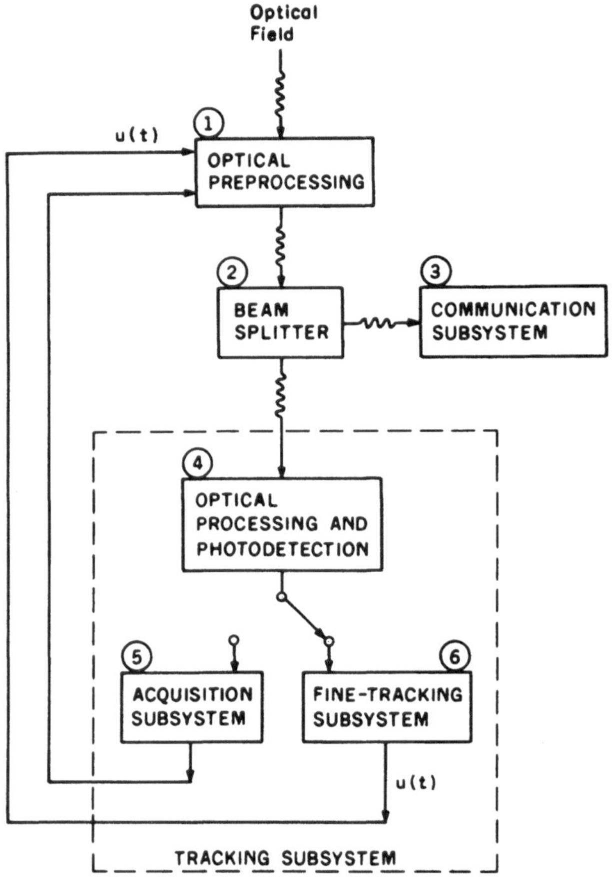

Figure 1. Optical-Communication Receiver

differential-equations. This does not seem to be a restrictive assumption for the limited range of operation associated with the "fine-tracking" mode of operation that we shall emphasize. Also, it is consistent with the linear, ordinary differential-equation models found in textbooks for most electromechanical devices. It is also an implicit assumption in our consideration of turbulence-induced effects that the receiving operture is about the size of one coherence-area of the received optical field or smaller.

2. beam splitter. Following preprocessing, the optical field is split into two components with, for example, a half-silvered mirror. One component is then routed to the communication subsystem where any information bearing signals modulating the optical field are detected and provided to the information sink. The other component is routed to the tracking subsystem where signals are generated and fedback to the electromechanical tracking elements described above. The communication and tracking subsystems are decoupled with both operating on their optical field independently. While there are presumably practical reasons for this, it is not evident that both communication and tracking performance are as good as they might be if coupling were permitted in the design. We shall return to this issue below. To conserve on photodetectors, the signals for the communication and tracking subsystems might alternatively be derived subsequent to photodetection with each subsystem operating independently on the photodetector output.

4. optical preprocessing. The purpose of the optical preprocessing in the tracking subsystem is to convert variations in the angle of arrival of the incident beam of light into variations in the position of a spot of light on the active surface of a photodetector. This can be accomplished with a lens. The photodetector then converts the spot of light into a position-sensitive electrical signal.

5,6. acquisition and fine-tracking subsystems. The tracking subsystem operates in two modes. The first is one of global search to seek and acquire the optical source initially. This is effected by the acquisition subsystem, which itself might operate in a coarse and a fine mode. Once the source is within a prescribed field of view, the mode of operation switches to that of fine-tracking to keep the optical source within the field of view. In the following discussion, we assume that the source has already been acquired, so the fine-tracking mode is in effect. Our concern is with the design of the fine-tracking subsystem.

For the design of the fine-tracking controller, we must: (1), characterize the spot of light on the photodetector's surface; (2), characterize the motions of this spot; (3), characterize the controlled electromechanical elements; (4), characterize the intensity fluctuations of the spot of light; (5), characterize the position-sensitive measurements available to the fine-tracking controller; and (6), specify the performance measure by which the fine-tracking controller is to be judged.

In the absence of any angle of arrival errors and intensity fluctuations, we assume that the spot of light on the active surface of the photodetector has a Gaussian intensity-profile given by

$$I(t,\vec{r}) = I_0(t)\exp\{-\tfrac{1}{2}\vec{r}'R^{-1}(t)\vec{r}\}, \tag{5}$$

where $I_0(t)$ is the maximum light intensity and $R(t)$ characterizes the shape of the light spot; $R(t)$ of the form ρI, where I is a 2×2 identity matrix, corresponds to a circularly symmetric spot. Propagation of the beam through a turbulent region of the atmosphere, vibration, and other effects cause the light spot to dance about in a random fashion and to fluctuate randomly in amplitude. Then, the light intensity becomes a random process described by

$$I(t,\vec{r}) = I_0(t)\exp\{-\tfrac{1}{2}(\vec{r} - y_m(t))'R^{-1}(t)(\vec{r} - y_m(t))\}, \tag{6}$$

where $y_m(t)$ is the random position of the spot at time t and $I_0(t)$ is now a random process due to the intensity fluctuations. We assume that $\{y_m(t);\ t\geq 0\}$ is derived from a Gaussian diffusion process $\{x_m(t);\ t\geq 0\}$ according to the linear stochastic-differential equation

$$dx_m(t) = F_m(t)x_m(t)dt + V_m(t)dv_m(t),\quad x_m(0) = x_{m0}$$

$$y_m(t) = H_m(t)x_m(t), \tag{7}$$

where $\{v_m(t);\ t\geq 0\}$ is a standard Wiener process. The peak-intensity process $\{I_0(t);\ t\geq 0\}$ is discussed below.

The purpose of the fine-tracking controller is to compensate for the random motions in order to maintain optical alignment. Thus, in the presence of a controller to position telescopes, mirrors, and other tracking elements, the light intensity becomes

$$I(t,\vec{r}) = I_0(t)\exp\{-\tfrac{1}{2}(\vec{r}-y_m(t) + y_p(t))'R^{-1}(t)(\vec{r}-y_m(t) + y_p(t))\}, \tag{8}$$

where $y_p(t) - y_m(t)$ is the tracking error at time t. Ideally, this error should be zero, but this cannot be accomplished for two reasons: the position error $y_m(t)$ is unknown and must be estimated from any position-sensitive measurements that are available, and the tracking elements will have some inertia so that $y_m(t)$ cannot be tracked instantaneously even if it were known. We model the tracking elements by a linear stochastic-differential equation

$$dx_p(t) = F_p(t)x_p(t)dt + G_p(t)u(t)dt + V_p(t)dv_p(t), \quad x_p(0) = x_{p0}$$

$$y_p(t) = H_p(t)x_p(t), \qquad (9)$$

where u(t) is the input to the tracking elements from the fine-tracking controller, and $\{v_p(t);\ t\geq 0\}$ is a standard Wiener process modeling local disturbances such as those due to vibration.

We can write (7), (8), and (9) in more concise notation by defining vectors $x' = [x_m' \ \ x_p']$ and $v' = [v_m' \ \ v_p']$ and defining matrices F, G, H, and V in an obvious way in terms of F_m, F_p, G_m, G_p, H_m, H_p, V_m, and V_p. Then

$$I(t,\vec{r}) = I_0(t)\exp\{-\tfrac{1}{2}(\vec{r} - H(t)x(t))'R^{-1}(t)(\vec{r} - H(t)x(t))\}, \qquad (10)$$

where

$$dx(t) = F(t)x(t)dt + G(t)u(t)dt + V(t)dv(t), \quad x(0) = x_0. \qquad (11)$$

Experimental evidence suggests that in the absence of any modulation by an information-bearing signal, the peak-intensity process $\{I_0(t);\ t\geq 0\}$ can be modeled as a lognormal process, a Rayleigh process, or a Ricean process depending on the propagation medium [3,4]. Modulation by analog or digital data results in a requirement for more complicated models. In the face of these potential complications, it is a remarkably fortunate circumstance that we find, subject to our assumptions, the design of the fine-tracking controller to be insensitive to any of these details of the peak-intensity process. All we shall require is that $I_0(t)$ be almost surely positive, that the expectation $E\int_0^T I_0(t)dt$ be finite for T finite, and that $\{I_0(t);\ t\geq 0\}$ be statistically independent of x_0 and $\{v(t);\ t\geq 0\}$. A practical requirement for the validity of this last assumption is that the peak intensity $I_0(t)$ and position x(t) processes be independent for u deterministic. For a receiving aperture about the size of one coherence area, $I_0(t)$ and x(t) result largely from amplitude and phase-gradiant fluctuations of the incident optical field, respectively [10]. Theoretical models for propagation effects indicate that, for lognormal fading, the logamplitude and phase-gradiant processes are uncorrelated, and hence independent since they are jointly normal, for a sufficiently small receiving aperture [11]†. Thus, our independence assumption seems consistent with other theoretical studies.

Next, we discuss the position-sensitive measurements available to the fine-tracking controller. These consist of a point-process component and a continuous component. The point-process component is derived as follows.

The moving spot of light on a photodetecting surface results in photoelectron conversions at a rate proportional to the light intensity; thus, from (3) and (10) the conversion rate within the surface will be

$$\lambda(t,\vec{r}) = \mu(t)\gamma(t,\vec{r},x(t)), \tag{12}$$

where we define

$$\mu(t) = 2\pi\alpha[\det(R(t))]^{\frac{1}{2}} I_0(t) \tag{13}$$

and

$$\gamma(t,\vec{r},x(t)) = (2\pi)^{-1}[\det(R(t))]^{-\frac{1}{2}}\exp\{-\tfrac{1}{2}(r-H(t)x(t))' R^{-1}(t)(\vec{r}-H(t)x(t))\}. \tag{14}$$

The point-process component of the fine-tracking controller input is derived from these photoelectron conversions in some manner. A quadrant photomultiplier is one implementation used in practice to accomplish this. With it, the photosensitive surface is, in effect, partitioned into four subregions, and photoelectron conversions are counted in each subregion without regard to their location within the subregion. This results in a decreased spatial resolution that can be improved with a finer partition. The trend in technology is toward a capability to implement such a finer partition [12], so the idealized extension we now make in our model to an infinitely fine partition seems not too unreasonable even though it is impractical today. We feel the controller design obtained with this idealization still provides useful insights into practical designs. In any case, the tracking performance we derive with this assumption underbounds that which can be achieved with a finite partition including that for a quadrant photomultiplier. To avoid edge

---

†This was elucidated for me by Professor J. Shapiro of M.I.T. who motivates the assertion by noting from (11) of his paper [11] that $\text{grad}_\rho[C_{\chi,\phi}(\rho,0)] = 0$ at $\rho=0$.

effects, we make the further idealization that the photosensitive surface is infinitely large; that is, $A = R^2$. This does not seem to be an overly impractical assumption for the fine-tracking mode which is our concern. Thus, the position sensitive measurements available from the photodetector form a time-space point process on $[0,\infty) \times R^2$, where the time and location coordinates of a point identifies the occurrence time and location of a corresponding photoelectron conversion.

The continuous component of the position-sensitive measurements is derived from sensors used to measure the state x_p of the tracking elements. For example, a strain gauge might be used to monitor a bender bimorph. We model these measurements as

$$dz(t) = C(t)x(t)dt + dw(t), \tag{15}$$

where $\{w(t);\ t\geq 0\}$ is a standard Wiener process and $C(t)$ is such that $C(t)\bar{x}(t)$ depends only on $x_p(t)$. Here, w models sensor noise, which we assume to be statistically independent of the peak-intensity process I_0. We denote by $\mathcal{Z}_t$ the continuous component of the observations up to time t; that is, $\mathcal{Z}_t = \{z(\sigma);\ 0\leq\sigma<t\}$.

Thus, the position-sensitive measurements available to the fine-tracking controller consist at time $t\geq 0$ of a point process component $\mathcal{N}_t$ and a continuous component $\mathcal{Z}_t$. We denote these pooled measurements by $\mathcal{B}_t = \mathcal{Z}_t \vee \mathcal{N}_t$. More precisely, let $(\Omega,\mathcal{F},P)$ be the underlying probability space for our model; all processes are defined on this space. Then, $\mathcal{Z}_t$ is the sub-σ-algebra of $\mathcal{F}$ generated by the continuous process z over $[0,t)$, $\mathcal{N}_t$ is the sub-σ-algebra of $\mathcal{F}$ generated by the discontinuous, time-space point process N over $[0,t)$, and $\mathcal{B}_t$ is the smallest sub-σ-algebra of $\mathcal{F}$ containing both $\mathcal{Z}_t$ and $\mathcal{N}_t$. We restrict attention to fine-tracking controller designs such that $u(t)$ is $\mathcal{B}_t$ measurable and the solution to (11) is well defined; controls of this type will be termed *admissible*. Admissible controls can depend on at most past measurements of z and N.

The following problems are solved in [7] by using stochastic calculus:

a. Find the conditional-mean estimates $\hat{x}(t) = E[x(t)|\mathcal{B}_t]$ and $\hat{\mu}(t) = E[\mu(t)|\mathcal{B}_t]$. Minimum mean square error estimates of $y_m(t)$ and $I_0(t)$ are linear, algebraic functions of $\hat{x}(t)$ and $\hat{\mu}(t)$.

b. Find the corresponding conditional covariances $\hat{\Sigma}(t) = \text{cov}[x(t)|\mathcal{B}_t]$ and $\hat{\Gamma}(t) = \text{cov}[\mu(t)|\mathcal{B}_t]$.

c. Find the admissible control $\{u(t);\ 0 \leq t < T\}$ that minimizes the average quadratic cost functional

$$J = E\{\int_0^T [u'(t)P(t)u(t) + x'(t)Q(t)x(t)]dt + x'(T)Sx(T)\}, \tag{16}$$

where P, Q, and S are symmetric, uniformly bounded matrix-valued time functions of appropriate dimensions with Q(t) and S nonnegative definite and P(t) positive definite.

d. Find precommutable upper and lower bounds on the estimation and control performances, $E[\hat{\Sigma}(t)]$ and J, respectively.

The solutions to these problems may be summarized as follows. Detailed derivations are in [7].

1. While x(t) is not normally distributed in general due to its dependence on u, and hence N, it is conditionally normal given $\mathcal{B}_t$ with mean $\hat{x}(t)$ and covariance $\hat{\Sigma}(t)$ which satisfy the following nonlinear, finite-dimensional, stochastic differential-equation for any causal control u:

$$d\hat{x}(t) = F(t)\hat{x}(t)dt + G(t)u(t)dt + \hat{\Sigma}(t)C'(t)[dz(t) - C(t)\hat{x}(t)dt] + \int_{R^2} M(t)[\vec{r} - H(t)\hat{x}(t)]N(dt \times d\vec{r}), \quad \hat{x}(0) = E(x_0) \tag{17}$$

$$d\hat{\Sigma}(t) = \{F(t)\hat{\Sigma}(t) + \hat{\Sigma}(t)F'(t) + V(t)V'(t) - \hat{\Sigma}(t)C'(t)C(t)\hat{\Sigma}(t)\}dt - M(t)H(t)\hat{\Sigma}(t)dN(t), \quad \hat{\Sigma}(0) = \text{cov}(x_0), \tag{18}$$

where

$$M(t) = \hat{\Sigma}(t)H'(t)[H(t)\hat{\Sigma}(t)H'(t) + R(t)]^{-1}. \tag{19}$$

2. The admissible control $\{u(t);\ 0 \leq t < T\}$ that minimizes J is the certainty-equivalent control given by

$$\hat{u}(t) = - P^{-1}(t)G'(t)K(t)\hat{x}(t) \triangleq - L(t)\hat{x}(t), \tag{20}$$

where K(t) is the symmetric nonnegative-definite matrix satisfying the Riccati equation

$$dK(t)/dt = - K(t)F(t) - F'(t)K(t) - Q(t) + K(t)G(t)P^{-1}(t)G'(t)K(t), \quad K(T) = S. \tag{21}$$

The corresponding minimum value of J is

$$J_{min} = E[x_0'K(0)x_0]$$

$$+\int_0^T tr[K(t)G(t)P^{-1}(t)G(t)K(t)E\{\hat{\Sigma}(t)\} + K(t)V(t)V'(t)]dt, \quad (22)$$

where tr[·] denotes the trace operator. Thus, according to (20), the fine-tracking controller minimizing J separates into an estimator to generate x(t) causally from the position-sensitive measurements and a controller having precisely the same design as if x(t) were known and not a random process.

3. Let $\mathcal{T}_t$ be the sub-σ-algebra of $\mathcal{F}$ generated by the counting process $N(t) = N([0,t) \times R^2)$; that is, $\mathcal{T}_t = \{t_1, t_2, \ldots, t_{N(t)}; N(t)\}$, where t_i is the occurrence time of the *ith* photoelectron conversion regardless of where it occurs in the photosensitive surface. Then, in [7], we find that the conditional distribution of μ(t) given $\mathcal{B}_t$ is the same as the conditional distribution given $\mathcal{T}_t$ and, further, that μ(t) and x(t) are conditionally independent given $\mathcal{B}_t$. This means that estimation procedures developed for purely temporal point processes, for example in [2, Chapters 6,7], can be used to characterize $\hat{\mu}(t)$ and $\hat{\Gamma}(t)$. Furthermore, there is the important practical implication that the problems of fine-tracking and of estimation of μ(t) can be treated separately without any loss of performance for either problem. Thus, for example, there would be no benefit in introducing any coupling between the communication and fine-tracking subsystems in Figure 1 if the assumptions on our model hold.

4. It is seen from (18) that $\hat{\Sigma}(t)$, the conditional covariance of $\hat{x}(t)$, is a stochastic process because of the dependence on N(t) in the last term. It follows that in contradistinction to the situation leading to the Kalman-Bucy filter, the conditional covariance is not precomputable. We are led, therefore, to consider $E[\hat{\Sigma}(t)]$ as an ensemble measure of the performance in estimating x(t) given $\mathcal{B}_t$. This measure is also of interest because it determines the control performance in (22). However, while $E[\hat{\Sigma}(t)]$ is in principle precomputable, this calculation is infinite dimensional. One way to see this is to examine the complications that arise upon taking the expectation of both sides of (18). The last term requires the calculation of $E\{\hat{\Sigma}H'[H\hat{\Sigma}H' + R]^{-1}H\hat{\Sigma}\}$. While a differential equation for this quantity can be developed, it, in turn, requires expectations of even more complicated nonlinear functions of $\hat{\Sigma}(t)$, and so on *ad infinitum* in an ever mushrooming manner. For this reason, we have in [7] developed the following upper and lower matrix-

ordering bounds on $E[\hat{\Sigma}(t)]$:

$$\Sigma_*(t) \leq E[\hat{\Sigma}(t)] \leq \Sigma^*(t), \tag{23}$$

where, with t-dependence suppressed,

$$d\Sigma_*/dt = F\Sigma_* + \Sigma_* F' + VV' - \Sigma_*[C'C + \bar{\mu}H'R^{-1}H]\Sigma_*, \tag{24}$$

$$d\Sigma^*(t)/dt = F\Sigma^* + \Sigma^* F' + VV' - \Sigma^* C'C\Sigma^* - \bar{\mu}\Sigma^* H'[H\Sigma^* H' + R]^{-1}H\Sigma^*, \tag{25}$$

where $\bar{\mu}(t) = E[\mu(t)]$, $\Sigma_*(0) = \Sigma^*(0) = \mathrm{cov}(x_0)$, $L(t) = \Sigma^*(t)C(t)$, and

$$M(t) = \Sigma^*(t)H'(t)[H(t)\Sigma^*(t)H'(t) + R(t)]^{-1}.$$

The upper bound in (23) is derived in [7] by evaluating exactly the performance of a parameterized family of suboptimum designs; one of these is identified as having a smaller mean square error than any other, thus providing a minimal upper bound within this family. The lower bound in (23) is derived in [7] directly by calculations involving inequalities. The bounds in (23) immediately imply corresponding bounds on the minimal control-performance functional J_{min}. If $f[E\{\Sigma(t)\}]$ denotes the right side of (22), then

$$J_* \leq J_{min} \leq J^*, \tag{26}$$

where $J_* = f[\Sigma_*(t)]$ and $J^* = f[\Sigma^*(t)]$. Thus, (24) and (25) provide precomputable upper and lower bounds on estimation and control performance. By comparison of (24) and (25), it is seen that both the estimation and control upper- and lower-bounds will be close to each other, and, hence, to $E[\hat{\Sigma}(t)]$ and J_{min}, if $H(t)\ \Sigma^*(t)H'(t)$ is small compared with $R(t)$. This precomputable condition will hold when the upper-bound tracking error is small compared to the size of the light spot on the photodetector surface. It is evident that the estimation and control lower bounds derived for the idealized observation of each photoelectron conversion are also lower bounds for both optimal and suboptimal trackers that employ observations obtained by temporal or spatial averaging as would be obtained using photon counting and a quadrant photomultiplier.

SEQUENTIAL DECISION MAKING

The work we now outline was performed in collaboration with R. E. Morley, Jr., of Micro-Term Inc., St. Louis.

In this section, we presume that the pointing and tracking issues discussed above have been solved, and we now turn our attention to certain aspects in the design of the communication subsystem. Imagine an information source that sequentially produces binary-valued digits with a new digit produced each T seconds. These digits modulate the intensity of a narrow beam of light used as a carrier. The received optical field is subsequently converted into an electrical signal $\{y(t); 0 \leq t \leq NT\}$ by the photodetection process described above. Suppose $s(t; i_1, i_2, \ldots, i_N)$ is the signal component of this waveform at time $t \in [0,NT]$, where $i_1, i_2, \ldots, i_N$ is a sequence of N binary-valued information digits. We suppose, further, that s is the result of the transmitted optical field having propagated through a randomly dispersive channel before arriving at the receiver. This occurs, as we have mentioned above, when the transmitted optical field propagates through a turbulent region of the clear atmosphere. Propagation through haze, fog, or a cloud similarly introduces randomly time-dependent dispersion. In the presence of such random dispersion, $s(t; i_1, i_2, \ldots, i_N)$ will be a random process even when the information sequence $i_1, i_2, \ldots, i_N$ is known.

The problem we address is that of designing the communication subsystem to form the maximum-likelihood estimate of the information sequence in terms of measurements $\{y(t); 0 \leq t \leq NT\}$ influenced by s. The form of these measurements depends upon the preprocessing used to convert the optical field into an electrical signal. For hetrodyne detection, we assume that the measurements are defined by

$$dy(t) = s(t; i_1, i_2, \ldots, i_N)dt + dw(t), \quad 0 \leq t \leq NT, \quad (27)$$

where $\{w(t); 0 \leq t \leq NT\}$ is a standard Wiener process. For direct detection, we assume that the measurements are defined by

$$dy(t) = dN(t), \quad 0 \leq t \leq NT, \quad (28)$$

where $\{N(t); 0 \leq t \leq NT\}$ is a doubly stochastic Poisson process with rate process $\{s(t; i_1, i_2, \ldots, i_N); 0 \leq t \leq NT\}$.

The separation theory associated with detecting stochastic signals from noisy observations [13,2] indicates that the following procedure can be used to form the maximum-likelihood estimate of $i_1, i_2, \ldots, i_N$ given measurements $\{y(t); 0 \leq t \leq NT\}$. For the hetrodyne-detection model, evaluate the generalized loglikelihood

$$\ell(i_1, i_2, \ldots, i_N) = \int_0^{NT} \hat{s}(t; i_1, i_2, \ldots, i_N)dy(t)$$

$$-\tfrac{1}{2}\int_0^{NT} \hat{s}^2(t; i_1, i_2, \ldots, i_N)dt \qquad (29)$$

for each of the 2^N possible information sequences. The desired maximum-likelihood sequence is the one with the largest loglikelihood. For the direct-detection model, this same procedure is used with the generalized loglikelihood given by

$$\ell(i_1, i_2, \ldots, i_N) = \int_0^{NT} \log \hat{s}(t; i_1, i_2, \ldots, i_N)dy(t)$$

$$-\tfrac{1}{2}\int_0^{NT} \hat{s}(t; i_1, i_2, \ldots, i_N)dt. \qquad (30)$$

In each of these expressions for ℓ,

$$\hat{s}(t; i_1, i_2, \ldots, i_N) = E[s(t; i_1, i_2, \ldots, i_N) | \mathcal{Y}_{0,t}; \; i_1, i_2, \ldots, i_N]$$

is the causal minimum mean square error estimate of s in terms of the past measurements $\mathcal{Y}_{0,t} = \{y(\tau);\ 0 \leq \tau \leq t\}$. The fundamental tool for deriving these expressions for ℓ is the stochastic calculus for continuous and discontinuous processes [13,2].

It is evident that the number of loglikelihoods to be evaluated, and hence the receiver complexity, grows exponentially with the message length N. We show in [8] that this rapid growth in complexity can be obviated when the channel dispersion processes have a finite memory L in the sense that

$$\hat{s}(t; i_1, i_2, \ldots, i_N) =$$
$$E[s(t; i_1, i_2, \ldots, i_N) | \mathcal{Y}_{\max(0,t-LT),t}; \; i_{\max(1,k-L)}, \ldots, i_k \qquad (31)$$

for all information sequences, and for each t such that $(k-1)T<t<kT$ and all $k=1,2,\cdots$. In this case, we show from the above loglikelihood expressions that for either form of measurements, the maximum likelihood estimate of $i_1, i_2, \ldots, i_N$

can be determined by employing a collection of 2^{L+1} signal estimators and a Viterbi algorithm having 2^L states [8]. Thus, the complexity depends exponentially on the channel memory regardless of the message length.

The above conclusion holds for both singly and doubly dispersive channels. The intersymbol interference results of Forney and others are included as a special case when the channel has deterministic time-dispersion. We also extend the above conclusion in [8] to include an m-ary information-symbol alphabet and trellis encoding at the transmitter.

CONCLUSION

What we have outlined is the theoretical solution of two design issues arising in optical communication systems. Stochastic calculus for continuous and discontinuous processes has been the basic mathematical tool for our approach to these issues. At the present time, we are unaware of any alternative mathematical tools that allow the same generality in our conclusions.

ACKNOWLEDGEMENTS

As noted in the text, the work reported herein was performed in collaboration with I. Rhodes, E. Hoversten, and R. Morley, Jr. Discussions with J. Shapiro were very helpful and are gratefully acknowledged.

REFERENCES

1. S. Karp, E. L. O'Neill, and R. M. Gagliardi, "Communication Theory for the Free-Space Optical Channel," *Proc. IEEE,* vol. 58, pp. 1611-1626, Oct. 1970.

2. D. L. Snyder, *Random Point Processes,* Wiley, New York, 1975.

3. E. V. Hoversten, R. O. Harger, and S. J. Halme, "Communication Theory for the Turbulent Atmosphere," *Proc. IEEE,* vol. 58, pp. 1626-1650, Oct. 1970.

4. R. S. Kennedy, "Communication Through Optical Scattering Channels: An Introduction," *Proc. IEEE,* vol. 58, pp. 1651-1665, Oct. 1970.

5. P. M. Fishman and D. L. Snyder, "The Statistical Analysis of Space-Time Point Processes," IEEE Trans. on Inform. Theory, vol. IT-22, May 1976.

6. D. L. Snyder, I. B. Rhodes, and E. V. Hoversten, "A Separation Theorem for Stochastic Control Problems with Point Process Observations," *Automatica*, vol. 13, Jan. 1977.

7. I. B. Rhodes and D. L. Snyder, "Estimation and Control Performance for Space-Time Point-Process Observations," *IEEE Trans. on Automatic Control*, vol. AC-22, No. 3, pp. 338-346, June 1977.

8. R. E. Morley, Jr. and D. L. Snyder, "Maximum Likelihood Sequence Estimation for Random Dispersive Channels," *IEEE Trans. on Inform. Theory*, submitted for publication.

9. K. Furutsu and Y. Furuhama, "Spot-Dancing and Related Saturation Phenomenon of Irradiance Scintillation of Optical Beams in a Random Medium," *Optica-Acta*, vol. 20, No. 9, pp. 707-719, 1973.

10. D. L. Fried, "Statistics of a Geometric Representation of Wavefront Distortion," *J. of the Optical Society of America*, vol. 55, No. 11, pp. 1427-1435, Nov. 1965.

11. J. H. Shapiro, "Point-Ahead Limitation on Reciprocity Tracking," *J. Optical Society of America*, vol. 65, No. 1, Jan. 1975.

12. B. R. Sandel and A. L. Broadfoot, "Photoelectron Counting with an Image Intensifier Tube and a Self-Scanned Photodiode Array," *Applied Optics*, vol. 15, No. 12, pp. 3111-3114, Dec. 1976.

13. T. Kailath, "A General Likelihood-Ratio Formula for Random Signals in Gaussian Noise," *IEEE Trans. on Inform. Theory*, 15, pp. 350-361, May 1969.

STATISTICAL MODELLING AND ANALYSIS FOR ADAPTIVE ROUTING IN COMPUTER NETWORKS*

Adrian Segall

Department of Electrical Engineering,
Technion, Israel Institute of Technology
Haifa, Israel

ABSTRACT. A model is developed for routing in data-communication networks. The model gives rise to a routing algorithm that can be implemented in a distributed fashion and furthermore has the properties of being loop-free, of reducing the delay in the network at each step and converging to optimal minimum delay routing.

1. INTRODUCTION

A data communication network is a facility which interconnects a number of data devices (such as computers, terminals, display units, etc.) by communication channels for the purpose of transmission of data between them. Each device can use the network to access some or all of the resources available throughout the network. These resources consist primarily of computational power (CPU time), memory capacity, data bases and specialized hardware and software. With the rapidly expanding role being played by data processing in today's society (from calculating interplanetary trajectories to issuing electric bills), it is clear that the sharing of computer resources is a desirability. In fact, the distinguished futurist Herman Kahn of the Hudson Institute has forecast that the "marriage of the telephone and the computer" will be one of the most socially significant technological achievements of the next two hundred years.

One of the most important problems in the area of data-com-

* Research supported by the Advanced Research Project Agency of The US Department of Defense (monitored by ONR) under contract Nr. N00014-75-C-1183.

munication networks is the design of efficient routing procedures for fast delivery of messages to their destinations. It is clear that the efficiency with which messages are transferred from source to destination determines to a great extent the desirability of networking data devices.

The data travelling along the links of the network is organized into *messages*, which are groups of bits which convey some information. One categorization of networks differentiates those which have message storage at the nodes from those which do not. Those with storage are known as *store-and-forward* networks. Another classification is made according to the manner in which the messages are sent through the network. In a *circuit-switching* network, one or more connected chains of links is set up from the node of origin to the destination node of the message, and certain proportions of data traffic between the origin and destination are then transmitted along these chains. The other category includes both message switching and packet switching networks. In *message-switching*, only one link at a time is used for the transmission of a given message. Starting at the source node,the message is stored in the node until its time comes to be transmitted on an outgoing link to a neighboring node. Having arrived at that node, it is once again stored until being transmitted to the next node. The message continues to traverse links and wait at nodes until it finally reaches its destination. *Packet-switching* is fundamentally the same as message switching, except that a message is decomposed into smaller pieces of maximum length called packets. These packets are properly identified and work their way through the network in the fashion of message switching. Once all packets belonging to a given message arrive at the destination node, the message is reassembled and delivered to the user.

For the purpose of this paper, we roughly classify routing procedures according to their dynamics into static, quasi-static and dynamic routing. Static routing is mostly of interest at the stage of designing the layout, topology and capacities of the network. For this design it is necessary to have some idea of the expected link flows in the network, and for this purpose one has to establish a routing algorithm. The analytical design is usually done by assuming some possible topology and capacity assignment, optimizing the routing and then redesign the topology to improve performance. This procedure is iteratively repeated, until a reasonable performance is obtained. In each stage of the algorithm, a static routing problem has to be solved, when topology, link capacities and traffic requirements are given and the routing that minimizes the expected delay has to be found. Using various statistical assumptions,it is possible to calculate the average expected delay in the network as a function of the link flows and then the situation reduces to a mathematical programming problem. Various algorithms for the solution of this problem have been previously proposed in the literature [1]-[2]. Although stating routing is very unreliable because it does not allow adaptivity to changing situations, it has and is being used in various operating networks, mainly because of its simplicity. The amount of overhead and protocols is minimal when working with static routing and it is relatively easy to implement and use.

Quasi-static routing refers to the situation when the network is operating, but over time the load is changing, new conversations are established and old ones are terminated. In addition, links or nodes may fail or go out of operation and new nodes and links may be added to the network. Clearly, every topological change requires rerouting of some of the traffic and in addition, it may be of importance to perform such rerouting in response to load variations. One possibility is to have a central computer that periodically receives status information from the network, performs a static routing optimization and returns to the network operating commands. This is referred to as centralized routing. Alternatively, a distributed algorithm may be implemented, where each node updates its routing procedure based on information that it receives periodically from neighboring nodes. The centralized algorithm is conceptually simple, but several difficulties arise in its implementation. First, the status messages from nodes to center and the commands from the center have to be routed themselves, so that a special routing procedure has to be designed for these particular messages. Secondly, the routes over which information on topological changes have to travel might in fact be destroyed by these same changes. Also, the links around the center will evidently be highly loaded with status and command messages and the network will be highly vulnerable because failure of the center will destroy the routing in the entire network. Finally, the question of synchronization of transmitting status information by the nodes to the center is a serious and unsolved problem.

Apparently centralized routing should have a better performance than distributed one because the center bases its decision on information it received from the entire network, whereas in a distributed algorithm each node has only local information. However, this is also not necessarily true, because of delays in transmission of status and command messages. The purpose of the present paper is to present a quasi-static routing algorithm that can be implemented in a distributed fashion and has the further property of converging to the optimal routing procedure. In addition, in each stage of the algorithm a loop-free network routing is obtained and every single algorithm step reduces the average delay in the network. These properties can be obtained by careful design of the propagation of updating and rerouting in the network.

Dynamic routing refers to the situation when messages are routed while taking into consideration the instantaneous length of the queues at the nodes of the network. An analytical model for description of the behaviour of the network under dynamic routing was developed in [3]. The mathematical model gives rise to an optimal control problem for which the dynamics, cost and constraints on the states and controls are all linear. A combination of techniques taken from dynamic programming, Pontryagin's maximum principle and linear programming are used to develop an algorithm for the construction of a feedback solution to this problem.

In this paper we are particularly concerned with quasi-static routing. In Section II the model is introduced and Section III presents the routing algorithm. Several final remarks are given in Section IV.

II QUASI-STATIC ROUTING — THE MODEL

The following model for quasi-static routing is introduced in [4]. Consider a data-communication network consisting of N nodes $\{1,2,\dots N\}$. The directed link connecting nodes i and k will be denoted by (i,k) and the collection of links by $\mathcal{L}$. We shall assume throughout the paper that all lines are byplex, namely if $(i,k) \in \mathcal{L}$, then $(k,i) \in \mathcal{L}$ and for each node i, denote by $Z(i)$ the collection of its neighbors.

Let $r_i(j) \geq 0$ be the average traffic entering the network at node i and destined for node j, $f_{ik}(j)$ be the flow in link (i,k) of messages destined for node j and C_{ik} be the capacity of link (i,k). Then the flows $f_{ik}(j)$ must satisfy

$$\sum_{k\in Z(i)} f_{ik}(j) - \sum_{\substack{\ell\in Z(i)\\ \ell\neq j}} f_{\ell i}(j) = r_i(j) \quad \text{all } i,j,\ i\neq j. \tag{1}$$

$$f_{ik}(j) \geq 0 \quad \text{all } i,j,k,\ i\neq j. \tag{2}$$

$$\sum_j f_{ik}(j) < C_{ik} \quad \text{all } (i,k) \in \mathcal{L}. \tag{3}$$

The objective of the routing is to minimize the average delay in the network. Let D_{ik} be the total delay per unit time of all traffic passing through link (i,k). Explicitly, D_{ik} is the average delay per unit of traffic multiplied by the amount of traffic per unit time transmitted over link (i,k). We shall assume here that D_{ik} is only a function of the total traffic flow $f_{ik} = \Sigma f_{ik}(j)$ passing through link (i,k). Then the total delay in the network per unit time is given by

$$D_T(f) = \sum_{(i,k)\in\mathcal{L}} D_{ik}(f_{ik}) \tag{4}$$

and since the total traffic in the network is independent of the routing procedure, we can minimize the average delay in the network by minimizing D_T. The main purpose of the following section is to indicate an iterative algorithm for performing this minimization.

Before proceeding, we should point out that the algorithm requires no explicit knowledge of the function $D_{ik}(f_{ik})$. Formulas for this function for various traffic models and assumptions have been previously obtained [5], [6], but here we shall need to assume only the following reasonable properties of the functions $D_{ik}(\cdot)$:

- D_{ik} is a non negative continuous increasing function of f_{ik}, with continuous first and second derivatives. (5a)
- D_{ik} is convex $\cup$. (5b)
- $\lim_{f_{ik} \uparrow C_{ik}} D_{ik}(f_{ik}) = \infty$. (5c)
- $D'_{ik}(f_{ik}) > 0$ for all f_{ik}, where D'_{ik} is the derivative of D_{ik}. (5d)

The following theorem whose proof appears in [4] provides the basis for the routing algorithm.

<u>Theorem 1</u> Assume that the set of flows satisfying (1), (2), (3) is nonempty. If the delay functions have properties given in (5), the flow $f^* = \{f^*_{ik}(j)\}$ minimizes D_T under constraints (1), (2), (3) if and only if there exists a set of numbers (Lagrange multipliers) $\lambda^* = \{\lambda^*_i(j)\}$ such that the Kuhn-Tucker conditions

$$D'_{ik}(f^*_{ik}) + \lambda^*_k(j) \begin{cases} = \lambda^*_i(j) & \text{if } f^*_{ik}(j) > 0 \\ \geqslant \lambda^*_i(j) & \text{if } f^*_{ik}(j) = 0 \end{cases} \qquad (6)$$

$$i \neq j,\ k \in Z(i)$$

are satisfied. Here

$$\lambda^*_j(j) = 0 \qquad (6a)$$

and D'_{ik} is the derivative of $D_{ik}(f_{ik})$.

It is well known [7, p.231] that the Lagrange multipliers λ^* are the sensitivity coefficients of the optimal cost with respect to the level of the contraint. In our situation, if the input flow $r_i(j)$ is increased by an incremental quantity $\delta r_i(j)$ and everything is held fixed, then the incremental increase in minimum delay will be $\lambda^*_i(j) \cdot \delta r_i(j)$. Consequently, we can give an interesting interpretation to the optimality conditions (6). Consider a given destination j and an arbitrary node i in the network. Look at all neighbors k of i and calculate the sum of their incremental delay coefficient $\lambda_k^*(j)$ and the incremental delay coefficient D'_{ik} on the line connecting i to k. Optimality requires that for all neighbors to which i sends traffic destined for j, this sum will be the same and no larger than the sum corresponding to neighbors to which i sends no traffic with final destination j. If and only if this is the situation for all nodes and all destinations in the network, the corresponding routing f^* is optimal.

Another fact we may note before proceeding is that in the optimality conditions (6), λ's corresponding to different destinations are not related. It is expected therefore that we shall be able to develop a rerouting algorithm that will evolve independently from one destination to another. Another interesting fact is that the flow requirements $r_i(j)$ do not enter explicitly in the optimality conditions. We shall see that these quantities do not enter in the algorithm either, so that it will not be necessary to know or estimate them.

III THE DISTRIBUTED ROUTING ALGORITHM

We shall indicate here a routing algorithm that was developed in [4] for line-switched data networks. A similar algorithm for networks working under packet or message switching regime was developed in [8].

Looking at the optimality conditions it is clear that generally speaking, the algorithm should be such that nodes will increase traffic on links with small incremental delay $D'_{ik} + \lambda_k(j)$ and decrease traffic on those with large incremental delay. In order to perform these actions, each node i will need the incremental delays D'_{ik} over each outgoing link (i,k) and the incremental delay $\lambda_k(j)$ of each neighbor k.

The quantity D'_{ik} can be obtained by node i by estimating f_{ik} and using one of the formulas for $D_{ik}(f_{ik})$. Alternatively, and probably preferably, node i can estimate D'_{ik} directly, thereby avoiding assumptions on the flow that are not always reasonable. Clearly both procedures will depend on the particular schemes for sending messages through the lines. An algorithm for estimating D'_{ik} for a virtual line-switched character multiplexing network was developed in [3, Eq.(30)-(33)].

The node incremental delays $\lambda_k(j)$ will have to be sent by the neighbors. This immediately brings up the question of a potential deadlock: in order to calculate $\lambda_i(j)$, node i needs the numbers $\lambda_k(j)$ from all the neighbors k, but to calculate its own $\lambda_k(j)$, a neighbor k needs the numbers from all its own neighbors, i included. It is therefore necessary to break this deadlock at the outset, and realize that in each step of the algorithm, each node will have to use only a subset of its neighbors to establish its number $\lambda_i(j)$.

Each step of the algorithm at every node will consist of two parts: updating of λ and then rerouting. In order to see how these operations progress through the network, we need several definitions. The discussion will refer to a given destination j. For a node i that has any flow passing through it destined for j, all neighbors k such that $f_{ik}(j) > 0$ are called its *real sons* and node i is called their *father* (a node can have more than one father). A node i such that $f_{ik}(j) = 0$ for all neighbors k has no real sons, but has exactly one *adopted son*; this is its prefered neighbor to which it would send any traffic destined for j if such traffic comes in. A node k is said to be a *son* of i, if it is either its real son or its adopted son. We denote by $S_i^n(j)$ the list of sons of node i for destination j at step n of the algorithm. If there is a sequence of nodes $i_1, i_2, \ldots i_m$ such that i_{r+1} is the son (for destination j) of i_r for $r = 1,2,\ldots (m-1)$, then we say that i_1 is *upstream* from i_m (for destination j) and i_m is *downstream* from i_1. The network is said to be *loop-free* if there are no two nodes that are each upstream from each other, and is said to have *loops* otherwise. If the network is loop-free for a given destination, then the downstream relationship forms a partial ordering of the nodes in the network.

A step of the algorithm will proceed such that the updating of λ's propagates from the destination upstream and the rerouting proper will propagate downstream from the peripheries towards the destination. We see therefore that maintaining loop-freedom in the network at each step of the algorithm not only saves resources, but is also essential to provide a natural sequencing for propagation of updating and rerouting signals in the network.

Before indicating the algorithm, it will be useful to discuss several special points connected with updating, loop-freedom and rerouting. We are still refering to a given destination j. Regarding updating, in order to be sure to prevent loops, we shall need the concept of *blocking* introduced in [8]. Briefly, if $f_{ik}(j) > 0$ and $\lambda_k(j) \leq \lambda_i(j)$, then there is danger of producing a loop in the next step. Therefore if, because of the constraints on the step-size, node i is not sure that in one step it can reroute all of $f_{ik}(j)$, then it declares itself blocked and so do all nodes upstream from it. If a node k was not the son of a node i at stage n and node k is blocked, then k cannot become its son at stage $(n+1)$. The exact procedure and proof that blocking prevents looping appear in the algorithm and the subsequent theorems in [4].

Another issue to be raised is connected with routing. Since we are dealing with (actual or virtual) line-switched networks, if a node decides to initiate the rerouting of a line, the entire portion of the old line from that node to the destination will

have to be cancelled and a new line established. This procedure will be performed in a distributed fashion, but it requires that a node will do its own rerouting only after all of its fathers — and in fact all nodes upstream from it — have completed their rerouting. In fact at each stage, the routing procedure at each node will consist of three possible parts: cancel those outgoing lines corresponding to lines that were previously coming in, but have been cancelled by fathers, establish outgoing lines corresponding to new incoming lines, and finally initiate rerouting.
A basic assumption for our analysis is that the lines are of small enough size, so that they can be reasonably approximated by a continuum. Each step of the algorithm is started by the destination, which sends a signal $\lambda_j(j) = 0$ to its neighbors. Each node i in the network collects numbers $\lambda_k(j)$ received from neighbors {k} until all its sons have send a number to i. Then node i establishes what nodes are *blocked* and chooses among nonblocked neighbors the one k_o with minimum $\lambda_k(j) + D'_{ik}$. This will be its prefered node for the present iteration. It sets

$$\lambda_i(j) = \lambda_{k_o}(j) + D'_{ik_o} \tag{7}$$

and sends $\lambda_i(j)$ to all neighbors *except sons*. At this stage the updating step is completed for i. It then waits until it receives $\lambda_k(j)$ from *all* neighbors {k} and at this stage it sends $\lambda_i(j)$ to the sons and at the same time performs the rerouting. This is done by transfering transmission lines from links (i,k) for which $f_{ik}(j) > 0$ and $k \neq k_o$ to line (i,k_o). The amount of transfered flow is proportional to the difference between $\lambda_k(j) + D'_{ik}$ and $\lambda_{k_o}(j) + D'_{ik_o}$.

The algorithm just indicated has the following properties proven in [4]

1) At all times there are no loops in the network
2) There exists a small enough step size such that, if the topology is fixed and the inputs stationary, the average delay in the network is strictly reduced at each step.
3) The routing in the network converges to the minimum delay routing.

IV DISCUSSION

The use of the algorithm for quasi-static routing should now be clear. Periodically, or immediately after the previous update has been completed, each destination starts a new update that propagates upstream through the network and then the rerouting propagates downstream. The nodes estimate the incremental delays D'_{ik} and use them in the update. The destination node knows that the update is completed as soon as it receives numbers λ from all neighbors. Old connections that are terminated are cancelled together with the rerouting, while the algorithm propagates downstream. New connections can be established at any time, but it may be preferable to wait for the next rerouting stage.

It can be seen that the algorithm as presented above does not have provisions to take into consideration topological changes. For example if a link fails, a node may never receive a number λ without which it cannot complete the updating-rerouting step and this may produce a deadlock. Similarly, if a link is added to the network, an unexpected message may arrive. Clearly such provisions are necessary in order to allow the algorithm to progress under all combinations of links failing and comming up.

The extension of the above algorithm to insure coverage of topological changes is a problem for future research. However, inspired by this algorithm, a similar simpler algorithm (in which each node has only one son) has been recently developed [9] where topological changes are indeed taken into account. It was rigorously proved [9] that in addition to the loop-free property, the algorithm allows the network to recover in finite time from an arbitrary number and timing of topological changes.

REFERENCES

1. D.G. Cantor & M. Gerla: Optimal routing in a packet-switched computer network, IEEE Trans. Comput., Vol. C-23, pp. 1062-1069, October 1974.
2. L. Fratta, M. Gerla & L. Kleinrock: The flow deviation method: An approach to store-and-forward communication network design, Networks, Vol.3, pp. 97-133, 1973.
3. A. Segall: The modelling of adaptive routing in data-communication networks, IEEE Trans. on Comm., Vol. COM-25, pp. 85-95, January 1977.
4. A. Segall: Optimal distributed routing for line switched data networks, submitted to IEEE Trans. on Comm.
5. L. Kleinrock: Communication Nets: Stochastic message flow and delay, Mc.Graw Hill, 1964.
6. L. Kleinrock: Analytic and simulation methods in computer network design, in 1970 Spring Joint Computer Conference, AFIPS Conf. Proc., Vol. 36, AFIPS Press, pp. 569-579, 1970.
7. D.G. Luenberger: Introduction to linear and nonlinear programming, Addison Wesley, 1973.
8. R.G. Gallager: A minimum delay routing algorithm using distributed computation, IEEE Trans. on Comm., Vol. COM-25, pp. 73-85, January 1977.
9. A. Segall & P. Merlin: A failsafe distributed routing algorithms, in preparation for IEEE Trans. on Comm.

Part XI

THE SHAPE TECHNICAL CENTRE SESSION

organised by

Dr. A. N. Ince,
Chief,
Communications Division

chaired by

Mr. D. W. Brown,
Communications Division,
SHAPE Technical Centre,
P.O. Box 174,
The Hague,
2076 Netherlands

RESEARCH ACTIVITIES ON COMMUNICATIONS AT THE SHAPE TECHNICAL CENTRE

A.N. Ince

Chief, Communications Division
SHAPE Technical Centre
The Hague, Netherlands

1. GENERAL

The SHAPE Technical Centre (STC) is involved in the development of a comprehensive, long-term, communications plan which embraces automatic switching and all forms of line and radio communications for operations within NATO. As part of this task the Centre proposes new systems and techniques that provide reliable means of meeting the stringent requirements that derive from modern command and control concepts. This work is carried out in close cooperation with a large number of national organizations, since the new systems have to interconnect with most of the communications systems that already exist in the Alliance, including national telegraph and telephone networks. Feasibility trials and simulations of new systems and techniques are performed in the Centre's laboratories and, when necessary, field experiments are conducted in the NATO countries concerned. In parallel with this work, effort is devoted to the improvement of currently-available communications with the aim of obtaining greater efficiency in circuit utilization and message handling.

In the course of developing the overall plan referred to above the Centre conducts studies related to the NATO Satellite Communications System, special radio systems for the interconnection of selected users, terminal equipment configurations, and switching systems - which together represent the essential components of an integrated, survivable communications system for NATO. The problems of optimizing the operational availability of this and similar systems in a wartime scenario involve the Centre in studies of electromagnetic countermeasures and counter-counter-measures.

Theoretical and experimental work (including speech coding and fibre optics), to increase the competence of the staff in areas in which the Centre may be required to render advice in the future, are undertaken in addition to the programme outlined above.

2. STUDIES ON RADIO SYSTEMS

For almost twenty years, STC has been involved in research and development work, interalia in the field of radio communications, including investigations into systems, techniques, and propagation. In frequency, the systems studied cover low frequency (LF) to superhigh frequency (SHF) and optical frequencies, and, in range, extend from a few meters to intercontinental distances. Examples of systems and propagation studies carried out at STC and reported in the literature are as follows.

(1) An adaptive meteor-burst system which makes use of automatic request for repetition (ARQ) and diversity reception to provide a highly reliable and efficient radio channel. A complete description of the system is given in [1].

(2) A compact and transportable ionoscatter system providing four telegraph channels by employing ARQ techniques and diversity [2].

(3) A study of propagation from buried antennas [3] and the development of a communication system using such antennas, together with a modulation system matching the LF-band channel. The system uses pseudonoise waveforms and correlation detection and achieves a processing gain of about 30 dB.

(4) The application of digital techniques and statistical concepts to communications using LF transmissions [4]. The system studies involved consideration of ground and sky waves, fading, noise distribution, and the use of a quarter-wave vertical wire raised by a captive helicopter.

(5) Theoretical and experimental studies for reliable air/ground communications in the VHF and UHF band, with particular emphasis on the effects of atomospheric refraction and surface roughness in the first Fresnel zone [5].

(6) Theoretical and experimental investigations of almost every aspect of satellite communications in the UHF, X band and above [6,7].

(7) Studies of adaptive techniques for increasing the transmission rate and reliability on dispersive channels such as troposcatter and HF.

In this Special Session we shall present papers on some of our studies related to the theme of this Institute. It will be appreciated that the presentations are based on work carried out by many scientists and engineers in addition to the authors, and their contributions are hereby gratefully acknowledged.

REFERENCES

1. P.J. Bartholomé and I.M. Vogt, COMET - A new meteor-burst system incorporating ARQ and diversity reception, IEEE Trans. Commun. Technol., vol COM-16, pp. 268-278, Apr. 1968.
2. ———, Design concepts for transportable ionoscatter systems and experimental results, IEEE Trans. Commun. Technol., vol. COM-15, pp. 839-847, Dec. 1967.
3. H.P. Williams, Buried antennas, Tidj. Ned. Elect. Rad. Gen., vol. 28, p. 271, 1963.
4. A.N. Ince and H.P. Williams, Range of LF transmissions using digital modulation, Proc. Inst. Elec. Eng., vol. 114, pp. 1391-1398, Oct. 1967.
5. ———, Design studies for reliable long-range ground-to-air communication, IEEE Trans. Commun. Technol., vol. COM-15, pp. 680-689, Oct. 1967.
6. A.N. Ince, Design and calibration of X-band satellite communication ground terminals, Ingenieur, vol. 84, pp. ET51-ET68, May 1972.
7. A.N. Ince, Design, testing and operation of an X-band satellite communications system, IEEE Trans. Com. Technol., Vol. COM-22, pp. 1338-1353, Sept. 1974.

CODE DIVISION MULTIPLEXING FOR SATELLITE SYSTEMS

A.N. Ince

Communications Division, SHAPE Technical Centre,
The Hague, Netherlands.

ABSTRACT. This paper discusses and compares various multiple access methods for satellite systems with particular emphasis on code-division multiplexing (CDM). It describes an extension of the CDMA technique to incorporate the message multiplexing functions normally performed in a separate time division multiplex (TDM) unit. In principle, each individual message signal (voice, data, telegraph, or any other type) modulates a subcarrier which is also modulated by a code that makes it near-orthogonal to all other subcarriers of that terminal and of all other terminals in the system. All subcarriers are combined in appropriate power ratios prior to transmission. Typical transmitter and receiver structures for the proposed CDM/CDMA system are described and some suitable codes are discussed together with the structure of the corresponding generator circuits. It is concluded that CDM/CDMA systems are eminently suitable for a range of applications where system flexibility and maximum throughput under varying conditions are primary requirements.

1. SPREAD SPECTRUM TECHNIQUES

Over the past decade STC has been studying a new class of communication systems which make use of a modulation technique known as "spread spectrum". The important characteristic of the technique is that the bandwidth of the signal transmitted is much greater than the information bandwidth and that the band-spread is determined by a function other than by the information to be sent. STC has designed, built and tested systems which employ spread spectrum modulation operating in the frequency bands of LF [1], HF [2,3], VHF/UHF and SHF [4] to obtain low-density output signals with resistance against interference and impulsive noise, to negate or circumvent degradation due to multipath, to provide message privacy and discrete addressing and finally to effect message multiplexing, multiple-access with resistance against jamming into a satellite as well as precise range measurement [4].

The basic principle of spectral spreading of the signal power can be implemented in several ways [5]:

(a) Frequency hopping (FH) in which the available frequency band is divided into a number of frequency slots and the carrier is shifted around them according to a pseudo-random pattern known at the receiver. It is possible to distinguish between fast and slow hopping depending on the time spent in each slot in relation to the inverse of the bandwidths of the slot.

(b) Phase hopping (PH), sometimes called "direct sequence" spread spectrum in which the carrier phase is modulated by a pseudo-random code.

(c) Time hopping in which the time slot in which the transmitted pulse occurs is controlled by a code generator.

(d) Pulse-FM or chirping in which the carrier is varied over a fixed frequency range in a known manner in order to convey the desired information.

A comparison between the two most commonly used SSM methods i.e. FH and PH above, is given in Appendix A. For most of the applications in which STC was interested the PH technique was found to have advantages; PH was therefore used for most of the applications mentioned above.

2. CODE DIVISION MULTIPLE ACCESS METHODS

In this paper we shall deal with the application of spread spectrum modulation to accessing of a communication satellite transponder by multiple jam-resistant links. This technique is known as spread spectrum multiple-access (SSMA) or alternatively as code-division multiple access (CDMA) where a certain allocated frequency band is shared by a number of users, each of whom occupies the whole band (as opposed to frequency division multiple access, FDMA, where each user is restricted to operate within some fraction of the band) with a duty cycle of 100% (as opposed to time division multiple-access, TDMA, where each user has exclusive use of the medium for a certain fraction of the time).

The capacity of a communication system using CDMA depends strongly on the characteristics of the transmission medium and the ground terminals. In [6] a comparison, the results of which are reproduced in Appendix B, is performed between FDMA, TDMA, and SSMA for a bandwidth-limited satellite communications system employing a hard limiting transponder.

This comparison shows that in general TDMA is the most efficient access method, and while the relative efficiencies of FDMA and SSMA depend on the number of accesses, FDMA is the more efficient when the number of accesses is small. Hence, the application of SSMA is hardly ever attractive from a capacity viewpoint alone and will normally only by considered when some of the other desirable features of SSMA have a decisive influence. This is certainly the case in a number of military applications where the resistance of spread spectrum modulated signals to intentional interference may be of vital importance; but SSMA has also found applications in mobile communication systems where multipath problems may be of great concern (maritime or urban area systems). The ability of an SSMA system to suppress multipath signals arises from the fact that a spreading signal can be designed to be uncorrelated with a time-shifted version of itself provided that the time shift exceeds one chip duration.

It is also noteworthy that an SSMA system provides a system flexibility superior to that of any other accessing technique. A TDMA system assumes a rigid and accurate system timing, and an FDMA system must always be based on an underlying frequency plan which is not easily changed to meet changing traffic requirements. With an SSMA system however, it is always possible to add another signal within the available band without greatly increasing the interference level experienced by the other signals. This "graceful" degradation of the system is very useful for applications where tight system control is not easily achieved.

Although an SSMA system does not require a rigid system timing, it is obviously necessary for receivers to be capable of synchronizing their reference waveforms exactly to the incoming signals. Naturally, the synchronization requirement implies increased circuit complexity for the receivers, but at the same time it makes life more difficult for potential jammers who, in order to perform selective jamming of a signal, must not only know the associated spreading code but also must be capable of timing their jamming signals correctly to within a fraction of a chip.

In addition to resistance to jamming, spread spectrum systems also provide some degree of protection against unauthorized interception. It must, however, be realized that interception is much easier to perform than selective jamming because the signals may be recorded for subsequent intensive off-line signal processing. The protection attainable by SSMA may nevertheless prove sufficient for many purposes if the codes used for spreading are chosen with a specific purpose in mind.

3. STC CDM/CDMA EQUIPMENT

3.1 Scope

In the following sections we shall describe a CDMA equipment which was developed at STC and which extends the CDMA technique to incorporate the message multiplexing function normally performed in a separate time division multiplex unit. The STC system will be called CDM/CDMA to distinguish it from the TDM/CDMA which has, so far, been considered in the literature and used in practical systems. The concept for CDM/CDMA was first described by G.R. Stette in [7]. This paper will describe the transmitter and receiver structures of the STC system and will discuss its performance in the presence of non-linearities and filtering in the ground terminal and satellite. The effects of non-linearities and filterings on PSK signals have been extensively studied in the literature [8 to 10] for TDM/CDMA systems. In the CDM/CDMA system of STC the sub-carriers transmitted from a gound terminal are coherent and this warrants special attention when they are subjected to heavy bandpass filtering and limiting.

3.2 Message multiplexing

While CDMA equipment in principle can be used for both analogue and digital message modulation, the latter has proved particularly suitable in connection with PSK as spread modulation. For many applications the digital information to be transmitted consists of several low-speed data streams from individually timed sources, isochronous data, and possibly also telegraph traffic of various speeds and character formats. This creates a multiplexing problem which can be solved in a number of ways, although the choice of method may have a large impact on cost, equipment complexity and ease of operation.

In terms of power utilization, the most efficient way of combining binary channels for transmission over a shared channel is Time Division Multiplexing (TDM), and this may be the only resort in severely bandlimited channels like 4 kHz voice channels when the data rates are a few kbit/s. The main problem with this method is that of synchronization. Bit integrity is usually required and speed equalizers must be incorporated at multiplexer and demultiplexer for each individual channel ("pulse stuffing"). One consequence of this is that restrictions are imposed upon the range of data rates which can be accommodated by the time derived channels. Another difficulty is that the channels may not be transparent with respect to character format, e.g., a time derived channel for isochronous traffic may not accept signals with varying element length. Certainly, there will be systems where these restrictions are of no concern, but there are also situations where this would impose severe operational problems.

Ideally, all the individual channels should be transparent, i.e., they should work properly for any modulation rate up to a given value. This maximum rate should also be greater than the message modulation rate, thus permitting forward error correcting codes to be applied.

As will be shown in the following the CDM/CDMA system avoids some of the inherent problems in TDM at the expense of a slight deterioration of the maximum obtainable performance.

3.3 Satellite system

The CDM/CDMA equipment to be described has been conceived to operate in a satellite communication system which has the basic features given below:

- There are many ground terminals (>10) with high interconnectivity.
- Each point-to-point link will be operated independently.
- There will be a wide range of link capacities (typically from 0.1 to more than 10 kbit/s).
- The traffic and the required link capacity will change during operation.
- The traffic composition is inhomogenous and the various types (start-stop and synchronous telegraphy isochronous data, analogue and digital voice etc.) have different performance requirements.

3.4 CDM/CDMA transmitter

The fundamental concept underlying CDM/CDMA is illustrated in the simplified block diagram shown in Fig. 1. Each individual message is fed to a separate four-phase modulator, together with a spread code which is unique for that channel. The carrier to be modulated is the same for all channels, but it may be applied with different phase to the various modulators. A separate two-phase modulator is used with spread modulation only to provide a useful tracking signal for the receiver. In fact, this tracking signal may also be used to carry a low-speed supervisory channel without causing any significant degree of degradation of the tracking performance of the receiver. All modulator outputs are combined in a suitable power ratio prior to transmission. A bandpass filter may be used to shape the transmitted spectrum if required.

In mathematical terms, the complete transmitter signal (before filtering) can be expressed as:

$$s(t) = \sum_{i=o}^{n} \sqrt{2P_i} \cos \{2\pi f_c t + m_i(t) + p_i(t) + \phi_i\} \qquad (1)$$

where P_i is the subcarrier power,

$m_i(t)$ is the message modulation, and is equal to either 0 or π depending on the time t,

$p_i(t)$ is the spread modulation and is equal to one of the following four values: 0, $\pi/2$, π and $3\pi/2$,

ϕ_i is the subcarrier phase,

for signal number i, while f_c is the carrier frequency. Note that only two-phase spread modulation is used for the tracking component since this choice facilitates the design of the receiver tracking circuits. Under the reasonable assumptions that the spread modulation bandwidth is much larger than the message modulation bandwidth and that the spread codes are uncorrelated and have perfect auto-correlation functions, the power spectrum of the aggregate signal s(t) can be expressed as:

$$s(f) = P\Delta \left(\frac{\sin \pi\Delta(f-f_c)}{\pi\Delta(f-f_c)} \right)^2 \qquad (2)$$

where the total power P is equal to the sum of the subcarrier powers:

$$P = \sum_{i=o}^{n} P_i \qquad (3)$$

and Δ denotes the chip duration. The power spectrum S(f) may be truncated by a transmitter bandpass filter. An analysis of the effect of filtering on message detection and code-tracking performance of SSMA signals shows that when the filter bandwidth is equal to the spread modulation rate carrier-to-noise ratio is reduced by about 1 dB while the code-tracking performance is hardly affected. This implies that the effect of the transmit filter may be compensated for at the receiver.

The main code generator may be a conventional feedback shift register from which a large number of mutually uncorrelated codes can be obtained by suitable combinations of output taps. Some examples of suitable codes for the purpose are given in Appendix C.

3.5 Demodulator and message separation

The way in which the messages are separated and recovered in the receiver is illustrated in Fig. 2. A code generator and a subcode selector identical to those of the transmitter provide the key for message separation. During normal operation the main code generator and consequently all codes in the receiver will be closely synchronized to the transmitter. The received aggregate signal plus interference is correlated against each of the codes used in the receiver for band-spreading, and the correlator outputs are demodulated coherently using a carrier reference derived from the tracking component of the signal.

Let the received signal be r(t) = s(t) + n(t) where s(t) is given by equation (1) and n(t) is a noise signal representing thermal noise and interference from unwanted signals. To recover the ith message the receiver first mixes r(t) with a local reference signal:

$$c_i(t) = \sqrt{2}\cos\{2\pi f_\ell t + p_i(t+\varepsilon) + \psi_i\} \tag{4}$$

where ε is the code tracking error, f_ℓ is the frequency, and ψ_i is the phase of the local oscillator. The mixer output is:

$$r(t)c_i(t) = \sum_{j=0}^{n} 2\sqrt{P_j}\cos\{2\pi f_c t + m_j(t) + p_j(t) + \phi_j\}.$$

$$.\cos\{2\pi f_\ell t + p_i(t+\varepsilon) + \psi_i\}$$

$$+ n(t)\sqrt{2}\cos\{2\pi f_\ell t + p_i(t+\varepsilon) + \psi_i\} \tag{5}$$

If account is taken of only terms centred around the difference frequency $f_c - f_\ell$, the mixer output becomes:

$$r(t)c_i(t) = \sum_{j=0}^{n} \sqrt{P_j} \cos\{2\pi(f_c - f_\ell)t + m_j(t) + p_j(t) - p_i(t+\varepsilon) + \phi_j - \psi_i\}$$
$$+ n_I(t) \cos\{2\pi(f_c - f_\ell)t + \phi_i - \psi_i\}$$
$$+ n_Q(t) \sin\{2\pi(f_c - f_\ell)t + \phi_i - \Psi_i\} \tag{6}$$

where the noise signal n(t) has been split into an in-phase component and a quadrature-phase component in accordance with the formula

$$n(t) = n_I(t)\sqrt{2} \cos(2\pi f_c t + \phi_i) + n_Q(t)\sqrt{2} \sin(2\pi f_c t + \phi_i) \tag{7}$$

In the coherent demodulator the signal r(t)c(t) is mixed with a second reference signal which, for perfect carrier tracking, is:

$$u_i(t) = \sqrt{2} \cos\{2\pi(f_c - f_\ell)t + \phi_i - \psi_i\} \tag{8}$$

So the demodulator output becomes:

$$d_i(t) = \sqrt{\frac{P_i}{2}} \cos\{m_i(t) + p_i(t) - p_i(t+\varepsilon)\} + \sum_{\substack{j=0 \\ j\neq i}}^{n} \sqrt{\frac{P_j}{2}} \cos\{m_j(t) + p_j(t) - p_i(t+\varepsilon) + \phi_j - \phi_i\} + n_I(t) \tag{9}$$

The first term in (9) represents the desired signal. If code tracking is not perfect, there will be a signal power loss proportional to the autocorrelation function of the spread modulation sequence, $p_i(t)$. For good pseudo-random sequences, the power reduction factor is $1 - |\varepsilon|/\Delta$ if $|\varepsilon| \leq \Delta$ and 0 otherwise. Therefore, if the tracking error exceeds Δ, the signal term is non-existent. The second term is interference from other signals from the same transmitter, and $n_I(t)$ is thermal noise plus interference from other sources. Provided that the codes $p_j(t)$ and $p_i(t+\varepsilon)$ are effectively uncorrelated for any value of ε, the second summation term in (9) will have only negligible effect.

The use of a separate tracking signal component for both code and carrier tracking is sub-optimum in the sense that only a fraction of the total received power is utilized for tracking. The method has, however, the advantage that it allows most of the tracking circuitry to be shared between all the message channels. The power requirement for the tracking signal is small, approximately the same as for a low-speed telegraph channel, and a generous margin could be maintained for this signal without greatly affecting the total power budget for the link.

3.6 Baseband interfaces

One of the important characteristics of the outlined CDM/CDMA system is that the message channels from the binary input of the four-phase modulators in the transmitter to the output of the coherent demodulator in the receiver may be described as "general-purpose", since they will accommodate any data rate up to a certain maximum determined by the equipment design (IF filter bandwidths, etc.). However, the binary detection circuits which follow the demodulators cannot be designed without some prior knowledge of the transmitted data rate (it is feasible to construct data detectors which will automatically choose the proper data rate from a selection of possible values). It must also be observed that the coherent phase-shift keying (CPSK) modulation scheme calls for differential data encoding at the transmitter because of the 180° ambiguity in the recovered phase reference. This is easy to achieve when the input signal comes with a clock signal, but otherwise the input interface in the transmitter will have to include a bit-timing circuit, which again requires a priori knowledge of the data rate. Thus, although the data channels provided by the CDM/CDMA modem are in principle completely transparent, the interface circuits will have to be designed for specific nominal data rates. The actual data rates can still be allowed to differ considerably from the design values without causing serious degradation.

For ordinary binary channels, the baseband interfaces will comprise differential encoders at the transmitter and bit conditioners and differential decoders at the receiver. Baseband interfaces may also include more complicated functions such as binary encoders and decoders for analogue voice (an efficient pulse duration modulation (PDM) modem for low-grade voice is described in [12]) or additional levels of multiplexing for low-rate signals. In Appendix D a scheme is described (Majority Logic Multiplexer) for combining a number of low-speed telegraph channels in code division multiplexing and transmitting the combined signal using a single general-purpose channel of the SSMA modem. Provision is made in the experimental equipment for multiplexing up to seven telegraph channels by using this method.

3.7 Code generation and tracking

The main code register consists of 64 stages with modulo-2 adders between successive stages to enable it to cater for any feedback configuration. The feedback connections are selectable via an octal thumbwheel switch array on the front panel of the code generator.

The output from the last stage of the code register is used to modulate the tracking signal component, this modulation being two-phase only. Each of the six sequences used to spread modulate the data signals (two for each of the three four-phase modulators) are obtained by modulo-2 adding up to four register outputs. If maximum length codes (m-sequences) are used, all the spread codes generated in this way will be time-displaced versions of each other (this follows from the shift-and-add property of m-sequences [14]).

The code generators have built-in clock generators that can be switched between clock rates of 5, 10, or 20 mega-bauds. A divider chain is provided for supplying clock rates which are sub-multiples of the code rates to various parts of the system. The multiplexer codes needed for the CDM telegraph multiplexer are derived by "decimation" of the main code, i.e., by sampling and holding every 2^L bit of the main code, where L is so chosen that the resulting "decimated" code rate becomes 611 Hz. The seven uncorrelated codes for the multiplexer are obtained as successive delays of that code. When the main code is of maximal length it is known that the "decimated" code is simply identical to the main code except for the translation in rate (see [14] for a proof of this statement).

A separate 13-bit shift register is provided for generating the short code to be used in the short/long code acquisition procedure described in Section 3.9.

The receiver code generator has all the features of the transmitter code generator, but is in addition equipped to track a received code. The code tracking loop employed is the 1-Δ delay-lock with envelope correlation described in [15].

3.8 Carrier Tracking

The received signals are down-converted in two steps from 70 MHz to 200 kHz, the spread modulation being removed in the first step of down-conversion. Coherent demodulation is then performed by mixing the down-converted signal with a fixed 200-kHz reference signal. The coherency is obtained by phase-controlling the second local oscillator signal at 19.3 MHz; the phase controller uses an error signal derived in the Costas-type demodulation loop for the tracking signal, see Fig. 3. Since the same carrier is used in the transmitter for generating all the sub-carriers, a reference signal locked to the tracking signal will bear a fixed phase relation to all the other signals in the receiver. However, small differences in the phase response of individual channels between a modulator in the transmitter and the corresponding demodulator in the receiver make it necessary to perform a phase calibration. This can be done simply by carefully adjusting critical cable lengths of each separate channel. An automatic phase-correcting circuit has been designed to avoid the need for such critical adjustment, see Fig. 4. According to this new technique the 200-kHz reference signal for each information channel is passed through a phase-shifter that is voltage-controlled in a closed-loop configuration. The control voltage is derived from a modified Costas loop demodulator for that channel. Some essential features of this approach are:

(a) Any phase difference between the tracking signal and an information-carrying signal at the receiver output is automatically compensated.

(b) The phase correcting loop can be extremely narrow because it is never required to track a frequency offset. As a result, it will contribute only insignificantly to the total reference phase jitter.

(c) The phase-correcting loops will always acquire the correct phase automatically, without the need for a search.

3.9 Code acquisition

The code acquisition circuitry in the STC CDM/CDMA equipment operates in three modes: RESET, SHORT, and LONG, and these are selectable via push-button switches on the front panel. In the RESET mode both of the shift registers are stopped so that no spread modulation is applied. This mode is useful for level calibrations with a CW signal. When the SHORT button is pressed on the transmitter the short code is applied to the tracking signal modulator, and pressing SHORT at the receiver causes a search along the short code to be initiated. After the short code has been acquired, a shift to long code occurs automatically when the LONG switch at the transmitter is activated, provided that both receiver and transmitter have been reset such that the two main code registers have been pre-loaded with the same initial contents. Resetting the code generators also causes the divider chains to return to their initial state. A fourth push-button on the front panel, LOAD, serves to load the auxiliary register with the contents of the thumbwheel switch array.

4. PERFORMANCE EVALUATION

4.1 Tracking loops

An important performance parameter of the tracking loop is the threshold which is defined as the lowest value of C/n at which both code tracking and carrier tracking can be maintained. With an unfiltered 20-megabaud CDMA signal the threshold was found to occur at C/n = 29 dBHz (corresponding to a signal-to-noise ratio of 1 dB out of a 400-Hz filter). Filtering the CDMA signal to a bandwidth of 20 MHz caused the threshold to increase by approximately 1 dB. It must be noted, however, that some uncertainty was involved in determining exactly where the threshold occurred because it was the carrier-tracking loop that lost lock first, causing the threshold to show as a gradually increasing rate of cycle skipping. The stated figures correspond to an average time between skips of about one minute. It was found that the code-tracking loop would maintain lock for a signal-to-noise ratio 3 dB below the overall threshold. An improvement of the threshold might therefore be obtained by decreasing the bandwidth of the carrier-tracking loop. It is finally to be noted that the stated thresholds were measured for zero doppler offset. When the VCO frequency was shifted to the limits of its range (+ 1.3 kHz and 2.5 kHz) a 3-dB increase in the threshold was observed.

4.2 Processing gain

The processing gain was measured for a CW interfering signal swept over the frequency range 60 to 80 MHz. With a spread rate of 20 megabauds a CW signal at the centre frequency is effectively spread over a 20-MHz bandwidth (i.e., the noise density created by the CW signal around the despread signal can be calculated by dividing the total power of the CW signal by 20 MHz). Consequently, the processing gain as defined by the ratio of the signal-to-noise ratio at the output of the pre-detection filter to the signal-to-noise ratio at the input to the CDMA receiver, can theoretically be calculated by dividing 20 MHz by the noise bandwidth of the pre-detection filter. If the CW signal is not at the centre frequency, the processing gain is increased inversely proportionally with the $(\sin x/x)^2$ - spectrum of the spread waveform. With the pre-detection filter 3-dB bandwidth set at 800 Hz, corresponding to a noise bandwidth of 1.26 kHz, the theoretical processing gain for a CW signal at 70 MHz is thus 42 dB. The theoretical curve showing the processing gain versus CW signal frequency is plotted in Fig. 5. Measured results obtained for unfiltered SSMA and for SSMA filtered to 20 MHz are also shown in Fig. 5. For the unfiltered case the processing gain at 70 MHz is only about 0.5 dB below the theoretical value, the difference being due partly to unavoidable residual filtering and partly to equipment imperfections and possibly also to the limited measuring accuracy. For filtered CDMA, an additional loss of about 1.0 dB is observed due to the mismatch between the filtered CDMA signal and the unfiltered reference signal.

4.3 Back-to-back transmission performance

CDMA carriers with the same centre frequency will to some extent interfere with each other. For a bandwidth-limited system, where the down-link thermal noise can be neglected, the predetection signal-to-noise ratio is given by $1/(N-1)$ assuming that the number of equal level carriers is N. Each carrier sees the other (N-1) carriers as interference. When Code Division Multiplex (CDM) is employed, each channel sees in addition all the other channels of its own carrier as interfering signals. Assuming that our carrier has M equal channels, this corresponds to an additional degradation of $10 \log_{10} [(N-1/M)/(N-1)]$. In a practical situation with thermal noise and other sources of interference the degradation will be even smaller. CDMA operation should therefore have a mutual interference performance not much worse than that to be expected with other mux methods.

Error-rate measurements were carried out on one of the 611-baud channels provided via the seven-channel majority-logic multiplexer described (Section 3.6 and Appendix D) in order to assess the total implementation losses. Differential encoding was used for the measurements. It should be noted that when six of the seven multiplexer channels are switched off, the remaining channel is just a normal differentially encoded coherent PSK (DCPSK) channel.

Measurements were done with 1, 3, 5, and 7 active channels and for multiplexing rates of 78 kilobauds and 1.25 megabauds. The CDMA signal was filtered in a bandpass filter that had a noise bandwidth of 20 MHz. No receiver filter was applied. Noise was added at the input of the 20-Mhz filter, and the signal-to-noise ratio was measured on a power meter at the output of the filter. The results are shown in Fig. 6.

The theoretical error-rate curve for 611-baud binary DCPSK is plotted in Fig. 6 for comparison. There is an apparent implementation loss of about 2 dB, but considering that the theoretical loss due to receiver mismatch is 1 dB when the SSMA signal is filtered to 20 MHz, the remaining actual implementation loss is about 1 dB, which was considered to be satisfactory.

With three, five, and seven channels active, the available power per channel is reduced by 4.8 dB, 7.0 dB, and 8.5 dB respectively. It is shown in [13] that the majority multiplexing technique implies an additional loss of $2n/[\pi(n-1)]$, where n is the number of channels, and for the three cases considered here the additional losses are 0.2 dB, 1.0 dB, and 1.3 dB, respectively. For the high multiplexing rate of 1.25 MHz, the total measured performance is about 1 dB worse than that predicted by these figures.

4.4 Non-linear suppression

If each carrier accessing the satellite transponder uses only a small fraction of the total transponder power, a hard-limiting transponder looks quasi-linear, except for a suppression of about 1 dB due to the creation of intermodulation. The non-constant-envelope properties of the carriers cause no further degradation. A difference between the CDM/CDMA system proposed here and a TDM/CDMA system could occur due to non-linear effects in ground terminals transmitting a small number of carriers. The suppression effects in this situation will be considered in some detail.

In its simplest form a CDM/CDMA carrier with bi-phase spread modulation will consist of an amplitude modulated signal with suppressed carrier and with possible amplitude values given by $\pm a_1 \pm a_2$ ---- $\pm a_N$. Let us consider the situation in which a single CDMA carrier with N channels is subjected to hard limiting in a ground terminal. The output level of a particular channel is then proportional to the probability that the amplitude of the composite signal has the same sign as a_i, the amplitude associated with the channel under consideration. It is immediately clear that if any $a_j, j=1,2,$ --, N is larger than the sum of the others, the whole power is associated with this channel and all the other channels are completely suppressed. This could for example happen for the combination of a delta-modulated voice channel plus a few telegraph channels. This situation is highly undesirable and although a pseudo-random, constant phase difference between the channel carriers would alleviate the problem, the case of binary spread modulation will not be considered further.

With four-phase modulation it is convenient to represent the signal by a sum of amplitude modulated sine and cosine waves

$$y(t) = \left(\sum_{i=0}^{N} a_i C_i W_i\right)\cos(\omega_o t+\phi_o)$$

$$+ \left(\sum_{i=0}^{N} a_i \overline{C}_i W_i\right)\sin(\omega_o t+\phi_o) \quad (10)$$

C_i (0 or 1) and W_i (+1 or -1) are binary pseudo-random sequences derived from the duo-binary sequence Z_i. $\overline{C}_i$ is the conjugate of C_i. a_i is an amplitude factor which represents the relative power level of carrier number i. The spectrum of this signal will have a $(\sin x/x)^2$ shape with a separation between the first spectral nulls equal to twice the spread modulation rate.

The instantaneous envelope is given by:

$$E = \left[\left(\sum_{i=o}^{N} a_i C_i W_i\right)^2 + \left(\sum_{i=o}^{N} a_i \overline{C}_i W_i\right)^2\right]^{\frac{1}{2}} \quad (11)$$

E is limited by Σa_i and for a large number of similar channels the distribution of E will approach a Rayleigh-distribution.

The total signal can be expressed as a sinusoid with amplitude and phase modulation

$$y(t) = E(t)\cos(\omega_o t+\psi(t)) \quad (12)$$

where

$$\psi(t) = \text{Arctan}\left[- \frac{\sum_{i=0}^{N} a_i \overline{C}_i W_i}{\sum_{i=0}^{N} a_i C_i W_i}\right] \quad (13)$$

The number of possible phase states is finite but increases rapidly with the number of channels.

After hard limiting, only the phase modulation is retained, i.e. the signal phasor has constant length, and the level associated with channel i is proportional with

$$< \cos(\psi-\phi_i)>_{ave} \tag{14}$$

where ϕ_i is the phase of sub-carrier i. Fig. 7 shows a typical phasor diagram.

For simplicity and without loss of generality, sub-carrier phasor i has been used as a reference ($\phi_i = 0$).

Power sharing between two channels of arbitrary level can be solved analytically. The wanted carrier has unity amplitude and the other carrier has amplitude a. Both carriers have ϕ=0. There are four possible values of the signal phasor with respect to the rotating reference system, each with a probability of 1/4.

For a <1 the average value of cos ($\psi-\phi_i$) is given by

$$<\cos(\psi-\phi_i)>_{ave} = 2\cdot\tfrac{1}{4}\cdot 1 + 2\cdot\tfrac{1}{4}\cdot\sqrt{\frac{1}{1 + a^2}}$$

$$= \tfrac{1}{2}\left(1 + \sqrt{\frac{1}{1 + a^2}}\right)$$

For a >1

$$<\cos(\psi-\phi_i)>_{ave} = \tfrac{1}{4}\cdot 1 - \tfrac{1}{4}\cdot 1 + 2\cdot\tfrac{1}{4}\sqrt{\frac{1}{1 + a^2}}$$

$$= \tfrac{1}{2}\left(\sqrt{\frac{1}{1 + a^2}}\right)$$

The non-linear suppression, defined as the difference in dB from proportional power sharing is shown in Fig. 8. Also shown is the non-linear suppression with two non-coherent sinusoidal carriers and it is noticed that the coherency results in a discontinuity for a =1 (0dB level difference). Similar calculations have been carried out for a higher number of channels and non-linear suppression of quadriphase modulated coherent channels represents no special problem.

It is seen that, as expected, weak signals tend to be suppressed in the ground terminal limiter, whereas strong signals are only slightly affected in this respect and may in fact even be enhanced. In other words, if the power allocated to each channel is proportional to the channel's data rate, the low speed channels must be expected to have a significantly higher error probability than the high-speed channels.

In the general case when N+1 carriers are present in a ground terminal limiter that is supporting data rates of R_o, R_1, ..., R_N, equal performance of the channel is obtained only when

$$\frac{\left[E^2\{\cos\psi\}\right]_o}{R_o} = \frac{\left[E^2\{\cos\psi\}\right]_1}{R_1} = \text{----} \frac{\left[E^2\{\cos\Psi\}\right]_N}{R_N} \tag{15}$$

where $\left[E^2\{\cos\Psi\}\right]_i$ denotes $E^2\{\cos\psi\}$ when the ith signal is taken to be the desired signal. Clearly, $\left[E^2\{\cos\psi\}\right]_i$ is a function of the relative signal amplitudes and, of course of the carrier phases. Given the rates and the phases, (15) must be used to determine the carrier levels. A calculation was carried out for the case N=7. Fig. 9 shows a plot of total power loss against R_N/R_0, and for comparison suppression in dB is plotted for one of the low-speed channels with proportional power allocation. When optimum power allocation is used, the worst loss encountered is about 1.3 dB. Fig. 9 assumes carriers of equal phases but it seems reasonable to expect the phase variations will have a minor influence on the total losses.

4.5 Comparison of TDM/CDMA with CDM/CDMA

From the calculations above the two systems can be compared under the conditions that both the satellite ground terminal and the satellite transponder are strongly non-linear, as they could be in a jamming situation. In such a case CDM/CDMA might be expected to perform slightly worse than TDM/CDMA because of the occurrence of mutual suppression of the individual sub-carriers in the ground terminal (SGT) amplifier. This type of suppression has two effects:

- a fraction of the total available output power is lost in the form of intermodulation products, and

- the relative powers of the subcarriers at the output of the klystron amplifiers differ from those assigned in the CDM/SSMA modem.

The intermodulation loss is known to be small, typically less than 1 dB. A similar loss would occur with TDM/CDMA whenever more than one modem were transmitting via the same amplifier. It appears that the most significant problem arising from limiting in the SGT is that of accurately controlling the subcarrier power levels at the output of the limiter.

The suppression of any one subcarrier can be expressed quantitatively by plotting the suppression in dB below proportional power-sharing against the interference-to-signal power ratio for various sub-carrier phase/amplitude relations. Such data can be used to evaluate the mapping from input power sharing to output power sharing (or vice versa).

It was found that for unfiltered CDM/CDMA signals in an ideal hard limiter the maximum suppression encountered is 6 dB. Because of the coherency of the sub-carriers, the suppression curves can exhibit abrupt discontinuities, and the size and position of these discontinuities can be very sensitive to small changes in the phases and amplitudes of the subcarriers. This observation indicates that the suppression effects may be difficult to predict when the system bandwidth is large and the limiter characteristic is close to that of an ideal hard limiter.

When the limiter is preceded by a bandpass filter which affects the transmitted waveform to a significant degree, the above-mentioned discontinuities can be shown to disappear while at the same time the maximum suppression is reduced to a value considerably less than 6 dB (Fig. 10). Both of these effects are explained by the introduction of amplitude modulation in the signal as a result of filtering. It is noted in particular, that two- and four-phase spread-modulated signals perform quite differently, the two-phase signal having a much weaker suppressing effect on other signals. This observation is confirmed by experiment.

Calculations carried out based on an accurate mathematical model of a klystron amplifier have shown (see Fig. 11) that the phase relations between the subcarriers were unimportant, even in the unfiltered case. This is mainly explained by the soft-limiting characteristic of the klystron, but the presence of AM/PM conversion also adds to this effect.

A difference between CDM/CDMA and TDM/CDMA which does not relate to SGT nonlinearity is the susceptibility of a subcarrier in a CDM/SSMA link to interference (in the form of an increased noise level) from not only other users (as with TDM/CDMA) but even from other subcarriers from the same modem. This effect can be shown to be small whenever each SGT only supplies a small fraction of the total input power to the satellite.

5. SUMMARY AND CONCLUSIONS

Satellite communication systems using CDMA operation with many ground terminals, large connectivity and inhomogeneous traffic composition may have a multiplexing problem of severe complexity. A system has been described and analysed in which the multiplexing process is an integrated part of the multiple access system. Each individual channel uses a separate wide-band carrier (code division message multiplexing) and all the carriers of a particular link are synchronized with a separate unmodulated carrier provided specifically for tracking purposes. This gives a system with a performance almost as good as TDM/CDMA, transparent channels with analogue and/or digital modulation, individual channel performance adjustment and a considerable flexibility of operation, in particular with respect to timing.

REFERENCES

1. A.N. Ince, H.P. Williams, Range of LF Transmission using Digital Modulation, Proc. IEE, Vol. 114, No. 10, 1967.
2. R. Schemel, A.N. Ince, Spread-Spectrum Modulation for LF ground-wave and HF sky-wave transmissions, A paper in preparation.
3. M. Darnell, New HF Data Transmission Techniques, presented at NATO Advanced Study Institute, Darlington, 8-20 August 1977.
4. A.N. Ince, Design, Testing and Operation of X-Band Satellite Communication System, IEEE Trans. on Communications, Vol. COM-22, No. 9, Sept. 1974.
5. AGARD Lecture Series No. 58, Spread Spectrum Communications, July 1973.
6. A.N. Ince, Design and Calibration of X-band satellite communications ground terminals, Electronics en Telecommuncatie (Royal Institute of Engineers, The Netherlands), No. 5, 15 May 1972.
7. G.R. Stette, Code Division Message Multiplexing in a satellite system, AIAA 5th Communications Satellite Systems Conference Paper 74-472, Los Angeles, Calif. April 1974.
8. J.M. Aein, Multiple Access to a hard-limiting communication satellite repeater, IEEE Trans. Space Elec. and Telemetry, Vol. SET-10, No. 4, pp 159-167, December 1964.
9. J.C. Springett and M.K. Simon, Analysis of the phase coherent-incoherent output of the bandpass limiter, IEEE Trans. Com. Technology, Vol. COM-19, No. 1, Feb. 1971.
10. J.R. Lesh, Signal-to-noise ratios in coherent soft limiters, IEEE Trans. Coms. Vol. COM-22, No. 6, June 1974.
11. R Gold, Optimal binary sequences for spread spectrum multiplexing, IEEE Trans. Inf. Th. Vol. IF 13, No. 4, Oct. 1967.

12. L. Jacobsen, R. Schoolcraft, P. Hooten, Highly efficient voice modulation for low C/N_o communications channels with hard-limiting repeaters, Proc. IEEE 1972, Int. Conf. on Com., Philadelphia, Pa., June 1972.
13. R. Titsworth, A Boolean-function-multiplexed telemetry system, IEEE Trans. Space Electr. and Telem., Vol. SET-9, No.2, June 1963.
14. S.W. Golomb, Shift Register Sequences, San Francisco Holden-Day, Inc., 1967.
15. W.J. Gill, A Comparison of binary delay-lock tracking-loop implementations, IEEE Trans. Aerosp. and Electron. Systems, Vol. AES-2, No. 4, July 1966.

APPENDIX A

A COMPARISON OF FREQUENCY AND PHASE HOPPING MODULATIONS

1. Processing gain : For the same occupied bandwidth there is about a 2-dB advantage in using FH due to its spectrum being flat while PH spectrum is $\left(\frac{\sin x}{x}\right)^2$.

2. Synchronization and timing : Timing requirements are more severe for PH systems as compared to FH by a factor which is approximately equal to the processing gain. Moreover, FH techniques can search out timing uncertainties faster by this same processing gain factor.

3. Demodulation efficiency : PH systems generally provide superior detection capability due to use of coherent techniques as compared to the non-coherent techniques usually employed in FH systems.

4. Complexity : FH systems (particularly receivers) tend to be more complex due to the use of frequency synthesizers and generally more complex decoders.

5. Ranging : PH systems provide more precise ranging capability than FH techniques by a factor approximately equal to the processing gain.

6. Multipath rejection : PH systems inherently exhibit multipath rejection capability due to a smaller correlation window.

APPENDIX B

COMPARISON OF MULTIPLE-ACCESS METHODS

The efficiencies of the access methods may be compared by consideration of:

(a) the percentage of the satellite EIRP which can be utilized for the transmission of information,

(b) the percentage of satellite bandwidth which could be utilized for the transmission of information if the satellite EIRP were unlimited,

(c) vulnerability to external interference.

Fig. B-1 and Fig. B-2 which are taken from [6] show the results with respect to the factors in (a) and (b) above for the three access methods.

APPENDIX C

CHOICE OF CODES

The CDM/CDMA systems require a "main code" (for tracking purposes) and a number of channel sub-codes, all of which should exhibit good auto- and cross-correlation properties. It is considered that regardless of the detailed choice of code the main code will be generated by a feedback shift register, and the sub-codes will be formed by a suitable combination of the output taps of this register. The following requirements influence the choice of code:

(a) The re-cycling period of the codes must be long.
This is necessary to ensure good "randomness" properties.

(b) The auto- and cross-correlation functions of the codes should be uniformly low.
The same should apply to the partial correlation functions measured over a period of time equal to the shortest expected bit duration of any input signal.

(c) The codes should be difficult to intercept.
For this reason a large class of uniformly good codes must be provided to permit frequent changes of codes.

(d) Acquisition of the main code must not be disturbed by the presence of other codes in the received signal.
As a consequence, no sub-code may be a shifted version of the main code unless the time shift is guaranteed to be large enough to prevent confusion.

The first important choice to be made is that between linear and non-linear codes. Non-linear codes have the definite advantage that they are difficult to intercept, since interception of a small part of the sequence does not generally identify the whole code. However, no analytical tool exists for working out the correlation properties of such codes, and because of the long recycling periods involved a trial-and-error approach to code selection is very difficult.

Among linear codes the maximum length sequences, often referred to as m-sequences, are probably best known. These codes have ideal two-valued auto-correlation but the cross-correlation function for two distinct codes of equal length may exhibit severe peaks, in particular if the re-cycling period contains small prime factors. Still, it may be feasible to use m-sequences in the proposed scheme if all codes are cyclic shifts of each other (in which case the cross-correlation properties are ideal), but care must be taken to avoid confusion during acquisition. The problem is complicated by the fact that even cross-correlation between codes in different modems is of significance.

Because of these deficiencies of m-sequences with regard to CDM-applications, efforts have been devoted to the task of finding classes of non-maximum length linear codes with superior performance. In [11] it is shown that for any n, it is possible to find a "preferred pair" of m-sequences with period $2^n - 1$ such that the class consisting of these two codes plus the $2^n - 1$ distinct codes formed by modulo-2 addition of one of the m-sequences with any cyclic shift of the other has the desired properties. For any pair of codes from this class the normalized cross-correlation function is shown to be numerically less than or equal to C, where

$$C = 2^{0.5(1-n)} + 1, \text{ n odd}$$

or $$C = 2^{0.5(2-n)} + 1, \text{ n even}$$

The auto-correlation of any code from the class satisfies the same bound.

Obviously, this class consisting of 2^{n+1} different codes of length 2^{n-1} may also be generated by a single 2n-bit shift register with suitable feedback connections. The particular code produced at, say, the last stage is determined by the initial contents of the register. Besides, different codes from the ensemble may be obtained simultaneously by suitable modulo-2 addition of selected output taps.

From this brief description it should be clear that the codes given in [11] possess many properties that are essential for the CDM/CDMA system described in this paper. Using a 62-bit shift register one may obtain a total of more than 10^9 distinct, uniformly good codes, each with a re-cycling period of about 100 seconds when a clock rate of 20 MHz is assumed.

APPENDIX D

MAJORITY-LOGIC MULTIPLEXER

The most distinctive property of the proposed CDMA system is the great flexibility obtained by supplying a separate modulator/demodulator for each message input. This approach could, however, become problematic if the required number of input channels turned out to be large. Two specific problems arise:

(a) The cost of modulator/demodulator equipment may become excessive.

(b) The complexity of the power control systems grows rapidly with the number of channels.

For applications where the main bulk of the traffic is carried by low-speed telegraph channels (50 or 75 bauds), it may prove advantageous to combine all low-speed telegraph channels into one bit-stream in a first stage of multiplexing and apply them to one single input of the CDMA modem. The low-speed channels may be combined using either TDM or CDM, and because of the low data rates involved the required transparency can be obtained simply by sampling the signals at, say, 5 to 10 times their maximum rate. TDM is easy to implement and has the advantage of making optimum use of power when all channels are busy.

A code division multiplexer featuring binary inputs and binary output has been suggested in [13]. The principle is illustrated in Fig D-1. Each of the n channels is allocated an L-bit codeword from a near-orthogonal code. For each time-slot of the input signals all channels send either their codeword or its inverse - depending on the data bit - to a logic device which performs a majority decision for each time-slot of the codewords. The output from the majority device is transmitted. In [13] it is shown that the single channels can be recovered from the received signal by a simple correlation with the corresponding codeword, provided that L is sufficiently large and the codewords are selected properly. This may be readily understood if it is realized that the majority device is equivalent to a summing network followed by a hard limiter. Reference [13] calculated the power loss due to this limiting action as close to 2 dB for most cases of practical interest.

An interesting feature of this CDM scheme is its inherent automatic power-sharing between active input channels. As a channel becomes idle the corresponding input to the majority function should be grounded and this immediately causes its power share to be distributed evenly among the remaining busy channels. It should be emphasized that this requires no corresponding action at the receiver.

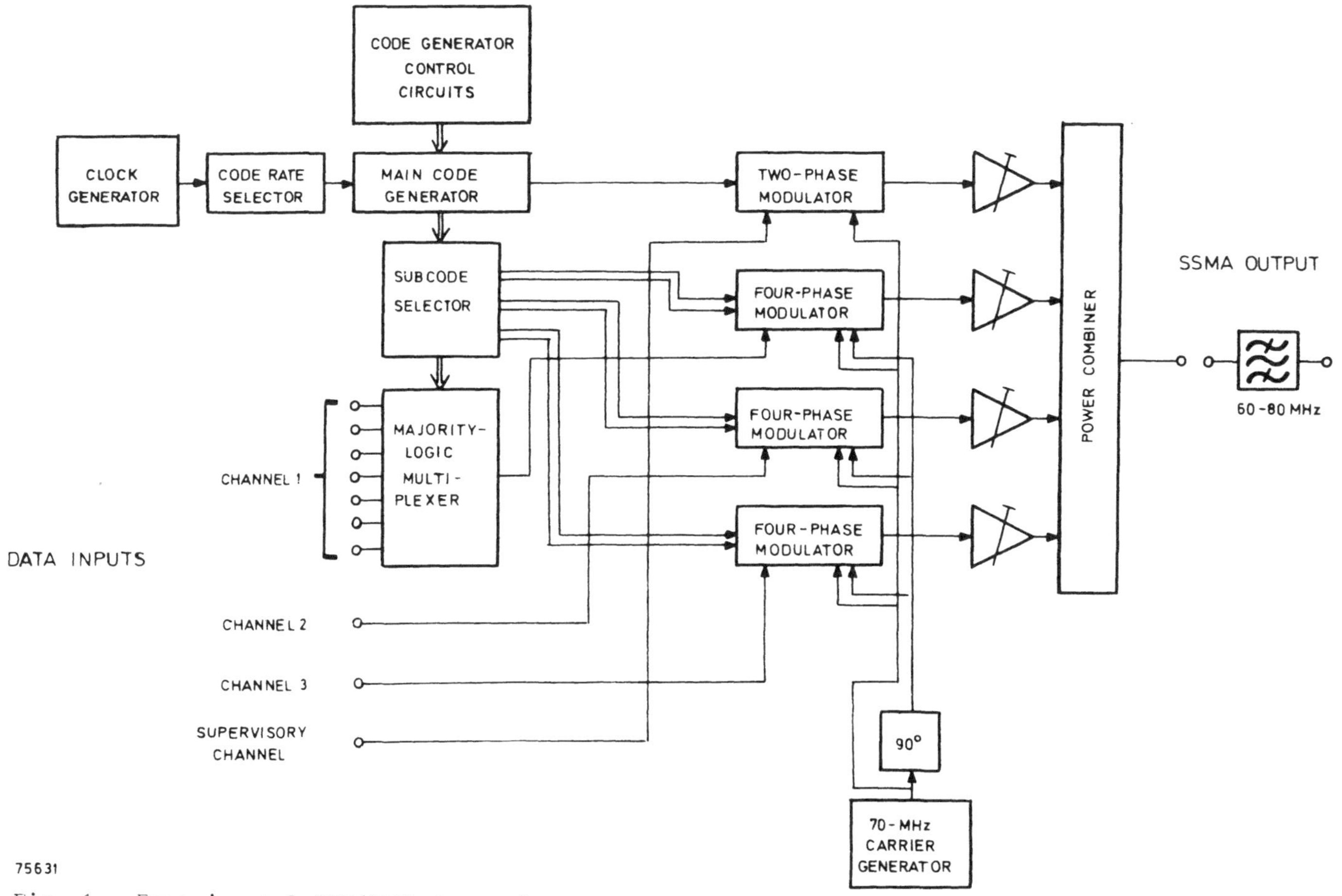

75631

Fig. 1 Experimental CDM/CDMA transmitter

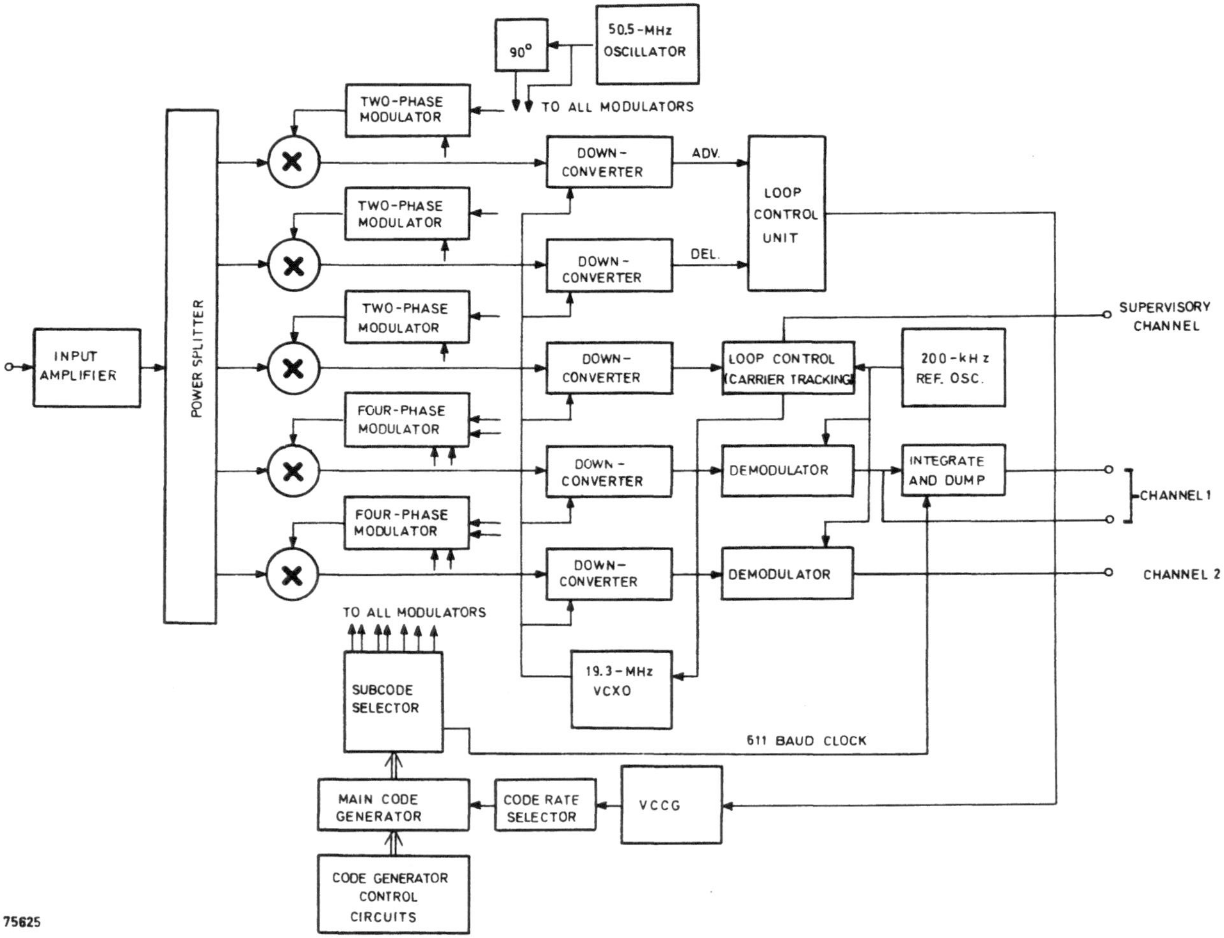

75625

Fig. 2 Experimental CDM/CDMA receiver

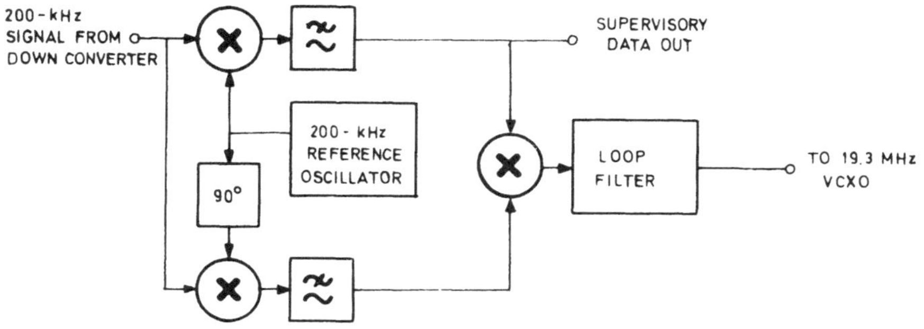

Fig. 3 Carrier tracking circuit (tracking channel)

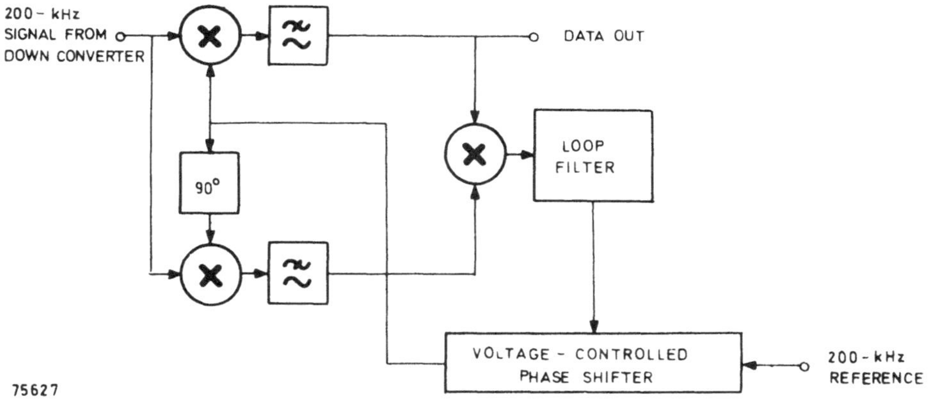

Fig. 4 Phase correcting circuit (data channel)

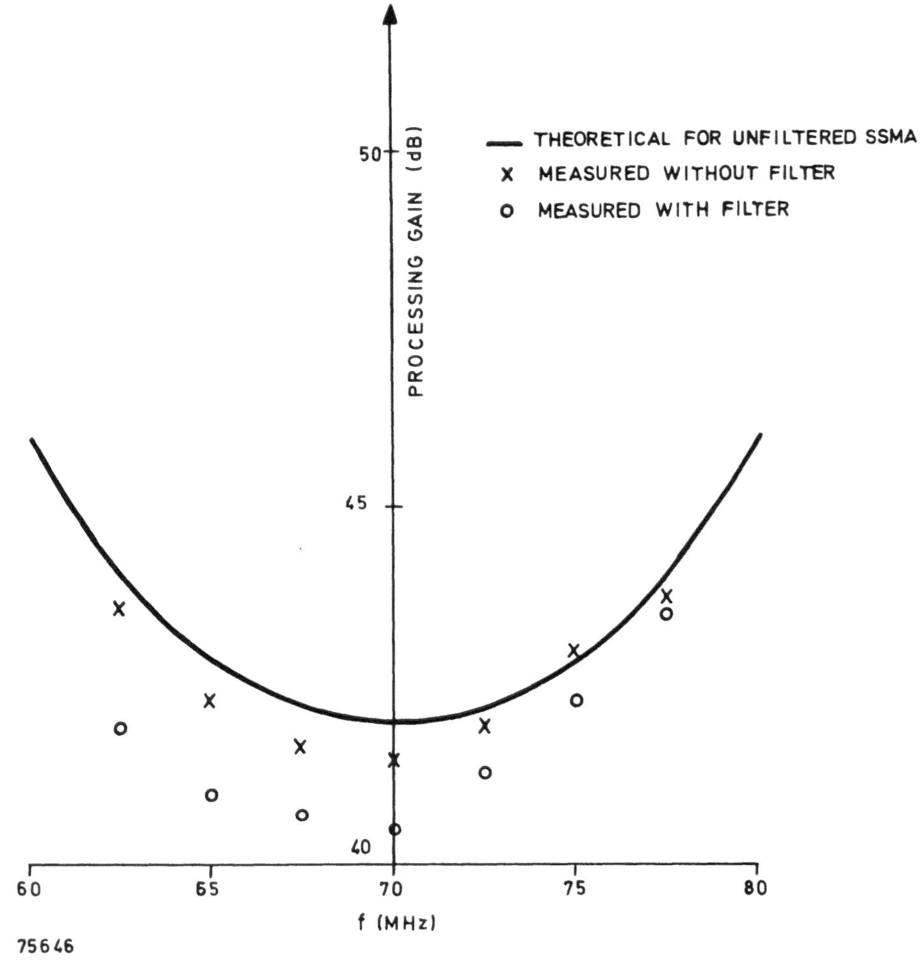

Fig. 5 Processing gain for CW interfering signal

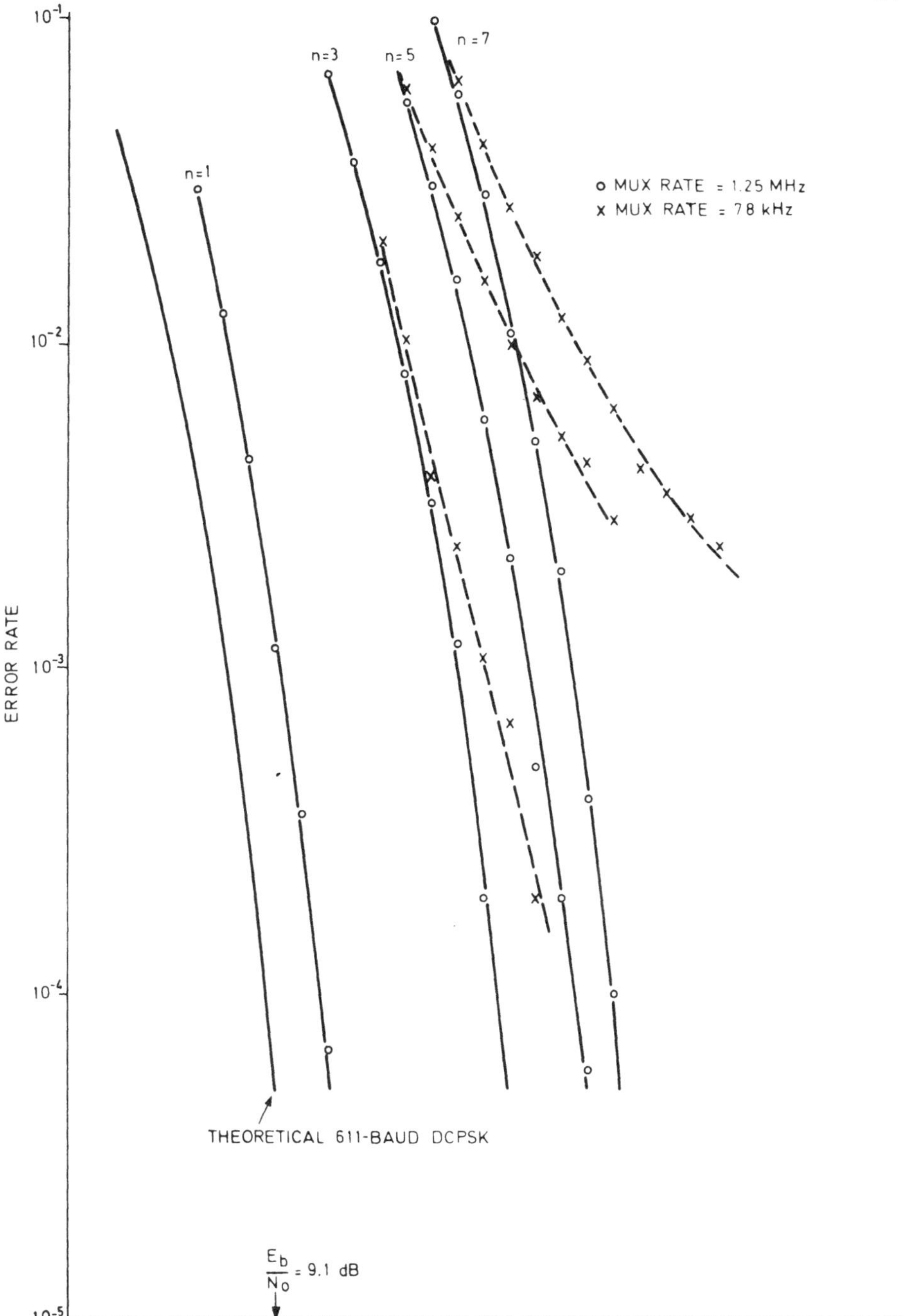

Fig. 6 Error rates for filtered CDMA

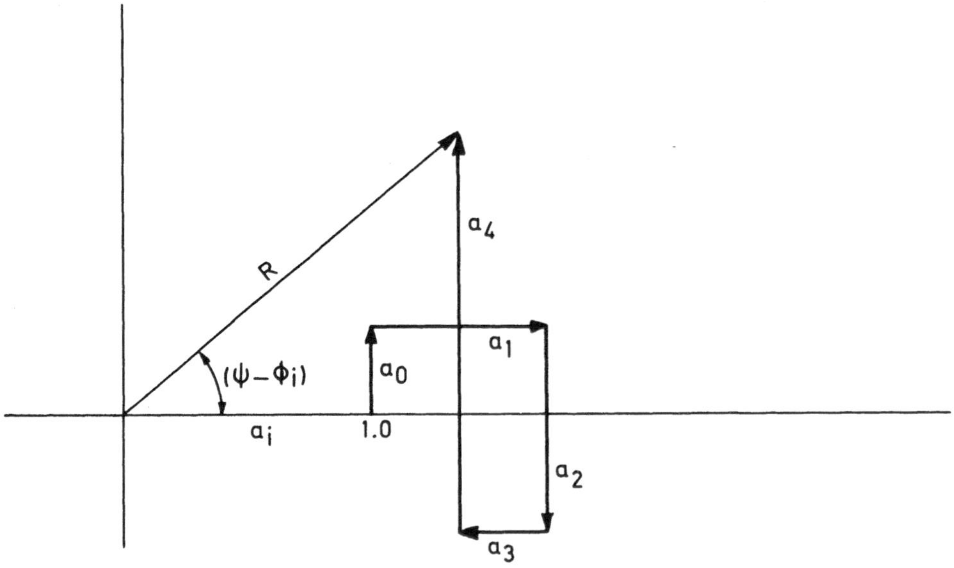

Fig. 7 Typical Phasor Diagram

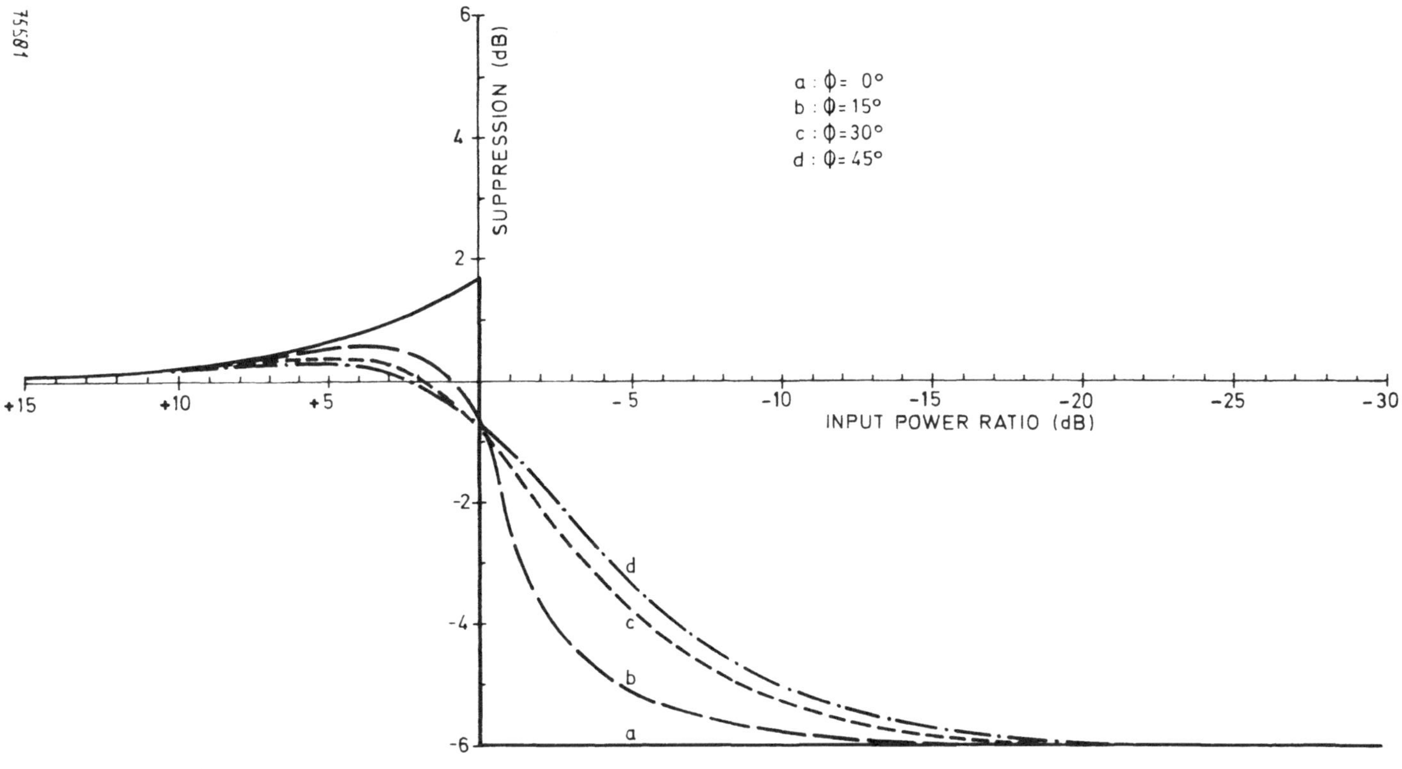

Fig. 8 Suppression of a CDMA subcarrier by one other subcarrier

75585

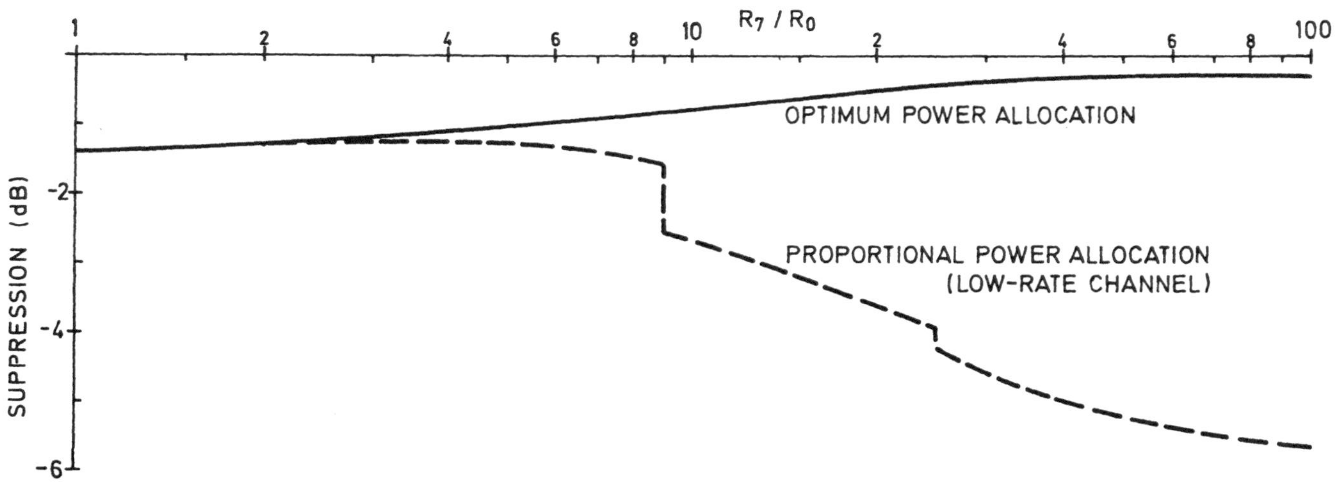

Fig. 9 Suppression of a CDMA subcarrier by seven other subcarriers with optimum and rate-proportional power assignment

75592

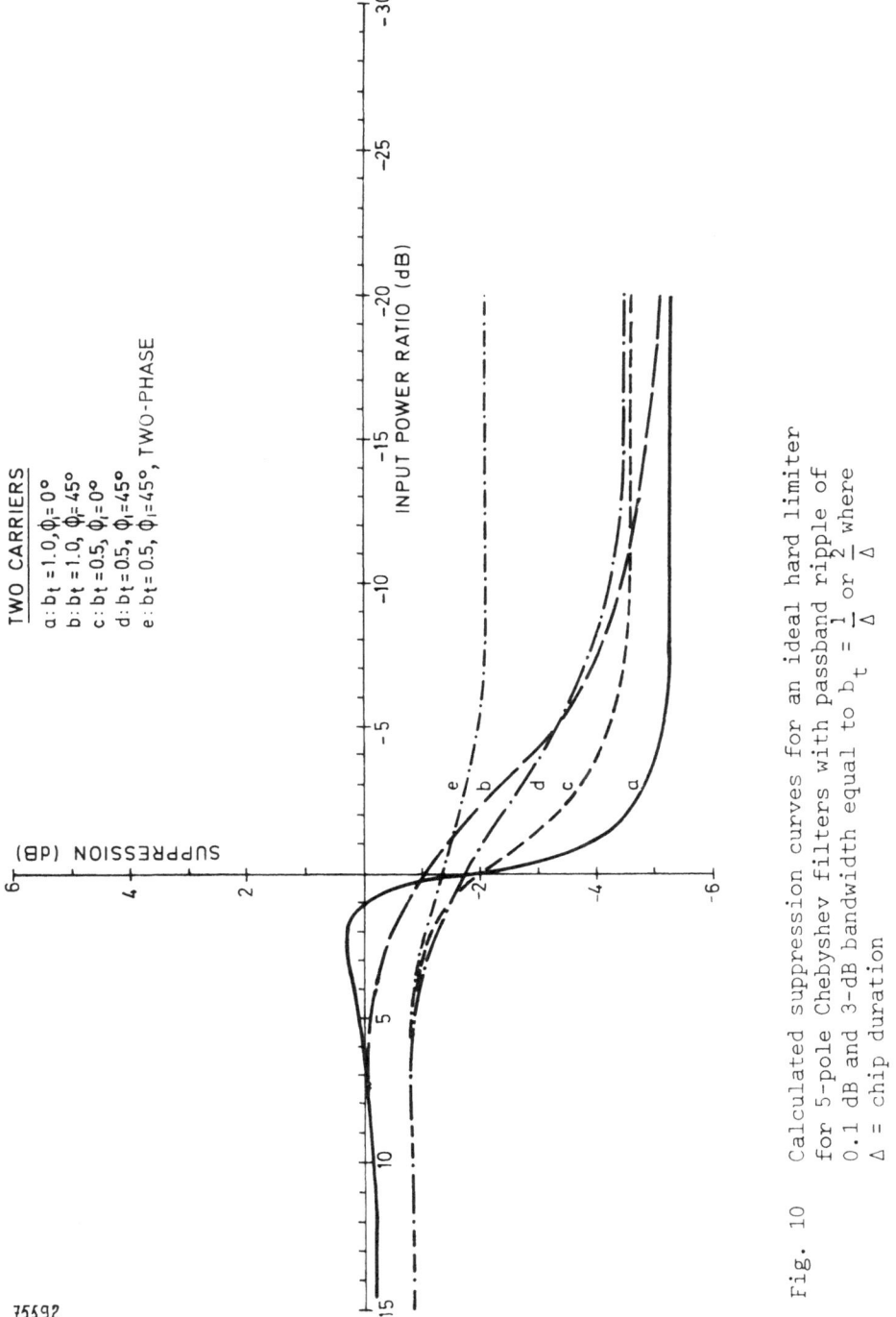

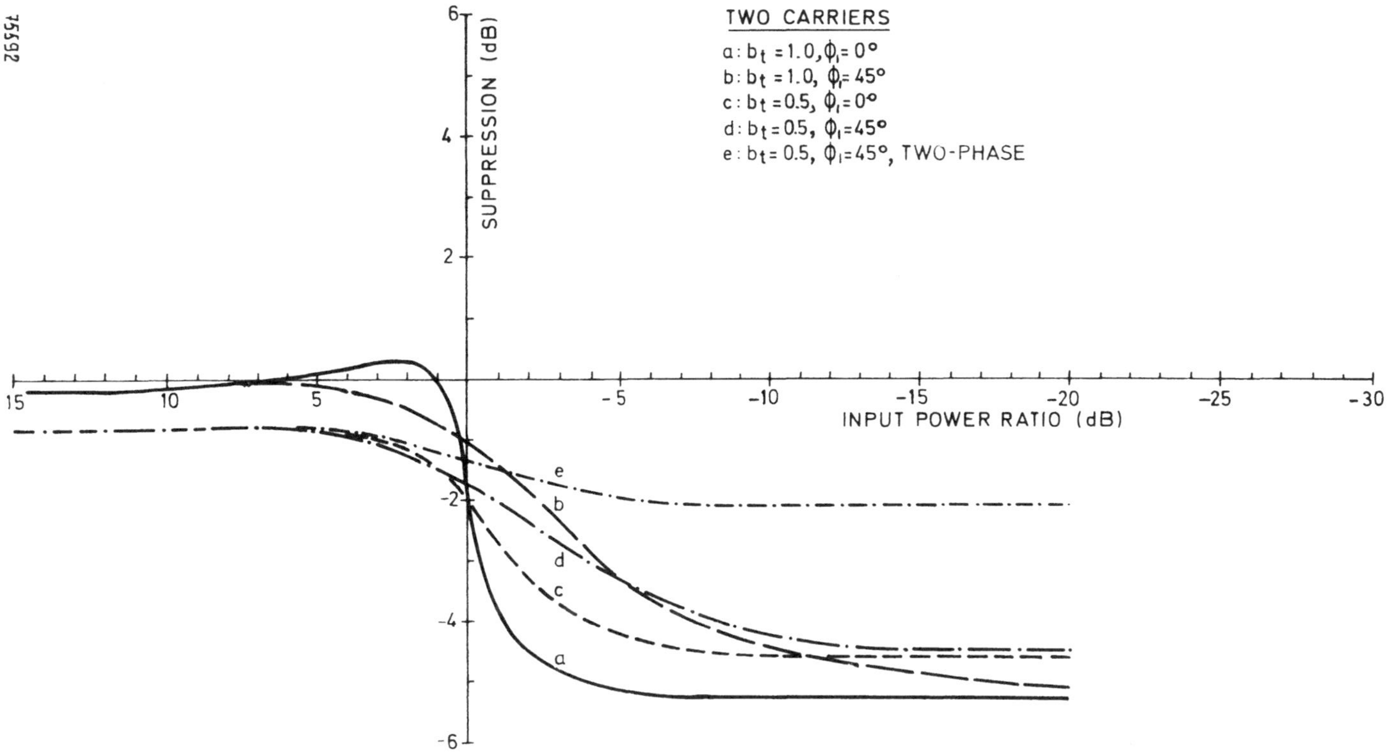

Fig. 10 Calculated suppression curves for an ideal hard limiter for 5-pole Chebyshev filters with passband ripple of 0.1 dB and 3-dB bandwidth equal to $b_t = \frac{1}{\Delta}$ or $\frac{2}{\Delta}$ where Δ = chip duration

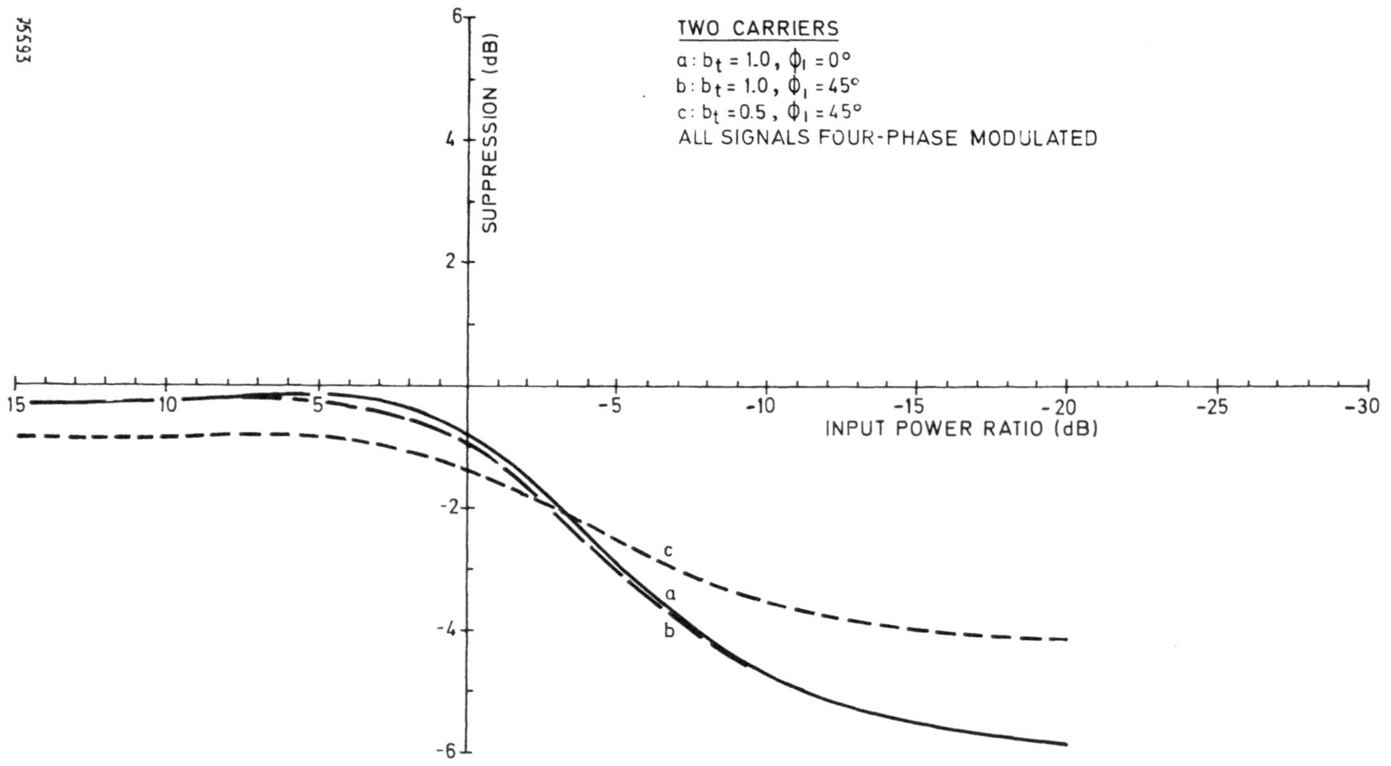

Fig. 11 Calculated suppression curves for a klystron amplifier

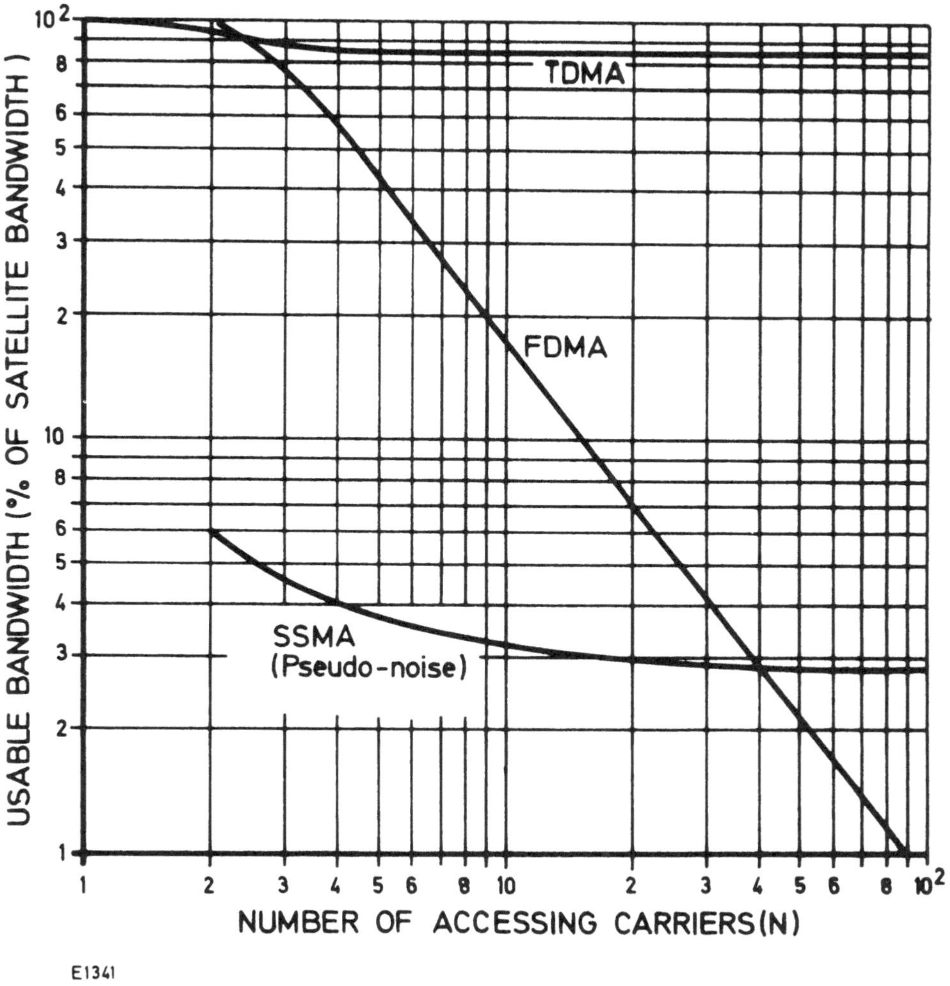

Fig. B-1 Usable satellite bandwidth for a hard-limiting transponder

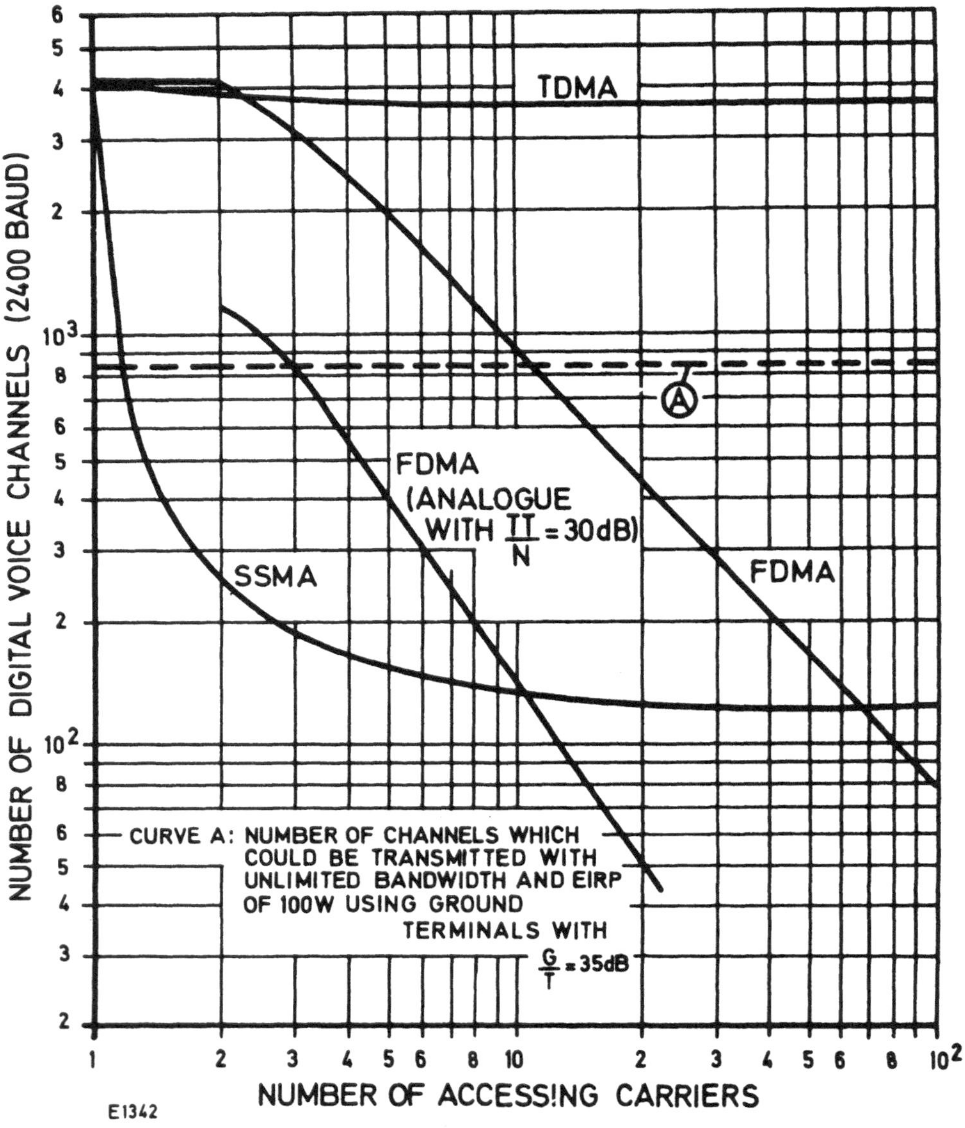

Fig. B-2 Maximum numbers of voice channels through a hard-limiting satellite of 20 MHz bandwidth.

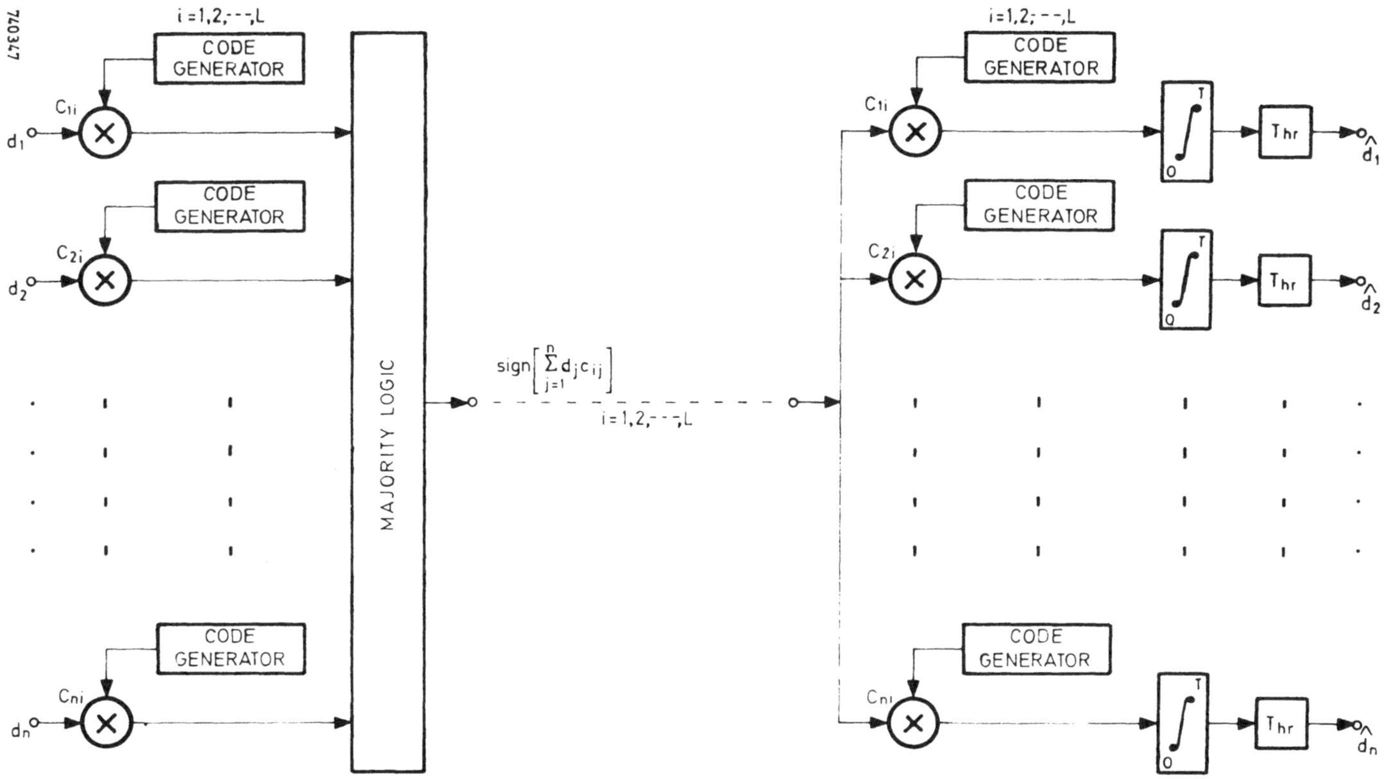

Fig. D-1 A majority-logic binary CDM system

THE POTENTIAL OF METEOR-BURST COMMUNICATIONS WITH PARTICULAR REFERENCE TO THE COMET SYSTEM

D.W.Brown

Communications Division, SHAPE Technical Centre,
The Hague, Netherlands

ABSTRACT. The paper reviews the phenomenon of propagation via meteor trails and describes the STC meteor burst communication system, COMET, which used ARQ and diversity reception to provide several telegraph channels on a 1000 km path. The results of extensive testing of COMET at 40 and 100 MHz are presented and it is shown that meteor-burst communications are highly reliable and that the STC system could be readily extended to support an average data rate of the order of 2400 bit/s on paths up to 2000 km using single frequency operation, simple equipment and antenna systems, and transmit powers less than 1000 watts. Alternatively very low power operation (e.g., 10 watts) would be possible if it were desired to return small quantities of data from a remote terminal, say, once per day. The immunity of meteor-burst propagation to auroral phenomena would make it an excellent candidate for operation in the Arctic.

1. INTRODUCTION

Meteor-burst communication has been tested at STC over a period of many years. Early testing of the Canadian Janet-B system at STC led to the development of a new system, later named COMET (COmmunication by MEteor Trails) [1], which makes use of a combination of diversity reception and the technique of automatic request for repetition (ARQ). It has the advantage over the earlier Janet-type systems, of making efficient use of the available duty cycle while at the same time keeping the error rate under control. At 40 MHz, the COMET system, using simple Yagi antennas and a power of only 200 watts, maintained an hourly average capacity of more than one 50-baud telegraph channel over a series of tests which covered all the seasons of the year [1]. The 24-hour average capacity was roughly equivalent to three 50-baud telegraph channels; for 90% of the time the error rate was better than 1:3000, with a worst recorded value of 1:1000.

For a military system, the inherent immunity of meteor-burst systems to jamming and interception is important. In general, an enemy who is not within, say, 200 km of either terminal of a meteor-burst circuit cannot jam or intercept that circuit, except during infrequent and short periods, which occur when there are two meteor trails available at the same time - one for reflection between ends of the circuit and the other for reflection between the circuit and the interception or jamming station. The meteors which support reflection between the terminals of a circuit will not normally support reflection from either end of the circuit to distant interception stations. Because the occurrence of meteors suitably oriented to support traffic on the circuit and the occurrence of those oriented for the interception station are independent events, the percentage of information intercepted will have the same value as the percentage duty cycle on the path to the interception station. Similarly, the maximum percentage of information that can be removed by jamming from a distance is equal to the duty cycle on the path to the jamming station.

Therefore, since duty cycle decreases with increasing frequency, and because of a desire to establish the capacity of meteor-burst systems at frequencies where sporadic-E and ionoscatter modes of propagation were totally absent, the STC COMET system was also tested at frequencies slightly above 100 MHz for a further period of a year. During certain parts of this time, testing simultaneously at 40 MHz and 100 MHz was carried out so that comparisons of performance on a burst-by-burst basis could be made.

Although the 100 MHz frequency did not prove particularly useful for meteor-burst communication, since 2000 watts at the transmitter produced an information duty cycle of only 0.5% at this frequency, the comparative results do nevertheless form the basis for an interpolation of performance at frequencies between 40 and 100 MHz. It was concluded that 2000 watts at 50 MHz would provide, with the 2000 bauds signalling rate used, an average capacity in excess of four 50-baud telegraph channels. Even at 70 MHz, with 2000 watts at 2000 bauds, an average capacity in excess of one 50-baud channel is to be expected.

STC's work on meteor-burst communications concluded with a very successful operational trial of the COMET system on an Allied Command Europe link in Northern Norway during 1969. The COMET system has not been further developed, nor is it now in operational service. It is perhaps in some sense a victim of the explosion in satellite communications which took place at about the time COMET was ready to be given to industry for commercial development. There are many applications which meteor-burst communications can satisfy very effectively and inexpensively, certainly at much lower cost than a satellite system. The purpose of this paper is to draw attention to the potential of meteor-burst communications systems and to present the results obtained with the STC COMET system.

The paper first addresses the propagation phenomenon exploited in meteor-burst communications, then briefly describes the COMET system and the performance achieved. It concludes by suggesting improvements in the system made possible by modern technology, and by indicating some applications.

2. METEORS AND THEIR IONIZATION TRAILS

Meteors are of two main types: those that seem to be solid particles and those that appear to be loose associations of dust and ice. The first type remains in one piece as it is ablated by the friction of the atmosphere. The second type, called a dustball, fragments as it enters the atmosphere and carries on as a group or cloud of smaller particles until it is consumed. Such meteors are assumed to be the remains of disintegrated comets. The solid variety may also be of cometary origin, but at least a few are in fact tiny asteroids.

The minimum speed of arrival is the same as the escape velocity from the earth, namely 11.2 km/sec. The maximum is determined by the earth's orbital velocity (30 km/sec) and that of the meteor itself. The latter velocity has an upper limit of 42 km/sec appropriate to a parabolic heliocentric orbit at the earth's distance from the sun. Thus, the range of possible velocities is from about 11 to 72 km/sec.

Experimental evidence shows that the distribution of meteor radiants in space is roughly uniform. It is easily seen, therefore, that a region on earth for which the local time is 06.00 hours would be rushing into and sweeping up the ambient meteors, whereas at 18.00 hours the same region would be receding from them. Radio measurements indicate a mean velocity of about 60 km/sec for the early morning and about half this value for the early evening. Correspondingly, the rate of incidence of meteors varies by about the same ratio or even more. This diurnal variation is greatest at the equator and least at the poles.

In addition to the diurnal variation in the number of meteors, there is an annual variation of roughly 2 to 1 in temperate latitudes with a minimum in April and a maximum in October in the Northern hemisphere. Unlike the diurnal variations, the seasonal variations are least near the equator and greatest near the poles.

All early work on meteors was performed using the same visual magnitude scale as that used in astronomy. On this scale, the light from the star Vega is zero magnitude, that from Venus -4.4 and that from the full moon -12.6. The faintest visible meteors are of about +5 magnitude. Figure 1 shows that three basic parameters can be represented by a single curve against a scale of visual magnitude. This unification is slightly idealized but accurate enough for the purpose. The values plotted are based on those given in [2] and [3].

Meteor-burst communication makes use of the short-lived columns of ionization formed as meteors disintegrate upon entering the earth's atmosphere. If their mass is greater than 10^{-7} grams, useful amounts of ionization are produced, smaller particles tending to float down to earth unchanged by their passage through the atmosphere [4]. More than 10^{12} meteors of this size or greater, enter the earth's atmosphere daily and those whose mass is greater than 10^{-3} grams produce visible traces in the night sky. It is likely that a casual observer would see only those meteors larger than 10^{-2} grams. It follows that, because the number of such bodies of a given mass is inversely proportional to their mass, meteors potentially useful for communication are more than 100,000 times as numerous as those seen by the casual observer on a normal night.

As meteors enter the denser air below about 120 km, they begin to collide frequently with air molecules. The resulting heat evaporates atoms from the surface of the particle and these, in turn, collide frequently before coming to rest. The meteor does not seem to lose much velocity in this process, which would imply that many of the atoms are lost after each collision with an air molecule, the final result being a long, thin column of heat, light and ionization.

The length of the column depends on the mass and velocity of the meteor, as well as the angle of entry to the atmosphere, the most probable length being some 15 km. The radius is determined by the number of collisions experienced by the meteor atoms after leaving the meteor and is therefore related to the mean free path. Typical initial radii for columns produced by single particles vary in exponential fashion from a few metres at a height of 120 km, to a fraction of a millimetre at 80 km. Meteors of the dustball variety presumably quickly disintegrate into a cloud of particles, each of which behaves in the above manner. If the cloud has itself some significant diameter, the radius of the trail will be determined by this rather than the length of the molecular mean free path, at least over a certain range of heights. This is borne out by radar studies [5] of initial radii, which indicates values of half a metre at a height of 80 km. If they were determined by the mean free path, these values would have been in tenths of millimetres. The radar values for 120 km, however, were of such a size as to indicate a causal relationship with the mean free path. The concept of an average dustball meteor breaking up into a cloud with a diameter of one metre would be consistent with the above results. At 80 km, the initial radius would be the same as that of the cloud, while at 120 km the mean free path is larger than the cloud radius and hence becomes the determining factor.

As soon as the meteor trail is formed, it begins to expand and dissipate through a process of diffusion. The rate of diffusion is also very dependent on height, with the result that trails or portions of trails at a high altitude expand much more rapidly than those at low altitudes. The maximum ionization height can vary from 120 km to 80 km, at which latter height the majority of meteors have been entirely consumed. Over this same range, the diffusion constant varies by a factor of more than 100.

3. RADIO REFLECTIONS FROM METEOR TRAILS

3.1 Categories of meteor trails

It is convenient to divide meteor trails into two categories, "underdense" and "overdense", on the basis of the electron line density in the column. Underdense trails have such low densities that each electron acts individually in scattering radio waves and, as a consequence, radio waves can pass through such trails without undue attenuation. Overdense trails, on the other hand, contain such a high density of electrons that a radio wave, after slight initial penetration, is completely reflected. These latter trails contain a core, inside which the dielectric constant is everywhere negative. Reflection can be thought of as taking place from the surface of this core in the same way as if it were a metal cylinder. The dividing line between underdense and overdense is roughly an electron line density of 2×10^{14} electrons per metre.

For an appreciable amount of energy to be returned to the transmitter from a meteor column, the latter must be tangent to a sphere with its centre at the transmitter. This is the "specular reflection" condition. The equivalent, in the case of a separate transmitter and receiver, is that the column should be tangent to an ellipsoid of revolution with foci at transmitter and receiver.

3.2 The underdense trail

Here, the signal at the receiver is simply the integral, over the signals scattered from all the electrons in the trail.

The major portion of the signal received comes from electrons in a length of the trail, known as the principal Fresnel zone, which lies on either side of the tangent point. This zone is defined as being that volume which is bounded by the surface of the reflection points at which the total path length, transmitter to receiver, exceeds the minimum path length by half a wavelength.

All the contributions from within the principal Fresnel zone, when added vectorially at the receiver, increase the signal strength. Other more remote Fresnel zones produce signal increases and decreases which more or less cancel each other out. The principal Fresnel zone on the plane tangent to the ellipsoid is roughly elliptically shaped, with dimensions in kilometres. The dimensions of the principal Fresnel zone are only of the order of the wavelength (a few metres for the frequencies of interest) in a direction perpendicular to that plane.

Each meteor-reflected signal will suffer attenuation as the result of destructive interference from electrons diffusing away from the trail axis and eventually leaving the principal Fresnel zone. In the same way, signals from trails formed with a large radius will suffer initial attenuation vis à vis the signal that would have pertained for an initial radius of zero.

The transmission equation takes the following form [4]:

$$\frac{P_R(t)}{P_T} = C_1 \lambda^2 A_1 A_2 (t) \tag{1}$$

where P_R and P_T are the received and transmitted powers,

C_1 is a constant determined by the antenna gains, the total scattering cross-section, and the geometry

λ is the wavelength

A_1 is the attenuation due to the initial radius of the trail and $A_2(t)$ is the time-varying attenuation due to the radial diffusion of electrons in the column.

The two attenuation factors expand as below:

$$A_1 = \exp\left(-\frac{8\pi^2 r_o^2}{\lambda^2 \sec^2\phi}\right) \quad (2)$$

$$A_2(t) = \exp\left(-\frac{32\pi^2 Dt}{\lambda^2 \sec^2\phi}\right) \quad (3)$$

where ϕ is half the angle between transmitter, tangent point and receiver

r_o is the initial radius,

and D is the diffusion constant at the centre of the principal Fresnel zone.

The underdense burst rises very quickly at first as the meteor traverses the principal Fresnel zone to a maximum, as given by the invariable time factors in the equation. The amplitude then decays exponentially at a rate determined by the position of the trail and the diffusion constant applicable at the trail height.

3.3 The overdense trail

The transmission equation for the overdense trail is developed in a fashion similar to the one applying to underdense trails, except that the concept of an expanding metallic cylinder replaces the summation over individual scattering electrons. The trail is assumed to have been formed with an initial radius, r_o, and a Gaussian distribution of electron density about the trail axis. The Gaussian distribution is maintained as the trail dissipates by ambipolar diffusion.

The transition from underdense to overdense behaviour could be defined as occurring when the core of the trail wherein the dialectric constant is negative, is of sufficient diameter to attenuate a radio wave to 1/e of its incident value. This implies that the radius of the critical core, r_c is greater than or equal to $\lambda/2\pi$. A trail is formed with a small value of r_c. As the trail diffuses, the radius of the core increases to a maximum and then begins to collapse, finally falling to zero very quickly. The signal from the overdense trail exhibits the same behaviour. The signal rises suddenly at the beginning, as the meteor traverses the principal Fresnel zone. The signal then increases slowly, as the critical diameter increases to a maximum and then decreases as the critical core collapses.

The transmission equation for the overdense case is:

$$\frac{P_R(t)}{P_T} = C_2\lambda^2 \left[\ln \left(\frac{r_e q\lambda^2 \sec^2\phi}{\pi^2\{4Dt+r_o^2\}} \right) \right]^{\frac{1}{2}} \tag{4}$$

and the duration of the overdense signal is:

$$T_{ov} = \frac{r_e q\lambda^2 \sec^2\phi}{4\pi^2 D} - \frac{r_o^2}{4D} \tag{5}$$

where C_2 is a constant depending on the antenna gains and the geometry

r_e is the classical radius of the electron

and q is the electron line density in the trail

Overdense duration consists of the difference of two terms. The first is the time taken for the critical radius to expand from zero to a maximum and then collapse to zero again. The second term is the fictitious time that r_o would take to expand to its initial value, if diffusion were the controlling mechanism.

Signals from overdense trails last much longer than do underdense signals and have very different shapes. Despite this, the peak heights of both types of signal vary as λ^3 and in both cases the duration varies as λ^2.

3.4 Where the trails which carry the bulk of the traffic are located

Some interesting spatial properties of meteor-burst propagation can now be seen. The attenuation occurring on the paths to and from the reflection point makes signals from trails in the neighbourhood of the two stations most numerous and highest in amplitude. The dependence of signal duration on $\sec^2\phi$ is the reason why from a communications point of view the centre of the path is the most important area. However, because meteors tangent to an ellipsoid at the path's mid-point must be travelling parallel to the ground, signals from this region are relatively rare. Instead, the areas that contribute the most to the duty cycle of a meteor-burst link are two large regions just off the great circle path on either side of the mid-point, as shown in Fig.2.

The heights of the reflection points are in the range 80-120 km, but because diffusion is so rapid at the higher end of this range, the greatest contribution to the duty cycle comes from meteors tangent in the lower half. This is particularly true at higher frequencies, where large initial trail diameters limit very severely all signals from above a height of 100 km. The fact that the low heights contribute most to the duty cycle, limits the practical length of meteor-burst circuits to slightly over 2000 km.

3.5 Frequency and time dispersion of signals from meteor trails

The frequency spread of a meteor-burst signal may be regarded as being negligible during the main portion of the echo. To start with a Doppler whistle is obtained from the head of the meteor and, depending on the velocity, this will be of the order of 1000 Hz decreasing within a fraction of a second to zero as the meteor passes the reflection point. Subsequently signals will be from the trail as a whole and then the "body Doppler " is generally only a fraction of 1 Hz and would rarely exceed 5 Hz. A more precise representation than the Doppler Whistle presented above is to consider the vector of the basic Cornu spiral pattern as the tip of the vector travels along the spiral until it finally simply joins the two end points of the spiral. Then atmosphere winds are represented by rotating the Cornu spiral pattern as a whole.

The time spread of a meteor reflected signal is also very small - less than 1 μsec during the main phase of the echo. In fact TV programmes have been received over long distances via meteor reflections [6] showing that coherent bandwidths of several MHz are possible. However, as winds begin to warp the trail, reflections begin to follow different total path lengths and the time spread increases to a few microseconds. Even then, the usable bandwidth is of the order of 100 kHz.

Thus, the dispersion for a single meteor echo is very small, both in time and frequency. Multiple echoes are, of course, possible, although if the duty cycle is low, so will be the chance of two echoes occurring simultaneously, since the probability of such an event is equal to the square of the duty cycle expressed as a fraction. In a communication system one must arrange that such multipath events are rejected by some form of error control. Even the tails of the meteor-burst may be used if bandwidths are restricted to less than 10 kHz. The successful use of these tails (where fading takes place) demands the use of diversity. At one time [7] the use of diversity for meteor scatter communication was thought to be unjustified, but this was largely because the tails were ignored.

Fig.3 shows a superimposed record of the same meteor trail reflection as received on three antennas spaced 4 wavelengths apart perpendicular to the great circle path. The lack of correlation in the tail portions is evident. The ability to use the tail portions, as in the STC COMET system, has two advantages, namely an extended duty period and the possibility of utilizing signals of ionoscatter origin when these are strong enough.

The STC meteor burst system, COMET, using error detection, ARQ, and diversity to provide a small number of telegraph channels was described in [1]. Only those general characteristics of the system necessary to interpret the results given in Section 5 are described below.

4. GENERAL DESCRIPTION OF THE STC COMET SYSTEM

The COMET system was designed to provide two-way transmission of telegraph information between two terminals at a maximum distance apart of 2000 km. The information was transmitted in frequency-shift keying with a total deviation of 6 kHz and at a signalling rate of 2000 bauds.

Transmitting and receiving antennas were five-element Yagi arrays used singly or in pairs. The receiving antenna system was four-fold and provided space and height diversity. The lower antennas were single Yagis mounted at a height of 1.2 λ, while the higher ones were twin Yagis at 2.6 λ above ground. Whereas the principle of frequency and space diversity is well known, it may be useful to recall that height diversity is a kind of angle diversity which takes advantage of the fact that antennas spaced vertically have different radiation patterns in the vertical plane.

The transmitting antennas consisted of twin Yagis and were mounted on a separate mast at a height of 2 λ.

Whenever a radio path was created by a meteor trail, the following process took place in the two directions. See Fig.4, which shows a block diagram of the equipment used at each terminal during the experimental phase (transmit and receive stores were used for the operational trial). The information was fed into the system in the form of a six-hole punched tape which was read at the speed of 285 characters per second. The five-element characters read from the tape were converted into the seven-element characters of the ARQ code and transmitted at 2000 bauds synchronously, i.e. without start and stop elements. At the receiver, the information was first checked character by character for the correct number of marks and spaces (3 and 4 respectively) peculiar to the ARQ code. When accepted, the information was converted back to the five element code. It could then be printed or punched at high speed or fed into a store of large capacity which acted as a buffer for printers or perforators working at lower speed. If a character was found to be in error, a request was transmitted in the reverse direction and a repetition cycle initiated. When the signal faded completely, the system remained in continual repetition.

Two different receiving stores were available. The first used magnetic cores as memory elements. It was divided into four compartments, each of a capacity of 2000 characters, from which messages could be read out by four teleprinters working simultaneously. It was therefore capable of handling the information of up to four 50-baud telegraph channels on a message-switching basis. The second store used a magnetic tape and had a capacity of some 20,000 characters; its output speed could be either 50 or 100 bauds. It was used for delivering messages at high speed by means of a tape perforator working at the rate of 150 characters per second.

In order to cope with unpredictable variations of path length, a means of quickly correcting the phase of the receiver clock had to be provided; this was made possible by the use of an RQ signal extended to 14 elements as described in [1].

$$L = 4c + 2t + d_1 + d_2 \qquad (6)$$

where $\underline{c}$ is the duration of a character, i.e. 3.5 ms

t is the longest propagation time over the radio path

d_1 is the delay between the detection of an error at station B and the beginning of the RQ1 sent by B

d_2 is the time delay between the detection of RQ1 at station A and the beginning of the RQ1 sent by A

To determine the value of L, one must observe that d_1 and d_2 can vary between zero and the duration of one character, depending on the phase relation between the clocks of the two terminals. However, d_1 and d_2 do not vary independently of each other; their sum remains constant and is included between c and 2c. Furthermore, it is clear that L must be a multiple of $\underline{c}$.

On the STC circuit the longest propagation time was 4 ms, so that $2c < 2t < 3c$. Consequently, the repetition cycle had to be at least 8 characters long.

In practice it is advisable to adopt a slightly longer repetition cycle to avoid loss of synchronization between the terminals when one of them fails to identify the request sent by the other, in which case a vicious circle can be created where both terminals repeat indefinitely. The repetition cycle used in COMET was 10 characters.

5. RESULTS OBTAINED DURING EXPERIMENTAL TRIALS WITH THE COMET SYSTEM

5.1 Traffic capacity

Because of the intermittent nature of the communications channel, the capacity for traffic must necessarily be expressed as an average value over a given period. In order to show short- and long-term variations, measurements were averaged over a period of different durations from a few minutes to 24 hours.

Examples of hourly average capacity measured with the COMET system on a 1000 km path are plotted against time of day in Figs.5 and 6 for 40 MHz and 100 MHz respectively (the exact frequencies were 36.59 MHz and 106.47 MHz in the direction to the measuring terminal). The data of Figs.5 and 6 were taken simultaneously using 200 watts of transmitted power at 40 MHz and 2000 watts at 100 MHz. These figures are representative of a wealth of data taken over a period of several years on the STC experimental link from Staalduinen near The Hague to La Crau in the South of France.

When the ratio of capacity at 40 MHz to that at 100 MHz was plotted, as in Fig.7, two interesting features were almost always observable: a minimum at the time of cosmic noise maximum at the receiving terminal (caused by the fact that the 40 MHz noise level was determined by cosmic noise and that at 100 MHz was set by receiver noise), and a maximum around noon due presumably to the prevalence at that time of 40 MHz propagation modes other than meteor-burst. The daily average ratio of capacity at 40 MHz to that at 100 MHz was observed to be 19 $\pm$ 1.5, except during two tests in the month of June, when meteor shower activity greatly increased the 100 MHz duty cycle. Fig.8 shows 100 MHz data taken during a shower and has a two peaked characteristic caused by the shower radiant being suitably oriented for the two hot spots on either side of the great circle paths (see Fig.2).

It can be seen from Figs.5 to 8 that the hourly average capacity is subject to large variations from day to day. Even more variable are the shorter-term average capacities. Fig.9 shows the variation for the 7-minute traffic capacity at 100 MHz as normalized to the pertinent hourly value. This variation is less at 40 MHz and was bounded by 0.5 and 2.0 times the hourly value in 90% of cases.

5.2 Burst statistics

The statistics presented below concern "information bursts" during which information is passed both ways on the circuit. In keeping with this definition, all bursts where less than 10 characters (the length of the repetition cycle) were received have been discarded. Similarly, short interruptions in the middle of bursts, where a repetition was requested were not counted as part of the burst but neither were they regarded as marking the end of one burst and the beginning of the next.

The data used to calculate burst-length and interval statistics were collected during the period 15 September to 22 December 1967. This extensive testing was necessary to obtain a large sample of "information bursts", which are relatively rare at 100 MHz.

Figure 10 shows the probability distribution of burst-lengths at 40 MHz and 100 MHz respectively. It can be seen that neither 40 nor 100 MHz burst-lengths fit an exponential probability distribution, except for short bursts of less than 0.5 seconds. Theory has it that the probability distribution of burst lengths for "underdense" bursts should be of the form $P(d) = e^{-d/\bar{d}}$ where $P(d)$ is the probability that the burst-lengths exceed d, and $\bar{d}$ is the average burst-length. This would seem to hold true in the underdense region, up to about 0.5 seconds duration. Bursts longer than about this value are overdense and should not be expected to fit an exponential distribution. In fact, the statistics for such long-duration bursts could be better characterized by a power law.

It should be noted that the average "information burst" at 100 MHz is only just a little shorter than at 40 MHz. The actual average burst-duration would be much less.

The distributions of burst-intervals shown in Fig.11 all fit exponential curves, as is to be expected when one considers that meteors occur at random. The difference between 40 MHz and 100 MHz is marked, 100-MHz bursts being much less frequent. This can also be seen in Fig.12, where the statistics for the number of bursts per 7-minute period for 40 and 100 MHz are compared.

5.3 Signal-level statistics

For all the tests, both the 40-100 MHz comparative tests and the purely 100-MHz ones, the percentage of time that two selected signals exceeded ten preset levels was recorded automatically by a digital recorder at Staalduinen. Duty-cycle vs signal-level results were derived from these data.

For the comparative tests, one signal from a 40-MHz and one signal from a 100-MHz receiver, were selected for signal-level recording. The signals chosen came from antennas with the same radiation pattern, i.e., the signal from a 40-MHz antenna at a height of 2.6 λ was compared with the signal from a 100-MHz antenna at the same relative height.

The results are shown in Fig.13 and represent the signal distribution, averaged for each frequency and test. The averages were carefully taken over three or four whole 24-hour periods for each test, in order to avoid bias with regard to any particular time of day. It can be seen that all the curves are approximately parallel. The values of the intercepts at a given signal level follow about the same increasing order as do the 24-hour average traffic capacities.

Figure 14 represents an average of signal-level statistics drawn from all the tests and is therefore useful for predicting the yearly average duty cycle for any system proposed. It can be seen in Fig.14 that the slopes of the two curves are equal in the expected operating regions. In fact, the curves are very closely parallel and have uniform slopes except at very high thresholds so that the curves can be approximated by the relation $D \simeq V^{-K}$, where D is the duty cycle, V is the voltage level of the received signal, and K is a constant with value 1.2. The value of the slope, K, indicates that it is very advantageous, in terms of information capacity, to increase the signalling rate and bandwidth.

The distance between the two curves is greater than is predictable from the simple λ^3 dependence of peak signal and λ^2 dependence of signal durations. The expected loss of signal amplitudes, 13.9 dB, and loss of duty cycle, a factor of 8.5, have been plotted in Fig.14. Note that the 13.9 dB is partially offset by a 10-dB increase in transmitter power. The predicted position of the 100-MHz curve differs from the actual one by a further 8.3 dB of signal level or a factor of 3.2 for the duty cycle. The real situation will be a combination of the two effects, i.e., some difference in peak heights and some difference in the duty cycle. This can be explained in terms of the large trail diameter compared to the 100 MHz wavelength at trail heights above 100 km and the inefficient usage by COMET of the shorter 100 MHz bursts.

5.4 Transmission delays

In a meteor burst system, where the signalling rate is relatively high, the electrical transmission time is very short but each message must necessarily wait for one or several signal bursts. The transmission delay, i.e. the time which lapses between the moment when the message becomes available for transmission and the moment when it is delivered in its entirety at the receiving end, depends on the rate of occurrence of bursts and on the length of the message. It may be shorter or longer than the transmission delay over a conventional link working at 50 bauds.

Delays were determined experimentally at 40 MHz for three lengths of message: 50, 150 and 350 characters and the results reproduced from Ref.1 are summarized in Tables I and II.

Message length, characters	50	150	350
Transmission time required at 50 bauds, seconds	7.5	22.5	52.5
Percentage of cases where delay measured is less than at 50 bauds	50	70	85
Average delay, seconds	10	17	30
Longest delay, minutes	about 1	about 2	about 2

Table I: Early morning

Message length, characters	50	150	350
Transmission time required at 50 bauds, seconds	7.5	22.5	52.5
Percentage of cases where delay measured is less than at 50 bauds	20	35	50
Average delay, seconds	30	50	80
Longest delay, minutes	between 2 and 3	between 3 and 4	between 4 and 5

Table II: Late afternoon

As the length of message increases, the average transmission delay tends to become more exactly proportional to that length divided by the average capacity of the circuit. With 200 watts of transmitter power at 40 MHz, the transmission delay for long messages would in the late afternoon be about the same as on a conventional 50-baud link; in the worst case it might be four to five minutes longer. In the early morning the delay would normally be four to eight times shorter. Delays could, of course, be shortened by raising the transmitter power. An increase from 200 watts to 1 or 2 kW would bring the evening performance to the level of the performance shown for the early morning.

6. CONCLUDING REMARKS

It has been demonstrated by a number of organizations that the characteristics of meteor-burst propagation offer a good solution to telegraph communications for distances up to 2000 km. Using frequencies in the region of 40 MHz, powers of about 1 kW and elevated 5-element Yagi antennas, it is possible to establish several 50-baud communication channels with very high reliability. The only drawback is that time delays of up to a minute or so can occur due to the intermittent nature of the propagation.

Probably the most advanced system developed to date is the STC COMET system, which makes use of diversity and ARQ. The COMET system worked highly successfully for the whole of a year's operational test period on a circuit in Norway. This carried encrypted telegraph traffic and used a signalling rate of 2000 baud during bursts to give an average transmission rate of two 50-baud channels.

However, the full potential and possibilities of meteor-burst communications have by no means been fully exploited. A tenfold increase in signalling rate to 20,000 baud can readily be used, even when the tails of the signal bursts are exploited (as is the case in diversity reception). Such rates would permit transmission of data and facsimile or a large number of multiplexed telegraph channels.

On the whole, it would seem that the most practical increase in transmission rate would be one which resulted in a minimum average rate of 2000 baud. This would require burst rates of between 20 to 80 kbit/s, depending on the other circuit parameters.

The STC tests at a carrier frequency of 106 MHz suggest that a frequency of about 70 MHz, if available, would offer a good compromise in case a system were required which was highly resistant to jamming. At such a frequency and with an intermittent signalling rate of 2000 baud, a power of some 2 kW would be required to give a capacity equivalent to a 50-baud channel. This is a very modest performance but the following advantages ensue:

(a) The system is virtually proof to jamming from a distance.

(b) The system would not only perform well in the auroral zone, but would even be highly resistant to polar cap absorption (PCA).

(c) The position of the terminals would be very difficult to locate by an enemy.

Yet another application of meteor-burst propagation is the use of a high HF frequency as a slow-speed order wire. One of the normally allocated frequencies could be used - for many circuits the highest allocated frequency is in the range 15 to 20 MHz. When normal HF contact is lost, the circuit could transfer to the meteor-burst mode using this either as a substitute or as a temporary means of communication to help re-establish HF communication. It has been found that the duty cycle at frequencies in the range 15 to 20 MHz exceeds even that given by the theoretical relationship which heavily favours such lower frequencies. The only adverse effect would be increased susceptibility to abnormal D-layer absorption, but this is likely to be accompanied by sporadic-E reflections by way of compensation.

It should perhaps be emphasized that meteor-burst communication does not require expensive or bulky equipment. Antennas are conventional, moderate transmitter power is sufficient, as are conventional receivers. Even the required buffer stores and the logic circuitry to control the ARQ process need not present problems if modern technology is exploited. These functions could all be handled by a mini-computer or even a microprocessor system.

Finally, it must be said that the use of meteor-burst propagation has been unjustly neglected during this last decade or so. It is hoped that a better understanding of the possibilities and limitations of meteor-burst propagation will help to restore this method of communication to its rightful place.

REFERENCES

1. P.J. Bartholomé and I.M. Vogt, COMET - A New Meteor-Burst System incorporating ARQ and Diversity Reception, IEEE Trans. Communication Technology, Vol. COM-16, pp. 268-278, April 1968.
2. D.W.R. McKinley, Meteor Science and Engineering, McGraw-Hill, New York, 1961

3. L.A. Manning and U.R. Eshleman, Meteors in the Ionosphere, Proc. IRE, Vol. 47, pp. 186-199, February 1957.
4. G.R. Sugar, Radio Propagation by Reflection from Meteor Trails, Proc. IEEE, Vol. 52, pp. 116-136, February 1964
5. J.S. Greenhow and J.E. Hall, The Importance of Initial Trail Radius on the Apparent Height and Number Distributions of Meteor Echoes, Monthly Notices Roy. Astron. Soc., Vol. 121, pp. 183-196, August 1960.
6. F. Akram, N.M.Sheikh, A. Javed and M.D. Grossi, Impulse Response of a Meteor-Burst Communication Channel determined by Ray-Tracing Techniques, IEEE Trans. COM-25,pp. 467-470, April 1977.
7. A.W. Ladd, Diversity Reception for Meteoric Communications, Trans. IRE, Vol. CS-9, pp. 145-148, June 1961.

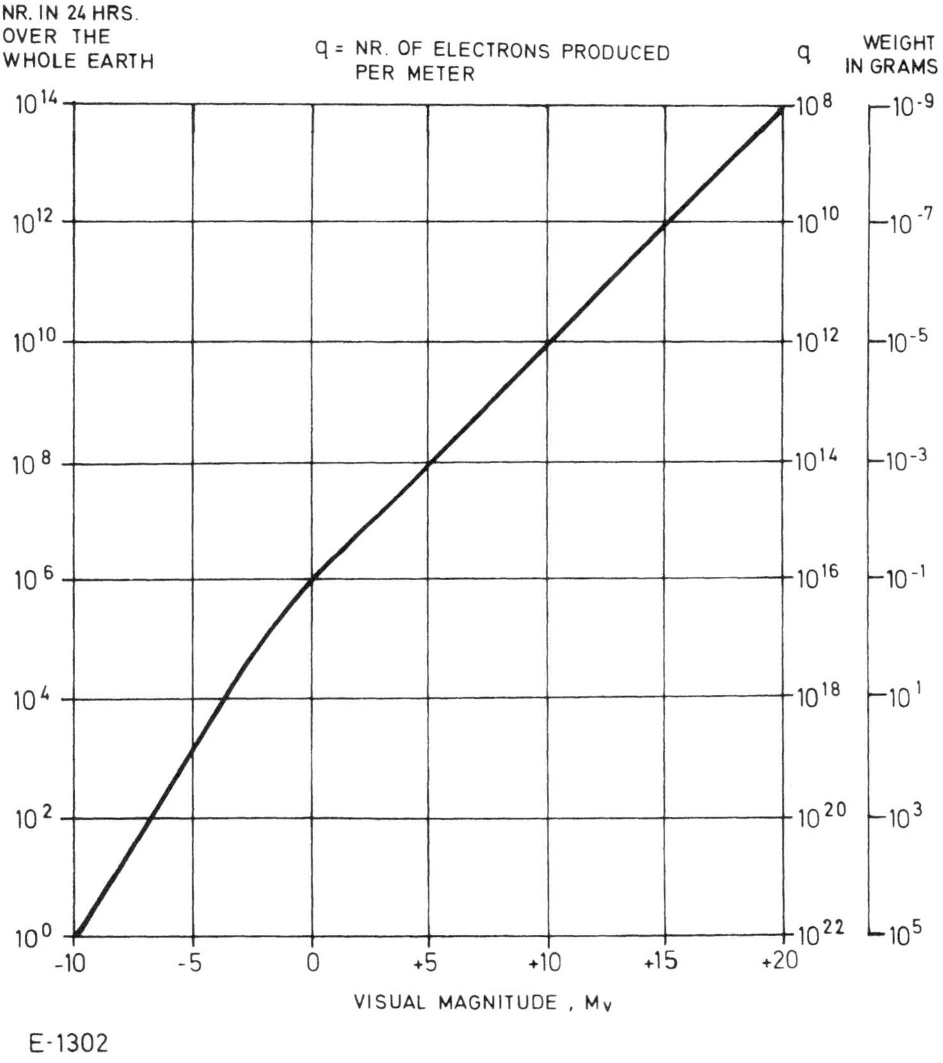

Fig. 1. Basic meteor statistics as a function of visual magnitude.

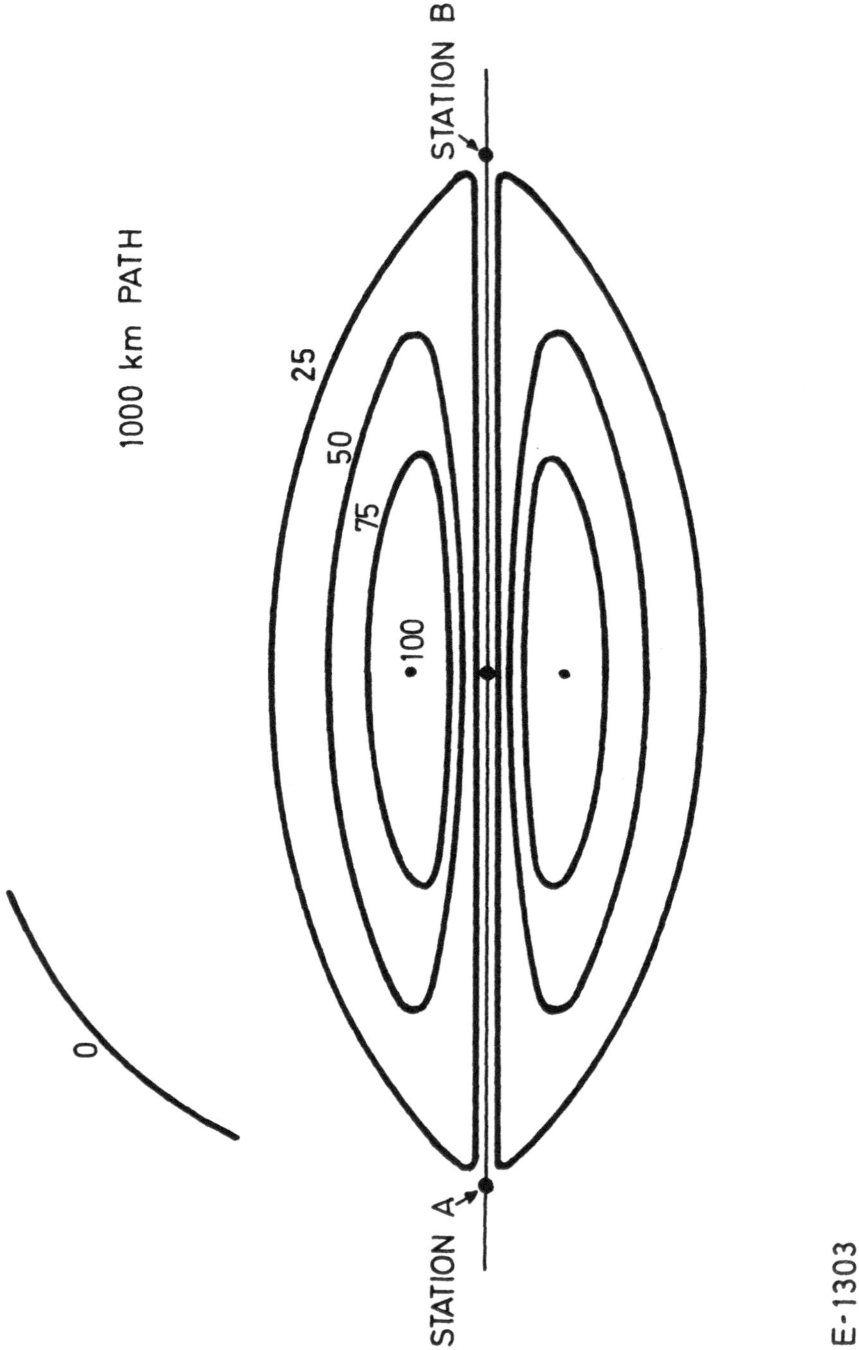

Fig. 2. Relative importance of geographical areas to meteor burst communications (sources of signal duration above threshold).

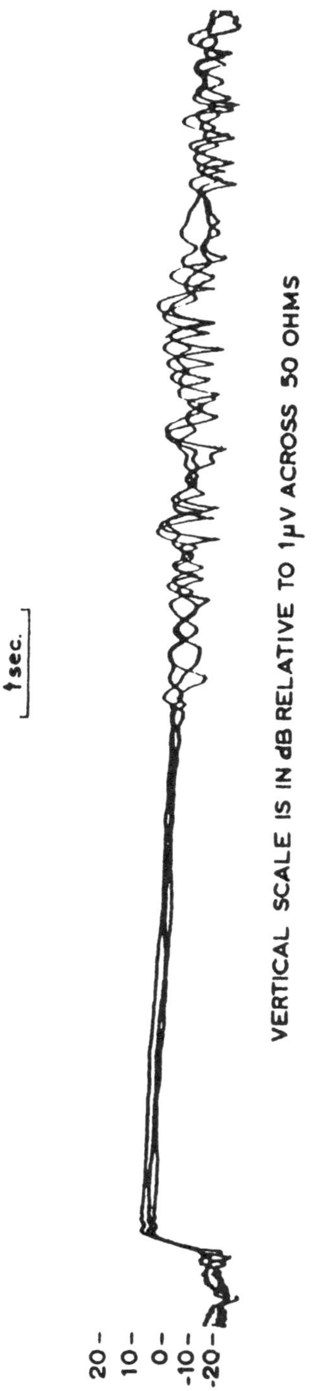

Fig. 3. Superimposed recordings of a meteor-trail reflection showing uncorrelation of fading in three diversity branches.

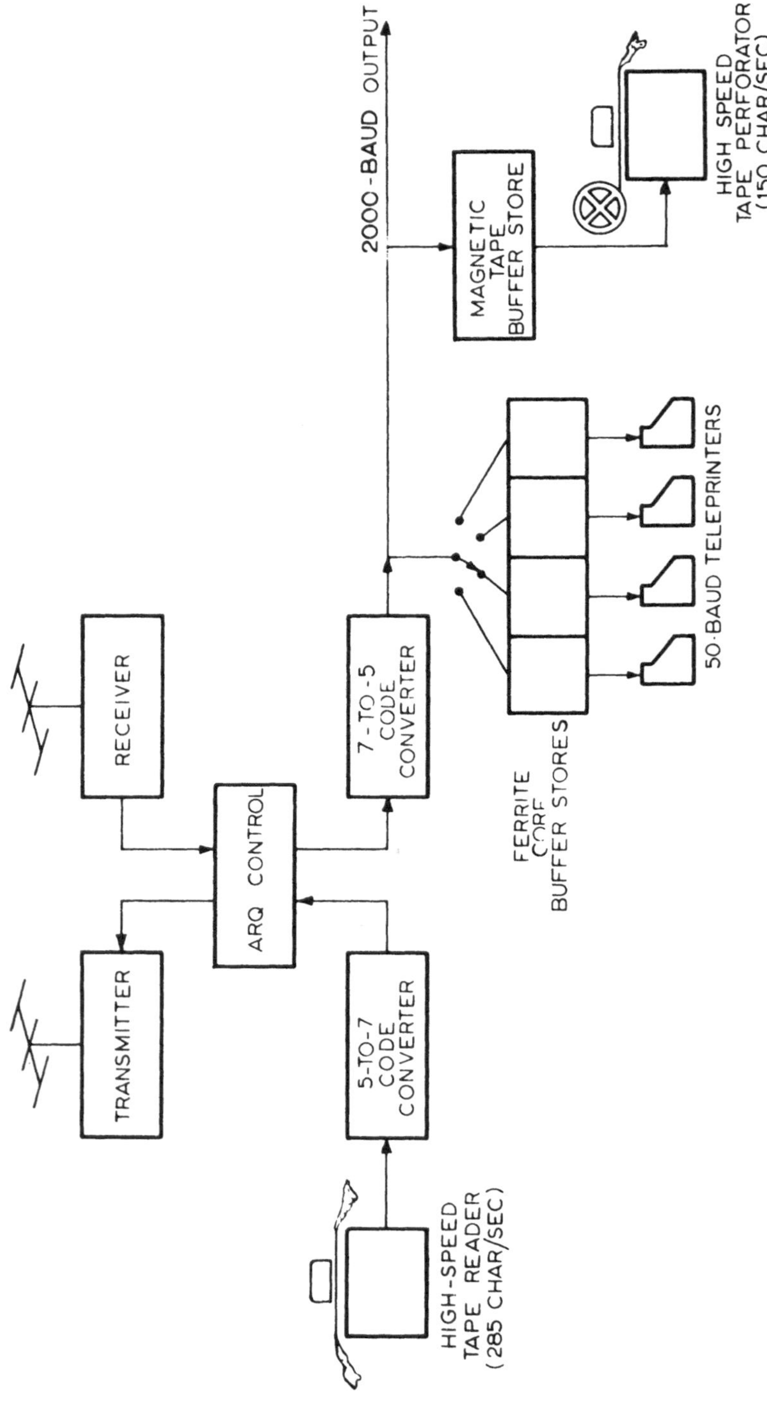

Fig. 4. Block diagram of terminal equipment used for the experimental trials.

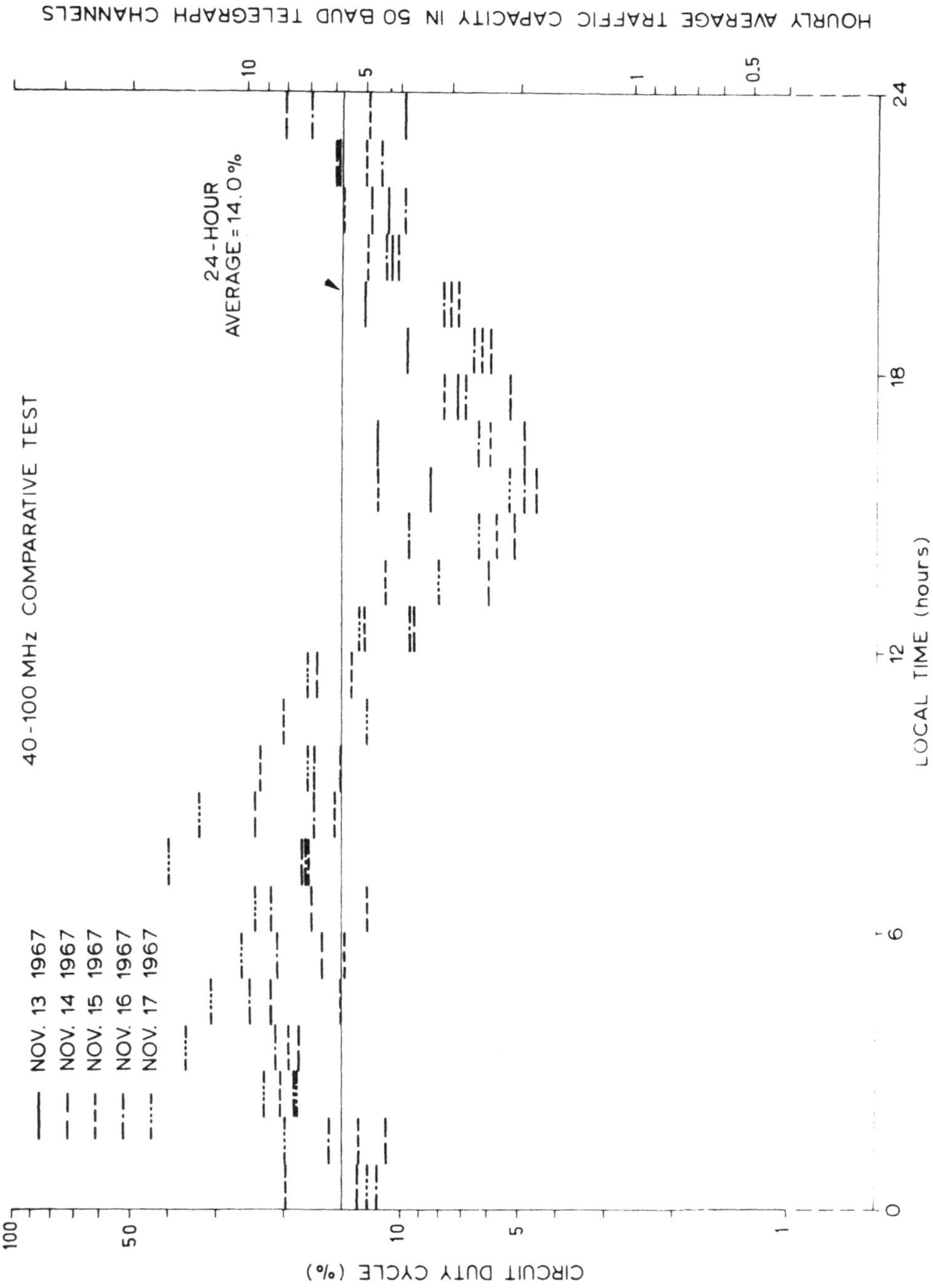

Fig. 5. Diurnal variation of 40-MHz traffic capacity (13-17 Nov. 1967).

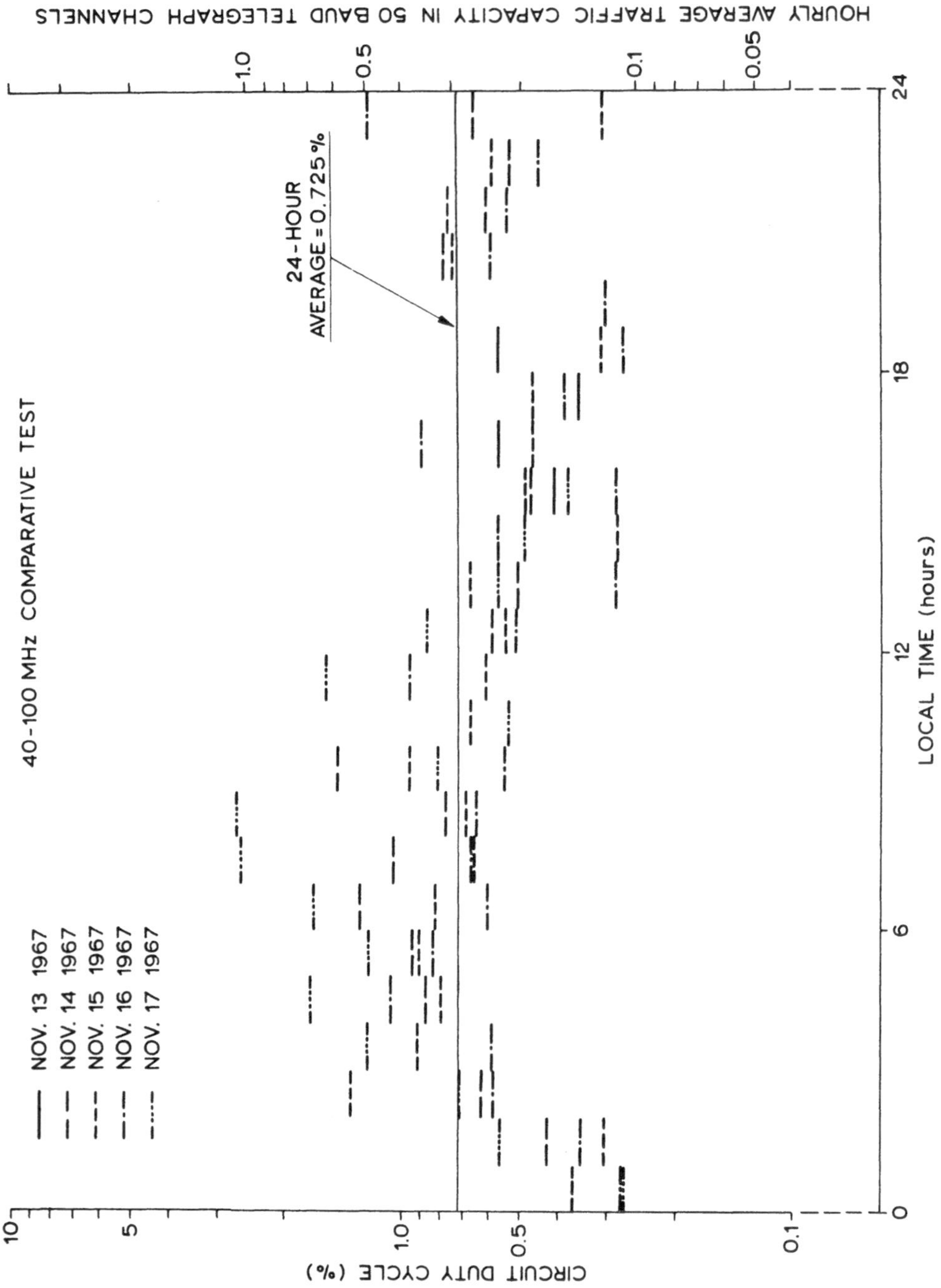

Fig. 6. Diurnal variation of 100-MHz traffic capacity (13-17 Nov. 1967).

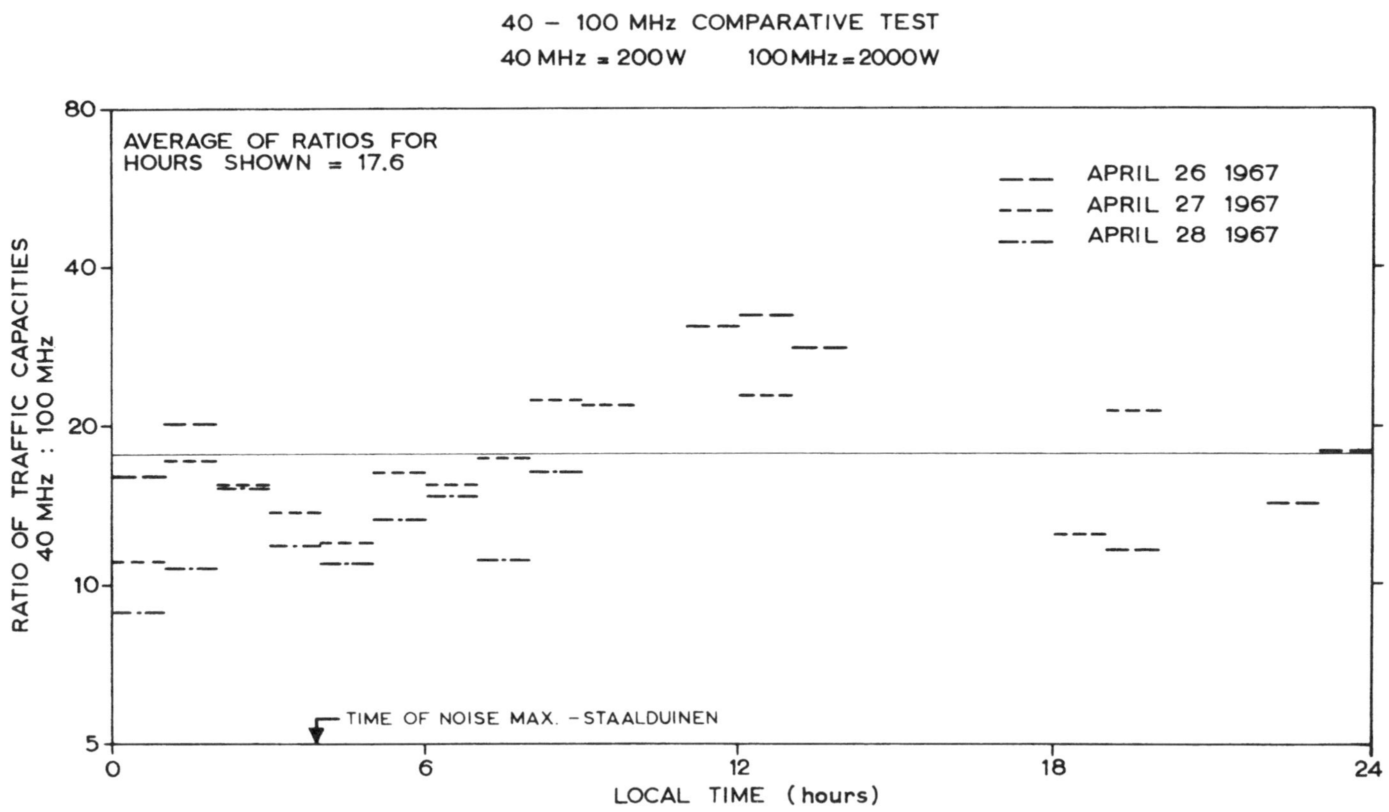

Fig. 7. Diurnal variation of the ratio of 40-MHz to 100-MHz traffic capacities (26-28 April 1967).

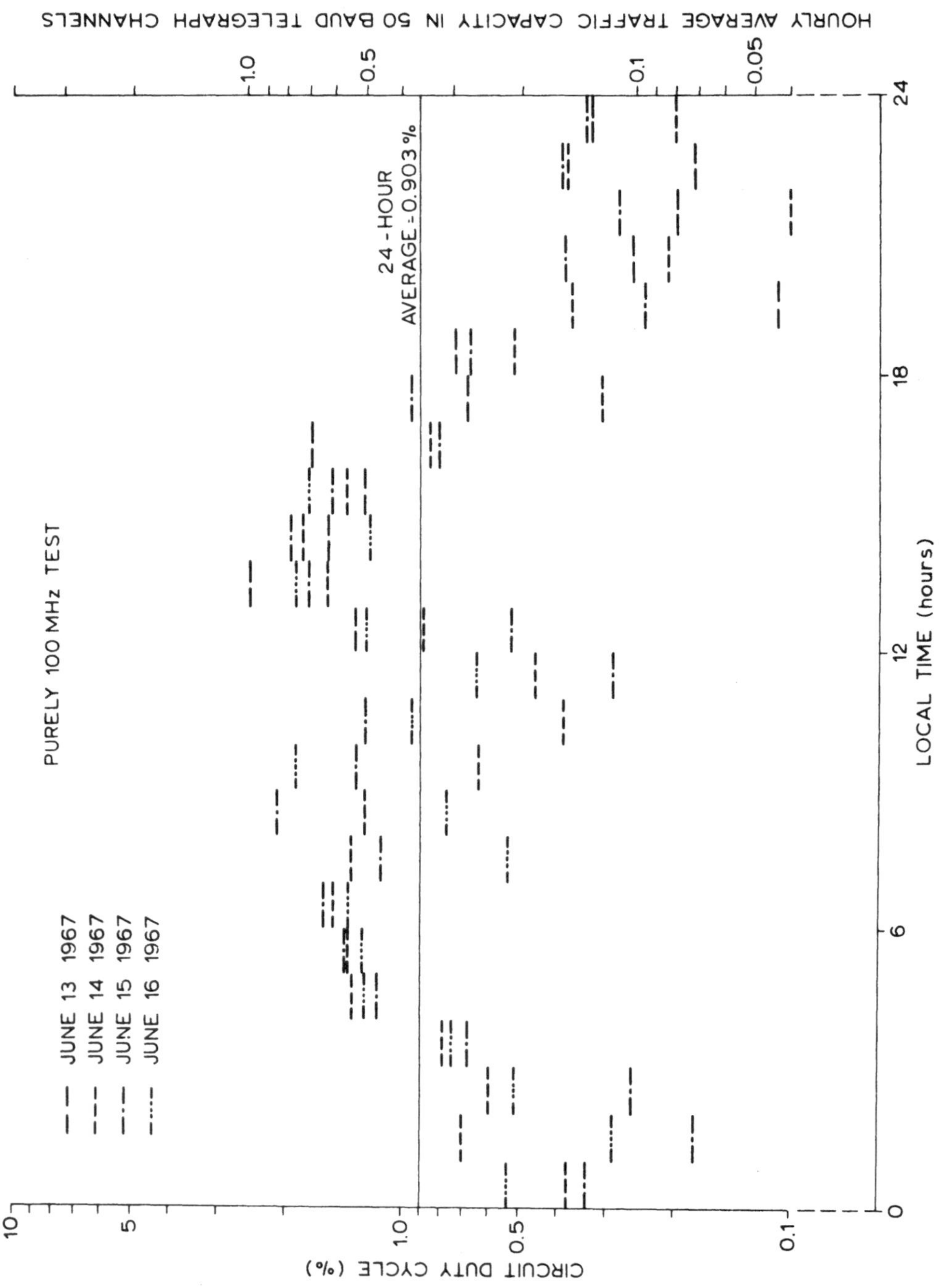

Fig. 8. Diurnal variation of 100-MHz traffic capacity (13-16 June 1967).

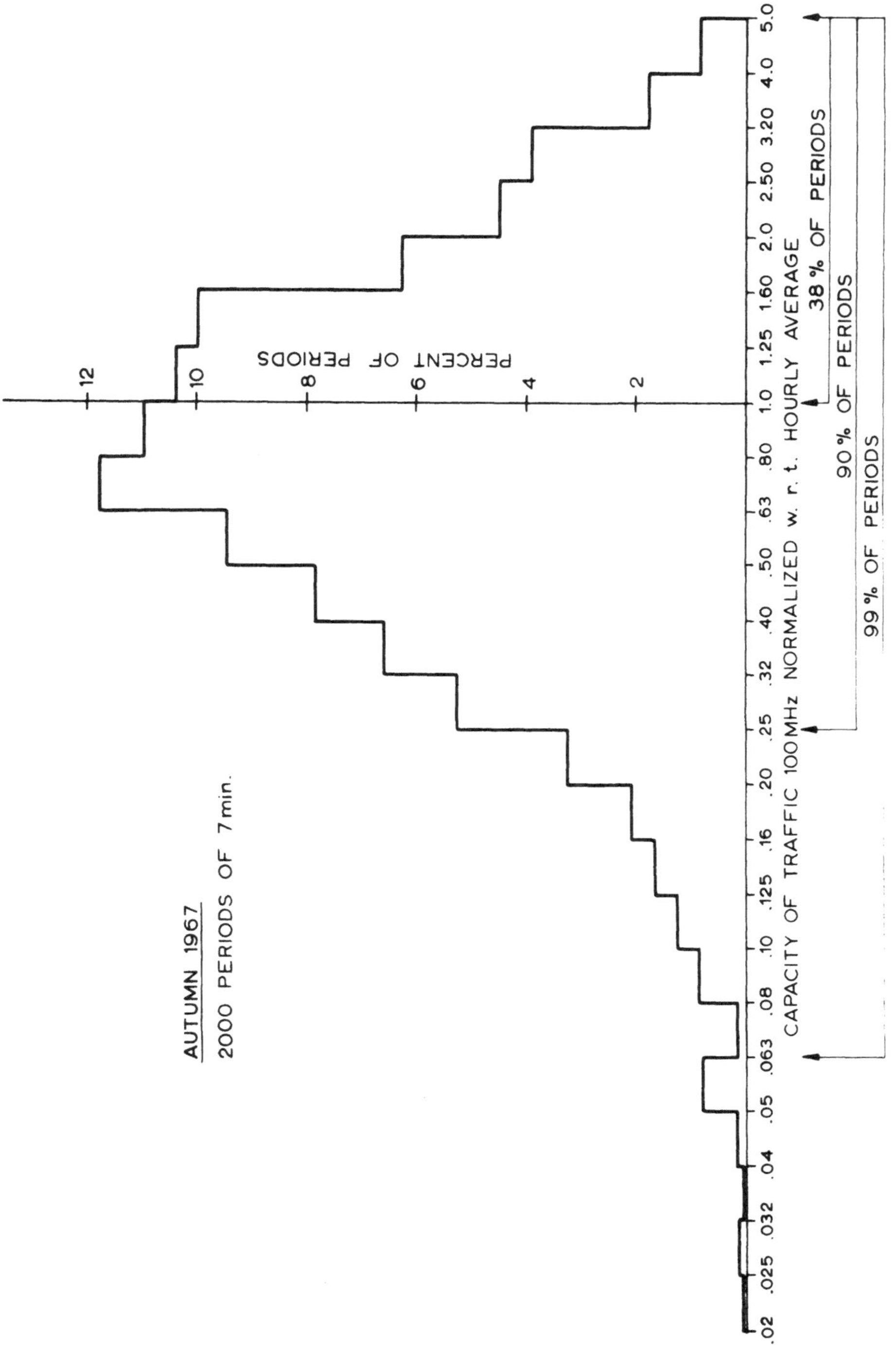

Fig. 9. The distribution of 7-minute 100-MHz average traffic capacities (normalized with respect to the hourly average).

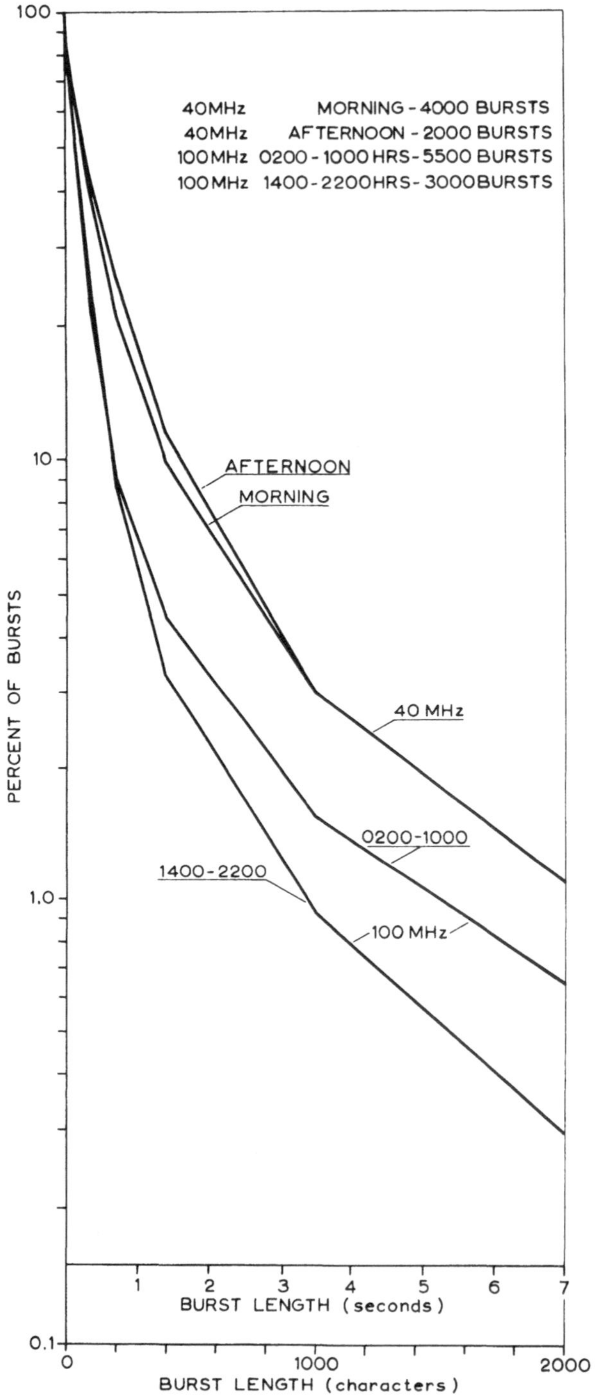

Fig.10. Probability distribution of burst-lengths.

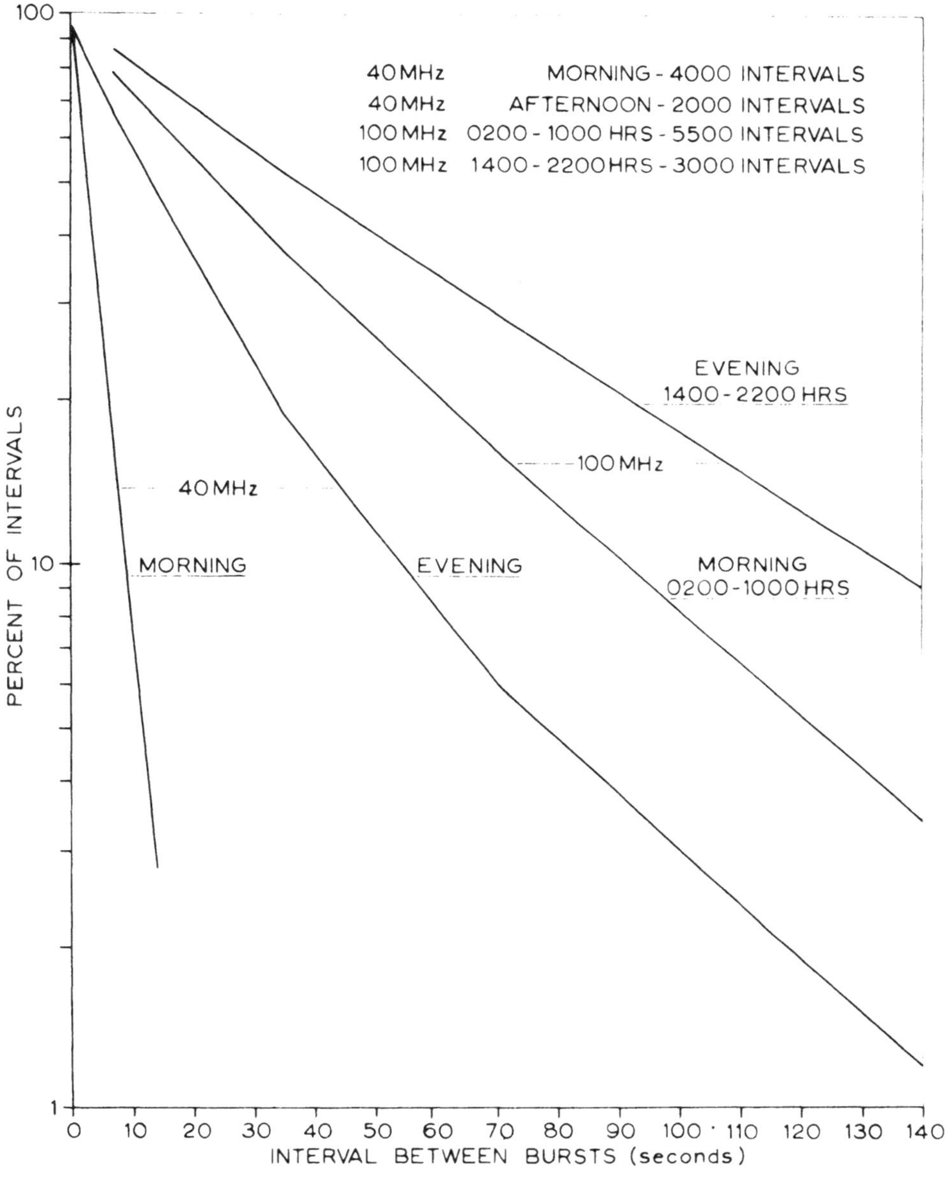

Fig.11. Probability distribution of burst-intervals.

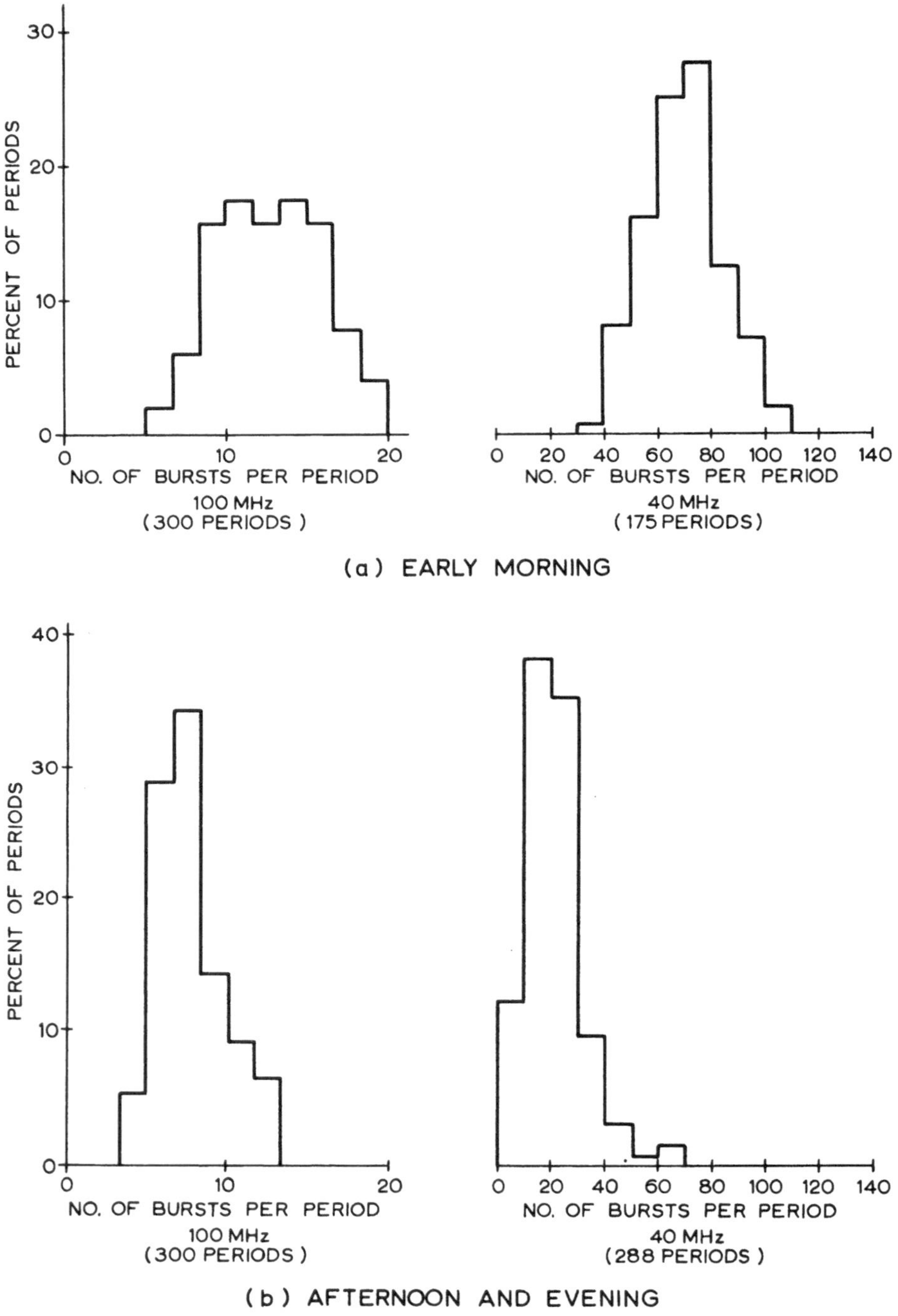

Fig.12. Number of transmission bursts per 7-minute period.

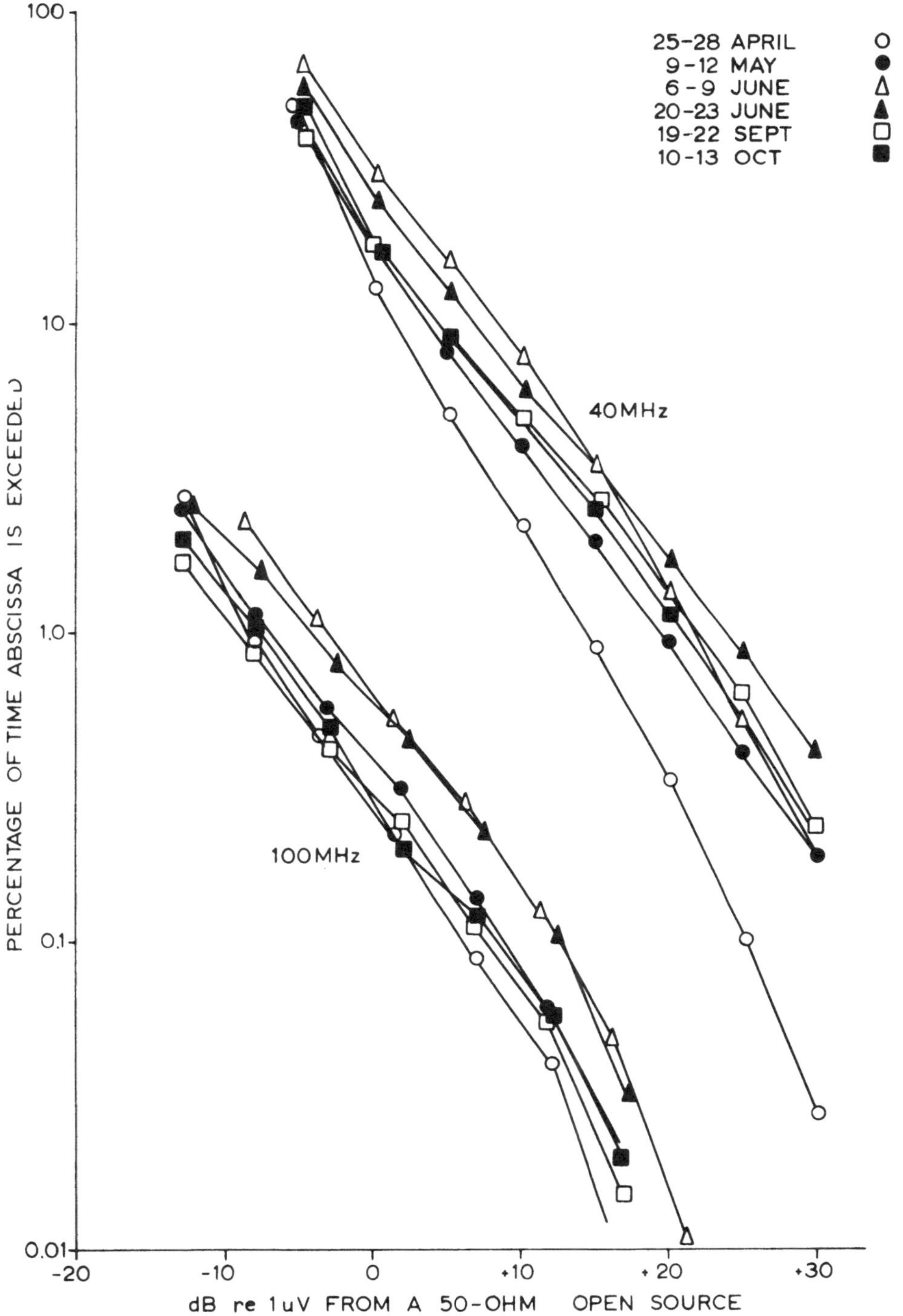

Fig.13. Distribution of signal levels (averages for each comparative test).

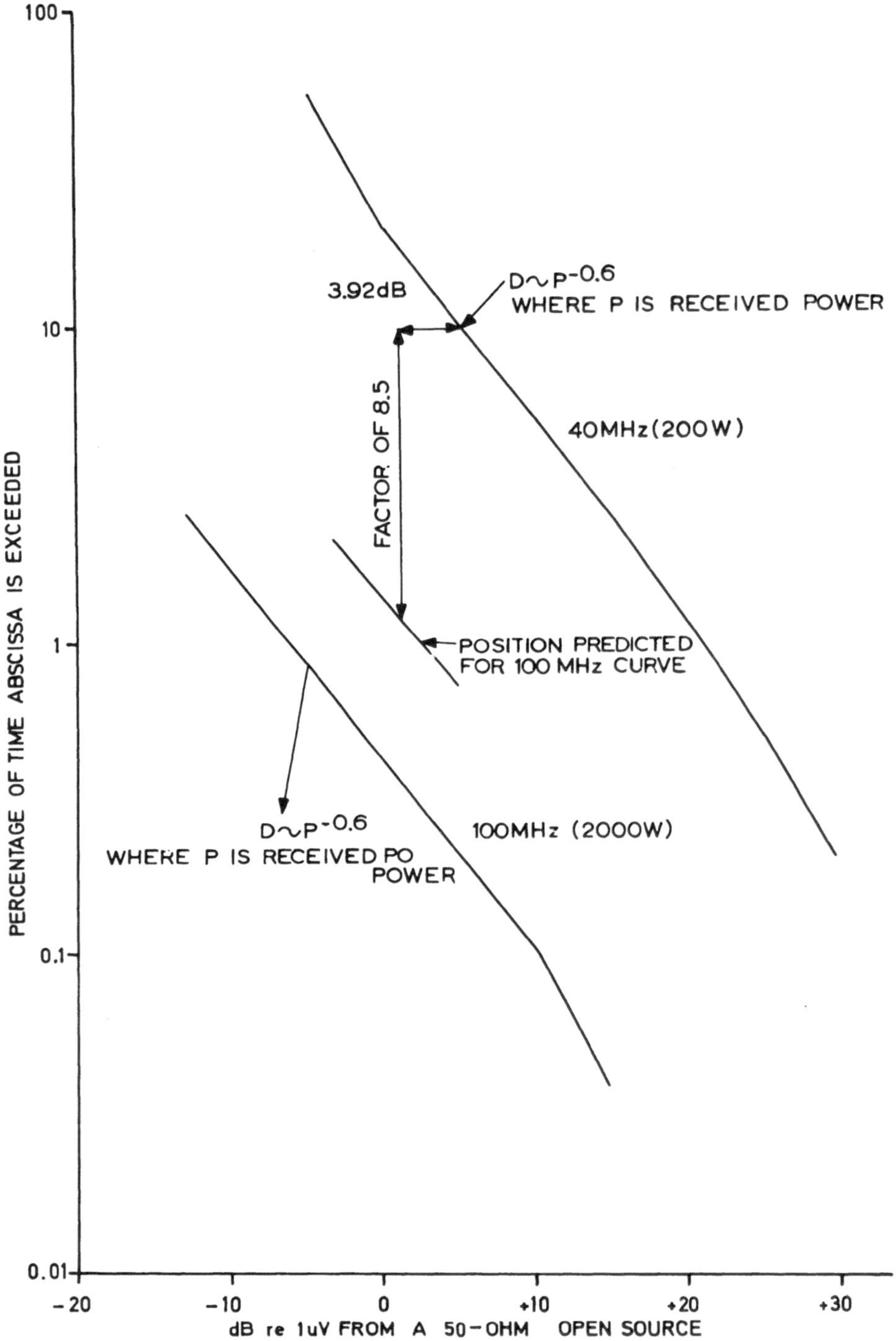

Fig.14. Yearly distribution of signal levels (average for each frequency over all comparative tests).

NEW HF DATA TRANSMISSION TECHNIQUES

M.Darnell

Communications Division, SHAPE Technical Centre,
The Hague, Netherlands

ABSTRACT. A survey of current types of HF data transmission systems is given, together with an indication of the performance levels achievable. The merits and disadvantages of these techniques are discussed by reference to a simplified physical model of the HF propagation path. Consideration of the propagation path structure allows conclusions to be drawn regarding the nature of data transmission schemes which might be better matched to the path properties than are the systems currently in use. One technique that has been investigated by STC is to employ spread-spectrum processing to combat multipath and interference effects. The design of a prototype system is outlined and performance results and future development discussed. A second technique is that of variable-rate HF data transmission. This would appear to provide a good match to the time-varying nature of the available HF channel capacity.

1. INTRODUCTION

In a companion paper [1] the essential nature of the high frequency (HF) propagation path is presented, together with a description of real-time channel evaluation (RTCE) techniques for selecting a channel, or channels, with certain properties from a set of alternative assigned channels. However, although RTCE schemes matched to different forms of transmission are described relatively little consideration is given in [1] to the suitability of the various types of information-bearing signals for HF paths. It is the purpose of this paper to discuss several aspects of HF signal design and processing, and to postulate transmission techniques which are better matched to the task of overcoming the effects of the propagation dispersion and noise/interference encountered with HF channels than are the majority of techniques currently employed.

1.1 HF communication: present status

To provide some background for a discussion of the status of HF communication, it is instructive to consider the major advantages and disadvantages of current HF systems; these are summarised in Table 1 below.

Advantages		Disadvantages	
(i)	Simple RF equipment	(i)	Channel capacity is time-varying and limited.
(ii)	Simple antennas.	(ii)	Path imposes time and frequency dispersion.
(iii)	Propagation medium is of common-user nature and is freely available.	(iii)	High levels of man-made interference.
(iv)	HF systems can be made readily transportable for mobile use.	(iv)	Manual system control required with current operating techniques.
(v)	No range limitation and relays not required.	(v)	Signal formats are not well matched to the characteristics of the propagation medium.
		(vi)	Propagation and noise predictions can be subject to large errors.

Table 1: Advantages and disadvantages of HF systems

The last period for significant development of HF techniques was in the mid-1950's: in this era, single-sideband (SSB) systems, high-precision frequency sources and modem technology all reached a relatively advanced state. It was only the "inertia" of these developments that carried HF systems through the period of disenchantment, in which satellites were expected to fulfill all long-distance civil and military communication requirements, to the mid-1970's when military users especially began to realize that complete reliance upon satellite systems could have significant operational disadvantages and that HF communication might indeed be an under-utilized resource with a valid operational role to play.

1.2 Outline of paper

In Section 2, a survey of HF data transmission techniques currently in use is given, whilst in Section 3 the mismatch between these techniques and the properties of the HF path is considered. Sections 4, 5 and 6 discuss data transmission techniques which potentially provide a better match to the path characteristics. The overall conclusions are given in Section 7.

2. BRIEF SURVEY OF CURRENT HF DATA TRANSMISSION TECHNIQUES

HF paths tend, for the most part, to be employed for low-speed (typically 50 - 75 bit/s) digital data transmission or for analogue speech circuits because of the multipath propagation limitation on channel capacity. For serial data transmission schemes, where the signalling element length is not less than about 4 ms, multipath effects over, say, a 1000 km path are normally not serious - even with simple signal formats. However, for higher data rates, more sophisticated signal generation and processing techniques must be adopted to combat multipath time dispersion.

Three important constraints on the operation of HF circuits which have influenced the developments of data transmission schemes to date are:

(a) Circuit planning and frequency selection are accomplished with the aid of off-line frequency prediction techniques.

(b) Channel bandwidth is normally limited to a maximum of 3 kHz for SSB systems operating in Europe, although larger bandwidths are used for multi-channel systems in the USA.

(c) Circuits are operated at a constant transmission rate with fixed signal generation and processing algorithms.

Current HF data transmission schemes are now described under two sub-headings: low-speed transmission systems with information rates in the range 20 - 250 bit/s, and medium-speed systems designed for information rates of the order of 1200 - 2400 bit/s.

2.1 Low-Speed Techniques

2.1.1 On-off-keying (OOK). OOK is perhaps the simplest of all HF data transmission schemes and, in some ways, one well-matched to the path characteristics. It is used primarily for manual morse transmission and involves only the interruption of a simple CW carrier in accordance with the morse symbols. Since the transmission rate is controlled entirely by a human operator it can be varied in response to path conditions as indicated by the receiving site. Also, at the receiver site, the signal detection bandwidth can be narrowed to a few tens of Hz if necessary, which has the effect of excluding much of the HF noise and interference that might be encountered if, say, a 3 kHz reception bandwidth were to be employed. The pattern recognition abilities of a trained morse operator are considerable, and he is capable of interpreting accurately very weak received signals in the presence of high levels of noise and man-made interference.

Therefore, although the transmission rate of an OOK manual morse system is low, its potential for adaptation in response to path conditions is great when used by skilled operators.

2.1.2. Binary frequency-shift-keying (FSK). The most widely used modulation scheme for low-speed data transmission is binary FSK. Originally, prior to the development of SSB techniques, FSK keying was carried out directly at RF; in addition, the frequency stability achievable with HF equipments in this area was poor and therefore a large shift had to be used to ensure that the two signalling frequencies were detected unambiguously - possibly in the presence of considerable frequency off-sets due to tuning instability and misalignments between transmitters and receivers. For these reasons wide-shift FSK, in which the two keying frequencies were separated by a nominal 850 Hz, was adopted. Clearly such a technique made inefficient use of the available bandwidth.

The high frequency stability resulting from the development of SSB technology led to frequency off-sets between transmitter and receiver being reduced to 1 Hz or less, rather than the tens of Hz with non-frequency-synthesized equipments. It was then feasible to reduce frequency shifts, typically to a value of 85 Hz for 50-bit/s traffic with generation and detection now taking place at baseband.

2.1.3 Multiple frequency-shift-keying (MFSK). The best known example of an HF MFSK system is Piccolo [2], which is designed for the transmission of 75 baud telegraphy. The set of 32 characters in say the CCITT No. 2 telegraph code are mapped directly into a set of 32 baseband tones, each separated by 10 Hz. An additional (33rd) tone is phase modulated at the baud rate for synchronisation purposes. Thus, the total bandwidth occupied by a Piccolo transmission is 330 Hz. The baud interval is 100 ms which allows the individual tones on the 10 Hz raster to be detected orthogonally by quenched resonators, i.e. the detection is a non-coherent process.

STC has carried out a detailed assessment of Piccolo, the results of which are described in [3]. It is concluded that Piccolo provides an improvement of between 6 & 12 dB in signal-to-noise density ratio for a given character error rate over a binary, narrow-shift, FSK system - both systems operating with dual space diversity. In general, Piccolo is capable of providing very high quality telegraphic copy over long HF paths using relatively low transmitter powers and simple antennas.

2.1.4 Binary frequency-exchange keying (FEK). Binary FEK or, as it is sometimes termed, two-tone keying, is a conceptually simple HF transmission technique particularly suited to telegraphic data rates. As in an FSK system, at the transmitter mark and space tones, with frequencies of f_m and f_s respectively, are keyed in response to a binary input data waveform so that at any instant either the mark or space tone will be present. At the receiver, however, detection is not carried out using a discriminator, as with FSK, but with the FEK filter-assessor arrangement illustrated in Fig. 1. The operation of this unit is now considered under both flat-fading and frequency-selective fading conditions. The term "flat-fading" is applied to the state where the variations in the energy levels of the two tones are highly correlated, wherease "frequency-selective" fading describes the state where the energy level variations are independent. For most practical HF paths, when the difference frequency between the two tones f_D, given by

$$f_D = |f_m - f_s| \tag{1}$$

is greater than a few hundreds of Hz, frequency-selective fading will be predominant.

As shown in Fig. 1, each of the OOK input tones is band-pass filtered, envelope detected and then low-pass filtered to remove tone components. Also, for each of the detected tones, the average energy level is measured by the energy detectors. Figures 2a and 2b show the corresponding waveforms under flat-fading and frequency-selective fading conditions respectively. In the flat-fading case, the average energy levels will tend to be similar and the decision threshold formed by their sum will correspond to the zero level. However, in the frequency-selective fading case, the decision threshold will tend to favour the tone with the greatest energy at any time, thus increasing the noise immunity of the detector as compared with a fixed threshold system. It is arranged that the averaging time-constant of the energy detectors, τ, is long compared with the baud interval of the binary data signal, Δt, but sufficiently short to track the level changes due to fading. Typically

$$\tau \simeq K.\ \Delta t$$

$$\text{where} \qquad 10 < K < 20 \tag{2}$$

Since the same information is transmitted on both mark and space sub-channels, filter-assessor detection of binary FEK can be considered as a form of dual in-band frequency diversity reception. In [3], it is concluded that the performance of such a system can be comparable with, or better than that of, Piccolo when the time differences between multipath propagation modes can be specified accurately and f_m and f_s optimised accordingly.

Fig. 3, derived from [4] and [5] illustrates the differences in performance between narrow-shift FSK (Section 2.1.2), Piccolo (Section 2.1.3) and binary FEK with filter-assessor detection, in terms of the average probability of character error, $\bar{P}_c$, as a function of

$$\frac{\text{Energy per character}}{\text{Noise spectral density}} \quad \left(\frac{E_c}{N_o}\right)$$

measured in dB. All three systems incorporated dual space diversity processing and the results were obtained using an HF path simulator with additive Gaussian noise producing two-path selective fading caused by path time delay differences of up to 4 ms. A fading rate of 40 fades per minute was employed. The advantages of Piccolo and FEK over FSK are clear and have been verified by tests over real HF paths.

An interesting point to note about the filter-assessor FEK detector is that it is essentially a soft-decision device in which the decision threshold is varied in accordance with a signal confidence parameter, namely the relative energy levels for the two tones. The further development of HF FEK transmission systems are discussed in Section 6 of this paper.

2.1.5 <u>Error detection and correction (EDC) techniques for HF low-speed data transmission</u>. In the four previous sections, the modulation schemes most commonly employed for HF low-speed data transmission have been discussed. The application of EDC to HF circuits has been somewhat limited, principally because of the difficulty in characterising a generalised channel model, and hence specifying an appropriate EDC coding scheme.

The EDC systems which have been applied to HF channels fall into two categories:

(a) Those requiring an automatic repeat request (ARQ) return channel for passing decision feedback information concerning errors detected on the forward channel and initiating retransmissions as necessary.

(b) Those in which forward error correction (FEC) only is used.

The best-known category (a) system, described in [6], takes the input data stream comprising 5-bit start-stop characters and converts them into 7-bit synchronous characters prior to transmission. Of the 2^7 possible characters, 35 comprise 3 mark and 4 space combinations and 32 of these are chosen to represent the original 5-bit alphabet. At the receiver, each character is checked for the 3 mark : 4 space ratio and, if this is found to be correct, the received 7-bit character is re-mapped to the original 5-bit start-stop code and passed to a teleprinter. Should a character be received incorrectly and the error detected, a repeat of that character is then requested.

A representative HF FEC scheme is described in [7]. In this case, the 5-bit input start-stop characters are mapped into 10-bit codewords, where the first 5 bits of the codewords are a replica of the character bits and the second 5 bits are parity elements. For example, if the 5-bit input character is $[\,I_1\ I_2\ I_3\ I_4\ I_5\,]$, the 10-bit codeword will be given by:

$$[\,I_1\ I_2\ I_3\ I_4\ I_5\ P_{\bar{1}}\ P_{\bar{2}}\ P_{\bar{3}}\ P_{\bar{4}}\ P_{\bar{5}}\,] \qquad (3)$$

Where $P_{\bar{1}}\ P_{\bar{2}}\ P_{\bar{3}}\ P_{\bar{4}}\ P_{\bar{5}}$ are the parity check bits formed as follows:

$$\left.\begin{aligned} P_{\bar{1}} &= I_2 \oplus I_3 \oplus I_4 \oplus I_5 \\ P_{\bar{2}} &= I_1 \oplus I_3 \oplus I_4 \oplus I_5 \\ P_{\bar{3}} &= I_1 \oplus I_2 \oplus I_4 \oplus I_5 \end{aligned}\right) \qquad (4)$$

$$P_{\bar{4}} = I_1 \oplus I_2 \oplus I_3 \oplus I_5$$
$$P_{\bar{5}} = I_1 \oplus I_2 \oplus I_3 \oplus I_4$$

When $\oplus$ denotes modulo -2 addition. At the receiver, inspection of the 5 parity checks allows all single errors per codeword to be detected and corrected. Multiple errors, when detected, give rise to a specific error character in the output copy.

STC has carried out trials with a version of this system that also incorporates bit interleaving whereby the elements of each 10-bit codeword can have interposed between them the corresponding elements taken from 10, 20 or 50 codewords. Fig. 4a shows an example of the teleprinter copy obtained using normal start-stop telegraphy and binary FEK modulation; Fig. 4b shows the corresponding teleprinter copy obtained using the FEC system with 20- bit interleaving. These tests were conducted over an 1100 km HF path between Oslo and The Hague. The one blank character marked in Fig. 4b is the special symbol indicating multiple detected, but uncorrected, errors. The degree of improvement obtained using FEC is typical of that measured during a comparative trial of the two systems over a wide range of path conditions.

2.2 Medium speed techniques

For simple modulation schemes the time dispersion due to multipath propagation tends to limit the serial transmission rate to about 250 bauds for, say, a 1000 km HF path. For this reason, medium-speed data transmission in the range 1200- 2400 bit/s is normally achieved by providing a modem having a multiplicity of low-speed parallel sub-channels.

2.2.1 Parallel sub-channel modems. The best known example of an HF parallel sub-channel modem is KINEPLEX [8]. There are several practical variants of this type of modem, but they all essentially employ multiple sub-channels (typically 16), each with quaternary differential phase-shift keying (DPSK) and orthogonal sub-channel frequency spacing. A typical modem accepts data at a rate of 2400 bit/s and formats this into 16 parallel sub-channels, each operating at a rate of 150 bit/s. These 16 data sub-channels then differentially phase-key 16 corresponding baseband sub-carriers; quaternary modulation allows the transmission of 2 bit/phase and thus the baud rate for each sub-channel is 75, i.e., a frame interval of 13.33 ms. Of this frame interval, 4.24 ms is employed as a "guard time" to allow the effects of inter-symbol interference to be rejected. The transmit and receive modems are synchronised prior to data transmission to allow only the middle (13.33 - 4.24) ms, or 9.09 ms, of the data frame to be used in the differential detection process. After detection, the data frames are reassembled into a 2400 bit/s serial data stream.

Such a modem will normally also accept data at a rate of 1200 bit/s and format this into a group of 8 sub-channels, or into two 8 sub-channel groups. In the latter mode, dual in-band frequency diversity operation is possible which assists in combatting the effects of the frequency-selective fading often encountered over HF paths.

The most significant development of the KINEPLEX principle is a modem termed CODEM [9]. This is a device with 25 parallel sub-channels employing a combination of modulation and coding to provide spectral redundancy by effectively spreading the energy of each data bit over the whole bandwidth available for transmission. CODEM uses soft-decision decoding in which certain confidence criteria, as measured at the demodulator, are employed to enhance the error detecting and correcting capability of the basic coding algorithm. The input data is first formatted into 16 parallel low-rate data streams, and each 16 digit data frame is transformed into a 25 sub-channel format by means of a (25, 16) product code using the row and column parity checks as shown below:

$$\begin{array}{cccc|c} I_1 & I_2 & I_3 & I_4 & P_{17} \\ I_5 & I_6 & I_7 & I_8 & P_{18} \\ I_9 & I_{10} & I_{11} & I_{12} & P_{19} \\ I_{13} & I_{14} & I_{15} & I_{16} & P_{20} \\ \hline P_{21} & P_{22} & P_{23} & P_{24} & P_{25} \end{array} \qquad (5)$$

Where $I_1 - I_{16}$ represent the information digits in the 16-digit data frame and $P_{17} - P_{25}$ are the parity check digits. The parity checks are formed by simple modulo -2 addition of row and column information digits, e.g.

$$P_{17} = I_1 \oplus I_2 \oplus I_3 \oplus I_4 \qquad (6)$$

and

$$P_{21} = I_1 \oplus I_5 \oplus I_9 \oplus I_{13} \qquad (7)$$

The parity digit P_{25} is a parity check on the complete frame of information digits. In practice, CODEM uses 4-phase modulation and thus each digit represents 2 bits of information. Therefore, two separate (25, 16) codes are used for in-phase and quadrature components. The digits of expression (5) are mapped into the 25 sub-channel format thus:

1	6	11	16	21	
22	2	7	12	17	
18	23	3	8	13	(8)
14	19	24	4	9	
10	15	20	25	5	

It will be noted that the information and parity digits are distributed across the raster of 25 sub-channels to overcome the fact that the fading characteristics of adjacent sub-channels will tend to be highly correlated. For each received bit, the values of tone amplitude and phase margin with respect to the demodulator reference are used to derive a soft-decision confidence level for that bit. The EDC decoding of the product code is carried out in the normal manner, except that when the parity checks fail, the soft-decision levels are invoked to determine which received bits have the highest probability of being in error. In this way, the polarities can be inverted systematically until all the parity checks are valid.

2.2.2 Serial modems. Two main serial HF medium-speed data transmission systems are discussed in the literature, namely ADAPTICOM [10] and ELSA [11]. ADAPTICOM is essentially a baseband adpative equalization scheme, requiring the transmission of training patterns from time to time in order to evaluate the multipath structure and thus adjust the parameters of the equalizer at the receiver.

ELSA also adapts to the time-varying structure of the propagation path by assuming that the multipath channel acts as a convolutional encoder. The decoder at the receiver then employs a sequential decoding algorithm to produce an estimate of the path encoding structure, and hence to recover the transmitted data.

In addition to the adaptive serial devices mentioned above, it is possible to use, say, a simple 4-phase DPSK serial data modem of the type normally employed in point-to-point data transmission networks of telephone circuits. Under single-mode propagation conditions, this latter class of modem can give a reasonable data transmission performance.

2.2.3 Comparison of parallel sub-channel and serial modem performance. In general, the serial modems mentioned above give rise to error rates significantly worse than those obtained using parallel sub-channel modems. STC has carried out practical tests in order to compare the performance of a modem with 16 parallel DPSK sub-channels with that of a simple serial DPSK modem, under similar path conditions, using both conventional frequency selection procedures and an RTCE channel selection system of the type described in [1].

Figure 5 shows the probability of achieving a given BER over an 1100 km path between Oslo and The Hague, using a 16 parallel sub-channel modem with dual in-band frequency diversity and the serial DPSK modem, both operating at 1200 bit/s. Two plots are given:

(a) For 2-frequency operation, as might be derived from off-line frequency prediction, and

(b) For RTCE frequency selection.

Similar tests at 1200 bit/s were also carried out over a 700 km path between Cornwall and The Hague in October/November 1976. Table 2 summarizes the results for these tests, together with those for the Oslo to The Hague path.

BER	Path	Parallel sub-channel modem availability(%)		Serial modem availability (%)	
		2-Frequency	RTCE	2-Frequency	RTCE
$\leqslant 10^{-2}$	Oslo-The Hague Cornwall-The Hague	75.0 45.8	100.0 75.0	62.5 37.5	79.2 70.8
$\leqslant 10^{-3}$	Oslo-The Hague Cornwall-The Hague	33.3 12.5	83.3 29.2	37.5 4.2	58.3 12.5
$\leqslant 10^{-4}$	Oslo-The Hague Cornwall-The Hague	0 0	29.2 4.2	33.3 0	45.8 8.3

Table 2: Results of medium-speed data transmission tests

3. FACTORS AFFECTING THE DESIGN OF HF DATA TRANSMISSION SYSTEMS

The inter-relationships between the various functional elements of a communication system can be conveniently identified by consideration of a generalized communication channel.

3.1 The generalized communications channel

The elements of a generalized communications channel are shown in Fig. 6. A data source produces a signal which often contains significant redundancy, i.e., it is only a partially random process. It is the function of the source encoder to determine which information should be transmitted by removing redundancy, in a deterministic manner, according to certain decision criteria. The channel encoder then specifies how the resulting data is to be transmitted by processing the "compact" data stream so as to introduce redundancy - but now in a form specifically designed to combat the particular types of distortion and interference which are created by the propagation path and other spectrum users. After transmission and reception, via appropriate antenna systems, the channel decoder makes use of the artificially-imposed redundancy to recover the transmitted data with a specified degree of fidelity. The source decoder then operates on the reconstituted compact data and re-inserts the original source redundancy, making use of a priori knowledge of the form of the source encoder, thus giving a replica of the data source output. In Fig. 6, a feedback link (dotted) is also shown: if such a link is available, even with a relatively low capacity, the power of the source and channel encoding/decoding algorithms can be enhanced and varied adaptively according to say channel state or received error rate criteria. Also, feedback data can be used to adapt transmitter power levels and antenna radiation patterns.

3.2 Channel encoding/decoding

This paper is not primarily concerned with source encoding/ decoding schemes, but rather with channel encoding/decoding techniques particularly suited to HF systems. In this context, the transmitting/receiving antennas as shown in Fig. 9 should be viewed as a component of the channel encoding/decoding process since they also can be optimized in response to path conditions. Certain conclusions regarding the most effective techniques to apply can be drawn from the discussion of existing HF data transmission systems given in the previous section. For example, the performance obtained with FEK telegraphy using a filter-assessor demodulator (Fig. 1) and the improved BER achieved using CODEM as opposed to a KINEPLEX-type modem tend to indicate the value of employing soft-decision data and diversity. For the HF path, which is both time and frequency variable, diversity of various types can perhaps provide the most effective means of improving communication system performance.

EDC techniques, particularly with time diversity or inter-leaving have been shown to provide worthwhile performance improvements over uncoded systems. It would seem that more attention could also be given to adaptive EDC schemes in which the type and power (error detection and correction capability) of the code are varied in accordance with the prevailing channel conditions and measured error characteristics. In cases where a feedback link from receiver to transmitter is available, such a scheme can be readily implemented and is sometimes termed "variable redundancy coding", e.g., [12].

A further point which should be made regarding the general application of EDC techniques: if the basic uncoded (unprotected) error rate for the channel is very high, or if the code is not reasonably suited to the type of error patterns occurring, it is unlikely that EDC will improve the error rate and it may well make the situation worse [13]. Therefore, it is most important that, wherever feasible, RTCE techniques are used prior to the application of EDC in order to identify channels which are likely to have low unprotected error rates and also to assess the nature of the error distributions so that maximum benefit can be obtained from this type of channel encoding.

In a discrete-state communications systems, it is usual for the channel decoder to make a "hard-decision" as to which state has been received from the library of possible states. This process inherently neglects much of the information present in the received signal, which potentially might be employed to enhance the reliability of the channel decoding decision. Methods which seek to make use of this additional information are termed collectively probabilistic, or "soft-decision", techniques. Instead of making a hard-decision on each discrete signal state received, the channel decoder initially decides whether the decoded signal state is above or below a decision threshold; it then computes a confidence factor which specifies the actual distance of the decoder output from the decision threshold. This number can in principle be an analogue quantity, but in practice would itself normally be quantized.

In general, care must be exercised in assessing the benefits of EDC and it should be considered whether or not a similar order of improvement might also have been achieved more economically by other means, e.g., frequency/space/polarization diversity, or increasing the transmitter power. This is particularly true in the case of the HF medium which is both time- and frequency-variable. It may be that for this type of path, it is more cost-effective to apply the majority of resources to developing techniques, such as RTCE, to allow the identification of the optimum working frequency. Having selected the best channel available, the complexity of an EDC scheme to yield a given circuit availability will be reduced in comparison with that required for an arbitrary HF channel.

3.3 The propagation path

RTCE techniques for the selection of the optimum member from a set of alternative assigned channels are discussed extensively in [1]. For HF data transmission, RTCE can be used for two main purposes:

(a) To allow the communication signal format and processing procedures to be adapted in response to the state of the channel. This entails the derivation of an appropriate channel model from RTCE measurements.

(b) To select a channel with characteristics appropriate to the type of communication signal format and processing procedures being used.

In the case of HF communication, (a) above can be a very complex process and (b) may well be a more attractive option leading to a relatively simple channel selection algorithm. Figure 7 illustrates one such selection algorithm when it is desired to find a channel having minimum time dispersion and maximum SNR. The display of propagation modes, or "ionogram", is analysed to find a region of single mode propagation with acceptable signal strength. The interference levels for the assigned channels in this region are then examined and the channel with the highest SNR selected for communication.

4. A PROTOTYPE HF SPREAD-SPECTRUM SYSTEM

4.1 Rationale

In an idealized memoryless continuous channel, perturbed by white Gaussian noise, the theoretical maximum error-free transmission capacity, C, is given by:

$$C = B \log_2 (1 + S/N) \text{ bit/s} \tag{9}$$

where B is the channel bandwidth in Hz, and S and N are the total signal and noise powers respectively within the bandwidth B. In dispersive channels, the quantity S/N is varying continuously with frequency and time and thus a more meaningful expression for the instantaneous value of C is:

$$C = \int_0^B \log_2 \left[1 + s(f)/n(f)\right] df \text{ bit/s} \tag{10}$$

where s(f) and n(f) are the one-sided signal and noise power density spectra, and n(f) can now take any form. Figure 8a shows typical plots of s(f) and n(f). It can be shown that, for a fixed total signal power S, where:

$$S = \int_0^B s(f)\, df \tag{11}$$

the capacity C is maximized if

$$s(f) + n(f) = \text{Constant} \tag{12}$$

wherever possible in the bandwidth B [14]. Figure 12b shows plots of s(f) and n(f) when the total signal power has been redistributed according to expression (12). The obvious results of this strategy are that regions of high noise level tend to be avoided and also that the signal energy is distributed relatively uniformly in regions of low noise level. To permit such an optimum distribution of signal power, the spectrum of the transmitted signal would have to be adapted continuously in response to the measured noise spectrum at the receiver. With rapid time variations of both signal level and noise/interference level at the receiver, as experienced on a typical HF link, such real-time adaptation can never be achieved because of the time delays involved in spectrum measurement and information feedback; a sub-optimum distribution of signal energy is therefore inevitable. Since interference in the HF band consists for the most part of high-level narrowband ($\leqslant$3 kHz) transmissions with relatively large regions of low noise between the high-level signals, it is postulated that it would be reasonable to distribute the available signal power approximately uniformly over a wide bandwidth B(>>3 kHz), such that most of the power would appear in low-noise regions and only small amounts would be completely masked by interference peaks. It is further postulated that such a system should be capable of working in any region of the HF band for which a propagation path exists, in contrast to a more conventional narrowband data transmission scheme where all the available signal power is concentrated within a 3 kHz pass-band and could be disrupted totally by a single high-level narrowband interfering signal.

STC is currently carrying out an experimental programme to investigate the performance of a prototype spread-spectrum HF (SSHF) modem and in particular its ability to:

(a) Randomize the interference background due to discrete interfering sources over a bandwidth of up to ±80 kHz in the HF band.

(b) Select, via correlation processing a single propagation mode from a multipath set.

(c) Operate simultaneously in the same frequency band as other spectrum users and produce low levels of interference, i.e., its EMC properties.

A description of the prototype SSHF system is now given, and the results of initial trials are presented.

4.2 System description

A functional block diagram of the SSHF modem is given in Fig. 9. The various sub-systems and the operation are now described individually.

4.2.1 Transmitter. The input data can be accepted in two main formats, either CCITT No 2 start/stop telegraph code, or in simple binary form. With start/stop telegraphy, the start and stop elements are removed leaving 5-bit characters; the simple binary input data stream is formatted into 5-bit groups. In both cases, a first level of M-ary encoding is applied whereby the 5-bit data groups are mapped into one of $M(=2^5)$ orthogonal states - a coding scheme conceptually similar to that used by Piccolo (Section 2.1.3). For the SSHF modem, however, the 32 orthogonal states are 32 63-bit codewords selected from the possible set of 63 phase-shifted versions of 2^6-1)-bit pseudonoise (PN) m sequence. In addition to 32 data code words, 4 of the remaining 31 states are used as control characters, e.g., for synchronization, as discussed in more detail in Section 4.2.3.

The first level encoded data is then added modulo-2 to a fast PN sequence, the clock rate of which can be set in the range 1.3 to 85 kbit/s. The composite signal then phase modulates a carrier prior to transmission.

4.2.2 Receiver. Turning now to the receiver section of Fig. 9, and considering only one path of the diversity arrangement, since the other path is identical: the 500 kHz IF waveform is split into in-phase (I) and quadrature (Q) channels by multiplying by local oscillator signals offset by 90° in phase. The use of a 500 kHz local oscillator also translates the signal to baseband. Prior to this, the received signal is multiplied by a correctly-timed version of the fast PN sequence used for the second level encoding at the transmitter. The outputs of the balanced modulators of the I and Q channels are therefore proportional to the sine and cosine of the phase difference between the input signal and the local reference. These outputs are integrated digitally over each bit interval of the 63-bit input codeword and, at the end of every bit interval, the output of the integrator is transferred to a memory prior to being reset to zero (7 bits + sign). Once the memory contains the number of samples corresponding to one codeword length, the input is transferred to a second memory and the contents of the first memory are processed at high rate whilst the second memory is being filled. The processing consists of recirculating the memory contents and correlating in turn with each member of the possible set of codewords with the moduli of the corresponding I and Q correlations being added. The two highest correlation values are then selected, the memory inputs switched over, and the process repeated for the next set of data. The ratio of the two highest correlation ordinates at the output of the correlation process is related to the SNR, and hence the confidence level, for the detection decision; a low value of the ratio may also be used to initiate repeat requests in an ARQ mode of working.

In the dual diversity mode, the best correlation outputs from the two channels are selected.

4.2.3 Synchronization. In the synchronization arrangements shown in Fig. 10 a standard delay-lock control loop procedure is employed with an initial drift search mode. In the acquisition phase, a distinct 63-bit synchronization codeword is transmitted continuously and, at the beginning of every codeword repeat, the fast PN spectrum-spreading sequence is reset to an initial state so that a truncated fast sequence of duration equal to the codeword period is actually transmitted. At the receiver, the delay-lock loop, or time discriminator, makes use of two signal detectors of the same type as described for data decoding, except that one pair of I and Q channels is multiplied by a PN sequence which is half a bit early and the other pair by a PN sequence which is half a bit late with respect to the wanted sequence.

Initially, the local fast PN sequence clock is drifted in one direction and when the received and local codewords are nearly synchronized the non-coherent detector outputs are approximately at dc. If the two detector outputs are connected in antiphase, the delay-lock characteristic shown results. The time discriminator output is then connected to a comparator which determines whether pulses should be added to or subtracted from the clock waveform. Once the incoming data and local reference waveforms are within one bit period of one another, this circuit maintains lock and data transmission can begin. At this point, the synchronization reference is replaced by the reconstituted detected information codeword and lock maintained in a similar manner.

One characteristic of this technique which should be noted is that, with the drift search mode, the system always locks to the first multipath component above some threshold value - this need not necessarily be the strongest or most stable component. If two separate processing paths are available, a different component can be selected for each and a mode diversity combining technique becomes a possibility.

4.2.4 Practical tests. The prototype SSHF modem is currently undergoing practical tests and the preliminary results are presented here.

Tests have also been carried out over two real HF paths to measure the circuit availability for 75 bit/s traffic in a non-diversity/non-ARQ mode with a PN sequence rate of 20 kbit/s. Transmitter radiated power levels were restricted to 10 W or less and selection of operating frequencies was based solely on predictions of available propagating frequency ranges; no attempt was made to deliberately avoid regions of high man-made interference. Figures 11a, b and c present sample 24-hour BER measurements, with BER values being averaged over 30-minute intervals.

Figure 11a shows the results obtained for a 200 km path where a day-time frequency, f_D, of about 5.5 MHz and a night-time frequency, f_N, of about 2.9 MHz were used. The propagation mode was via skywave, ionospheric F-layer refraction.

Figures 11b and 11c are similar plots for a short 21 km path. To obtain the results shown in Fig. 11b, the SSHF modem was locked to a skywave mode, as for Fig. 11a and, for the results of Fig. 11c, the modem was locked to a groundwave propagation mode. The improved performance resulting from the more stable groundwave mode is evident. In this latter case, no frequency changes were necessary during the 24-hour period and a single frequency, f_{DN}, of approximately 2.1 MHz was used throughout.

In Figs. 11a and 11b particularly, the diurnal variation of man-made interference level is apparent. This is shown by the higher BER values at night due to the available skywave frequency propagation range contracting and the interfering signals being compressed into a much smaller frequency band than during daytime. The diurnal effects are less pronounced with groundwave propagation, although here also BER values are on average higher at night.

The results shown are representative of those obtained over other 24-hour periods. Circuit availabilities for various BER thresholds are summarized in Table 3 below.

BER threshold	Circuit availability (%)		
	(a) 200 km Circuit (skywave)	(b) 21 km Circuit (skywave)	(c) 21 km Circuit (groundwave)
$\leq 10^{-2}$	72.9	75.0	100.0
$\leq 10^{-3}$	35.4	33.3	75.0
$\leq 10^{-4}$	10.4	25.0	56.3

Table 3: Circuit availability for SSHF modem

4.2.5 General conclusions. As a result of the tests carried out to date the following conclusions can be drawn:

(a) Under conditions where the HF propagation mode structure is changing relatively rapidly, e.g. at sunrise and sunset, the SSHF modem can exhibit synchronization loss due to the fact that the mode to which it was previously locked has disappeared and other propagation modes have been established with different path delay times.

(b) It is to be expected that the use of diversity techniques and ARQ operation will improve the circuit availability appreciably.

(c) RTCE techniques were not employed for the practical tests. Had they been used, for example, to select regions of single mode skywave propagation, it is probable that the system performance would have improved.

(d) The use of PN rates greater that 20 kbit/s was precluded by transmitter bandwidth limitations. It is planned to carry out tests at 80 kbit/s PN rate in the near future. In principle, the simple limit on maximum PN rate is set by the time dispersion of a single propagation mode. Measurements quoted [15] indicate that such dispersion (of an impulse) is less than 5 μs for 50% of daylight hours, that E-W geomagnetic paths were better than N-S paths and that minimum dispersion occurred at a frequency of about 70% of the measured maximum usable frequency (MUF).

(e) The SSHF system described represents an attempt to "live" with the current HF environment by removing the normal bandwidth constraint applied to HF channels. The initial trials indicate that the technique has sufficient potential to merit further development.

5. VARIABLE RATE HF TRANSMISSION

5.1 The nature of the HF path

In a multipath situation, the individual modes will tend to exhibit independent variations in amplitude and phase and therefore, at certain times, one propagation mode may be completely dominant. Similarly, interfering signal variations will, to a large extent be uncorrelated with those of the wanted signal; thus, even although on average an interfering signal may have a greater strength than a wanted signal, there will be short periods when the wanted signal will be dominant. Also, as different modes become dominant, the effective path propagation delay will change, causing synchronization problems of the type described for the SSHF modem in Section 4.

To attempt to give more insight into the characteristics of signals propagated via the HF medium, Table 4 lists BER samples taken using a radio frequency of approximately 7.7 MHz over the path between Oslo and The Hague. The test signal was the 1200 bit/s serial DPSK modem waveform transmitted at an average power of 1 kW. Off-line analysis was applied to the received data in such a way that the number of errors occurring in blocks of 100, 1000 or 10,000 data bits could be computed. Therefore, the same set of data was sampled and analysed three times - once for each of the specified block lengths.

Error counts per block							
Block length 100			Block length 1000			Block length 10,000	
45	0	0	17	139	4	516	215
5	0	0	21	0	70	597	5
12	0	3	112	4	0	809	28
16	0	11	0	6	0	0	587
6	27	0	0	42	87	720	656
13	0	11	0	0		359	224
0	0	0	215	2		627	159
0	1	0	130	4		95	
0	18	0	113	2			
7	0		0	573			

Table 4: Sampled error counts

Clearly, the probability of obtaining an error-free block of N bits, $P_{EF}(N)$, decreases as N increases. Hence:

$$P_{EF}(100) > P_{EF}(1000) > P_{EF}(10000) \qquad (13)$$

It is seen that most errors on the path occur in bursts, due either to signal fading or man-made interference. In the case of the channel of Table 4, the actual probability values are:

$$\left.\begin{aligned} P_{EF}(100) &\simeq 0.5 \\ P_{EF}(1000) &\simeq 0.25 \\ P_{EF}(10000) &\simeq 0.07 \end{aligned}\right\} \qquad (14)$$

Therefore, a medium-speed data transmission system in which the data are assembled and transmitted in short, high-rate blocks would appear to be better matched to the nature of the HF medium than the more conventional continuous transmission systems described in this paper. A block-formatted system could work in an ARQ mode at high serial transmission rates. An error detection and correction coding scheme applied to the blocks could be used to monitor the presence of errors, correct them where possible, and reject uncorrectable blocks. In this way, the operation of the transmission system could take account of the inherently time-varying capacity of the HF channel.

Although, the BER values quoted in Table 4 are a very small sample, they are reasonably representative since similar values of $P_{EF}(N)$ have been computed from extensive quantities of test data collected over several different HF paths.

5.2 Techniques for variable rate HF systems

As mentioned in Section 4, the simplest form of variable rate transmission technique to match the information rate to the path characteristics is ARQ. It was also shown in Section 4 that, with high rate serial transmission in a synchronous mode, rapid changes in path propagation time could cause synchronization problems.

Consider an SSHF signal received via a given propagation mode with propagation delay τ_1: if a sudden change in propagation conditions occurs, such that a new propagation mode with propagation delay τ_2 is established and the delay-lock loop is unable to track the change, then the system will only lock to the new mode if

$$|\tau_2-\tau_1| < \Delta t/2 \qquad (15)$$

where Δt is the clock interval of the fast PN sequence. This can be seen from the delay-lock loop characteristic shown in Fig. 10. For a 20 kbit/s PN rate, this represents a maximum delay change of 25 μs. Since time differences between HF propagation modes typically take values in the range 0.5 to 2 ms, it can be seen that the SSHF modem will require to be re-synchronized in such a situation. For short term fades of a given mode, the stability of the receiver clock is sufficient to allow automatic re-synchronization when the fade is over; the data transmitted during the fade may be lost however due to inadequate signal quality.

The weakness of the SSHF system described in Section 4 is that it makes use of PN sequences with periodic correlation properties. It is postulated that, in a situation where propagation time may vary rapidly, the design of the system should be based upon sequences with suitable aperiodic correlation properties, i.e. the autocorrelation function (acf) of a single period of the sequence should be impulsive. For an N-digit binary sequence, $[x_i]_{i=1}^{N}$, the aperiodic acf, $\emptyset_{xx}(r)$, is thus ideally given by:

$$\emptyset_{xx}(r) = \begin{cases} N; & r = 0 \\ 0; & \text{otherwise} \end{cases} \qquad (16)$$

where r is a discrete shift variable. Expression (16) will normally not describe the single period acf of a periodic PN sequence. In practice, several classes of sequences do have aperiodic acf's approximating to expression (16). The best known of these are Barker codes [16], Huffman sequences [17] and binary sequences found by computer search algorithms [18]. However, the type of signals described in [19] and termed "complementary sequences", appear to have greatest potential as data transmission waveforms. A survey of the theory and applications of complementary sequences is given in [20] and therefore only essential background is included here.

A pair of binary complementary sequences is defined as a pair of finite, equal-length, two-level sequences having the property that the number of pairs of like elements with any given spacing in one sequence is equal to the number of pairs of unlike elements with the same spacing in the other sequence. The sum of the individual aperiodic acf's of the two sequences is zero, except at the zero delay position, where it takes the value 2N, N being the number of digits in each sequence. If the two sequences are denoted by $[x_i]_{i=1}^{N}$ and $[y_i]_{i=1}^{N}$ and their respective aperiodic acf's are $\emptyset_{xx}(r)$ and $\emptyset_{yy}(r)$, then

$$\emptyset_{xx}(r) + \emptyset_{yy}(r) = \sum_{\ell=1}^{N=r} x_\ell x_{\ell+r} + \sum_{\ell=1}^{N=r} Y_\ell Y_{\ell+r} \qquad (17)$$

$$= \begin{cases} 2N; & r=0 \\ 0; & \text{otherwise} \end{cases} \qquad (18)$$

It should be noted, however, that the individual acf's, $\emptyset_{xx}(r)$ and $\emptyset_{yy}(r)$ will not themselves be of the form given in expression (18), but will have significant sidelobes.

In general $\quad N = 2^n$
where n is an integer $\geqslant 1$ (19)

Complementary sequences exhibit a basic "block" structure as follows:

Sequence 1:	$+\vec{A}$	$+\vec{B}$
Sequence 2:	$+\overleftarrow{B}$	$-\overleftarrow{A}$

(20)

Where A and B represent blocks of 2^{n-1} digits, the arrows indicate the orientation of the blocks, and the signs specify the relative polarity of all the digits in a given block. Complementary sequences can be synthesized recursively as shown below:

```
Initial Elements  :  +1  ) selected arbitrarily
                     +1  )

1st Expansion     :  +1 | +1 ) using expression (20)
                     +1 | -1 )
                                                        (21)
2nd Expansion     :  +1 +1 | -1 +1 )
                     +1 -1 | -1 -1 )

3rd Expansion     :  +1 +1 -1 +1 | -1 -1 -1 +1 )
                     +1 -1 -1 -1 | -1 +1 -1 -1 )
                     |
                     |
                     etc
```

It is seen that the first 2^{n-1} digits for each pair of sequences are identical with the sequences from the previous recursion.

It is also possible to synthesize orthogonal sets of complementary sequences for use as modulating waveforms. The two sets of sequences with block structures

1st sequence

2nd sequence

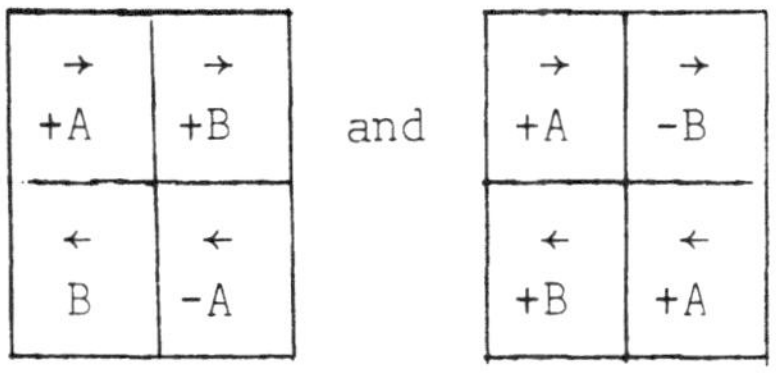

(22)

are orthogonal for all phase shifts. This can be demonstrated by computing the aperiodic crosscorrelation functions (ccfs) between the corresponding block sequences and summing for each phase shift:

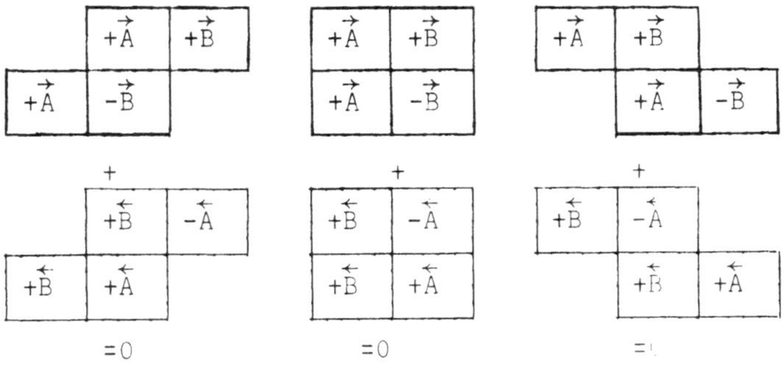

(23)

This simplification is valid since, as mentioned above, A and B themselves are a pair of complementary sequences and will only have a non-zero summed correlation when they are in phase. Thus four pairs of complementary sequences can be specified with block structures:

SET A		SET B		SET C		SET D	
$+\vec{A}$	$+\vec{B}$	$+\vec{A}$	$-\vec{B}$	$-\vec{A}$	$-\vec{B}$	$-\vec{A}$	$+\vec{B}$
$+\overleftarrow{B}$	$-\overleftarrow{A}$	$+\overleftarrow{B}$	$+\overleftarrow{A}$	$-\overleftarrow{B}$	$+\overleftarrow{A}$	$-\overleftarrow{B}$	$-\overleftarrow{A}$

(24)

which have aperiodic in-phase ccf values given by

IN-PHASE CCF	SET A	SET B	SET C	SET D
SET A	2N	0	-2N	0
SET B	0	2N	0	-2N
SET C	-2N	0	2N	0
SET D	0	-2N	0	2N

Elsewhere, all summed aperiodic values are zero.

Complementary sequences based upon the block structures given in (24) can therefore form the basis of an orthogonal state data transmission system. Other, more general, techniques for synthesizing orthogonal sets of more than two complementary sequences are described in [20].

It is proposed that sets of complementary sequences should be transmitted as shown in Fig. 12, such that there is no spectral overlap and therefore no interaction, and hence crosscorrelation, between the two sequences of a given set. At the receiver, the two sequences are processed independently, but in synchronism, via matched filters, or correlation detectors, and the outputs of the matched filters summed. Separate pairs of matched filters are necessary for each orthogonal set of sequences.

The block structure of the proposed coding scheme is particularly suited to ARQ operation. In addition, forward EDC coding could be applied over a group of blocks, with soft-decision data derived from the matched filter output levels being used to enhance the power of the coding.

It should also be noted that the summed matched filter outputs will provide a measure of the impulse response h(t) of the channel. Let the first complementary sequence be x(t) and the second y(t); the corresponding received signals are $x^1(t)$ and $y^1(t)$ respectively. Therefore, the individual input-output ccf's for the channel, $\phi_{xx^1}(\tau)$ and $\phi_{yy^1}(\tau)$, are given by [21]:

$$\phi_{xx^1}(\tau) = \int_{-\infty}^{\infty} h(u)\phi_{xx}(\tau-u)du \quad (25)$$

and

$$\phi_{yy^1}(\tau) = \int_{-\infty}^{\infty} h(u)\phi_{yy}(\tau-u)du \quad (26)$$

Summing (25) and (26) yields

$$\phi_{xx^1}(\tau) + \phi_{yy^1}(\tau) = \int_{-\infty}^{\infty} h(u) \left[\phi_{xx}(\tau-u) + \phi_{yy}(\tau-u)\right] du \quad (27)$$

In equation (27), the term on the RHS in square brackets is simply the summed acf of the complementary sequence pair, which is defined by expression (16). Hence, the summed input-output ccf is proportional to the channel impulse response. This information can be used for RTCE purposes and to optimize the processing algorithms.

5.3 Discussion

Complementary sequence encoding can potentially provide a high degree of protection for data transmitted over the dispersive HF medium. It provides redundancy simultaneously in both time and frequency and gives the optimum acf for matched filter reception purposes. Additionally, the system is essentially asynchronous and is thus directly compatible with use in variable rate transmission systems which can adapt in response to the state of the channel. As will be described in the following section, although the technique is particularly suited to spread-spectrum transmission, it can also be used for in-band signalling where the bandwidth is $\leqslant$3 kHz.

6. IN-BAND HF DATA TRANSMISSION

The data transmission systems described in the previous two sections assume that there are no specific restrictions on signal design and, in particular, bandwidth, apart from non-interference with other spectrum users. In this section, techniques applicable to HF data transmission within the limits of a 3 kHz bandwidth channel will be considered.

6.1 Flexible low-rate data modem design

STC is currently developing a flexible modem for reliable low-rate data transmission. A block diagram of the system is shown in Fig.

13. The modem is designed to operate in three main alternative modes:

(a) Mode 1: OOK, 1 to 4 channels;

(b) Mode 2: FEK, 1 or 2 channels.

(c) Mode 3: Multiple tone signalling - up to 4 from 16.

As indicated previously, much of the man-made interference in the HF band is narrowband in nature, due to the fact that the majority of the traffic is low-speed telegraphy requiring a bandwidth of only a few hundreds of Hz. In Fig. 13, the 16 bandpass filters can be used to monitor the interference levels continuously in 16 contiguous 170 Hz-wide sub-channels to provide in-band RTCE control information [22]. At the transmitter, between 1 and 4 oscillators can be selected from the library of 16 and OOK-modulated, either simultaneously or independently.

6.2 Mode 1 operation

At the receiver, up to 4 of the bandpass filters can be selected simultaneously for detection purposes. In general, if a feedback link to the transmitter is available, the corresponding transmitter oscillators can be positioned in the sub-channels exhibiting the lowest noise/interference levels, as monitored at the receiver.

OOK detectors can be applied to all 4 sub-channels and the oscillators can be keyed either independently or in parallel. Also soft-decision outputs indicating the average received energy levels in the keyed sub-channels can be used in confidence level assessments. This mode of operation might be used to implement, say, a time diversity modem with soft-decision majority logic detection as follows: 4 transmitter tones would be selected and keyed by 4 identical binary data streams, but each offset in time with respect to the others (see Section 2.1.5). At the receiver, the keyed tones are demodulated using envelope detectors, the time delays are removed so that all 4 sub-channels are now in synchronism, and a majority logic element assesses the transmitted binary states. The soft-decision energy levels can be used to allow sub-channels with low energy, due perhaps to selective fading, to be eliminated from the majority logic decision. Alternatively, the majority logic element inputs could be weighted in accordance with their average energy levels.

Simpler modes of operation include OOK with in-band frequency diversity.

6.3 Mode 2 operation

In Mode 2, either 2 or 4 sub-channels can be selected for 1- or 2-channel FEK transmission (see Section 2.1.4). At the receiver, filter-assessor demodulators can be applied after the selected filters. The same data can be transmitted simultaneously in the two channels to give quadruple in-band frequency diversity; alternatively, separate binary data streams can be transmitted on each channel. In this latter mode, the sets of complementary sequences discussed in Section 5 could be employed as modulating waveforms in an asynchronous, variable rate scheme. It should be noted that, at the lower data rates for which the modem is designed, multipath interference between successive sequence sets will no longer be significant.

By employing the inherent frequency agility of the modem, the frequency separation between the 2 tones of each of the FEK channels can be optimized for any given multipath time delay.

6.4 Mode 3 operation

The third mode of operation possible with the modem is a simple multi-tone signalling scheme in which 4 of 16 tones are selected to represent the data symbols. The total number of these selections is:

$$ {}_{16}C_4 = \frac{16!}{4!12!} = 1820 \qquad (28) $$

In practice, the number of combinations employed would tend to be limited to avoid an excessive decoding period. One convenient arrangement might ba a 4-state orthogonal encoding scheme in which the state-to-tone mapping is as follows:

$$ \left.\begin{array}{lllll} \text{State 1} \equiv \text{tones } 1 & 5 & 9 & 13 \\ \text{State 2} \equiv \text{tones } 2 & 6 & 10 & 14 \\ \text{State 3} \equiv \text{tones } 3 & 7 & 11 & 15 \\ \text{State 4} \equiv \text{tones } 4 & 8 & 12 & 16 \end{array}\right\} \qquad (29) $$

Clearly, many other state-to-tone mapping arrangements are also possible.

7. CONCLUSIONS

The main aim of this paper has been to present briefly the state-of-the-art in HF data transmission and to indicate what appear to be the most promising techniques for future exploitation. It is intended that it shall encourage the reader to think about possible solutions to the problems of transmitting data reliably over an operationally-important, but highly dispersive and time-variable path. The practical importance of the HF path is that it provides the only alternative to satellite communication for long-distance circuits involving mobiles, such as ships and aircraft.

In Sections 4 and 5, data transmission systems which are not constrained by the normal HF bandwidth limitations are discussed. Practical tests have shown that the prototype SSHF modem does not present an EMC problem since it causes negligible interference to other spectrum users. It would appear that the synchronization difficulties associated with the SSHF modem in a multipath environment can be overcome by the use of spectrum-spreading sequences with the appropriate aperiodic correlation properties.

The flexible in-band ($\leqslant$ 3 kHz bandwidth) modem described in Section 6 can be implemented simply and economically; it is capable of operating in a variety of modes, according to the state of the channel and the characteristics of the data source.

Two major considerations must be taken into account before efficient and effective transmission systems, matched to the characteristics of the HF channel, can be designed:

(a) The information transmission capacity of the HF channel can vary rapidly with time over a wide range; the constant-rate systems currently in use are thus fundamentally mismatched to the path properties.

(b) Efficient HF signal generation and processing requires a knowledge of the state of the channel at any time. Therefore, it is essential that RTCE techniques are incorporated into the communication system to provide the control data necessary for system adaptation in response to changes in path characteristics.

REFERENCES

1. M. Darnell, Channel Evaluation Techniques for Dispersive Communication Paths, presented at the NATO ASI on Communication Systems and Random Process Theory, Darlington, August 1977.
2. D. Bayley & J.D. Ralphs, Piccolo 32 - Tone Telegraph System in Diplomatic Communication, Proc IEE, Vol 119, No 9, September 1972, pp 1229-1236.
3. R.E. Schemel, An Assessment of Piccolo, a 32-Tone Telegraph System, SHAPE Technical Centre Memorandum STC TM-337, July 1972.
4. Research Report No 20879, The Performance of a 32-Tone Telegraph System subjected to Noise and Fading, Eng Dept, GPO(UK), 1964.
5. P.K. Ridout & L.K. Wheeler, Choice of Multi-Channel Telegraph Systems for Use on HF Radio Links, Proc IEE, Vol 110, No 8, August 1963, pp 1402-1410.
6. H.C.A. Van Duuren, Typendruktelegrafie over Radioverkindingen (TOR), Tijdschr Nederlands Radiogenoot, Vol 16, No 2, March 1951, pp 53-67.
7. P.R. Keller, An Automatic Error Correction System for Unidirectional HF Teleprinter Circuits, Point-to-point Telecommunications, Vol 7, No 3, June 1963, pp 14-29.
8. M.L. Doelz, E.T. Heald & D.L. Martin, Binary Data Transmission Techniques for Linear Systems, Proc IRE, Vol 45, No 5, Pt 1, May 1957, pp 656-661.
9. D. Chase, A Combined Coding and Modulation Approach for Communication over Dispersive Channels, IEE Trans, Vol COM-21, No 3, March 1973, pp 159-174.
10. B. Goldberg, M.J. Di Toro & J. Hanulec, Design and Performance of a Simulated Time-Variable Multipath HF Link, IEEE Annual Communications Conf Record, Boulder, June 1965, pp 769-773.
11. M.M. Goutmann & M.M. Gutman, High Rate Serial Transmission Experiment over the HF Channel, NTC, 1974, pp 750-754.
12. R.M.F. Goodman & P.G. Farrell, Data Transmission with Variable Redundancy Error Control over a High Frequency Channel, Proc IEE, Vol 122, No 2, February 1975.
13. J.D. Ralphs, The Limitations of Error Detection Coding at High Error Rates, Proc IEE, Vol 118, No 314, March/April 1971, pp 409-416.
14. S. Goldman, Information Theory, Constable, 1953, pp 158-161.
15. J.T. Lynch, R.B. Fenwick & O.G. Villard, Measurement of the Best Time Delay Resolution obtained along E-W and N-S Ionospheric Paths, J. Rad Sci, Vol 7, No 10, October 1972.
16. R.H.Barker, Group Synchronizing of Binary Digital Systems, in "Communication Theory", Butterworth, London, 1953 pp 273-287.
17. D.A. Huffman, The Generation of Impulse-Equivalent Pulse Trains, IRE Trans, Vol IT-8, September 1962, pp S10-S16.

18. H.B. Mann (Editor), Error Correcting Codes, Wiley, 1968, pp 195-225.
19. M.J.E. Golay, Complementary Series, IRE Trans, Vol IT-7, April 1961, pp 82-87.
20. M.Darnell, Principles and Applications of Binary Complementary Sequences, presented at Symposium on "Walsh and other Non-Sinusoidal Functions", Hatfield, July 1975.
21. Y.W.Lee, Statistical Theory of Communication, Wiley, 1960. pp 323-351.
22. Pye/RAE Combine for FSK Demodulator, Communications International, Vol.1, No.1, 1974, p.10.

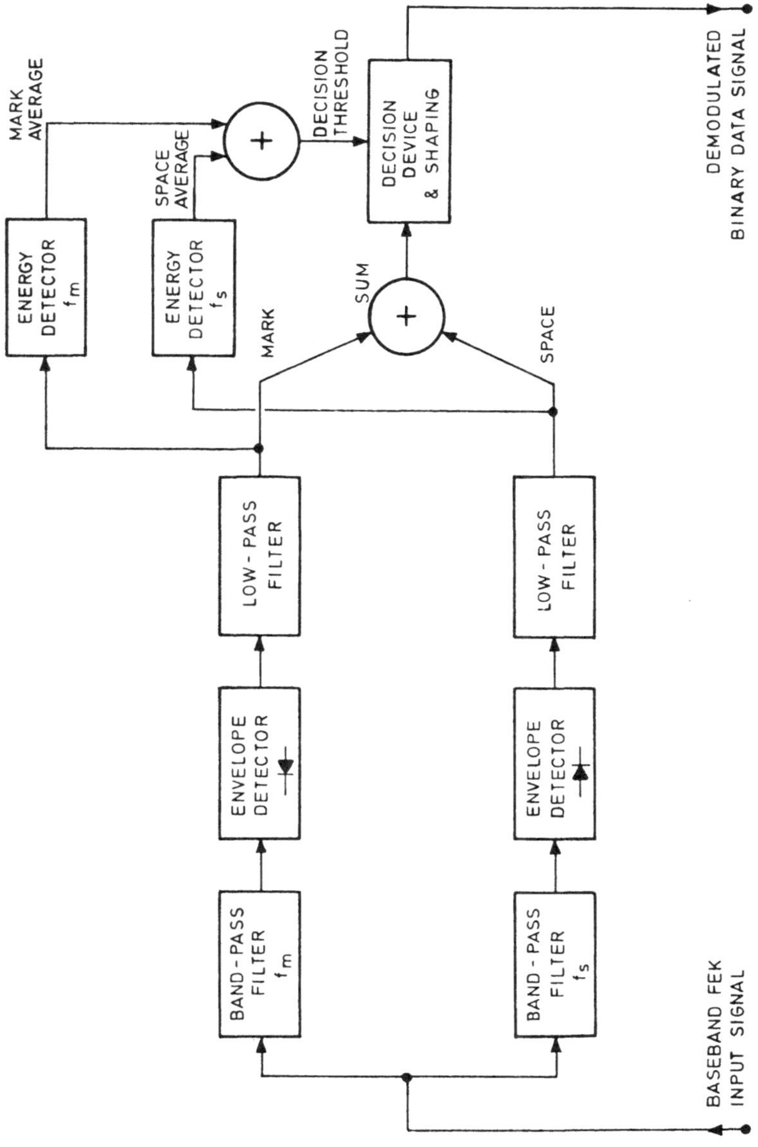

Fig. 1. Binary FEK detection using filter-assessor technique

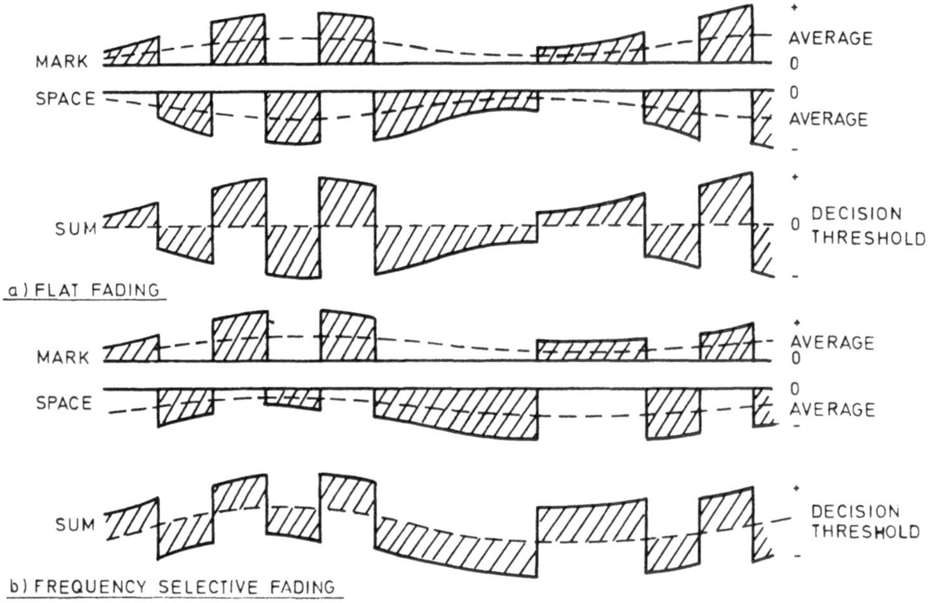

Fig. 2. Principle of operation of filter-assessor binary FEK detector

10^{-1}
10^{-2}
10^{-3}
10^{-4}
10^{-5}
$\bar{P}_c$
A: Piccolo (path time
difference up to 4 ms)
B1: Binary FEK (path time
differences 0 and 1 ms) -
5-unit synchronous
B2: Binary FEK (path time
differences 2 and 3 ms) -
5-unit synchronous
C: Binary FSK (path time
difference 1 ms) - 5-unit synchronous;
shift = 0.8 baud rate
B1
A
B2
C
10
15
20
25
30
35
40
E_c/N_o (dB)

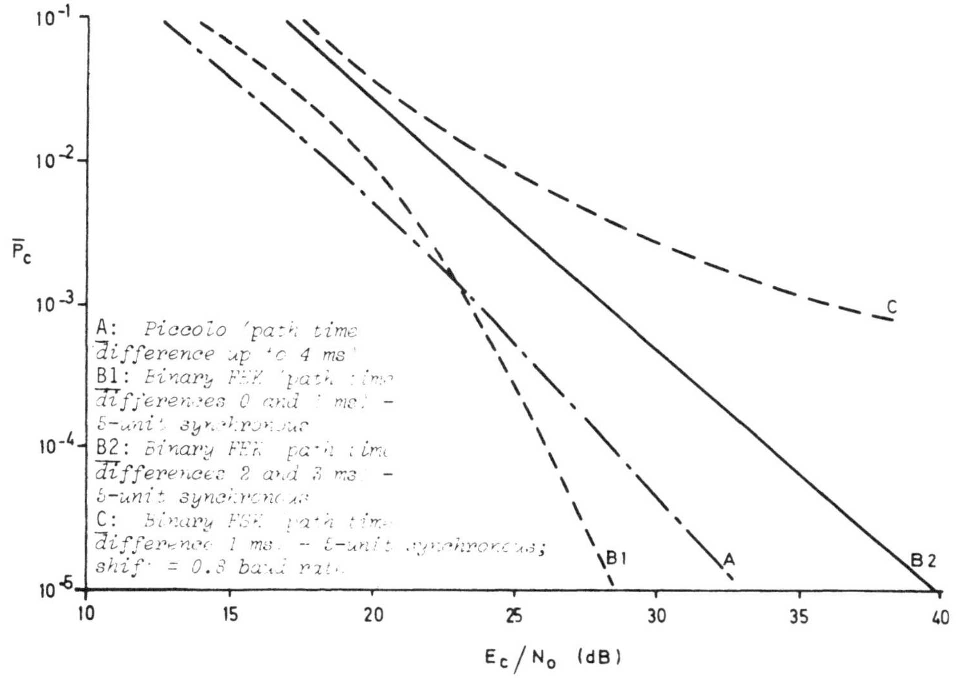

Fig. 3. Measured performance of Piccolo, binary FEK and binary FSK data transmission systems

```
-0007            5¾3 178:( ?492, ¼9/ ର7.0'8?eg hhenlazyndog01235¼?/
     fadtkaklsmhmapct, ¼./,-7=0:'9=-:'9¾3 )2?6 ⊕½½0123456789?
0008sadlctest hyepquick brogn koxijumps over thb lazy dgv0213¼56789
ttf09ujdmphestctrzlvfajgx
         ovej ope lazy dog0123456789
0010sadtctest thesquick brown fox jumps over the lazy dog0123456789
0011sadtctest the quick brown fox jumps over the lazy dog0123456789
0012sadtctest the quick brown fox jumps over thz ljzy hgljrqwertyuio
ppqesadtcozst tpe quick brown fox jumps over the lazy dog0123456789
0014sadtctest the quick brown fox jumps over the lazyhbog=:83496789
0015sadtctest the quick brown foxhjuvps over the lazy dog0123456789
0016sadtctest the quick brown fox jumps over the lazy dog0123456789
0017sadtctest the quick brown fox jumps over the lazy dog013456789
0018sadtctest zhe quick brown fox jumps xer the lazyidog0123456789
0019sadtctest the quick brown fox jumps bver oqe lazy dog0123456789
0000$sadddcest the quick brown fox jumps over the lazy dog0123456789
          s
           t lhdsquick brgwn foxmjuvps over hye lazy dog0123456789
0022sadtctesg the quick brown fox jumps bvzr ohe lazy dog012+ର56789
0023sadtktdsxntheiquukk brmwn fox jumps mv
                                         r ohxqlazy dgg012345/789
0024ର7dtctbst4t5¾3quick b42


                          ⊕9½012349670y
ppwtsjjtvlzst thdhquick brownifm jumpfcmf/¾⊕'½-+6 ?9½0¼(((¼:¼'
                          3                                    77¼/62?
07:7'::ର½09/,ର7=ifvqmymheip61:?¾½01234?677.¾/)0027sadtctesg bhvgqkla
```

a) START - STOP

```
0007sadtctest the quick brown fox jumps over the lazy dog0123456789
0008sadtctest the quick brown fox jumps over the lazy dog0123456789
0009sadtctest the quick brown fox jumps over the lazy dog0123456789
0010sadtctest the quick brown fox jumps over the lazy dog0123456789
0011sadtctest the quick brown fox jumps over the lazy dog0123456789
0012sadtctest the quick brown fox jumps over the lazy dog0123456789
0013sadtctest the quick brown fox jumps over the lazy dog0123456789
0014sadtctest the quick brown fox jumps over the lazy dog0123456789
0015sadtctest the quick brown fox jumps over the lazy dog0123456789
0016sadtctest the quick brown fox jumps over the lazy dog0123456789
0017sadtctest the quick brown fox jumps over the lazy dog0123456789
0018sadtctest the quick brown fox j_mps over the lazy dog0123456789
0019sadtctest the quick brown fox jumps over the lazy dog0123456789
0020sadtctest the quick brown fox jumps over the lazy dog0123456789
0021sadtctest the quick brown fox jumps over the lazy dog0123456789
0022sadtctest the quick brown fox jumps over the lazy dog0123456789
0023sadtctest the quick brown fox jumps over the lazy dog0123456789
0024sadtctest the quick brown fox jumps over the lazy dog0123456789
0025sadtctest the quick brown fox jumps over the lazy dog0123456789
0026sadtctest the quick brown fox jumps over the lazy dog0123456789
```

b) FEC

Fig. 4. Comparison of teleprinter copy obtained using start-stop telegraphy and FEC.
Oslo - The Hague, June 1976, FEK, 100 W.

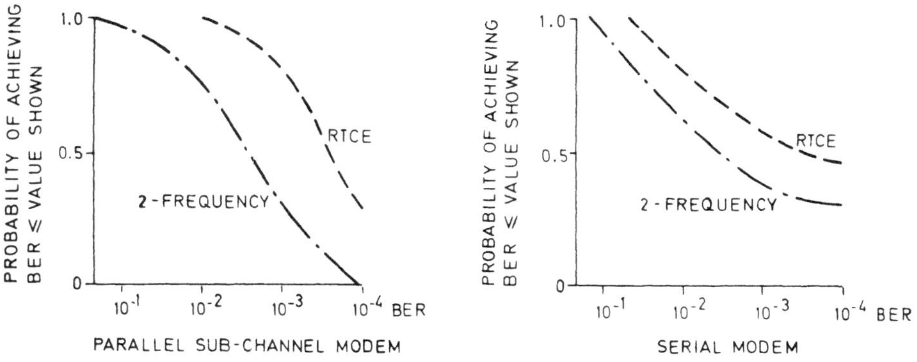

Fig. 5. Probabilities of achieving a given BER for parallel sub-channel and serial medium-speed data modems. Oslo - The Hague, 1200 bit/s.

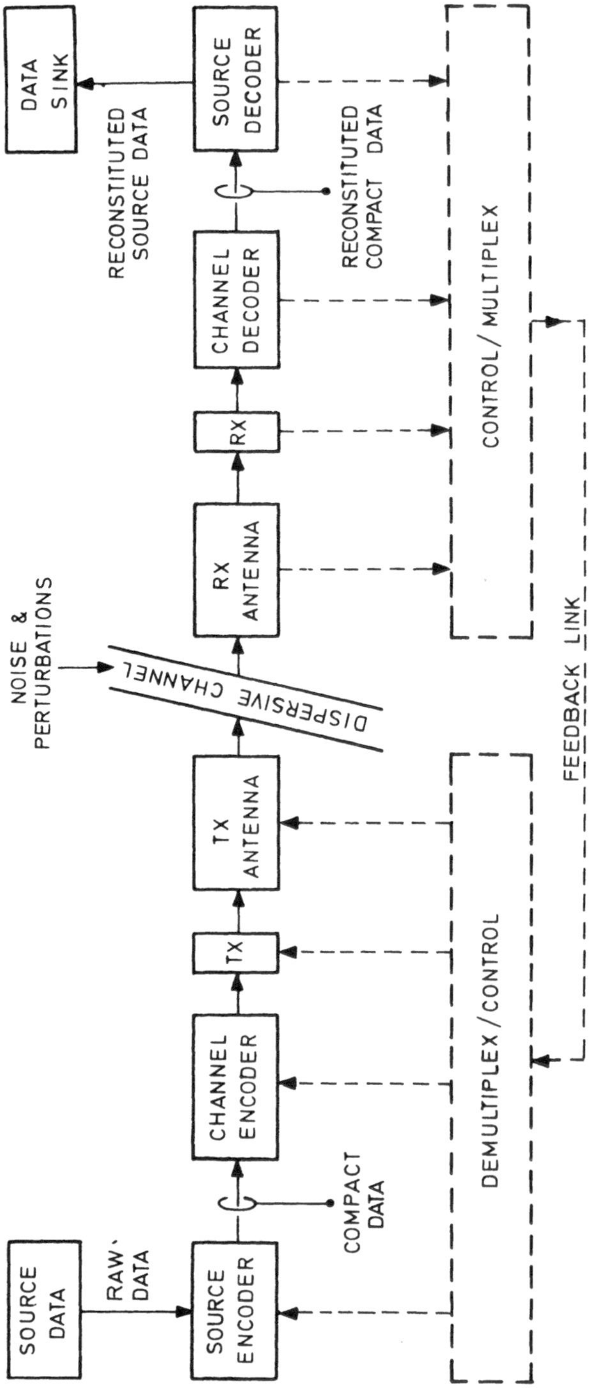

Fig. 6. Elements of a generalized communication channel

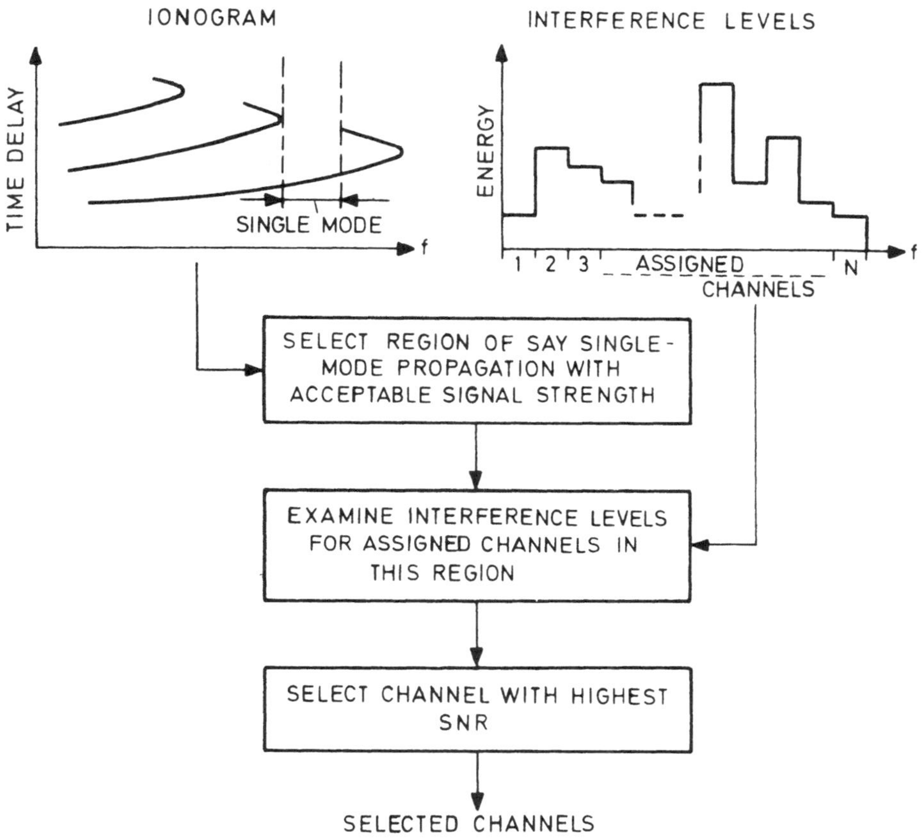

Fig. 7. Channel selection algorithm

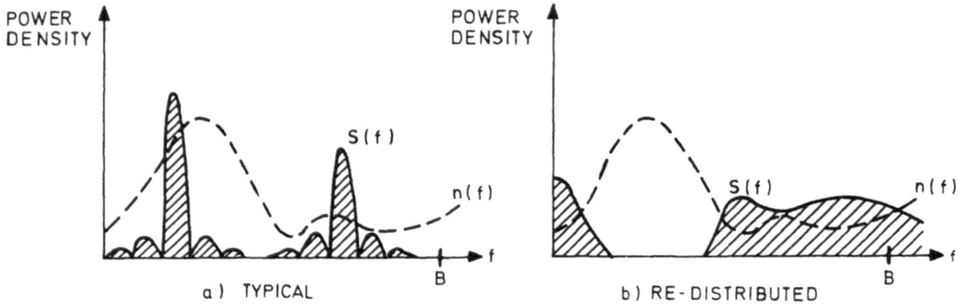

Fig. 8. Signal and noise power density spectra

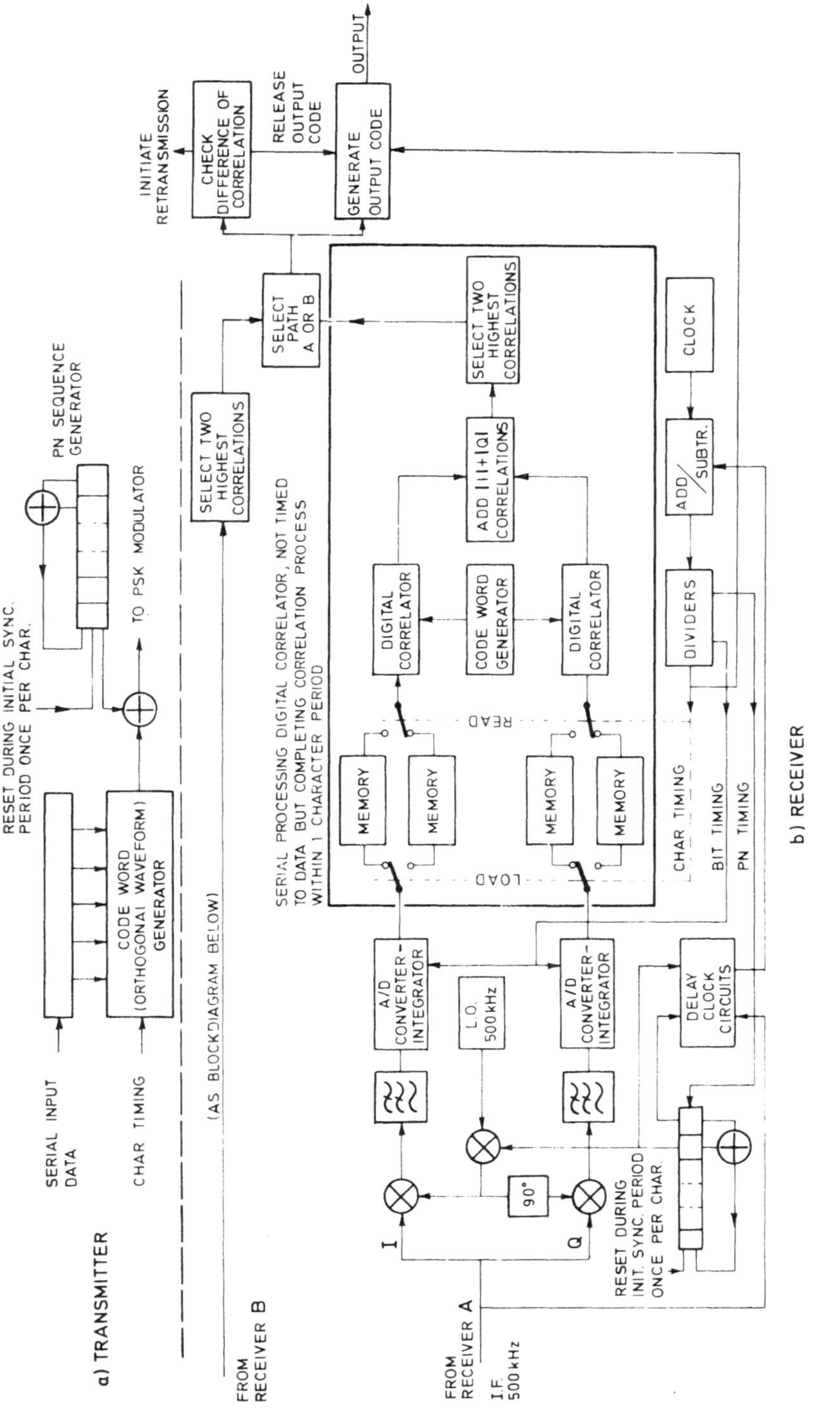

Fig. 9. Functional block diagram of equipment

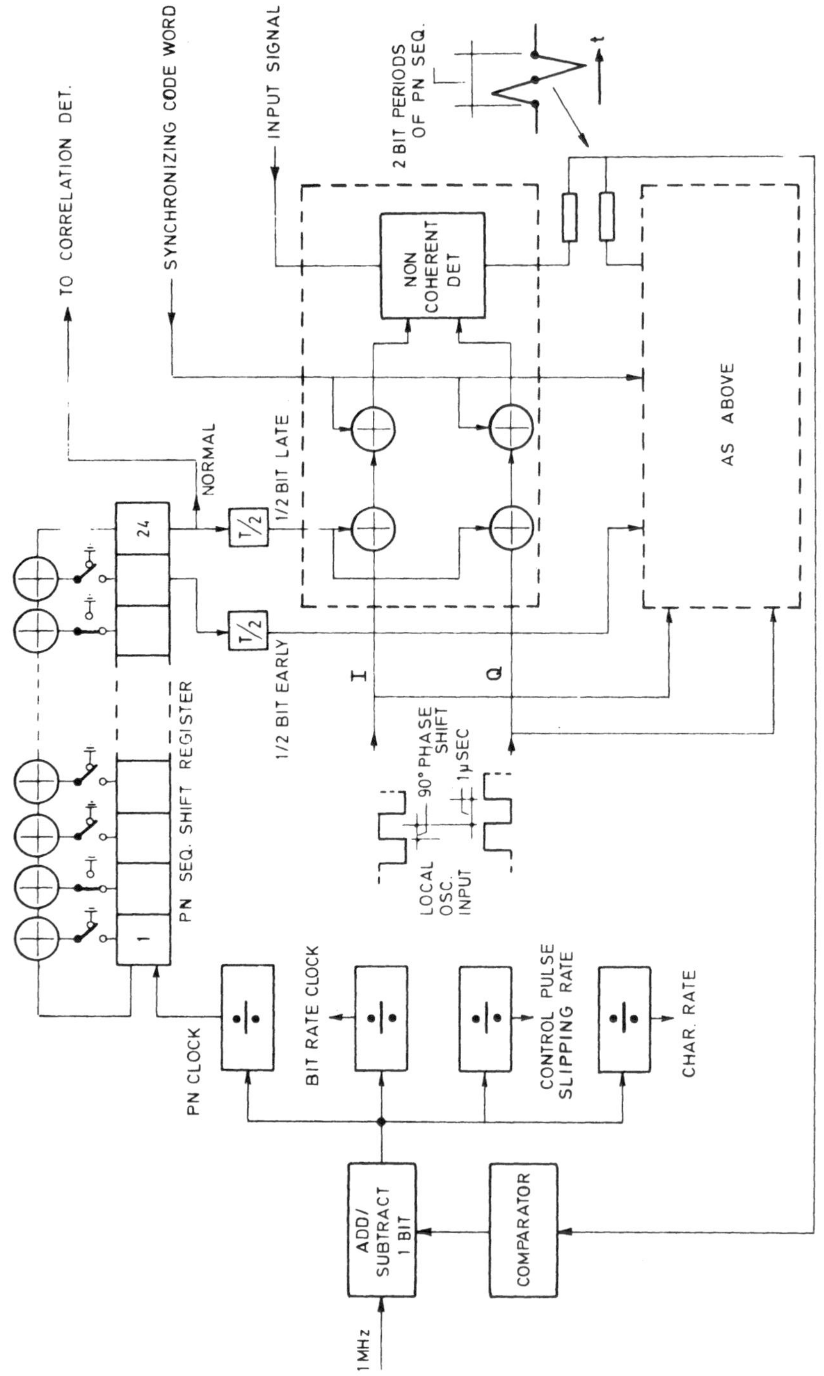

Fig. 10. Synchronizing circuits

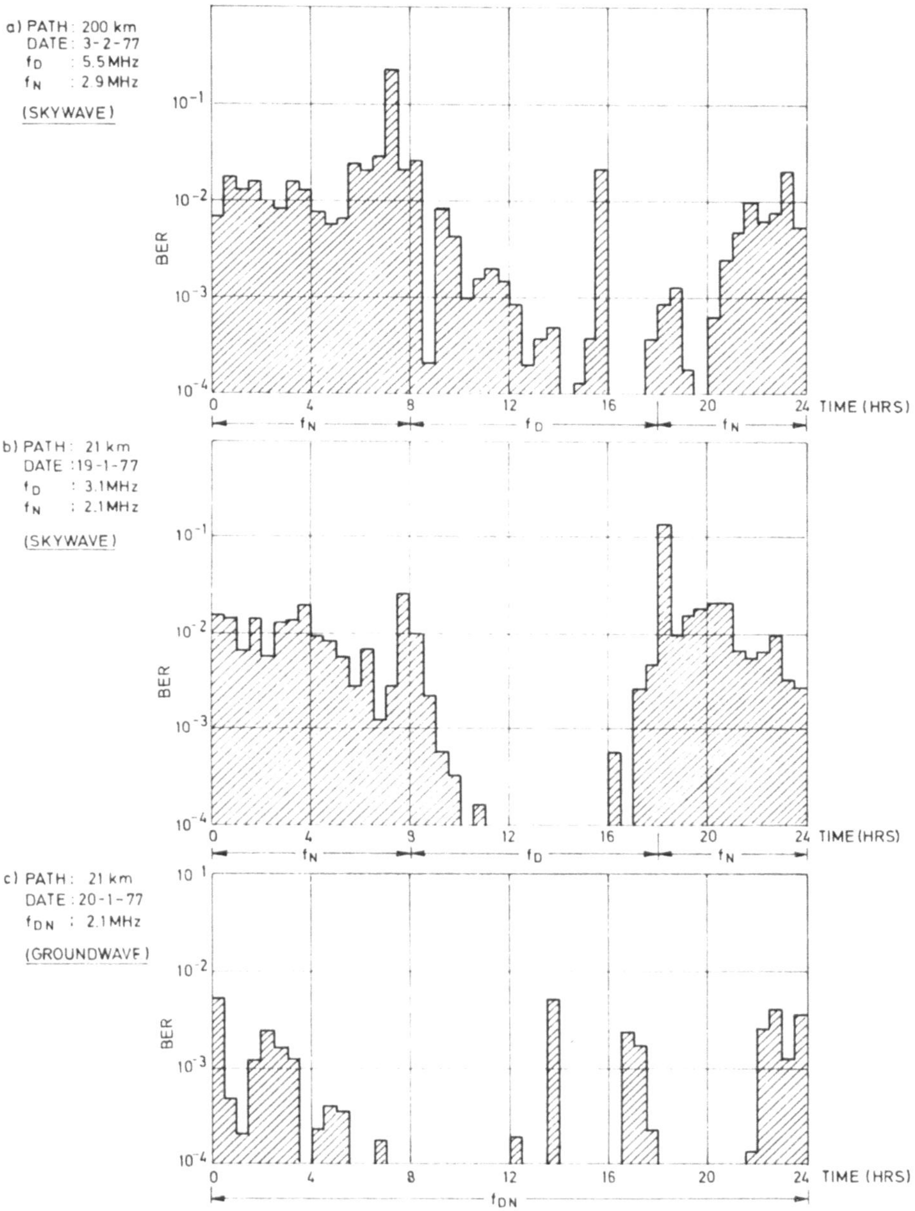

Fig. 11. BER performance of SSHF modem over 24-hour periods.
Date rate 75 bit/s PN rate 20 kbit/s;
Tx power 5 - 10 W.

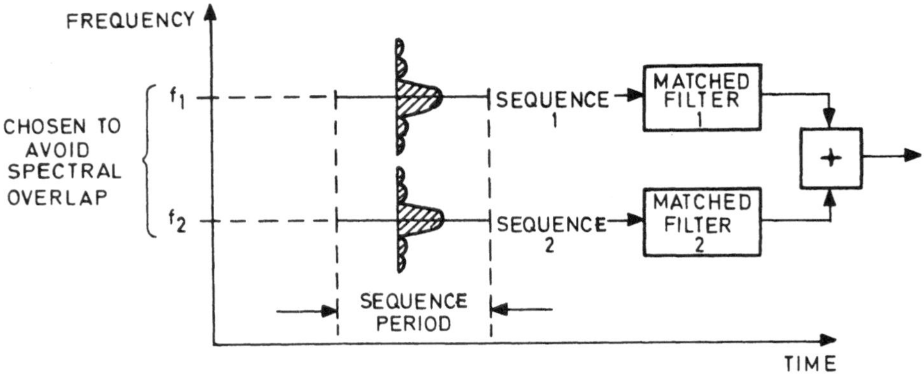

Fig. 12. Complementary sequence transmission mode

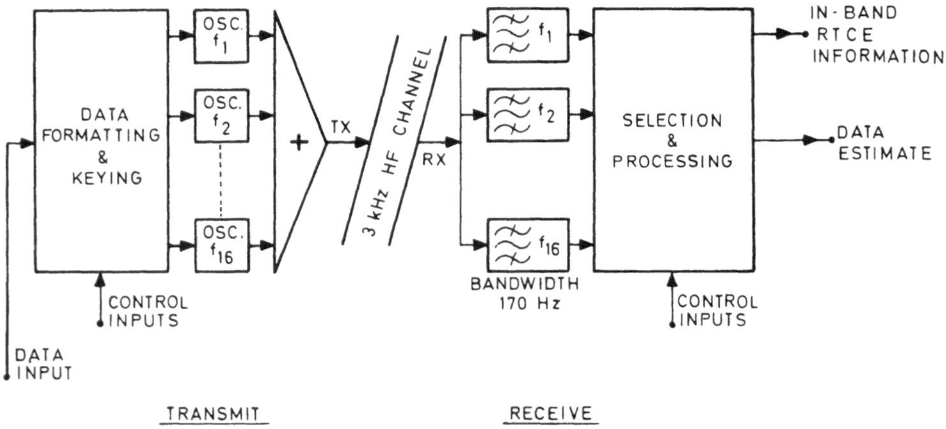

Fig. 13. Block diagram of flexible HF low-rate data modem

SPEECH PROCESSING FOR LOW DATA RATE DIGITAL VOICE COMMUNICATIONS

E.V. Stansfield

Communications Division, SHAPE Technical Centre,
The Hague, Netherlands.

ABSTRACT. The research being carried out at STC in the field of narrow band digital speech encoding is described. An important prerequisite for this research was an understanding of the results achieved to date by other workers, particularly those aspects which cause inferior performance in an operational environment. The paper firstly indicates the general trends of digital speech communications research, highlighting the problem areas, and then discusses the approaches being taken at STC to overcome these problems.

1. INTRODUCTION

The motivation for narrow band digital speech encoding research was originally the provision of secure voice communications for the diplomatic services and the armed forces. Today, the possibility of significant bandwidth economies in digital satellite and submarine cable communications links has provided additional motivation. Wideband techniques such as pulse code modulation (PCM) and delta modulation (DM) have not provided good enough speech quality for either military or civilian networks when operated at low data rates, whereas the speech quality obtained with narrow band techniques such as channel vocoding and linear predictive coding (LPC) has been accepted by the armed forces but not by civilian telephone administrations. In addition, wideband techniques have tended to involve much lower equipment costs than narrow band techniques, and economics have therefore dictated that wideband PCM or DM be employed in civilian digital networks. However, recent advances in narrow band techniques promise much improved speech quality, and the rapid pace of development in semiconductor technology promises cost-effective implementation for future systems.

In this paper, wideband encoding techniques are first surveyed, and their unsuitability for low data rates discussed. Narrow band encoding techniques are then described, with emphasis on the assumptions made in their derivation and on the problems encountered in practice. Finally, the speech quality versus data rate and cost trade-offs involved in hardware implementation are discussed, with reference to recent developments in semi-conductor technology.

The active areas of research at STC are indicated in the relevant sections. Briefly, these are the problems associated with reliable voiced-unvoiced-silence decisions and pitch detection, the development of new techniques in spectrum vocoding, and the elimination of assumptions currently made in linear prediction modelling of speech waveforms. Methods for improving the performance of speech coders in noisy acoustic environments, and in the presence of transmission errors, are also being investigated. A prime consideration in this research is cost-effective hardware implementation making maximum use of the features offered by current and future large scale integrated circuits, including micro-processors. The ultimate objective is to achieve good quality secure voice communications at data rates in the range from 8 kbit/s to 2.4 kbit/s and below.

2. METHODS OF SPEECH DIGITISATION

Wideband encoding techniques such as PCM and DM are essentially waveform coders, since they endeavour to reproduce as accurately as possible the amplitude-time characteristics of the original speech. They do not make use of any special speech characteristics in their operation, and can be applied to almost any kind of band-limited signal.

On the other hand, narrow band speech encoding techniques such as channel vocoding and linear predictive coding (LPC) depend on the presence of redundancy and do utilize the characteristics of speech for their operation. The speech signal is analysed at the transmitter into parameters requiring lower transmission bandwidth than the original speech, and synthesized from these parameters at the receiver. Analysis-synthesis telephony can thus only be applied to speech-like signals.

The important factors which need to be taken into consideration when comparing different encoding techniques are the speech quality achievable in the presence of both transmission errors and acoustic noise, the data rate required for transmission, the cost of implementation (a function of coder complexity), and the physical size of the equipment.

2.1 Waveform coding

The most basic type of waveform coding is pulse code modulation (PCM) consisting of sampling (usually at 8 kHz), quantising to a finite number of levels, and binary encoding. The quantiser can have either uniform or non-uniform steps giving rise to linear and logarithmic PCM respectively. Log-PCM has a much wider dynamic range than linear PCM for a given number of bits per sample, because low amplitude signals are better represented, and as a result logarithmic quantisation is nearly always used in wideband speech communications applications. A data rate of 56 to 64 kbit/s is required for commercial quality speech, and 40 to 48 kbit/s for military tactical quality.

There are many variations on the basic PCM idea, the most common being differential encoding and adaptive quantisation. Each variation has the object of reducing the data rate required for a given speech quality, a saving of approximately 1 bit per sample (8 kbit/s) being achieved when each is optimally employed. In differential PCM (DPCM) the sampled speech signal is compared with a locally decoded version of the previous sample prior to quantisation so that the transmitted signal is the quantised difference between samples. In adaptive PCM (APCM) the quantiser gain is adjusted to the prevailing signal amplitude, either on a short term basis or syllabically. By controlling the adaption logic from the quantiser output, the quantiser gain can be recovered at the receiver without the need for additional information to be transmitted. Adaptive differential PCM (ADPCM) is a combination of DPCM and APCM which saves approximately 2 bits per sample compared with PCM, thus requiring 40 to 48 kbit/s for commercial quality speech and 24 to 32 kbit/s for tactical quality.

When the number of quantisation levels in DPCM is reduced to two, delta modulation (DM) results. The sampling frequency in this case is equal to the data rate, but it has to be well above the Nyquist frequency to ensure that the binary quantisation of the difference signal does not produce excessive quantisation noise. Just as with PCM, there are many variations of DM, and the right hand side of Fig. 1 illustrates some of them. The most important form of DM used in digital speech communications is syllabically companded DM; there are a number of closely related versions of this, examples being continuously variable slope DM (CVSD) and digitally controlled DM (DCDM). The data rate requirements are a minimum of about 16 kbit/s for military tactical quality speech and about 48 kbit/s for commercial quality.

When operated at data rates of 12 kbit/s and lower, the speech quality obtained with PCM and DM coders is poor, and consequently they cannot be used as narrow band devices. However, the principles of operation of wideband coders are useful in analysis-synthesis telephony once significant redundancy has been removed from the speech waveform. Examples of this are digital encoding for the transmission of individual speech parameters, and the relationship between LPC and DPCM indicated in Fig. 1.

2.2 Analysis-synthesis telephony

Analysis-synthesis telephony techniques are based on a model of speech production. Figure 2a shows a lateral cross-section through the human head, and illustrates the various organs of speech production. Briefly, these are the vocal tract running from the vocal chords at the top of the larynx to the mouth opening at the lips, and the nasal tract branching off the vocal tract at the velum and running to the nose opening at the nostrils. The glottis (the space between the vocal chords) and the sub-glottal air pressure from the lungs together regulate the flow of air into the vocal tract, and the velum regulates the degree of coupling between the vocal and nasal tracts (i.e., the nasalisation).

There are two basic types of speech sound which can be produced, namely voiced and unvoiced sounds. Voiced sounds occur when the vocal chords are tightened in such a way that the sub-glottal air pressure forces them to open and close quasi-periodically, thereby generating "puffs" of air which acoustically excite the vocal cavities. The pitch of voiced sounds is simply the frequency at which the vocal chords vibrate. On the other hand, unvoiced sounds are produced by forced air turbulence at a point of constriction in the vocal tract, giving rise to a noise-like excitation, or "hiss".

A model of speech production often used for the design of analysis-synthesis vocoders is shown in Fig. 2b. In this model, a number of simplifications have been made, the most important ones being that the excitation source for both voiced and unvoiced sounds is located at the glottis, that the excitation waveform is not affected by the shape of the vocal tract, and that the nasal tract can be incorporated by suitably modifying the vocal tract. These simplifications lead to differing subjective effects, depending on the type of speech sound and the particular vocoder being used.

In channel vocoding the speech is analysed by processing through a bank of parallel band pass filters, and the speech amplitude in each frequency band is digitized using PCM techniques. For synthesis, the vocal and nasal tracts are represented by a set of controlled gain, lossy resonators, and either pulses or white noise are used to excite them. In pitch-excited vocoders, the excitation is explicitly derived in the analysis, whereas in voice-excited vocoders it is derived by non-linear processing of the speech signal in a few of the low frequency channels combined into one. Pitch-excited vocoders require data rates in the range from 1200 to 2400 bit/s and yield poor quality speech, whereas voice-excited vocoders will provide reasonable speech quality at 4800 bit/s and good quality at 9600 bit/s.

A formant vocoder is similar to a channel vocoder, but has the fixed filters replaced by formant tracking filters. The centre frequencies of these filters along with the corresponding speech formant amplitudes are the transmitted parameters. The main problem is in acquiring and maintaining lock on the relevant spectral peaks during vowel-consonant-vowel transitions, and also during periods where the formants become ill-defined. The data rate required for formant vocoders can be as low as 600 bit/s, but the speech quality is poor. The minimum data rate required to achieve good quality speech is about 1200 bit/s, but to date this result has only been obtained using semi-automated analysis with manually interpolated and corrected formant tracks.

The third method of analysis-synthesis telephony to have achieved some importance is linear predictive coding. In this technique the parameters of a linearised speech production model are estimated using mean square error minimisation procedures. The parameters estimated are not acoustic ones as in channel and formant vocoders, but articulatory ones related to the shape of the vocal tract. For a given speech quality, a transmission data rate reduction in comparison with acoustic parameter vocoding should be achieved because of the lower redundancy present. Just as with channel and formant vocoders, excitation for the synthesizer has to be derived from a separate analysis, the usual terminology being pitch-excited or residual excited, corresponding to pitch or voice excitation in a channel vocoder. LPC is a very active area of speech research, and new results appear regularly. At present, data rates as low as 2400 bit/s have been achieved for pitch-excited LPC with reasonable quality speech, and in the range from 9.6 kbit/s to 16 kbit/s for residual excited LPC with good speech quality. STC is currently working in this area also, as discussed in Section 4.

2.3 Discussion

The more important speech digitisation techniques have now been briefly surveyed, but it should be noted that many other possibilities exist (Fig. 1). In general, each basic technique will have a number of variations; for example, the adaption logic in APCM, ADPCM and companded DM could be implemented either pitch asynchronously or pitch synchronously. The possibilities are numerous, but there are a number of general conclusions which one can make. Firstly, wideband waveform coding techniques are cheap to implement compared with narrow-band analysis-synthesis techniques. Secondly, wideband techniques do not provide good speech quality when modified to operate at data rates below about 12 kbit/s. Thirdly, the speech quality achieved to date with analysis-synthesis techniques at data rates of 2400 bit/s or less is far from being commercially acceptable. These three basic conclusions are the main motivation for current research effort on low data rate speech encoding.

An important prerequisite for performing this research is an understanding of just why attempts to date have proved unsuccessful. Many explanations can be given as to why particular types of speech coder do not perform well at low data rates. With waveform coders, it is generally accepted that the main reason is excessive quantization noise despite companding and/or adaptive logic. With analysis-synthesis techniques, the main reasons are over-simplification of the vocal tract model, leading to imprecise spectral characterization, and unreliable pitch detection and voiced-unvoiced-silence decisions in the analyser which, coupled with an over-simplified excitation model in the synthesizer, lead to imprecise temporal characterisation and a lack of naturalness in the synthetic speech. STC is studying all these aspects of narrow-band techniques in an attempt to achieve improved performance. At the present time, our special interest is a new design of spectrum vocoder avoiding some of the above problems, and techniques for improving existing LPC vocoders.

3. REQUIREMENTS FOR FUTURE SYSTEMS

Present day switched communications networks tend to contain a mixture of FDM analogue voice channels and TDM digital voice channels. This state of affairs is due to the current evolutionary trend from the all-analogue networks of the past to the all-digital networks of the future. The requirements for digital speech encoding during the transition phase must therefore be such as to permit working with either digital or analogue transmission media. The civilian telecommunications authorities have tackled this problem by implementing 64 kbit/s companded PCM voice channels with space-division base-band switching for all their digital plant, since this gives performance comparable to the established FDM analogue 3 kHz voice channels plant. In the military arena, however, the voice channel data rates for digital transmission vary from nation to nation, examples being 16, 32 and 48 kbit/s. In the NATO environment, this poses interoperability problems, particularly when encryption is involved and parts of a circuit connection have to pass over FDM 3 kHz analogue channels as well as TDM digital channels.

A speech encoder data rate of 8000 bit/s will be compatible with both digital and analogue transmission media, but will involve fairly sophisticated modems for the analogue sections. On the other hand, a lower encoder data rate of, say, 2400 bit/s will enable relatively simple modems to be used for the analogue sections, and permit three analogue voice channels to be multiplexed into each available 8 kbit/s digital channel, thereby making economic use of the digital plant.

In the military environment there is a further important factor which needs to be considered, namely the use of HF radio links. Reliable digital transmission over HF radio requires the use of specially-designed modems, but even so the error rates tend to become unacceptable for data rates in excess of about 1200 to 1800 bit/s. From the point of view of logistics and equipment standardisation, the speech encoding technique chosen for future military systems should therefore be capable of operating not only over the switched network at the selected data rate, but also over HF radio links at data rates of 1200 bit/s or less with a reduced but acceptable speech quality.

In determining the cost-effectiveness of particular implementations, all costs must be taken into consideration and this includes both the required modem costs and the speech encoder costs. The cost of speech encoders tends to increase with decreasing data rate, whereas the opposite tends to be true for voice band modems. There is thus a trade-off between encoder and modem costs in deciding the most cost-effective data rate. Until recently, limitations in the state-of-the-art inhibited such trade-offs being made in practice, but new developments in LPC, in spectrum vocoding, in modems, and in semi-conductor technology have changed this. The current status is very fluid, however, and as a result the likely overall implementation costs must now be considered in parallel with theoretical developments and feasibility studies.

As has already been mentioned, STC is currently working on spectrum and LPC vocoding. A new design of spectrum vocoder is at a very early stage of development, and any definitive statements regarding its performance and costs would be premature. However, LPC vocoders are more established and our thinking in this area is covered next.

4. LINEAR PREDICTIVE CODING

Linear predictive coding of speech waveforms is based on an articulatory model of the speech production organs. The vocal tract is generally modelled as a concatenation of contiguous, lossless, constant cross-sectional area acoustic tubes. The driving source for voiced sounds is assumed to be equivalent to a piston vibrating at the pitch frequency in the glottal opening of the vocal chords; it is known that the volume velocity of this source has a gross spectral characteristic approximately proportional to the inverse of the square of the frequency. The lip termination is assumed to be equivalent to a piston vibrating in a spherical baffle, and to a first order this approximates to an inductive load having a spectral characteristic proportional to frequency. The speech pressure waveform at the lips thus has a gross spectral characteristic proportional to the vocal tract volume velocity transfer function weighted - 6 dB per octave. This fact is utilised in LPC to determine from the speech waveform the reflection coefficients at the junctions between the sections of the vocal tract model and it is these reflection coefficients which are the transmitted articulatory parameters in LPC vocoders.

4.1 Basic theory

Space does not permit a detailed treatment of LPC theory, but the following will indicate the essential points and the approximations which are generally made. By considering the propagation of forward and backward travelling components of volume velocity in the piecewise constant lossless acoustic tube model, the transmission matrix of the vocal tract can be derived in z-transform notation as

$$\begin{pmatrix} U_1(z) \\ V_1(z) \end{pmatrix} = H_M(z) \begin{pmatrix} A_M(z) & -z^{-M}B_M(z^{-1}) \\ -B_M(z) & z^{-M}A_M(z^{-1}) \end{pmatrix} \begin{pmatrix} U_{M+1}(z) \\ V_{M+1}(z) \end{pmatrix} \quad (1)$$

where $z = \exp(j\omega T)$ with T = twice the propagation time along each section of the model.

$U_1(z)$, $V_1(z)$ = forward, backward components of volume velocity at the glottis

$U_{M+1}(z)$, $V_{M+1}(z)$ = forward, backward components of volume velocity at the lips.

$H_M(z)$ = $z^{\frac{1}{2}M} / \prod_{i=1}^{M} (1-k_i)$

k_i = reflection coefficient at the ith junction

and $A_M(z)$ and $B_M(z)$ can be generated from the k_i i=1...M via

$$A_m(z) = A_{m-1}(z) + k_m z^{-m} B_{m-1}(z^{-1})$$

$$B_m(z) = B_{m-1}(z) + k_m z^{-m} A_{m-1}(z^{-1})$$

with initial conditions $A_o(z) = 1$ and $B_o(z) = 0$.

To derive the vocal tract transfer function G(z) from the transmission matrix, boundary conditions must be defined. At the glottis, continuity of volume velocity dictates that

$$U_1(z) + V_1(z) = U_G(z)$$

where $U_G(z)$ is the volume velocity of the excitation source. At the lips it is assumed that the backward component $V_{M+1}(z)$ is zero - a reasonable assumption, since in a normal talking environment the speech signals reflected from the floors and walls (i.e., those contributing to the backward component at the lips) will be small compared with the speech volume velocity $U_L(z) = U_{M+1}(z)$ radiated forward from the lips. With these boundary conditions, the volume velocity transfer function G(z) can be derived as

$$G(z) = U_L(z)/U_G(z) = 1/H_M(z)P_M(z) \tag{2}$$

where $P_M(z) = B_M(z)-A_M(z)$ can be recursively generated from the k_i i=1...M via

$$P_m(z) = P_{m-1}(z) - k_m z^{-m} P_{m-1}(z^{-1}) \tag{3}$$

with the initial condition $P_o(z) = -1$.

This expression for the vocal tract transfer function is next related to the speech pressure waveform s(t). With the radiation impedance at the lips defined as $Z_L(\omega)$, the Fourier spectrum $S'(\omega)$ of the differentiated speech waveform $s'(t) = d\{s(t)\}/dt$ is given by

$$S'(\omega) = j\omega S(\omega) = j\omega U_L(\omega)Z_L(\omega) = j\omega U_G(\omega)G(\omega)Z_L(\omega)$$

Defining a new function $E(\omega)$ as

$$E(\omega) = j\omega Z_L(\omega)U_G(\omega)/H_M(\omega) \tag{4}$$

the expression for $S'(\omega)$ becomes

$$S'(\omega) = E(\omega)G(\omega)H_M(\omega) \tag{5}$$

Reverting to z-transform notation and substituting for G(z) then yields

$$S'(z) = E(z)/P_M(z) \tag{6}$$

which can be expressed in the sampled time domain as

$$e(nT) = s'(nT) - \sum_{i=1}^{M} p_i^M \, s'(nT-iT) \tag{7}$$

where the parameters p_i^M i=1..M are the coefficients of $P_M(z)$.

Equation (7) is the fundamental equation of LPC analysis; it states that the value of the signal e(t) at a time t=nT is the difference (or "error") between the differentiated speech signal s'(t) at a time t=nT and its linearly predicted value calculated from the previous M samples. Now, the signal e(t) has a flat magnitude spectrum; this can be readily deduced from equation (4) when the gross spectral characteristics of $Z_L(\omega)$, $U_G(\omega)$ and $H_M(\omega)$ are considered. In fact, for voiced sounds, e(t) will ideally be a sequence of periodic impulses at the pitch frequency, with all samples of e(t) between the pitch pulses being zero. A minimum mean square error (MMSE) analysis based on e(t) as the error signal is thus an appropriate method for computing the coefficients p_i^M i=1...M from the speech waveform, and these can then be converted to the set of reflection coefficients k_i i=1...M by invoking the inverse of the recursion expressed in equation (3), namely

$$P_{m-1}(z) = \{P_m(z) + k_m z^{-m} P_m(z^{-1})\}/(1-k_m^2) \tag{8}$$

where $k_m = P_m^m$.

4.2 Digital filter representations

The foregoing section has briefly sketched the underlying theory involved in LPC analysis of speech waveforms. Greater insight into this theory can be obtained if the equations are studied in terms of digital filters. A ladder form exactly models the reflection mechanism at each junction, but requires the greatest number of multiplication and addition operations. On the other hand, a direct form is canonic, but does not have a simple physical interpretation for its coefficients. A lattice form falls halfway between these extremes, and is the one most frequently used in LPC studies. It is a direct representation of the vocal tract reflections mechanism when the transfer function has been normalized by the factor $H_M(z)$. Consideration of the quantisation, round-off and stability properties for the different forms has shown that the lattice form is the most desirable, since it is the least sensitive to quantisation and round-off errors, and, more important, its stability can be assured simply by ensuring that the reflection coefficients k_i are in the range -1 to +1.

The procedure for extracting the vocal tract parameters from the speech waveform can also be represented in digital filter form, the analysis filter simply having a transfer function which is the inverse of the vocal tract response with the delay of $\frac{1}{2}MT$ removed to permit realisability. When the input to the inverse filter is the speech waveform with the combined effects of the excitation spectrum and lip termination removed, the output corresponds to the error sequence e_n, whose mean square value is minimized. Again, the lattice form of the inverse filter is that most commonly encountered, firstly because of its superior quantisation and round-off error properties, and secondly because not only is the overall transfer function the inverse of the vocal tract response, but each stage is also the inverse of each stage in the lattice representation of the vocal tract.

4.3 General observations

Having now presented the basic theory of LPC and a variety of digital filter representations, some general observations are in order:

(a) Implicit in the theory is a unique relationship between the predictor order M and the sampling frequency 1/T. The sampling period T is equal to twice the propagation time along each section of the vocal tract, and because the total length of the vocal tract is essentially constant, the number of sections M in the model is directly proportional to the sampling frequency.

(b) In deriving a relationship between the vocal tract transfer function $G(\omega)$ and the speech pressure spectrum $S(\omega)$, it was assumed that the source spectrum $U_G(\omega)$ is inversely proportional to frequency squared. This is valid for voiced sounds, and results in spectral flattening in the form of time domain differentiation being performed prior to processing. The theory can also be applied to unvoiced sounds, however, if the different excitation characteristics are taken into account. The gross spectral characteristic of unvoiced excitation is essentially flat, and hence by integrating rather than differentiating prior to processing, the theory will be equally applicable. The error sequence e_n at the output of the inverse filter in this case will ideally be sampled random noise.

(c) The vocal tract model on which LPC is based assumes that the vocal tract is lossless, whereas in the real vocal tract heat conduction, soft flesh, moisture and the viscosity of air all introduce losses. In fitting boundary conditions to the model, it was further assumed that the sum of the forward and backward components of volume velocity just inside the glottis was equal to the source volume velocity (effectively equivalent to a closed glottis condition), and also that the backward component of volume velocity at the lips $V_{M+1}(z)$ was zero (effectively equivalent to terminating the lips with a matched resistive load). In the analysis, therefore, all vocal tract losses are effectively transferred to the load at the lips, and this gives rise to an apparent anomaly, because in relating the speech spectrum $S(\omega)$ to the model vocal tract transfer function $G(\omega)$, it was assumed that the lip termination is a pure inductance. The net effect of these assumptions is that the LPC procedure as described models the frequencies of the formants fairly accurately but not their bandwidths, and results in inferior speech quality. STC is currently studying methods for overcoming these effects.

(d) From equation (2) the vocal tract transfer function $G(\omega)$ is seen to be all-pole. The presence of spectral zeros due to (a) the nasal tract branching off the vocal tract at the velum and (b) the excitation source being forward of the glottis are thus ignored in LPC theory. However, it is possible to approximate zeros to any specified degree of accuracy by including a sufficient number of extra poles in the spectrum, and this is what is done in practice. The first four formants alone require only an eight order predictor, and for a sampling frequency of about 8 kHz this agrees with the order required based on vocal tract length considerations. In most LPC implementations reported to date, the predictor order is chosen to be between 10 and 14 to take account of spectral zeros. However, once zeros appear in the spectrum and extra poles are utilised in the analysis to approximate them, the acoustic tube model representation for the extracted parameters can no longer be directly related to the human vocal tract. Exploitation of the lower redundancy properties of articulatory parameters compared with their acoustic correlates may therefore not now be applicable, leading to imprecise temporal characteristics in the synthesized speech. The approximation of zeros by a limited number of poles also manifests itself in the bandwidths of the synthesized formants being too wide, leading to the characteristic "buzziness" of LPC synthetic speech. This is in addition to the bandwidth increases caused by the resistive lip termination mentioned earlier. Possible methods for incorporating zeros in the model to overcome the speech quality degradations resulting from these effects are being investigated at STC.

(e) The model of speech production used to derive the LPC analysis procedure is also a model which can be employed to generate synthetic speech at the receiver from the transmitted parameters. The important considerations for the synthesizer are the stability of the speech production model, the effect of round-off errors on the synthetic speech, and the excitation function used to drive the synthesizer. The lattice representations of the vocal tract model and its inverse are the most desirable because of their superior properties with respect to stability, quantisation and round-off errors. Stability is particularly important, and is assured if the reflection coefficients k_i computed in the analysis have a magnitude of less than unity. It is for these reasons, among others, that the lattice form tends to be used for analysis and synthesis in LPC vocoders.

The excitation waveform used to drive the synthesizer is an important consideration for speech quality. For voiced sounds, it will ideally be a periodic pulse, and for unvoiced sounds random noise. However, because the bandwidths of the synthetic speech formants tend to be too wide, for the reasons discussed earlier, the nature of the excitation affects the speech quality more than it should. Techniques are being devised at STC to overcome this problem.

4.4 Discussion

Many different approaches to LPC analysis can be found in the literature, usually differing only in the method of implementing the MMSE procedure. Among the more important are Autocorrelation, Covariance, Partial Correlation (PARCOR) and Sequential Estimation. Space does not permit a detailed discussion of these, but the following points are worth mentioning. The Autocorrelation approach results when the speech waveform is assumed to be both stationary and of finite duration; it guarantees a reliable, stable solution. The Covariance approach does not need the stationarity assumption, and does not guarantee a stable solution. However, better results can be achieved with the Covariance method compared with the Autocorrelation, particularly when the analysis is performed pitch synchronously. The PARCOR approach was one of the earliest methods of LPC analysis, and will produce results identical to the Autocorrelation method when the same assumptions are made. However, even if the speech waveform is not stationary, the PARCOR method does guarantee a stable solution. Figure 3 shows a complete PARCOR vocoder; it is based on the lattice representations of the vocal tract and its inverse, and consequently has the desirable properties discussed earlier.

All the approaches to LPC investigated to date are based on the vocal tract model presented earlier, and depend on the validity of the assumptions made therein. The effect of the assumption of a lossless vocal tract and the boundary conditions imposed have already been mentioned, as have the problems resulting from the approximation of spectral zeros by poles. However, there are in addition a number of other problem areas which STC is currently investigating. These primarily relate to pitch detection, the effects of acoustic noise on synthetic speech quality and the degradations resulting from the transmission of narrowband vocoder signals over wideband systems.

5. OUTSTANDING PROBLEMS

One of the major outstanding problems in all analysis-synthesis vocoders is the analysis of the pitch frequency for voiced sounds and the most appropriate form of excitation to use in the synthesizer. In relation to LPC, it has already been mentioned that if the speech production model were perfect, then the "error" signal E(z) at the output of the inverse filter would be synonymous with the idealised vocal tract excitation signal (i.e., periodic pulses for voiced sounds and sampled random noise for unvoiced sounds). The properties of the error signal E(z) can be exploited to provide a data base from which the excitation parameters are determined, but more established techniques such as spectrum analysis or zero-crossings could also be employed. None of the techniques tried to date has been found to be 100% reliable in practice, however, because temporal effects and closely-spaced formants tend to cause a pitch detection algorithm to find either a harmonic or a sub-harmonic of the true pitch. Heuristic methods can be employed to ensure that rapid changes in pitch do not occur, but there are occasions when the pitch frequency really does vary quite rapidly, and when this happens the correction procedure inevitably introduces errors.

The search for the most suitable forms of synthesizer excitation waveform for pitch-excited vocoders is also still a problem. The idealised periodic pulse form for voiced sounds is theoretically valid, but in practice it results in the synthetic speech waveform being "peaky" in appearance and "rough" in subjective quality. This is because the excitation energy is applied over a much shorter interval of time than is the case in the real vocal tract. The adverse effects can be reduced somewhat by spreading the excitation energy in time, any resulting spectral differences being compensated for at the synthesizer output if necessary. However, such a procedure does not completely eliminate the machine-like synthetic quality, firstly because the synthesizer does not precisely represent the human vocal tract for the reasons discussed earlier, and secondly because the pitch in real speech is not smoothly varying but has a variable "jitter" component. Attempts to analyse this and/or compensate for it by deliberately introducing jitter have met with only partial success.

In all analysis-synthesis vocoding techniques, the presence of acoustic noise in the talker environment has a detrimental effect on synthetic speech quality. A particularly bad example of this occurs when narrow-band secure-voice communications are operated between a helicopter pilot and air-traffic control. In the case of vocoders employing autocorrelation or power spectrum processing, one possible approach to overcome the problem is to estimate the noise statistics during silence periods and to assume that they remain unchanged during speech periods. The autocorrelation approach to LPC with interfering acoustic noise present will serve to illustrate the principles. The speech waveform plus noise enters the analysis only via its autocorrelation function and, since it can reasonably be assumed that the noise and speech are uncorrelated, the effect of the noise will be to produce an autocorrelation which is the sum of the speech and noise autocorrelations. If the noise is random, only its zero delay autocorrelation value will be finite and its effect could be eliminated fairly easily. In tanks, helicopters and aircraft cockpits, however, the noise is certainly not random and the autocorrelation is likely to be affected at all values of delay. The effects of such interfering noise could be reduced simply by subtracting the noise statistics from those of the speech in the autocorrelation domain. There are problems with this technique, however, firstly because the noise characteristics may not be short-term stationary, and secondly because it requires a speech-silence decision to be made, a difficult task in the presence of interfering acoustic noise. Furthermore, the noise will complicate the voiced-unvoiced decision necessary in pitch-excited LPC vocoders and, if the noise has a periodic component, it will complicate the pitch detection problem also.

Interfering noise in the transmission medium affects synthetic speech quality because of errors in the received parameters. Some speech coding techniques are more sensitive to transmission errors than others, a particularly striking example being the relative immunity of delta modulation to channel errors compared with pulse code modulation. With all coding techniques, the effects of errors can be reduced by deliberately adding redundancy in the form of forward error correction. This inevitably increases the data rate and offsets the bandwidth compression advantages of source encoding. However, by choosing the encoder frame structure and error correction codes judiciously, and by matching them to the transmission channel statistics, the increase in data rate required to achieve a given performance criterion can be minimized.

The final problem area of narrow-band vocoding concerns its performance in analogue tandem with delta modulated (e.g., CVSD) tactical communications systems. An assessment of the degradation in synthetic speech quality caused by tandem connections is difficult to ascertain from individual assessments of the constituent coder degradations because the human ear is tolerant of many types of severe distortions in the speech waveform, but quite intolerant of even mild distortions of other types, whereas narrow-band speech coders tend to be intolerant of nearly all distortions. Vocoded speech tends to be "peaky" and thus contains quite high rates of change. CVSD is a slope-limited waveform following technique, and the high slope portions of vocoded speech will not be properly followed by a CVSD, leading to an overall degradation in speech quality which is more than would be expected from separate assessments of the individual vocoder and CVSD performances. One method for overcoming this problem is to use an all-pass network to reduce the "peaky" nature of vocoder speech without affecting its amplitude spectrum. If this is done at the synthesizer input, it can be considered as part of the excitation function, and as such has a theoretical basis. The speech production model is minimum phase because of its all-pole characteristics, but there is no guarantee that the glottal excitation waveform in the real vocal tract possesses this property. An all-pass phase shift network will effectively transform the synthetic excitation into a more representative mixed-phase waveform, but the most suitable choice of all-pass characteristics remains a problem. In the reverse direction, namely a CVSD followed by a vocoder, the delta coded synthetic speech can effectively be considered as the sum of the original speech and quantisation noise, resulting in the subsequent vocoder voiced-unvoiced decision algorithm making more errors than it otherwise would. In both directions, therefore, the synthetic speech quality of vocoder-CVSD analogue tandem connections is degraded by more than might reasonably be expected from individual performance assessments, and further investigations are being undertaken at STC to find ways of reducing the adverse effects.

6. HARDWARE IMPLEMENTATION CONSIDERATIONS

In this final section, the factors involved in deciding the most cost-effective means for providing secure voice communications will be considered. Advances in the ever-changing field of semi-conductor technology are continually shifting the balance of trade-offs in speech quality, data rate and cost, and in addition they have resulted in the overall weight, size and power consumption requirements being reduced.

Narrow band speech coding techniques such as spectrum and LPC vocoding require a considerable amount of real time complex signal processing to be performed. The devices currently available for performing such processing operations are many and varied, ranging from discrete component circuits to large scale intergrated (LSI) circuits, including the relatively recent innovation of high-speed microprocessors. In addition, specialized analogue signal processing elements such as charge-coupled devices (CCD) and surface acoustic wave (SAW) devices have also appeared on the market within the last few years. Due to the continued improvement of semi-conductor manufacturing techniques, the cost per function achieved with these devices is continually decreasing and, in order that the cost-effectiveness of future military communications be optimum, it is necessary that the hardware designs make maximum use of the features and advantages offered.

Once the theory of a speech coding technique has been established, the usual approach to hardware design is to employ a simulation on a general purpose digital computer to validate the theory, and to determine the optimum operating parameters and configuration. However, there are two problems with this; firstly the architecture of the general purpose computer will usually differ from that of the equipment being simulated (e.g., in arithmetic manipulation, word formats, rounding, etc.), and secondly, real-time simulations will not be possible unless the computer system is equipped with high speed peripheral processors. As far as the author is aware, the complex programming problems involved in ensuring that a simulation precisely matches a given hardware configuration have not yet been solved, and final optimisation using a simulation is not possible. Completion of the hardware construction is thus likely to be the first occasion when the performance of a technique can be properly assessed under real-time operating conditions. With a programmable microcomputer type hardware, the capability for modifying the algorithm is available and the overall performance can be optimized relatively easily compared with hardwired equipments.

Having decided that a programmable microcomputer is the most convenient approach to speech coder hardware design, there are a variety of technologies and corresponding architectures which can be employed. As a general rule, the technology will determine both the speed of operation and the amount of LSI which can be undertaken, because high speed technologies cannot be integrated to the same extent as slower technologies. Highly integrated single chip microprocessors with powerful instruction sets thus tend to be slow in computation and control compared with equivalent but less highly integrated units constructed from microprocessing elements using a faster technology. However, because the slower technology devices are produced in large quantities, they tend to be less expensive, and because fewer chips are required for a given number of functions, they also tend to be more reliable and consume less power.

In speech coding, a considerable proportion of the complex signal processing needs to be done at high speed in order to meet real-time operating constraints. The processing circuitry also needs to be highly integrated in order to keep the size and power consumption requirements to a minimum. It is unlikely, however, that a highly integrated single-chip microprocessor architecture will be able to meet the speed requirements, and alternatives must be sought. A suitable architecture might be one which uses a powerful slow speed microprocessor for the overall control function (management), a less flexible but higher speed processor to organize array processing, and a fast multiply-add peripheral interfacing directly to the high-speed processor. Such a structure is not completely flexible but is likely to be optimum in terms of processing speed, size, power consumption and reliability.

The foregoing discussion has indicated the important factors involved in the hardware design of narrow band voice coding devices. It can be seen that modern semi-conductor technology has opened up the possibility for producing, in a reasonably quick and straightforward manner, designs which are optimum in terms of cost, performance, size, weight, power consumption and reliability. To take full advantage of this possibility, the hardware design engineers need to be conversant with the theoretical aspects of speech coding, and conversely the speech coding experts need to be conversant with the hardware aspects of implementation. This is so because speech coding techniques such as LPC require state-of-the-art digital hardware for implementation, and the optimisation of a design for a given data rate in terms of the program, the hardware configuration and the speech quality requires knowledge which covers the specialist fields of both digital hardware design and speech coding techniques. The existence of project teams having such dual capabilities means that new ideas can be tested and engineered within a matter of months instead of the years previously required. The design of future communications systems thus needs to be flexible so that the advantages offered by new developments can be exploited to the full in a timely manner. This, in essence, is the impact which modern technology is having today on world-wide civil and military communications systems.

The speech processing group at STC consists of specialists in both hardware design and theoretical techniques. The areas discussed in this paper, and particularly those aspects indicated as requiring further investigation, are the subject of our continuing research to achieve high quality, low data rate secure voice communications cost effectively.

To assist in this research, we have built up a flexible digital signal processing system consisting of a fairly fast control processor together with its associated memory and other peripheral equipment. An important feature of the system is a facility for the rapid selection and display of data. For simulation development, a signal processing interpreter is being used, since this permits program changes to be made quickly and the effects to be assessed almost immediately. The iterative process of checking and optimising speech coder designs is therefore accomplished with relative ease. The disadvantage of large computation times being required for long samples of speech will be overcome at a future date by adding high-speed array processors to perform critical tasks, and by using a compiler to translate the interpreter-written code into non-interpreter object code. Any particularly critical and/or time-consuming sections of a simulation can, of course, always be written directly in assembler language to achieve faster operation.

In parallel with the simulation work, we are also studying developments in the microprocessor field with particular emphasis on speech coding applications. Experience in microprocessor use is already available within the team, and will be extended by designing and building high-speed array processors. The experience and lessons learned from this will then enable optimum architectures to be selected for both real-time speech coder hardware implementations, and for the in-house simulation system.

7. ACKNOWLEDGEMENTS

The author would like to thank all those who have contributed to the ideas presented in this paper, and particularly his colleague Herr. W. Grooteboer for many stimulating discussions.

SELECTED BIBLIOGRAPHY

1. K.W. Cattermole, Principles of Pulse Code Modulation, Iliffe, London, 1970.
2. C.G.M. Fant, Acoustic Theory of Speech Production, Mouton, The Hague, 1970.
3. J.L. Flanagan, Speech Analysis Synthesis and Perception, Spring-Verlag, Berlin, 1972.
4. J.D. Markel and A.H. Gray, Linear Prediction of Speech, Spring-Verlag, Berlin, 1976.
5. R.Steele, Delta Modulation Systems, Pentech Press, 1975.

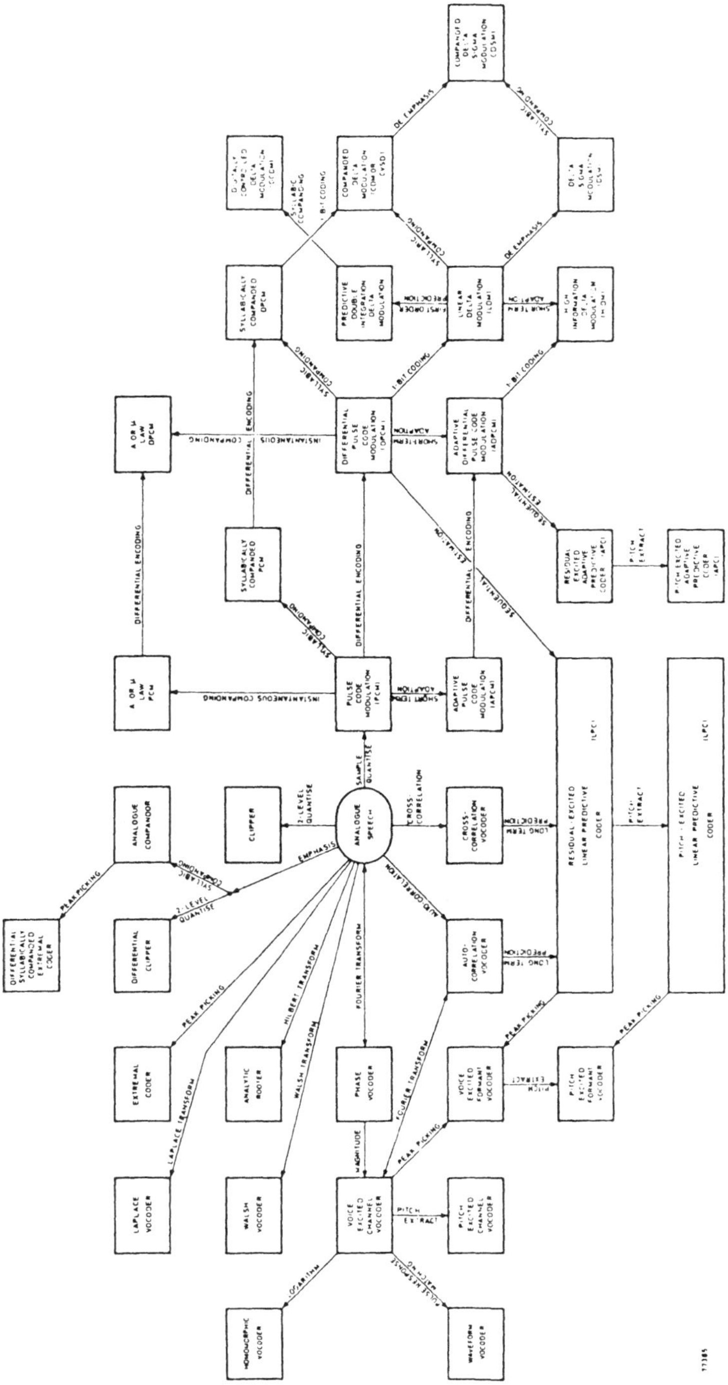

Fig.1 Relationship between different speech bandwidth compression and coding techniques.

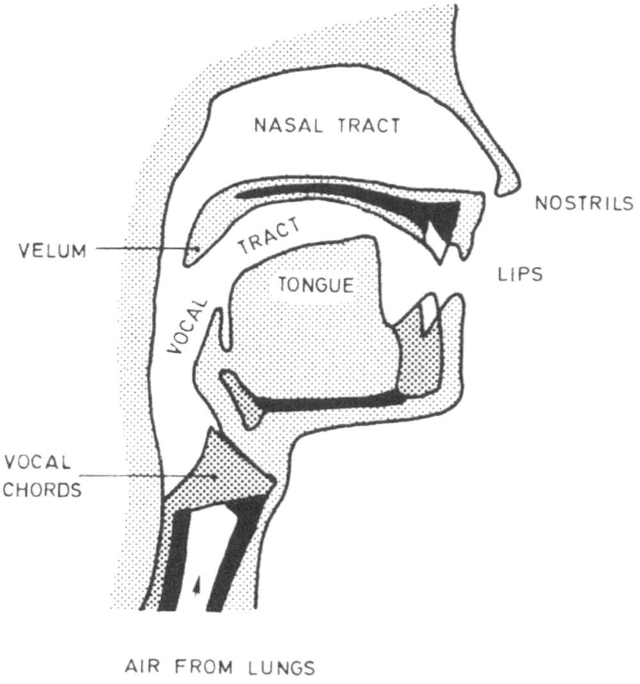

Fig.2 Principles of speech production.

a. Lateral cross-section of human head

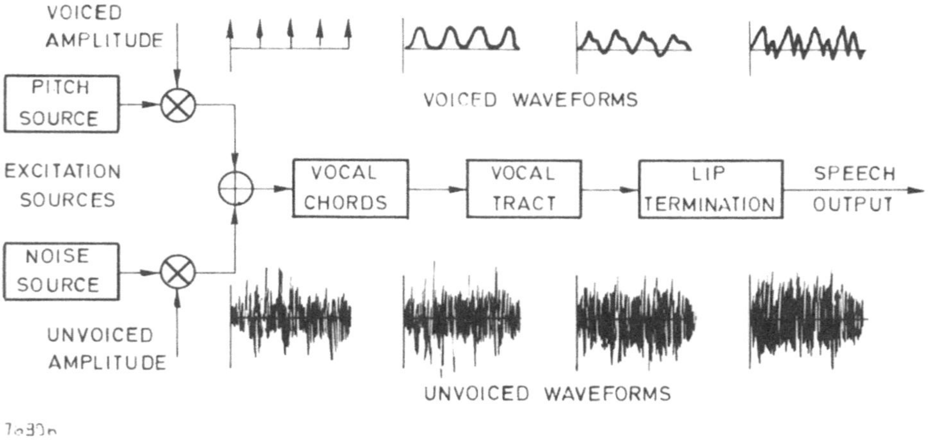

b. Simplified model of speech production

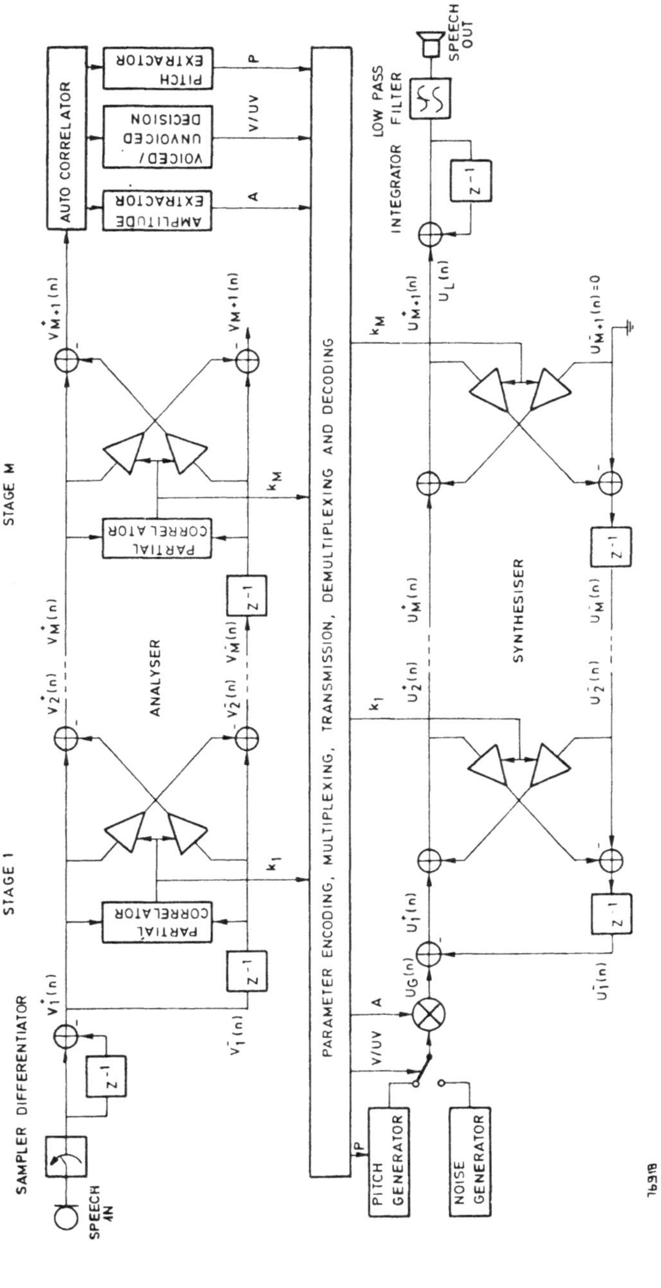

Fig.3 PARCOR LPC vocoder.

CHARACTERIZATION OF TROPOSCATTER CHANNELS FOR DIGITAL TRANSMISSION

P.T. Nielsen

Communications Division, SHAPE Technical Centre,
The Hague, Netherlands

ABSTRACT. The achievable data rates for digital transmission over troposcatter channels are restricted by intersymbol interference caused by dispersion. Adaptive signal processing techniques are required for high-speed transmission over such channels where the reciprocal of the transmission rate is comparable with the differential path delay. A fairly detailed statistical model of the troposcatter channel is a necessary tool for the design of efficient adaptive modems. Preferably, such a model should moreover be usable for predicting digital troposcatter performance.

The most widely accepted model for describing the short-term behaviour of troposcatter channels is the Wide Sense Stationary Uncorrelated Scattering (WSSUS) channel. This model and its range of applicability is discussed together with methods for accounting for long term variations and special propagation effects not included in the WSSUS model.

1. BACKGROUND AND SCOPE

Over the past two decades, troposcatter communications systems have come into widespread use for a range of military and commercial applications. Microwave line-of-sight (LOS) relay systems are limited by the need for suitably spaced relay stations, a need which - cost considerations aside - may sometimes be impossible to meet, for instance if communications are to be over water or over territory which for political reasons is inaccessible. Satellite communications on the other hand, may be too costly for certain high-capacity, medium-range communications links. Thus, in a sense there exists a gap between microwave relays and satellite links, in which troposcatter systems occasionally turn out to provide the most attractive solution. A recent example is the provision of troposcatter communications to British Oil rigs in the North Sea.

While most existing troposcatter links employ FDM/FM modulation, the development of high speed digital transmission techniques for LOS and satellite communications has recently led to an interest in considering similar techniques for troposcatter channels. Driving factors have been the need for privacy and the desire to be able to integrate troposcatter circuits into larger digital nets. Due to the fading and dispersive nature of the troposcatter channel, however, high speed digital transmission over such channels is much more difficult than for LOS or satellite, and generally requires the use of adaptive receiver techniques. The challenging problem of developing good adaptive receivers and the question as to how to determine the performance of digital troposcatter links have caused workers in the field to take a new, closer look at the basic propagation mechanisms in troposcatter and to develop more accurate mathematical models for describing the statistical fluctuations of such channels.

STC's current interest in troposcatter communications is intimately related to the future of ACE High, NATO's strategic troposcatter network, which extends from the most northern flank of NATO in Norway to the easternmost flank in Turkey (Fig. 1). If, at some point in time, ACE High is going to be converted from the present FDM/FM mode of operation to digital transmission, STC will provide systems engineering assistance to the NATO agencies responsible for systems planning and implementation. For this reason, STC has engaged in a detailed study of suitable systems engineering methods for digital troposcatter.

Although STC's work in the digital troposcatter field is driven primarily by the need for ACE High planning, the problems associated with high-speed digital transmission over dispersive channels are also being addressed under the more research-oriented part of the STC programme of work. In this way, the many basic commonalities with other problems of high speed digital transmission and adaptive systems can be identified and utilized in the solutions.

The present paper reviews from a system engineer's point of view the recent developments in digital troposcatter transmission with special emphasis on the channel characterization problem. Section 2 introduces the concept of digital troposcatter transmission and considers some alternative techniques. Section 3 addresses the problem of RF channel modelling, while the evaluation of digital troposcatter performance is discussed in Section 4. In order to be able to determine quantitatively the achievable digital performance of a given link, it is necessary to determine certain parameters of the RF channel model; to this end, techniques for prediction, measurement and simulation of troposcatter channels are considered in Section 5. Finally, Section 6 outlines STC current and planned activities in the field.

2. DIGITAL TRANSMISSION OVER TROPOSCATTER CHANNELS

2.1 Bit rate requirements

For typical applications, the required traffic capacity of a digital troposcatter link may range between 30 and 120 voice channels. If standard 64 kbit/s PCM is used for A/D conversion, this corresponds to aggregate data rates between 2 and 8 Mbit/s, approximately. The discussion in this paper assumes data rates of this order of magnitude.

2.2 Digital modulation techniques for troposcatter

The same criteria that apply for the choice of modulation techniques for LOS systems are valid a fortiori for troposcatter systems. High power efficiency calls for the use of near constant envelope signals, and bandwidth is particularly at a premium because troposcatter links require frequency coordination over extremely large geographical areas. A third criterion which is unique to troposcatter is that the modulation technique (with suitable receiver processing) must be usable under severe intersymbol interference conditions: typical average multipath dispersion for a long troposcatter link is 250 ns which is comparable to the symbol duration for the data rates mentioned in Section 2.1. The use of conventional PSK modems, for example, would result in high or very high irreducible error rates due to intersymbol interference.

A number of detailed studies have been performed in the past to examine and compare a number of candidate modulation schemes, see [1] to [4]. The general conclusion has been that QPSK is a suitable technique; optionally QPSK may be comined with transmitter time gating to simplify the receiver design at the expense of larger bandwidth requirements.

2.3 Receiver processing techniques

A receiver for high rate digital troposcatter signals must perform two basic functions. Firstly, it must recombine the energy in the received, randomly distorted pulses in a coherent manner so as to obtain a high signal-to-noise ratio for detection. Secondly, it must supress interference from adjacent pulses to an acceptably small level. If the former function is performed well, the channel is, in fact, utilized as additional diversity: although channel dispersion and fading causes the received waveform to vary, the probability of a fade affecting scatter returns at all delays simultaneously is small, if dispersion is significant compared to the symbol duration. This phenonmenon is usually referred to as implicit diversity.

Some approaches to adaptive receiver design are described in [1] to [4]. However, two specific techniques deserve special mention here because they have been used in the development of prototype modems. One is the use of adaptive matched filtering based on correlation of the received signal with a stored reference continuously updated in a recycling delay line. When used in connection with transmitter time gating, this principle offers reasonable performance for links which are only moderately dispersive, as has been demonstrated by Raytheon Corporation in their development of the DAR (distortion adaptive receiver) modem [5]. The other approach involves the use of adaptive equalization techniques similar to those employed in voice frequency data modems. This method, which offers higher potential performance at the expense of greater complexity, has been used by GTE Sylvania in their development of the MDTS (megabit digital troposcatter subsystem) modem [6].

2.4 Bandwidth requirements

With current techniques, digital troposcatter transmission requires an RF bandwidth of at least 1 Hz per bit/s. Assuming the use of 64 kbit/s PCM, this leads to a bandwidth occupancy of approximately twice that normally used to support the same number of voice channels using FDM/FM. Due to the scarcity of frequency allocations, therefore, there is a real need to direct future efforts in this area towards the achievement of higher bandwidth efficiency.

3. TROPOSCATTER CHANNEL MODELS

3.1 General

After the very brief introduction to digital troposcatter techniques given in Section 2, the remainder of the paper concentrates on the statistical behaviour of the troposcatter channel and its impact on digital troposcatter performance.

The received signal level (RSL) in a troposcatter channel usually exhibits Rayleigh fading with a median that varies slowly, following approximately a log-normal distribution. For the prediction of analogue troposcatter performance, it is usually sufficent to consider the RSL statistics, supplemented perhaps by the differential transmission delay if intermodulation noise may become a limiting factor. However, for digital transmission where waveform distortion plays a predominant role, it is desirable to obtain more detailed statistical descriptions of the troposcatter channel, with emphasis on the time domain characterization.

3.2 The WSSUS model

From the nature of the troposcatter propagation mechanism it is reasonable to assume that the troposcatter channel (excluding transmit and receive equipment) for all practical purposes can be modelled as a time varying linear filter. In its most general form, such a filter can be modelled as a tapped delay line filter as shown in Fig. 2. The tap gains are complex stochastic processes. By making the two additional assumptions that, on a short-term basis:

- a received carrier can be characterized as a stationary narrow-band Gaussian process, and

- for narrow-band signals, scatter contributions with different delays exhibit uncorrelated fading,

one is led to the generally accepted so-called Wide Sense Stationary Uncorrelated Scattering (WSSUS) model, in which all tap gains are independent, stationary, Gaussian processes. This model is described by Bello in his classical paper of 1969 [7]; an excellent textbook reference is Kennedy [8]. While the statistical behaviour of the WSSUS channel is completely described by the channel scattering function [8], it is often more convenient for practical purposes to work with the delay power spectrum, or multipath profile, which can be envisioned with the aid of Fig. 2 as the variance of the complex tap gains as a function of delay. It can be shown that the delay power spectrum is related to the more generally known frequency correlation function by the Fourier transform. To take account of the Doppler information contained in the scattering function, the delay power spectrum must be supplemented with information about fade rate.

3.3 Limitations and generalisations of the WSSUS model

It is important to emphasize the limitations of the WSSUS model for practical troposcatter applications. First of all, it is a stationary model which does not account for long-term variations caused by changing refractivity and meteorological conditions. Moreover, it does not in its basic form account for special propagation effects such as diffraction, ducting or reflections caused by aircraft traversing the common volume. To the extent that such effects can be adequately modelled by the presence of a constant signal in addition tc the fading troposcatter signal, it is possible to generalize the tapped delay line model to include them. Figure 3 shows an example of a mixed propagation mode consisting of double knife-edge diffraction and troposcatter. If the two component signals are of comparable powers, the multipath handling capability required for a digital modem to operate over such a channel may need to be considerably larger than would be predicted based on the troposcatter mode alone.

3.4 Troposcatter channel models: periods of validity

It was noted above in Section 3.3 that the WSSUS is a "short-term" model only. However, it is important to acquire a quantitative feeling for what short-term really means.

Apart from aircraft fading, typical fade rates for troposcatter channels range from 0.1 Hz to 10 Hz. This means that for periods not exceeding, say 0.1 s, the channel could be modelled as a time-invariant linear channel (this period is still very long compared to the symbol duration). The term "frozen channel" representation has been coined for this situation. For longer periods, perhaps up to about a quarter of an hour, the stationary WSSUS model applies. Any description to be valid over periods exceeding $\frac{1}{4}$ hour must necessarily be non-stationary in nature. These observations are summarized in Table 1, which also serves as a definition of terms for the remainder of this paper.

Term	Statistical description	Max. safe period of validity
Frozen channel	Time-invariant	0.1 s
Short-term Rayleigh	Time-varying stationary	15 min
Long-term	Time-varying non-stationary	NA

Table 1: Approximate periods of validity for troposcatter channel models

4. CHARACTERIZATION OF DIGITAL TROPOSCATTER PERFORMANCE

4.1 General

Before addressing the question of how to apply the channel models described in Section 3 to the problem of quantitatively determining the performance of digital troposcatter systems, it is useful to consider briefly some methods available for characterizing overall digital system performance and also to look at suitable ways of specifying the performance of a digital modem per se.

4.2 Digital troposcatter systems specification

For analogue systems, it is customary to specify the required performance in terms of a certain average circuit quality and a worst-case quality to be exceeded for all but a certain, small percentage of time. Special rules are applied to translate user requirements into requirements for individual links, taking account of the maximum number of links to occur in tandem.

Digital modems tend to exhibit very rapid performance degradation as the signal-to-noise ratio (SNR) decreases below a certain, critical value. This threshold effect combined with the large amount of fading experienced on troposcatter channels tends to make a specification of long term average performance less interesting than for an analogue system. It is therefore almost certain that a digital system would be specified in terms of certain performance limits and the percentages of time during which these limits may be violated. The detailed method of specification, however, would depend on the detailed user requirements, network structure, etc. Specifications may, for example, be in terms of:

(a) Rate and duration of outages, an outage being defined as an interval exceeding, say, 0.1 s or 1.0 s, during which the error rate remains above a certain threshold, for example, 10^{-4}.

(b) Average bit error rate (BER), as measured over, say, a 15-min. interval, not to be exceeded for more than a certain percentage of time.

A user requirement would often be stated in a fashion similar to (a), while a specification such as (b) has the advantage of lending itself more readily to measurement. In some cases, it may be allowable to transform a user requirement of the form (a) into a systems specification of the form (b) on the basis of past experience with digital troposcatter performance. A certain margin would then be necessary to cover the uncertainties in such a transformation.

There is an important performance measure for digital communications systems which has no counterpart in the analogue world: statistics for loss of bit count integrity (BCI). Loss of BCI may occur in the timing recovery circuits within the modem, but it may also result from excessive error bursts causing loss of frame syncrhonization in the associated TDM equipment, in which latter case the loss of BCI statistics is closely related to the outage statistics (a) described above.

4.3 Characterization of digital modem performance

A digital troposcatter modem can be characterized in a pseudo-deterministic manner in terms of its performance on a frozen channel. Given the shape of the channel impulse response, the modem performance can be described by plotting BER as a function of SNR, just as for a conventional PSK modem. However, due to the multiplicity of impulse responses experienced on a troposcatter channel, it is necessary to resort to some kind of simplified description in which the multipath dispersion enters as a parameter.

It is reasonable to expect that for a well-designed adaptive modem, the performance will be relatively insensitive to the detailed "multipath delay spread", MDS, defined in some suitable manner, e.g., as the two-sigma duration of the power impulse response. A useful frozen channel characterization of a modem can therefore be obtained by plotting curves of constant BER in a coordinate system with SNR and MDS along the axes (see Fig. 4a). The type of curve shown is believed to be typical: as long as the MDS remains within the multipath handling capability of the modem, performance is virtually unaffected, but beyond some critical value of MDS, a rapid degradation occurs as a result of intersymbol interference and reduced received power utilization. It should be noted that for diversity reception, SNR and MDS are to be taken as the post-combining values.

An alternative approach is to express the performance of a modem in terms of its average BER when operating over a perfect WSSUS channel. A typical performance description of this type is shown in Fig. 4b, where MDS is to be taken to mean the two-sigma duration of the delay power spectrum. Due to the effect of implicit diversity, the performance is generally optimum for a non-zero value of MDS, and the degradation at higher values of MDS does not occur as abruptly because of the averaging process. For diversity reception, a separate set of curves is required for each order of explicit diversity.

4.4 Prediction of digital troposcatter link performance

Existing methods for predicting troposcatter propagation characteristics only provide expected values or distributions of short-term average parameters. However, as will be seen in Section 5, these statistics can be used in conjunction with a short-term channel model such as the WSSUS model to derive predicted sample statistics (i.e. frozen channel statistics). Depending on how the link performance requirements have been stated, there are at least two distinct approaches to the link prediction problem, as illustrated in Fig. 5. If outage statistics are desired, the predicted sample statistics (joint SNR and MDS statistics) are used with the frozen channel modem description (Fig. 5a). To obtain average BER statistics, the predicted short-term average parameters are used directly with the description of average modem performance on a WSSUS channel (Fig. 5b).

Section 5 addresses the problem of how to predict the joint SNR and MDS statistics required for the two procedures in Fig. 5. It should be noted, however, that the required accuracy of this prediction depends to a great extent on the quantitative modem performance characteristics. This important question of how sensitive the resulting prediction of digital link performance is to the accuracy of the various parts of the RF predictions is currently being studied by STC.

5. PREDICTION, MEASUREMENT AND SIMULATION OF TROPOSCATTER CHANNEL CHARACTERISTICS

5.1 Prediction techniques

It was shown in Section 4 that, at least in principle, the prediction of troposcatter link performance requires the prediction of joint statistics of path loss and multipath dispersion. Depending on the approach taken, see Fig. 5, the statistics required are either sample statistics or statistics of short-term average values.

Techniques for prediction of troposcatter path loss distribution are well established (see, for example [9]), and have been standardized by CCIR and NBS. The statistics obtained by these methods describe the long-term distribution of one-hour median path loss. Reference [10] provides a comparison of measured and predicted path loss statistics for almost 800 links in all parts of the world, the agreement generally being good.

One aspect of troposcatter path loss prediction which may still require further investigations is that of estimating aperture-to-medium coupling loss statistics. The dependency of this gain loss on the scatter angle and the height of the common volume has not yet been satisfactorily established. A recent reference is [11].

Bello [7] has proposed a simple method of predicting the delay power spectrum for a link, given antenna sizes, line geometry, and effective earth radius. More recently, this method has been slightly modified to provide a better fit with experimental data for short links [12]. No general method exists for predicting the long-term distribution of MDS, except that knowledge of the distribution of refractive index can be used with the Bello model to obtain an estimate of the MDS distribution. There is a reasonable amount of measured MDS data that appears to suggest that the distribution of MDS is approximately Gaussian and that the variation of MDS is largest for short links, ([13] and [14]) but more work appears necessary in this area. From a systems engineering viewpoint, the main interest is on predicting the worst case multipath not exceeded for more than a small percentage of time. Reference [15] provides a discussion of this issue.

There are theoretical considerations ([13] and [16]) which suggest that there should be a positive correlation between path loss and dispersion, and this conjecture is, at least partially, supported by experiments (e.g. [14]). However, experimental data exhibits a large amount of scatter (see Fig. 6), and it is doubtful whether a positive correlation can be relied upon to give an additional implicit diversity advantage during high path loss conditions.

As discussed in Section 4, given an accurate short-term stationary channel model, long-term and short-term statistics can be combined to yield predictions of sample (frozen channel) statistics. To be of interest for the prediction of link performance, however, the frozen channel statistics must be post-combining statistics and take account of both explicit and implicit diversity gains. The appropriate question is therefore: given the link, the order of explicit diversity and the data rate, how does one predict the joint distribution of frozen-channel RSL and MDS? This problem needs to be solved, at least approximately, if the link performance assessment method illustrated in Fig. 5 is to be useful. Computer simulations coupled with hardware for analysis and simulation of troposcatter channels will probably be required to develop frozen channel prediction methods.

The prediction of fade rate distributions is also an issue of some interest, since it is closely related to the distribution of outage duration on a digital link. Some experimental results are reported in [13], but predictions are at the present time usually restricted to giving an estimate on the "likely range" of fade rates, based on general experience.

5.2 Channel measurement techniques

While RSL measurements are straightforward to carry out, the measurement of dispersion and Doppler characteristics requires specialized instrumentation. Pulses of very short duration could, in principle, be used to probe the medium for its impulse response but, due to the limited available peak power in the transmitter, the received pulse energy would be too small for the pulse shape to be detected. As in radar, therefore, one resorts to channel probing using pseudo-random pulse sequences with good autocorrelation properties. The basic principle of this approach, sometimes referred to as the RAKE method, is illustrated in Fig. 7: a transmitted biphase coded carrier is correlated at the receiver with a number of time-displaced replicas of the code used for modulation, and the sampled impulse response emerges. By processing a large number of consecutive measurements, the scattering function can be estimated. An example of such a measured scatter function is shown in Fig. 8. Perhaps a more illustrative presentation is that shown in Fig.9, which shows the evolution of the impulse response with time. A detailed description of a practical RAKE equipment is given in [17].

The impulse response measurements can be processed in real time to provide a measure of the MDS parameter. Bello [18] has proposed an alternative method for obtaining one-parameter measures of multipath and Doppler, based on the observation of two slightly separated carriers or on one FM-modulated carrier.

5.3 Channel simulation techniques

The tapped delay line channel model shown in Fig. 2 can be implemented in hardware or software to provide an accurate realization of a troposcatter channel. A hardware simulator relying on this principle is described in detail in [19]. It is particularly useful as a tool for laboratory testing of adaptive modems, but it can also be used to examine experimentally the detailed statistical properties of various channel models, including the WSSUS channel.

6. CURRENT AND PLANNED STC ACTIVITIES RELATING TO TROPOSCATTER CHANNEL CHARACTERIZATION

During the first half of 1977, STC has, in cooperation with the US Defense Communications Engineering Centre, conducted a series of digital troposcatter tests over two links in ACE High. The purpose of the trials, which employed US prototype modems, was inter alia, to evaluate the validity of existing performance prediction methods and to provide a large data base for use in future work in this area. The data obtained during the tests are now being processed, and a report is due later this year.

After the completion of the combined US/NATO tests over ACE High, STC will continue the collection of troposcatter channel statistics relevant for digital systems engineering. This work will be a cooperative effort with a German research establishment, the Forschungsinstitut für Hochfrequenzphysik in Werthhoven, and will make use of a test link owned by that Institute. Simultaneous path loss and dispersion data will be collected at two frequencies (900 MHz and 4.5 GHz) over an extended period of time. Parts of that programme will be devoted to the investigation of aperture-to-medium coupling loss and of angle diversity in conjunction with digital transmission.

In parallel with the channel measurement programme, STC plans to engage in detailed theoretical and experimental studies of adaptive modems with special emphasis on troposcatter applications. A troposcatter channel simulator is being constructed now in preparation for these studies. An important part of the modem work will be directed towards the problem of finding more bandwidth-efficient modulation schemes.

REFERENCES

1. P. Monsen and S.H. Richman, Adaptive Data Transmission Study, RADC-TR-73-268, August 1973.
2. P.J. Crepeau, Digital Troposcatter:State-of-the-Art Modulation Techniques, Technical Report ECCM-4132, US Army Electronics Command, Fort Monmouth, New Jersey, June 1973.
3. P.A. Bello et al., Troposcatter Multichannel Digital Systems Study, RADC-TR-67-218, May 1967.
4. P.A. Bello, Selection of Multichannel Digital Data Systems for Troposcatter Channels, IEEE Trans. Comm. Techn. Vol. COM-17, No. 2, pp. 138-161, April 1969
5. Final report for the Hybrid Tropo Transmission System, prepared for RADC by Raytheon Corporation, 30 March 1976.
6. Final report for the Megabit Digital Troposcatter Subsystem, prepared for ECOM by GTE Sylvania In. and Signatron Inc., April 1977.
7. P.A. Bello, A Troposcatter Channel Model, IEEE Trans. Comm. Techn., Vol. COM-17, No. 2, pp. 130-137, April 1969.
8. Robert S. Kennedy, Fading Dispersive Communication Channels, New York: John Wiley and Sons, 1969.
9. Philip F. Panter, Communications Systems Design, New York: McGraw-Hill, 1972.
10. A.G. Longeley et al., Measured and Predicted Long-Term Distributions of Tropospheric Transmission Loss, Report No. OT/TRER 16, US Department of Commerce, Office of Telecommunications, July 1971.

11. D.T. Gjessing and K.S. McCormich, On the Prediction of the Characteristic Parameters of Long-Distance Tropospheric Communication Links", IEE Trans. Comm., Vol. COM-22, No. 9, pp. 1325-1331, September 1974.

12. L.D. Daniel and R.A. Reinman, Performance Prediction for Short-Range Troposcatter Links, IEE Trans. Comm., Vol. COM-24, No. 6, pp. 670-672, June 1976.

13. L. Ehrman and J.W. Graham, Digital Troposcatter Experiments, report produced by Signatron Inc. (SIG-A-125) for RADC (RADC-TR-72-350), January 1973.

14. A.R. Sherwood and I.A. Fantera, Multipath Measurements over Troposcatter Paths with Application to Digital Transmission, NTC 75 Conference Record, pp. 28-1 to 28-9, December 1975.

15. P. Monsen, Troposcatter Multipath - Availability Considerations, NTC 74 Conference Record, pp. 845-851, December 1974.

16. P.A. Bello et al., Signal Distortion and Intermodulation with Tropospheric Scatter, AGARD Conf. Proc. No. 70, pp. 36-1 to 36-17 February 1971.

17. R. Gallant, Troposcatter Multipath Analyzer, report produced by Sylvania Electronic Systems for RADC (RADC-TR-70-155), August 1970.

18. P.A. Bello, Some Techniques for the Instantaneous Real-Time Measurement of Multipath and Doppler Spread, IEEE Trans. Comm. Techn., Vol. 13, No. 3, pp. 285-292, September 1965.

19. Tropo Diversity Simulator Development, RADC-TR-74-188, July 1974.

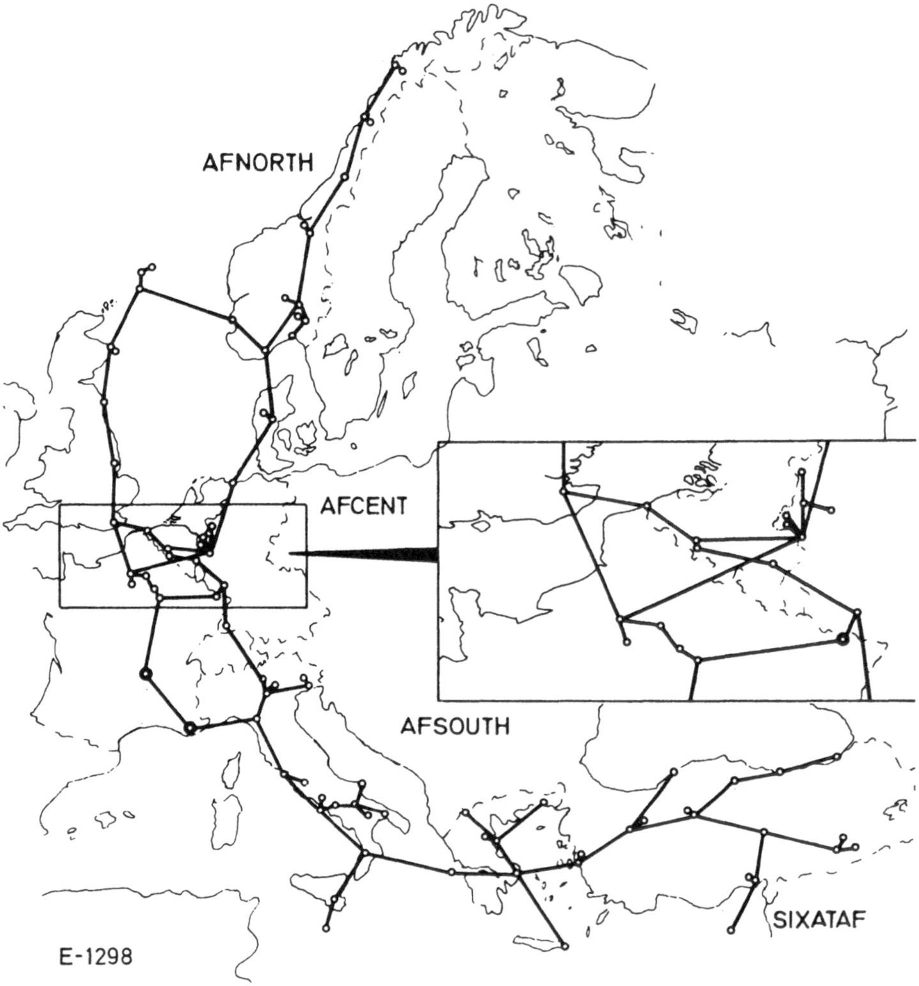

Fig. 1. Outline of the NATO ACE High troposcatter system

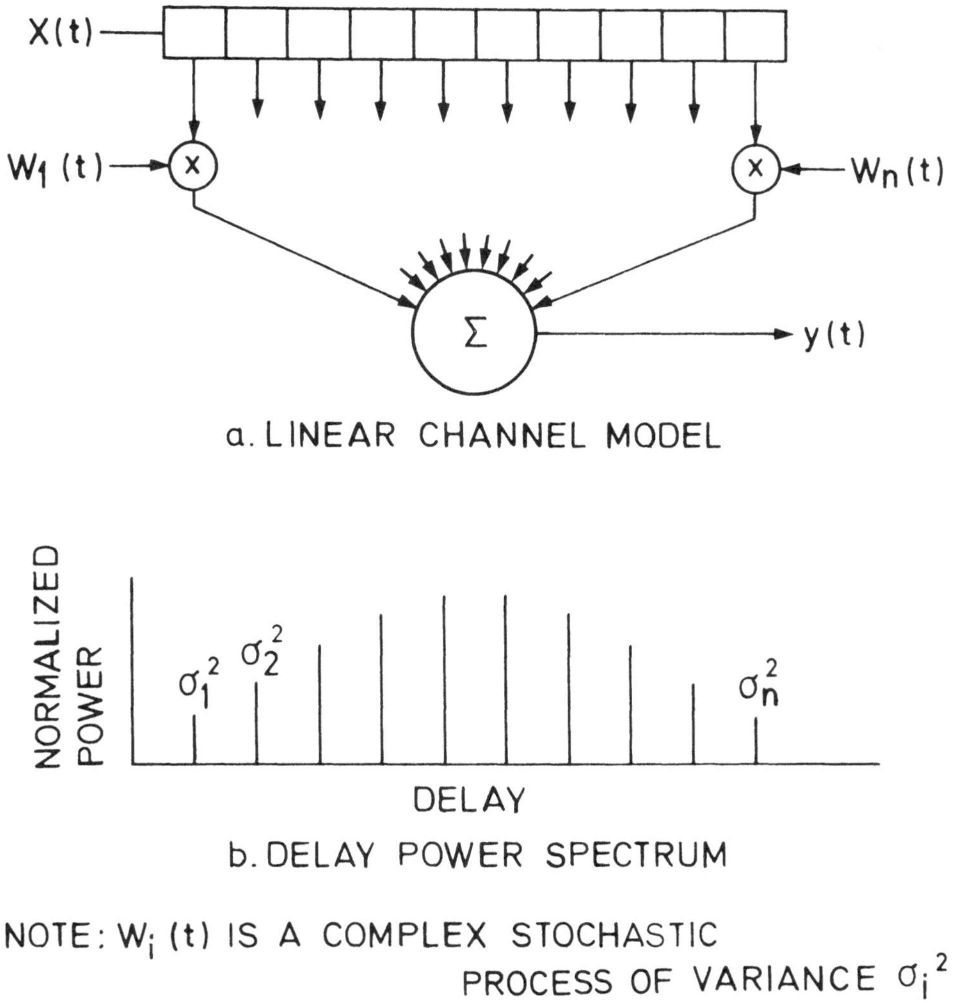

E 1291

Fig. 2. Troposcatter channel model and associated delay power spectrum

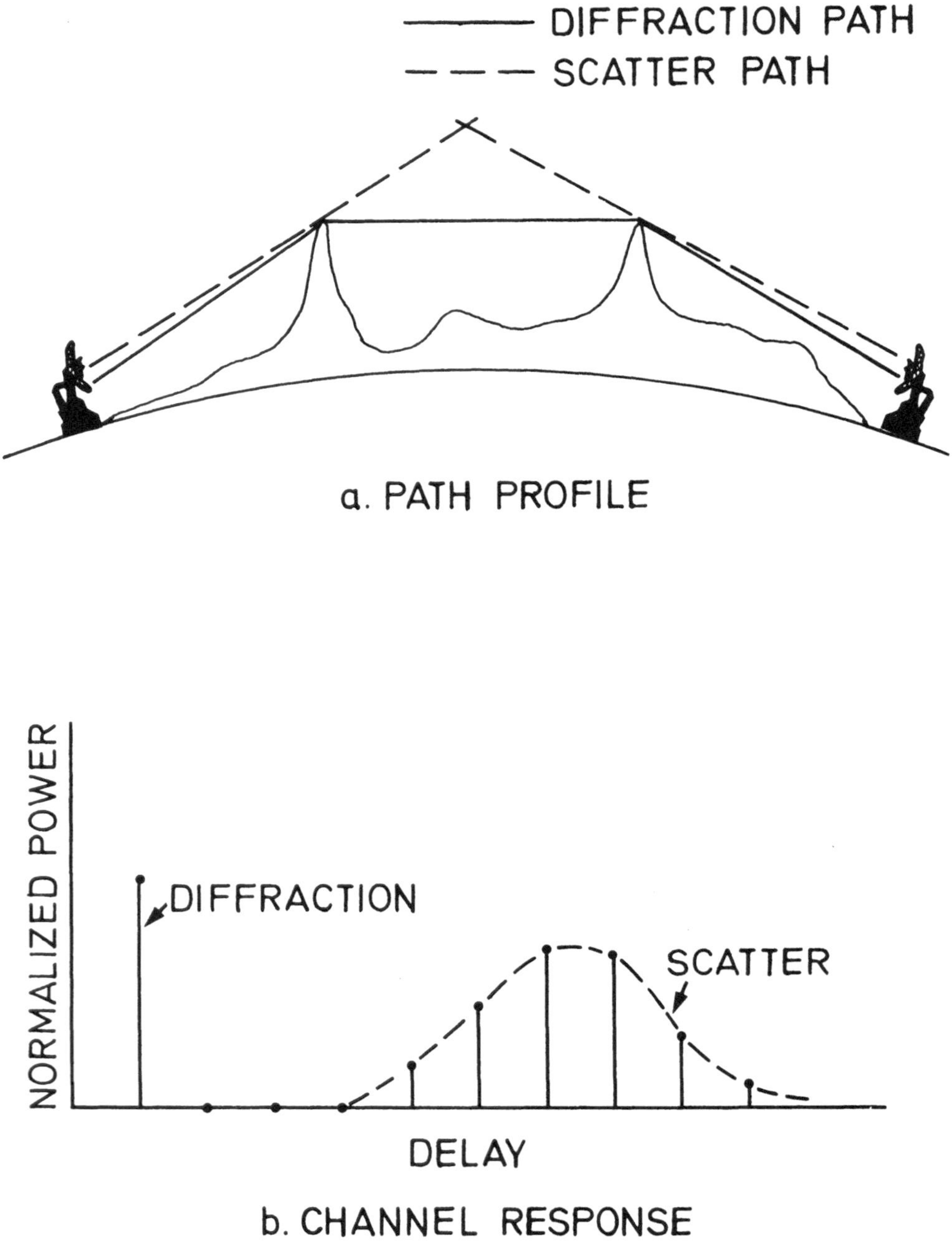

E 1293

Fig. 3. Example of mixed propagation mode

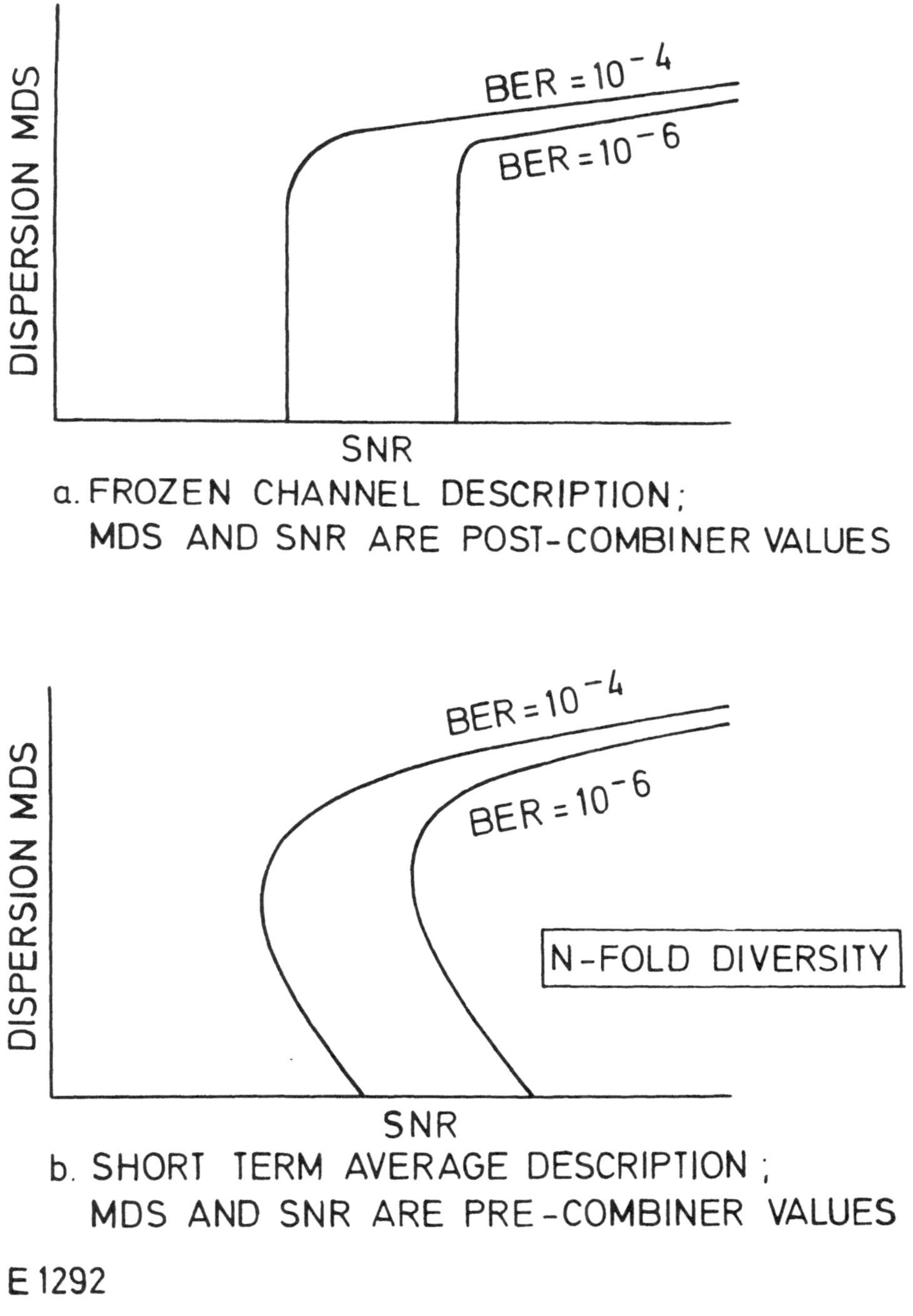

Fig. 4. Digital modem characterizations

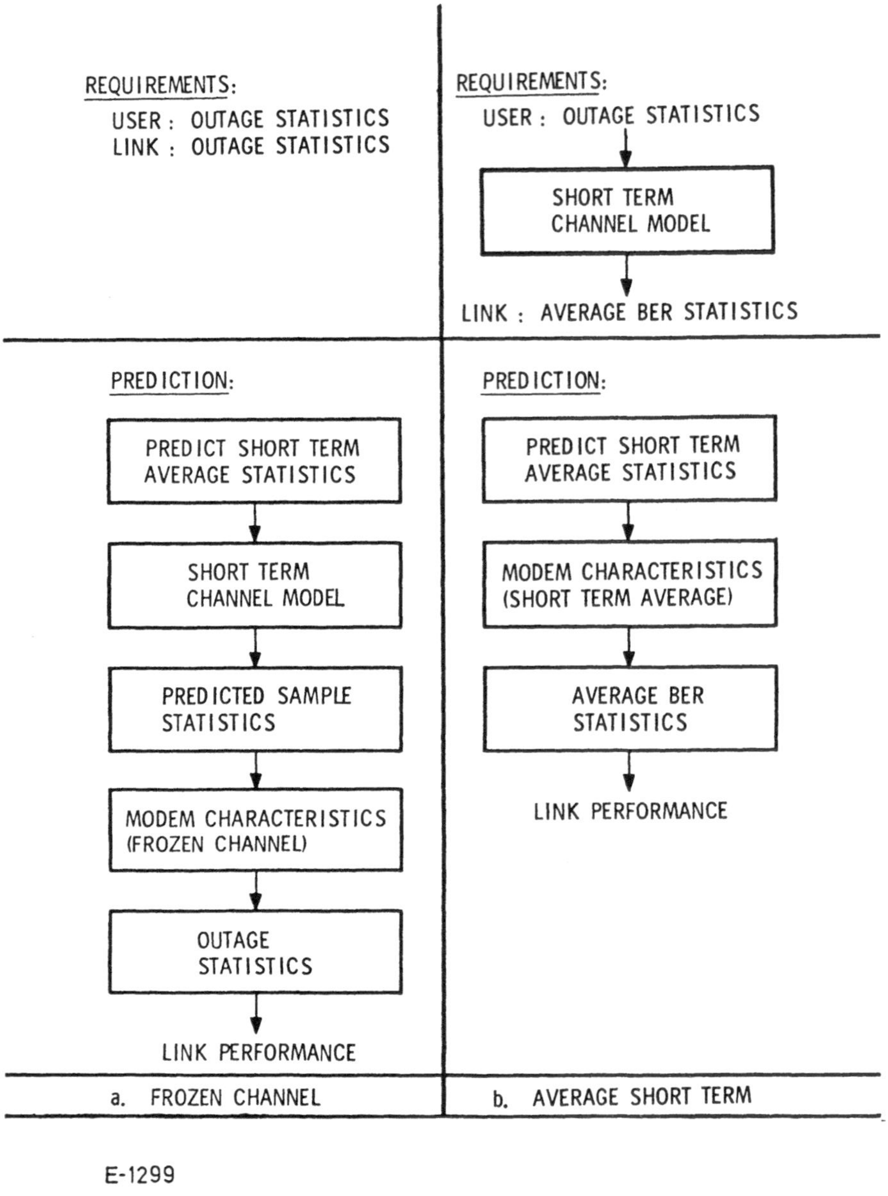

Fig. 5. Link prediction approaches

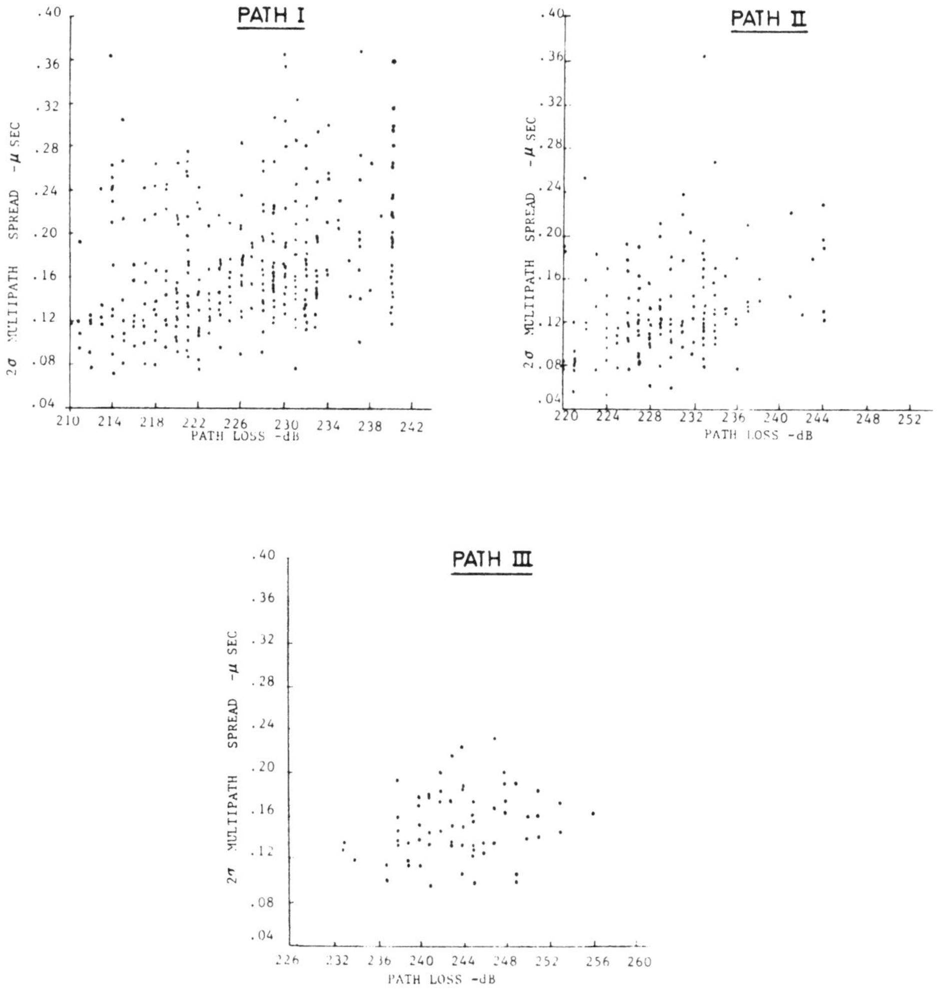

E-1296

Fig. 6. Typical examples of multipath/path loss scatter plots (from [14])

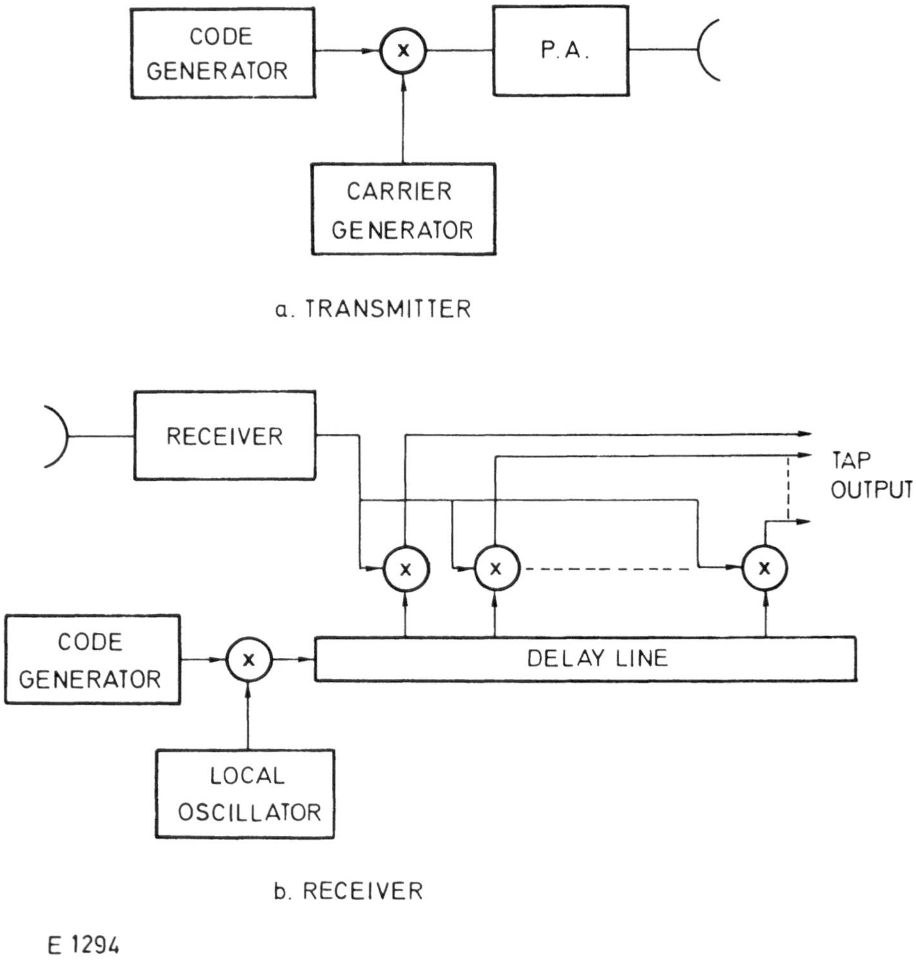

Fig. 7. Simplified principle of multipath analyzer (RAKE)

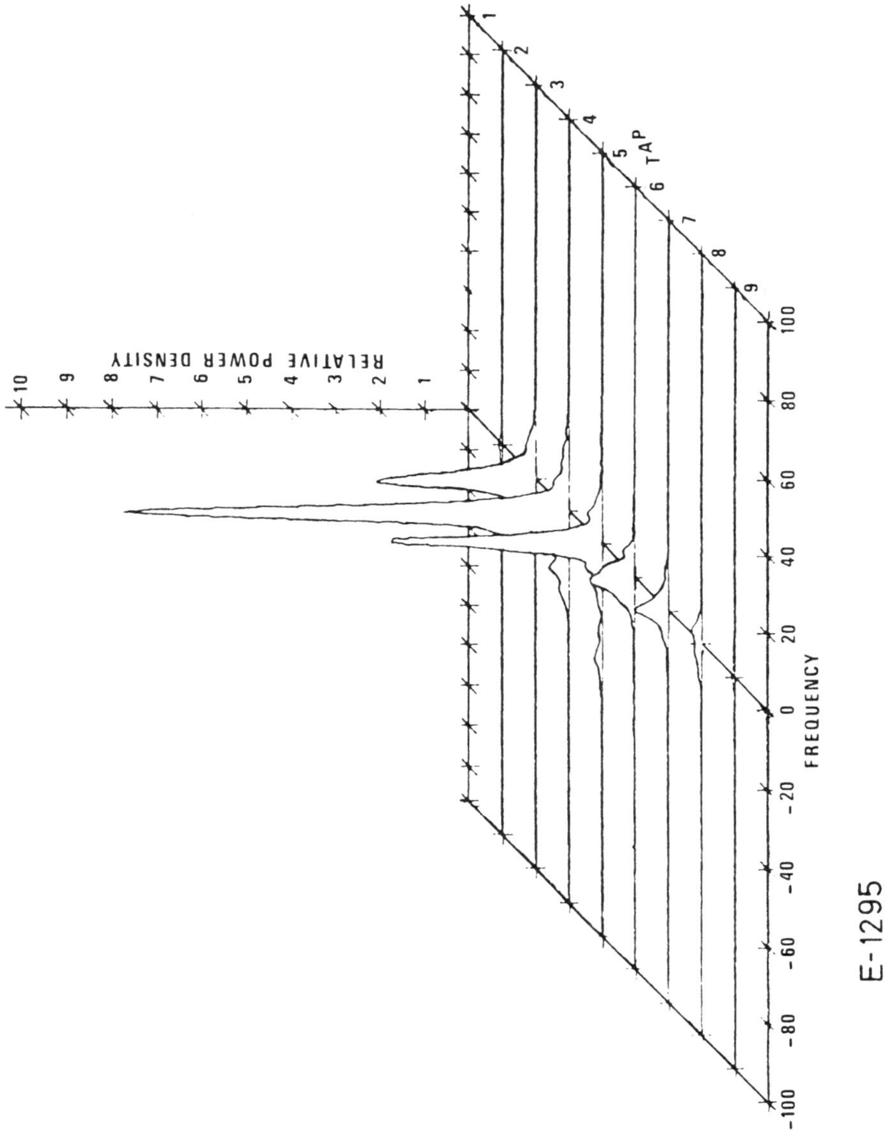

Fig. 8. Example of measured scatter function

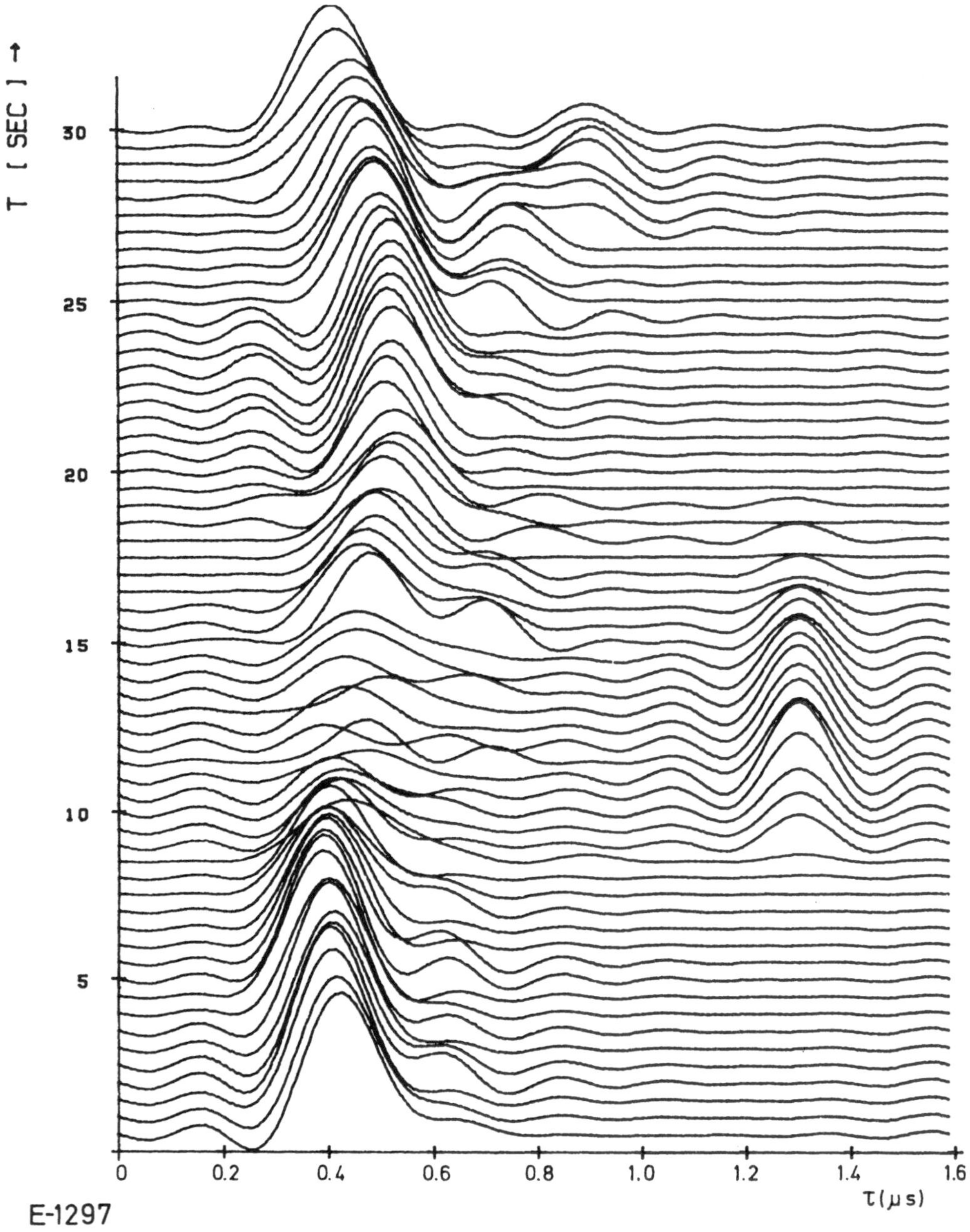

Fig. 9. Example of evolution of channel impulse response with time

LIST OF LECTURERS & DELEGATES

AGRESTI, A.C., Ansoldo S.G.E., Genova, Italy
ALLEN, K.C., Longmont, Col., USA
ALTES, R.A., Dr., Science Applications Inc., La Jolla, Cal., USA
ANDERSON, J.B., Dr., McMaster University, Hamilton, Ont., Canada
ANNECKE, K.H., TH Aachen Institut für Elecktr. Nachrichtentechnik, Aachen, BRD
BASAR, T., Dr., Marmara Research Institute, Istanbul, Turkey
BASAR, T., Mrs, Marmara Research Institute, Istanbul, Turkey
BENEDETTO, S., Prof., Politecnico di Torino, Italy
BIGLIERI, E., Dr., Politecnico di Torino, Italy
BLACHMAN, N.M., Dr., US Navy, London, U.K.
BOEL, R.K., Dr., State University of Gent, Belgium
BREMAUD, P., Dr., IRIA/ LABORIA, Le Chesnay, France
BRITTON, R.L., Prof., California State University at Chico, Cal.,USA
BROOKS, G., GEC Marconi Electronics Ltd., Chelmsford, Essex, U.K.
BROUWER, J.A.M. de, Eindhoven Polytechnic, The Netherlands
BROWN, D.W., SHAPE, The Hague, The Netherlands
BRUYLAND, I.L., Dr., State University of Gent, Belgium
CATTERMOLE, K.W., Prof., University of Essex, Colchester, U.K.
CIFTCIOGLU, O., Dr., Technical University of Istanbul, Turkey
CLARK, J.M.C., Dr., Imperial College, London, U.K.
COCHRANE, P., Post Office Research Centre, Ipswich, Suffolk, U.K.
CONTE, E., Prof., Instituto Elettrotecnico, Naples, Italy
COOPER, A.B., Dr., US Army, Maryland, USA
COOPER, D.C., Dr., University of Birmingham, Edgbaston, U.K.
DARNELL, M., Dr., SHAPE, The Hague, The Netherlands
DAVIES, B.H., Dr., Royal Signals & Radar Establishment, Christchurch, Dorset, U.K.
DAVIS, M.H.A., Dr., Imperial College, London, U.K.
DECOUVELAERE, M., Mme., E.N.S. Telecommunications, Paris, France
DEIGHTON, P.D., Dr., Post Office Research Centre, Ipswich, U.K.
DELGADO-PENIN, J.A., Dr.Ing., Escuela Tecnica Superior Ingenieros de Telecommunicacion, Barcelona, Spain
DEL RE, E., Dr., Instituto di Elettronica, Firenze, Italy
DERIN, H., Prof., Middle East Technical University, Ankara, Turkey
DIETERICH, G., Dr.Ing., Standard-Elektrik Lorenz AG, Stuttgart, BRD
DOLIVO, F.B., Dr., IBM Research Laboratory, Rueschlikon, Switzerland
DURRANI, T.S., Dr., University of Strathclyde, Glasgow, Scotland, U.K.
EPHREMIDES, A., Prof., University of Maryland, College Park, Mar., USA
ERBACH, D.W., Dr., Concordia University, Montreal, Quebec, Canada
FARRELL, P.G., Dr., The University of Kent at Canterbury, U.K.
FRANKS, L.E., Prof., University of Massachusetts, Amherst, Mass., USA
GARDNER, W.A., Dr., University of California, Davies, Cal., USA
GOBIEN, J.O., Prof., AFIT/ENG, Wright-Patterson AFB, Ohio, USA
GODLEWSKI, P., E.N.S.T., Paris, France
GOODMAN, R.M.F., Dr., University of Hull, U.K.

GRONEVELD, E.W., Prof., Twente Polytechnic, Enschede, The Netherlands
GRUPP, W., Universität Karlsruhe, BRD
GUALTIEROTTI, A.F., Ecole Polytechnique Fédérale, Lausanne, Switzerland
HABER, F., Prof., University of Pennsylvania, Philadelphia, Pa., USA
HACCOUN, D., Prof., University of Montreal, Quebec, Canada
HADJUFOTIOU, A., Dr., STL, Ltd., Harlow, Essex, U.K.
HAGENAUER, H.J., Dr., Ing., DFVLR, Wessling, BRD
HAYRE, H.S., Prof., University of Houston, Texas, USA
HAZER, S., Dr., Ege University, Bornova-Izmir, Turkey
HEALY, T.J., Dr., University of Santa Clara, USA
HERMSTRUWER, G., Krupp-Atlas-Elektronik, Bremen, BRD
INGRAM, D.G.W., University of Cambridge, U.K.
JOHANNESSON, R., Dr., Tekniska Högskolan, Lund, Sweden
JONES, E.V., Dr., University of Essex, Colchester, U.K.
JOURDAIN, G., Dr., CEPHAG, Grenoble, France
KASTNER, M.P., Dr., Harvard University, Cambridge, Mass., USA
KAWAS-KALEH, G., E.N.S.T., Paris, France
LACOUME, J.L., Prof., CEPHAG, Grenoble, France
LONGO, G., Prof., University of Trieste, Italy
MAARSEVEEN VAN, M.F.A.M., Twente Polytechnic, Enschede, The Netherlands
MADAMS, C.J., A.S.W.E., Portsmouth, U.K.
MARKO, H., Prof., Dr.Ing., Technische Universität, Munich, BRD
MARSAN, M.A., Prof., Politecnico di Torino, Italy
MASSEY, J., Prof., M.I.T., Cambridge, Mass., USA
MEIER, L., Dr., SACLANT, San Bartolomeo, La Spezia, Italy
MILBOURN, A.J., Dr., Brunel University, Uxbridge, Middlesex, U.K.
MODESTINO, J.W., Dr., Rensselaer Polytechnic Institute, Troy, New York, USA
MORLEY, R.E., Washington University, St. Louis, Miss., USA
MORRIS, J.M., Dr., US Navy, Washington DC, USA
NEUHOFF, D.L., Dr., University of Michigan, Ann Arbor, Mich., USA
NIELSON, P.T., SHAPE Technical Centre, The Hague, The Netherlands
NORTON-WAYNE, L., Dr., The City University, London, U.K.
OLGAYTO, E., Dr., University of Strathclyde, Glasgow, Scotland, U.K.
PAEPE, A.J.P. de, Eindhoven Polytechnic, The Netherlands
PANAYIRCI, E., Prof., Technical University of Istanbul, Turkey
PARZEN, P., Prof., University of Illinois at Chicago, USA
PEARCE, J. LeRoy, Dr., Communications Research Centre, Ottawa, Ont., Canada
PENNINGTON, J., A.S.W.E., Portsdown, Cosham, U.K.
PICINBONO, B., Prof., Centre National de la Recherche Scientifique, Gif-sur-Yvette, France
PIRANI, G., Dr., Eng., CSELT, Torino, Italy
PLAISANT, A., SACLANT, San Bartolomeo, La Spezia, Italy
PRABHAKAR, J.C., Dr., Texas Technical University, Lubbock, Texas, USA
PROAKIS, J.G., Dr., North Eastern University, Boston, Mass., USA

PROTONOTARIOS, E.N., Prof., National Technical University of Athens, Greece
REITBERGER, P., Technische Universität Munich, BRD
RICHARDS, G.A., Dr., GEC-Marconi Electronics, Ltd., Great Baddow, Chelmsford, Essex, U.K.
ROBINSON, S.R., Prof., AFIT/ENG, Wright-Patterson AFB, Ohio, USA
ROOT, W.L., Prof., University of Michigan, Ann Arbor, Mich., USA
ROOYACKERS, J.E., Eindhoven Polytechnic, The Netherlands
SADLER, M., Miss, GEC-Marconi Electronics, Ltd., Great Baddow, Chelmsford, Essex, U.K.
SAKRISON, D., Prof., University of California, Berkeley, Cal., USA
SANKUR, B., Dr., Bogazici University, Istanbul, Turkey
SCHALKWIJK, J.P.M., Prof., Eindhoven Polytechnic, The Netherlands
SCHARF, L.L., Prof., Honeywell Inc., Seattle, Washington, USA
SCHENDEL, A.P.C. van, Eindhoven Polytechnic, The Netherlands
SEGALL, A., Prof., Technion IIT, Haifa, Israel
SELBUZ, H., Marmara Scientific & Industrial Research Institute, Kadiköy, Istanbul, Turkey
SKWIRZYNSKI, J.K., GEC-Marconi Electronics, Ltd., Great Baddow, Chelmsford, Essex, U.K.
SMITH, G.K., E.S.T.E.C., Noordwijk, The Netherlands
SNYDER, D.L., Prof., Washington University, St. Louis, Miss., USA
STANSFIELD, E.V., SHAPE Technical Centre, The Hague The Netherlands
STEENAART, W., Dr., University of Ottawa, Ontario, Canada
STUMPERS, F.L., Prof., IEEE, Eindhoven, The Netherlands
SUNDBERG, G.E.W., Dr., University of Lund, Sweden
TACCONI, G., Prof., Instituto de Elettrotecnica, Genova, Italy
THOMAS, J.B., Prof., Princeton University, New Jersey, USA
VENETSANOPOULOS, A.N., Prof., University of Toronto, Ontario, Canada
VINCK, A.J., Eindhoven Polytechnic, The Netherlands
WEST, B., GEC-Marconi Electronics, Ltd., Great Baddow, Chelmsford, Essex, U.K.
WILLIAMS, C.S., Dr., Stanford University, California, USA
WILSON, S.G., Prof., University of Virginia, Charlottesville, Virginia, USA
WOLF, J.K., Prof., University of Massachusetts, Amherst, Mass., USA
WONG, E., Prof., University of California, Berkeley, Cal., USA
WRIGHT, P., GEC-Marconi Electronics, Ltd., Great Baddow, Chelmsford, Essex, U.K.
YAO, K., Prof., University of California, Los Angeles, Cal., USA
YAVUZ, D., Dr., Middle East Technical University, Ankara, Turkey